Marcus Laurel, Marius Mitrea

Weighted Morrey Spaces

De Gruyter Studies in Mathematics

Volume 99

Marcus Laurel, Marius Mitrea

Weighted Morrey Spaces

Calderón-Zygmund Theory and Boundary Problems

DE GRUYTER

Mathematics Subject Classification 2020
Primary: 31B10, 35C15, 35J25, 35J57; Secondary: 42B20, 42B37, 46E30

Authors
Dr. Marcus Laurel
Baylor University
Department of Mathematics
Waco, TX 76706, USA
Marcus_Laurel@baylor.edu

Prof. Dr. Marius Mitrea
Baylor University
Department of Mathematics
Waco, TX 76706, USA
Marius_Mitrea@baylor.edu

ISBN 978-3-11-145816-8
e-ISBN (PDF) 978-3-11-145827-4
e-ISBN (EPUB) 978-3-11-146145-8
ISSN 0179-0986

Library of Congress Control Number: 2024938450

Bibliographic information published by the Deutsche Nationalbibliothek
The Deutsche Nationalbibliothek lists this publication in the Deutsche Nationalbibliografie; detailed bibliographic data are available on the Internet at http://dnb.dnb.de.

Typesetting: VTeX UAB, Lithuania
Printing and binding: CPI books GmbH, Leck

www.degruyter.com

Contents

1 Introduction

Morrey spaces are ubiquitous in a number of areas of mathematics, including harmonic analysis and partial differential equations. The work undertaken in this monograph has three principal aims:

1. *to develop a functional analytic framework for Muckenhoupt-weighted Morrey spaces and their pre-duals in general geometric settings;*
2. *to develop a Calderón–Zygmund theory on uniformly rectifiable sets for singular integral operators acting on Muckenhoupt-weighted Morrey spaces and their pre-duals;*
3. *to develop boundary layer methods as a means of solving a variety of boundary value problems with data arbitrarily prescribed in Muckenhoupt-weighted Morrey spaces and their pre-duals.*

This places our project at the crossroads between harmonic analysis, partial differential equations, functional analysis, and geometric measure theory. The three basic objectives outlined above interface tightly with one another. To elaborate on this aspect, we first reexamine the general philosophy of boundary value problems.

The theory of function spaces, which is concerned with measuring the size and smoothness of functions in various geometric settings, is intimately linked with the area of boundary value problems. In its most basic form (which, for now, lacks quantitative aspects), the latter attempts to answer questions such as:

$$\text{when is a function, } f, \text{ defined on } \partial\Omega \text{ (the boundary of a given open set } \Omega \subseteq \mathbb{R}^n\text{), extendable to a null-solution, } u, \text{ of a given elliptic, second-order operator } L \text{ (such as, e.g., the Laplacian) in } \Omega? \tag{1.1}$$

As opposed to considering a single function at a time, placing the boundary datum f in a suitable function space, call it $\mathscr{X}(\partial\Omega)$, is a powerful idea which opens the door for bringing in results and techniques from functional analysis. In turn, this allows for the simultaneous treatment of a large class of boundary data, greatly extending the scope and applicability of the question posed in (1.1). Once this point of view is adopted, it is only natural that one identifies a companion function space, say $\mathscr{Y}(\Omega)$, serving as an ambient in which a solution, u, is sought. The membership $u \in \mathscr{Y}(\Omega)$, typically quantified as $\|u\|_{\mathscr{Y}(\Omega)} < +\infty$, should be thought of as a size condition imposed on u. The above discussion then leads to the consideration of the following boundary value problem:

$$\begin{cases} u \in \mathscr{Y}(\Omega), \\ Lu = 0 \quad \text{in } \Omega, \\ \text{“}u|_{\partial\Omega}\text{”} = f \in \mathscr{X}(\partial\Omega), \end{cases} \tag{1.2}$$

where the restriction to the boundary “$u|_{\partial\Omega}$” is suitably understood in the context of the space $\mathscr{Y}(\Omega)$. This is a framework within which one now has a chance to also prove uniqueness for the solution u.

https://doi.org/10.1515/9783111458274-001

The choices made for the space of boundary data, $\mathscr{X}(\partial\Omega)$, and its companion space, $\mathscr{Y}(\Omega)$, in which a solution is sought, decidedly affect the nature of the question posed in (1.1). There are in fact infinitely many possible choices for both spaces, and as witnessed by the very essence of the question formulated in (1.1), there is no "canonical" or all-encompassing space of boundary data $\mathscr{X}(\partial\Omega)$. This is why any successful resolution of a boundary value problem like (1.2) writes a very distinct chapter in the quest for fully understanding the multifaceted question posed in (1.1). Additionally, even when a good understanding of (1.2) exists, it is likely that additional features of f, such as membership to other function spaces, entail extra properties of the extension u that might be of interest and would otherwise be overlooked by a singular focus on $\mathscr{X}(\partial\Omega)$ and $\mathscr{Y}(\Omega)$.

To summarize, the answer to (1.1) is akin to a book with infinitely many chapters, each focusing on a specific functional analytic setting, i. e., an explicit choice for the space of boundary data $\mathscr{X}(\partial\Omega)$ and a naturally accompanying space $\mathscr{Y}(\Omega)$ in which to find a solution. While these chapters may be closely related and tied to one another in principle, the actual technical challenges for solving the corresponding boundary value problems are very much specific to the analytic/geometric setting one chooses to consider. Overcoming these mathematical hurdles requires continually developing new, ever more potent and flexible methods and techniques.

The goal here is to write a new chapter in this metaphorical encyclopedia of boundary value problems, in which we focus on $\mathscr{X}(\partial\Omega) := M^{p,\lambda}(\partial\Omega, w)$, a Muckenhoupt-weighted Morrey space on $\partial\Omega$, and with $\mathscr{Y}(\Omega)$ taken to be the collection of Lebesgue-measurable functions u defined in Ω, whose nontangential maximal operator (with aperture κ), denoted by $\mathcal{N}_\kappa u$, belongs to $M^{p,\lambda}(\partial\Omega, w)$, and whose pointwise nontangential boundary trace $u\big|_{\partial\Omega}^{\kappa-\mathrm{n.t.}}$ exists $\mathcal{H}^{n-1}$-a. e. on $\partial\Omega$ (where $\mathcal{H}^{n-1}$ is the $(n-1)$-dimensional Hausdorff measure in $\mathbb{R}^n$). We shall amply elaborate on these aspects later in the narrative. For now, to place matters into a broader perspective, we take a look at the scale of Morrey spaces in the simplest geometric setting, namely that of the entire Euclidean space.

Morrey spaces have a long history in functional and harmonic analysis. There exist many different types and variations (e. g., classical, weighted, generalized Morrey spaces, etc.[1]), and there are several research monographs dedicated solely to their study, e. g., [2, 130, 131]. From their inception, Morrey spaces have been closely related to the field of partial differential equations. Charles B. Morrey, Jr. first introduced the notion of a function having finite Morrey norm in 1938 while studying the regularity of solutions to second-order, quasilinear, elliptic partial differential equations in two variables [107]. There, Morrey shows that functions belonging to a certain Morrey space must also satisfy some type of Hölder regularity, a fact frequently referred to as Morrey's lemma (or Morrey's inequality) in the literature. To see how the notion of a Morrey space naturally arises, consider a function u in the (local) Sobolev space $W^{1,1}_{\mathrm{loc}}(\mathbb{R}^n)$. Denote by $Q \subseteq \mathbb{R}^n$ any

1 We shall attempt a classification later.

cube with sides parallel to the coordinate axes in $\mathbb{R}^n$, which has side-length $\ell(Q)$, and set $u_Q := ⨍_Q u\,d\mathcal{L}^n$, the integral average of u over Q with respect to the n-dimensional Lebesgue measure $\mathcal{L}^n$. Then for any $\alpha, \lambda \in \mathbb{R}$ and $p \in [1,\infty)$, a combination of Poincaré's inequality and Hölder's inequality yields the following estimate, valid for some purely dimensional constant $C_n \in (0,\infty)$:

$$\begin{aligned}\frac{1}{\ell(Q)^\alpha} ⨍_Q |u - u_Q|\,d\mathcal{L}^n &\leq C_n \frac{\ell(Q)}{\ell(Q)^\alpha} ⨍_Q |\nabla u|\,d\mathcal{L}^n \\ &\leq C_n \ell(Q)^{1-\alpha}\Big(⨍_Q |\nabla u|^p\,d\mathcal{L}^n\Big)^{1/p} \\ &= C_n \ell(Q)^{1-\alpha+(\lambda-n)/p}\Big(\frac{1}{\ell(Q)^\lambda}\int_Q |\nabla u|^p\,d\mathcal{L}^n\Big)^{1/p}.\end{aligned} \tag{1.3}$$

Choosing $\alpha := 1 - (n-\lambda)/p$ eliminates the power of $\ell(Q)$ outside of the parentheses, a scenario in which we simply obtain

$$\frac{1}{\ell(Q)^\alpha} ⨍_Q |u - u_Q|\,d\mathcal{L}^n \leq C_n\Big(\frac{1}{\ell(Q)^\lambda}\int_Q |\nabla u|^p\,d\mathcal{L}^n\Big)^{1/p}. \tag{1.4}$$

Let us now require that $\alpha \in (0,1)$, which forces both $\lambda < n$ and $p > n-\lambda$. The membership of α to the interval $(0,1)$ is assumed in order to ensure the applicability of N. Meyers's integral characterization of Hölder continuity, to the effect that

$$\|u\|_{\dot{\mathscr{C}}^\alpha(\mathbb{R}^n)} := \sup_{\substack{x,y\in\mathbb{R}^n \\ x\neq y}} \frac{|u(x)-u(y)|}{|x-y|^\alpha} \leq \sup_{\substack{Q\subseteq\mathbb{R}^n \\ Q \text{ cube}}} \frac{C_{n,\alpha}}{\ell(Q)^\alpha} ⨍_Q |u - u_Q|\,d\mathcal{L}^n; \tag{1.5}$$

cf. [83]. Granted this, we see from (1.4) that having

$$\sup_{\substack{Q\subseteq\mathbb{R}^n \\ Q \text{ cube}}} \Big(\frac{1}{\ell(Q)^\lambda}\int_Q |\nabla u|^p\,d\mathcal{L}^n\Big)^{1/p} < +\infty \tag{1.6}$$

guarantees the membership of u to the homogeneous Hölder space $\dot{\mathscr{C}}^\alpha(\mathbb{R}^n)$.

The finiteness condition in (1.6) serves as motivation for the norm that defines a classical Morrey space. In the definition that we choose to introduce, we consider balls instead of cubes, though this distinction is inconsequential as it leads to an equivalent norm. Specifically, given $p \in (0,\infty)$ and $\lambda \in [0,\infty)$, the *classical* Morrey space, $M^{p,\lambda}(\mathbb{R}^n, \mathcal{L}^n)$, is defined as the set of all Lebesgue measurable functions f on $\mathbb{R}^n$ for which the following quantity is finite:

$$\|f\|_{M^{p,\lambda}(\mathbb{R}^n,\mathcal{L}^n)} := \sup_{\substack{x\in\mathbb{R}^n\\ r\in(0,\infty)}} \left(\frac{1}{r^\lambda}\int_{B(x,r)} |f|^p\, d\mathcal{L}^n\right)^{1/p}. \tag{1.7}$$

It is common, however, to restrict λ to the interval $(0, n)$, since Lebesgue's differentiation theorem implies that $M^{p,\lambda}(\mathbb{R}^n, \mathcal{L}^n) = \{0\}$ whenever $\lambda > n$, and that $M^{p,n}(\mathbb{R}^n, \mathcal{L}^n)$ coincides with $L^\infty(\mathbb{R}^n, \mathcal{L}^n)$. It is also clear from (1.7) that $M^{p,0}(\mathbb{R}^n, \mathcal{L}^n) = L^p(\mathbb{R}^n, \mathcal{L}^n)$, making Lebesgue spaces an endpoint of the scale of Morrey spaces, corresponding to the case when $\lambda = 0$.

Put differently, having $f \in M^{p,\lambda}(\mathbb{R}^n, \mathcal{L}^n)$ comes down to the existence of some universal constant $C \in (0, \infty)$ with the property that

$$\left(\int_{B(x,r)} |f|^p\, d\mathcal{L}^n\right)^{1/p} \leq C r^{\lambda/p} \quad \text{for all } (x, r) \in \mathbb{R}^n \times (0, \infty), \tag{1.8}$$

and the best constant above is precisely $\|f\|_{M^{p,\lambda}(\mathbb{R}^n,\mathcal{L}^n)}$. From this point of view, *the membership of a function to a Morrey space amounts to having power-type growth* (resp., *vanishing*) *control of the L^p-styled norms of said function over Euclidean balls, in a uniform manner with respect to the center, as the radius goes to infinity* (resp., *goes to zero*).

Because the condition of having finite Morrey norm originally arose as a means to grant a priori regularity for solutions to systems of PDE, the notion of a Morrey *space* was not quite realized as an object of study at the epoch of its inception. It is not until the 1960s in the pioneering works of S. Campanato, G. Stampacchia, and J. Peetre [14, 139, 113] that we begin to see the emergence of a treatment of Morrey spaces as function spaces proper (through the consideration of such topics as boundedness of operators and interpolation). The late 1980s saw Morrey spaces more firmly solidified in the realm of harmonic analysis. As one instance of this, F. Chiarenza and M. Frasca [16] established the boundedness of the Hardy–Littlewood maximal operator on classical Morrey spaces. In a separate development in 1986, C. Zorko [153] identified the pre-dual of Morrey spaces as consisting of functions created by gluing together certain bump-functions (characterized in terms of localization and size) via ℓ^1 sequences of numerical coefficients. This is a function space now commonly referred to as a Block space,[2] denoted by $\mathcal{B}^{q,\lambda}(\mathbb{R}^n, \mathcal{L}^n)$, which for $\lambda \in (0, n)$ satisfies

$$\begin{gathered}(\mathcal{B}^{q,\lambda}(\mathbb{R}^n, \mathcal{L}^n))^* = M^{p,\lambda}(\mathbb{R}^n, \mathcal{L}^n)\\ \text{for } p, q \in (1, \infty) \text{ with } 1/p + 1/q = 1,\end{gathered} \tag{1.9}$$

where equality above is understood as an isomorphism between spaces. See [130, Section 9] and the references therein for more information on Block spaces in the Euclidean setting. A collection of results concerning classical Morrey spaces may be found in [2]

2 The label Zorko space is also occasionally used in the literature.

with emphasis on the boundedness of Riesz potentials in such a functional analytic setting. See also [41] for an account of boundedness results of commutators on classical Morrey spaces, and [119] for a discussion of classical Calderón–Zygmund operators on Morrey spaces. In more geometrically rough settings, work in [137] considers boundary value problems for elliptic systems with data in Morrey spaces on Lipschitz domains. Moving beyond the Euclidean setting, there is work concerning Morrey spaces on spaces of homogeneous type in [40]. Bensoussan and Frehse [10] employ Morrey spaces in the realm of nonlinear elliptic systems, and in a similar vein, [45, 145] consider the Navier–Stokes equation with data belonging to Morrey spaces.

Since classical Morrey spaces are more of a template rather than an end product (as far as considering function spaces whose norms are inspired by (1.7)), it is of interest to present a brief zoology of Morrey spaces.

Z-I. Weighted Morrey spaces of the first kind. One way to add flexibility in (1.7) is to inject a weight. That is, for $p \in (0,\infty)$, $\lambda \in [0,\infty)$, and any weight function[3] w on $\mathbb{R}^n$, define $M^{p,\lambda}(\mathbb{R}^n, w)$ as the collection of $\mathcal{L}^n$-measurable functions f in $\mathbb{R}^n$ satisfying

$$\|f\|_{M^{p,\lambda}(\mathbb{R}^n,w)} := \sup_{\substack{x\in\mathbb{R}^n\\ r\in(0,\infty)}} \left(\frac{1}{r^\lambda}\int_{B(x,r)} |f|^p w\, d\mathcal{L}^n\right)^{1/p} < +\infty. \tag{1.10}$$

Morrey spaces of this nature were introduced by N. Samko in [123] for a certain choice of w, and expounded upon in [125], studying necessary conditions for the boundedness of the Hardy–Littlewood maximal operator on these spaces. See also [131, Section 15.2] and the references therein, as well as [144] for two-weight norm inequalities involving this type of weighted Morrey space. This is the brand of Morrey spaces that will be the focus of this monograph, albeit in a geometrically rough setting, working with a closed Ahlfors regular set in place of $\mathbb{R}^n$.

Z-II. Weighted Morrey spaces of the second kind. Weights can be introduced in a way that affect not just in the measure of integration, but also in how they relate to the extra parameter λ. Specifically, for $p \in (0,\infty)$, $\lambda \in (0,\infty)$, and any weight function w on $\mathbb{R}^n$, define $\widetilde{M}^{p,\lambda}(\mathbb{R}^n, w)$ as the collection of $\mathcal{L}^n$-measurable functions f in $\mathbb{R}^n$ such that

$$\|f\|_{\widetilde{M}^{p,\lambda}(\mathbb{R}^n,w)} := \sup_{\substack{x\in\mathbb{R}^n\\ r\in(0,\infty)}} \left(\frac{1}{w(B(x,r))^{\lambda/n}}\int_{B(x,r)} |f|^p w\, d\mathcal{L}^n\right)^{1/p} < +\infty. \tag{1.11}$$

This type of weighted Morrey space was introduced by Komori and Shirai in [70], wherein they prove boundedness of the Hardy–Littlewood maximal operator and operators of Calderón–Zygmund type. For a general treatment, see [131, Section 15.1] and the

3 i. e., a measurable function which is strictly positive and finite almost everywhere.

references within. See also [143] for the consideration of fractional integral operators on these types of weighted Morrey spaces, and [138] for applications of these weighted Morrey spaces of the second kind to elliptic equations in nondivergence form.

Z-III. Hybrid weighted Morrey spaces. A hybrid version of (1.10) and (1.11) can be constructed as follows. Fix two parameters $\lambda_1, \lambda_2 \in \mathbb{R}$, and set $\lambda := (\lambda_1, \lambda_2)$. Having defined

$$\|f\|_{\widehat{M}^{p,\lambda}(\mathbb{R}^n,w)} := \sup_{\substack{x\in\mathbb{R}^n \\ r\in(0,\infty)}} \left(\frac{1}{r^{\lambda_1} \cdot w(B(x,r))^{\lambda_2/n}} \int_{B(x,r)} |f|^p w \, d\mathcal{L}^n \right)^{1/p}, \tag{1.12}$$

we then let $\widehat{M}^{p,\lambda}(\mathbb{R}^n, w)$ denote the space of $\mathcal{L}^n$-measurable functions for which (1.12) is finite. A cursory inspection reveals that when $\lambda_1 = 0$ this brand of space reduces precisely to that of Komori and Shirai, whose norm is given in (1.11), while corresponding to $\lambda_2 = 0$ we recover the weighted Morrey spaces of Samko type, whose norm is as in (1.10). Thus, (1.12) blends both (1.10) and (1.11) into a more flexible, and more inclusive scale. This kind of Morrey space has been considered by Duoandikoetxea and Rosenthal in [30], where the authors characterize various boundedness properties of the Hardy–Littlewood maximal operator on such a scale. In particular, the relevance of the Muckenhoupt class of weights is quite apparent in their work, as the authors show that boundedness of the Hardy–Littlewood maximal operator from $\widehat{M}^{p,\lambda}(\mathbb{R}^n, w)$ into a weak version of $\widehat{M}^{p,\lambda}(\mathbb{R}^n, w)$ requires that the weight w belongs to a certain Muckenhoupt class. See also [31, 125, 112] for the relevance of Muckenhoupt weights to weighted Morrey spaces of the first and second kinds.

The reader is reminded that a nonnegative $\mathcal{L}^n$-measurable function $w : \mathbb{R}^n \to [0, +\infty]$, which satisfies $0 < w < +\infty$ at $\mathcal{L}^n$-a.e. point on $\mathbb{R}^n$, is said to belong to the Muckenhoupt class $A_p(\mathbb{R}^n, \mathcal{L}^n)$ with $1 \leq p < \infty$ provided

$$[w]_{A_p} := \sup_{\substack{Q\subseteq\mathbb{R}^n \\ Q\,\text{cube}}} \left(⨍_Q w \, d\mathcal{L}^n \right) \left(⨍_Q w^{1-p'} \, d\mathcal{L}^n \right)^{p-1} < +\infty \tag{1.13}$$

if $p > 1$ and $p' := (1 - 1/p)^{-1}$, while corresponding to $p = 1$ the demand is

$$[w]_{A_1} := \sup_{\substack{Q\subseteq\mathbb{R}^n \\ Q\,\text{cube}}} \left(\operatorname*{ess\,inf}_{x\in Q} w(x) \right)^{-1} ⨍_Q w \, d\mathcal{L}^n < +\infty. \tag{1.14}$$

In this vein, we wish to observe that the Muckenhoupt class of weights also allows for a version of Morrey's lemma for the aforementioned hybrid Morrey spaces. To elaborate, fix a Muckenhoupt weight $w \in A_p(\mathbb{R}^n, \mathcal{L}^n)$ with $p \in (1, \infty)$ and, as usual, abbreviate $w(Q) := \int_Q w \, d\mathcal{L}^n$ for each cube $Q \subseteq \mathbb{R}^n$. Then, given any function $u \in W^{1,1}_{\mathrm{loc}}(\mathbb{R}^n)$, for any $a, \lambda_1 \in \mathbb{R}$ we may write

$$\frac{1}{\ell(Q)^{\alpha}} \fint_Q |u - u_Q| \, d\mathcal{L}^n \tag{1.15}$$
$$\le C_n \ell(Q)^{1-\alpha} \fint_Q |\nabla u| \, d\mathcal{L}^n$$
$$\le C_{n,p,w} \ell(Q)^{1-\alpha} \Big(\frac{1}{w(Q)} \int_Q |\nabla u|^p w \, d\mathcal{L}^n \Big)^{1/p}$$
$$\le C_{n,p,w} \ell(Q)^{1-\alpha+\lambda_1/p} \Big(\frac{1}{\ell(Q)^{\lambda_1} \cdot w(Q)} \int_Q |\nabla u|^p w \, d\mathcal{L}^n \Big)^{1/p},$$

for some constant $C_{n,p,w} \in (0, \infty)$. Above, the first line is due to Poincaré's inequality, while the second line is a consequence of properties of Muckenhoupt weights (cf. Lemma 2.34). Choosing $\alpha := 1 + \lambda_1/p$ and asking that α belongs to $(0, 1)$ forces $-p < \lambda_1 < 0$. In such a scenario, (1.15), (1.12), and N. Meyer's integral characterization of Hölder continuity recalled in (1.5) imply that membership of u to $\dot{\mathscr{C}}^{\alpha}(\mathbb{R}^n)$ is guaranteed whenever $|\nabla u|$ belongs to $\widehat{M}^{p,\lambda}(\mathbb{R}^n, w)$, where $\lambda := (\lambda_1, n)$.

Z-IV. Generalized Morrey spaces. Pushing the idea of a Morrey space farther still, we have the notion of a generalized Morrey space. The idea is that the integral is now modulated by some function

$$\varphi : \mathbb{R}^n \times (0, \infty) \to (0, \infty), \tag{1.16}$$

allowing a generic dependency on both the location $x \in \mathbb{R}^n$ and the scale $r \in (0, \infty)$. Concretely, the generalized Morrey space $\mathcal{M}^{p,\varphi}(\mathbb{R}^n, \mathcal{L}^n)$ is the collection of $\mathcal{L}^n$-measurable functions f on $\mathbb{R}^n$ such that

$$\|f\|_{\mathcal{M}^{p,\varphi}(\mathbb{R}^n, \mathcal{L}^n)} := \sup_{\substack{x \in \mathbb{R}^n \\ r \in (0,\infty)}} \Big(\frac{1}{\varphi(x, r)} \int_{B(x,r)} |f|^p \, d\mathcal{L}^n \Big)^{1/p} < +\infty. \tag{1.17}$$

In the case that $\varphi(x, r) = r^{\lambda}$ for all $(x, r) \in \mathbb{R}^n \times (0, \infty)$ for some fixed $\lambda \in [0, \infty)$, we recover from (1.17) the classical Morrey norm introduced in (1.7). A broad overview of these spaces may be found in [131, Section 12.1] as well as the references therein. See [105] for boundedness results for the Hardy–Littlewood maximal operator on this scale of spaces in the case that $\varphi(x, r) = \varphi(r)$ satisfies a certain growth condition (the reader is also referred to [110] for a generalization of this result). Work in [106] looks at commutators on this scale of spaces, and [72] contains results for singular integral operators on these generalized Morrey spaces. In [51], a version of (1.17) is studied in which the exponent p is variable. Works such as [76, 34, 127] consider a more abstract formulation of (1.17) in the setting of metric/quasimetric measure spaces. There are even variable exponent generalized Morrey spaces, as described in [132]. In a slightly different direction,

[47] considers an anisotropic version of generalized Morrey spaces and proves boundedness of anisotropic singular integral operators, a subject whose motivation is rooted in the realm of parabolic equations.

Z-V. Weighted generalized Morrey spaces. It is also of interest to define a weighted generalized Morrey space in a similar way by injecting a weight into the integral. That is, given a weight function w on $\mathbb{R}^n$, the space $\mathcal{M}^{p,\varphi}(\mathbb{R}^n, w)$ is described as the collection of $\mathcal{L}^n$-measurable functions f in $\mathbb{R}^n$ for which

$$\|f\|_{\mathcal{M}^{p,\varphi}(\mathbb{R}^n,w)} := \sup_{\substack{x\in\mathbb{R}^n \\ r\in(0,\infty)}} \left(\frac{1}{\varphi(x,r)} \int_{B(x,r)} |f|^p w\, d\mathcal{L}^n\right)^{1/p} < +\infty. \tag{1.18}$$

Note that taking $\varphi(x,r) := r^\lambda$ for all $(x,r) \in \mathbb{R}^n \times (0,\infty)$ for some fixed $\lambda \in [0,\infty)$ in the above norm recovers (1.10). If instead one now takes $\varphi(x,r) := r^{\lambda_1} \cdot w(B(x,r))^{\lambda_2/n}$ for some fixed $\lambda_1, \lambda_2 \in \mathbb{R}$, these generalized weighted Morrey spaces become the type defined by the norm introduced in (1.12). See [122, 73, 132] for work pertaining to this scale.

Our zoology of Morrey spaces may be further augmented by considering scales of spaces obtained via the “Morreyfication” of other known spaces.

M-I. Sobolev–Morrey spaces. The usual notion of a Sobolev space can be “Morreyfied” as exhibited in the following definition. For $p \in (1,\infty)$ and $\lambda \in (0,n)$, introduce the *inhomogenous* Morrey-based Sobolev space (of order one) in $\mathbb{R}^n$ as

$$M_1^{p,\lambda}(\mathbb{R}^n, \mathcal{L}^n) := \{f \in M^{p,\lambda}(\mathbb{R}^n, \mathcal{L}^n) : \partial_j f \text{ exists in a weak sense and } \partial_j f \in M^{p,\lambda}(\mathbb{R}^n, \mathcal{L}^n) \text{ for } 1 \le j \le n\}, \tag{1.19}$$

equipped with the natural norm. For an overview of these kinds of spaces, see [131, Section 14.2] and the references therein. In spite of attributes which are strongly linked with properties of the Euclidean space (viz. weak derivatives), Morrey-based Sobolev spaces are resilient and have been considered in much more general settings (cf. [77, 79, 87]).

A *homogeneous* version of the above Sobolev space could be constructed according to

$$\dot{M}_1^{p,\lambda}(\mathbb{R}^n, \mathcal{L}^n) := \{f \in L^1_{\text{loc}}(\mathbb{R}^n, \mathcal{L}^n) : \partial_j f \in M^{p,\lambda}(\mathbb{R}^n, \mathcal{L}^n) \text{ for } 1 \le j \le n\}. \tag{1.20}$$

Interestingly, because we are working in Euclidean space, there is a global integrability property “hidden” in (1.20), in the sense that the above definition self-improves to

$$\dot{M}_1^{p,\lambda}(\mathbb{R}^n, \mathcal{L}^n) = \left\{f \in L^1\left(\mathbb{R}^n, \frac{\mathcal{L}^n(x)}{1+|x|^{n+1}}\right) : \partial_j f \in M^{p,\lambda}(\mathbb{R}^n, \mathcal{L}^n) \text{ for } 1 \le j \le n\right\}. \tag{1.21}$$

Indeed, taking $a = 0$ in the version of (1.3) written for balls in place of cubes implies that we can control the growth of the mean oscillation of any given $f \in W^{1,1}_{\mathrm{loc}}(\mathbb{R}^n)$ as follows:

$$\fint_{B(0,t)} |f - f_{B(0,t)}|\, d\mathcal{L}^n \leq C_n t^{1-(n-\lambda)/p} \|\nabla f\|_{[M^{p,\lambda}(\mathbb{R}^n,\mathcal{L}^n)]^n}, \tag{1.22}$$

for all $t \in (0, \infty)$, where $f_{B(0,t)} := \fint_{B(0,t)} f\, d\mathcal{L}^n$. Granted this, we may invoke [86, Lemma 7.4.14, p. 603] (presently used with $X := \mathbb{R}^n$, $\mu := \mathcal{L}^n$, $\rho := |\cdot - \cdot|$, $d = n$, $x_0 := 0 \in \mathbb{R}^n$, $\varepsilon := 1$, $p := 1$, $q := 1$, and $r := 1$), which controls the global integrability of a function with respect to the measure $\frac{dx}{1+|x|^n}$ in terms of the growth of the mean oscillation of f with respect to the radius, to estimate

$$\begin{aligned}\Big(\int_{\mathbb{R}^n} \frac{|f(x) - f_{B(0,1)}|^p}{1 + |x|^{n+1}}\, dx\Big)^{1/p} &\leq C_n \|\nabla f\|_{[M^{p,\lambda}(\mathbb{R}^n,\mathcal{L}^n)]^n} \int_1^\infty t^{1-(n-\lambda)/p} \frac{dt}{t^2} \\ &\leq C_{n,p} \|\nabla f\|_{[M^{p,\lambda}(\mathbb{R}^n,\mathcal{L}^n)]^n},\end{aligned} \tag{1.23}$$

for some $C_{n,p} \in (0, \infty)$ (with the finiteness of $C_{n,p}$ guaranteed by the assumption that $\lambda < n$). From (1.23), we readily obtain the inclusion

$$\dot{M}^{p,\lambda}_1(\mathbb{R}^n, \mathcal{L}^n) \subseteq L^1\Big(\mathbb{R}^n, \frac{\mathcal{L}^n(x)}{1 + |x|^{n+1}}\Big), \tag{1.24}$$

justifying (1.21). The usefulness of the alternative characterization of $\dot{M}^{p,\lambda}_1(\mathbb{R}^n, \mathcal{L}^n)$ presented in (1.21) is not fully appreciated until trying to go beyond the Euclidean ambient. For instance, identifying a versatile notion of homogeneous Morrey-based Sobolev space on a closed Ahlfors regular set is a delicate task, especially if this is to be designed to provide a natural environment for the kind of singular integral operators we have in mind. Naïvely trying to adapt (1.20) in as direct a manner as possible to a more geometrically abstract setting does not yield a reasonable version since singular integral operators cannot be applied to generic functions in $L^1_{\mathrm{loc}}(\mathbb{R}^n, \mathcal{L}^n)$. By way of contrast, the space $L^1(\mathbb{R}^n, \frac{\mathcal{L}^n(x)}{1+|x|^{n+1}})$ as opposed to $L^1_{\mathrm{loc}}(\mathbb{R}^n, \mathcal{L}^n)$ is a much friendlier environment from the perspective of Calderón–Zygmund singular integral operators. See our treatment of homogeneous Morrey-based Sobolev spaces on Ahlfors regular domains in Chapter 8.

M-II. BMO-Morrey spaces. Another space to "Morreyfy" is the John–Nirenberg space of bounded mean oscillations (BMO). S. Campanato in 1964 was the first to introduce such a scale of spaces, now called Morrey–Campanato spaces [14]. See also [130, Section 5] and [131, Section 14.3] as well as the references therein for a more modern treatment of this scale of spaces. Fix $k \in \mathbb{N}_0$, $p \in (0, \infty)$, $\lambda \in \mathbb{R}$, and let $\mathcal{P}_k(\mathbb{R}^n)$ denote the set of all polynomials of degree k in $\mathbb{R}^n$. Then the Morrey–Campanato space $\mathcal{L}^{p,\lambda}_k(\mathbb{R}^n, \mathcal{L}^n)$ is defined as the set of all $\mathcal{L}^n$-measurable functions, f, for which

$$\|f\|_{\mathcal{L}_k^{p,\lambda}(\mathbb{R}^n,\mathcal{L}^n)} := \sup_{\substack{x\in\mathbb{R}^n\\ r\in(0,\infty)}} \left(\inf_{P\in\mathcal{P}_k(\mathbb{R}^n)} \frac{1}{r^\lambda} \fint_{B(x,r)} |f-P|^p \, d\mathcal{L}^n \right)^{1/p} < \infty. \tag{1.25}$$

One important aspect of this space is its close relationship to many other familiar function spaces. For instance, it is immediate from the definition that in the case $k = 0$ and $\lambda = 0$ we recover the space of functions of bounded mean oscillations; cf. (2.81). That is, $\mathcal{L}_0^{p,0}(\mathbb{R}^n,\mathcal{L}^n) = \mathrm{BMO}(\mathbb{R}^n,\mathcal{L}^n)$. This is a consequence of the John–Nirenberg inequality, which says for any $f \in L^1_{\mathrm{loc}}(\mathbb{R}^n,\mathcal{L}^n)$ and $p \in [1,\infty)$ there exists a constant $C \in (0,\infty)$, depending only on the dimension n and p, such that

$$\sup_{\substack{x\in\mathbb{R}^n\\ r\in(0,\infty)}} \left(\fint_{B(x,r)} |f - f_{B(x,r)}|^p \, d\mathcal{L}^n \right)^{1/p} \leq C\|f\|_{\mathrm{BMO}(\mathbb{R}^n,\mathcal{L}^n)}, \tag{1.26}$$

where $f_{B(x,r)}$ denotes the integral average of f over the ball $B(x,r)$ with respect to $\mathcal{L}^n$. The connection between classical Morrey spaces and Hölder regularity has been mentioned earlier and, given the close relationship between Morrey–Campanato spaces and classical Morrey spaces, it should not come as a surprise that Morrey–Campanato spaces are themselves related to spaces of Hölder continuous functions. To elaborate, assume $k = 0$, $\lambda \in (0,\infty)$, and $p > \lambda$, then introduce $\alpha := \lambda/p \in (0,1)$. Upon observing that (1.5), (1.25), and Hölder's inequality imply

$$\begin{aligned} \|f\|_{\dot{\mathscr{C}}^\alpha(\mathbb{R}^n)} &\leq \sup_{\substack{x\in\mathbb{R}^n\\ r\in(0,\infty)}} \frac{C_{n,\alpha}}{r^\alpha} \fint_{B(x,r)} |f - f_{B(x,r)}| \, d\mathcal{L}^n \\ &\leq C_{n,\lambda,p} \cdot \sup_{\substack{x\in\mathbb{R}^n\\ r\in(0,\infty)}} \left(\frac{1}{r^\lambda} \fint_{B(x,r)} |f - f_{B(x,r)}|^p \, d\mathcal{L}^n \right)^{1/p} \\ &\leq C_{n,\lambda,p} \cdot \|f\|_{\mathcal{L}_0^{p,\lambda}(\mathbb{R}^n,\mathcal{L}^n)} \end{aligned} \tag{1.27}$$

for some $C_{n,\lambda,p} \in (0,\infty)$, we conclude that the Morrey–Campanato space $\mathcal{L}_0^{p,\lambda}(\mathbb{R}^n,\mathcal{L}^n)$ embeds continuously into the homogeneous Hölder space $\dot{\mathscr{C}}^\alpha(\mathbb{R}^n)$. In the opposite direction, based on definitions and Hölder's inequality we have

$$\begin{aligned} \|f\|_{\mathcal{L}_0^{p,\lambda}(\mathbb{R}^n,\mathcal{L}^n)} &\leq \sup_{\substack{x\in\mathbb{R}^n\\ r\in(0,\infty)}} \left(\frac{1}{r^\lambda} \fint_{B(x,r)} |f - f_{B(x,r)}|^p \, d\mathcal{L}^n \right)^{1/p} \\ &\leq \sup_{\substack{x\in\mathbb{R}^n\\ r\in(0,\infty)}} \left(\frac{1}{r^\lambda} \fint_{B(x,r)} \fint_{B(x,r)} |f(y) - f(z)|^p \, dy \, dz \right)^{1/p} \\ &\leq 2^\lambda \|f\|_{\dot{\mathscr{C}}^\alpha(\mathbb{R}^n)} \end{aligned} \tag{1.28}$$

so that, ultimately, $\mathcal{L}_0^{p,\lambda}(\mathbb{R}^n,\mathcal{L}^n) = \dot{\mathscr{C}}^\alpha(\mathbb{R}^n)$ (cf. [13]).

Another way in which Morrey spaces are related to BMO is in the structure of their duality (we refer to the survey [120] for results concerning Morrey spaces and their pre-duals in Euclidean space). Recall the duality relationship between the Hardy space $H^1(\mathbb{R}^n, \mathcal{L}^n)$, the John–Nirenberg space of bounded mean oscillations $\mathrm{BMO}(\mathbb{R}^n, \mathcal{L}^n)$, and the Sarason space of vanishing mean oscillations $\mathrm{VMO}(\mathbb{R}^n, \mathcal{L}^n)$ (cf. [140]):

$$(\mathrm{VMO}(\mathbb{R}^n, \mathcal{L}^n))^* = H^1(\mathbb{R}^n, \mathcal{L}^n), \quad (H^1(\mathbb{R}^n, \mathcal{L}^n))^* = \mathrm{BMO}(\mathbb{R}^n, \mathcal{L}^n). \tag{1.29}$$

In the context of classical Morrey spaces, there exists a triplet of spaces, $\mathring{M}^{p,\lambda}(\mathbb{R}^n, \mathcal{L}^n)$, $\mathcal{B}^{p',\lambda}(\mathbb{R}^n, \mathcal{L}^n)$, $M^{p,\lambda}(\mathbb{R}^n, \mathcal{L}^n)$ with

$$(\mathring{M}^{p,\lambda}(\mathbb{R}^n, \mathcal{L}^n))^* = \mathcal{B}^{p',\lambda}(\mathbb{R}^n, \mathcal{L}^n) \quad (\mathcal{B}^{p',\lambda}(\mathbb{R}^n, \mathcal{L}^n))^* = M^{p,\lambda}(\mathbb{R}^n, \mathcal{L}^n), \tag{1.30}$$

which is in analogy with (1.29). Here, $\mathring{M}^{p,\lambda}(\mathbb{R}^n, \mathcal{L}^n)$ is yet another version of a Morrey space, called a *vanishing* Morrey space, which first appeared in [151]. Functions f belonging to $\mathring{M}^{p,\lambda}(\mathbb{R}^n, \mathcal{L}^n)$ must have finite Morrey norm and satisfy the vanishing condition

$$\lim_{R\to 0^+} \sup_{\substack{x\in\mathbb{R}^n \\ r\in(0,R)}} \left(\frac{1}{r^\lambda} \int_{B(x,r)} |f|^p \, d\mathcal{L}^n \right)^{1/p} = 0. \tag{1.31}$$

This is in close analogy with functions f belonging to the Sarason space $\mathrm{VMO}(\mathbb{R}^n, \mathcal{L}^n)$, which have $\|f\|_{\mathrm{BMO}(\mathbb{R}^n,\mathcal{L}^n)} < \infty$ and satisfy the vanishing condition

$$\lim_{R\to 0^+} \sup_{\substack{x\in\mathbb{R}^n \\ r\in(0,R)}} \fint_{B(x,r)} |f - f_{B(x,r)}| \, d\mathcal{L}^n = 0. \tag{1.32}$$

Equivalently,

$$\begin{aligned}&\text{when } p \in (1,\infty) \text{ and } \lambda \in (0,n), \text{ the vanishing Morrey space } \mathring{M}^{p,\lambda}(\mathbb{R}^n, \mathcal{L}^n)\\ &\text{is the closure of the Lebesgue space } L^s(\mathbb{R}^n, \mathcal{L}^n) \text{ in the classical Morrey}\\ &\text{space } M^{p,\lambda}(\mathbb{R}^n, \mathcal{L}^n), \text{ where } s := \tfrac{pn}{(n-\lambda)(p-1)} \in (1,\infty).\end{aligned} \tag{1.33}$$

What makes this characterization possible is the validity of the following continuous embedding:

$$\begin{aligned}&L^s(\mathbb{R}^n, \mathcal{L}^n) \hookrightarrow M^{p,\lambda}(\mathbb{R}^n, \mathcal{L}^n) \quad \text{whenever}\\ &p \in (1,\infty), \quad \lambda \in (0,n), \quad s := \frac{pn}{(n-\lambda)(p-1)} \in (1,\infty),\end{aligned} \tag{1.34}$$

which is itself a consequence of definitions and Hölder's inequality. Since basic singular integral operators (of convolution type), such as the Riesz transforms in $\mathbb{R}^n$, are bounded both on the Morrey scale and the Lebesgue scale, we conclude from (1.33) that said singular integral operators induce well-defined mappings from the vanishing Morrey space $\mathring{M}^{p,\lambda}(\mathbb{R}^n, \mathcal{L}^n)$ into itself (as long as $1 < p < \infty$ and $0 < \lambda < n$). For work on classical vanishing Morrey spaces, see [117, 6, 53]. For work on generalized vanishing Morrey spaces, there is a wealth of recent results, e. g., [114, 124, 126].

In this work, we develop a suitable notion of Muckenhoupt-weighted vanishing Morrey space on rough sets that extends work done in [87, 79] in the unweighted case. The novel difficulty is the fact that, in the presence of a weight, there is no direct analogue of (1.34) so we are forced to adopt an alternative approach.

M-III. Morreyfying other scales. There is also progress in the "Morreyfication" of Besov and Triebel–Lizorkin spaces. Succinctly, the usual expressions defining the Besov and Triebel–Lizorkin norms contain an L^p norm, which is now replaced by a Morrey norm as in (1.7). We refer to [135, 128, 152] wherein an atomic and molecular theory is developed for these "Morreyfied" scales of spaces. For Orlicz spaces of Morrey flavor, see [111, 134, 52, 68, 57], and [131, Section 13] as well as the references therein. The notion of a Lorentz space with a Morrey character appears in [1, 3], and has been considered more in depth in [84, 118, 7]. For more recent results concerning these spaces, the reader is referred to [55, 147, 15, 56]. There are even instances of blending Hardy spaces with Morrey spaces (cf. [65]), and more recently a blend of Herz and Morrey spaces (cf. [116] and [130, Section 1.3.2]).

In this monograph, we are concerned with Muckenhoupt-weighted Morrey spaces defined on closed Ahlfors regular sets (cf. Definition 2.2). Let $\Sigma \subseteq \mathbb{R}^n$ be such a set with "surface measure" $\sigma := \mathcal{H}^{n-1}\lfloor\Sigma$. Fix an integrability exponent $p \in (1,\infty)$, and some threshold $\lambda \in [0,\infty)$. Also, fix a Muckenhoupt weight w in $A_p(\Sigma,\sigma)$, a class defined by imposing a size condition analogously to (1.13) (cf. (2.47)). The Muckenhoupt-weighted Morrey space $M^{p,\lambda}(\Sigma, w)$ is then introduced as the space of all σ-measurable functions $f : \Sigma \to \mathbb{R}$ for which

$$\|f\|_{M^{p,\lambda}(\Sigma,w)} := \sup_{\substack{x\in\Sigma \\ r\in(0,2\,\mathrm{diam}(\Sigma))}} \left(\frac{1}{r^\lambda}\int_{\Delta(x,r)} |f|^p w\,d\sigma\right)^{1/p} < +\infty. \tag{1.35}$$

Here and elsewhere, $\Delta(x,r) := B(x,r) \cap \Sigma$ is the "surface ball" centered at $x \in \Sigma$ and of radius $r \in (0,\infty)$. Muckenhoupt-weighted Morrey spaces can be seen as an extension of the classical, unweighted Morrey spaces by taking the weight to be constant. Indeed, (nonzero) constant weights belong to $A_p(\Sigma,\sigma)$ for any choice of $p \in (1,\infty)$ in view of the definition of the A_p characteristic (cf. (2.47)). In the unweighted case, by taking $\lambda = 0$, we readily obtain that $M^{p,0}(\Sigma,\sigma) = L^p(\Sigma,\sigma)$ for all $p \in (1,\infty)$, so that Lebesgue spaces fit within the scale of classical Morrey spaces as an endpoint. In this way, the threshold λ can be thought of as a measure of how far $M^{p,\lambda}(\Sigma,\sigma)$ is away from $L^p(\Sigma,\sigma)$. A natural question is how far from $L^p(\Sigma,\sigma)$ can we go? It turns out that the natural range for λ is the interval $(0, n-1)$, making it apparent that the range for λ is dependent solely on the dimension of the Euclidean ambient. See (4.4)–(4.9) for a more detailed discussion. When introducing a Muckenhoupt weight $w \in A_p(\Sigma, \sigma)$ for some $p \in (1, \infty)$, we also have $M^{p,0}(\Sigma, w) = L^p(\Sigma, w)$, so we can likewise think of λ as a measure of how far $M^{p,\lambda}(\Sigma, w)$ is from $L^p(\Sigma, w)$. We can then ask the same question: how far from $L^p(\Sigma, w)$ can we go? In principle, quite far. In fact, it makes sense in the definition to consider any $\lambda \in (0, \infty)$. However, we do not want to go so far that we start losing useful properties of $M^{p,\lambda}(\Sigma, w)$, e. g., the boundedness of the Hardy–Littlewood maximal operator and other operators of Calderón–Zygmund type, certain embedding results, etc. Throughout this work, we find that a natural range for the threshold λ is the interval

$$\begin{aligned}&(0, (n-1)/q'), \text{ where } q' := q/(q-1) \in (1, \infty) \text{ and } q \in (1, \infty)\\ &\text{is such that the weight } w \text{ belongs to } \mathrm{RH}_q(\Sigma, \sigma), \text{ the reverse}\\ &\text{Hölder class with exponent } q \text{ (cf. (2.63)).}\end{aligned} \tag{1.36}$$

Note that such a $q \in (1, \infty)$ is guaranteed by the classical theory of Muckenhoupt weights; cf. item (2) of Proposition 2.33. Thus, in the weighted case, the “good” range for λ is determined by not only the dimension of the Euclidean ambient, but also the reverse Hölder exponent associated with the weight. The more restrictive a reverse Hölder class that the weight belongs to (i. e., the larger q is), the larger the good range for λ. Indeed, taking w to be constant, say $w \equiv 1$ on Σ, we find $w \in \mathrm{RH}_q(\Sigma, \sigma)$ for all $q \in (1, \infty)$. Subsequently, we can take

$$\lambda \in \bigcup_{q \in (1,\infty)} (0, (n-1)/q') = (0, n-1), \tag{1.37}$$

which coincides with the good range for λ in the unweighted case, as expected.

Muckenhoupt weights have a rich history in analysis. By introducing the ubiquitous class of weights now bearing his name, G. Muckenhoupt sparked a revolution in harmonic analysis. Historically, if one wanted a weighted norm inequality, essentially, the original proof in the unweighted setting would have to be entirely redone, but now trying to accommodate the presence of a specific weight. As a glimpse into this era of mathematical development, we recall E. M. Stein’s 1957 result [141] in which, following work in the one-dimensional setting by Hardy and Littlewood [54] and Babenko [8], Stein initiated the study of weighted norm inequalities for higher-dimensional singular integral operators. Stein achieved boundedness results for singular integral operators of Calderón–Zygmund type on $L^p(\mathbb{R}^n, |\cdot|^\beta \mathcal{L}^n)$ with $n \in \mathbb{N}$, $p \in (1, \infty)$, and $-n/p < \beta < n/p$. Remarkably, though Muckenhoupt weights had not yet been discovered (and would not be for another 15 years in [109]), the permissible scale of powers deduced by Stein is the sharp range of powers ensuring that the weights belong to $A_p(\mathbb{R}^n, \mathcal{L}^n)$; cf. (1.41). It was not until 1985 with the work of J. García-Cuerva and J. Rubio de Francia in [43] that we see the first monograph systematically treating weighted norm inequalities.

There are a myriad of good reasons for focusing on Muckenhoupt weights. To highlight their relevance, bring in the Hardy–Littlewood maximal operator on Σ, which we denote by $\mathcal{M}_\Sigma$, whose action on each σ-measurable function $f : \Sigma \to \mathbb{R}$ is described according to

$$(\mathcal{M}_\Sigma f)(x) := \sup_{r>0} \fint_{\Delta(x,r)} |f|\,d\sigma \quad \text{at each } x \in \Sigma. \tag{1.38}$$

Consider a weight function w on Σ (i. e., $w : \Sigma \to [0,\infty]$ is σ-measurable and satisfies $0 < w < \infty$ at σ-a. e. point on Σ). Henceforth, identify the function w with the measure $w\sigma$, i. e., set $w(E) := \int_E w\,d\sigma$ for each σ-measurable set $E \subseteq \Sigma$. Finally, fix an arbitrary integrability exponent $p \in (1,\infty)$. A basic result (due to B. Muckenhoupt in the entire Euclidean setting; cf. [109]) then states that $\mathcal{M}_\Sigma$ is a bounded (sublinear) operator from the space $L^p(\Sigma, w)$ into itself if and only if $w \in A_p(\Sigma, \sigma)$. In view of the fact that, as noted earlier, corresponding to $\lambda = 0$ we have $M^{p,0}(\Sigma, w) = L^p(\Sigma, w)$, this points to the significance of Muckenhoupt weights in relation to Morrey spaces. A more precise result, proved in Proposition 5.2, states that for $p \in (1,\infty)$ and $\lambda \in [0, n-1)$, if for some arbitrary weight function $w : \Sigma \to [0,\infty]$, it happens that the Hardy–Littlewood maximal operator $\mathcal{M}_\Sigma$ is bounded from $M^{p,\lambda}(\Sigma, w)$ into itself, then necessarily the weight function w belongs to the Muckenhoupt class $A_{p+\frac{\lambda}{n-1}}(\Sigma, \sigma)$.

Another striking result involving Muckenhoupt weights with far reaching consequences is Rubio de Francia's extrapolation theorem, cf. [43], which itself (or at the very least the idea of it) is of great import to our work as it relates to boundedness properties of singular integral operators. Historically, singular integral operators, their existence and boundedness properties, were studied one at a time. It was not until the early 1950s in the work of A. P. Calderón and A. Zygmund (cf. [12]) that we start to see an entire class of singular integral operators treated all at once. Coupled with the development of Muckenhoupt weights, this led to a burgeoning in the field of singular integral operators and their applications to integral equations. Speaking on these historical roots, A. Lonseth notes in [74, p. 241]:

> By contrast with differential equations, which got off to a flying start with Isaac Newton's second law of motion (1687, in his Principia), integral equations arrived late; they made their first appearance, sporadically and even furtively, during the third and fourth decades of the nineteenth century. Sometimes they were seen only as alternative formulations of differential problems, sometimes not. General recognition of their independent importance came slowly. Even the name "integral equation" was not suggested until the late 1880s, and it was adopted only in the early 1900s. In the last 4 or 5 years of the nineteenth century, Vito Volterra and Ivar Fredholm succeeded in working out fundamental linear theories of the two types, which have since carried their names. Suddenly, integral equations blazed forth in the mathematical heavens, a supernova heralding the analysis of the twentieth century.

Classically, Rubio de Francia's extrapolation theorem takes estimates which hold on Muckenhoupt-weighted Lebesgue spaces for a *fixed* exponent $p_0 \in (1, \infty)$ but arbitrary weights $w \in A_{p_0}$ and turns them into estimates which hold on Muckenhoupt-weighted Lebesgue space corresponding to *any* integrability exponent $p \in (1, \infty)$ and *any* weight $w \in A_p$. Inspired by the work in [29] in the classical Euclidean setting $\mathbb{R}^n$, we jump-start this project by proving in Theorem 4.19 a similar result (now working in the class of closed Ahlfors regular sets), which serves as a portal between Muckenhoupt-weighted Lebesgue spaces and Muckenhoupt-weighted Morrey spaces. Intriguingly, the additional threshold λ in the definition of a Morrey space is placed in a suitable range, influenced by the exponent of a reverse Hölder class to which the Muckenhoupt weight belongs.

With this powerful tool in hand, we are able to transfer results which provide estimates at the level of Muckenhoupt-weighted Lebesgue spaces for singular integral operators of Calderón–Zygmund type (the likes of which have been recently developed in [79]) to estimates at the level of Muckenhoupt-weighted Morrey spaces, allowing us to develop a cohesive Calderón–Zygmund theory for this brand of Morrey space in the optimal setting of uniformly rectifiable sets. Of course, since $w \equiv 1$ is a Muckenhoupt weight, we actually recover many results of a Calderón–Zygmund theoretic nature already developed for classical Morrey spaces on Ahlfors regular sets in [86]. There has also been much work for singular integral operators on Morrey spaces of various types in the Euclidean setting, for instance, in [29, 30, 31, 48, 49, 50, 70, 129, 133]. In contrast, our work here aims to develop results for singular integral operators in a more general geometric environment to suit our needs later down the road when studying boundary value problems in rough domains. That being said, the work done here on singular integral operators both at the level of Muckenhoupt-weighted Morrey and Block spaces is of inherent interest.

The relevance of Muckenhoupt weights to PDE is also apparent when employing the method of harmonic measures to solve the classical Dirichlet problem for the Laplacian on Lebesgue spaces. This is concerned with finding a harmonic function u in a domain $\Omega \subseteq \mathbb{R}^n$ (an open, connected set), which is continuous up to and including the boundary of Ω, with the additional requirement that its restriction to $\partial\Omega$ is equal to some given arbitrary function (which, necessarily, is itself continuous on $\partial\Omega$). Specifically, given a domain $\Omega \subseteq \mathbb{R}^n$, this classical Dirichlet problem reads

$$\begin{cases} u \in \mathscr{C}^\infty(\Omega) \cap \mathscr{C}^0(\overline{\Omega}), \\ \Delta u = 0 \quad \text{in } \Omega, \\ u|_{\partial\Omega} = f \in \mathscr{C}^0(\partial\Omega). \end{cases} \tag{1.39}$$

Regarding the solvability of (1.39), the following fundamental result (cf. [69]) holds. Suppose Ω is a nontangentially accessible (NTA) domain, in the sense of D. Jerison and C. Kenig, and set $\sigma := \mathcal{H}^{n-1}\lfloor\partial\Omega$. Then the following conditions are equivalent:

(1) the harmonic measure ω for Ω is a Muckenhoupt weight on $\partial\Omega$, i. e., satisfies $\omega \in A_\infty(\partial\Omega,\sigma) := \bigcup_{1\le p<\infty} A_p(\partial\Omega,\sigma)$ (cf. (2.50)).
(2) There exist $p \in (1,\infty)$ and $C \in (0,\infty)$ such that if u is the solution to (1.39), then

$$\|\mathcal{N}_\kappa u\|_{L^p(\partial\Omega,\sigma)} \le C\|f\|_{L^p(\partial\Omega,\sigma)} \tag{1.40}$$

for some (or any) aperture parameter $\kappa \in (0,\infty)$.
(3) $\omega \ll \sigma$ (i. e., the harmonic measure ω is absolutely continuous with respect to the surface measure σ on $\partial\Omega$) and the Radon–Nikodym derivative $k := \frac{d\omega}{d\sigma}$, known as the *Poisson kernel* in this context, belongs to the reverse Hölder class $\mathrm{RH}_q(\partial\Omega,\sigma)$ (cf. (2.63)), where $q \in (1,\infty)$ is such that $\frac{1}{p}+\frac{1}{q}=1$ and p is as in item (2).

For a closely related result in the class of AR domains satisfying a 2-sided corkscrew condition, cf. [108, Theorem 9.2, pp. 59–60].

Another natural instance where Muckenhoupt weights appear in PDE is in the context of boundary value problems on polygonal domains in $\mathbb{R}^2$. Such boundary problems with boundary data in weighted Sobolev spaces have been considered (cf., e. g., [71] and the references therein) for weights of the form $|\cdot - x_0|^\alpha$ where x_0 is a corner of the polygonal domain, and α a parameter that models the power-like growth of solutions near the corner. The fact that power weights are Muckenhoupt weights for certain powers is encapsulated in the following fact (cf. [44, Proposition 1.5.9, p. 42]):

$$\begin{aligned}&\text{if } \Sigma \subseteq \mathbb{R}^n \text{ is a closed Ahlfors regular set, } \sigma := \mathcal{H}^{n-1}\lfloor\Sigma, \text{ and } p \in (1,\infty),\\ &\text{then the power weight } |\cdot - x_0|^\alpha \text{ belongs to } A_p(\Sigma,\sigma) \text{ if and only if}\\ &\alpha \in (1-n,(p-1)(n-1)).\end{aligned} \tag{1.41}$$

This naturally generalizes further to higher dimensions (cf. [32]):

$$\begin{aligned}&\text{if } \Sigma \subseteq \mathbb{R}^n \text{ is a closed Ahlfors regular set and } E \subseteq \Sigma \text{ is a “}d\text{-dimen-}\\ &\text{sional” set for some } d \in (0,n-1), \text{ i. e., there exists a Borel outer}\\ &\text{measure } \mu \text{ on } E \text{ for which } \mu(B(x,r)\cap E) \approx r^d \text{ uniformly in } x \in E,\\ &\text{then } \operatorname{dist}(\cdot,E)^\alpha \text{ belongs to } A_p(\Sigma,\sigma) \text{ if } \alpha \in (d+1-n,(n-1)(n-1-d)).\end{aligned} \tag{1.42}$$

Such a result is, e. g., useful when dealing with boundary value problems in polyhedra. As opposed to polygons, corners are no longer the only types of singularity on the boundary, as now there are edges to take into consideration. It is therefore natural to consider weights involving powers of the distance function to such lower-dimensional singular subsets of the boundary (cf. [82]), and (1.42) links this to Muckenhoupt weights.

Continue to assume that $\Sigma \subseteq \mathbb{R}^n$ is a closed Ahlfors regular set and, as before, abbreviate $\sigma := \mathcal{H}^{n-1}\lfloor\Sigma$. A category of weights closely related to Muckenhoupt classes are the reverse Hölder classes, denoted $\mathrm{RH}_q(\Sigma,\sigma)$ for $q \in (1,\infty)$. Specifically, a nonnegative function $w \in L^1_{\mathrm{loc}}(\Sigma,\sigma)$ is said to belongs to $\mathrm{RH}_q(\Sigma,\sigma)$ for some $q \in (1,\infty)$ provided there exists $C \in (0,\infty)$ such that

$$\left(\fint_{\Delta} w^q\,d\sigma\right)^{1/q} \leq C \fint_{\Delta} w\,d\sigma \quad \text{for each surface ball } \Delta \subseteq \Sigma. \tag{1.43}$$

In relation to this, we make two observations. First, it is part of the classical theory that any weight $w \in A_p(\Sigma,\sigma)$ with $p \in (1,\infty)$ belongs to a reverse Hölder class $\mathrm{RH}_q(\Sigma,\sigma)$ for *some* $q \in (1,\infty)$. Second, having fixed a reference point $x_0 \in \Sigma$, we claim that

$$|\cdot - x_0|^a \in \mathrm{RH}_q(\Sigma,\sigma) \quad \text{for each } q \in (1,\infty) \text{ and } a > (1-n)/q. \tag{1.44}$$

This comes down to showing that for each $q \in (1,\infty)$ and $a > (1-n)/q$ there exists a constant $C \in (0,\infty)$ such that for all $x \in \Sigma$ and $r \in (0, \operatorname{diam}\Sigma)$ we have

$$\left(\fint_{\Delta(x,r)} |y - x_0|^{aq}\,d\sigma(y)\right)^{1/q} \leq C \fint_{\Delta(x,r)} |y - x_0|^a\,d\sigma(y). \tag{1.45}$$

In turn, (1.45) is seen by analyzing three cases. First, in the scenario when $|x - x_0| < r/2$, we have $\Delta(x,r) \subseteq \Delta(x_0, 3r/2)$ so the left side above is bounded by

$$\begin{aligned}
&\left(\fint_{\Delta(x,r)} |y - x_0|^{aq}\,d\sigma(y)\right)^{1/q} \\
&\quad\leq C\left(r^{1-n} \int_{\Delta(x_0,3r/2)} |y - x_0|^{aq}\,d\sigma(y)\right)^{1/q} \\
&\quad= C\left(r^{1-n} \sum_{j=0}^{\infty} \int_{\Delta(x_0,3r/2^j)\setminus\Delta(x_0,3r/2^{j+1})} |y - x_0|^{aq}\,d\sigma(y)\right)^{1/q} \\
&\quad\leq C\left(r^{1-n} \sum_{j=0}^{\infty} (r/2^j)^{aq} (r/2^j)^{n-1}\right)^{1/q} \\
&\quad\leq Cr^a \left(\sum_{j=0}^{\infty} 2^{-jaq} 2^{-j(n-1)}\right)^{1/q} \leq Cr^a,
\end{aligned} \tag{1.46}$$

since the geometric series converges given that $aq + n - 1 > 0$. At the same time, when $|x - x_0| < r/2$ we have $\Delta(x_0, r/2) \subseteq \Delta(x,r)$ so the right side in (1.45) bounds

$$\begin{aligned}
\fint_{\Delta(x,r)} |y - x_0|^a\,d\sigma(y) &\geq cr^{1-n} \int_{\Delta(x_0,r/2)} |y - x_0|^a\,d\sigma(y) \\
&= cr^{1-n} \sum_{j=0}^{\infty} \int_{\Delta(x_0,r/2^j)\setminus\Delta(x_0,r/2^{j+1})} |y - x_0|^a\,d\sigma(y)
\end{aligned} \tag{1.47}$$

$$
\begin{aligned}
&\geq cr^{1-n}\sum_{j=0}^{\infty}(r/2^j)^a(r/2^j)^{n-1}\\
&= cr^a\sum_{j=0}^{\infty}2^{-ja}2^{-j(n-1)} = cr^a,
\end{aligned}
$$

since having $a+n-1>[aq+n-1]/q>0$ makes the geometric series converge. Thus, (1.45) holds whenever $|x-x_0|<r/2$. Second, in the case that $r/2\leq|x-x_0|\leq 2r$ we have $\Delta(x,r)\subseteq\Delta(x_0,3r)$ so the left side in (1.45) is bounded by

$$
\begin{aligned}
\Big(\fint_{\Delta(x,r)}|y-x_0|^{aq}\,d\sigma(y)\Big)^{1/q} &\leq C\Big(r^{1-n}\int_{\Delta(x_0,3r)}|y-x_0|^{aq}\,d\sigma(y)\Big)^{1/q}\\
&\leq Cr^a,
\end{aligned}
\tag{1.48}
$$

much as in (1.46). At the same time, when $r/2\leq|x-x_0|\leq 2r$ we have $|y-x_0|\approx r$ uniformly for $y\in\Delta(x,r/4)$ so the right side in (1.45) bounds

$$
\begin{aligned}
\fint_{\Delta(x,r)}|y-x_0|^{a}\,d\sigma(y) &\geq cr^{1-n}\int_{\Delta(x,r/4)}|y-x_0|^{a}\,d\sigma(y)\\
&\geq cr^{1-n}r^ar^{n-1}=r^a,
\end{aligned}
\tag{1.49}
$$

so (1.45) holds whenever $r/2\leq|x-x_0|\leq 2r$. Third, in the case when $|x-x_0|>2r$ we have $|y-x_0|\approx|x-x_0|$ uniformly for $y\in\Delta(x,r)$, so both sides in (1.45) are $\approx|x-x_0|^a$. As such, the desired estimate holds in this case as well, completing the proof of (1.44).

In closing, we note that the result in (1.44) is optimal since, generally speaking, the membership of any nonnegative function $w\in L^1_{\mathrm{loc}}(\Sigma,\sigma)$ to $\mathrm{RH}_q(\Sigma,\sigma)$ for some $q\in(1,\infty)$ implies $w\in L^q_{\mathrm{loc}}(\Sigma,\sigma)$.

The previous discussion shows that Muckenhoupt weights have a penchant for showing up in boundary value problems when rough geometry (permitting singularities) is involved. For a more specific example of this phenomena, let us illustrate how the use of conformal mappings permits one to transfer the job of solving the Dirichlet problem for the Laplacian, with boundary data prescribed in plain Lebesgue spaces, in the infinite planar sector

$$
\Omega_\theta := \Big\{(x,y)\in\mathbb{R}^2\setminus\{(0,0)\} : \frac{y}{\sqrt{x^2+y^2}}>\cos(\theta/2)\Big\},
\tag{1.50}
$$

(of a given aperture $\theta\in(0,2\pi)$), to the more tractable task of proving the solvability of the Dirichlet problem for the Laplacian in geometrically tamer setting of the upper half-plane, but now with allowing the boundary data to belong to a Muckenhoupt-weighted Lebesgue space. To elaborate, abbreviate $\sigma_\theta:=\mathcal{H}^1\lfloor\partial\Omega_\theta$, fix an aperture parameter $\kappa \in (0,\infty)$, and pick an integrability exponent $p \in (1,\infty)$. The L^p Dirichlet problem in Ω_θ for the Laplacian reads

$$\begin{cases} u \in \mathscr{C}^\infty(\Omega_\theta), \\ \Delta u = 0 \quad \text{in } \Omega_\theta, \\ \mathcal{N}_\kappa u \in L^p(\partial\Omega_\theta, \sigma_\theta), \\ u\big|_{\partial\Omega_\theta}^{\kappa-\text{n.t.}} = f \in L^p(\partial\Omega_\theta, \sigma_\theta). \end{cases} \tag{1.51}$$

The function

$$\Phi : \mathbb{C}_+ \to \Omega_\theta, \quad \Phi(z) := i\big[(-i)z\big]^{\theta/\pi} \quad \text{for each } z \in \mathbb{C}_+ \tag{1.52}$$

is a conformal mapping of the upper half-plane $\mathbb{R}^2_+ \equiv \mathbb{C}_+$ onto the sector Ω_θ, which extends to a homeomorphism $\Phi : \overline{\mathbb{R}^2_+} \to \overline{\Omega_\theta}$ with $\Phi(\partial\mathbb{R}^2_+) = \partial\Omega_\theta$. Thanks to (1.41) we see that

$$\begin{aligned} &\big|\Phi'(x)\big| = (\theta/\pi)|x|^{(\theta/\pi)-1} \in A_p(\mathbb{R}, \mathcal{L}^1) \\ &\quad \text{if and only if} \quad p > \theta/\pi. \end{aligned} \tag{1.53}$$

The relevance of this membership to the Muckenhoupt class is made clear by work in [90], which shows that under the transformation

$$u \mapsto \widetilde{u} := u \circ \Phi, \tag{1.54}$$

the L^p Dirichlet problem in (1.51) is equivalent to the following weighted Dirichlet Problem in the upper half-plane:

$$\begin{cases} \widetilde{u} \in \mathscr{C}^\infty(\mathbb{R}^2_+), \\ \Delta \widetilde{u} = 0 \quad \text{in } \mathbb{R}^2_+, \\ \mathcal{N}_\kappa \widetilde{u} \in L^p(\mathbb{R}, |\Phi'(x)|\,dx), \\ \widetilde{u}\big|_{\partial\mathbb{R}^2_+}^{\kappa-\text{n.t.}} = \widetilde{f} \in L^p(\mathbb{R}, |\Phi'(x)|\,dx). \end{cases} \tag{1.55}$$

In turn, this boundary value problem is known (cf. [79]) to be well posed whenever $|\Phi'| \in A_p(\mathbb{R}, \mathcal{L}^1)$. Thus, recalling (1.53), we see that the boundary value problem in (1.51) is well posed whenever

$$\theta \in (0, 2\pi) \quad \text{and} \quad p \in \big(\max\{1, \theta/\pi\}, \infty\big). \tag{1.56}$$

This is yet another instance where the theory of Muckenhoupt weights plays a basic role in the solution of a problem, even though weights are not present in the original formulation of said problem.

The results in (1.41) and (1.42) offer but a few examples of Muckenhoupt weights, but the class of Muckenhoupt weights is quite rich. One way to see this is to recall that A_1 weights can be constructed using the Hardy–Littlewood maximal operator, $\mathcal{M}_\Sigma$. Namely, given any function f belonging to $L^1_{\mathrm{loc}}(\Sigma,\sigma)$, not identically zero, and satisfying $\mathcal{M}_\Sigma f < +\infty$ at σ-a. e. point on Σ, it follows that

$$(\mathcal{M}_\Sigma f)^\theta \in A_1(\Sigma,\sigma) \quad \text{for each } \theta \in (0,1). \tag{1.57}$$

Because the Hardy–Littlewood maximal operator is bounded on $L^p(\Sigma,\sigma)$ for $p \in (1,\infty)$, any function belonging to a Lebesgue space with integrability exponent $p \in (1,\infty)$ can be used to create a Muckenhoupt weight. In fact, in this work we show that the Hardy–Littlewood maximal operator is bounded on Morrey spaces and their pre-duals, Block spaces, which give many more examples of functions, which can be used to construct A_1 weights from the Hardy–Littlewood maximal operator. Going further, for any $p \in (1,\infty)$ one can construct Muckenhoupt weights belonging to $A_p(\Sigma,\sigma)$ from any two weights $w_0, w_1 \in A_1(\Sigma,\sigma)$, namely

$$w_0 \cdot w_1^{1-p} \in A_p(\Sigma,\sigma) \quad \text{with } [w_0 \cdot w_1^{1-p}]_{A_p} \le [w_0]_{A_1} \cdot [w_1]_{A_1}^{p-1}. \tag{1.58}$$

This fact along with (1.57) points to the abundance of Muckenhoupt weights. In fact, the class of Muckenhoupt weights is a log-convex cone, which contains all positive constant functions. That is, given any $w_0, w_1 \in A_p(\Sigma,\sigma)$ with $p \in [1,\infty)$, and $\lambda \in (0,\infty)$, it follows that

$$\begin{aligned} &1 \in A_p(\Sigma,\sigma), \quad w_0 + w_1 \in A_p(\Sigma,\sigma), \\ &\text{and} \quad \lambda w_0^\theta \cdot w_1^{1-\theta} \in A_p(\Sigma,\sigma) \quad \text{for each } \theta \in [0,1]. \end{aligned} \tag{1.59}$$

This gives yet another way to create other Muckenhoupt weights from previously known Muckenhoupt weights. There are other ways still to build Muckenhoupt weights from other Muckenhoupt weights, which we can see by combining some of the facts we have already seen. For instance, using the conclusions in (1.57) and (1.58) and identifying both $w_0 \equiv 1$ and $w_1 \equiv (\mathcal{M}_\Sigma f)^\theta$, we can see for any function $f \in L^1_{\mathrm{loc}}(\Sigma,\sigma)$ not identically zero with $\mathcal{M}_\Sigma f < +\infty$ at σ-a. e. point on Σ, that $(\mathcal{M}_\Sigma f)^{\theta(1-p)}$ belongs to $A_p(\Sigma,\sigma)$ for any $\theta \in (0,1)$ and $p \in (1,\infty)$ arbitrary. More generally, using (1.57) and (1.58) again, given any two functions $f_1, f_2 \in L^1_{\mathrm{loc}}(\Sigma,\sigma)$ both not identically zero and with $\mathcal{M}_\Sigma f_1 < +\infty$ and $\mathcal{M}_\Sigma f_2 < +\infty$ at σ-a. e. point on Σ, and any $\theta_1, \theta_2 \in (0,1)$, one has $(\mathcal{M}_\Sigma f)^{\theta_1}(\mathcal{M}_\Sigma f_2)^{\theta_2(p-1)} \in A_p(\Sigma,\sigma)$. As a final interesting property about Muckenhoupt weights that can used to obtain more such weights, there is an interpolation result (cf. [86, Lemma 7.7.4, pp. 645–646]) of the following flavor:

$$\begin{aligned} &\text{Let } p_0, p_1 \in (1,\infty),\ w_0 \in A_{p_0}(\Sigma,\sigma), \text{ and } w_1 \in A_{p_1}(\Sigma,\sigma). \text{ For each } \theta \in [0,1], \\ &\text{define } p := \big(\tfrac{1-\theta}{p_0} + \tfrac{\theta}{p_1}\big)^{-1} \in (1,\infty). \text{ Then the weight } w := \big(w_0^{\frac{1-\theta}{p_0}} \cdot w_1^{\frac{\theta}{p_1}}\big)^p \\ &\text{belongs to } A_p(\Sigma,\sigma). \end{aligned} \tag{1.60}$$

To review, Muckenhoupt weights are abundant, have many interesting properties, work well in relation to a wealth of topics in harmonic analysis and partial differential equations, and many natural weights that arise in practice are actually Muckenhoupt weights. This justifies our focus on *Muckenhoupt-weighted* Morrey spaces in the broader context of boundary value problems.

When working with boundary value problems, there are three things to consider. Namely, the differential equation, the functional analytic setting, and the geometry of the domain. We have discussed several aspects of the functional analytic setting of interest for us here, and we now turn our attention to geometry. In the classical case, i. e., (1.39), it is known precisely how irregular a boundary can become before (1.39) fails to be solvable. Concretely, (1.39) is solvable if and only if every point in $\partial\Omega$ satisfies the Wiener criterion; equivalently, all boundary points are what are known as regular points (cf., e. g., [46]). However, matters are not so clear-cut when relaxing the requirement of continuity up to and including the boundary.

Consider the Dirichlet problem for the Laplacian $\Delta := \sum_{j=1}^{n} \partial_j^2$ in some (nonempty) open set $\Omega \subseteq \mathbb{R}^n$ with the corresponding "surface measure" denoted $\sigma := \mathcal{H}^{n-1}\lfloor\partial\Omega$, with boundary data arbitrarily prescribed in a Lebesgue space with integrability exponent $p \in (1, \infty)$:

$$\begin{cases} u \in \mathscr{C}^{\infty}(\Omega), \\ \Delta u = 0 \quad \text{in } \Omega, \\ \mathcal{N}_{\kappa} u \in L^p(\partial\Omega, \sigma), \\ u\big|_{\partial\Omega}^{\kappa-\text{n.t.}} = f \in L^p(\partial\Omega, \sigma), \end{cases} \tag{1.61}$$

where $\kappa > 0$ is a fixed parameter (regulating the aperture of the nontangential approach regions used to define the nontangential maximal operator $\mathcal{N}_{\kappa} u$ as well as the nontangential boundary trace $u\big|_{\partial\Omega}^{\kappa-\text{n.t.}}$).

It is of interest to know what is the largest class of domains for which (1.61) is well posed. Colloquially, how rough can $\partial\Omega$ become, while still allowing (1.61) to be well posed, or at least admit a solution? In 1977, Bjorn Dahlberg [21] used harmonic measure techniques to prove that when Ω is a bounded Lipschitz domain, then (1.61) is well posed for $p \in (2-\varepsilon_\Omega, \infty)$, where $\varepsilon_\Omega \in (0, 1)$ depends on the geometry of Ω (one should think of ε_Ω as an embodiment of the worst type of "corner singularity" that Ω exhibits). Not long after, Dahlberg was also able to show (cf. [22]) that if Ω is a bounded $\mathscr{C}^1$ domain then (1.61) is well posed for all $p \in (1, \infty)$. While effective in the case of Dirichlet boundary conditions and scalar operators with real coefficients, the method of harmonic measures is not directly relevant in the context of the Neumann and regularity problems, operators with complex coefficients, or systems of PDEs, so new techniques are needed in those situations. In 1980, D. Jerison and C. Kenig [62] recover Dahlberg's 1977 result concerning the well posedness of the L^p Dirichlet problem in Lipschitz domains using an integral identity involving the Radon–Nikodym derivative $k = \frac{d\omega}{d\sigma}$ of the harmonic measure ω with respect to the surface measure σ. Shortly thereafter in 1981 [63], Jerison and Kenig employ a Rellich type identity in order to extend Dahlberg's work so as to accommodate the case of the L^2 Neumann and L_1^2 regularity boundary value problems for the Laplacian in Lipschitz domains.

In 1979, E. Fabes, M. Jodeit, and N. Rivière employed the method of singular integral operators of boundary layer type in [37] to prove the well posedness of (1.61) when Ω is a bounded $\mathscr{C}^1$ domain and $p \in (1, \infty)$ is arbitrary. Under the same geometric assumptions and the same range of p, they also proved the well posedness of the Neumann problem, and the so-called regularity problem. The breakthrough of R. Coifman, A. McIntosh, and Y. Meyer, who in 1982 completed A. P. Calderón's program aimed at establishing boundedness properties of singular integral operators on Lipschitz surfaces (cf. [18]), opened the door for employing boundary integral methods in Lipschitz domains.

Following important work by G. Verchota in 1984 (who first used Jerison and Kenig's Rellich identity in connection with boundary layer potentials [149]), the theory developed by Fabes, Jodeit, and Rivière was subsequently augmented in 1987 by B. Dahlberg and C. Kenig, who used singular integrals of boundary layer type to not only recover Dahlberg's earlier results pertaining to the well posedness of (1.61) when Ω is a bounded Lipschitz domain, but also to prove the well posedness of the Neumann problem and the regularity problem when Ω is an arbitrary bounded Lipschitz domain and for sharp ranges of p's (cf. [23]). The theory of boundary value problems in Lipschitz domains is incredibly intricate and nuanced, with many other operators and boundary conditions successfully treated in such a geometric setting, including the case of the Lamé system of elasticity by B. Dahlberg, C. Kenig, and G. Verchota [25], the Stokes system of hydrodynamics by E. Fabes, C. Kenig, and G. Verchota [38], as well as M. Mitrea and M. Taylor [103] and M. Mitrea and M. Wright [104], the heat operator by R. Brown [11], the parabolic versions of the Lamé and Stokes systems by Z. Shen [136], the Maxwell system of electromagnetism by M. Mitrea and B. Jawerth [96, 61], D. Mitrea, M. Mitrea, and J. Pipher [93], and later in the manifold setting by M. Mitrea and D. Mitrea [92], the Laplace–Beltrami operator on Lipschitz subdomains of Riemannian manifolds by M. Mitrea and M. Taylor [97, 98, 99, 100, 101, 102], the Transmission problem for the Laplacian by L. Escauriaza and M. Mitrea [35], the bi-Laplacian in L^2 by B. Dahlberg, C. Kenig, and G. Verchota in [24] and in L^p by J. Pipher and G. Verchota [115], the polyharmonic operator by Verchota [150], higher-order systems by I. Mitrea and M. Mitrea [95], the Poisson problem for the Laplacian with Dirichlet boundary conditions by D. Jerison and C. Kenig [64] and with Neumann boundary conditions by E. Fabes, O. Mendez, and M. Mitrea [39], the Poisson problem for higher-order systems by V. Maz'ya, M. Mitrea, and T. Shaposhnikova [81], the mixed problem by I. Mitrea and M. Mitrea [94], the Hodge Laplacian by D. Mitrea, M. Mitrea, and M. Taylor [91], among others. See, e. g., C. Kenig's book [69] for additional information and references.

Subsequently, efforts have been focused on the issue of proving boundedness results for singular integral operators in geometric settings more general than Lipschitz surfaces. The breakthrough came in the early 1990s when G. David and S. Semmes identified the optimal geometric context in which singular integral operators of convolution type are bounded on Lebesgue spaces. This involves the concept of uniformly rectifiable set (cf. [28]), meaning that at all scales and locations the set in question contains big pieces of (images of) Lipschitz surfaces, in a uniform fashion. Simply put, the largest class of sets for which there exists a robust Calderón–Zygmund theory are those that are uniformly rectifiable. David and Semmes thus established a two-way street between analysis and geometry, something notoriously elusive. Such progress once again brought into focus the method of singular integrals of boundary layer type as a means of solving boundary value problems in very general domains.

Calderón–Zygmund theory is useful for methods involving singular integrals of boundary layer type, because those operators have integral kernels manufactured by taking the conormal derivative of the fundamental solution of the operator (or system of PDEs) in question, which give them the right smoothness, homogeneity, and parity, so that said theory applies. We consider layer potentials on uniformly rectifiable domains, an area which has seen significant development through the works undertaken in [86] and [79]. In fact, thanks to such progress, there now exists a Calderón–Zygmund theory on Muckenhoupt-weighted Lebesgue spaces in [79], which is an ideal launching point from which to develop a Calderón–Zygmund theory at the level of Muckenhoupt-weighted Morrey spaces. Here is where our extrapolation result in Theorem 4.19 plays a decisive role. Placing the work undertaken here in a broader context, we have been motivated by the comments of C. Kenig in his 1991 CBMS book [69] who, at the bottom of page 116, writes "*it would be of interest to study the regularity and Neumann problems on* [*classes of* UR *domains*]." As such, our work provides a partial answer to the question posited by Kenig in [69, problem 3.2.2., p. 117], which asks

> if $\Omega \subseteq \mathbb{R}^n$ is a NTA domain such that $\partial\Omega$ is Ahlfors regular, does there exist $p > 1$ such that the Neumann and regularity problems for the Laplacian are solvable? Are layer potentials invertible on $L^p(\partial\Omega, \mathcal{H}^{n-1}\lfloor\partial\Omega)$ and $L^p_1(\partial\Omega, \mathcal{H}^{n-1}\lfloor\partial\Omega)$? (1.62)

This question is actually raised again by T. Toro in [146, question 2.2, p. 1491]. While we work in Muckenhoupt-weighted Morrey spaces instead of Lebesgue spaces, the nature of the question is very much relevant to what we are trying to achieve. Such a question is heavily influenced by both geometry and analysis. The Calderón–Zygmund theory developed in [79] contains a key result, which goes beyond the usual boundedness of singular integral operators of boundary layer type acting on Muckenhoupt-weighted Lebesgue spaces, by producing a bound which for certain classes of singular integral operators is sensitive to flatness: the flatter the underlying set, the smaller the bound on the operator norm in question. We shall call these estimates, which bring in the geometry of the domain in such a specific fashion, "geometric estimates." Fundamentally important for the present work, we are able to extrapolate this result to our functional analytic setting of interest, and produce geometric estimates for certain classes of singular integral operators acting on Muckenhoupt-weighted Morrey spaces.

To elaborate on this crucial result further, fix a uniformly rectifiable domain $\Omega \subseteq \mathbb{R}^n$ with surface measure $\sigma := \mathcal{H}^{n-1}$, along with two integrability exponents $p, q \in (1, \infty)$, a Muckenhoupt weight $w \in A_p(\partial\Omega, \sigma)$ belonging to the reverse Hölder class $\mathrm{RH}_q(\partial\Omega, \sigma)$, and a threshold $\lambda \in (0, (n-1)/q')$, where $q' := q/(q-1)$. Finally, pick an aperture parameter $\kappa > 0$ and consider the Dirichlet problem for the Laplacian with boundary data in $M^{p,\lambda}(\partial\Omega, w)$, stated as

$$\begin{cases} u \in \mathscr{C}^\infty(\Omega), \\ \Delta u = 0 \quad \text{in } \Omega, \\ \mathcal{N}_\kappa u \in M^{p,\lambda}(\partial\Omega, w), \\ u\big|_{\partial\Omega}^{\kappa-\mathrm{n.t.}} = f \in M^{p,\lambda}(\partial\Omega, w). \end{cases} \tag{1.63}$$

Recall that the boundary-to-domain harmonic double layer potential operator acts on any function $g \in L^1(\partial\Omega, \frac{d\sigma(x)}{1+|x|^{n-1}})$ according to

$$(\mathcal{D}_\Delta g)(x) := \frac{1}{\omega_{n-1}} \int_{\partial\Omega} \frac{\langle \nu(y), y - x\rangle}{|x-y|^n} g(y)\, d\sigma(y) \quad \text{for all } x \in \Omega, \tag{1.64}$$

where ν is the geometric measure theoretic outward unit normal to Ω, and ω_{n-1} is the surface area of the unit sphere in $\mathbb{R}^n$. The double layer is a good place to start for the solvability of (1.63), because we know that for any g as above,

$$\mathcal{D}_\Delta g \in \mathscr{C}^\infty(\Omega) \quad \text{and} \quad \Delta(\mathcal{D}_\Delta g) = 0 \quad \text{in } \Omega. \tag{1.65}$$

Moreover, $\mathcal{D}_\Delta$ is an integral operator covered by Theorem 5.7, which means it satisfies nontangential estimates that put $\mathcal{N}_\kappa(\mathcal{D}_\Delta g)$ in $M^{p,\lambda}(\partial\Omega, w)$ whenever g originally belongs to $M^{p,\lambda}(\partial\Omega, w)$. In addition, work in [86] identifies the nontangential boundary trace of $\mathcal{D}_\Delta g$ as

$$(\mathcal{D}g)\big|_{\partial\Omega}^{\kappa-\mathrm{n.t.}} = \Big(\frac{1}{2}I + K_\Delta\Big)g \quad \text{at } \sigma\text{-a. e. point on } \partial\Omega, \tag{1.66}$$

where I is the identity on $M^{p,\lambda}(\partial\Omega, w)$, and K_Δ is the boundary-to-boundary harmonic double layer potential, acting on each $g \in L^1(\partial\Omega, \frac{\sigma(x)}{1+|x|^{n-1}})$ according to

$$(K_\Delta g)(x) := \lim_{\varepsilon\to 0^+} \frac{1}{\omega_{n-1}} \int_{\substack{y\in\partial\Omega \\ |x-y|>\varepsilon}} \frac{\langle \nu(y), y - x\rangle}{|x-y|^n} g(y)\, d\sigma(y) \tag{1.67}$$

at σ-a. e. point $x \in \partial\Omega$. Thanks to these inherent properties of the harmonic double layer, solving the problem formulated in (1.63) reduces to the task of determining a function

$g \in M^{p,\lambda}(\partial\Omega, w)$ such that $(\frac{1}{2}I + K_\Delta)g = f$. This brings into focus the issue of whether the operator $\frac{1}{2}I + K_\Delta$ is invertible on the Muckenhoupt-weighted Morrey space $M^{p,\lambda}(\partial\Omega, w)$. Inspired by the Calderón–Zygmund theory developed in [86] for Muckenhoupt-weighted Lebesgue spaces, we show that for each $m \in \mathbb{N}$ there exists some constant $C_m \in (0,\infty)$ such that

$$\|K_\Delta\|_{M^{p,\lambda}(\partial\Omega,w)\to M^{p,\lambda}(\partial\Omega,w)} \leq C_m \|\nu\|_{\mathrm{BMO}(\partial\Omega,\sigma)}^{\langle m\rangle}, \tag{1.68}$$

where the notation $t^{\langle m\rangle}$ for $t \geq 0$ is detailed in Definition 5.9 (essentially, it stands for the m-fold logarithmic multiplicative perturbation of t itself). Importantly, $t^{\langle m\rangle} \to 0$ as $t \to 0^+$. Notably, (1.68) shows that if $\|\nu\|_{\mathrm{BMO}(\partial\Omega,\sigma)}$ is small, then so is the operator norm of K_Δ on $M^{p,\lambda}(\partial\Omega, w)$. Since the latter space is Banach (cf. Proposition 4.5), we may then invert the operator $\frac{1}{2}I + K_\Delta$ via a Neumann series (cf. Lemma 11.2) provided the operator norm of K_Δ on $M^{p,\lambda}(\partial\Omega, w)$ is strictly smaller than $\frac{1}{2}$.

The geometric estimate in (1.68) is the linchpin of this work. Such an estimate brings into play all three guiding pillars of this monograph as outlined at the beginning of this Introduction. It is a beautiful result as it is an instance of geometry subtly controlling analysis. Specifically, (1.68) shows that the operator norm of the boundary-to-boundary harmonic double layer potential operator in $M^{p,\lambda}(\partial\Omega, w)$, an analytic quantity, is controlled by the BMO seminorm of the measure theoretic outward unit normal to Ω, a purely geometric quantity, which is sensitive to flatness. There is a clear interplay between functional analysis, geometry, and partial differential equations, perfectly encapsulated in the kind of estimate given in (1.68). Indeed, the functional analytic environment needs to support a robust Calderón–Zygmund theory, the geometry of the domain under consideration needs to lend itself to a small BMO seminorm, and the PDE should lead to a double layer potential whose integral kernel is of the right type to be covered by the Calderón–Zygmund theory that produces these kinds of small-norm estimates.

The estimate in (1.68) leads us to consider a class of open sets $\Omega \subseteq \mathbb{R}^n$, which are sufficiently flat in the sense that the bounded mean oscillation of the outward unit normal to Ω is small. These are the so-called δ-AR domains (cf. Definition 2.37), where $\delta \in (0,1)$ is a small parameter that controls the BMO seminorm of the geometric measure theoretic outward unit normal to Ω. Specifically, given $\delta \in (0,1)$, following [79] we say $\Omega \subseteq \mathbb{R}^n$ is a δ-AR domain provided Ω is an Ahlfors regular domain (cf. Definition 2.5) whose geometric measure theoretic outward unit normal ν satisfies

$$\|\nu\|_{\mathrm{BMO}(\partial\Omega,\sigma)} < \delta. \tag{1.69}$$

The requirement that δ belongs to the interval $(0,1)$ is an artifact of the observation that $0 \leq \|\nu\|_{\mathrm{BMO}(\partial\Omega,\sigma)} \leq 1$ for any AR domain $\Omega \subseteq \mathbb{R}^n$. While the presence of a single (inner or outer) cusp forces $\|\nu\|_{\mathrm{BMO}(\partial\Omega,\sigma)} = 1$, the class of δ-AR domains is quite an inclusive environment, which even goes beyond the class of domains of upper-graph type (we explore this in more detail later). Heuristically, δ-AR domains (for δ small) are the geometric measure theoretically sharp version of the class of domains lying above the graphs of real-valued Lipschitz functions with small Lipschitz constants.

The simplest example of a δ-AR domain is the upper half-space $\mathbb{R}^n_+$, which happens to be a δ-AR domain for any choice of $\delta \in (0,1)$. This follows from the simple observation that the outward unit normal to $\mathbb{R}^n_+$ is constant, and hence $\|\nu\|_{\mathrm{BMO}(\partial\Omega,\sigma)} = 0$, which is of course less than any $\delta \in (0,1)$. It has also been noted in [79] that, given any $\delta \in (0,1)$, an infinite cone $C_\theta := \{x \in \mathbb{R}^n \setminus \{0\} : \frac{x_n}{|x|} > \cos(\theta/2)\}$ with aperture angle $\theta \in (0,2\pi)$ is a δ-AR domain provided $|\cos(\theta/2)| < \delta$, i. e., θ is sufficiently close to π. Thus, in this case, the "flatness" condition quantifies how close the cone is to becoming a half-space.

Next, upper-graph Lipschitz domains with small enough Lipschitz constants are δ-AR domains for a given $\delta \in (0,1)$. Specifically, starting with a choice of $\delta \in (0,1)$, an upper-graph Lipschitz domains with Lipschitz constant less than $2^{-3/2}\delta$, is a δ-AR domain (cf. [79]). Even though they are gently sloped, such domains can develop as tall peaks and as deep valleys as one wants; see Figure 1.1 below.

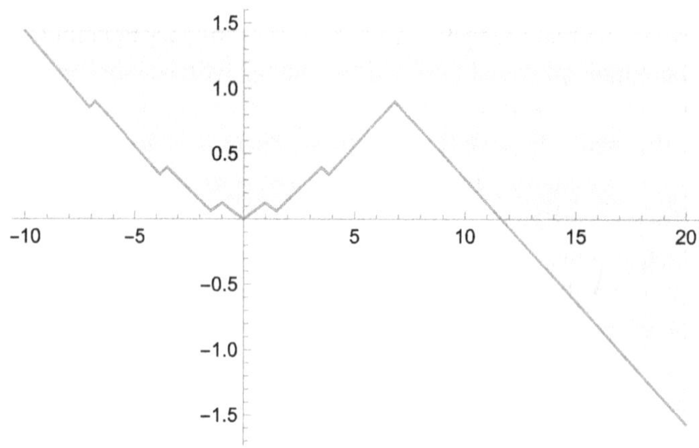

Figure 1.1: A prototype of an upper-graph Lipschitz domain; this is a δ-AR domain if the Lipschitz constant is $\leq 2^{-3/2}\delta$.

It was mentioned earlier that the class of δ-AR domains goes beyond the class of upper-graph domains. As one striking instance of this, the class of δ-AR domains can accommodate logarithmic spirals provided the spiral branches chase one another in a sufficiently gentle fashion; see Figure 1.2 below.

Having considered the functional analytic context and the geometric nature of our boundary value problem, we now turn our attention to partial differential equations and the types of differential operators/systems we can consider. As a first step, we recognize that the method of layer potentials is robust enough to consider weakly elliptic, second-order, homogeneous, constant complex coefficient $M \times M$ systems in $\mathbb{R}^n$,

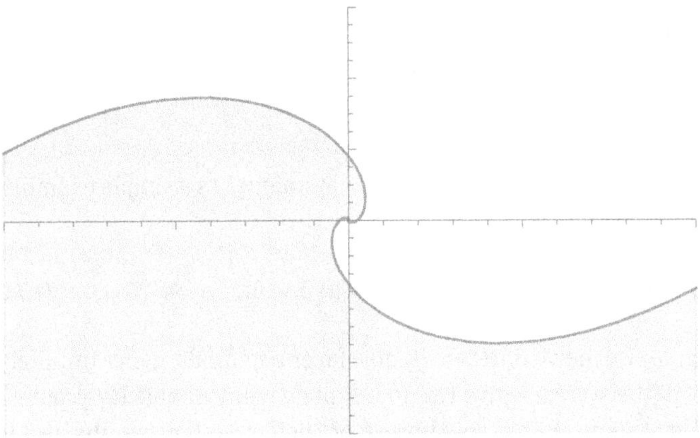

Figure 1.2: A δ-AR domain not of upper-graph type. Indeed, the spiral point at the origin does not past the vertical line test, no matter how one rotates the coordinate axes.

with $M \in \mathbb{N}$ arbitrary. The weak ellipticity assumption ensures the existence of a well-behaved (matrix-valued) fundamental solution $E = (E_{\alpha\beta})_{1\leq\alpha,\beta\leq M}$ of L. In turn, this is used to construct boundary layer potential operators, which enjoy vastly similar properties to $\mathcal{D}_\Delta$ and K_Δ associated with the Laplacian as in (1.64) and (1.67), with the possible exception of the geometric estimate recorded in (1.68).

It turns out that such a geometric estimate is unique to integral operators possessing what has been termed in [86] as a *chord-dot-normal* structure for their kernels. More specifically, given an Ahlfors regular domain $\Omega \subseteq \mathbb{R}^n$ with geometric measure theoretic outward unit normal ν and "surface measure" $\sigma := \mathcal{H}^{n-1}\lfloor\partial\Omega$, call a singular integral operator T on $\partial\Omega$ of chord-dot-normal type if its integral kernel is of the form $\langle x-y, \nu(y)\rangle k(x-y)$ for some sufficiently smooth function k that is even and positive homogeneous of degree $-n$ in $\mathbb{R}^n \setminus \{0\}$. For such chord-dot-normal integral operators we are able to show (again inspired and guided by the work in [86, 79]) that for each $m \in \mathbb{N}$ there exists a constant $C_m \in (0, \infty)$ such that (cf. Definition 5.9)

$$\|T\|_{M^{p,\lambda}(\partial\Omega,w)\to M^{p,\lambda}(\partial\Omega,w)} \leq C\|\nu\|^{\langle m\rangle}_{\mathrm{BMO}(\partial\Omega,\sigma)}. \tag{1.70}$$

The question then becomes, when does a system of partial differential equations have a double layer potential operator of chord-dot-normal type? One basic example that we have already seen is the Laplacian, $\Delta := \sum_{j=1}^n \partial_j^2$, whose boundary-to-boundary double layer, written in (1.67), clearly exhibits such a chord-dot-normal structure.

The quest for addressing this question starts from the realization that the double layer potential operator for a given system depends heavily on the specific manner one chooses to write the system in question. A second-order, homogeneous $M \times M$ system L with constant complex coefficients in $\mathbb{R}^n$ can be written according to

$$L = (a_{jk}^{\alpha\beta}\partial_j\partial_k)_{1\leq\alpha,\beta\leq M} \tag{1.71}$$

(with summation over repeated indices), where

$$A := (a_{jk}^{\alpha\beta})_{\substack{1\leq\alpha,\beta\leq M\\ 1\leq j,k\leq n}} \in \mathbb{C}^{M^2\cdot n^2} \tag{1.72}$$

is referred to as a coefficient tensor for L. Such an A is not unique. As a simple example, the Laplacian in $\mathbb{R}^2$ may be written as

$$\Delta = \partial_1^2 + \partial_2^2 \quad \text{or} \quad \Delta = \partial_1^2 + i\partial_1\partial_2 - i\partial_2\partial_1 + \partial_2^2. \tag{1.73}$$

Two different writings give rise to two different double layer potentials, hence infinitely many different writings of the system L give rise to infinitely many double layer potentials associated with said system. Within this library of coefficient tensors, the aim is then to identify those (if any) which yield a double layer potential operator exhibiting a chord-dot-normal structure. Such a coefficient tensor, introduced and characterized in [86], is referred to as being a *distinguished* coefficient tensor for the system L. The set of all distinguished coefficient tensors for a system L is denoted by $\mathfrak{A}_L^{\mathrm{dis}}$. Hence, demanding that $\mathfrak{A}_L^{\mathrm{dis}} \neq \varnothing$ means that *the system L may be written as in* (1.71) *for a choice of a coefficient tensor A as in* (1.72), *which gives rise to a double layer potential operator of chord-dot-normal type.* Following work in [86], we document the characterization of these distinguished coefficient tensors in Chapter 3.

Since the norm of a principal-value double layer potential operator of chord-dot-normal type gives rise to geometric estimates of similar flavor as in (1.68), a key ingredient in solving the boundary integral equation to which the Dirichlet problem is reduced by the method of boundary layers (cf. (1.66)), the above discussion begs the question: what kinds of weakly elliptic systems fit under the scope of this theory by actually possessing a distinguished coefficient tensor?

Recent progress in this regard has been made in [86]. For instance, it has been shown that

> every scalar, second-order, homogeneous, constant complex coefficient, strongly elliptic operator in $\mathbb{R}^n$ with $n \geq 2$ has a distinguished coefficient tensor. (1.74)

Recall that a scalar, second-order, homogeneous, constant complex coefficient operator $L := \sum_{j,k=1}^n a_{jk}\partial_j\partial_k$ with $a_{jk} \in \mathbb{C}$ for each $j,k \in \{1,\dots,n\}$ is called strongly elliptic (in the sense of Legendre–Hadamard) provided there exists some $c \in (0,\infty)$ such that

$$\mathrm{Re}\left[\sum_{j,k=1}^{n} a_{jk}\xi_j\xi_k\right] \geq c|\xi|^2 \quad \text{for all } \xi = (\xi_1,\dots,\xi_n) \in \mathbb{R}^n. \tag{1.75}$$

Famously, the work of Agmon, Douglis, and Nirenberg in [4, 5] shows that a strongly elliptic system in the upper half-space possess a Poisson kernel. Additionally, in [80] it is shown that such a Poisson kernel must be unique. While the existence of a Poisson kernel eventually leads to well-posedness results for the Dirichlet problem in the upper half-space, this method loses its efficacy when domains of a more general nature (than $\mathbb{R}^n_+$) are considered. By way of contrast, a large variety of boundary value problems are solvable for weakly elliptic systems possessing a distinguished coefficient tensor in the inclusive class of δ-AR domains (with $\delta \in (0,1)$ appropriately small). Interestingly, when $n \geq 3$, weak ellipticity alone is enough to guarantee existence of a distinguished coefficient tensor when dealing with *scalar* operators. That is to say (cf. [86]),

$$\text{for } n \geq 3 \text{ any scalar, second-order, weakly elliptic, homogeneous, constant complex coefficient operator in } \mathbb{R}^n \text{ has a distinguished coefficient tensor.} \tag{1.76}$$

Generic second-order systems are considerably more difficult to categorize. The example of the (complex) Lamé system of elasticity,

$$L_{\mu,\lambda} := \mu\Delta + (\mu + \lambda)\nabla \operatorname{div} \quad \text{where } \lambda, \mu \in \mathbb{C} \tag{1.77}$$

is, however, well understood. For one thing, this system is weakly elliptic if and only if $\mu \neq 0$ and $2\mu + \lambda \neq 0$. It can be checked that each coefficient tensor in the one-parameter family given by

$$\begin{aligned} A(\zeta) &:= \big(a^{\alpha\beta}_{jk}(\zeta)\big)_{\substack{1\leq j,k,\leq n\\ 1\leq\alpha,\beta\leq n}} \quad \text{where} \\ a^{\alpha\beta}_{jk} &:= \mu\delta_{jk}\delta_{\alpha\beta} + (\mu + \lambda - \zeta)\delta_{jk}\delta_{k\beta} + \zeta\delta_{j\beta}\delta_{k\alpha} \quad \text{for } \zeta \in \mathbb{C}, \end{aligned} \tag{1.78}$$

may be used to express the Lamé system (again, with the summation convention over repeated indices) as

$$L_{\mu,\lambda} = \big(a^{\alpha\beta}_{jk}(\zeta)\partial_j\partial_k\big)_{1\leq\alpha,\beta\leq M} \quad \text{for each } \zeta \in \mathbb{C}. \tag{1.79}$$

It is shown in [86, 79] that $A(\zeta)$ is a distinguished coefficient tensor for $L_{\mu,\lambda}$ precisely when $3\mu + \lambda \neq 0$ and $\zeta := \frac{\mu(\mu+\lambda)}{3\mu+\lambda}$. To summarize,

$$\text{the Lamé system of elasticity } L_{\mu,\lambda} := \mu\Delta+(\mu+\lambda)\nabla \operatorname{div} \text{ where } \mu, \lambda \in \mathbb{C} \text{ are such that } \mu \neq 0,\ 2\mu + \lambda \neq 0, \text{ and } 3\mu + \lambda \neq 0 \text{ has the property that } \mathfrak{A}^{\mathrm{dis}}_{L_{\mu,\lambda}} = \mathfrak{A}^{\mathrm{dis}}_{L^{\top}_{\mu,\lambda}} \neq \varnothing. \tag{1.80}$$

The fact that we are able to accommodate a system such as the Lamé system of elasticity in our main well-posedness results for various boundary value problems, including the Dirichlet problem, Neumann problem, regularity problem, and the transmission problem, formulated in δ-AR domains, and with boundary data in Muckenhoupt-weighted Morrey spaces (as well as other related function spaces) is a testament to the power and versatility of the theory developed here.

The aforementioned theory weaves together geometric measure theory, functional analysis, harmonic analysis, and partial differential equations into a beautiful tapestry and leads to results of the following nature.

Let $\Omega \subseteq \mathbb{R}^n$ be a δ-AR domain. Denote by ν its geometric measure theoretic outward unit normal, and regard $\sigma := \mathcal{H}^{n-1}\lfloor\partial\Omega$ as the "surface measure" on its topological boundary. Fix any two exponents $p, q \in (1,\infty)$, along with a Muckenhoupt weight $w \in A_p(\partial\Omega,\sigma) \cap \mathrm{RH}_q(\partial\Omega,\sigma)$, and select some arbitrary threshold $\lambda \in (0,(n-1)/q')$, where $q' := q/(q-1)$. Finally, select an aperture parameter $\kappa > 0$. With $M_1^{p,\lambda}(\partial\Omega, w)$ denoting the Muckenhoupt-weighted Morrey-based Sobolev space of order one on $\partial\Omega$ (cf. Definition 8.8), consider the Dirichlet and regularity problems for a weakly elliptic, second-order, homogeneous, constant complex coefficient, $M \times M$ system L in $\mathbb{R}^n$, formulated in Ω as indicated below:

$$\begin{cases} u \in [\mathscr{C}^\infty(\Omega)]^M, \\ Lu = 0 \quad \text{in } \Omega, \\ \mathcal{N}_\kappa u \in M^{p,\lambda}(\partial\Omega, w), \\ u\big|_{\partial\Omega}^{\kappa-\text{n.t.}} = f \in [M^{p,\lambda}(\partial\Omega, w)]^M, \end{cases} \qquad \begin{cases} u \in [\mathscr{C}^\infty(\Omega)]^M, \\ Lu = 0 \quad \text{in } \Omega, \\ \mathcal{N}_\kappa u,\ \mathcal{N}_\kappa(\nabla u) \in M^{p,\lambda}(\partial\Omega, w), \\ u\big|_{\partial\Omega}^{\kappa-\text{n.t.}} = f \in [M_1^{p,\lambda}(\partial\Omega, w)]^M. \end{cases} \tag{1.81}$$

Then *these problems are well posed provided both L and its (real) transpose* $L^\top$ *have distinguished coefficient tensors, i. e.,* $\mathfrak{A}_L^{\mathrm{dis}} \neq \varnothing$, $\mathfrak{A}_{L^\top}^{\mathrm{dis}} \neq \varnothing$, *and assuming* $\delta \in (0,1)$ *is sufficiently small.* See Theorem 11.9 and Theorem 11.29 for precise formulations of these problems.

A more nuanced statement asserts that L having a distinguished coefficient tensor leads to solvability, and interestingly, it turns out that the requirement of $L^\top$ having a distinguished coefficient tensor ensures uniqueness in either of the boundary value problems in (1.81). To establish uniqueness, we use an argument involving a Green function. The construction of a Green function requires solving a suitable regularity problem for the transpose system, $L^\top$, with boundary data in a space which pairs well, in an integral sense, with functions from $M^{p,\lambda}(\partial\Omega, w)$.

This motivates us to develop a parallel theory for Muckenhoupt-weighted Block spaces, a brand of function spaces serving as pre-duals to our Muckenhoupt-weighted Morrey spaces. To be more specific, work on an closed Ahlfors regular set $\Sigma \subseteq \mathbb{R}^n$ and abbreviate $\sigma := \mathcal{H}^{n-1}\lfloor\Sigma$. Fix two exponents $p, q \in (1,\infty)$ and denote their Hölder conjugates by p' and q', respectively. Also, pick some $w \in A_p(\Sigma,\sigma) \cap \mathrm{RH}_q(\Sigma,\sigma)$ and write $w' := w^{-p'/p}$, which belongs to $A_{p'}(\Sigma,\sigma)$, for the conjugate weight to w. Finally, choose $\lambda \in (0,(n-1)/q')$. In this setting, we then have (cf. Theorem 6.29)

$$\big(\mathcal{B}^{p',\lambda}(\Sigma, w')\big)^* = M^{p,\lambda}(\Sigma, w), \tag{1.82}$$

where $\mathcal{B}^{p',\lambda}(\Sigma, w')$ denotes the Muckenhoupt-weighted Block space introduced and discussed at length in Chapter 6. Moreover, the aforementioned duality is realized in an integral sense, and a natural Hölder inequality holds in this context (cf. Theorem 6.28).

Along the way, we also develop all the necessary tools to solve boundary value problems with data in Muckenhoupt-weigthed Block spaces, akin to (1.81), starting with the functional analytic foundations, and going through the Calderón–Zygmund theory all the way to the theory of boundary layer potentials. The reward is a very satisfactory theory of boundary problems similar to those posited in (1.81), but formulated for Muckenhoupt-weighted Block spaces.

Significantly, the body of results and techniques developed here is able to accommodate the Neumann problem formulated on both Muckenhoupt-weighted Morrey spaces and Muckenhoupt-weighted Block spaces. In the context of a δ-AR domain, and a weakly elliptic, second-order, homogeneous, constant complex coefficient, $M \times M$ system L in $\mathbb{R}^n$, the Neumann problem on a Muckenhoupt-weighted Morrey space reads

$$\begin{cases} u \in [\mathscr{C}^\infty(\Omega)]^M, \\ Lu = 0 \quad \text{in } \Omega, \\ \mathcal{N}_\kappa(\nabla u) \in M^{p,\lambda}(\partial\Omega, w), \\ \partial_\nu^A u = f \in [M^{p,\lambda}(\partial\Omega, w)]^M, \end{cases} \tag{1.83}$$

where $\partial_\nu^A u$ denotes the conormal derivative of u with respect to the coefficient tensor A (chosen to express the system L as in (1.71)–(1.72); cf. Definition 3.18). In Theorem 11.22, we establish the well posedness of (1.83) in the case when $A \in \mathfrak{A}_L^{\text{dis}}$ and $A^\top \in \mathfrak{A}_{L^\top}^{\text{dis}}$, and δ is sufficiently small. Opposite to the case of the Dirichlet and regularity problems, now the requirement that $L^\top$ has a distinguished coefficient tensor ensures existence, while L having a distinguished coefficient tensor ensures uniqueness.

It should be noted that, in the well-posedness results stated so far, the requirement that the system and its transpose have distinguished coefficient tensors are structural hypotheses (in the sense that the conclusions may fail if said conditions are dropped) and not mere artifacts of the method of proof we employ. Indeed, in [86] there exist examples of systems, which do not possess distinguished coefficient tensors for which both (1.81) and (1.83) fail to be well posed. In fact, such failure can be quite catastrophic, in that the space of "admissible" boundary data (i. e., the space of all boundary data which lead to a solvable problem) has infinite-dimensional cokernel in the "full space" of boundary data, and the space of null-solutions for said problems may be infinite-dimensional, even when the underlying domain is the upper half-space. Ultimately, this shows that the boundary value problems in the tame geometric setting of the upper half-space but corresponding to such systems without distinguished coefficient tensors may not even be Fredholm solvable.

As the ultimate litmus test for the potency of our theory of layer potentials developed here, we are able to treat the transmission problem, stated as follows. Given an NTA domain $\Omega \subseteq \mathbb{R}^n$, denote $\Omega_+ := \Omega$ and $\Omega_- := \mathbb{R}^n \setminus \overline{\Omega}$. Fix a weakly elliptic, second-order, homogeneous, constant complex coefficient, $M \times M$ system L in $\mathbb{R}^n$, and consider

$$\begin{cases} u^{\pm} \in [\mathscr{C}^{\infty}(\Omega)]^M, \\ Lu^{\pm} = 0 \quad \text{in } \Omega_{\pm}, \\ \mathcal{N}_{\kappa}(\nabla u^{\pm}) \in M^{p,\lambda}(\partial\Omega, w), \\ u^{+}\big|_{\partial\Omega}^{\kappa-\text{n.t.}} - u^{-}\big|_{\partial\Omega}^{\kappa-\text{n.t.}} = g \in [\dot{M}_1^{p,\lambda}(\partial\Omega, w)]^M, \\ \partial_{\nu}^{A} u^{+} - \eta \cdot \partial_{\nu}^{A} u^{-} = f \in [M^{p,\lambda}(\partial\Omega, w)]^M, \end{cases} \tag{1.84}$$

where $\eta \in \mathbb{C}$ is the transmission (or coupling) parameter, and $\dot{M}_1^{p,\lambda}(\partial\Omega, w)$ denotes a Muckenhoupt-weighted homogeneous Morrey-based Sobolev space (cf. Definition 8.19). Then (1.84) is well posed whenever $\eta \neq -1$ and

$$\begin{aligned} &\left|\frac{1-\eta}{1+\eta}\right| \|\nu\|_{\mathrm{BMO}(\partial\Omega,\sigma)} \quad \text{is sufficiently small and} \\ &\text{either } A^{\top} \in \mathfrak{A}_{L^{\top}}^{\mathrm{dis}} \text{ with } \eta \in \mathbb{C} \setminus \{-1\}, \\ &\text{or } A \in \mathfrak{A}_{L}^{\mathrm{dis}} \text{ with } \eta \in \mathbb{C} \setminus \{-1, 0\}. \end{aligned} \tag{1.85}$$

In this case, it is remarkable that no flatness assumption on the underlying domain is imposed outright. Indeed, as seen from (1.85), the usual δ-AR flatness condition on $\partial\Omega$ can be relaxed, provided the coupling parameter η is correspondingly close to one. This is an interesting feature, because it allows for the consideration of, say, Lipschitz domains with an arbitrarily large Lipschitz constant in the case when the transmission parameter η is sufficiently close to 1. This is yet another subtle way in which the solvability of a boundary value problem is intricately tied up with both geometry and the partial differential equation itself.

While historically Morrey spaces have always been close to the realm of partial differential equations, it still begs the question, why consider Morrey spaces as a space of boundary data for, say, the Dirichlet problem? Moreover, why further complicate matters by considering a weighted version of this scale of spaces? The importance of considering a multitude of functional analytic settings for boundary value problems stems from the realization that, in applications, the boundary datum is determined by nature, so there is no specification regarding its membership to a particular function space.

As an example, fix some power $b \in \mathbb{R}$ and consider the following Dirichlet problem for the Laplacian on $\mathbb{R}^n_+$ (naturally identifying $\partial\mathbb{R}^n_+ \equiv \mathbb{R}^{n-1}$):

$$\begin{cases} u \in \mathscr{C}^{\infty}(\mathbb{R}^n_+), \\ \Delta u = 0 \quad \text{in } \mathbb{R}^n_+, \\ u\big|_{\partial\mathbb{R}^n_+}^{\kappa-\text{n.t.}} = |\cdot|^{-b} \quad \text{at } \mathcal{L}^{n-1}\text{-a. e. point on } \mathbb{R}^{n-1}. \end{cases} \tag{1.86}$$

This is a boundary value problem that, no matter how robust a Lebesgue space theory there is (and it is quite robust at this point), could never fall under that theory's scope for the simple fact that the power function $|\cdot|^{-b}$ does not belong to *any* Lebesgue space in $\mathbb{R}^{n-1}$ no matter the choice of b. However, we can place $|\cdot|^{-b}$ into a Muckenhoupt-weighted Morrey space in $\mathbb{R}^{n-1}$ provided[4] $b \in (0, n-1)$. Specifically, we claim that for any fixed power $b \in (0, n-1)$ and integrability exponent $p \in (1, \infty)$ there exist

$$\begin{aligned}&a \in (1-n, (p-1)(n-1)) \text{ together with } q \in (1, \infty) \text{ such that } a > (n-1)/q,\\ &\text{as well as } \lambda \in (0, (n-1)/q') \text{ where } q' \text{ is the Hölder conjugate exponent}\\ &\text{of } q,\end{aligned} \tag{1.87}$$

with the property that

$$\begin{aligned}&|\cdot|^{-b} \in M^{p,\lambda}(\mathbb{R}^{n-1}, |\cdot|^a \mathcal{L}^{n-1}) \quad \text{and}\\ &|\cdot|^a \in A_p(\mathbb{R}^{n-1}, \mathcal{L}^{n-1}) \cap \mathrm{RH}_q(\mathbb{R}^{n-1}, \mathcal{L}^{n-1}).\end{aligned} \tag{1.88}$$

Indeed, the membership in the second line of (1.88) comes from (1.87), (1.41), and (1.44). As for the membership in the first line of (1.88), a simple continuity/connectivity argument shows that for any given power $b \in (0, n-1)$ there exist a, q, λ as in (1.87) such that b may be expressed $b = (n - 1 - \lambda + a)/p$. Assuming this is the case, we then note that

$$\begin{aligned}\big\| |\cdot|^{-b} \big\|_{M^{p,\lambda}(\mathbb{R}^{n-1}, |\cdot|^a \mathcal{L}^{n-1})} &= \big\| |\cdot|^{-b} \times |\cdot|^{a/p} \big\|_{M^{p,\lambda}(\mathbb{R}^{n-1}, \mathcal{L}^{n-1})}\\ &= \big\| |\cdot|^{-(n-1-\lambda)/p} \big\|_{M^{p,\lambda}(\mathbb{R}^{n-1}, \mathcal{L}^{n-1})} < +\infty,\end{aligned} \tag{1.89}$$

with the final inequality implied by [79, Example 7.1, p. 434]. This justifies the membership in the first line of (1.88). Given the scope of our theory (which includes a well-posedness result for the Dirichlet problem for the Laplacian in the upper half-space with boundary datum in a Muckenhoupt-weighted Morrey space), we conclude that we may attach a functional analytic context (involving Muckenhoupt-weighted Morrey spaces) as to turn (1.86) into a well-posed boundary value problem. This is a new and nontrivial result, which never could have come out of a theory with a singular focus on Lebesgue spaces.

To sum up, we use a large spectrum of tools and techniques from across areas like geometric measure theory, harmonic analysis, functional analysis, and partial differential equations, to prove well-posedness results for a large variety of boundary value problems, formulated in very general geometric settings. All our main results are new and nontrivial even in such geometrically simple settings like that of the upper half-space. The interdisciplinary work undertaken in this monograph is predicated upon recent progress registered in [79, 86, 87, 88, 89, 90].

The organization of this monograph is as follows. Chapter 2 is concerned with reviewing aspects from geometric measure theory and harmonic analysis, which serve as a foundation for much of our work. We introduce notions from geometric measure theory that underpin not only our functional analytic framework, but also the kinds of

---

4 in view of the fact that any Muckenhoupt-weighted Morrey space in $\mathbb{R}^{n-1}$ embeds into the space $L^1(\mathbb{R}^{n-1}, \frac{dx'}{1+|x'|^{n-1}})$, it follows that the range $b \in (0, n-1)$ is optimal.

domains and geometries we would like to consider later in the formulation of our boundary value problems. Chapter 2 is also concerned with fundamentals from harmonic analysis including the Hardy–Littlewood maximal operator, Muckenhoupt weights, the John–Nirenberg space of functions of bounded mean oscillations, and dyadic grid structures. Each of these tools are used to build up the main pillars of this monograph as mentioned at the beginning of the introduction.

Chapter 3 details a treatment of boundary layer potentials in as accommodating a geometric and functional analytic environment as possible, more specifically, in the setting of uniformly rectifiable domains. We present several known trace results, differentiability properties, and a theory for modified versions of the usual double and single layer potentials. It is here where we are also introduced to the notion of a distinguished coefficient tensor, a crucial concept in both the solvability and uniqueness of the boundary value problems we consider.

In Chapter 4, we present a thorough treatment of Muckenhoupt-weighted Morrey spaces from examples to embedding results, leading up eventually to an extrapolation theorem in the spirit of Rubio de Francia's famous extrapolation for Muckenhoupt-weighted Lebesgue spaces. While each result in this chapter is important in its own right, the significance of our Morrey space extrapolation cannot be understated. In fact, Chapter 5 is a testament to the utility of such a result as we develop a Calderón–Zygmund theory on uniformly rectifiable sets for singular integral operators acting on Muckenhoupt-weighted Morrey spaces. Of particular significance, we establish geometric estimates (involving the BMO seminorm of the outward unit normal) for the operator norm of convolution-type singular integrals which have a certain chord-dot-normal structure in their integral kernel.

We change directions in Chapter 6 to develop a functional analytic framework for the pre-duals of Muckenhoupt-weighted Morrey spaces, what we call Muckenhoupt-weighted Block spaces. This includes a survey of basic properties as well as a myriad of useful embedding results, until ultimately establishing the duality between Muckenhoupt-weighted Morrey spaces and Muckenhoupt-weighted Block spaces. With the functional analytic preliminaries addressed, we move on to Chapter 7 where a powerful and versatile Calderón–Zygmund theory for this flavor of Block spaces is developed, including geometric estimates (involving the BMO seminorm of the outward unit normal) for chord-dot-normal convolution-type singular integral operators acting Muckenhoupt-weighted Block spaces.

We continue our foray into functional analysis in Chapter 8 by considering Sobolev spaces, which have the flavor of Muckenhoupt-weighted Morrey and Block spaces. In this chapter, we make an effort to consider suitable notions of both homogeneous and inhomogeneous versions of Muckenhoupt-weighted Morrey and Block-based Sobolev spaces in geometrically rough environments.

Moving closer to our final goal of solving boundary value problems, Chapter 9 is concerned with applying our Calderón–Zygmund theory for Muckenhoupt-weighted Morrey and Block spaces to the boundary layer potentials introduced throughout Chapter 3.

If Chapter 3 is the foundation for solving boundary value problems, then Chapter 9 is the scaffolding. The developments here put us in a prime position to solve the kinds of boundary value problems we have in mind.

In Chapter 10, we diver further into functional analysis, now in the direction of generalized Banach function spaces. This newer notion of function space, which has many remarkable properties such as a nice duality theory (duality here is not strictly in the sense of a topological dual), and notably an extrapolation theory, originated in [86]. We successfully place Muckenhoupt-weighted Morrey and Block spaces into this abstract framework. This perspective yields an extrapolation result involving Muckenhoupt-weighted Block spaces, and while this can be used as an effective tool to develop a Calderón–Zygmund theory for this scale of spaces, following this route does not entirely recover the results recorded throughout Chapter 7 as those results remain slightly more flexible regarding the parameters involved in the definition of a Muckenhoupt-weighted Block space; see Remark 10.56 for a more nuanced explanation. Placing Muckenhoupt-weighted Morrey and Block spaces into the framework of generalized Banach function spaces is highly nontrivial and such a feat requires us to develop a suitable notion of a "Muckenhoupt-weighted *vanishing* Morrey space," a scale of spaces we show serve as a pre-dual to Muckenhoupt-weighted Block spaces. This is a delicate matter. We also explain how many of our previous results concerning our Calderón–Zygmund theory on Muckenhoupt-weighted Morrey spaces can (with some care) be adapted to these Muckenhoupt-weighted vanishing Morrey spaces.

The monograph finishes with results that were once the lofty goals that started this journey. In Chapter 11, we examine a large assortment of boundary value problems in rough domains with boundary data in Muckenhoupt-weighted Morrey and vanishing Morrey spaces, as well as Muckenhoupt-weighted Block spaces. At this late stage in our work, we enjoy much success in employing the method of boundary layer potentials to establish genuine well-posedness results (in the sense of Legendre–Hadamard) for such problems including the Dirichlet, Neumann, (homogeneous and inhomogeneous) Regularity, and Transmission problems. We also provide several specific and nontrivial examples to which our abstract theory applies. Finally, in Chapter 12, we examine boundary problems in domains with compact boundaries, including bounded domains and exterior domains. This makes full use of the results in Chapter 10 that show Muckenhoupt-weighted Morrey and Block spaces are generalized Banach function spaces as well as recent work in [90], which treat boundary problems on compact domains in this much more general functional analytic setting.

2 Preliminaries

2.1 Basic notation

We list some basic notation that will be adopted throughout this work:

$\mathbb{N} := \{1, 2, 3, \dots\}$ and $\mathbb{N}_0 := \mathbb{N} \cup \{0\}$.
$\mathbb{Z}$ the set of integers.
$\mathbb{Q}$ the set of rational numbers.
$\mathbb{R}$ the set of real numbers.
$\mathbb{C}$ the set of complex numbers.
$\mathcal{L}^n$ n-dimensional Lebesgue measure in $\mathbb{R}^n$.
$\mathcal{H}^s$ s-dimensional Hausdorff measure in $\mathbb{R}^n$.

Starting with an open set $\Omega \subseteq \mathbb{R}^n$, the space $\mathscr{C}^0(\Omega)$ denotes the collection of all continuous functions on Ω. For $n \in \mathbb{N}$, $\mathscr{C}^n(\Omega)$ denotes the space of functions on Ω which are continuously differentiable up to and including order n. Lastly, $\mathscr{C}^\infty(\Omega)$ denotes the space of infinitely differentiable functions on Ω.

Given a measure space $(X, \mathfrak{M}, \mu)$ and some $p \in (0, \infty)$, denote by $L^p(X, \mu)$ the set of all μ-measurable functions such that

$$\int_X |f|^p \, d\mu < \infty.$$

Given an arbitrary topological space X and a Borel measure μ on X, the space $L^\infty_{\mathrm{comp}}(X, \mu)$ denotes the set of all μ-measurable functions, which are essentially bounded with compact support. That is, $f \in L^\infty_{\mathrm{comp}}(X, \mu)$ provided f is μ-measurable, there exists some compact set $K \subseteq X$ for which $\operatorname{supp} f \subseteq K$, and $\operatorname{ess\,sup}_{x \in X} f(x) < +\infty$.

Starting with a Banach space $(X, \|\cdot\|_X)$, for any $M \in \mathbb{N}$ we construct a new Banach space $([X]^M, \|\cdot\|_{[X]^M})$, where $[X]^M := \underbrace{X \times \cdots \times X}_{M \text{ times}}$ and for $f = (f_\alpha)_{1 \le \alpha \le M} \in [X]^M$,

$$\|f\|_{[X]^M} := \max_{1 \le \alpha \le M} \|f_\alpha\|_X \tag{2.1}$$

Let $(X, \mathfrak{M}, \mu)$ be a measure space, and $E \in \mathfrak{M}$ with $0 < \mu(E) < +\infty$. Then we denote the integral average of a μ-measurable function f on X by

$$\fint_E f \, d\mu := \frac{1}{\mu(E)} \int_E f \, d\mu. \tag{2.2}$$

Write $B(x, r) := \{y \in \mathbb{R}^n : |x - y| < r\}$ to denote an open ball in $\mathbb{R}^n$, centered at $x \in \mathbb{R}^n$ with radius $r \in (0, \infty)$.

The standard inner product on $\mathbb{R}^n$ is denoted by $\langle \cdot, \cdot \rangle$. That is, for two arbitrary points $x = (x_1, \dots, x_n) \in \mathbb{R}^n$, $y = (y_1, \dots, y_n) \in \mathbb{R}^n$, one has $\langle x, y \rangle := \sum_{j=1}^n x_j y_j$.

We say that $A \approx B$ provided there exist two constants $c, C \in (0, \infty)$ such that $cA \le B \le CA$.

https://doi.org/10.1515/9783111458274-002

2.2 Results from geometric measure theory

Our main goal of this work is to solve boundary value problems in as rough a setting as possible. The mathematical discipline of geometric measure theory provides a framework to describe an entire zoo of different types of sets. As such, there are many technical definitions involved, which quantify the different types of "bad behavior" a set can have.

We will frequently study sets of some lower dimension (with respect to Hausdorff measure) within the Euclidean ambient, $\mathbb{R}^n$. Our first definition helps to define the notion of a ball on these lower-dimensional sets.

Definition 2.1. Given a closed set $\Sigma \subseteq \mathbb{R}^n$, we write $\Delta(x,r) := \Sigma \cap B(x,r)$ to denote a `surface ball` centered at $x \in \Sigma$ with radius $r \in (0, 2\,\mathrm{diam}(\Sigma))$. If omitting the center is unambiguous, we may write Δ_r, and if omitting both the center and radius is unambiguous, we will simply write Δ.

We use the term surface ball, because typically, Σ behaves as some type of $(n-1)$-dimensional "surface" in a certain measure theoretic way, described below.

Definition 2.2. A closed set $\Sigma \subseteq \mathbb{R}^n$ with surface measure $\sigma := \mathcal{H}^{n-1}\lfloor\Sigma$ is called `Ahlfors regular` or `Ahlfors-David regular` if there exist constants $c_\Sigma, C_\Sigma \in (0,\infty)$ such that for each $x \in \Sigma$ and each $r \in (0, 2\,\mathrm{diam}(\Sigma))$,

$$c_\Sigma r^{n-1} \leq \sigma(\Delta(x,r)) \leq C_\Sigma r^{n-1}. \tag{2.3}$$

We express the above relationship more succinctly by $\sigma(\Delta(x,r)) \approx r^{n-1}$ uniformly in $x \in \Sigma$. Collectively, the constants c_Σ and C_Σ are referred to as the ADR `character` (or constants) of Σ. The first inequality in (2.3) constitutes what is known as `lower Ahlfors regularity`, while the second inequality in (2.3) constitutes `upper Ahlfors regularity`.

The topological boundary of an open set $\Omega \subseteq \mathbb{R}^n$ is insufficient in many ways when trying to develop some sort of calculus on it. One of the most glaring shortcomings is the inability to define a good notion of outward unit normal. This next definition provides a more suitable notion of boundary from a measure theoretic perspective on which we can define some type of outward unit normal.

Definition 2.3. Let $\Omega \subseteq \mathbb{R}^n$ be a $\mathcal{L}^n$-measurable set. Define the `geometric measure theoretic` boundary $\partial_*\Omega$ as follows:

$$\partial_*\Omega := \left\{ x \in \partial\Omega : \limsup_{r\to 0^+} \frac{\mathcal{L}^n(B(x,r)\cap\Omega)}{r^n} > 0, \right. \tag{2.4}$$
$$\left. \limsup_{r\to 0^+} \frac{\mathcal{L}^n(B(x,r)\setminus\Omega)}{r^n} > 0 \right\}.$$

This is a subset of the topological boundary consisting of all points with sufficient mass at infinitesimal scales both in the interior of Ω and in the complement of Ω. For example, the geometric measure theoretic boundary excludes points located at cusps, or the points on the slit in a slit disk.

Recall that a set $\Omega \subseteq \mathbb{R}^n$ of locally finite perimeter is one whose characteristic function has finite variation in all open bounded subsets of $\mathbb{R}^n$. Specifically, for each relatively compact subset $\mathcal{O} \subseteq \mathbb{R}^n$ one has

$$\sup\left\{\left|\int_{\Omega} \operatorname{div} \vec{\varphi}\, d\mathcal{L}^n : \vec{\varphi} \in [\mathscr{C}_c^{\infty}(\mathcal{O})]^n \text{ with } \sup_{\mathbb{R}^n} |\vec{\varphi}| \leq 1\right|\right\} < \infty. \tag{2.5}$$

Equivalently (cf. [36, Theorem 5.1, pp. 194–195] in this regard), there exist a locally finite Borel-regular measure σ_* in $\mathbb{R}^n$ and some $\nu \in [L^{\infty}(\mathbb{R}^n, \sigma_*)]^n$ with $|\nu(x)| = 1$ at σ_*-a. e. point $x \in \mathbb{R}^n$ and

$$\nabla 1_{\Omega} = -\nu\sigma_* \quad \text{in } [\mathcal{D}'(\mathbb{R}^n)]^n. \tag{2.6}$$

Definition 2.4. The function ν appearing in (2.6) is referred to as the (De Giorgi–Federer) `geometric measure theoretic outward unit normal` to Ω.

Concerning the measure σ_* in the context of (2.6), it turns out (combining [36, Theorem 5.15(iii), p. 232] and [36, Lemma 5.5, p. 235]) that

$$\sigma_* = \mathcal{H}^{n-1} \lfloor \partial_* \Omega. \tag{2.7}$$

Thus, the geometric measure theoretic boundary is a good environment in which to speak of some notion of outward unit normal at σ_*-a. e. point on the boundary, which of course is sufficient from the perspective of integration.

Definition 2.5. Let $\Omega \subseteq \mathbb{R}^n$ be an open, nonempty, and proper subset of $\mathbb{R}^n$. We call Ω an AR `domain` (a.k.a `Ahlfors regular domain`) provided $\partial\Omega$ is an Ahlfors regular set and $\mathcal{H}^{n-1}(\partial\Omega \setminus \partial_*\Omega) = 0$.

The requirement that the geometric measure theoretic boundary has full measure in the topological boundary with respect to $(n-1)$-dimensional Hausdorff measure ensures that the geometric measure theoretic outward unit normal exists $\mathcal{H}^{n-1}$-a. e. on $\partial\Omega$ (see the discussion surrounding (2.7)).

Definition 2.6. A closed set $\Sigma \subseteq \mathbb{R}^n$ is called `uniformly rectifiable` (sometimes UR for short) if Σ is an Ahlfors regular set (cf. Definition 2.2) and there exist two constants $\varepsilon, M \in (0, \infty)$ with the following significance. For each $x \in \Sigma$ and each $R \in (0, 2\operatorname{diam}(\Sigma))$, there exists a Lipschitz map $\varphi : B_R^{n-1} \to \mathbb{R}^n$ with Lipschitz constant $\leq M$ and such that

$$\mathcal{H}^{n-1}(\Sigma \cap B(x, R) \cap \varphi(B_R^{n-1})) \geq \varepsilon R^{n-1}. \tag{2.8}$$

Collectively, ε and M are referred to as the UR `character` (or constants) of Σ.

This definition encodes the fact that uniformly rectifiable sets contain big pieces of Lipschitz images at all points and at all scales in a uniform fashion. The significance of uniform rectifiability is not immediately clear from the definition. However, the remarkable property of uniform rectifiability, owed to David and Semmes [28], is the if and only if L^2-boundedness of the Cauchy operator on uniformly rectifiable sets. Subsequently, from the perspective of Calderón–Zygmund theory and singular integral operators of Calderón–Zygmund type, uniformly rectifiable domains are as rough as the geometry can be to still accommodate the various boundedness properties and size estimates that one would want for these kinds of operators.

Definition 2.7. An open, nonempty, proper subset Ω of $\mathbb{R}^n$ is called a UR `domain` provided that $\partial\Omega$ is a UR set and $\mathcal{H}^{n-1}(\partial\Omega \setminus \partial_*\Omega) = 0$.

As in the definition of an Ahlfors regular domain (cf. Definition 2.5), the requirement that $\mathcal{H}^{n-1}(\partial\Omega \setminus \partial_*\Omega) = 0$ ensures that the geometric measure theoretic outward unit normal to Ω is defined σ-a. e. on $\partial\Omega$.

There is actually quite a variety of different ways to describe the regularity of the boundary, and we list several definitions now, so that various terms are familiar as they appear in this work.

Definition 2.8. Fix $R \in (0,\infty]$ and $c \in (0,1)$. A nonempty proper subset Ω of $\mathbb{R}^n$ is said to satisfy the (R,c)-`corkscrew condition`, or simply a `corkscrew condition` if the particular values of R and c are not important, if for each location $x \in \partial\Omega$ and each scale $r \in (0,R)$ there exists a point $z \in \Omega$, called a corkscrew point relative to x and r, satisfying $B(z,cr) \subseteq B(x,r)\cap\Omega$. A nonempty proper subset Ω of $\mathbb{R}^n$ is said to satisfy a `two-sided corkscrew condition` provided both Ω and $\mathbb{R}^n \setminus \Omega$ satisfy a corkscrew condition.

The above definition implies that $\partial_*\Omega = \partial\Omega$ whenever Ω satisfies a two-sided corkscrew condition. This highlights the extent of the kinds of irregularities a domain satisfying a two-sided corkscrew condition permits. In fact (cf. [26, Theorem 1, p. 840]),

> any proper, nonempty, open subset $\Omega \subseteq \mathbb{R}^n$, which satisfies a two-sided corkscrew condition and is such that $\partial\Omega$ is an Ahlfors regular set, is actually a uniformly rectifiable domain. (2.9)

The following definition may be thought of as a certain kind of connectivity condition for an open set.

Definition 2.9. Fix $R \in (0,\infty]$ and $N \in \mathbb{N}$. An open set $\Omega \subseteq \mathbb{R}^n$ is said to satisfy the (R,N)-`Harnack chain condition`, or simply a `Harnack chain condition` if the particular values of R, N are unimportant, provided whenever

$$\begin{aligned}&\varepsilon > 0,\quad k \in \mathbb{N},\quad z \in \partial\Omega,\quad r \in (0,R),\quad \text{and}\quad x,y \in B(z,r/4)\cap\Omega\\ &\text{satisfy } |x-y| \le 2^k\varepsilon \text{ and } \min\{\operatorname{dist}(x,\partial\Omega),\, \operatorname{dist}(y,\partial\Omega)\} \ge \varepsilon,\end{aligned} \tag{2.10}$$

one may find a chain of balls $B_1, B_2, \dots, B_K$ with $K \leq Nk$, such that $x \in B_1, y \in B_K$, $B_i \cap B_{i+1} \neq \varnothing$ for every $i \in \{1, \dots, K-1\}$, and

$$\begin{aligned} &N^{-1} \cdot \operatorname{diam}(B_i) \leq \operatorname{dist}(B_i, \partial\Omega) \leq N \cdot \operatorname{diam}(B_i), \\ &\operatorname{diam}(B_i) \geq N^{-1} \cdot \min\{\operatorname{dist}(x, B_i), \operatorname{dist}(y, B_i)\}, \end{aligned} \tag{2.11}$$

for every $i \in \{1, \dots, K\}$.

As observed in [79, (2.52), p. 38], an open set which satisfies a (∞, N) Harnack chain condition for some $N \in \mathbb{N}$ must actually be pathwise connected, and hence connected.

It turns out that a corkscrew condition and a Harnack chain condition are the right definition to introduce the notion of a NTA domain, the higher-dimensional analogue of a chord-arc domain.

Definition 2.10. Fix $R \in (0, \infty]$ and $N \in \mathbb{N}$. An open, nonempty, proper subset Ω of $\mathbb{R}^n$ is said to be an (R, N)-nontangentially accessible domain (or simply an NTA domain if the particular values of R, N are not important) if Ω satisfies both the (R, N^{-1})-two-sided corkscrew condition and the (R, N)-Harnack chain condition.

Call Ω a (R, N)-two-sided nontangentially accessible domain (or, simply, a two-sided NTA domain if the particular values of R, N are not relevant) provided Ω is an open, nonempty, proper subset of $\mathbb{R}^n$ with the property that both Ω and $\mathbb{R}^n \setminus \overline{\Omega}$ are (R, N)-nontangentially accessible domains.

A set $\Omega \subseteq \mathbb{R}^n$ is said to be an (R, N)-one-sided NTA domain provided Ω satisfies the (R, N)-Harnack chain condition and the (R, N^{-1})-corkscrew condition (once again, with the convention that the parameters R, N are dropped if theirs values are not relevant).

Finally, it is agreed that, in all cases, one takes $R = \infty$ if and only if $\partial\Omega$ is unbounded.

This concludes our brief list of the types of sets and domains we will consider. Now, we turn to more quantitative aspects of geometric measure theory.

Definition 2.11. Let $\Omega \subseteq \mathbb{R}^n$ be an open set and fix an aperture parameter $\kappa \in (0, \infty)$. We define the nontangential approach regions

$$\Gamma_\kappa(x) := \{y \in \Omega : |y - x| < (1 + \kappa) \operatorname{dist}(y, \partial\Omega)\} \quad \text{for each } x \in \partial\Omega. \tag{2.12}$$

In the case that Ω is a half-space, the nontangential approach regions $\Gamma_\kappa(x)$ are genuine one-sided (upright) cones with vertex at x and aperture angle controlled by κ.

Definition 2.12. Define the nontangential maximal operator, $\mathcal{N}_\kappa$, which acts on each $\mathcal{L}^n$-measurable function $u : \Omega \to \mathbb{C}$ according to

$$(\mathcal{N}_\kappa u)(x) := \|u\|_{L^\infty(\Gamma_\kappa(x), \mathcal{L}^n)} \quad \text{at each } x \in \partial\Omega, \tag{2.13}$$

with the convention that $(\mathcal{N}_\kappa u)(x) = 0$ whenever $\Gamma_\kappa(x) = \varnothing$. If u is as before and $E \subseteq \Omega$ is a $\mathcal{L}^n$-measurable set, $\mathcal{N}_\kappa^E u : \partial\Omega \longrightarrow [0, \infty]$ denotes the nontangential maximal function of u restricted to E, defined via

$$(\mathcal{N}_\kappa^E u)(x) := \|u\|_{L^\infty(\Gamma_\kappa(x)\cap E,\, \mathcal{L}^n)} \quad \text{for each } x \in \partial\Omega. \tag{2.14}$$

Remark 2.13. For each $\mathcal{L}^n$-measurable function u, defined in an open set $\Omega \subseteq \mathbb{R}^n$ with surface measure $\sigma := \mathcal{H}^{n-1}\lfloor\partial\Omega$, and every aperture parameter $\kappa \in (0,\infty)$ the function $\mathcal{N}_\kappa u : \partial\Omega \to [0,\infty]$ is lower semicontinuous (see [86, Proposition 8.2.3, p. 684]). In particular, this implies that the function $\mathcal{N}_\kappa u : \partial\Omega \to [0,\infty]$ is σ-measurable.

Definition 2.14. Given an open set $\Omega \subseteq \mathbb{R}^n$, and a Lebesgue-measurable function u on Ω, the `nontangential limit` of u at $x \in \partial\Omega$ from within $\Gamma_\kappa(x)$ exists and its value is $a \in \mathbb{C}$, provided

$$\begin{aligned}&\text{for every } \varepsilon > 0 \text{ there exists some } r > 0 \text{ such that } |u(y)-a| < \varepsilon\\ &\text{for } \mathcal{L}^n\text{-a. e. point } y \in \Gamma_\kappa(x) \cap B(x,r).\end{aligned} \tag{2.15}$$

When the nontangential limit (or trace) of u at x from within $\Gamma_\kappa(x)$ exists, we denote this limit by $(u\big|_{\partial\Omega}^{\kappa-\text{n.t.}})(x)$.

Definition 2.15. Given an open, nonempty, proper subset $\Omega \subseteq \mathbb{R}^n$, define the subset of the boundary

$$A_\kappa(\partial\Omega) := \{x \in \partial\Omega : x \in \overline{\Gamma_\kappa(x)}\}. \tag{2.16}$$

That is, the set of all points on the topological boundary that are accessible in a nontangential sense, or more specifically from within $\Gamma_\kappa(x)$. We also agree to denote the `nontangentially accessible boundary` as $\partial_{\text{nta}}\Omega$, which is defined as

$$\partial_{\text{nta}}\Omega := \bigcap_{\kappa>0} A_\kappa(\partial\Omega). \tag{2.17}$$

That is, $\partial_{\text{nta}}\Omega$ consists of all points $x \in \partial\Omega$, which are accumulation points of the nontangential approach region $\Gamma_\kappa(x)$ for every aperture parameter $\kappa > 0$.

As remarked in [86, (8.9.8), p. 788], for a Lebesgue measurable function u on some open, nonempty, proper subset $\Omega \subseteq \mathbb{R}^n$, whenever $x \in A_\kappa(\partial\Omega)$ for some $\kappa \in (0,\infty)$, then

$$\big|(u\big|_{\partial\Omega}^{\kappa-\text{n.t.}})(x)\big| \le (\mathcal{N}_\kappa u)(x). \tag{2.18}$$

Interestingly (cf. [86, Proposition 8.8.6, p. 782]), working on Ω, an open, nonempty, proper subset of $\mathbb{R}^n$,

$$\begin{aligned}&\text{if } \partial\Omega \text{ is lower Ahlfors regular and } \sigma := \mathcal{H}^{n-1}\lfloor\partial\Omega \text{ is a}\\ &\text{doubling measure, then } \sigma(\partial_*\Omega \setminus \partial_{\text{nta}}\Omega) = 0.\end{aligned} \tag{2.19}$$

In particular, working in an AR domain (as we often do when considering boundary value problems in Chapter 11), both the topological boundary and the nontangentially accessible boundary differ by only a null set (with respect to the $(n-1)$-dimensional Hausdorff measure). This is due to (2.19) and the requirement in Definition 2.5 that for an AR domain, the topological boundary and the geometric measure theoretic boundary differ by only a null set (with respect to the $(n-1)$-dimensional Hausdorff measure).

As a consequence of the sublinearity of the nontangential maximal operator, we arrive at the following lemma.

Lemma 2.16. *Fix an aperture parameter $\kappa \in (0,\infty)$. Given some open set $\Omega \subseteq \mathbb{R}^n$ and two $\mathcal{L}^n$-measurable functions $u, v : \Omega \to \mathbb{R}$, then*

$$|(\mathcal{N}_\kappa u)(x) - (\mathcal{N}_\kappa v)(x)| \le (\mathcal{N}_\kappa(u-v))(x) \tag{2.20}$$

at each $x \in \partial\Omega$ for which $\min\{(\mathcal{N}_\kappa u)(x), (\mathcal{N}_\kappa v)(x)\} < \infty$.

Proof. Note by definition that $\mathcal{N}_\kappa$ is sublinear, and hence

$$\mathcal{N}_\kappa(u+v)(x) \le (\mathcal{N}_\kappa u)(x) + (\mathcal{N}_\kappa v)(x) \quad \text{for all } x \in \partial\Omega. \tag{2.21}$$

Replacing u by $u - v$ in (2.21) and assuming $\min\{(\mathcal{N}_\kappa u)(x), (\mathcal{N}_\kappa v)(x)\} < \infty$ implies

$$(\mathcal{N}_\kappa u)(x) - (\mathcal{N}_\kappa v)(x) \le \mathcal{N}_\kappa(u-v)(x). \tag{2.22}$$

Similarly, under our assumption that $\min\{(\mathcal{N}_\kappa u)(x), (\mathcal{N}_\kappa v)(x)\} < \infty$, replacing v by $v - u$ in (2.21) implies

$$(\mathcal{N}_\kappa v)(x) - (\mathcal{N}_\kappa u)(x) \le \mathcal{N}_\kappa(v-u)(x). \tag{2.23}$$

Noting that $\mathcal{N}_\kappa(u-v)$ is symmetric in u and v, we see that (2.22) and (2.23) imply the desired result. □

A closely related object to the nontangential maximal function is the so-called *tangential* maximal function.

Definition 2.17. Let Ω be an open, nonempty, proper subset of $\mathbb{R}^n$. Also, fix a power $M \in [0,\infty)$. For a Lebesgue measurable function $u : \Omega \to \mathbb{R}$, the `tangential maximal function` of u (with exponent M) is the function

$$\begin{aligned} &u_M^{\max} : \partial\Omega \to [0,+\infty] \quad \text{defined by} \\ &u_M^{\max}(x) := \left\| u(y)\left(\frac{\operatorname{dist}(y,\partial\Omega)}{|x-y|}\right)^M \right\|_{L_y^\infty(\Omega,\mathcal{L}^n)} \quad \text{for each } x \in \partial\Omega, \end{aligned} \tag{2.24}$$

where the subscript y indicates that the essential supremum is taken with respect to the variable $y \in \Omega$.

The interest of this maximal function is that it avoids any reference to nontangential approach regions. That is, the essential supremum that appears is taken over the entire open set as opposed to over a particular nontangential approach region. In the context of Muckenhoupt-weighted Morrey and Block spaces, which is the primary functional analytic focus of this work, it turns out that both the *nontangential* maximal function and *tangential* maximal function are quantitatively equivalent; cf. Proposition 7.7. As such, it is a matter of preference whether to describe the size of a function in terms of its nontangential maximal function versus its tangential maximal function.

This next lemma, whose proof is routine, and hence omitted, will be used frequently as part of many geometric arguments.

Lemma 2.18. *Fix an arbitrary* $N \in (0, \infty)$.

(1) *One has* $(1 + |x|)^N \approx 1 + |x|^N$ *uniformly for* $x \in \mathbb{R}^n$ *with implicit constants depending only on* N.
(2) *Fix any* $x_0 \in \mathbb{R}^n$, *then* $1 + |x - x_0|^N \approx 1 + |x|^N$ *uniformly for* $x \in \mathbb{R}^n$ *with implicit constants depending only on* N *and* x_0.
(3) *One has* $1 \approx 1 + |x|^N$ *uniformly for* x *in a bounded subset of* $\mathbb{R}^n$.
(4) *Fix any* $x_0 \in \mathbb{R}^n$ *and* $R \in (0, \infty)$. *Then* $1+|x-x_0| \approx |x-x_0|$ *uniformly for* $x \in \mathbb{R}^n \backslash B(x_0, R)$ *with implicit constants depending only on* R.

2.3 Results from harmonic analysis

One of the most important objects in harmonic analysis is the Hardy–Littlewood maximal operator. Originally studied in $\mathbb{R}^n$, the theory has been stripped to its bare bones so that we can discuss a version of this operator in as abstract a setting as possible, namely spaces of homogeneous type. These are quasimetric spaces with a background doubling measure.

Definition 2.19. Let (X, ρ) be a metric space. We say that a Borel measure μ is `doubling` provided there exists some constant $C \in (0, \infty)$ such that

$$0 < \mu(B_\rho(x, r)) \le C\mu(B_\rho(x, r/2)) < +\infty \tag{2.25}$$

for all $x \in X$ and all $r \in (0, \infty)$. The constant C appearing in (2.25) is referred to as the `doubling character` of μ.

It follows from Definition 2.2 that given an Ahlfors regular set $\Sigma \subseteq \mathbb{R}^n$, the measure $\sigma := \mathcal{H}^{n-1}\lfloor\Sigma$ is doubling. This is useful to us, because it means that an Ahlfors regular set can be thought of as a space of homogeneous type, and thus an environment in which we cannot only make sense of the Hardy–Littlewood maximal operator, but also retain many of its properties, e. g., L^p boundedness for $p \in (1, \infty]$.

Definition 2.20. Let $\Sigma \subseteq \mathbb{R}^n$ be a closed set with a Borel measure σ on Σ with the property that $0 < \sigma(\Delta) < +\infty$ for all surface balls $\Delta \subseteq \Sigma$. The Hardy–Littlewood maximal operator acts on each σ-measurable function $f : \Sigma \to \mathbb{C}$ according to

$$(\mathcal{M}_\Sigma f)(x) := \sup_{r\in(0,2\,\mathrm{diam}(\Sigma))} \fint_{\Delta(x,r)} |f|\, d\sigma \quad \text{for all } x \in \Sigma. \tag{2.26}$$

The following is a version of the Lebesgue differentiation theorem, abstracted to our setting of interest; cf. [86, Proposition 7.4.4, p. 585] for a more general result, which holds for spaces of homogeneous type.

Proposition 2.21. *Let $\Sigma \subseteq \mathbb{R}^n$ be a closed set and let $\sigma := \mathcal{H}^{n-1}\lfloor\Sigma$ be doubling. Then σ is a complete, locally finite, Borel-regular measure on Σ and for each $f \in L^1_{\mathrm{loc}}(\Sigma,\sigma)$ one has*

$$\lim_{r\to 0^+} \fint_{\Delta(x,r)} f\, d\sigma = f(x) \tag{2.27}$$

for σ-a. e. point $x \in \Sigma$. Equivalently,

$$\lim_{r\to 0^+} \fint_{\Delta(x,r)} |f(x) - f(y)|\, d\sigma(y) = 0 \tag{2.28}$$

for σ-a. e. point $x \in \Sigma$.

A simple corollary is that the Hardy–Littlewood maximal operator applied to a function dominates the absolute value of said function in a pointwise sense.

Corollary 2.22. *Let $\Sigma \subseteq \mathbb{R}^n$ be a closed set, and let $\sigma := \mathcal{H}^{n-1}\lfloor\Sigma$ be doubling. Then for each $f \in L^1_{\mathrm{loc}}(\Sigma,\sigma)$ one has*

$$\mathcal{M}_\Sigma f \geq |f| \quad \textit{at } \sigma\textit{-a. e. point on } \Sigma. \tag{2.29}$$

Proof. This is clear from (2.26) and (2.27). □

Here is another immediate consequence of Lebesgue's differentiation theorem, which gives a criterion for when a function is zero in an almost everywhere sense.

Corollary 2.23. *Let $\Sigma \subseteq \mathbb{R}^n$ be a closed set and let $\sigma := \mathcal{H}^{n-1}\lfloor\Sigma$ be doubling. If $g \in L^1_{\mathrm{loc}}(\Sigma,\sigma)$ is such that*

$$\int_\Delta g\, d\sigma = 0 \quad \textit{for each bounded surface ball } \Delta \subseteq \Sigma, \tag{2.30}$$

then $g = 0$ at σ-a. e. point on Σ.

This next lemma is a kind of triangle inequality involving an infinite series and the Hardy–Littlewood maximal operator.

Lemma 2.24. *Let $\Sigma \subseteq \mathbb{R}^n$ be a closed set with σ, a Borel measure on Σ with the property that $0 < \sigma(\Delta) < +\infty$ for all surface balls $\Delta \subseteq \Sigma$. If $\{f_j\}_{j\in\mathbb{N}}$ is a sequence of σ-measurable functions, then*

$$\mathcal{M}_\Sigma\left(\sum_{j=1}^{\infty} f_j\right) \leq \sum_{j=1}^{\infty} \mathcal{M}_\Sigma(f_j). \tag{2.31}$$

Proof. To begin, notice that $\sum_{j=1}^{N} |f_j|$ increases pointwise on Σ to $\sum_{j=1}^{\infty} |f_j|$ as $N \to \infty$. It then follows that $\mathcal{M}_\Sigma(\sum_{j=1}^{N} |f_j|)$ increases pointwise on Σ to $\mathcal{M}_\Sigma(\sum_{j=1}^{\infty} |f_j|)$ as $N \to \infty$. By the sublinearity of $\mathcal{M}_\Sigma$, we have

$$\mathcal{M}_\Sigma\left(\sum_{j=1}^{N} |f_j|\right) \leq \sum_{j=1}^{N} \mathcal{M}_\Sigma(f_j) \leq \sum_{j=1}^{\infty} \mathcal{M}_\Sigma(f_j) \quad \text{for all } N \in \mathbb{N}. \tag{2.32}$$

Consequently, $\mathcal{M}_\Sigma(\sum_{j=1}^{\infty} |f_j|) \leq \sum_{j=1}^{\infty} \mathcal{M}_\Sigma(f_j)$, and hence

$$\mathcal{M}_\Sigma\left(\sum_{j=1}^{\infty} f_j\right) \leq \mathcal{M}_\Sigma\left(\sum_{j=1}^{\infty} |f_j|\right) \leq \sum_{j=1}^{\infty} \mathcal{M}_\Sigma(f_j) \tag{2.33}$$

as wanted. □

We now present a lemma, stated and proved in [79, (8.45)–(8.51), pp. 506–507]. For completeness, we record the proof here as well. See also [86, Proposition 7.6.8, p. 632] for a more general result of this flavor. Remarkably, we have a quantitative description of how the Hardy–Littlewood maximal operator acts on characteristic functions of surface balls.

Lemma 2.25. *Let $\Sigma \subseteq \mathbb{R}^n$ be a closed Ahlfors regular set and abbreviate $\sigma := \mathcal{H}^{n-1}\lfloor\Sigma$. Fix an arbitrary reference point $x_0 \in \Sigma$. Then*

$$\mathcal{M}_\Sigma(\mathbf{1}_{\Delta(x_0,1)})(x) \approx \frac{1}{1+|x-x_0|^{n-1}} \quad \textit{uniformly in } x \in \Sigma, \tag{2.34}$$

with implicit constants that depend only on n and the ADR *character of* Σ.

Proof. Fix $x \in \Sigma$. Then

$$x \in \Delta(x_0, 1+|x-x_0|) \quad \text{and} \quad \Delta(x_0,1) \subseteq \Delta(x_0, 1+|x-x_0|). \tag{2.35}$$

As such, we can estimate

$$\begin{aligned}\mathcal{M}_\Sigma(\mathbf{1}_{\Delta(x_0,1)})(x) &\geq \fint_{\Delta(x_0,1+|x-x_0|)} \mathbf{1}_{\Delta(x_0,1)}\,d\sigma \\ &= \frac{\sigma(\Delta(x_0,1))}{\sigma(\Delta(x_0,1+|x-x_0|))} \\ &\geq \frac{C}{1+|x-x_0|^{n-1}},\end{aligned} \tag{2.36}$$

wherein the last inequality above follows from the Ahlfors regularity of Σ as well as Lemma 2.18, and the constant $C \in (0,\infty)$ depends only on the ADR character of Σ and n. This takes care of one of the inequalities in (2.34).

It remains to show the opposite inequality in (2.34) holds. Note that

$$\mathcal{M}_\Sigma(\mathbf{1}_{\Delta(x_0,1)})(x) \leq \|\mathbf{1}_{\Delta(x_0,1)}\|_{L^\infty(\Sigma,\sigma)} = 1. \tag{2.37}$$

Now, consider the case that $x \in \Delta(x_0,2)$. Then because of (2.37) we can estimate

$$\mathcal{M}_\Sigma(\mathbf{1}_{\Delta(x_0,1)})(x) \leq 1 = \frac{1+|x-x_0|^{n-1}}{1+|x-x_0|^{n-1}} \leq \frac{1+2^{n-1}}{1+|x-x_0|^{n-1}}. \tag{2.38}$$

Assume now $x \in \Sigma \setminus \Delta(x_0,2)$. Choose some radius $r \in (0, 2\,\mathrm{diam}(\Sigma))$ and a location $y \in \Sigma$. If $\Delta(y,r) \cap \Delta(x_0,1) = \varnothing$, then

$$\fint_{\Delta(y,r)} \mathbf{1}_{\Delta(x_0,1)}\,d\sigma = 0, \tag{2.39}$$

and the desired inequality is trivially satisfied. Thus, work under the assumption that $\Delta(y,r) \cap \Delta(x_0,1) \neq \varnothing$. Then, in fact,

$$2 \leq |x-x_0| \leq |x-y| + |y-x_0| \leq r + (r+1) = 2r+1. \tag{2.40}$$

This implies $1 \leq 2r$, which inserting back into (2.40) shows $|x-x_0| \leq 4r$. It follows that

$$\begin{aligned}\fint_{\Delta(y,r)} \mathbf{1}_{\Delta(x_0,1)}\,d\sigma &\leq \frac{\sigma(\Delta(x_0,1))}{\sigma(\Delta(y,r))} \leq \frac{C}{r^{n-1}} \\ &\leq \frac{C}{|x-x_0|^{n-1}} \leq \frac{C}{1+|x-x_0|^{n-1}},\end{aligned} \tag{2.41}$$

for some constant $C \in (0,\infty)$ depending only on the ADR character of Σ and n. The second inequality in (2.41) is a consequence of the Ahlfors regularity of Σ, the third inequality uses that $|x-x_0| \leq 4r$, and the last inequality follows from our present assumption that $x \in \Sigma \setminus \Delta(x_0,2)$ and item (4) of Lemma 2.18. We now have for $x \in \Sigma \setminus \Delta(x_0,2)$,

$$
\begin{aligned}
\mathcal{M}_\Sigma(\mathbf{1}_{\Delta(x_0,1)})(x) &= \sup_{\Delta \ni x} \fint_\Delta \mathbf{1}_{\Delta(x_0,1)}\, d\sigma \\
&= \sup_{\substack{r\in(0,2\operatorname{diam}(\Sigma))\\ y\in\Sigma,\, \Delta(y,r)\ni x}} \fint_{\Delta(y,r)} \mathbf{1}_{\Delta(x_0,1)}\, d\sigma \\
&\le \frac{C}{1+|x-x_0|^{n-1}}.
\end{aligned} \tag{2.42}
$$

The suprema are taken over all surface balls containing x, and the last inequality is owed to (2.41). Thus, (2.38) and (2.42) establish the opposite inequality in (2.34) as wanted. In view of (2.36), this completes the proof of Lemma 2.25. □

Definition 2.26. Let $\Sigma \subseteq \mathbb{R}^n$ be a closed set and write $\sigma := \mathcal{H}^{n-1}\lfloor\Sigma$. A weight on Σ is a σ-measurable function $w : \Sigma \to [0,\infty]$ such that $0 < w < \infty$ at σ-a.e.point in Σ.

We often tacitly identify the weight function w with the measure $w\sigma$. With this in mind, we see that σ and w share the same sigma-algebra of sets. Consequently, a property holds w-a. e. if and only if it holds σ-a. e. Additionally, for any σ-measurable set $E \subseteq \Sigma$, we often suggestively write $w(E) := \int_E w\, d\sigma$.

Let $\Sigma \subseteq \mathbb{R}^n$ be a closed set such that $\sigma := \mathcal{H}^{n-1}\lfloor\Sigma$ is doubling. Weighted Lebesgue spaces, written $L^p(\Sigma, w) := L^p(\Sigma, w\sigma)$, are of interest in this work, and Muckenhoupt weights are of particular import, because they have the property that (cf. [86, Lemma 7.7.1, pp. 641–643] for a more general result which holds for spaces of homogeneous type)

for $p \in (1,\infty]$ the Hardy–Littlewood maximal operator $\mathcal{M}_\Sigma$ (cf. (2.26)) is bounded on $L^p(\Sigma, w)$ if and only if w belongs to the Muckenhoupt class $A_p(\Sigma, \sigma)$. (2.43)

Analogously, when $p = 1$, we have that $\mathcal{M}_\Sigma$ satisfies the weak-(1, 1) estimate

$$
\sup_{a\in(0,\infty)} a \cdot w(\{x \in \Sigma : \mathcal{M}_\Sigma f(x) > a\}) \le C\|f\|_{L^1(\Sigma,w)} \tag{2.44}
$$

(where C is independent of f), if and only if $w \in A_1(\Sigma, \sigma)$. That is to say, the mapping

$$
\mathcal{M}_\Sigma : L^1(\Sigma, w) \to L^{1,\infty}(\Sigma, w) \tag{2.45}
$$

is well-defined, sublinear, and bounded whenever w belongs to $A_1(\Sigma, \sigma)$.

Definition 2.27. Let $\Sigma \subseteq \mathbb{R}^n$ be a closed set such that σ is a Borel measure on Σ with the property that $0 < \sigma(\Delta) < +\infty$ for all surface balls $\Delta \subseteq \Sigma$. Fix $p \in [1,\infty)$. The Muckenhoupt class $A_p(\Sigma, \sigma)$ is defined as the collection of weight functions

$$
A_p(\Sigma, \sigma) := \{w : \Sigma \to [0,\infty] : w \text{ is a weight on } \Sigma \text{ and } [w]_{A_p} < \infty\}, \tag{2.46}
$$

where for $p \in (1, \infty)$ and $p' := p/(p-1)$, the A_p characteristic $[w]_{A_p}$ is

$$[w]_{A_p} := \sup_{\Delta \subseteq \Sigma} \Big(⨍_{\Delta} w \, d\sigma \Big) \Big(⨍_{\Delta} w^{1-p'} \, d\sigma \Big)^{p-1} \tag{2.47}$$
$$= \sup_{\Delta \subseteq \Sigma} \Big(⨍_{\Delta} w \, d\sigma \Big) \Big(⨍_{\Delta} w^{-1/(p-1)} \, d\sigma \Big)^{p-1}$$
$$= \sup_{\Delta \subseteq \Sigma} \Big(⨍_{\Delta} w \, d\sigma \Big) \Big(⨍_{\Delta} w' \, d\sigma \Big)^{p-1},$$

where $w' := w^{-p'/p} = w^{-\frac{1}{p-1}} = w^{1-p'}$ is referred to as the conjugate weight to w. Parenthetically, notice how the second writing in (2.47) makes it clear that

$$[\lambda w]_{A_p} = [w]_{A_p} \quad \text{for all } \lambda > 0. \tag{2.48}$$

For $p = 1$, the A_1 characteristic is

$$[w]_{A_1} := \sup_{\Delta \subseteq \Sigma} \big(\operatorname*{ess\,inf}_{x \in \Delta} w(x) \big)^{-1} ⨍_{\Delta} w \, d\sigma. \tag{2.49}$$

Note that the suprema in (2.47) and (2.49) are taken over all surface balls. When $p = \infty$, define

$$A_\infty(\Sigma, \sigma) := \bigcup_{p \in [1,\infty)} A_p(\Sigma, \sigma). \tag{2.50}$$

The notation in (2.50) is perhaps suggestive of the property that the range of Muckenhoupt classes is nested. That is (cf. [86, Lemma 7.7.1, pp. 641–643]),

$$\begin{aligned} &\text{given any } p, q \in [1, \infty) \text{ with } p < q \text{ one has} \\ &A_p(\Sigma, \sigma) \subseteq A_q(\Sigma, \sigma). \end{aligned} \tag{2.51}$$

Remark 2.28. As is apparent from definitions, $A_p(\Sigma, \sigma) \subseteq L^1_{\text{loc}}(\Sigma, \sigma)$ for all $p \in [1, \infty)$, and hence also $A_\infty(\Sigma, \sigma) \subseteq L^1_{\text{loc}}(\Sigma, \sigma)$.

A useful estimate for weights can be derived in the following way. Fix some $p \in (1, \infty)$, and denote by p' the Hölder conjugate of p. Then for each $w \in A_p(\Sigma, \sigma)$ and any surface ball $\Delta \subseteq \Sigma$, Hölder's inequality yields

$$1 = ⨍_{\Delta} 1 \, d\sigma = ⨍_{\Delta} w^{1/p} w^{-1/p} \, d\sigma \tag{2.52}$$

$$\leq \Big(\fint_{\Delta} w \, d\sigma \Big)^{1/p} \Big(\fint_{\Delta} w^{1-p'} \, d\sigma \Big)^{1/p'}$$
$$= \frac{w(\Delta)^{1/p} w^{1-p'}(\Delta)^{1/p'}}{\sigma(\Delta)}.$$

Furthermore, using (2.47) with (2.52) we can say that

$$\frac{w(\Delta)^{1/p} w^{1-p'}(\Delta)^{1/p'}}{\sigma(\Delta)} \leq [w]_{A_p}^{1/p} \quad \text{for all surface balls } \Delta \subseteq \Sigma, \tag{2.53}$$

and in particular $[w]_{A_p} \geq 1$ for all $w \in A_p(\Sigma, \sigma)$.

The following is a summary of useful properties of Muckenhoupt weights on Ahlfors regular sets (cf. [86, Lemma 7.7.1, pp. 641–643], which summarizes many other properties in the context of spaces of homogeneous type).

Proposition 2.29. *Let $\Sigma \subseteq \mathbb{R}^n$ be a closed Ahlfors regular set, and define $\sigma := \mathcal{H}^{n-1}\lfloor\Sigma$.*

(1) *Fix $p \in (1, \infty)$ and let $w \in A_p(\Sigma, \sigma)$. Then there exist $q \in (1, \infty)$, $s \in (1, p)$, and $\varepsilon \in (0, p-1)$, each of which depends only on the Ahlfors regularity character of Σ, n, p, and $[w]_{A_p}$, such that*

$$w^q \in A_p(\Sigma, \sigma), \quad w \in A_{p/s}(\Sigma, \sigma), \quad \text{and} \quad w \in A_{p-\varepsilon}(\Sigma, \sigma). \tag{2.54}$$

(2) *Fix $p \in (1, \infty)$ with Hölder conjugate p', and let $w \in A_p(\Sigma, \sigma)$. Then the conjugate weight $w' := w^{-p'/p}$ belongs to $A_{p'}(\Sigma, \sigma)$. Moreover,*

$$[w']_{A_{p'}} = [w]_{A_p}^{1/(p-1)}. \tag{2.55}$$

(3) *Let $f \in L^1_{\mathrm{loc}}(\Sigma, \sigma)$, not identically zero, and $(\mathcal{M}_\Sigma f)(x) < +\infty$ for σ-a. e. $x \in \Sigma$. Then for each $s \in (1, \infty)$, one has $(\mathcal{M}_\Sigma f)^{1/s} \in A_1(\Sigma, \sigma)$ and there exists a constant $C \in (0, \infty)$ depending only on Σ and n such that*

$$[(\mathcal{M}_\Sigma f)^{1/s}]_{A_1} \leq \frac{Cs}{s-1}. \tag{2.56}$$

(4) *Fix $p \in (1, \infty)$ and $w \in A_p(\Sigma, \sigma)$. Let $\Delta \subseteq \Sigma$ be a surface ball and let $E \subseteq \Delta$ be σ-measurable. Then*

$$\left(\frac{\sigma(E)}{\sigma(\Delta)} \right)^p \leq [w]_{A_p} \cdot \frac{w(E)}{w(\Delta)}. \tag{2.57}$$

In particular, this implies the measure w is doubling.

(5) *Let $w_1, w_2 \in A_1(\Sigma, \sigma)$. For any $p \in (1, \infty)$, one has*

$$w_1 \cdot w_2^{1-p} \in A_p(\Sigma, \sigma) \quad \text{with } [w_1 \cdot w_2^{1-p}] \leq [w_1]_{A_1} [w_2]_{A_1}^{p-1}. \tag{2.58}$$

(6) *Fix $p \in [1,\infty)$ and any $\theta \in [0,1]$. If $w_1, w_2 \in A_p(\Sigma,\sigma)$, then $w_1^\theta \cdot w_2^{1-\theta}$ belongs to $A_p(\Sigma,\sigma)$ with $[w_1^\theta \cdot w_2^{1-\theta}]_{A_p} \le [w_1]_{A_p}^\theta [w_2]_{A_p}^{1-\theta}$. In particular, corresponding to the case that w_2 is constant, $w \in A_p(\Sigma,\sigma)$ implies $w^\theta \in A_p(\Sigma,\sigma)$ for any $\theta \in [0,1]$.*

This next lemma details a sufficient condition for being an A_1 weight.

Lemma 2.30. *Let $\Sigma \subseteq \mathbb{R}^n$ be a closed Ahlfors regular set, and denote $\sigma := \mathcal{H}^{n-1}\lfloor\Sigma$. Let $w : \Sigma \to [-\infty,+\infty]$ be σ-measurable with $w \neq 0$ for σ-a. e. point on Σ and $w < +\infty$ at σ-a. e. point on Σ. Additionally, suppose there exists some $C_0 \in (0,\infty)$ such that $(\mathcal{M}_\Sigma w)(x) \le C_0 w(x)$ for σ-a. e. point $x \in \Sigma$.*

Then, in fact, $w \in A_1(\Sigma,\sigma)$ and there exists some $C_\Sigma \in (0,\infty)$, depending only on the Ahlfors regularity constants of Σ, such that

$$[w]_{A_1} \le C_0 C_\Sigma. \tag{2.59}$$

Proof. Fix an arbitrary surface ball $\Delta \subseteq \Sigma$. By the definition of $\mathcal{M}_\Sigma$ in (2.26), we see that there exists some $C_\Sigma \in (0,\infty)$, independent of Δ, such that

$$\fint_\Delta w\,d\sigma \le C_\Sigma \cdot (\mathcal{M}_\Sigma w)(x) \quad \text{for } \sigma\text{-a. e. point } x \in \Delta. \tag{2.60}$$

By assumption, we find that $C_\Sigma(\mathcal{M}_\Sigma w)(x) \le C_0 C_\Sigma w(x)$ for σ-a. e. point $x \in \Sigma$. Taking an essential infimum over all $x \in \Delta$, we obtain from (2.60) that

$$\fint_\Delta w\,d\sigma \le C_0 C_\Sigma \cdot \operatorname*{ess\,inf}_{x\in\Delta} w(x). \tag{2.61}$$

Since the constant $C_0 C_\Sigma$ is independent of Δ, the definition of the A_1 characteristic (2.49) shows that indeed, $[w]_{A_1} < C_0 C_\Sigma < \infty$ and $w \in A_1(\Sigma,\sigma)$ as wanted. □

Remark 2.31. As it turns out, any function w satisfying the hypotheses of Lemma 2.30 is strictly positive almost everywhere. Indeed, let w be as in the statement of Lemma 2.30. If $w = 0$ at σ-a. e. point on Σ, we are done, so assume otherwise. Then necessarily $(\mathcal{M}_\Sigma w)(x) > 0$ for all $x \in \Sigma$. To see this, let $x \in \Sigma$ and suppose for the sake of a contradiction that $(\mathcal{M}_\Sigma w)(x) = 0$. Then

$$\sup_{r>0} \fint_{\Delta(x,r)} |w|\,d\sigma = 0 \implies \int_{\Delta(x,r)} |w|\,d\sigma = 0 \quad \text{for all } r > 0. \tag{2.62}$$

This further implies $w = 0$ at σ-a. e. point in $\Delta(x,r)$ for any $r > 0$, and hence $w = 0$ at σ-a. e. point on Σ, a contradiction.

Closely related to the Muckenhoupt class is the reverse Hölder class.

Definition 2.32. Let $\Sigma \subseteq \mathbb{R}^n$ be a closed Ahlfors regular set, and write $\sigma := \mathcal{H}^{n-1}\lfloor\Sigma$. Fix $q \in (1, \infty)$. A weight $w : \Sigma \to [0, \infty]$ belongs to the `reverse Hölder class`, $\mathrm{RH}_q(\Sigma, \sigma)$ provided there exists a $C \in (0, \infty)$ such that

$$\Big(\fint_{\Delta} w^q \, d\sigma\Big)^{1/q} \leq C \fint_{\Delta} w \, d\sigma, \tag{2.63}$$

for all surface balls $\Delta \subseteq \Sigma$. The infimum over all such constants C for which (2.63) holds is denoted by $[w]_{\mathrm{RH}_q}$, and referred to as the `reverse Hölder characteristic` of w. Given any σ-measurable function f on Σ, any integrability exponent $q \in (1, \infty)$, and any weight $w \in \mathrm{RH}_q(\Sigma, \sigma)$, as a consequence of Hölder's inequality and (2.63) (with $C := [w]_{\mathrm{RH}_q}$), one has

$$\begin{aligned} \fint_{\Delta} |f| \, dw &\leq \frac{1}{w(\Delta)} \Big(\int_{\Delta} |f|^{q'} \, d\sigma\Big)^{1/q'} \Big(\int_{\Delta} w^q \, d\sigma\Big)^{1/q} \\ &= \frac{\sigma(\Delta)^{1/q}}{w(\Delta)} \Big(\int_{\Delta} |f|^{q'} \, d\sigma\Big)^{1/q'} \Big(\fint_{\Delta} w^q \, d\sigma\Big)^{1/q} \\ &\leq \frac{\sigma(\Delta)^{1/q}}{w(\Delta)} \Big(\fint_{\Delta} |f|^{q'} \, d\sigma\Big)^{1/q'} [w]_{\mathrm{RH}_q} \fint_{\Delta} w \, d\sigma \\ &= [w]_{\mathrm{RH}_q} \Big(\fint_{\Delta} |f|^{q'} \, d\sigma\Big)^{1/q'} \end{aligned} \tag{2.64}$$

for every surface ball $\Delta \subseteq \Sigma$, where $q' := q/(q-1)$.

The following is a summary of useful properties of the reverse Hölder class. The proofs of (1), (2), and (3) may be found in [86, Section 7.7]. The result in (4) may be traced back to [142], and the proof in [66, Lemma 3.1, p. 264] naturally adapts to our more general setting of interest. An equivalent version of (5) in the one-dimensional Euclidean setting may be found in [67, (P6), p. 641] (see also [20]), which says for $p, \sigma \in (1, \infty)$ that $w^\sigma \in A_p$ if and only if $w \in A_{1+(p-1)/\sigma} \cap \mathrm{RH}_\sigma$. We choose to express this result differently to better highlight the alternative characterization of $A_p \cap \mathrm{RH}_q$.

Proposition 2.33. *Let $\Sigma \subseteq \mathbb{R}^n$ be a closed Ahlfors regular set, and define $\sigma := \mathcal{H}^{n-1}\lfloor\Sigma$.*

(1) *The reverse Hölder classes are monotonic. That is, if $1 < q_1 \leq q_2 < \infty$, then*

$$\mathrm{RH}_{q_2}(\Sigma, \sigma) \subseteq \mathrm{RH}_{q_1}(\Sigma, \sigma). \tag{2.65}$$

(2) *Fix $p \in [1, \infty)$ and some $w \in A_p(\Sigma, \sigma)$. There exist $q \in (1, \infty)$ and $C \in (0, \infty)$, both depending on the* AR *character of Σ, n, p, and $[w]_{A_p}$ such that*

$$\left(\fint_{\Delta} w^q \, d\sigma\right)^{1/q} \le C \fint_{\Delta} w \, d\sigma \quad \textit{for each surface ball } \Delta \subseteq \Sigma. \tag{2.66}$$

That is, $w \in \mathrm{RH}_q(\Sigma, \sigma)$. Moreover, for any given σ-measurable function $f : \Sigma \to \mathbb{C}$, and any surface ball $\Delta \subseteq \Sigma$, one has

$$\fint_{\Delta} |f| \, dw \le C\left(\fint_{\Delta} |f|^{q'} \, d\sigma\right)^{1/q'}, \tag{2.67}$$

where q' is the Hölder conjugate of q and C is as in (2.66).

(3) *Fix $q \in (1, \infty)$ and let $w \in \mathrm{RH}_q(\Sigma, \sigma)$, then for any surface ball $\Delta \subseteq \Sigma$ and any σ-measurable $E \subseteq \Delta$, one has*

$$\frac{w(E)}{w(\Delta)} \le [w]_{\mathrm{RH}_q}\left(\frac{\sigma(E)}{\sigma(\Delta)}\right)^{1/q'}, \tag{2.68}$$

where q' denotes the Hölder conjugate exponent of q. In fact, starting with any $w \in A_p(\Sigma, \sigma)$ for some $p \in [1, \infty)$, there exists $q \in (1, \infty)$ with Hölder conjugate q' such that (2.68) holds with $[w]_{\mathrm{RH}_q}$ replaced by the constant $C \in (0, \infty)$ appearing in (2.66).

(4) *Fix $q \in (1, \infty)$. Then $w \in \mathrm{RH}_q(\Sigma, \sigma)$ if and only if $w^q \in A_\infty(\Sigma, \sigma)$.*

(5) *Fix $p \in [1, \infty)$ and some $q \in (1, \infty)$. Then*

$$w \in A_p(\Sigma, \sigma) \cap \mathrm{RH}_q(\Sigma, \sigma) \quad \textit{if and only if} \quad w^q \in A_{q(p-1)+1}(\Sigma, \sigma). \tag{2.69}$$

(6) *If $w \in A_p(\Sigma, \sigma)$ for some $p \in (1, \infty)$, then for any $\theta \in (0, 1)$, the weight w^θ belongs to $\mathrm{RH}_{1/\theta}(\Sigma, \sigma)$. Moreover, $w^\theta \in A_p(\Sigma, \sigma) \cap \mathrm{RH}_{1/\theta}(\Sigma, \sigma)$.*

Proof. To prove (5), we start with the forward implication in the case that $p = 1$. Fix $q \in (1, \infty)$ and suppose $w \in A_1(\Sigma, \sigma) \cap \mathrm{RH}_q(\Sigma, \sigma)$. Let $\Delta \subseteq \Sigma$ be an arbitrary surface ball. Then there exists some constant $C \in (0, \infty)$ such that

$$\begin{aligned}
\left(\operatorname*{ess\,inf}_{x \in \Delta} w(x)^q\right)^{-1} \fint_{\Delta} w^q \, d\sigma &= \left(\operatorname*{ess\,inf}_{x \in \Delta} w(x)\right)^{-q} \fint_{\Delta} w^q \, d\sigma \\
&\le C\left(\operatorname*{ess\,inf}_{x \in \Delta} w(x)\right)^{-q}\left(\fint_{\Delta} w \, d\sigma\right)^q \\
&\le C[w]_{A_1}^q.
\end{aligned} \tag{2.70}$$

The first inequality follows from the fact that $w \in \mathrm{RH}_q(\Sigma, \sigma)$, and the last inequality uses the fact that $w \in A_1(\Sigma, \sigma)$. Thus, from (2.49) we find that $w^q \in A_1(\Sigma, \sigma)$.

Now we prove the forward implication when $p > 1$. Fix $p \in (1,\infty)$ and $q \in (1,\infty)$, and define $s := q(p-1)+1$. Let p', q', and s' denote the Hölder conjugates of p, q, and s, respectively. Notice that $q(1-s') = 1-p'$. Indeed,

$$\frac{1}{s'} = 1 - \frac{1}{q(p-1)+1} \iff s' = 1 + \frac{1}{q(p-1)}, \tag{2.71}$$

so that

$$q(1-s') = -\frac{1}{p-1} = 1-p'. \tag{2.72}$$

Subsequently, since $w \in \mathrm{RH}_q(\Sigma,\sigma)$, for an arbitrary surface ball $\Delta \subseteq \Sigma$ there exists $C \in (0,\infty)$ such that

$$\begin{aligned}\Big(\fint_\Delta w^q\,d\sigma\Big)\Big(\fint_\Delta w^{q(1-s')}\Big)^{s-1} &\le C\Big(\fint_\Delta w\,d\sigma\Big)^q\Big(\fint_\Delta w^{1-p'}\,d\sigma\Big)^{q(p-1)}\\ &\le C[w]_{A_p}^q,\end{aligned} \tag{2.73}$$

where the last inequality follows bearing in mind (2.47) and that w belongs to $A_p(\Sigma,\sigma)$ by assumption. Thus, since $\Delta \subseteq \Sigma$ was arbitrary, (2.73) and (2.47) imply that w^q belongs to $A_{q(p-1)+1}(\Sigma,\sigma)$ as desired.

For the reverse implication, suppose $w^q \in A_{q(p-1)+1}(\Sigma,\sigma)$. Then by (2.50), we know that $w^q \in A_\infty(\Sigma,\sigma)$, and hence $w \in \mathrm{RH}_q(\Sigma,\sigma)$ by (4). It remains to show that w belongs to $A_p(\Sigma,\sigma)$. First, consider the case that $p=1$, and let $w^q \in A_1(\Sigma,\sigma)$. For any surface ball $\Delta \subseteq \Sigma$, we can estimate via (2.49) and Hölder's inequality,

$$\begin{aligned}[w]_{A_1} &\ge \big(\operatorname*{ess\,inf}_{x\in\Delta} w^q(x)\big)^{-1}\fint_\Delta w^q\,d\sigma\\ &\ge \big(\operatorname*{ess\,inf}_{x\in\Delta} w(x)\big)^{-q}\Big(\fint_\Delta w\,d\sigma\Big)^q.\end{aligned} \tag{2.74}$$

Given $\Delta \subseteq \Sigma$ is arbitrary, this implies $[w]_{A_1} \le [w^q]_{A_1}^{1/q} < +\infty$ (where finiteness is by assumption), showing that $w \in A_1(\Sigma,\sigma)$ as required. Next, suppose $p \in (1,\infty)$ and define $s := q(p-1)+1$ as before, which has Hölder conjugate $s' = 1 + \frac{1}{q(p-1)}$ (cf. (2.71)). We find for any surface ball $\Delta \subseteq \Sigma$ that

$$\begin{aligned}[w^q]_{A_s} &\ge \Big(\fint_\Delta w^q\,d\sigma\Big)\Big(\fint_\Delta w^{1-p'}\,d\sigma\Big)^{q(p-1)}\\ &\ge \Big(\fint_\Delta w\,d\sigma\Big)^q\Big(\fint_\Delta w^{1-p'}\,d\sigma\Big)^{q(p-1)}.\end{aligned} \tag{2.75}$$

Notice the use of (2.72) and (2.47) for the first inequality, and Hölder's inequality for the second. Given that $\Delta \subseteq \Sigma$ was arbitrary, (2.75) implies that $[w]_{A_p} \leq [w^q]_{A_s}^{1/q} < +\infty$ (where finiteness is by assumption), so in fact, $w \in A_p(\Sigma, \sigma)$ as well, completing the proof of item (5).

The proof of (6) follows from the observation that by (4) (and since $\theta \in (0,1)$ implies $1/\theta \in (1,\infty)$), we have $w^\theta \in \mathrm{RH}_{1/\theta}(\Sigma,\sigma)$ if and only if $w = (w^\theta)^{1/\theta} \in A_\infty(\Sigma,\sigma)$ if any only if $w \in A_p(\Sigma,\sigma)$ for some $p \in (1,\infty)$. The fact that $w^\theta \in A_p(\Sigma,\sigma)$ as well follows from item (6) of Lemma 2.29. □

The following lemma, taken from [79, Lemma 2.12, p. 133], shows that an average integral with respect to a "surface" measure can be estimated by an average integral involving a Muckenhoupt weight acting as the measure of integration. This is a highly useful result, because it allows one to inject a Muckenhoupt weight into a given inequality in which such a weight may be missing.

Lemma 2.34. *Suppose $\Sigma \subseteq \mathbb{R}^n$ is a closed set, which is Ahlfors regular, and abbreviate $\sigma := \mathcal{H}^{n-1}\lfloor\Sigma$. Then for each $p \in (1,\infty)$, each Muckenhoupt weight $w \in A_p(\Sigma,\sigma)$, and each σ-measurable function f on Σ one has*

$$\fint_\Delta |f|\,d\sigma \leq [w]_{A_p}^{1/p}\Big(\fint_\Delta |f|^p\,dw\Big)^{1/p}, \quad \text{for each surface ball } \Delta \subseteq \Sigma. \tag{2.76}$$

Conversely, if $p \in (1,\infty)$ and w is a weight on Σ with the property that there exists a constant $C \in (0,\infty)$ such that

$$\fint_\Delta |f|\,d\sigma \leq C\Big(\fint_\Delta |f|^p\,dw\Big)^{1/p} \quad \text{for each function } f \in L^p_{\mathrm{loc}}(\Sigma,w) \text{ and surface ball } \Delta \subseteq \Sigma, \tag{2.77}$$

then actually $w \in A_p(\Sigma,\sigma)$ and $[w]_{A_p} \leq C^p$.

From Lemma 2.34, we obtain the following corollary, which allows for membership of a Muckenhoupt-weighted local Lebesgue space into an unweighted local Lebesgue space.

Corollary 2.35. *Suppose $\Sigma \subseteq \mathbb{R}^n$ is a closed Ahlfors regular set and write $\sigma := \mathcal{H}^{n-1}\lfloor\Sigma$. Fix any integrability exponent $p \in (1,\infty)$ and any Muckenhoupt weight $w \in A_p(\Sigma,\sigma)$. Then there exists $p_0 \in (1,\infty)$ such that*

$$L^p_{\mathrm{loc}}(\Sigma,w) \subseteq L^{p_0}_{\mathrm{loc}}(\Sigma,\sigma). \tag{2.78}$$

Proof. Let f be a σ-measurable function on Σ. By item (1) of Proposition 2.29, there exists $\varepsilon \in (0,p-1)$ such that $w \in A_{p-\varepsilon}(\Sigma,\sigma)$. Now, fix an arbitrary $p_0 \in (1,\infty)$ to be determined later. Using (2.76) with $p - \varepsilon$ playing the role of p, and $|f|^{p_0}$ playing the role of f, we find upon taking p_0-th roots that

$$\left(\fint_\Delta |f|^{p_0}\,d\sigma\right)^{1/p_0} \le [w]_{A_{p-\varepsilon}}^{1/(p_0(p-\varepsilon))}\left(\fint_\Delta |f|^{p_0(p-\varepsilon)}\,dw\right)^{1/(p_0(p-\varepsilon))} \tag{2.79}$$

for any surface ball $\Delta \subseteq \Sigma$. We now choose $p_0 := p/(p-\varepsilon) \in (1,\infty)$, and (2.79) becomes

$$\left(\fint_\Delta |f|^{p_0}\,d\sigma\right)^{1/p_0} \le [w]_{A_{p-\varepsilon}}^{1/p}\left(\fint_\Delta |f|^{p}\,dw\right)^{1/p}. \tag{2.80}$$

Thus, if $f \in L^p_{\text{loc}}(\Sigma, w)$, then necessarily $f \in L^{p_0}_{\text{loc}}(\Sigma,\sigma)$, which gives the desired inclusion advertised in (2.78). □

Definition 2.36. Let $\Sigma \subseteq \mathbb{R}^n$ be a closed set, and let μ be a doubling Borel measure on Σ. The space of `bounded mean oscillations`, $\mathrm{BMO}(\Sigma,\mu)$, is the collection of functions $f \in L^1_{\text{loc}}(\Sigma,\mu)$ such that

$$\|f\|_{\mathrm{BMO}(\Sigma,\mu)} := \sup_{\Delta\subseteq\Sigma} \fint_\Delta \left| f - \fint_\Delta f\,d\mu\right| d\mu < \infty, \tag{2.81}$$

where the supremum is taken over all surface balls $\Delta \subseteq \Sigma$.

Fix an integrability exponent $p \in (1,\infty)$. We also define the L^p-based space of bounded mean oscillations $\mathrm{BMO}_p(\Sigma,\mu)$ as the collection of functions $f \in L^1_{\text{loc}}(\Sigma,\mu)$ such that

$$\|f\|_{\mathrm{BMO}_p(\Sigma,\mu)} := \sup_{\Delta\subseteq\Sigma}\left(\fint_\Delta \left| f - \fint_\Delta f\,d\mu\right|^p d\mu\right)^{1/p} < +\infty, \tag{2.82}$$

where again the supremum is taken over all surface balls $\Delta \subseteq \Sigma$.

Note that both $\|\cdot\|_{\mathrm{BMO}(\Sigma,\mu)}$ and $\|\cdot\|_{\mathrm{BMO}_p(\Sigma,\mu)}$ constitute seminorms, since for any surface ball $\Delta \subseteq \Sigma$ we have $f = \fint_\Delta f\,d\sigma$ whenever f is constant.

With these definitions in hand, we can define the class of δ-AR domains.

Definition 2.37. Given some $\delta \in (0,1)$, an Ahlfors regular domain $\Omega \subseteq \mathbb{R}^n$ is called a `δ-AR domain` provided $\|\nu\|_{[\mathrm{BMO}(\partial\Omega,\sigma)]^n} < \delta$. Here, $\sigma := \mathcal{H}^{n-1}\lfloor\partial\Omega$ and ν denotes the geometric measure theoretic outward unit normal to Ω.

In fact, (cf. [79, Lemma 2.8, p. 52]) for any Ahlfors regular domain $\Omega \subseteq \mathbb{R}^n$ with $\sigma := \mathcal{H}^{n-1}\lfloor\partial\Omega$, it is always the case that

$$\|\nu\|_{[\mathrm{BMO}(\partial\Omega,\sigma)]^n} \le 1. \tag{2.83}$$

Moreover, if $\partial\Omega$ is bounded, then

$$\|\nu\|_{[\mathrm{BMO}(\partial\Omega,\sigma)]^n} = 1, \tag{2.84}$$

and consequently,

$$\text{when } \|\nu\|_{[\mathrm{BMO}(\partial\Omega,\sigma)]^n} < 1, \text{ then } \partial\Omega \text{ is unbounded.} \tag{2.85}$$

We now present several lemmas that will aid in obtaining estimates involving BMO seminorms. This first lemma shows the remarkable fact, owed to the John–Nirenberg inequality, that the BMO seminorm and L^p-based BMO seminorm are equivalent. For justification, see [79, (2.98), p. 47].

Lemma 2.38. *Let $\Sigma \subseteq \mathbb{R}^n$ be a closed set, and let μ be a doubling Borel measure on Σ. For each $f \in L^1_{\mathrm{loc}}(\Sigma,\mu)$ and integrability exponent $p \in (1,\infty)$, one has*

$$\|f\|_{\mathrm{BMO}(\Sigma,\mu)} \approx \|f\|_{\mathrm{BMO}_p(\Sigma,\mu)}. \tag{2.86}$$

Moreover, the implicit constants depend only on the doubling character of μ and p.

This next lemma presents another remarkable fact, namely that the BMO seminorm with respect to the surface measure is equivalent to the BMO seminorm with respect to any Muckenhoupt weight. For a proof, see [79, Lemma 2.14, p. 143].

Lemma 2.39. *Let $\Sigma \subseteq \mathbb{R}^n$ be a closed ADR set, and write $\sigma := \mathcal{H}^{n-1}\lfloor\Sigma$. Fix an integrability exponent $p \in [1,\infty)$. For each $f \in L^1_{\mathrm{loc}}(\Sigma,\sigma)$, and Muckenhoupt weight $w \in A_p(\Sigma,\sigma)$, one has*

$$\|f\|_{\mathrm{BMO}(\Sigma,\sigma)} \approx \|f\|_{\mathrm{BMO}(\Sigma,w)}. \tag{2.87}$$

Moreover, the implicit constants depend only on the ADR character of Σ, n, p, and $[w]_{A_p}$.

The following lemma and proof can be found in [79, Proposition 2.15, p. 100].

Lemma 2.40. *Let $\Omega \subseteq \mathbb{R}^n$ be an AR domain, abbreviate $\sigma := \mathcal{H}^{n-1}\lfloor\partial\Omega$, and denote by ν the geometric measure theoretic outward unit normal to Ω. Let $z \in \partial\Omega$ and $R > 0$, and define $\nu_{\Delta(z,R)} := ⨍_{\Delta(z,R)} \nu\, d\sigma$. Then there exists a constant $C \in (0,\infty)$ (depending only on n and $\partial\Omega$) such that for each dilation parameter $\lambda \in [1,\infty)$,*

$$\begin{aligned} \sup_{z\in\partial\Omega}\sup_{R>0}\sup_{x,y\in\Delta(z,R)} & R^{-1}\left|\langle x-y, \nu_{\Delta(z,\lambda R)}\rangle\right| \\ & \leq C\lambda(1+\log_2\lambda)\|\nu\|_{[\mathrm{BMO}(\partial\Omega,\sigma)]^n}. \end{aligned} \tag{2.88}$$

A fundamental result in the theory of Muckenhoupt weights is Rubio de Francia's extrapolation theorem, which in the context of Ahlfors regular sets reads as follows (cf. [19], [59, Proposition 2.17, p. 2603] for a more general statement in the context of spaces of homogeneous type).

Proposition 2.41. *Let $\Sigma \subseteq \mathbb{R}^n$ be a closed ADR set and write $\sigma := \mathcal{H}^{n-1}\lfloor\Sigma$. Let f and g be two nonnegative σ-measurable functions on Σ with the property that there exists $p_0 \in [1,\infty)$ such that for every Muckenhoupt weight $\omega \in A_{p_0}(\Sigma,\sigma)$ one has*

$$\|g\|_{L^{p_0}(\Sigma,\omega)} \leq C_\omega \|f\|_{L^{p_0}(\Sigma,\omega)}, \tag{2.89}$$

where the constant $C_\omega \in (0,\infty)$ depends only on $[\omega]_{A_{p_0}}$ in a manner that renders C_ω bounded if $[\omega]_{A_{p_0}}$ stays bounded.

Then for each integrability exponent $p \in (1,\infty)$ (and also for $p = 1$ if $p_0 = 1$) and each weight $w \in A_p(\Sigma,\sigma)$ there exists a constant $K \in (0,\infty)$, depending only on Σ, n, p, and $[w]_{A_p}$, and a constant $W \in (0,\infty)$, depending only on Σ, n, and p, such that

$$\|g\|_{L^p(\Sigma,w)} \leq C_w \|f\|_{L^p(\Sigma,w)}, \tag{2.90}$$

where

$$C_w := K \cdot \Big(\sup_{\substack{\omega \in A_{p_0}(\Sigma,\sigma) \\ [\omega]_{A_{p_0}} \leq W}} C_\omega \Big) \in (0,\infty). \tag{2.91}$$

In Chapter 4, we will develop a result with a similar philosophy, which will allow us to extrapolate from Muckenhoupt-weighted Lebesgue spaces to Muckenhoupt-weighted Morrey spaces.

To finish this chapter, we introduce the following lemma, which details a dyadic grid structure on closed Ahlfors regular sets. This lemma is a specific case of the result recorded in [86, Proposition 7.5.4, pp. 612–615], which details a dyadic grid structure on geometrically doubling quasimetric spaces. The result below can be traced to M. Christ [17] (see also [27, Theorem 3.2, p. 143]), and [58, Proposition 2.11, p. 19], which slightly refines the result due to Christ.

Lemma 2.42. *Let $\Sigma \subseteq \mathbb{R}^n$ be a closed ADR set, abbreviate $\sigma := \mathcal{H}^{n-1}\lfloor\Sigma$, and fix some $\kappa_\Sigma \in \mathbb{Z} \cup \{\infty\}$ with the property that*

$$2^{-\kappa_\Sigma - 1} \leq \operatorname{diam}(\Sigma) \leq 2^{-\kappa_\Sigma}. \tag{2.92}$$

Then there are finite constants $a_1 \geq a_0 > 0$ such that for each $m \in \mathbb{Z}$ with $m \geq \kappa_\Sigma$ there exists a collection

$$\mathbb{D}_m(\Sigma) := \{Q_\alpha^m\}_{\alpha \in I_m} \tag{2.93}$$

of subsets of Σ indexed by a nonempty, at most countable set of indices I_m, as well as a family $\{x_\alpha^m\}_{\alpha \in I_m}$ of points in Σ, for which the collection of all dyadic cubes in Σ, i. e.,

$$\mathbb{D}(\Sigma) := \bigcup_{m \in \mathbb{Z},\, m \geq \kappa_\Sigma} \mathbb{D}_m(\Sigma), \tag{2.94}$$

has the following properties:

(1) [All dyadic cubes are open]. *For each $m \in \mathbb{Z}$ with $m \geq \kappa_\Sigma$ and each $\alpha \in I_m$ the set Q_α^m is relatively open in Σ.*

(2) [Dyadic cubes are mutually disjoint within the same generation]. *For each $m \in \mathbb{Z}$ with $m \geq \kappa_\Sigma$ and each $\alpha, \beta \in I_m$ with $\alpha \neq \beta$ there holds $Q_\alpha^m \cap Q_\beta^m = \varnothing$;*

(3) [No partial overlap across generations]. *For each $m, \ell \in \mathbb{Z}$, which satisfy $\ell > m \geq \kappa_\Sigma$ and each $\alpha \in I_m$, $\beta \in I_\ell$, either $Q_\beta^\ell \subseteq Q_\alpha^m$ or $Q_\alpha^m \cap Q_\beta^\ell = \varnothing$.*

(4) [Any dyadic cube has a unique ancestor in any earlier generation]. *For each integers $m, \ell \in \mathbb{Z}$ with $m > \ell \geq \kappa_\Sigma$ and each $\alpha \in I_m$ there is a unique $\beta \in I_\ell$ such that $Q_\alpha^m \subseteq Q_\beta^\ell$. In particular, for each $m \in \mathbb{Z}$ and each $\alpha \in I_m$ there exists a unique $\beta \in I_{m-1}$ such that $Q_\alpha^m \subseteq Q_\beta^{m-1}$ (a scenario in which Q_β^{m-1} is referred to as the parent of Q_α^m).*

(5) [The size is dyadically related to the generation]. *For each $m \in \mathbb{Z}$ with $m \geq \kappa_\Sigma$ and each $\alpha \in I_m$, one has*

$$\Delta(x_\alpha^m, a_0 2^{-m}) \subseteq Q_\alpha^m \subseteq \Delta(x_\alpha^m, a_1 2^{-m}). \tag{2.95}$$

(6) [Control of the number of children]. *There exists an integer $M \in \mathbb{N}$ with the property that for each $m \in \mathbb{Z}$ with $m \geq \kappa_\Sigma$ and each $\alpha \in I_m$ one has*

$$\#\{\beta \in I_{m+1} : Q_\beta^{m+1} \subseteq Q_\alpha^m\} \leq M. \tag{2.96}$$

Also, this integer may be chosen such that for each $m \in \mathbb{Z}$ with $m \geq \kappa_\Sigma$, each $x \in \Sigma$, and each $r \in (0, 2^{-m})$ the number of Q's in $\mathbb{D}_m(\Sigma)$ that intersect $\Delta(x, r)$ is at most M.

(7) [Any generation covers a dense subset of the entire space]. *For each integer $m \in \mathbb{Z}$ with $m \geq \kappa_\Sigma$, the set $\bigcup_{\alpha \in I_m} Q_\alpha^m$ is dense in Σ. Additionally, there exist $b_0, b_1 \in (0, \infty)$, depending only on the* ADR *character of Σ with the property that for each $x \in \Sigma$ and each $r \in (0, 2\,\mathrm{diam}(\Sigma))$ there exist $m \in \mathbb{Z}$ with $m \geq \kappa_\Sigma$ and $\alpha \in I_m$ such that $Q_\alpha^m \subseteq \Delta(x, r)$ and $b_0 r \leq 2^{-k} \leq b_1 r$.*

(8) [Each generation covers the space σ-a. e.]. *For each $m \in \mathbb{Z}$ with $m \geq \kappa_\Sigma$, one has*

$$\sigma\Big(\Sigma \setminus \bigcup_{\alpha \in I_m} Q_\alpha^m\Big) = 0. \tag{2.97}$$

In particular,

$$N := \bigcup_{m \in \mathbb{Z}} \Big(\Sigma \setminus \bigcup_{\alpha \in I_m} Q_\alpha^m\Big) \Longrightarrow \sigma(N) = 0, \tag{2.98}$$

and for each $m \in \mathbb{Z}$ with $m \geq \kappa_\Sigma$ and each $\alpha \in I_m$ one has

$$\sigma\Big(Q_\alpha^m \setminus \bigcup_{\beta \in I_{m+1},\, Q_\beta^{m+1} \subseteq Q_\alpha^m} Q_\beta^{m+1}\Big) = 0. \tag{2.99}$$

(9) [Dyadic cubes have thin boundaries]. *There exist constants, some small* $\vartheta \in (0,1)$ *along with some large* $C \in (0,\infty)$, *such that for each* $m \in \mathbb{Z}$ *with* $m \geq \kappa_\Sigma$, *each* $\alpha \in I_m$, *and each* $t > 0$ *one has*

$$\sigma(\{x \in Q_\alpha^m : \operatorname{dist}(x, \Sigma \setminus Q_\alpha^m) \leq t \cdot 2^{-m}\}) \leq Ct^\vartheta \cdot \sigma(Q_\alpha^m). \tag{2.100}$$

Definition 2.43. In the context of Lemma 2.42, for each $k \in \mathbb{Z}$, the collection $\mathbb{D}_k(\Sigma)$ will be called the dyadic cubes of generation k. For each $Q \in \mathbb{D}_k(\Sigma)$, define the side length of Q to be $\ell(Q) := 2^{-k}$, and the center of Q to be the point $x_Q := x_\alpha^k \in \Sigma$ if $Q = Q_\alpha^k$.

3 Layer potentials on UR domains

This section serves to set the stage for our discussion on the solvability of boundary value problems in later sections. Much of the machinery we develop in this chapter closely follows the exposition found in [79, Sections 3.1–3.3]. See [89, 90] for a more in-depth treatment and complete proofs.

Fix $n \in \mathbb{N}$ with $n \geq 2$ along with some $M \in \mathbb{N}$. Denote the collection of all homogeneous constant complex coefficient second-order $M \times M$ systems L in $\mathbb{R}^n$ by $\mathfrak{L}$. As such, one may write any system L in $\mathfrak{L}$ as a matrix of differential operators of the form $L = (a_{jk}^{\alpha\beta} \partial_j \partial_k)_{1 \leq \alpha,\beta \leq M}$, where $a_{jk}^{\alpha\beta}$ belongs to $\mathbb{C}$ for each $j, k \in \{1, \ldots, n\}$ and $\alpha, \beta \in \{1, \ldots, M\}$. Note that here and throughout, we adopt the standard convention of summation over repeated indices. It follows that the action of L on any given vector-valued distribution $u = (u_\beta)_{1 \leq \beta \leq M}$ is given via

$$Lu = (a_{jk}^{\alpha\beta} \partial_j \partial_k u_\beta)_{1 \leq \alpha \leq M}. \tag{3.1}$$

Also, denote by $L^\top := (a_{kj}^{\beta\alpha} \partial_j \partial_k)_{1 \leq \alpha,\beta \leq M}$ the (real) transpose of the system L.

Definition 3.1. The `characteristic matrix` of L is defined according to

$$L(\xi) := -\big[(a_{jk}^{\alpha\beta} \xi_j \xi_k)_{1 \leq \alpha,\beta \leq M}\big] \quad \text{for each } \xi = (\xi_i)_{1 \leq i \leq n} \in \mathbb{R}^n. \tag{3.2}$$

Definition 3.2. A system $L \in \mathfrak{L}$ will be called `weakly elliptic` provided the characteristic matrix is invertible for all nonzero vectors; more specifically, $\det[L(\xi)] \neq 0$ for all $\xi \in \mathbb{R}^n \setminus \{0\}$.

Definition 3.3. A more demanding condition on the system L is that of `Legendre–Hadamard (strong) ellipticity` which is characterized by the existence of some $c > 0$ for which

$$\operatorname{Re}\big\langle -L(\xi)\zeta, \overline{\zeta} \big\rangle \geq c|\xi|^2|\zeta|^2 \quad \text{for all } \xi \in \mathbb{R}^n \text{ and } \zeta \in \mathbb{C}^M. \tag{3.3}$$

At this point, we introduce the collection (or library) of all coefficient tensors with constant complex entries,

$$\mathfrak{A} := \{A = (a_{jk}^{\alpha\beta})_{\substack{1 \leq \alpha,\beta \leq M \\ 1 \leq j,k \leq n}} : \text{ each } a_{jk}^{\alpha\beta} \text{ belongs to } \mathbb{C}\}. \tag{3.4}$$

Under the natural operations of componentwise addition and multiplication by scalars, the set $\mathfrak{A}$ is realized as a finite-dimensional vector space (over $\mathbb{C}$), which we endow with the norm

$$\|A\| := \sum_{\substack{1 \leq \alpha,\beta \leq M \\ 1 \leq j,k \leq n}} |a_{jk}^{\alpha\beta}| \quad \text{for each } A = (a_{jk}^{\alpha\beta})_{\substack{1 \leq \alpha,\beta \leq M \\ 1 \leq j,k \leq n}} \in \mathfrak{A}. \tag{3.5}$$

https://doi.org/10.1515/9783111458274-003

Notice that if the transpose of each given $A = (a_{jk}^{\alpha\beta})_{\substack{1\le\alpha,\beta\le M\\1\le j,k\le n}} \in \mathfrak{A}$ is the coefficient tensor $A^\top := (a_{kj}^{\beta\alpha})_{\substack{1\le\alpha,\beta\le M\\1\le j,k\le n}}$, then $\mathfrak{A} \ni A \mapsto A^\top \in \mathfrak{A}$ is an isometry. With each coefficient tensor $A = (a_{jk}^{\alpha\beta})_{\substack{1\le\alpha,\beta\le M\\1\le j,k\le n}} \in \mathfrak{A}$ associate the system $L_A \in \mathfrak{L}$ according to

$$L_A := (a_{jk}^{\alpha\beta}\partial_j\partial_k)_{1\le\alpha,\beta\le M}. \tag{3.6}$$

The following map

$$\mathfrak{A} \ni A \longmapsto L_A \in \mathfrak{L} \tag{3.7}$$

is seen to be both linear and surjective, though it drastically fails to be injective. That is, multiple coefficient tensors may represent the same system, and any given coefficient tensor A associated with the system L is not intrinsic to the system itself. Contrast this with the characteristic matrix $L(\xi)$, which despite requiring an explicit choice of coefficient tensor to write, is independent of the choice of coefficient tensor. In fact, $L_A = L_B$ if and only if $A - B$ is antisymmetric in the lower indices. To see this, given any $(c_{jk})_{1\le j,k\le n}$ which is antisymmetric, i. e., $c_{jk} = -c_{kj}$, we have that $c_{jk}\partial_j\partial_k = 0$ since

$$\begin{aligned}\sum_{j,k=1}^{n} c_{jk}\partial_j\partial_k &= \frac{1}{2}\sum_{j,k=1}^{n} c_{jk}\partial_j\partial_k + \frac{1}{2}\sum_{j,k=1}^{n} c_{kj}\partial_k\partial_j \\ &= \frac{1}{2}\sum_{j,k=1}^{n} c_{jk}\partial_j\partial_k - \frac{1}{2}\sum_{j,k=1}^{n} c_{jk}\partial_j\partial_k = 0.\end{aligned} \tag{3.8}$$

As such, for $A := A = (a_{jk}^{\alpha\beta})_{\substack{1\le\alpha,\beta\le M\\1\le j,k\le n}}$ and $B := B = (b_{jk}^{\alpha\beta})_{\substack{1\le\alpha,\beta\le M\\1\le j,k\le n}}$ such that $A - B$ is antisymmetric in the lower indices, we find that

$$L_A - L_B = ((a_{jk}^{\alpha\beta} - b_{jk}^{\alpha\beta})\partial_j\partial_k)_{1\le\alpha,\beta\le M} = 0, \tag{3.9}$$

so that $L_A = L_B$. The space of all coefficient tensors associated with the system $L \in \mathfrak{L}$ is defined as

$$\mathfrak{A}_L := \{A \in \mathfrak{A} : L = L_A\}, \tag{3.10}$$

and our previous discussion shows that $\mathfrak{A}_L$ is not a singleton set.

One important aspect of differential equations is the notion of a *fundamental solution*. The following theorem proved in [85] demonstrates that a canonical choice of fundamental solution exists for any given weakly elliptic system. It is worth remarking that the fundamental solution described below is intrinsic to the system itself and is therefore independent of the choice of coefficient tensor associated to the system.

Proposition 3.4. *Let L be a homogeneous, second-order, constant complex coefficient, $M \times M$ system in $\mathbb{R}^n$, which is weakly elliptic (cf. Definition 3.2). Then there exists a $M \times M$ matrix-valued function $E = (E_{\alpha\beta})_{1\leq\alpha,\beta\leq M}$, canonically associated with the given system L, such that the following properties are true:*

1. *For each $\alpha, \beta \in \{1, \dots, M\}$, one has that $E_{\alpha\beta} \in \mathscr{C}^\infty(\mathbb{R}^n \setminus \{0\})$ and $E_{\alpha\beta}(x) = E_{\alpha\beta}(-x)$ for every $x \in \mathbb{R}^n \setminus \{0\}$.*
2. *For each fixed point $y \in \mathbb{R}^n$, one has $L[E(\cdot - y)] = \delta_y I_{M\times M}$ in the sense of distributions in $\mathbb{R}^n$, where $I_{M\times M}$ is the $M \times M$ identity matrix and δ_y denotes the Dirac distribution with mass at y in $\mathbb{R}^n$. That is, using the standard Kronecker delta notation,*

$$a^{\alpha\beta}_{jk} \partial_{x_j} \partial_{x_k} [E_{\beta\gamma}(x - y)] = \delta_{\alpha\gamma} \delta_y(x), \quad x \in \mathbb{R}^n, \tag{3.11}$$

   in the sense of distributions, for every $\alpha, \gamma \in \{1, \dots, M\}$.
3. *The transpose of E, i. e., $E^\top = (E_{\beta\alpha})_{1\leq\alpha,\beta\leq M}$, is a fundamental solution for the transpose system $L^\top$. In other words, for each fixed point $y \in \mathbb{R}^n$ one has $L^\top[E^\top(\cdot - y)] = \delta_y I_{M\times M}$ in the sense of distributions in $\mathbb{R}^n$, i. e.,*

$$a^{\beta\alpha}_{kj} \partial_{x_j} \partial_{x_k} [E_{\gamma\beta}(x - y)] = \delta_{\alpha\gamma} \delta_y(x), \quad x \in \mathbb{R}^n, \tag{3.12}$$

   in the sense of distributions, for every $\alpha, \gamma \in \{1, \dots, M\}$.
4. *For every multiindex $\alpha \in \mathbb{N}_0^n$ with $n + |\alpha| > 2$, the function $\partial^\alpha E$ is positive homogeneous of degree $2 - n - |\alpha|$ and there exists a constant $C_\alpha \in (0, \infty)$ with the property that*

$$|(\partial^\alpha E)(x)| \leq C_\alpha |x|^{2-n-|\alpha|} \quad \textit{for all } x \in \mathbb{R}^n \setminus \{0\}. \tag{3.13}$$

   Finally, corresponding to $n = 2$ and $\alpha = (0, \dots, 0)$, there exists a constant $C \in (0, \infty)$ such that $|E(x)| \leq C(1 + |\ln|x||)$ for every $x \in \mathbb{R}^2 \setminus \{0\}$.
5. *Let "hat" denote the Fourier transform in $\mathbb{R}^n$ (originally defined on Schwartz functions, then extended to tempered distributions via duality). Then $\widehat{E}$ is a tempered distribution in $\mathbb{R}^n$ (which is positive homogeneous of degree -2 if $n \geq 3$), whose restriction to $\mathbb{R}^n \setminus \{0\}$ is a (matrix-valued) function of class $\mathscr{C}^\infty$. In fact,*

$$\widehat{E}(\xi) = -[L(\xi)]^{-1} \quad \textit{for every } \xi \in \mathbb{R}^n \setminus \{0\}. \tag{3.14}$$

   More generally, given any $\gamma \in \mathbb{N}_0^n$ it follows that the tempered distribution $\widehat{\partial^\gamma E}$ is a function of class $\mathscr{C}^\infty$ when restricted to $\mathbb{R}^n \setminus \{0\}$ which, regarded as such, satisfies

$$\widehat{\partial^\gamma E}(\xi) = -i^{|\gamma|} \xi^\gamma [L(\xi)]^{-1} \quad \textit{for every } \xi \in \mathbb{R}^n \setminus \{0\}, \tag{3.15}$$

   and

$$\begin{gathered}\textit{if } \gamma \in \mathbb{N}_0^n \textit{ then } \widehat{\partial^\gamma E} = -i^{|\gamma|} \xi^\gamma [L(\xi)]^{-1} \textit{ as tempered distributions in } \mathbb{R}^n \\ \textit{when either } |\gamma| > 0 \textit{ or } n \geq 3.\end{gathered} \tag{3.16}$$

6. *Writing E_L in place of E to emphasize the dependence on L, matters may be arranged so that*

$$\begin{aligned}(E_L)^\top = E_{L^\top}, \quad \overline{(E_L)} = E_{\overline{L}}, \quad (E_L)^* = E_{L^*},\\ \text{as well as} \quad E_{\lambda L} = \lambda^{-1} E_L \quad \text{for each } \lambda \in \mathbb{C}\setminus\{0\},\end{aligned} \tag{3.17}$$

where $\top$, $\bar{\cdot}$, $*$ denote, respectively, transposition, complex conjugation, and complex (or Hermitian) adjunction.

We continue by recording a number of basic definitions and results pertaining to boundary layer potentials. From the perspective of Calderón–Zygmund theory, the class of UR domains is the largest geometric environment in which we could hope to make sense of the following double and single layer potentials while retaining nice boundedness properties.

Definition 3.5. Assume $\Omega \subseteq \mathbb{R}^n$ is a UR domain. Abbreviate $\sigma := \mathcal{H}^{n-1}\lfloor\partial\Omega$ and denote by $\nu = (\nu_1, \dots, \nu_n)$ the geometric measure theoretic outward unit normal to Ω. In addition, consider a homogeneous, second-order, constant complex coefficient, weakly elliptic $M \times M$ system L in $\mathbb{R}^n$ and consider the matrix-valued fundamental solution $E = (E_{\alpha\beta})_{1\le\alpha,\beta\le M}$ associated with L as in Proposition 3.4. Finally, fix a coefficient tensor $A = (a_{jk}^{\alpha\beta})_{\substack{1\le\alpha,\beta\le M\\ 1\le j,k\le n}} \in \mathfrak{A}_L$. In this setting, define the action of the `boundary-to-domain double layer potential operator`, $\mathcal{D}_A$, on each $f \in [L^1(\partial\Omega, \frac{\sigma(x)}{1+|x|^{n-1}})]^M$ via

$$\mathcal{D}_A f(x) := \Big(-\int_{\partial\Omega} \nu_k(y) a_{jk}^{\beta\alpha} (\partial_j E_{\gamma\beta})(x-y) f_\alpha(y)\, d\sigma(y) \Big)_{1\le\gamma\le M} \tag{3.18}$$

at each point $x \in \Omega$. Also, define the action of the `boundary-to-boundary double layer potential operator`, K_A, on each $f \in [L^1(\partial\Omega, \frac{\sigma(x)}{1+|x|^{n-1}})]^M$ by setting

$$K_A f(x) := \Big(-\lim_{\varepsilon\to 0^+} \int_{\substack{y\in\partial\Omega\\ |x-y|>\varepsilon}} \nu_k(y) a_{jk}^{\beta\alpha} (\partial_j E_{\gamma\beta})(x-y) f_\alpha(y)\, d\sigma(y) \Big)_{1\le\gamma\le M} \tag{3.19}$$

at σ-a. e. point $x \in \partial\Omega$. Another singular integral operator which is closely related to (3.19) is the so-called `transpose boundary-to-boundary double layer potential operator`, $K_A^\#$, defined for each $f \in [L^1(\partial\Omega, \frac{\sigma(x)}{1+|x|^{n-1}})]^M$ according to

$$K_A^\# f(x) := \Big(\lim_{\varepsilon\to 0^+} \int_{\substack{y\in\partial\Omega\\ |x-y|>\varepsilon}} \nu_k(x) a_{jk}^{\beta\alpha} (\partial_j E_{\gamma\beta})(x-y) f_\gamma(y)\, d\sigma(y) \Big)_{1\le\alpha\le M} \tag{3.20}$$

at σ-a. e. point $x \in \partial\Omega$.

From (3.13), we see that $[L^1(\partial\Omega, \frac{\sigma(x)}{1+|x|^{n-1}})]^M$ is the most general functional analytic setting in which the integral in (3.18) is absolutely convergent. The double layer operator $\mathcal{D}_A$ may be regarded as a mechanism for generating many null-solutions for the given system L in Ω since, as is apparent from (3.18) and Proposition 3.4,

$$\begin{aligned} &\text{for each function } f \in \Big[L^1\Big(\partial\Omega, \frac{\sigma(x)}{1+|x|^{n-1}}\Big)\Big]^M \text{ we have} \\ &\mathcal{D}_A f \in [\mathscr{C}^\infty(\Omega)]^M \quad \text{and} \quad L(\mathcal{D}_A f) = 0 \quad \text{in } \Omega. \end{aligned} \tag{3.21}$$

The above facts are particularly useful in solving the Dirichlet problem. Since we are presently assuming that Ω is a UR domain, work in [89] also guarantees that (3.19) and (3.20) are well defined in a σ-a. e. pointwise fashion for each function belonging to the space $[L^1(\partial\Omega, \frac{\sigma(x)}{1+|x|^{n-1}})]^M$.

For each coefficient tensor $A \in \mathfrak{A}_L$, we can relate the boundary-to-domain double layer potential operator $\mathcal{D}_A$ to the boundary-to-boundary double layer potential operator K_A via the following jump-formula (proved in [86]). Namely, if I denotes the identity operator and $\kappa \in (0, \infty)$ is an arbitrary aperture parameter, then

$$\begin{aligned} &(\mathcal{D}_A f)\big|_{\partial\Omega}^{\kappa-\text{n.t.}} = \Big(\frac{1}{2}I + K_A\Big)f \quad \text{at } \sigma\text{-a. e. point on } \partial\Omega, \\ &\text{for each given function} \quad f \in \Big[L^1\Big(\partial\Omega, \frac{\sigma(x)}{1+|x|^{n-1}}\Big)\Big]^M. \end{aligned} \tag{3.22}$$

It is worth remarking that the layer potentials are highly sensitive to the choice of coefficient tensor, and our notation is suggestive of this fact. This crucial observation allows for the possibility of choosing a coefficient tensor that creates a "nice" layer potential, i. e., a layer potential with a certain algebraic structure which, given a suitable functional analytic setting, ensures small operator norm estimates of the double layer in terms of the BMO seminorm of the outward unit normal.

Example 3.6. The standard fundamental solution for the Laplacian in $\mathbb{R}^n$ is defined (cf., e. g., [85]) for each $x \in \mathbb{R}^n \setminus \{0\}$ by

$$E_\Delta(x) := \begin{cases} \frac{1}{\omega_{n-1}(2-n)} \frac{1}{|x|^{n-2}}, & \text{if } n \geq 3, \\ \frac{1}{2\pi} \ln|x|, & \text{if } n = 2, \end{cases} \tag{3.23}$$

where, as usual, ω_{n-1} denotes the surface area of the unit sphere in $\mathbb{R}^n$. Given an Ahlfors regular domain $\Omega \subseteq \mathbb{R}^n$, abbreviate $\sigma := \mathcal{H}^{n-1}\lfloor\partial\Omega$ and denote by ν the geometric measure theoretic outward unit normal to Ω. Set $a_{jk}^{\alpha\beta} := a_{jk} := \delta_{jk}$ in (3.1) so that $L = \Delta$, and refer to $\mathcal{D}_\Delta, K_\Delta$ (constructed as in (3.18), (3.19) for this choice of coefficient tensor, i. e., for $A := I_{n\times n}$, the identity matrix) as being the (classical) harmonic double layer potentials. Concretely, for each function $f \in L^1(\partial\Omega, \frac{\sigma(x)}{1+|x|^{n-1}})$ we have (writing in this case, $\mathcal{D}_\Delta$, K_Δ,

$K^{\#}_{\Delta}$ in place of $\mathcal{D}_{I_{n\times n}}, K_{I_{n\times n}}, K^{\#}_{I_{n\times n}}$)

$$\mathcal{D}_{\Delta}f(x) = \frac{1}{\omega_{n-1}} \int_{\partial\Omega} \frac{\langle \nu(y), y - x\rangle}{|x-y|^n} f(y)\, d\sigma(y), \quad \text{for all } x \in \Omega, \tag{3.24}$$

and, at σ-a. e. point $x \in \partial\Omega$,

$$K_{\Delta}f(x) = \lim_{\varepsilon\to 0^+} \frac{1}{\omega_{n-1}} \int_{\substack{y\in\partial\Omega \\ |x-y|>\varepsilon}} \frac{\langle \nu(y), y - x\rangle}{|x-y|^n} f(y)\, d\sigma(y), \tag{3.25}$$

$$K^{\#}_{\Delta}f(x) = \lim_{\varepsilon\to 0^+} \frac{1}{\omega_{n-1}} \int_{\substack{y\in\partial\Omega \\ |x-y|>\varepsilon}} \frac{\langle \nu(x), x - y\rangle}{|x-y|^n} f(y)\, d\sigma(y). \tag{3.26}$$

In the above double layer potentials for the Laplacian, notice the inner product between the chord $x - y$ and ν, the geometric measure theoretic outward unit normal to Ω. Such an algebraic structure grants the above double layers some remarkable properties. For example, when Ω is any half-space, this forces the inner products to be zero and subsequently we see that $K_{\Delta} = K^{\#}_{\Delta} = 0$. The nontangential trace of $\mathcal{D}_{\Delta}$ is then quite simple in view of the jump-formula in (3.22). Further on in this chapter, we will focus on this "chord-dot-normal" structure, which appears in the integral kernel of the double layer potentials for the Laplacian (cf. (3.24)–(3.26)), and use it to try and characterize other weakly elliptic operators and systems, which possess a specific coefficient tensor that leads to a similar algebraic structure.

This next proposition is illustrative of the fact that the boundary-to-domain double layer potential enjoys remarkable regularity properties. More specifically, spacial derivatives of the double layer can be absorbed all the way inside the integral to the boundary function on which the operator acts, assuming said boundary function has reasonable weak derivatives. A proof may be found in [89].

Proposition 3.7. *Let $\Omega \subseteq \mathbb{R}^n$ be an AR domain. Define $\sigma := \mathcal{H}^{n-1}\lfloor\partial\Omega$, and denote by $\nu = (\nu_1, \dots, \nu_n)$ the geometric measure theoretic outward unit normal to Ω. Also, for some $M \in \mathbb{N}$, consider a weakly elliptic, homogeneous, second-order, constant complex coefficient, $M \times M$ system L in $\mathbb{R}^n$, written as in* (3.1) *for some choice of a coefficient tensor $A = (a^{\alpha\beta}_{rs})_{\substack{1\le r,s\le n \\ 1\le\alpha,\beta\le M}}$. Finally, associate with A and Ω the double layer potential operator $\mathcal{D}_A$ as in* (3.18), *and consider a function*

$$\begin{aligned} &f = (f_\alpha)_{1\le\alpha\le M} \in \Big[L^1\Big(\partial\Omega, \frac{\sigma(x)}{1+|x|^{n-1}}\Big)\Big]^M \quad \textit{with the property that} \\ &\partial_{\tau_{jk}} f_\alpha \in L^1\Big(\partial\Omega, \frac{\sigma(x)}{1+|x|^{n-1}}\Big) \quad \textit{for } 1 \le j, k \le n \textit{ and } 1 \le \alpha \le M. \end{aligned} \tag{3.27}$$

Then for each index $\ell \in \{1, \dots, n\}$ and each point $x \in \Omega$ one has

$$\partial_\ell(\mathcal{D}_A f)(x) = \Big(\int_{\partial\Omega} a^{\beta\alpha}_{rs}(\partial_r E_{\gamma\beta})(x-y)(\partial_{\tau_{\ell s}} f_\alpha)(y)\,d\sigma(y)\Big)_{1\le\gamma\le M}. \tag{3.28}$$

In particular, for each aperture parameter $\kappa \in (0,\infty)$, *the nontangential boundary trace*

$$(\nabla\mathcal{D}_A f)\big|^{\kappa-\mathrm{n.t.}}_{\partial\Omega} \quad \text{exists (in } \mathbb{C}^{n\cdot M}) \text{ at } \sigma\text{-a. e. point on } \partial\Omega. \tag{3.29}$$

This next proposition demonstrates how weak tangential derivatives commute with the boundary-to-boundary double layer potential. This can be done, but at the cost of picking up commutators between convolution-type singular integral operators of Calderón–Zygmund type and the operation of pointwise multiplication by components of the geometric measure theoretic outward unit normal.

Proposition 3.8. *Suppose* $\Omega \subseteq \mathbb{R}^n$ *is a* UR *domain, write* $\sigma := \mathcal{H}^{n-1}\lfloor\partial\Omega$, *and denote by* $\nu = (\nu_1,\dots,\nu_n)$ *the geometric measure theoretic outward unit normal to* Ω. *Let* L *be a weakly elliptic, second-order, homogeneous, constant complex coefficient,* $M\times M$ *system in* $\mathbb{R}^n$ *and consider the fundamental solution* $E = (E_{\alpha\beta})_{1\le\alpha,\beta\le M}$ *associated with* L *as in Proposition 3.4. Also, pick a coefficient tensor* $A = (a^{\alpha\beta}_{jk})_{\substack{1\le\alpha,\beta\le M\\ 1\le j,k\le n}} \in \mathfrak{A}_L$ *and bring in* K_A, *the boundary-to-boundary double layer potential operator associated with* Ω *and* A *as in (3.19). In addition, for each* $j,k \in \{1,\dots,n\}$ *define the singular integral operator* U_{jk} *acting on each given matrix-valued function* $F = (F_{\alpha s})_{\substack{1\le\alpha\le M\\ 1\le s\le n}}$ *that belongs to* $[L^1(\partial\Omega, \frac{\sigma(x)}{1+|x|^{n-1}})]^{n\cdot M}$ *as* $U_{jk}F = ((U_{jk}F)_\gamma)_{1\le\gamma\le M}$ *where, for each index* $\gamma \in \{1,\dots,M\}$,

$$\begin{aligned}(U_{jk}F)_\gamma(x) := &-\lim_{\varepsilon\to0^+}\int_{\substack{y\in\partial\Omega\\ |x-y|>\varepsilon}}[\nu_k(x)-\nu_k(y)]\nu_j(y)a^{\beta\alpha}_{rs}(\partial_r E_{\gamma\beta})(x-y)F_{\alpha s}(y)\,d\sigma(y)\\ &+\lim_{\varepsilon\to0^+}\int_{\substack{y\in\partial\Omega\\ |x-y|>\varepsilon}}[\nu_j(x)-\nu_j(y)]\nu_k(y)a^{\beta\alpha}_{rs}(\partial_r E_{\gamma\beta})(x-y)F_{\alpha s}(y)\,d\sigma(y)\\ &+\lim_{\varepsilon\to0^+}\int_{\substack{y\in\partial\Omega\\ |x-y|>\varepsilon}}[\nu_k(y)-\nu_k(x)]\nu_s(y)a^{\beta\alpha}_{rs}(\partial_r E_{\gamma\beta})(x-y)F_{\alpha j}(y)\,d\sigma(y)\\ &-\lim_{\varepsilon\to0^+}\int_{\substack{y\in\partial\Omega\\ |x-y|>\varepsilon}}[\nu_j(y)-\nu_j(x)]\nu_s(y)a^{\beta\alpha}_{rs}(\partial_r E_{\gamma\beta})(x-y)F_{\alpha k}(y)\,d\sigma(y)\end{aligned} \tag{3.30}$$

at σ*-a. e. point* $x\in\partial\Omega$. *Finally, fix some integrability exponents* $p,q\in(1,\infty]$ *and consider a function*

$$\begin{aligned} f &\in \Big[L^1\Big(\partial\Omega, \frac{\sigma(x)}{1+|x|^{n-1}}\Big)\cap L^p_{\mathrm{loc}}(\partial\Omega,\sigma)\Big]^M \quad \text{with the property that}\\ \partial_{\tau_{jk}} f &\in \Big[L^1\Big(\partial\Omega, \frac{\sigma(x)}{1+|x|^{n-1}}\Big)\cap L^q_{\mathrm{loc}}(\partial\Omega,\sigma)\Big]^M \quad \text{for all } j,k\in\{1,\dots,n\}.\end{aligned} \tag{3.31}$$

Then for each $j, k \in \{1, \dots, n\}$ one has

$$\partial_{\tau_{jk}}(K_A f) = K_A(\partial_{\tau_{jk}} f) + U_{jk}(\nabla_{\tan} f) \quad \text{at } \sigma\text{-a. e. point on } \partial\Omega, \tag{3.32}$$

where $\nabla_{\tan} f$ is the $M \times n$ matrix-valued function $F = (F_{\alpha s})_{\substack{1\le\alpha\le M\\ 1\le s\le n}}$ whose entry $F_{\alpha s}$ is the s-th component of $\nabla_{\tan} f_\alpha$.

Remark 3.9. Working in the context of Proposition 3.8, it is worth mentioning that in any half-space the outward unit normal is necessarily constant (cf. [86, Proposition 5.10.12, p. 470]). Therefore, in view of (3.30), we see that when Ω is a half-space, then $U_{jk} = 0$ for each $j, k \in \{1, \dots, n\}$. Subsequently, the identity in (3.32) simplifies to

$$\partial_{\tau_{jk}}(K_A f) = K_A(\partial_{\tau_{jk}} f) \quad \text{at } \sigma\text{-a. e. point on } \partial\Omega. \tag{3.33}$$

Sometimes we will find it useful to work with *modified* versions of the double layers. The following theorem, which is taken from [89, Section 1.5], defines and catalogs numerous useful properties of these so-called modified double layer potentials.

Proposition 3.10. *Let $\Omega \subseteq \mathbb{R}^n$ be a UR domain and write $\sigma := \mathcal{H}^{n-1}\lfloor\partial\Omega$. Denote by $\nu = (\nu_1, \dots, \nu_n)$ the geometric measure theoretic outward unit normal to Ω. Also, let $L = (a_{rs}^{\alpha\beta}\partial_r\partial_s)_{1\le\alpha,\beta\le M}$ be a homogeneous, weakly elliptic, constant (complex) coefficient, second-order $M \times M$ system in $\mathbb{R}^n$.*

Consider the following `modified boundary-to-domain double layer potential operator` *acting on each $f = (f_\alpha)_{1\le\alpha\le M} \in [L^1(\partial\Omega, \frac{\sigma(x)}{1+|x|^n})]^M$ according to (keeping in mind the summation convention over repeated indices):*

$$(\mathcal{D}_{A,\mathrm{mod}} f)(x) := \Big(-\int_{\partial\Omega} \nu_s(y) a_{rs}^{\beta\alpha}\{(\partial_r E_{\gamma\beta})(x-y) - k_1^{(r\gamma\beta)}(-y)\} f_\alpha(y)\, d\sigma(y)\Big)_{1\le\gamma\le M} \tag{3.34}$$

at each point $x \in \Omega$, where

$$k_1^{(r\gamma\beta)} := (\partial_r E_{\gamma\beta}) \cdot \mathbf{1}_{\mathbb{R}^n \setminus \overline{B(0,1)}} \quad \text{for every } r, \gamma, \beta, \tag{3.35}$$

and $E = (E_{\gamma\beta})_{1\le\gamma,\beta\le M}$ is the matrix-valued fundamental solution associated with L as in Proposition 3.4. Then the following properties hold:

(1) *The operator $\mathcal{D}_{A,\mathrm{mod}}$ is meaningfully defined, and satisfies*

$$\begin{aligned} &\mathcal{D}_{A,\mathrm{mod}} f \in [\mathscr{C}^\infty(\Omega)]^M \quad \text{and} \quad L(\mathcal{D}_{A,\mathrm{mod}} f) = 0 \quad \text{in } \Omega, \\ &\text{for each } f \in \Big[L^1\Big(\partial\Omega, \frac{\sigma(x)}{1+|x|^n}\Big)\Big]^M. \end{aligned} \tag{3.36}$$

In addition, the operator $\mathcal{D}_{A,\mathrm{mod}}$ is compatible with $\mathcal{D}_A$ from (3.18), in the sense that for each function f belonging to the smaller space $[L^1(\partial\Omega, \frac{\sigma(x)}{1+|x|^{n-1}})]^M$ the difference

$$C_f := \mathcal{D}_{A,\mathrm{mod}} f - \mathcal{D}_A f \quad \text{is a constant (belonging to } \mathbb{C}^M\text{) in } \Omega. \tag{3.37}$$

As a consequence,

$$\nabla \mathcal{D}_{A,\mathrm{mod}} f = \nabla \mathcal{D}_A f \quad \text{in } \Omega \text{ for each } f \in \Big[L^1\Big(\partial\Omega\,, \frac{\sigma(x)}{1+|x|^{n-1}}\Big)\Big]^M. \tag{3.38}$$

Moreover,

$$\mathcal{D}_{A,\mathrm{mod}} \text{ maps constant } (\mathbb{C}^M\text{-valued) functions on } \partial\Omega \text{ into constant } (\mathbb{C}^M\text{-valued) functions in } \Omega. \tag{3.39}$$

In addition, at each point $x \in \Omega$ one may express

$$\partial^\mu(\mathcal{D}_{A,\mathrm{mod}} f)(x) = \Big(-\int_{\partial\Omega} \nu_s(y) a_{rs}^{\beta\alpha}(\partial^\mu \partial_r E_{\gamma\beta})(x-y) f_\alpha(y)\,d\sigma(y)\Big)_{1\le\gamma\le M} \tag{3.40}$$

for each $\mu \in \mathbb{N}_0^n$ with $|\mu| > 0$ and $f = (f_\alpha)_{1\le\alpha\le M} \in [L^1(\partial\Omega\,, \frac{\sigma(x)}{1+|x|^n})]^M$.
Finally, given any function

$$\begin{aligned} &f = (f_\alpha)_{1\le\alpha\le M} \in \Big[L^1\Big(\partial\Omega\,, \frac{\sigma(x)}{1+|x|^n}\Big)\Big]^M \quad \text{with} \\ &\text{the property that } \partial_{\tau_{jk}} f_\alpha \in L^1\Big(\partial\Omega\,, \frac{\sigma(x)}{1+|x|^{n-1}}\Big) \\ &\text{for all } j,k \in \{1,\dots,n\} \text{ and } \alpha \in \{1,\dots,M\}, \end{aligned} \tag{3.41}$$

it follows that for each index $\ell \in \{1,\dots,n\}$ and each point $x \in \Omega$ one has

$$\partial_\ell(\mathcal{D}_{A,\mathrm{mod}} f)(x) = \Big(\int_{\partial\Omega} a_{rs}^{\beta\alpha}(\partial_r E_{\gamma\beta})(x-y)(\partial_{\tau_{\ell s}} f_\alpha)(y)\,d\sigma(y)\Big)_{1\le\gamma\le M}. \tag{3.42}$$

(2) *Fix an aperture parameter $\kappa \in (0,\infty)$. Then the nontangential boundary trace*

$$(\partial_\ell \mathcal{D}_{A,\mathrm{mod}} f)\big|_{\partial\Omega}^{\kappa-\mathrm{n.t.}} \text{ exists (in } \mathbb{C}^M\text{) at } \sigma\text{-a.e. point on } \partial\Omega \text{ for each function } f \text{ as in (3.41) and each index } \ell \in \{1,\dots,n\}. \tag{3.43}$$

Moreover, for each $\varepsilon > 0$ and each $p \in (1,\infty)$,

$$\begin{aligned} &\mathcal{N}_\kappa^\varepsilon(\nabla(\mathcal{D}_{A,\mathrm{mod}} f)) \in L^p_{\mathrm{loc}}(\partial\Omega,\sigma) \quad \text{for each function} \\ &f = (f_\alpha)_{1\le\alpha\le M} \in \Big[L^1\Big(\partial\Omega, \frac{\sigma(x)}{1+|x|^n}\Big)\Big]^M \quad \text{such that} \\ &\partial_{\tau_{jk}} f_\alpha \in L^1\Big(\partial\Omega\,, \frac{\sigma(x)}{1+|x|^{n-1}}\Big) \cap L^p_{\mathrm{loc}}(\partial\Omega,\sigma) \\ &\text{for all } j,k \in \{1,\dots,n\} \text{ and all } \alpha \in \{1,\dots,M\}. \end{aligned} \tag{3.44}$$

In addition, for each truncation parameter $\varepsilon \in (0,\infty)$ *one has*

$$\begin{aligned}&\mathcal{N}_\kappa^\varepsilon(\mathcal{D}_{A,\mathrm{mod}}f) \in L^p_{\mathrm{loc}}(\partial\Omega,\sigma) \quad \textit{for each function}\\ &f \in \Big[L^1\Big(\partial\Omega, \frac{\sigma(x)}{1+|x|^n}\Big) \cap L^p_{\mathrm{loc}}(\partial\Omega,\sigma)\Big]^M \quad \textit{with } p\in(1,\infty).\end{aligned} \tag{3.45}$$

Furthermore, for each function f as in (3.41) *one has*

$$\begin{aligned}&(\partial_\nu^A(\mathcal{D}_{A,\mathrm{mod}}f))(x) =\\ &\Big(\lim_{\varepsilon\to0^+} \int\limits_{\substack{y\in\partial\Omega\\ |x-y|>\varepsilon}} \nu_i(x)a_{ij}^{\mu\gamma}a_{rs}^{\beta\alpha}(\partial_r E_{\gamma\beta})(x-y)(\partial_{\tau_{js}}f_\alpha)(y)\,d\sigma(y)\Big)_{1\le\mu\le M}\end{aligned} \tag{3.46}$$

where the conormal derivative ∂_ν^A *is considered as in* (3.74).

Next, let $K_{A,\mathrm{mod}}$ *denote the* modified boundary-to-boundary double layer potential operator *acting on each* $f = (f_\alpha)_{1\le\alpha\le M} \in [L^1(\partial\Omega, \frac{\sigma(x)}{1+|x|^n})]^M$ *at* σ*-a. e. point* $x\in\partial\Omega$ *according to*

$$\begin{aligned}K_{A,\mathrm{mod}}f(x) := \Big(-\lim_{\varepsilon\to0^+}\int\limits_{\partial\Omega} \nu_s(y)a_{rs}^{\beta\alpha}\times\\ \times\{k_\varepsilon^{(r\gamma\beta)}(x-y) - k_1^{(r\gamma\beta)}(-y)\}f_\alpha(y)\,d\sigma(y)\Big)_{1\le\gamma\le M}\end{aligned} \tag{3.47}$$

where

$$k_\varepsilon^{(r\gamma\beta)} := (\partial_r E_{\gamma\beta})\cdot \mathbf{1}_{\mathbb{R}^n\setminus\overline{B(0,\varepsilon)}} \quad \textit{for each } \varepsilon>0. \tag{3.48}$$

Then the operator $K_{A,\mathrm{mod}}$ *is compatible with* K_A *(acting on functions belonging to the space* $[L^1(\partial\Omega, \frac{\sigma(x)}{1+|x|^{n-1}})]^M$ *as in* (3.19)) *in the sense that*

$$\begin{aligned}&\textit{for each function } f\in[L^1(\partial\Omega, \tfrac{\sigma(x)}{1+|x|^{n-1}})]^M \textit{ the difference } c_f := K_{A,\mathrm{mod}}f - K_Af \textit{ is}\\ &\textit{a constant (belonging to } \mathbb{C}^M) \textit{ on } \partial\Omega \textit{ such that } |c_f| \le C_{\partial\Omega,L}\cdot\|f\|_{[L^1(\partial\Omega,\frac{\sigma(x)}{1+|x|^{n-1}})]^M}\\ &\textit{for some finite constant } C_{\partial\Omega,L}>0.\end{aligned} \tag{3.49}$$

Furthermore, the following jump-formula holds:

$$\begin{aligned}&(\mathcal{D}_{A,\mathrm{mod}}f)\big|_{\partial\Omega}^{\kappa-\mathrm{n.t.}} = \Big(\frac{1}{2}I + K_{A,\mathrm{mod}}\Big)f \quad \textit{at } \sigma\textit{-a. e. point on } \partial\Omega,\\ &\textit{for each given function } f\in\Big[L^1\Big(\partial\Omega, \frac{\sigma(x)}{1+|x|^n}\Big)\Big]^M,\end{aligned} \tag{3.50}$$

where I is the identity operator. Lastly,

the operator $K_{A,\mathrm{mod}}$ maps constant ($\mathbb{C}^M$-valued) functions on $\partial\Omega$ into constant ($\mathbb{C}^M$-valued) functions on $\partial\Omega$. (3.51)

Remark 3.11. A slightly more general version of Proposition 3.8 continues to hold if instead we work with the modified boundary-to-boundary double layer potential. Namely, retaining the context of Proposition 3.8, for each $j, k \in \{1, \dots, n\}$ one has

$$\partial_{\tau_{jk}}(K_{A,\mathrm{mod}} f) = K_A(\partial_{\tau_{jk}} f) + U_{jk}(\nabla_{\tan} f) \quad \text{at } \sigma\text{-a. e. point on } \partial\Omega. \tag{3.52}$$

At this point, we would like to discuss the kinds of coefficient tensors that are "good" in the sense that they lead to an integral kernel in the double layer potential that has chord-dot-normal structure (see the discussion following Example 3.6). The next result from [86] that we include contains a number of equivalent properties for the coefficient tensor of a weakly elliptic system. This result paves the way for introducing the notion of a "distinguished" coefficient tensor for a given weakly elliptic system; cf. Definition 3.13.

Proposition 3.12. *Let L be a homogeneous, second-order, constant complex coefficient, weakly elliptic $M \times M$ system in $\mathbb{R}^n$, and consider the matrix-valued function defined for each $\xi \in \mathbb{R}^n \setminus \{0\}$ as*

$$(\mathcal{E}_{\gamma\beta}(\xi))_{1\le\gamma,\beta\le M} := -[L(\xi)]^{-1} \in \mathbb{C}^{M\times M} \tag{3.53}$$

(recall that the characteristic matrix $L(\xi)$ of L has been defined in (3.2)). Also, let $E = (E_{\alpha\beta})_{1\le\alpha,\beta\le M}$ be the fundamental solution associated with the given system L as in Proposition 3.4.

Then for each coefficient tensor $A = (a_{jk}^{\alpha\beta})_{\substack{1\le\alpha,\beta\le M\\ 1\le j,k\le n}} \in \mathfrak{A}_L$ (cf. (3.10)) the following conditions are equivalent:

(a) *For each $k, k' \in \{1, \dots, n\}$ and each $\alpha, \gamma \in \{1, \dots, M\}$ there holds*

$$\begin{aligned}(x_{k'} a_{jk}^{\beta\alpha} - x_k a_{jk'}^{\beta\alpha})(\partial_j E_{\gamma\beta})(x) &= 0\\ \text{for all } x = (x_i)_{1\le i\le n} &\in \mathbb{R}^n \setminus \{0\}.\end{aligned} \tag{3.54}$$

(b) *For each $s, s' \in \{1, \dots, n\}$ and each $\alpha, \gamma \in \{1, \dots, M\}$, in the sense of tempered distributions in $\mathbb{R}^n$ one has*

$$[a_{rs}^{\beta\alpha}\partial_{\xi_{s'}} - a_{rs'}^{\beta\alpha}\partial_{\xi_s}][\xi_r \mathcal{E}_{\gamma\beta}(\xi)] = 0. \tag{3.55}$$

(c) *For each $k, k' \in \{1, \dots, n\}$ and each $\alpha, \gamma \in \{1, \dots, M\}$, one has*

$$(a_{k'k}^{\beta\alpha} - a_{kk'}^{\beta\alpha} + \xi_j a_{jk}^{\beta\alpha}\partial_{\xi_{k'}} - \xi_j a_{jk'}^{\beta\alpha}\partial_{\xi_k})\mathcal{E}_{\gamma\beta}(\xi) = 0 \tag{3.56}$$

for all $\xi \in \mathbb{R}^n \setminus \{0\}$, and also

$$\int_{S^1} (a^{\beta\alpha}_{jk}\xi_{k'} - a^{\beta\alpha}_{jk'}\xi_k)\xi_j\,\mathcal{E}_{\gamma\beta}(\xi)\,d\mathcal{H}^1(\xi) = 0 \quad \textit{if } n = 2. \tag{3.57}$$

(d) *One has*

$$\begin{gathered}\xi_r\xi_j[a^{\beta\alpha}_{rs'}(a^{\lambda\mu}_{sj} + a^{\lambda\mu}_{js}) - a^{\beta\alpha}_{rs}(a^{\lambda\mu}_{s'j} + a^{\lambda\mu}_{js'})]\mathcal{E}_{\mu\beta}(\xi) + a^{\lambda\alpha}_{s's} - a^{\lambda\alpha}_{ss'} = 0\\ \textit{for all } \xi \in S^{n-1}, \textit{ all } s, s' \in \{1,\dots,n\}, \textit{ and all } \alpha, \lambda \in \{1,\dots,M\},\end{gathered} \tag{3.58}$$

with the cancellation condition

$$\begin{gathered}\int_{S^1} (a^{\beta\alpha}_{rs}\xi_{s'} - a^{\beta\alpha}_{rs'}\xi_s)\xi_r\mathcal{E}_{\lambda\beta}(\xi)\,d\mathcal{H}^1(\xi) = 0\\ \textit{for all } s, s' \in \{1,\dots,n\} \textit{ and } \alpha, \lambda \in \{1,\dots,M\},\end{gathered} \tag{3.59}$$

additionally imposed in the case when $n = 2$.

(e) *For each $\xi \in S^{n-1}$ and each $\alpha, \lambda \in \{1,\dots,M\}$,*

$$\begin{gathered}\textit{the expression } (a^{\lambda\mu}_{sj} + a^{\lambda\mu}_{js})\mathcal{E}_{\mu\beta}(\xi)\xi_j\xi_r a^{\beta\alpha}_{rs'} + a^{\lambda\alpha}_{s's}\\ \textit{is symmetric in the indices } s, s' \in \{1,\dots,n\},\end{gathered} \tag{3.60}$$

with the condition that for each $\alpha, \lambda \in \{1,\dots,M\}$

$$\begin{gathered}\textit{the expression } \int_{S^1} a^{\beta\alpha}_{rs}\xi_{s'}\xi_r\mathcal{E}_{\lambda\beta}(\xi)\,d\mathcal{H}^1(\xi)\\ \textit{is symmetric in the indices } s, s' \in \{1,2\},\end{gathered} \tag{3.61}$$

also imposed in the case when $n = 2$.

(f) *There exists a matrix-valued function*

$$k = \{k_{\gamma\alpha}\}_{1\le\gamma,\alpha\le M} : \mathbb{R}^n \setminus \{0\} \longrightarrow \mathbb{C}^{M\times M} \tag{3.62}$$

with the property that for each $\gamma, \alpha \in \{1,\dots,M\}$ and $s \in \{1,\dots,n\}$ one has

$$a^{\beta\alpha}_{rs}(\partial_r E_{\gamma\beta})(x) = x_s k_{\gamma\alpha}(x) \quad \textit{for all } x \in \mathbb{R}^n \setminus \{0\}. \tag{3.63}$$

The following definition plays an important role in this work, especially as it relates to the well posedness of the boundary value problems we consider throughout Chapter 11.

Definition 3.13. Given a second-order, weakly elliptic, homogeneous, $M \times M$ system L in $\mathbb{R}^n$, with constant complex coefficients, call

$$A = (a^{\alpha\beta}_{rs})_{\substack{1\le r,s\le n\\ 1\le\alpha,\beta\le M}} \in \mathfrak{A}_L \tag{3.64}$$

a distinguished coefficient tensor for the system L provided any of the conditions (a)–(f) in Proposition 3.12 holds. Also, denote by $\mathfrak{A}_L^{\mathrm{dis}}$ the family of such distinguished coefficient tensors for L, more explicitly,

$$\mathfrak{A}_L^{\mathrm{dis}} := \{A = (a_{rs}^{\alpha\beta})_{\substack{1\le r,s\le n\\ 1\le\alpha,\beta\le M}} \in \mathfrak{A}_L : \text{condition (3.54) holds} \quad \text{for each } k,k' \in \{1,\dots,n\} \text{ and } \alpha,\gamma \in \{1,\dots,M\}\}. \tag{3.65}$$

Remark 3.14. In view of [90, (h) of Proposition 1.2.1, pp. 11–13] and Definition 3.16, an alternative characterization of the space of distinguished coefficient tensors is as follows:

$$\mathfrak{A}_L^{\mathrm{dis}} = \{A \in \mathfrak{A}_L : \partial_\xi^{A^\top} E_{L^\top}(x) = 0 \text{ for all } x \in \mathbb{R}^n \setminus \{0\} \text{ and } \xi \in \mathbb{R}^n \text{ such that } x\cdot\xi = 0\}, \tag{3.66}$$

where, generally speaking, given $B = (b_{jk}^{\alpha\beta})_{\substack{1\le\alpha,\beta\le M\\ 1\le j,k\le n}} \in \mathbb{C}^{M^2\times n^2}$ and $\xi \in \mathbb{R}^n$,

$$\partial_\xi^B := (\xi_j b_{jk}^{\alpha\beta}\partial_k)_{1\le\alpha,\beta\le M} \tag{3.67}$$

denotes the directional derivative operator.

Remark 3.15. According to [90, (g) of Proposition 1.2.1, pp. 11–13, (1.2.14)–(1.2.15), p. 13], having $\mathfrak{A}_L^{\mathrm{dis}} \neq \varnothing$ is equivalent to the existence of some function $k \in [\mathscr{C}^\infty(\mathbb{R}^n \setminus \{0\})]^{M\times M}$ which is even, positive homogeneous of degree $-n$, and satisfies

$$\int_{S^{n-1}} k\, d\mathcal{H}^{n-1} = I_{M\times M}, \quad \text{and} \quad L(x_j k(x)) = 0\cdot I_{M\times M} \quad \text{for all } x \in \mathbb{R}^n \setminus \{0\} \text{ and all } j \in \{1,\dots,n\}. \tag{3.68}$$

We record a result from [86] elaborating on the significance of the class of distinguished coefficient tensors from the perspective of double layer potentials. Of particular importance is item (iii), which introduces the notion of an integral kernel having a specific algebraic structure which we refer to as an integral kernel of chord-dot-normal type.

Proposition 3.16. *Let L be a homogeneous, second-order, constant complex coefficient, weakly elliptic $M \times M$ system in $\mathbb{R}^n$, and suppose $A \in \mathfrak{A}_L$. Then the following statements are equivalent:*

(i) *The coefficient tensor A belongs to $\mathfrak{A}_L^{\mathrm{dis}}$.*
(ii) *Whenever Ω is a half-space in $\mathbb{R}^n$, the boundary-to-boundary double layer potential K_A associated with A and Ω as in (3.19) is the zero operator.*
(iii) *There exists a matrix-valued function $k \in [\mathscr{C}^\infty(\mathbb{R}^n \setminus \{0\})]^{M\times M}$ which is even, positive homogeneous of degree $-n$, and with the property that for each* UR *domain $\Omega \subseteq \mathbb{R}^n$*

the matrix-valued integral kernel of the double layer potential operator K_A associated with A and Ω as in (3.19) *has the chord-dot-normal structure*

$$\begin{aligned}&\langle \nu(y), x-y\rangle k(x-y)\\ &\text{for each } x \in \partial\Omega \text{ and } \mathcal{H}^{n-1}\text{-a. e. } y \in \partial\Omega,\end{aligned} \tag{3.69}$$

where ν is the geometric measure theoretic outward unit normal to Ω.

(iv) *Whenever Ω is a half-space in $\mathbb{R}^n$, the "transpose" double layer potential $K_A^\#$ associated with A and Ω as in* (3.19) *is the zero operator.*

(v) *There exists a matrix-valued function $k^\# \in [\mathscr{C}^\infty(\mathbb{R}^n\setminus\{0\})]^{M\times M}$, which is even, positive homogeneous of degree $-n$, and with the property that for each* UR *domain $\Omega \subseteq \mathbb{R}^n$ the (matrix-valued) integral kernel of the "transpose" double layer potential operator $K_A^\#$ associated with A and Ω as in* (3.20) *has the chord-dot-normal structure*

$$\begin{aligned}&\langle \nu(x), y-x\rangle k^\#(x-y)\\ &\text{for } \mathcal{H}^{n-1}\text{-a. e. } x \in \partial\Omega \text{ and each } y \in \partial\Omega,\end{aligned} \tag{3.70}$$

where ν is the geometric measure theoretic outward unit normal to Ω.

Moreover, whenever either (hence all) of the above conditions holds, the matrices k, $k^\#$ in items (iii), (v) *above are related to each other via $k^\# = k^\top$, where the superscript $\top$ indicates transposition.*

Another useful tool for solving elliptic systems is the single layer potential. In order that the single layer potential has the right kind of integral kernel to fit into the framework of Calderón–Zygmund theory, we consider the following modified version.

Definition 3.17. Let $\Omega \subseteq \mathbb{R}^n$ be a UR domain and write $\sigma := \mathcal{H}^{n-1}\lfloor\partial\Omega$. Recall that in such a setting that ν, the geometric measure theoretic outward unit normal to Ω, is defined σ-a. e. on $\partial\Omega$. Let L be a weakly elliptic, second-order, homogeneous, constant (complex) coefficient, $M \times M$ system in $\mathbb{R}^n$. Let E denote the fundamental solution associated to L as in Proposition 3.4. We define as in [89] the `modified boundary-to-domain single layer potential operator`

$$\begin{aligned}&\mathscr{S}_{\text{mod}} f(x) := \int_{\partial\Omega} \{E(x-y) - E_*(-y)\} f(y)\, d\sigma(y) \quad \text{for each } x \in \Omega,\\ &\text{for each } f \in \Big[L^1\Big(\partial\Omega, \frac{\sigma(x)}{1+|x|^{n-1}}\Big)\Big]^M, \text{ where } E_* := E \cdot \mathbf{1}_{\mathbb{R}^n\setminus B(0,1)}.\end{aligned} \tag{3.71}$$

Notice that the integral kernel appearing in (3.71), via the mean value theorem, enhances the decay of what ordinarily (in the standard definition of the single layer potential) would be just the fundamental solution. Thus, recalling (d) of Proposition 3.4, we see that $\mathscr{S}_{\text{mod}} f$ is defined via an absolutely convergent integral for each

$f \in [L^1(\partial\Omega, \frac{\sigma(x)}{1+|x|^{n-1}})]^M$. Note also that unlike the boundary-to-domain double layer (cf. (3.18)), $\mathscr{S}_{\mathrm{mod}}$ is independent of any coefficient tensor and is intrinsic to the system L. The modified single layer also has nice differentiability properties as witnessed below (cf. [89]). For each multiindex $\alpha \in \mathbb{N}_0^n$ with $|\alpha| \geq 1$ and any function $f \in [L^1(\partial\Omega, \frac{\sigma(x)}{1+|x|^{n-1}})]^M$, one has

$$\partial^\alpha(\mathscr{S}_{\mathrm{mod}} f)(x) = \int_{\partial\Omega} (\partial^\alpha E)(x-y) f(y)\, d\sigma(y) \quad \text{for each } x \in \Omega. \tag{3.72}$$

In particular, for any $f \in [L^1(\partial\Omega, \frac{\sigma(x)}{1+|x|^{n-1}})]^M$, the modified single layer potential operator satisfies

$$\mathscr{S}_{\mathrm{mod}} f \in [\mathscr{C}^\infty(\Omega)]^M \quad \text{and} \quad L(\mathscr{S}_{\mathrm{mod}} f) = 0 \quad \text{in } \Omega. \tag{3.73}$$

The above facts demonstrate that the modified single layer potential is also an excellent mechanism for producing many null solutions to a given weakly elliptic system L. Contrasting with the double layer, the single layer potential operator is important in the context of solving the Neumann problem. To elaborate more on that idea, we introduce the notion of conormal derivative.

Definition 3.18. Let $\Omega \subseteq \mathbb{R}^n$ be a UR domain and write $\sigma := \mathcal{H}^{n-1}\lfloor\partial\Omega$, and denote by ν the geometric measure theoretic outward unit normal to Ω. Let L be a weakly elliptic, second-order, homogeneous, constant complex coefficient, $M \times M$ system in $\mathbb{R}^n$. Next, fix an arbitrary coefficient tensor $A = (a_{jk}^{\alpha\beta})_{\substack{1\leq j,k\leq n\\ 1\leq\alpha,\beta\leq M}} \in \mathfrak{A}_L$ and an aperture parameter $\kappa \in (0,\infty)$. Given any function $u = (u_\beta)_{1\leq\beta\leq M} \in [W^{1,1}_{\mathrm{loc}}(\partial\Omega,\sigma)]^M$ for which $(\nabla u)\big|^{\kappa-\mathrm{n.t.}}_{\partial\Omega}$ exists σ-a. e. on $\partial\Omega$, we define the `conormal derivative` $\partial_\nu^A u$ as the $\mathbb{C}^M$-valued function

$$\partial_\nu^A u := \big(\nu_j a_{jk}^{\alpha\beta} (\partial_k u_\beta)\big|^{\kappa-\mathrm{n.t.}}_{\partial\Omega}\big)_{1\leq\alpha\leq M} \quad \text{at } \sigma\text{-a. e. point on } \partial\Omega. \tag{3.74}$$

Retaining the geometric context of Definition 3.18, it turns out (as proved in [89]) that for any function $f \in [L^1(\partial\Omega, \frac{\sigma(x)}{1+|x|^{n-1}})]^M$ one has the jump-relation

$$\partial_\nu^A(\mathscr{S}_{\mathrm{mod}} f) = \Big(-\frac{1}{2}I + K^\#_{A^\top}\Big) f \quad \text{at } \sigma\text{-a. e. point on } \partial\Omega. \tag{3.75}$$

Here, I is the identity and $K^\#_{A^\top}$ is the transpose boundary-to-boundary double layer associated with Ω and $A^\top$ as in (3.20).

Definition 3.19. Working in a UR domain $\Omega \subseteq \mathbb{R}^n$, write $\sigma := \mathcal{H}^{n-1}\lfloor\partial\Omega$, and let L be a weakly elliptic, second-order, homogeneous, constant, complex coefficient, $M \times M$ system in $\mathbb{R}^n$. Analogously define the `modified boundary-to-boundary single layer potential operator`

$$S_{\text{mod}}f(x) := \int_{\partial\Omega} \{E(x-y) - E_*(-y)\} f(y)\, d\sigma(y) \quad \text{at } \sigma\text{-a. e. } x \in \partial\Omega,$$
$$\text{for each } f \in \Big[L^1\Big(\partial\Omega, \frac{\sigma(x)}{1+|x|^{n-1}}\Big)\Big]^M, \text{ where } E_* := E \cdot \mathbf{1}_{\mathbb{R}^n \setminus B(0,1)}, \tag{3.76}$$

and E is the fundamental solution to L in the sense of Proposition 3.4.

Some useful facts about S_{mod}, proved in [89], are as follows: for each $\varepsilon > 0$, the operator

$$S_{\text{mod}} : L^1\Big[\Big(\partial\Omega, \frac{\sigma(x)}{1+|x|^{n-1-\varepsilon}}\Big)\Big]^M \to \Big[L^1\Big(\partial\Omega, \frac{\sigma(x)}{1+|x|^{n}}\Big)\Big]^M \tag{3.77}$$
is well-defined, linear, and bounded.

Also, for each $f \in [L^1(\partial\Omega, \frac{\sigma(x)}{1+|x|^{n-1}})]^M$ and aperture parameter $\kappa \in (0,\infty)$, one has

$$((\mathscr{S}_{\text{mod}}f)\big|^{\kappa-\text{n.t.}}_{\partial\Omega})(x) = (S_{\text{mod}}f)(x) \quad \text{at } \sigma\text{-a. e. point } x \in \partial\Omega. \tag{3.78}$$

We conclude this section with the following integral representation formula for null solutions to weakly elliptic systems on Ahlfors regular domains (as opposed to the more restrictive class of UR domains). This result is a showcase of the usefulness of the modified (boundary-to-domain) double and single layer potentials. Its proof may be found in [89, Section 1.8].

Proposition 3.20. *Let $\Omega \subseteq \mathbb{R}^n$ (where $n \in \mathbb{N}$, $n \geq 2$) be an Ahlfors regular domain which is either bounded or unbounded. Denote $\sigma := \mathcal{H}^{n-1}\lfloor\partial\Omega$, and let ν be the geometric measure theoretic outward unit normal to Ω. Suppose L is a weakly elliptic, second-order, homogeneous, constant complex coefficient, $M \times M$ system in $\mathbb{R}^n$, and choose some $A = (a^{\alpha\beta}_{rs})_{\substack{1\le r,s\le n\\ 1\le\alpha,\beta\le M}} \in \mathfrak{A}_L$. In this setting, recall the modified version of the double layer operator $\mathcal{D}_{A,\text{mod}}$ from* (3.34)*, and the modified version of the single layer operator $\mathscr{S}_{\text{mod}}$ from* (3.71)*. Finally, fix an arbitrary aperture parameter $\kappa \in (0,\infty)$, a truncation parameter $\varepsilon \in (0,\infty)$, and consider a function $u : \Omega \to \mathbb{C}^M$ satisfying*

$$\begin{aligned} &u \in [\mathscr{C}^\infty(\Omega)]^M, \quad Lu = 0 \quad \text{in } \Omega, \quad \mathcal{N}^{\varepsilon}_{\kappa} u \in L^1_{\text{loc}}(\partial\Omega, \sigma),\\ &u\big|^{\kappa-\text{n.t.}}_{\partial\Omega} \quad \text{exists } \sigma\text{-a. e. on } \partial\Omega \text{ and } u\big|^{\kappa-\text{n.t.}}_{\partial\Omega} \in \Big[L^1\Big(\partial\Omega, \frac{\sigma(x)}{1+|x|^{n}}\Big)\Big]^M,\\ &(\nabla u)\big|^{\kappa-\text{n.t.}}_{\partial\Omega} \quad \text{exists } \sigma\text{-a. e. on } \partial\Omega \text{ and } \mathcal{N}_\kappa(\nabla u) \in L^1\Big(\partial\Omega, \frac{\sigma(x)}{1+|x|^{n-1}}\Big). \end{aligned} \tag{3.79}$$

Then there exists some $\mathbb{C}^M$-valued locally constant function c_u in Ω with the property that

$$u = \mathcal{D}_{A,\text{mod}}\big(u\big|^{\kappa-\text{n.t.}}_{\partial\Omega}\big) - \mathscr{S}_{\text{mod}}(\partial^A_\nu u) + c_u \quad \text{in } \Omega. \tag{3.80}$$

4 Weighted Morrey spaces

This chapter is concerned with a certain brand of weighted Morrey space. Though we start with arbitrary weights, we soon focus on Muckenhoupt-weighted Morrey spaces, as the additional structure of the weight allows for more to be said about the space in terms of things like embeddings, and even a Rubio de Francia-type extrapolation result, which is the pinnacle of results in this chapter.

4.1 Definitions and basic examples

There are several notions of "weighted" Morrey space that can be found in the literature. We start by defining the brand of weighted Morrey space that we will be working with from here all the way through Chapter 11 on boundary value problems.

Definition 4.1. Start with a closed set $\Sigma \subseteq \mathbb{R}^n$ with $\sigma := \mathcal{H}^{n-1}\lfloor\Sigma$, fix an integrability exponent $p \in (0,\infty)$, along with a weight function $w : \Sigma \to [0,\infty]$, and some threshold $\lambda \in [0,\infty)$. The `weighted Morrey space` is defined by

$$M^{p,\lambda}(\Sigma, w) := \{f : \Sigma \to \mathbb{C} : f \text{ is } \sigma\text{-measurable and } \|f\|_{M^{p,\lambda}(\Sigma,w)} < +\infty\}, \tag{4.1}$$

where

$$\|f\|_{M^{p,\lambda}(\Sigma,w)} := \sup_{\substack{x\in\Sigma \\ r\in(0,2\,\mathrm{diam}(\Sigma))}} \left(\frac{1}{r^\lambda}\int_{\Delta(x,r)} |f|^p \, dw\right)^{1/p}, \tag{4.2}$$

for each σ-measurable function f on Σ.

As a consequence of this definition, given any $f : \Sigma \to \mathbb{C}$, a σ-measurable function, any $p_0 \in (0,\infty)$ and any $\lambda \in (0,\infty)$, one has

$$\|f^{p_0}\|_{M^{p/p_0,\lambda}(\Sigma,w)} = \|f\|^{p_0}_{M^{p,\lambda}(\Sigma,w)}. \tag{4.3}$$

In the case that $\lambda = 0$, we see (4.1) and (4.2) imply

$$M^{p,0}(\Sigma, w) = L^p(\Sigma, w) \quad \text{with coincidence of norms}, \tag{4.4}$$

so that L^p spaces are an endpoint on the scale of Morrey spaces. Since the latter scale is well understood, we often restrict λ to be strictly positive. On the other hand, we will often restrict $\lambda < n-1$. When $\lambda > n-1$ and in the unweighted case (i. e., $w \equiv 1$, or in our notation convention, $w = \sigma$), the Lebesgue differentiation theorem (cf. Proposition 2.21) implies for any $x_0 \in \Sigma$,

https://doi.org/10.1515/9783111458274-004

$$
\begin{aligned}
\|f\|_{M^{p,\lambda}(\Sigma,\sigma)} &= \sup_{\substack{x\in\Sigma\\ r\in(0,2\operatorname{diam}(\Sigma))}} \left(\frac{1}{r^{\lambda}} \int_{\Delta(x,r)} |f|^p \, d\sigma\right)^{1/p} \qquad (4.5)\\
&\geq \lim_{r\to 0^+} \left(\frac{C}{r^{\lambda+1-n}} \fint_{\Delta(x_0,r)} |f|^p \, d\sigma\right)^{1/p}\\
&= C|f(x_0)| \lim_{r\to 0^+} r^{(n-1-\lambda)/p} = \infty,
\end{aligned}
$$

where $C \in (0,\infty)$ depends only on the ADR character of Σ and p. As such, $M^{p,\lambda}(\Sigma,\sigma)$ is trivial whenever $\lambda > n-1$. A similar computation as in (4.5) reveals for the case $\lambda = n-1$ that

$$
M^{p,n-1}(\Sigma,\sigma) \hookrightarrow L^{\infty}(\Sigma,\sigma) \quad \text{continuously.} \qquad (4.6)
$$

It is also the case that

$$
f \in M^{p,n-1}(\Sigma,\sigma) \quad \text{implies} \quad f \in L^{\infty}(\Sigma,\sigma). \qquad (4.7)
$$

This is seen from the fact that if $f \in M^{p,n-1}(\Sigma,\sigma)$, there exists a constant $C \in (0,\infty)$, depending only on the ADR character of Σ and p, such that

$$
\left(\fint_{\Delta} |f|^p \, d\sigma\right)^{1/p} \leq C\|f\|_{M^{p,\lambda}(\Sigma,\sigma)} \quad \text{for all surface balls } \Delta \subseteq \Sigma. \qquad (4.8)
$$

The Lebesgue differentiation theorem (cf. Proposition 2.21) then implies (4.7). Together, (4.6) and (4.7) show that

$$
M^{p,n-1}(\Sigma,\sigma) = L^{\infty}(\Sigma,\sigma) \quad \text{for all } p \in (0,\infty). \qquad (4.9)
$$

Since $L^{\infty}(\Sigma,\sigma)$ is well understood, this is why we restrict λ to be *strictly* less than $n-1$.

Remark 4.2. Note that for all $f \in M^{p,\lambda}(\Sigma,w)$ and for all $g \in L^{\infty}_{\text{comp}}(\Sigma,\sigma)$ one has $fg \in L^p(\Sigma,w)$. Indeed, if $g \in L^{\infty}_{\text{comp}}(\Sigma,\sigma)$, then there exists a surface ball $\Delta \subseteq \Sigma$ with radius $R \in (0,\infty)$ on which $\operatorname{supp} g \subseteq \Delta$. Thus, by (4.2),

$$
\begin{aligned}
\left(\int_{\Sigma} |fg|^p \, dw\right)^{1/p} &= \left(\int_{\Delta} |fg|^p \, dw\right)^{1/p} \qquad (4.10)\\
&\leq \|g\|_{L^{\infty}(\Sigma,\sigma)} R^{\lambda/p} \left(\frac{1}{R^{\lambda}} \int_{\Delta} |f|^p \, dw\right)^{1/p}\\
&\leq \|g\|_{L^{\infty}(\Sigma,\sigma)} R^{\lambda/p} \|f\|_{M^{p,\lambda}(\Sigma,w)} < \infty,
\end{aligned}
$$

which places fg in $L^p(\Sigma,w)$ as wanted.

Remark 4.3. As is apparent in (4.2), any σ-measurable function f on Σ for which $\|f\|_{M^{p,\lambda}(\Sigma,w)} < +\infty$ forces $|f| < +\infty$ at σ-a. e. point on Σ.

Remark 4.4. It is also clear from (4.2) that the scale of weighted Morrey spaces are stable under multiplication by an essentially bounded function. That is, if $f \in L^\infty(\Sigma,\sigma)$ and $g \in M^{p,\lambda}(\Sigma,w)$, then $fg \in M^{p,\lambda}(\Sigma,w)$ with

$$\|fg\|_{M^{p,\lambda}(\Sigma,w)} \le \|f\|_{L^\infty(\Sigma,\sigma)}\|g\|_{M^{p,\lambda}(\Sigma,w)}. \tag{4.11}$$

As one might expect, (4.2) defines a genuine norm that turns $M^{p,\lambda}(\Sigma,w)$ into a Banach space whenever $p \ge 1$.

Proposition 4.5. *Let $\Sigma \subseteq \mathbb{R}^n$ be a closed set and abbreviate $\sigma := \mathcal{H}^{n-1}\lfloor\Sigma$. Fix an integrability exponent $p \in [1,\infty)$ along with an arbitrary weight function $w : \Sigma \to [0,\infty]$, and let $\lambda \in (0,n-1)$.*

Then the weighted Morrey space $M^{p,\lambda}(\Sigma,w)$, equipped with the norm $\|\cdot\|_{M^{p,\lambda}(\Sigma,w)}$, defined in (4.2), is a Banach space.

Proof. The fact that $\|\cdot\|_{M^{p,\lambda}(\Sigma,w)}$ is nondegenerate, homogeneous, and satisfies a triangle inequality is clear from (4.2). As such, it is clear that $(M^{p,\lambda}(\Sigma,w), \|\cdot\|_{M^{p,\lambda}(\Sigma,w)})$ is a normed vector space. To show that it is in fact complete with respect to the norm, let $\{f_j\}_{j\in\mathbb{N}}$ be a Cauchy sequence in $M^{p,\lambda}(\Sigma,w)$. For each fixed $x \in \Sigma$ and $r \in (0, 2\,\mathrm{diam}(\Sigma))$, we have for the surface ball $\Delta := \Delta(x,r)$ that

$$\|f_j - f_k\|_{L^p(\Delta,w)} \le r^{\lambda/p}\|f_j - f_k\|_{M^{p,\lambda}(\Sigma,w)} \quad \text{for all } j,k \in \mathbb{N}. \tag{4.12}$$

Hence, $\{f_j|_\Delta\}_{j\in\mathbb{N}}$ is a Cauchy sequence in $L^p(\Delta,w)$. Since $L^p(\Delta,w)$ is a Banach space, there exists some $f^\Delta \in L^p(\Delta,w)$ such that $f_j|_\Delta \to f^\Delta$ as $j \to \infty$ in $L^p(\Delta,w)$.

Note that if Δ' is another surface ball on Σ with $\Delta' \subseteq \Delta$, then in fact $f^\Delta|_{\Delta'} = f^{\Delta'}$. Indeed, $(f_j|_\Delta)|_{\Delta'} = f_j|_{\Delta'}$ and $f_j|_{\Delta'} \xrightarrow[j\to\infty]{L^p(\Delta',w)} f^{\Delta'}$. Also,

$$f_j|_\Delta \xrightarrow[j\to\infty]{L^p(\Delta,w)} f^\Delta \quad \text{implies} \quad (f_j|_\Delta)|_{\Delta'} \xrightarrow[j\to\infty]{L^p(\Delta',w)} f^\Delta|_{\Delta'}, \tag{4.13}$$

and hence we find that $f^\Delta|_{\Delta'} = f^{\Delta'}$ via uniqueness of the limits.

Now, define $f : \Sigma \to \mathbb{C}$ by setting $f := f^\Delta$ on the surface ball $\Delta \subseteq \Sigma$. Note that per the previous discussion this is an unambiguous definition. Pick some $x \in \Sigma$ and $r \in (0, 2\,\mathrm{diam}(\Sigma))$, both arbitrary. Let $\varepsilon > 0$, then since $\{f_j\}_{j\in\mathbb{N}}$ is assumed to be Cauchy in $M^{p,\lambda}(\Sigma,w)$, we find that

$$\left(\frac{1}{r^\lambda}\int_{\Delta(x,r)} |f_j - f_k|^p\,dw\right)^{1/p} \le \|f_j - f_k\|_{M^{p,\lambda}(\Sigma,w)} < \varepsilon \tag{4.14}$$

provided, $j,k \geq N_\varepsilon$, for sufficiently large $N_\varepsilon \in \mathbb{N}$. Sending k to infinity in (4.14) implies

$$r^{-\lambda/p}\|f_j - f\|_{L^p(\Delta,w)} < \varepsilon, \tag{4.15}$$

and so

$$\left(\frac{1}{r^\lambda}\int_{\Delta(x,r)} |f_j - f|^p\,dw\right)^{1/p} < \varepsilon \quad \text{whenever } j > N_\varepsilon. \tag{4.16}$$

In view of (4.2), taking a supremum over all $x \in \Sigma$ and $r \in (0, 2\,\mathrm{diam}(\Sigma))$ yields

$$\|f_j - f\|_{M^{p,\lambda}(\Sigma,w)} < \varepsilon \quad \text{for all } j > N_\varepsilon, \tag{4.17}$$

and hence $f \in M^{p,\lambda}(\Sigma, w)$ with $f_j \to f$ as $j \to \infty$ in $M^{p,\lambda}(\Sigma, w)$, demonstrating that $(M^{p,\lambda}(\Sigma, w), \|\cdot\|_{M^{p,\lambda}(\Sigma,w)})$ is complete. □

Corollary 4.6. *Let $\Sigma \subseteq \mathbb{R}^n$ be a closed set and abbreviate $\sigma := \mathcal{H}^{n-1}\lfloor\Sigma$. Fix an integrability exponent $p \in (0,1)$ along with an arbitrary weight function $w : \Sigma \to [0,\infty]$, and let $\lambda \in [0,\infty)$.*

Then the weighted Morrey space $M^{p,\lambda}(\Sigma, w)$, equipped with the quasinorm $\|\cdot\|_{M^{p,\lambda}(\Sigma,w)}$, defined in (4.2) is a quasi-Banach space.

Proof. We may argue as in Proposition 4.5, noting that $\|\cdot\|_{M^{p,\lambda}(\Sigma,w)}$ satisfies a quasitriangle inequality whenever $p \in (0,1)$. □

The scale of weighted Morrey spaces is similar to the scale of Lebesgue spaces in several ways. As this next proposition states, weighted Morrey spaces inherit a lattice property from the latter scale of spaces.

Proposition 4.7. *Let $\Sigma \subseteq \mathbb{R}^n$ be a closed set and write $\sigma := \mathcal{H}^{n-1}\lfloor\Sigma$. Fix an integrability exponent $p \in (0,\infty)$, any weight function $w : \Sigma \to [0,\infty]$, and some threshold $\lambda \in (0,\infty)$.*

Then for any $g \in M^{p,\lambda}(\Sigma, w)$ and any σ-measurable function f, which satisfy $|f| \leq g$ at σ-a. e. point on Σ, one has $f \in M^{p,\lambda}(\Sigma, w)$ with $\|f\|_{M^{p,\lambda}(\Sigma,w)} \leq \|g\|_{M^{p,\lambda}(\Sigma,w)}$.

Proof. This follows from the fact that $L^p(\Delta, w)$ is a lattice of functions for any surface ball $\Delta \subseteq \Sigma$. Therefore, for any σ-measurable function f, and some $g \in M^{p,\lambda}(\Sigma, w)$ such that $|f| \leq g$ at σ-a. e. point on Σ, we have that

$$\|f|_{\Delta(x,r)}\|_{L^p(\Delta(x,r),w)} \leq \|g|_{\Delta(x,r)}\|_{L^p(\Delta(x,r),w)} \tag{4.18}$$

for each surface ball $\Delta(x,r) \subseteq \Sigma$ with $x \in \Sigma$ and $r \in (0, 2\,\mathrm{diam}(\Sigma))$. Multiplying by $r^{-\lambda/p}$ and taking a supremum over all $x \in \Sigma$ and $r \in (0, 2\,\mathrm{diam}(\Sigma))$ on both sides of (4.18) yields

$$\|f\|_{M^{p,\lambda}(\Sigma,w)} \leq \|g\|_{M^{p,\lambda}(\Sigma,w)}, \tag{4.19}$$

on account of (4.2), completing the proof. □

From this point on in our development of weighted Morrey spaces, we require a little more geometric structure on the underlying set Σ, as well as more analytic structure on the weight function w. Namely, we ask that Σ is Ahlfors regular in the sense of Definition 2.2, and we require that w belongs to the Muckenhoupt class of weights (cf. Definition 2.27). We refer to such a Morrey space as a `Muckenhoupt-weighted Morrey space`.

Remark 4.8. Bring in an Ahlfors regular set $\Sigma \subseteq \mathbb{R}^n$ and $\sigma := \mathcal{H}^{n-1}\lfloor\Sigma$. Define $M^{p,\lambda}(\Sigma, w)$ as in Definition 4.1 for $p \in (1,\infty)$, a weight function w, and $\lambda \in (0,\infty)$. In many of our results involving Muckenhoupt-weighted Morrey spaces, we ask that not only the weight w belongs to $A_p(\Sigma,\sigma)$, but also that we know a value $q \in (1,\infty)$ such that w satisfies a reverse Hölder inequality with exponent q. Such a q is always guaranteed to exist by the theory of Muckenhoupt weights (see item (2) of Proposition 2.33), though in principle the value of q need not be the one guaranteed by the theory. In turn q modulates the "good" range for λ, with a larger value of q meaning a larger interval for λ. Specifically, we often restrict λ to the interval $(0,(n-1)/q')$ where $q' := q/(q-1)$ is the Hölder conjugate to q. This range of λ is consistent with the standard range for λ in the unweighted case, i. e., the interval $(0, n-1)$ (see the discussion surrounding (4.5)–(4.9)). Indeed, the unweighted case corresponds to a weight function that is identically one. From (2.63), it is clear that all positive constants belong to $\mathrm{RH}_q(\Sigma,\sigma)$ for any choice of $q \in (1,\infty)$. As such, writing $q' := q/(q-1)$ for each $q \in (1,\infty)$, we may take

$$\lambda \in \bigcup_{q\in(1,\infty)} (0,(n-1)/q') = (0, n-1). \tag{4.20}$$

The next series of results present specific examples of functions that belong to a Muckenhoupt-weighted Morrey space, and serve to demonstrate that these function spaces are rich. Our first result in this direction simply states that the characteristic function of a surface ball belongs to a Muckenhoupt-weighted Morrey space.

Lemma 4.9. *Let $\Sigma \subseteq \mathbb{R}^n$ be a closed Ahlfors regular set, and abbreviate $\sigma := \mathcal{H}^{n-1}\lfloor\Sigma$. Fix two integrability exponents $p, q \in (1,\infty)$, along with an arbitrary Muckenhoupt weight $w \in A_p(\Sigma,\sigma) \cap \mathrm{RH}_q(\Sigma,\sigma)$ and some threshold $\lambda \in (0,(n-1)/q')$, where $q' := q/(q-1)$.*

Then for any reference point $x_0 \in \Sigma$ and any $R \in (0, 2\operatorname{diam}(\Sigma))$, one has

$$\|\mathbf{1}_{\Delta(x_0,R)}\|_{M^{p,\lambda}(\Sigma,w)} \approx \frac{w(\Delta(x_0,R))^{1/p}}{R^{\lambda/p}} \quad \textit{uniformly in } x_0 \textit{ and } R. \tag{4.21}$$

In particular, one has the membership

$$\mathbf{1}_{\Delta(x_0,R)} \in M^{p,\lambda}(\Sigma, w). \tag{4.22}$$

Proof. To set the stage, pick $x_0 \in \Sigma$ and $R \in (0, 2\operatorname{diam}(\Sigma))$. Consider an arbitrary surface ball $\Delta(x,r) \subseteq \Sigma$ with $x \in \Sigma$ and $r \in (0, 2\operatorname{diam}(\Sigma))$. Additionally, assume that $\Delta(x_0,R) \cap \Delta(x,r) \neq \varnothing$. We consider two cases.

Case 1: *Assume r ≥ R.* Then

$$\left(\frac{1}{r^\lambda}\int_{\Delta(x,r)}|\mathbf{1}_{\Delta(x_0,R)}|^p\,dw\right)^{1/p} \le \left(\frac{1}{R^\lambda}\int_{\Delta(x,r)}|\mathbf{1}_{\Delta(x_0,R)}|^p\,dw\right)^{1/p} \le \left(\frac{w(\Delta(x_0,R))}{R^\lambda}\right)^{1/p}. \tag{4.23}$$

Case 2: *Assume r < R.* In this case, since we are assuming that $\Delta(x_0,R)$ and $\Delta(x,r)$ are disjoint, we have the inclusion $\Delta(x,r)\subseteq\Delta(x_0,3R)$. Indeed, if $y\in\Delta(x,r)$, then

$$|y-x_0|\le|y-x|+|x-x_0|<r+2R<3R, \tag{4.24}$$

placing $y\in\Delta(x_0,3R)$, demonstrating the desired inclusion. By item (2) of Proposition 2.33, we may write

$$\frac{w(\Delta(x,r))}{w(\Delta(x_0,3R))}\le C\left(\frac{\sigma(\Delta(x,r))}{\sigma(\Delta(x_0,3R))}\right)^{1/q'} \le C_\Sigma\frac{r^{(n-1)/q'}}{R^{(n-1)/q'}}, \tag{4.25}$$

where the last inequality follows via the Ahlfors regularity of Σ, and the constant $C_\Sigma\in(0,\infty)$ depends only on n, p, q, $[w]_{A_p}$, and the ADR character of Σ. Thus, we find

$$w(\Delta(x,r))\le Cw(\Delta(x_0,3R))\frac{r^{(n-1)/q'}}{R^{(n-1)/q'}}. \tag{4.26}$$

With this estimate available to us and bearing in mind the assumption that $r\in(0,R)$, we can write

$$\begin{aligned}\left(\frac{1}{r^\lambda}\int_{\Delta(x,r)}|\mathbf{1}_{\Delta(x_0,R)}|^p\,dw\right)^{1/p} &= \left(\frac{w(\Delta(x,r)\cap\Delta(x_0,R))}{r^\lambda}\right)^{1/p}\\ &\le \left(\frac{w(\Delta(x,r))}{r^\lambda}\right)^{1/p}\\ &\le \left(C_\Sigma\frac{w(\Delta(x_0,3R))}{R^{(n-1)/q'}}\cdot\frac{r^{(n-1)/q'}}{r^\lambda}\right)^{1/p}\\ &\le C\left(\frac{w(\Delta(x_0,R))}{R^\lambda}\right)^{1/p}\end{aligned} \tag{4.27}$$

for all $r\in(0,R)$ and all $x\in\Sigma$. Note that the third line above follows from (4.26), and for the last line, we used the fact that we are working in the regime $r<R$, our assumption that $(n-1)/q'-\lambda>0$, and the doubling nature of w (cf. item (4) of Proposition 2.29).

Since the quantities appearing on the right-hand sides of (4.23) and (4.27) are independent of x and r, we ultimately find that

$$\begin{aligned}\|\mathbf{1}_{\Delta(x_0,R)}\|_{M^{p,\lambda}(\Sigma,w)} &:= \sup_{\substack{x\in\Sigma\\ r\in(0,2\,\mathrm{diam}(\Sigma))}} \Big(\frac{1}{r^\lambda}\int_{\Delta(x,r)} \mathbf{1}_{\Delta(x_0,R)}\,dw\Big)^{1/p} \\ &\leq C\Big(\frac{w(\Delta(x_0,R))}{R^\lambda}\Big)^{1/p} < +\infty \end{aligned} \tag{4.28}$$

for all $x_0 \in \Sigma$ and $R \in (0, 2\,\mathrm{diam}(\Sigma))$, where the constant $C \in (0,\infty)$ depends exclusively on $n, p, q, [w]_{A_p}$, and the ADR character of Σ. Recall Remark 2.28 from which we guarantee finiteness in (4.28). This also gives the left-pointing inequality in (4.21), while the right-pointing inequality is a consequence of the first line in (4.28). This completes the proof of Lemma 4.9. □

Remark 4.10. Note that in the case that $\lambda = 0$ and $w = 1$, the estimate appearing in (4.28) is exactly what we would expect, since in this case $M^{p,\lambda}(\Sigma, w) = L^p(\Sigma, \sigma)$.

As a consequence of Lemma 4.9, we find that the space of essentially bounded functions with compact support is contained in a Muckenhoupt-weighted Morrey space.

Corollary 4.11. *Let $\Sigma \subseteq \mathbb{R}^n$ be a closed ADR set with $\sigma := \mathcal{H}^{n-1}\lfloor\Sigma$. Fix two arbitrary integrability exponents $p, q \in (1,\infty)$, a Muckenhoupt weight $w \in A_p(\Sigma,\sigma)\cap \mathrm{RH}_q(\Sigma,\sigma)$, and some threshold $\lambda \in (0, (n-1)/q')$, where $q' := q/(q-1)$. Then*

$$L^\infty_{\mathrm{comp}}(\Sigma,\sigma) \subseteq M^{p,\lambda}(\Sigma, w). \tag{4.29}$$

Proof. Let $f \in L^\infty_{\mathrm{comp}}(\Sigma,\sigma)$. As such, there exists some $C \in (0,\infty)$ such that $|f(x)| \leq C$ for σ-a. e. point $x \in \Sigma$. Fix a reference point $x_0 \in \Sigma$. Because f has compact support, there exists some $R \in (0, 2\,\mathrm{diam}(\Sigma))$ such that $\mathrm{supp}\, f \subseteq \Delta(x_0, R)$. It follows that

$$|f| \leq C\mathbf{1}_{\Delta(x_0,R)} \quad \text{at } \sigma\text{-a. e. point } x \in \Sigma. \tag{4.30}$$

Since f is σ-measurable by assumption, the lattice property of $M^{p,\lambda}(\Sigma, w)$ (cf. Proposition 4.7) along with Lemma 4.9 implies that $f \in M^{p,\lambda}(\Sigma, w)$ as well. The arbitrary nature of f ensures the desired inclusion. □

Throughout our development of a Calderón–Zygmund theory in Chapter 5, we often encounter the weighted Lebesgue space $L^1(\Sigma, \frac{\sigma(x)}{1+|x|^{n-1}})$ (cf. Remark 5.14). As it turns out, the weight function $\frac{1}{1+|\cdot|^{n-1}}$ belongs to a Muckenhoupt-weighted Morrey space.

Lemma 4.12. *Let $\Sigma \subseteq \mathbb{R}^n$ be a closed ADR set and write $\sigma := \mathcal{H}^{n-1}\lfloor\Sigma$. Fix two arbitrary integrability exponents $p, q \in (1,\infty)$, a Muckenhoupt weight $w \in A_p(\Sigma,\sigma)\cap \mathrm{RH}_q(\Sigma,\sigma)$, and some threshold $\lambda \in (0, (n-1)/q')$, where $q' := q/(q-1)$.*

Then the function $f : \Sigma \to \mathbb{R}$ given by

$$f(x) := \frac{1}{1 + |x|^{n-1}} \quad \textit{for all } x \in \Sigma \tag{4.31}$$

belongs to $M^{p,\lambda}(\Sigma, w)$.

As a corollary of this and the lattice property for Morrey spaces (Proposition 4.7), any σ-measurable function $f : \Sigma \to \mathbb{R}$ with the property that there exists $C \in (0, \infty)$ such that $|f(x)| \le C(1 + |x|)^{1-n}$ for σ-a. e. point $x \in \Sigma$ belongs to $M^{p,\lambda}(\Sigma, w)$.

The following proof relies on Proposition 5.1, which states that the Hardy–Littlewood maximal operator

$$\mathcal{M}_\Sigma : M^{p,\lambda}(\Sigma, w) \to M^{p,\lambda}(\Sigma, w) \tag{4.32}$$

is bounded. We assure the concerned reader that despite the fact that this result is proved later in the work, there is no circular reasoning.

Proof of Lemma 4.12. Fix an arbitrary reference point $x_0 \in \Sigma$. First, observe thanks to Lemma 2.18 item (2) that

$$\frac{1}{1 + |x|^{n-1}} \approx \frac{1}{1 + |x - x_0|^{n-1}} \quad \text{uniformly in } x \in \Sigma. \tag{4.33}$$

Recall by Lemma 2.25 that

$$\mathcal{M}_\Sigma(\mathbf{1}_{\Delta(x_0,1)})(x) \approx \frac{1}{1 + |x - x_0|^{n-1}}, \quad \text{uniformly in } x \in \Sigma. \tag{4.34}$$

By Proposition 5.1 and the lattice property of $M^{p,\lambda}(\Sigma, w)$ (cf. Proposition 4.7), we find that

$$\begin{aligned}\left\| \frac{1}{1 + |\cdot|^{n-1}} \right\|_{M^{p,\lambda}(\Sigma,w)} &\le C \left\| \frac{1}{1 + |\cdot - x_0|^{n-1}} \right\|_{M^{p,\lambda}(\Sigma,w)} \\ &\le C \left\| \mathcal{M}_\Sigma(\mathbf{1}_{\Delta(x_0,1)}) \right\|_{M^{p,\lambda}(\Sigma,w)} \\ &\le C \|\mathbf{1}_{\Delta(x_0,1)}\|_{M^{p,\lambda}(\Sigma,w)} < \infty.\end{aligned} \tag{4.35}$$

Note in addition to the lattice property, the first inequality uses (4.33), the second inequality uses (4.34), and the third inequality uses the boundedness of $\mathcal{M}_\Sigma$ on $M^{p,\lambda}(\Sigma, w)$ and Lemma 4.9, which ensures that $\mathbf{1}_{\Delta(x_0,1)}$ belongs to $M^{p,\lambda}(\Sigma, w)$. This yields the desired result. □

4.2 Embedding results

Having seen several examples of functions belonging to a Morrey space, we would now like to present several embedding results involving weighted Morrey spaces. The embeddings throughout this section while interesting in their own right, actually have far-reaching consequences to areas such as extrapolation theory, Calderón–Zygmund theory, and boundary value problems.

The first result is a lemma, which will be useful in ascertaining other embedding results as well as in aiding the proof of a Rubio de Francia-type extrapolation result for Muckenhoupt-weighted Morrey spaces. More specifically, we find that a certain class of weighted Morrey space beyond the Muckenhoupt range continuously embeds into a weighted local Lebesgue spaces with the same integrability exponent.

Lemma 4.13. *Let $\Sigma \subseteq \mathbb{R}^n$ be a closed set, and define $\sigma := \mathcal{H}^{n-1}\lfloor\Sigma$. Fix an integrability exponent $p \in (0,\infty)$ and a threshold $\lambda \in [0,\infty)$. Suppose w is a weight function on Σ and let $\alpha \in (-\infty,\lambda)$. Now, fix a reference point $x_0 \in \Sigma$, and abbreviate $w_\alpha := |\cdot - x_0|^\alpha w$, a weight function on Σ.*

Then for every $r \in (0, 2\operatorname{diam}(\Sigma))$, there exists a constant $C \in (0,\infty)$, depending only on p, λ, and α such that for each σ-measurable $f : \Sigma \to \mathbb{C}$, one has

$$\Big(\int_{\Delta(x_0,r)} |f|^p w\,d\sigma\Big)^{1/p} \le C r^{(\lambda-\alpha)/p} \|f\|_{M^{p,\lambda}(\Sigma,w_\alpha)}. \tag{4.36}$$

As an immediate corollary, $M^{p,\lambda}(\Sigma,w_\alpha)$ embeds continuously into the space $L^p_{\mathrm{loc}}(\Sigma,w)$.

Proof. Let $r \in (0, 2\operatorname{diam}(\Sigma))$. For each $j \in \mathbb{N}$, define

$$\Delta_j := \Delta(x_0, 2^{-j+1}r) \setminus \Delta(x_0, 2^{-j}r). \tag{4.37}$$

If $y \in \Delta_j$, then we have $2\cdot 2^{-j}r \ge |y-x_0| \ge 2^{-j}r$. In particular, for any $\alpha \in (-\infty,\lambda)$,

$$\Big(\frac{|y-x_0|}{2^{-j}r}\Big)^\alpha \ge \min\{1, 2^\alpha\} =: C_\alpha. \tag{4.38}$$

Note also that $\Delta(x_0,r)\setminus\{x_0\} = \bigcup_{j=1}^\infty \Delta_j$ is a disjoint union. Hence, by (4.2) and the fact that $\sigma(\{x_0\}) = 0$, we can estimate

$$\begin{aligned}\int_{\Delta(x_0,r)} |f|^p w\,d\sigma &\le C_\alpha^{-1}\sum_{j=1}^\infty \int_{\Delta_j} |f(y)|^p w(y)\Big(\frac{|y-x_0|}{2^{-j}r}\Big)^\alpha d\sigma(y) \\ &\le C_\alpha^{-1}\sum_{j=1}^\infty \frac{(2^{-j+1}r)^\lambda}{(2^{-j+1}r)^\lambda(2^{-j}r)^\alpha}\times \\ &\quad\times \int_{\Delta(x_0,2^{-j+1}r)} |f(y)|^p w(y)|y-x_0|^\alpha\,d\sigma(y) \\ &\le C_\alpha^{-1}2^\lambda \sum_{j=1}^\infty (2^{-j}r)^{\lambda-\alpha}\|f\|^p_{M^{p,\lambda}(\Sigma,w_\alpha)},\end{aligned} \tag{4.39}$$

where the first inequality is implied by (4.38), and the last inequality is owed to (4.2). The above geometric series converges since $\alpha < \lambda$ by assumption, and hence upon taking pth roots in (4.39) we arrive at (4.36) as intended. □

As this next lemma shows, Muckenhoupt-weighted Morrey spaces embed into the space of locally integrable functions.

Lemma 4.14. *Let Σ be a closed* ADR *set and write $\sigma := \mathcal{H}^{n-1}\lfloor\Sigma$. Fix an integrability exponent $p \in (1,\infty)$, a Muckenhoupt weight $w \in A_p(\Sigma,\sigma)$, a threshold $\lambda \in (0,\infty)$, and a power $\alpha \in (-\infty,\lambda)$. Choose a reference point $x_0 \in \Sigma$ and define $w_\alpha := |\cdot - x_0|^\alpha w$, a weight function on Σ. Then*

$$M^{p,\lambda}(\Sigma, w_\alpha) \hookrightarrow L^1_{\mathrm{loc}}(\Sigma,\sigma) \quad \textit{continuously.} \tag{4.40}$$

Proof. This is an immediate consequence of (2.76) and (4.36). □

The following embedding result for Muckenhoupt-weighted Morrey spaces amounts to showing that functions in a Muckenhoupt-weighted Morrey space also belong to the most general class of functions for which convolution type singular integral operators of Calderón–Zygmund type are well-defined (cf. Remark 5.4 for more detail on the relevance of the weighted Lebesgue space to which we are alluding).

Proposition 4.15. *Let $\Sigma \subseteq \mathbb{R}^n$ be a closed Ahlfors regular set and write $\sigma := \mathcal{H}^{n-1}\lfloor\Sigma$. Fix two integrability exponents $p, q \in (1,\infty)$, along with an arbitrary Muckenhoupt weight $w \in A_p(\Sigma,\sigma) \cap \mathrm{RH}_q(\Sigma,\sigma)$, and some threshold $\lambda \in (0,(n-1)/q')$, where $q' := q/(q-1)$. Having fixed $\alpha \in [0,\lambda)$ and a reference point $x_0 \in \Sigma$, abbreviate $w_\alpha := |\cdot - x_0|^\alpha w$, a weight on Σ.*

There exists $\varepsilon_* \in (0,\infty)$ such that for all $\varepsilon \in [0,\varepsilon_*)$ the following continuous embedding holds:

$$M^{p,\lambda}(\Sigma, w_\alpha) \hookrightarrow L^1\Big(\Sigma, \frac{\sigma(x)}{1+|x|^{n-1-\varepsilon}}\Big). \tag{4.41}$$

Proof. First, fix $x_0 \in \Sigma$ and observe that for any $\varepsilon \in [0, n-1)$, we have the quantitative equivalence $1+|x|^{n-1-\varepsilon} \approx 1+|x-x_0|^{n-1-\varepsilon}$ uniformly in $x \in \Sigma$ by Lemma 2.18. Consequently, for any $\varepsilon \in [0,n-1)$ and given any $f \in M^{p,\lambda}(\Sigma, w_\alpha)$, we have

$$\begin{aligned}
\int_\Sigma \frac{|f(x)|}{1+|x|^{n-1-\varepsilon}}\,d\sigma(x) &\le C \int_\Sigma \frac{|f(x)|}{1+|x-x_0|^{n-1-\varepsilon}}\,d\sigma(x) \\
&\le C \int_{\Delta(x_0,1)} |f(x)|\,d\sigma(x) \\
&\quad + C\sum_{j=0}^\infty \int_{\Delta(x_0,2^{j+1})\setminus\Delta(x_0,2^j)} \frac{|f(x)|}{|x-x_0|^{n-1-\varepsilon}}\,d\sigma(x) \\
&\le C\sum_{j=0}^\infty 2^{j\varepsilon} \fint_{\Delta(x_0,2^j)} |f(x)|\,d\sigma(x)
\end{aligned} \tag{4.42}$$

for some constant $C = C(n, \varepsilon, x_0) \in (0, \infty)$, where we have enlarged the domain of integration and used Ahlfors regularity to create the integral average in the last inequality. By Lemma 2.34, we may further estimate

$$\sum_{j=0}^{\infty} 2^{j\varepsilon} \fint_{\Delta(x_0,2^j)} |f|\,d\sigma \le [w]_{A_p}^{1/p} \sum_{j=0}^{\infty} 2^{j\varepsilon} \Big(\fint_{\Delta(x_0,2^j)} |f|^p\,dw \Big)^{1/p} \tag{4.43}$$

Unpacking the average integral on the right-hand side of (4.43), multiplying and dividing by $w(\Delta(x_0,1))^{1/p}$, and appealing to Lemma 4.13, we see that (4.42) and (4.43) imply

$$\begin{aligned} &\int_{\Sigma} \frac{|f(x)|}{1+|x|^{n-1}}\,d\sigma(x) \\ &\quad \le \frac{C}{w(\Delta(x_0,1))^{1/p}} \sum_{j=0}^{\infty} 2^{j\varepsilon} \Big(\frac{w(\Delta(x_0,1))}{w(\Delta(x_0,2^j))} \Big)^{1/p} (2^j)^{(\lambda-a)/p} \|f\|_{M^{p,\lambda}(\Sigma,w_a)}. \end{aligned} \tag{4.44}$$

By item (2) of Proposition 2.33 applied to $E := \Delta(x_0,1)$ and $\Delta := \Delta(x_0,2^j)$, we may write for each $j \in \mathbb{N}_0$,

$$\Big(\frac{w(\Delta(x_0,1))}{w(\Delta(x_0,2^j))} \Big)^{1/p} \le \Big(\frac{\sigma(\Delta(x_0,1))}{\sigma(\Delta(x_0,2^j))} \Big)^{1/(pq')}. \tag{4.45}$$

By Ahlfors regularity, there exists a constant $C \in (0,\infty)$ such that

$$\frac{\sigma(\Delta(x_0,1))}{\sigma(\Delta(x_0,2^j))} \le C 2^{j(1-n)}. \tag{4.46}$$

Inserting the estimates from (4.46) and (4.45) into (4.44), we conclude that

$$\begin{aligned} &\int_{\Sigma} \frac{|f(x)|}{1+|x|^{n-1}}\,d\sigma(x) \\ &\quad \le \frac{C}{w(\Delta(x_0,1))^{1/p}} \sum_{j=0}^{\infty} 2^{j[-(n-1)/(pq')+(\lambda-a)/p+\varepsilon]} \|f\|_{M^{p,\lambda}(\Sigma,w_a)}. \end{aligned} \tag{4.47}$$

The above geometric series will converge whenever ε is small enough, because our assumptions that $\lambda < (n-1)/q'$ and $0 \le a < \lambda$ ensure that

$$-\frac{n-1}{pq'} + \frac{\lambda - a}{p} < 0 \tag{4.48}$$

Thus, there exists $\varepsilon_* \in (0, n-1)$ such that

$$-\frac{n-1}{pq'} + \frac{\lambda - a}{p} + \varepsilon < 0 \quad \text{for all } \varepsilon \in [0,\varepsilon_*). \tag{4.49}$$

Ultimately, from (4.47), we can conclude that there exists a constant $C \in (0,\infty)$, independent of f, such that for any $\varepsilon \in [0, \varepsilon_*)$,

$$\|f\|_{L^1(\Sigma, \frac{\sigma(x)}{1+|x|^{n-1-\varepsilon}})} \leq C\|f\|_{M^{p,\lambda}(\Sigma, w_a)}. \tag{4.50}$$

Given the arbitrary nature of f, this establishes the desired continuous embedding. □

The next embedding result involving weighted Morrey spaces that we would like to present states that Morrey spaces continuously embed into local Lebesgue spaces for some small range of integrability exponents. The proof is adapted from [86, Lemma 7.7.13, p. 659]. Note that this is a slightly stronger result than what is presented in Lemma 4.14, but this comes at the cost of a more restricted range for the threshold λ.

Proposition 4.16. *Let $\Sigma \subseteq \mathbb{R}^n$ be a closed ADR set and write $\sigma := \mathcal{H}^{n-1}\lfloor\Sigma$. Fix two integrability exponents $p, q \in (1,\infty)$, along with an arbitrary Muckenhoupt weight $w \in A_p(\Sigma,\sigma) \cap \mathrm{RH}_q(\Sigma,\sigma)$, and a threshold $\lambda \in (0, (n-1)/q')$, where q' is the Hölder conjugate of q.*

Then there exists some $p_* \in (1,\infty)$ such that for all $p_0 \in (0, p_*)$ one has the continuous embedding

$$M^{p,\lambda}(\Sigma, w) \hookrightarrow L^1\Big(\Sigma, \frac{\sigma(x)}{1+|x|^{n-1}}\Big) \cap L^{p_0}_{\mathrm{loc}}(\Sigma, \sigma). \tag{4.51}$$

Proof. Thanks to Proposition 4.15, it remains only to show that there exists $p_* \in (1,\infty)$ such that for any $p \in (0, p_*)$, we have

$$M^{p,\lambda}(\Sigma, w) \hookrightarrow L^{p_0}_{\mathrm{loc}}(\Sigma, \sigma) \quad \text{continuously.} \tag{4.52}$$

To set the stage, we first prove a useful estimate involving Muckenhoupt weights. Let $s \in (1,\infty)$ and let $\omega \in A_s(\Sigma,\sigma)$. The openness property of Muckenhoupt weights (item (1) of Proposition 2.29) ensures that there exists some small $\varepsilon \in (0, s-1)$ such that $\omega \in A_{s-\varepsilon}(\Sigma,\sigma)$. Let $x_0 \in \Sigma$. Then there exists some $C = C(x_0) \in (0,\infty)$ such that $1 + |x - x_0| \leq C(1 + |x|)$ for all $x \in \mathbb{R}^n$. Therefore,

$$\begin{aligned} \int_\Sigma \frac{d\omega(x)}{(1+|x|^{n-1})^s} &\leq C \int_\Sigma \frac{d\omega(x)}{(1+|x-x_0|^{n-1})^s} \\ &\leq C \int_{\Delta(x_0,1)} d\omega + C \sum_{j=0}^\infty \int_{\Delta(x_0,2^{j+1})\setminus\Delta(x_0,2^j)} \frac{d\omega(x)}{|x-x_0|^{(n-1)s}} \\ &\leq C\omega(\Delta(x_0,1)) + C\sum_{j=0}^\infty \frac{1}{2^{j(n-1)s}}\omega(\Delta(x_0,2^{j+1})) \\ &\leq C\omega(\Delta(x_0,1)) + C\sum_{j=0}^\infty \frac{2^{j(n-1)(s-\varepsilon)}}{2^{j(n-1)s}}\omega(\Delta(x_0,1)) \end{aligned} \tag{4.53}$$

$$\leq C\left(1+\sum_{j=0}^{\infty}2^{-j(n-1)\varepsilon}\right)\omega(\Delta(x_0,1)) < +\infty.$$

Note that the fourth inequality above follows via doubling (Proposition 2.29 item (4)). Indeed, since $w \in A_{s-\varepsilon}(\Sigma,\sigma)$, the Ahlfors regularity of Σ and (2.57) (with $s-\varepsilon$, $\Delta(x_0,1)$, and $\Delta(x_0,2^{j+1})$ in place of p, E, and Δ, respectively) yield

$$\begin{aligned}\omega(\Delta(x_0,2^{j+1})) &\leq [\omega]_{A_{s-\varepsilon}}\omega(\Delta(x_0,1))\left(\frac{\sigma(\Delta(x_0,2^{j+1}))}{\sigma(\Delta(x_0,1))}\right)^{s-\varepsilon}\\ &\leq [\omega]_{A_{s-\varepsilon}}\omega(\Delta(x_0,1))\frac{C_\Sigma 2^{(j+1)(n-1)(s-\varepsilon)}}{c_\Sigma}\\ &\leq C\omega(\Delta(x_0,1))2^{j(n-1)(s-\varepsilon)},\end{aligned} \tag{4.54}$$

where c_Σ and C_Σ are the lower and upper Ahlfors regularity constants of Σ, respectively, and $C \in (0,\infty)$ depends only on the ADR character of Σ.

Moving to the task of proving (4.52), the openness and monotonicity properties of the Muckenhoupt class (cf. item (1) of Proposition 2.29 and (2.51)) guarantee that there exists some $r_0 \in [1,p)$ such that $w \in A_r(\Sigma,\sigma)$ for each $r \in (r_0,\infty)$. Define $p_* := p/r_0$ and choose any $p_0 \in (0,p_*)$. Set $r := p/p_0 \in (r_0,\infty)$ and denote $r' := r/(r-1) \in (1,\infty)$. Then by item (2) of Proposition 2.29, we have that $\omega := w^{-r'/r}$ belongs to $A_{r'}(\Sigma,\sigma)$. Now, fix some σ-measurable function $f:\Sigma\to\mathbb{C}$ and let $\Delta\subseteq\Sigma$ be an arbitrary surface ball. We estimate

$$\begin{aligned}\int_\Delta |f|^{p_0}\,d\sigma &\leq C_\Delta\int_\Delta \frac{|f(x)|^{p_0}}{1+|x|^{n-1}}\,d\sigma(x)\\ &= C_\Delta\int_\Delta \big(|f(x)|^{p_0}w(x)^{p_0/p}\big)\frac{w(x)^{-p_0/p}}{1+|x|^{n-1}}\,d\sigma(x)\\ &\leq C_\Delta\left[\int_\Delta \big(|f|^{p_0}w^{p_0/p}\big)^r\,d\sigma\right]^{1/r}\left[\int_\Delta \frac{w(x)^{-(p_0/p)r'}}{(1+|x|^{n-1})^{r'}}\,d\sigma(x)\right]^{1/r'}\\ &= C_\Delta\|f\|^{p_0}_{L^p(\Delta,w)}\left[\int_\Delta \frac{1}{(1+|x|^{n-1})^{r'}}\,d\omega(x)\right]^{1/r'}\\ &\leq C_\Delta\|f\|^{p_0}_{L^p(\Delta,w)}\end{aligned} \tag{4.55}$$

for some constant $C_\Delta \in (0,\infty)$, which depends on Δ. The first line above follows from (3) of Lemma 2.18, the third line follows from Hölder's inequality, the fourth line follows because $w^{-(p_0/p)r'} = w^{-r'/r} = \omega$, and the last line is a consequence of (4.53) with r' in place of s. With this estimate in hand, we are now ready to complete the proof.

Let $f \in M^{p,\lambda}(\Sigma,w)$. Then f is σ-measurable and as such, (4.55) implies for any $p_0 \in (0,p_*)$, and any surface ball $\Delta\subseteq\Sigma$,

$$\|f\|_{L^{p_0}(\Delta,\sigma)} \le C_\Delta \|f\|_{L^p(\Delta,w)} \le C_\Delta \|f\|_{M^{p,\lambda}(\Sigma,w)}, \tag{4.56}$$

whereby the last inequality is a consequence of Lemma 4.13 and $C_\Delta \in (0,\infty)$ depends on Δ, but is independent of f. Given that $f \in M^{p,\lambda}(\Sigma, w)$ was arbitrary, this yields the desired continuous embedding in (4.52), thereby completing the proof. □

Our last embedding result for this section presents a very general embedding result for a certain weighted Lebesgue space into a weighted Morrey space. It is worth calling attention to the fact that the weight can be arbitrary in this case, not necessarily Muckenhoupt.

Proposition 4.17. *Let $\Sigma \subseteq \mathbb{R}^n$ be a closed ADR set and write $\sigma := \mathcal{H}^{n-1}\lfloor\Sigma$. Fix an integrability exponent $p \in (1,\infty)$, a threshold $\lambda \in (0, n-1)$, and an arbitrary weight function $w : \Sigma \to [0,\infty]$. Define $s := \frac{p(n-1)}{n-1-\lambda}$. Then the following continuous embedding holds:*

$$L^s(\Sigma, w^{s/p}) \hookrightarrow M^{p,\lambda}(\Sigma, w). \tag{4.57}$$

Proof. Begin by fixing a location $x \in \Sigma$ and a scale $r \in (0, 2\operatorname{diam}(\Sigma))$. Suppose that $f : \Sigma \to \mathbb{C}$ is a σ-measurable function. We can apply Hölder's inequality with the exponent s/p and its conjugate,

$$\left(1 - \frac{p}{s}\right)^{-1} = \left(1 - \frac{p}{\frac{p(n-1)}{n-1-\lambda}}\right)^{-1} = \left(1 - \frac{n-1-\lambda}{n-1}\right)^{-1} = \left(\frac{\lambda}{n-1}\right)^{-1}, \tag{4.58}$$

to estimate

$$\begin{aligned} \frac{1}{r^\lambda} \int_{\Delta(x,r)} |f|^p w \, d\sigma &\le \frac{1}{r^\lambda} \Big(\int_{\Delta(x,r)} |f|^{p(s/p)} w^{s/p} \, d\sigma \Big)^{p/s} \sigma(\Delta(x,r))^{\lambda/(n-1)} \\ &\le \frac{1}{r^\lambda} \Big(\int_\Sigma |f|^s w^{s/p} \, d\sigma \Big)^{p/s} (C r^{n-1})^{\lambda/(n-1)} \\ &\le C \|f\|^p_{L^s(\Sigma, w^{s/p})}. \end{aligned} \tag{4.59}$$

Note that we used the Ahlfors regularity of Σ to obtain the second inequality, and the constant $C \in (0,\infty)$ appearing in the final line depends only on the ADR character of Σ, n, and λ. Taking pth roots on both sides of the inequality, followed by a supremum over all $x \in \Sigma$ and $r \in (0, 2\operatorname{diam}(\Sigma))$ in view of (4.2) yields

$$\|f\|_{M^{p,\lambda}(\Sigma,w)} \le C \|f\|_{L^s(\Sigma, w^{s/p})}, \tag{4.60}$$

where the constant C now depends only on Σ, n, λ, and p. This implies the desired embedding, given that f was arbitrary. □

Remark 4.18. It is worth remarking that in the context of Proposition 4.17, the case that w is identically one becomes

$$L^s(\Sigma,\sigma) \hookrightarrow M^{p,\lambda}(\Sigma,\sigma) \quad \text{continuously with } s := \frac{p(n-1)}{n-1-\lambda}. \tag{4.61}$$

This turns out to be a significant embedding result, because it helps usher in the notion of a "vanishing" Morrey space; cf. [87, Section 6.2].

We will discuss a weighted version of vanishing Morrey spaces in Chapter 10. We will see that even with the embedding in Proposition 4.17, it is not immediately clear how to incorporate a weight. For instance, when working with Muckenhoupt-weighted Morrey spaces, there is a "natural" environment from which to select a weight. Namely, starting with some $p \in (1,\infty)$, the weight in question should belong not only to $A_p(\Sigma,\sigma)$ but also $\mathrm{RH}_q(\Sigma,\sigma)$ for some additional choice of $q \in (1,\infty)$ (such a q is guaranteed by item (2) of Proposition 2.33, though in principle q could differ from the one guaranteed by the theory). However, for $s := \frac{p(n-1)}{n-1-\lambda}$, it is not expected that the weight $w^{s/p}$ belongs to $A_s(\Sigma,\sigma)$, the "natural" class of Muckenhoupt weights for the Lebesgue space $L^s(\Sigma, w^{s/p})$ from the perspective of the boundedness of the Hardy–Littlewood maximal operator. As such, the task of coming up with a suitable version of a *weighted* vanishing Morrey space is more nebulous, and we hold off on that task for now.

4.3 Extrapolation theory

The following theorem is the main result of this section. We seek to generalize the extrapolation result of Duoandikoetxea and Rosenthal [29] to our geometric setting of interest. Namely, we obtain a Rubio de Francia-type extrapolation result on the class of closed Ahlfors regular sets for Muckenhoupt-weighted Morrey spaces. Loosely, Muckenhoupt-weighted L^p estimates for a *single* exponent p and *all* A_p weights imply analogous estimates at the level of Muckenhoupt-weighted Morrey spaces for *all* integrability exponents and *all* corresponding Muckenhoupt weights.

Theorem 4.19. *Let $n \in \mathbb{N}$ with $n \geq 2$. Let $\Sigma \subseteq \mathbb{R}^n$ be a closed ADR set and write $\sigma := \mathcal{H}^{n-1}\lfloor\Sigma$. Fix a reference point $x_0 \in \Sigma$ along with an arbitrary integrability exponent $p_0 \in [1,\infty)$. Let $\mathcal{F}$ be a collection of pairs of σ-measurable functions on Σ such that for all $(f,g) \in \mathcal{F}$ one has*

$$\|g\|_{L^{p_0}(\Sigma,\omega)} \leq C_\omega \|f\|_{L^{p_0}(\Sigma,\omega)} \tag{4.62}$$

for each Muckenhoupt weight $\omega \in A_{p_0}(\Sigma,\sigma)$, where the constant $C_\omega \in (0,\infty)$ is independent of f and g and depends on $[\omega]_{A_{p_0}}$ in such a way that C_ω stays bounded provided $[\omega]_{A_{p_0}}$ stays bounded. More precisely, for all $W \in (0,\infty)$,

$$\sup_{\substack{\omega\in A_{p_0}(\Sigma,\sigma)\\ [\omega]_{A_{p_0}}\leq W}} C_\omega < +\infty. \tag{4.63}$$

For any given $\alpha \in [0,\infty)$ abbreviate $w_\alpha := |\cdot - x_0|^\alpha w$, a weight function on Σ.

Then for all integrability exponents $p, q \in (1,\infty)$ (and also for $p = 1$ in the case that $p_0 = 1$) and all Muckenhoupt weights $w \in A_p(\Sigma,\sigma) \cap \mathrm{RH}_q(\Sigma,\sigma)$, there exist finite positive constants K and W, both depending only on the ADR character of Σ, n, p, q, λ, α, $[w]_{A_p}$, and $[w]_{\mathrm{RH}_q}$ such that

$$\|g\|_{M^{p,\lambda}(\Sigma,w_\alpha)} \leq C_W \|f\|_{M^{p,\lambda}(\Sigma,w_\alpha)}, \tag{4.64}$$

for all $(f,g) \in \mathcal{F}$, provided

$$0 \leq \lambda < (n-1)/q' \quad \text{and} \quad 0 \leq \alpha < \lambda, \tag{4.65}$$

where $q' := q/(q-1)$, and

$$C_W := K \cdot \Big(\sup_{\substack{\omega\in A_{p_0}(\Sigma,\sigma)\\ [\omega]_{A_{p_0}}\leq W}} C_\omega \Big). \tag{4.66}$$

Proof. Without loss of generality, we can assume both f and g are nonnegative functions. Otherwise, work with $|f|$ and $|g|$ in place of f and g. There are two main cases to distinguish, and the second case largely follows from the classical Rubio de Francia extrapolation theorem (cf. Proposition 2.41) and work done in the first.

Case 1: *Suppose* $p_0 = 1$. Fix $p \in [1,\infty)$, and pick an arbitrary Muckenhoupt weight $w \in A_p(\Sigma,\sigma)$. Assume that (4.62) holds for $p_0 = 1$. Let $\Delta(x,r) \subseteq \Sigma$ be a surface ball with center $x \in \Sigma$ and radius $r \in (0, 2\operatorname{diam}(\Sigma))$. Fix some $(f,g) \in \mathcal{F}$, then

$$\begin{aligned} &\int_{\Delta(x,r)} g(y)^p |x_0 - y|^\alpha w(y)\, d\sigma(y) \\ &\quad \leq \begin{cases} r^\alpha \int_{\Delta(x_0,r)} g^p w\, d\sigma, & \text{if } x = x_0, \\ (5/4)^\alpha |x - x_0|^\alpha \int_{\Delta(x,r)} g^p w\, d\sigma, & \text{if } 0 < r \leq |x - x_0|/4, \end{cases} \end{aligned} \tag{4.67}$$

where in the second case we use the fact that when $r \leq |x - x_0| < 4$ and for $y \in \Delta(x,r)$, we have

$$|x_0 - y| \leq |x_0 - x| + |x - y| \leq (5/4)|x - x_0|. \tag{4.68}$$

Subsequently, from (4.67) we obtain

$$\left(\frac{1}{r^{\lambda}}\int_{\Delta(x,r)} g(y)^p|y-x_0|^a w(y)\,d\sigma(y)\right)^{1/p} \leq \begin{cases} r^{(a-\lambda)/p}\|g\|_{L^p(\Delta(x_0,r),w)}, & \text{if } x = x_0, \\ (5/4)^{a/p}|x-x_0|^{a/p}r^{-\lambda/p}\|g\|_{L^p(\Delta(x,r),w)}, & \text{if } 0 < r \leq |x-x_0|/4. \end{cases} \tag{4.69}$$

At this point, we find it useful to distinguish when $p \in (1,\infty)$ and $p = 1$.

Subcase 1A: *Suppose* $p \in (1,\infty)$. Denote by $p' := p/(p-1)$ the Hölder conjugate of p. Fix some nonnegative

$$h \in L^{p'}(\Delta(x,r),w) \quad \text{with } \|h\|_{L^{p'}(\Delta(x,r),w)} = 1. \tag{4.70}$$

Observe by item (2) of Proposition 2.29 that $w \in A_p(\Sigma,\sigma)$ implies $w^{1-p'}$ belongs to $A_{p'}(\Sigma,\sigma)$, and hence by item (1) of Proposition 2.29 there exists $s \in (1,p)$, depending on Σ, n, p, and $[w]_{A_p}$, such that $w^{1-p'}$ belongs to $A_{p'/s}(\Sigma,\sigma)$. Write $s' := s/(s-1)$. By Hölder's inequality with the exponent $p'/s \in (1,\infty)$ and its Hölder conjugate $(p'/s)' := p'/(p'-s)$, we can estimate

$$\begin{aligned}
\left(\int_{\Sigma} h^s w^s \mathbf{1}_{\Delta(x,r)}\,d\sigma\right)^{1/s} &= \left(\int_{\Delta(x,r)} h^s w^{s/p'} w^{s/p}\,d\sigma\right)^{1/s} \\
&\leq \left(\int_{\Delta(x,r)} h^{p'} w\,d\sigma\right)^{1/p'}\left(\int_{\Delta(x,r)} w^{(s/p)(p'/s)'}\,d\sigma\right)^{(1/s)/(p'/s)'} \\
&= \sigma(\Delta(x,r))^{1/s-1/p'}\left(\fint_{\Delta(x,r)} w^{(p'-1)[1-(p'/s)']}\,d\sigma\right)^{1/s-1/p'} \\
&\leq \sigma(\Delta(x,r))^{1/s-1/p'}[w^{1-p'}]_{A_{p'/s}}^{1/p'}\left(\fint_{\Delta(x,r)} w^{1-p'}\,d\sigma\right)^{-1/p'} \\
&= [w^{1-p'}]_{A_{p'/s}}^{1/p'}\sigma(\Delta(x,r))^{1/s} w^{1-p'}(\Delta(x,r))^{-1/p'}
\end{aligned} \tag{4.71}$$

where $C = C(\Sigma,n,p,[w]_{A_p}) \in (0,\infty)$. The third line above follows from (4.70) and the fact that $(p'/s)' = p'/(p'-s)$. Since $w^{1-p'} \in A_{p'/s}(\Sigma,\sigma)$, the fourth line above holds because of (2.47) (with p'/s and $w^{1-p'}$ playing the roles of p and w, respectively). Then notice that by the lower Ahlfors regularity condition on Σ, and (2.52), we can estimate

$$C_\Sigma r^{n-1} \leq \sigma(\Delta(x,r)) \leq w(\Delta(x,r))^{1/p} w^{1-p'}(\Delta(x,r))^{1/p'}. \tag{4.72}$$

Dividing the above inequality by $w^{1-p'}(\Delta(x,r))^{1/p'} c_\Sigma r^{n-1}$ in combination with (4.71) and the Ahlfors regularity of Σ implies

$$\Big(\int_\Sigma h^s w^s \mathbf{1}_{\Delta(x,r)}\,d\sigma\Big)^{1/s} \le C_\Sigma [w^{1-p'}]^{1/p'}_{A_{p'/s}} w(\Delta(x,r))^{1/p} r^{-(n-1)+(n-1)/s}$$
$$\le C w(\Delta(x,r))^{1/p} r^{-(n-1)/s'} \tag{4.73}$$

where the constant $C \in (0,\infty)$ depends only on Σ, n, p, and $[w]_{A_p}$.

Note that (4.71) implies $h^s w^s \mathbf{1}_{\Delta(x,r)} \in L^1(\Sigma,\sigma)$, so recalling our discussion in (2.45), we find that $\mathcal{M}_\Sigma(h^s w^s \mathbf{1}_{\Delta(x,r)}) \in L^{1,\infty}(\Sigma,\sigma)$, and hence $\mathcal{M}_\Sigma(h^s w^s \mathbf{1}_{\Delta(x,r)})(x) < +\infty$ at σ-a. e. point $x \in \Sigma$. Subsequently, item (3) of Proposition 2.29 gives

$$\begin{gathered}(\mathcal{M}_\Sigma(h^s w^s \mathbf{1}_{\Delta(x,r)}))^{\frac{1}{s}} \in A_1(\Sigma,\sigma) \quad \text{and}\\ [(\mathcal{M}_\Sigma(h^s w^s \mathbf{1}_{\Delta(x,r)}))^{\frac{1}{s}}]_{A_1} \le \frac{Cs}{s-1} =: W\end{gathered} \tag{4.74}$$

for some constant $C \in (0,\infty)$, depending only on n and the ADR character of Σ. Observe that for

$$C_* := \sup_{\substack{\omega\in A_1(\Sigma,\sigma)\\ [\omega]_{A_1}<W}} C_\omega \in (0,\infty), \tag{4.75}$$

we may estimate

$$\begin{aligned}\int_{\Delta(x,r)} ghw\,d\sigma &= \int_\Sigma ghw\mathbf{1}_{\Delta(x,r)}\,d\sigma = \int_\Sigma g(h^s w^s \mathbf{1}_{\Delta(x,r)})^{1/s}\,d\sigma \\ &\le \int_\Sigma g(\mathcal{M}_\Sigma(h^s w^s \mathbf{1}_{\Delta(x,r)}))^{1/s}\,d\sigma \\ &\le C_* \int_\Sigma f(\mathcal{M}_\Sigma(h^s w^s \mathbf{1}_{\Delta(x,r)}))^{1/s}\,d\sigma,\end{aligned} \tag{4.76}$$

where the first inequality uses Corollary 2.22, and the last inequality follows by assumption (4.62) with $p_0 = 1$ and for the weight $(\mathcal{M}_\Sigma(h^s w^s \mathbf{1}_{\Delta(x,r)}))^{1/s} \in A_1(\Sigma,\sigma)$ (cf. (4.74)–(4.75)).

We will split the last integral of (4.76) into an integral over $\Delta(x,2r)$ and an integral over $\Sigma \setminus \Delta(x,2r)$, so define

$$\begin{aligned}\mathrm{I} &:= \int_{\Delta(x,2r)} f(\mathcal{M}_\Sigma(h^s w^s \mathbf{1}_{\Delta(x,r)}))^{1/s}\,d\sigma,\\ \mathrm{II} &:= \int_{\Sigma\setminus\Delta(x,2r)} f(\mathcal{M}_\Sigma(h^s w^s \mathbf{1}_{\Delta(x,r)}))^{1/s}\,d\sigma.\end{aligned} \tag{4.77}$$

To treat the case near x, we use Hölder's inequality to estimate

$$
\begin{aligned}
\mathrm{I} &= \int_{\Delta(x,2r)} f(\mathcal{M}_\Sigma(h^s w^s \mathbf{1}_{\Delta(x,r)}))^{1/s} w^{1/p} w^{-1/p}\, d\sigma \\
&\le \Big(\int_{\Delta(x,2r)} f^p w\, d\sigma\Big)^{1/p} \Big(\int_{\Delta(x,2r)} (\mathcal{M}_\Sigma(h^s w^s \mathbf{1}_{\Delta(x,r)}))^{p'/s} w^{1-p'}\, d\sigma\Big)^{1/p'}.
\end{aligned}
\tag{4.78}
$$

To estimate the second integral in the last line of (4.78), note that $\mathcal{M}_\Sigma$ is bounded on $L^{p'/s}(\Sigma, w^{1-p'})$, because $p'/s > 1$ and $w^{1-p'} \in A_{p'/s}(\Sigma, \sigma)$ (see the remark in (2.43)). Hence,

$$
\begin{aligned}
&\Big(\int_{\Delta(x,2r)} (\mathcal{M}_\Sigma(h^s w^s \mathbf{1}_{\Delta(x,r)}))^{p'/s} w^{1-p'}\, d\sigma\Big)^{1/p'} \\
&\quad \le \big\|\mathcal{M}_\Sigma(h^s w^s \mathbf{1}_{\Delta(x,r)})\big\|_{L^{p'/s}(\Sigma, w^{1-p'})}^{1/s} \\
&\quad \le C\big\|h^s w^s \mathbf{1}_{\Delta(x,r)}\big\|_{L^{p'/s}(\Sigma, w^{1-p'})}^{1/s} \\
&\quad = C\|h\|_{L^{p'}(\Delta(x,r),w)}^{1/s} = C,
\end{aligned}
\tag{4.79}
$$

where the last line follows from (4.70), and $C \in (0, \infty)$ depends only on Σ, n, p, and $[w]_{A_p}$. Concerning the first integral in the last line of (4.78), we find it easier to distinguish the case $x = x_0$ from $0 < r \le |x - x_0|/4$. Observe by Lemma 4.13 that there exists a constant $C = C(\alpha, \lambda) \in (0, \infty)$ such that

$$
\int_{\Delta(x,2r)} f^p w\, d\sigma \le C(2r)^{\lambda-\alpha} \|f\|_{M^{p,\lambda}(\Sigma, w_\alpha)}^p \quad \text{if } x = x_0.
\tag{4.80}
$$

Now, when $0 < r \le |x - x_0|/4$, note for $y \in \Delta(x, 2r)$ one has

$$
|x - x_0| \le |x - y| + |y - x_0| \le |x - x_0|/2 + |y - x_0|,
\tag{4.81}
$$

which implies $|x - x_0| \le 2|y - x_0|$, and hence (4.2) and $\alpha \ge 0$ imply

$$
\int_{\Delta(x,2r)} f^p w\, d\sigma \le 2^\alpha |x - x_0|^{-\alpha} (2r)^\lambda \|f\|_{M^{p,\lambda}(\Sigma, w_\alpha)}^p
\tag{4.82}
$$

if $0 < r \le |x - x_0|/4$. Bringing the estimates in (4.79), (4.80), and (4.82) together in (4.78) yield

$$
\mathrm{I} \le \begin{cases} C(2r)^{(\lambda-\alpha)/p} \|f\|_{M^{p,\lambda}(\Sigma, w_\alpha)} & \text{if } x = x_0, \\ 2^{\alpha/p} |x - x_0|^{-\alpha/p} (2r)^{\lambda/p} \|f\|_{M^{p,\lambda}(\Sigma, w_\alpha)} & \text{if } 0 < r \le |x - x_0|/4. \end{cases}
\tag{4.83}
$$

We now move on to the treatment of II as defined in (4.77). Let $j \in \mathbb{N}$ and define $\Delta_j := \Delta(x, 2^{j+1}r) \setminus \Delta(x, 2^j r)$, so that $\Sigma \setminus \Delta(x, 2r) = \bigcup_{j=1}^\infty \Delta_j$. Fix $j \in \mathbb{N}$ along with $y \in \Delta_j$ arbitrary, and assume $R > 0$ is such that $\Delta(y, R) \cap \Delta(x, r) \ne \varnothing$ (otherwise, the estimate we endeavor to obtain in (4.85) is trivially true by (2.26)). Then for $z \in \Delta(y,R) \cap \Delta(x,r)$ we have

$$2^j r \le |y-x| \le |y-z| + |z-x| < R + r. \tag{4.84}$$

In particular, this forces $2^{j-1}r \le (2^j - 1)r < R$ and it follows that

$$\begin{aligned}
\mathcal{M}_\Sigma(h^s w^s \mathbf{1}_{\Delta(x,r)})(y) &= \sup_{R>0} \fint_{\Delta(y,R)} h^s w^s \mathbf{1}_{\Delta(x,r)}\, d\sigma \\
&\le \sup_{\substack{R>0 \text{ such that} \\ \Delta(y,R)\cap\Delta(x,r)\neq\varnothing}} \fint_{\Delta(y,R)} h^s w^s \mathbf{1}_{\Delta(x,r)}\, d\sigma \\
&\le \sup_{\substack{R>0 \text{ such that} \\ \Delta(y,R)\cap\Delta(x,r)\neq\varnothing}} \frac{C}{R^{n-1}} \int_{\Delta(x,r)} h^s w^s\, d\sigma \\
&\le \frac{C}{(2^{-j}r)^{n-1}} \int_{\Delta(x,r)} h^s w^s\, d\sigma \\
&\le \frac{C}{2^{j(n-1)}} w(\Delta(x,r))^{s/p} r^{-(n-1)[(s/s')+1]},
\end{aligned} \tag{4.85}$$

for some constant $C = C(\Sigma, n, p, [w]_{A_p}) \in (0,\infty)$. The third line above follows from Ahlfors regularity of Σ, the fourth line from the fact that $2^{j-1}r \le R$, and the last line above is a consequence of the bound in (4.71) upon raising both sides to the sth power. With the same C as in (4.85), it follows (bearing in mind (4.77)) that

$$\begin{aligned}
\mathrm{II} &\le C \sum_{j=1}^{\infty} \frac{w(\Delta(x,r))^{1/p}}{2^{j(n-1)/s} r^{(n-1)[(s/s')+1]/s}} \int_{A_j} f\, d\sigma \\
&\le C \sum_{j=1}^{\infty} \frac{w(\Delta(x,r))^{1/p} w^{1-p'}(\Delta(x,2^{j+1}r))^{1/p'}}{2^{j(n-1)/s} r^{n-1}} \Big(\int_{\Delta(x,2^{j+1}r)} f^p w\, d\sigma \Big)^{1/p},
\end{aligned} \tag{4.86}$$

where the second inequality follows from multiplication by $w^{1/p}w^{-1/p}$, an application of Höder's inequality, and recalling that $1/s + 1/s' = 1$. Since we are assuming $w \in \mathrm{RH}_q(\Sigma,\sigma)$, item (3) of Proposition 2.33 applied to the surface balls $\Delta(x,r)$ and $\Delta(x,2^{j+1}r)$, together with the Ahlfors regularity of Σ implies

$$\begin{aligned}
w(\Delta(x,r))^{1/p} &\le [w]_{\mathrm{RH}_q}^{1/p} w(\Delta(x,2^{j+1}r))^{1/p} \Big(\frac{\sigma(\Delta(x,r))}{\sigma(\Delta(x,2^{j+1}r))} \Big)^{1/pq'} \\
&\le C w(\Delta(x,2^{j+1}r))^{1/p} 2^{-(j+1)(n-1)/pq'}
\end{aligned} \tag{4.87}$$

with constant $C = C(\Sigma, n, p, [w]_{\mathrm{RH}_q}) \in (0,\infty)$. Recalling (2.53) and using the upper Ahlfors regularity of Σ, we can estimate

$$\begin{aligned} & w(\Delta(x, 2^{j+1}r))^{1/p} w^{1-p'}(\Delta(x, 2^{j+1}r))^{1/p'} \\ & \quad \le [w]_{A_p}^{1/p} \sigma(\Delta(x, 2^{j+1}r)) \le C 2^{j(n-1)} r^{n-1}, \end{aligned} \tag{4.88}$$

with $C = C(\Sigma, p, [w]_{A_p}) \in (0, \infty)$. Combining (4.86)–(4.88) yields

$$\begin{aligned} & \int_{\Sigma\setminus\Delta(x,2r)} f(\mathcal{M}(h^s w^s \mathbf{1}_{\Delta(x,r)}))^{1/s}\, d\sigma \\ & \quad \le C \sum_{j=1}^{\infty} \Big(\int_{\Delta(x,2^{j+1}r)} f^p w\, d\sigma \Big)^{1/p} 2^{j(n-1)/s'} 2^{-j(n-1)/(pq')}, \end{aligned} \tag{4.89}$$

where $C = C(\Sigma, n, p, [w]_{A_p}, [w]_{\mathrm{RH}_q}) \in (0, \infty)$. Here, we again distinguish the case when $x = x_0$ and when $0 < r \le |x - x_0|/4$.

If $x = x_0$, then by Lemma 4.13,

$$\begin{aligned} & \sum_{j=1}^{\infty} \Big(\int_{\Delta(x_0,2^{j+1}r)} f^p w\, d\sigma \Big)^{1/p} 2^{j(n-1)/s'} 2^{-j(n-1)/(pq')} \\ & \quad \le C \Big\{ \sum_{j=1}^{\infty} 2^{j[(1/p)(\lambda-a-(n-1)/q')+(n-1)/s']} \Big\} r^{(\lambda-a)/p} \|f\|_{M^{p,\lambda}(\Sigma,w_a)}, \end{aligned} \tag{4.90}$$

with $C = C(\Sigma, n, p, \lambda, a, [w]_{A_p}, [w]_{\mathrm{RH}_q}) \in (0, \infty)$. Working under the assumptions in (4.65), we find that

$$\frac{1}{p}\Big(\lambda - a - \frac{n-1}{q'}\Big) < -\frac{a}{p}. \tag{4.91}$$

At this point, choose s sufficiently close to 1 so that $(n-1)/s' < a/p$, which ensures

$$\Big(\lambda - a - \frac{n-1}{q'}\Big)\frac{1}{p} + \frac{n-1}{s'} < 0. \tag{4.92}$$

Ultimately, (4.92) implies that the sum in (4.90) is a convergent geometric series. Subsequently, (4.90), (4.89), and (4.77) imply that

$$\mathrm{II} \le C r^{(\lambda-a)/p} \|f\|_{M^{p,\lambda}(\Sigma,w_a)} \quad \text{if } x = x_0, \tag{4.93}$$

where $C \in (0, \infty)$ depends only on Σ, n, p, q, λ, a, $[w]_{A_p}$, and $[w]_{\mathrm{RH}_q}$.

Turning our attention to the case that $0 < r \le |x - x_0|/4$, there exists $N \in \mathbb{N}$, with $N \ge 2$, such that

$$2^N r < |x - x_0| \le 2^{N+1} r. \tag{4.94}$$

We distinguish the cases $j \le N-2$ and $j \ge N-1$. If $j \le N-2$, then for $y \in \Delta(x, 2^{j+1}r)$, we have (bearing in mind that $N \ge 2$ and $r < |x - x_0|/4$)

$$|y - x| < 2^{j+1}r \leq 2^{N-1}r < 2^{N-3}|x - x_0| < \frac{1}{2}|x - x_0|, \tag{4.95}$$

and hence

$$|x - x_0| \leq |x - y| + |y - x_0| \quad \text{implies} \quad |x - x_0| \leq 2|y - x_0|. \tag{4.96}$$

Thus,

$$\begin{aligned} \int\limits_{\Delta(x,2^{j+1}r)} f^p w \, d\sigma &\leq 2^a |x - x_0|^{-a} \int\limits_{\Delta(x,2^{j+1}r)} f(y)^p |y - x_0|^a w(y) \, d\sigma(y) \\ &\leq 2^{\lambda+a} |x - x_0|^{-a} (2^j r)^\lambda \|f\|^p_{M^{\lambda,p}(\Sigma,w_a)}, \end{aligned} \tag{4.97}$$

where the second inequality is due to (4.2). When $j \geq N - 1$, we have the set inclusion $\Delta(x, 2^{j+1}r) \subseteq \Delta(x_0, 2^{j+3}r)$. Indeed, if $y \in \Delta(x, 2^{j+1}r)$, then keeping in mind (4.94),

$$\begin{aligned} |y - x_0| \leq |y - x| + |x - x_0| &\leq 2^{j+1}r + 2^{N+1}r \\ &\leq 2^{j+1}r + 2^{j+2}r \leq 2^{j+3}r. \end{aligned} \tag{4.98}$$

Appealing to Lemma 4.13 gives

$$\int\limits_{\Delta(x,2^{j+1}r)} f^p w \, d\sigma \leq \int\limits_{\Delta(x_0,2^{j+3}r)} f^p w \, d\sigma \leq C(2^j r)^{\lambda-a} \|f\|^p_{M^{\lambda,p}(\Sigma,w_a)} \tag{4.99}$$

for some constant $C = C(a, \lambda) \in (0, \infty)$. Inserting the results from (4.97) and (4.99) into (4.89) yields

$$\begin{aligned} &\int\limits_{\Sigma\setminus\Delta(x,2r)} f(\mathcal{M}_\Sigma(h^s w^s \mathbf{1}_{\Delta(x,r)}))^{1/s} \, d\sigma \\ &\quad \leq C \sum_{j=1}^{N-2} |x - x_0|^{-a/p} r^{\lambda/p} 2^{j[(\lambda-(n-1)/q')(1/p)+(n-1)/s']} \|f\|_{M^{\lambda,p}(\Sigma,w_a)} \\ &\qquad + C \sum_{j=N-1}^{\infty} r^{(\lambda-a)/p} 2^{[(\lambda-a-(n-1)/q')(1/p)+(n-1)/s']} \|f\|_{M^{\lambda,p}(\Sigma,w_a)}, \end{aligned} \tag{4.100}$$

with constant $C = C(\Sigma, n, p, \lambda, a, [w]_{A_p}, [w]_{\mathrm{RH}_q}) \in (0, \infty)$. In a similar manner as (4.91)–(4.92), when (4.65) is satisfied and s is chosen sufficiently close to 1, we have that the exponent of 2 in the first sum in (4.100) is negative, and hence, being a geometric sum, is bounded independently of N. Likewise, with this choice of s the exponent of 2 in the second sum of (4.100) is also negative, and hence, a convergent geometric series with

$$\begin{aligned}\sum_{j=N+1}^{\infty} & r^{(\lambda-a)/p} 2^{j[(\lambda-a-(n-1)/q')(1/p)+(n-1)/s']} \\ & \leq C r^{\frac{\lambda-a}{p}} 2^{N[(\lambda-a-(n-1)/q')(1/p)+(n-1)/s']} \\ & \leq C|x-x_0|^{-\frac{a}{p}} r^{\frac{\lambda}{p}},\end{aligned} \tag{4.101}$$

where the constant $C \in (0,\infty)$ depends only on p, λ, and a. Note that the last inequality above is a consequence of the fact that

$$2^{N[(\lambda-a-(n-1)/q')(1/p)+(n-1)/s']} \leq 2^{a/p}|x-x_0|^{-a/p} r^{a/p}. \tag{4.102}$$

Indeed, recall by (4.94) that $r|x-x_0|^{-1} \geq 2^{-N-1}$, which implies

$$r^{a/p}|x-x_0|^{-a/p} \geq 2^{-a/p} \cdot 2^{-(Na)/p}. \tag{4.103}$$

Since our choice of s ensures $(\lambda - \frac{n-1}{q'})\frac{1}{p} + \frac{n-1}{s'} < 0$, we find that $-\frac{a}{p} > (\lambda - a - \frac{n-1}{q'})\frac{1}{p} + \frac{n-1}{s'}$, making it clear how (4.102) follows from (4.103). Thus, (4.100) and (4.77) imply

$$\mathrm{II} \leq C|x-x_0|^{-a/p} r^{\lambda/p} \|f\|_{M^{p,\lambda}(\Sigma,w_a)} \quad \text{if } 0 < r \leq |x-x_0|/4, \tag{4.104}$$

for some constant $C \in (0,\infty)$ depending on Σ, n, p, q, λ, a, $[w]_{A_p}$, and $[w]_{\mathrm{RH}_q}$.

At this point, we insert the estimates in (4.83), (4.93), and (4.104) into (4.76) (while recalling (4.77)) to obtain

$$\begin{aligned}\int_{\Delta(x,r)} ghw\, d\sigma &\leq C_*(\mathrm{I}+\mathrm{II}) \\ &\leq \begin{cases} KC_* r^{(\lambda-a)/p}\|f\|_{M^{p,\lambda}(\Sigma,w_a)} & \text{if } x = x_0, \\ KC_*|x-x_0|^{-a/p} r^{\lambda/p}\|f\|_{M^{\lambda,p}(\Sigma,w_a)} & \text{if } 0 < r \leq |x-x_0|/4,\end{cases}\end{aligned} \tag{4.105}$$

where $K = K(\Sigma, n, p, q, \lambda, a, [w]_{A_p}, [w]_{\mathrm{RH}_q}) \in (0,\infty)$, and C_* is as in (4.75).

For the purposes of establishing (4.64), there is no loss of generality in assuming that $f \in M^{p,\lambda}(\Sigma, w_a)$, otherwise the conclusion in (4.64) is trivially true. Keeping this in mind and specializing (4.105) to the case where h is a constant,[1] we see that $g \in L^1(\Delta(x,r), w)$ in either of the specified cases. Appealing to [42, Theorem 6.14, p. 189], we have by duality that there exists $h \in L^{p'}(\Delta(x,r), w)$ with $\|h\|_{L^{p'}(\Delta(x,r),w)} = 1$ such that

$$\|g\|_{L^p(\Delta(x,r),w)} = \int_{\Delta(x,r)} ghw\, d\sigma. \tag{4.106}$$

1 e. g., one could take $h := \sigma(\Delta(x,r))^{-1/p'} \in L^{p'}(\Delta(x,r),\sigma)$.

Consequently, (4.69), (4.106), and (4.105) imply that whenever it is the case that $x = x_0$ or instead if $0 < r \leq |x - x_0|/4$, there exists a positive, finite constant K, depending exclusively on Σ, n, p, q, λ, a, $[w]_{A_p}$, and $[w]_{\mathrm{RH}_q}$, such that

$$\left(\frac{1}{r^\lambda}\int_{\Delta(x,r)} g(y)^p|y - x_0|^a w(y)\,d\sigma(y)\right)^{1/p} \leq KC_*\|f\|_{M^{p,\lambda}(\Sigma,w_a)}, \tag{4.107}$$

where C_* is as in (4.75).

The estimate in (4.107) only holds when $x = x_0$ or $0 < r \leq |x - x_0|/4$. However, this is sufficient, for if $\Delta(x,r)$ is such that $r > |x - x_0|/4$, then $\Delta(x,r) \subseteq \Delta(x_0, 5r)$, and as such

$$\begin{aligned}
&\left(\frac{1}{r^\lambda}\int_{\Delta(x,r)} |g(y)|^p|y - x_0|^a w(y)\,d\sigma(y)\right)^{1/p} \\
&\leq \left(\frac{1}{r^\lambda}\int_{\Delta(x_0,5r)} |g(y)|^p|y - x_0|^a w(y)\,d\sigma(y)\right)^{1/p} \\
&\leq \left(\frac{5^\lambda}{(5r)^\lambda}\int_{\Delta(x_0,5r)} |g(y)|^p|y - x_0|^a w(y)\,d\sigma(y)\right)^{1/p} \\
&\leq 5^{\lambda/p}KC_*\|f\|_{M^{p,\lambda}(\Sigma,w_a)},
\end{aligned} \tag{4.108}$$

where the last inequality holds by (4.107) used in the case that $x = x_0$ and with $5r$ in place of r. Therefore, recalling (4.2), taking a supremum on both sides of (4.107) over all $x \in \Sigma$ and $r \in (0, 2\,\mathrm{diam}(\Sigma))$ yields the desired result for subcase 1A.

Subcase 1B: *Suppose p* = 1. By assumption, $w \in A_1(\Sigma,\sigma) \cap \mathrm{RH}_q(\Sigma,\sigma)$, which implies that $w^q \in A_1(\Sigma,\sigma)$ by item (5) in Proposition 2.33. For any exponent $s \in (1,q)$, the monotonicity of the reverse Hölder class (cf. item (1) of Proposition 2.33) ensures that $w \in A_1(\Sigma,\sigma) \cap \mathrm{RH}_s(\Sigma,\sigma)$, and hence $w^s \in A_1(\Sigma,\sigma)$ again by item (5) of Proposition 2.33. By the A_1 condition on w^s (cf. (2.49)) and the definition of $\mathcal{M}_\Sigma$ (cf. (2.26)), we have

$$\begin{aligned}
&(\mathcal{M}_\Sigma(w^s\mathbf{1}_{\Delta(x,r)}))^{1/s} \leq (\mathcal{M}_\Sigma(w^s))^{1/s} \leq w < +\infty \quad \text{at } \sigma\text{-a. e. point on } \Sigma, \\
&\text{and} \quad (\mathcal{M}_\Sigma(w^s\mathbf{1}_{\Delta(x,r)}))^{1/s} \in A_1(\Sigma,\sigma).
\end{aligned} \tag{4.109}$$

Furthermore, we have the following estimate due to item (3) of Proposition 2.29:

$$[(\mathcal{M}_\Sigma(w^s\mathbf{1}_{\Delta(x,r)}))^{1/s}]_{A_1} \leq \frac{Cs}{s-1} =: W \tag{4.110}$$

where $C \in (0,\infty)$ depends only on n and Σ. Observe that for

$$C_* := \sup_{\substack{\omega\in A_1(\Sigma,\sigma)\\ [\omega]_{A_1} < W}} C_\omega, \tag{4.111}$$

and by assumption (4.62) in the case $p_0 = 1$, we have (thanks to Corollary 2.22) that

$$\begin{aligned}\|g\|_{L^1(\Delta(x,r),w)} &= \int_{\Delta(x,r)} gw\,d\sigma \leq \int_{\Sigma} g\cdot\big(\mathcal{M}_{\Sigma}(w^s\mathbf{1}_{\Delta(x,r)})\big)^{1/s}\,d\sigma \\ &\leq C_* \int_{\Sigma} f\cdot\big(\mathcal{M}_{\Sigma}(w^s\mathbf{1}_{\Delta(x,r)})\big)^{1/s}\,d\sigma.\end{aligned} \tag{4.112}$$

We proceed in a similar fashion as before where we split the last integral in (4.112) into an integral over $\Delta(x,2r)$ and an integral over $\Sigma\setminus\Delta(x,2r)$. That is, write

$$\begin{aligned}\mathrm{I} &:= \int_{\Delta(x,2r)} f\big(\mathcal{M}_{\Sigma}(w^s\mathbf{1}_{\Delta(x,r)})\big)^{1/s}\,d\sigma, \\ \mathrm{II} &:= \int_{\Sigma\setminus\Delta(x,2r)} f\big(\mathcal{M}_{\Sigma}(w^s\mathbf{1}_{\Delta(x,r)})\big)^{1/s}\,d\sigma.\end{aligned} \tag{4.113}$$

To treat I recall (4.109), then observe that there exists a constant $C = C(a,\lambda) \in (0,\infty)$ such that

$$\begin{aligned}\mathrm{I} &\leq \int_{\Delta(x,2r)} fw\,d\sigma \\ &\leq \begin{cases} C(2r)^{\lambda-a}\|f\|_{M^{1,\lambda}(\Sigma,w_a)} & \text{if } x = x_0, \\ 2^a|x-x_0|^{-a}(2r)^{\lambda}\|f\|_{M^{1,\lambda}(\Sigma,w_a)} & \text{if } 0 < r \leq |x-x_0|/4, \end{cases}\end{aligned} \tag{4.114}$$

where the first case follows from Lemma 4.13, and the second case follows from reasoning as in (4.81)–(4.82).

We now move on to the treatment of II. Define

$$\Delta_j := \Delta(x, 2^{j+1}r)\setminus\Delta(x,2^j r) \quad \text{for each } j \in \mathbb{N}. \tag{4.115}$$

Fix $j \in \mathbb{N}$ and let $y \in \Delta_j$. Then $\operatorname{dist}(y,\Delta(x,r)) \geq (2^j - 1)r \geq 2^{j-1}r$, so that $\Delta(x,r)\cap\Delta(y,R) = 0$ whenever $R < 2^{j-1}r$. As such,

$$\begin{aligned}\big(\mathcal{M}_{\Sigma}(w^s\mathbf{1}_{\Delta(x,r)})(y)\big)^{1/s} &= \Big(\sup_{R>0} \fint_{\Delta(y,R)} w^s\mathbf{1}_{\Delta(x,r)}\,d\sigma\Big)^{1/s} \\ &\leq \Big(\sup_{R\geq 2^{j-1}r} \fint_{\Delta(y,R)} w^s\mathbf{1}_{\Delta(x,r)}\,d\sigma\Big)^{1/s} \\ &\leq C\Big(\frac{w^s(\Delta(x,r))}{(2^j r)^{n-1}}\Big)^{1/s},\end{aligned} \tag{4.116}$$

where the last inequality above makes use of the Ahlfors regularity of Σ. Because we are presently assuming $s < q$, we know $w \in \mathrm{RH}_s(\Sigma, \sigma)$ by the monotonicity of the reverse Hölder class (cf. Proposition 2.33). Additionally, we estimate using (2.63),

$$\begin{aligned} w^s(\Delta(x,r))^{1/s} &= \sigma(\Delta(x,r))^{1/s}\Big(\fint_{\Delta(x,r)} w^s \, d\sigma \Big)^{1/s} \\ &\leq [w]_{\mathrm{RH}_s} \cdot \sigma(\Delta(x,r))^{1/s} \fint_{\Delta(x,r)} w \, d\sigma \\ &= [w]_{\mathrm{RH}_s} \sigma(\Delta(x,r))^{-1/s'} w(\Delta(x,r)) \\ &\leq Cr^{-(n-1)/s'} w(\Delta(x,r)), \end{aligned} \tag{4.117}$$

where $s' := s/(s-1)$ is the Hölder conjugate of s. The last inequality above follows by the upper Ahlfors regularity of Σ, and the constant $C \in (0, \infty)$ depends only on Σ, n, and $[w]_{\mathrm{RH}_s}$, which itself is controlled by $[w]_{\mathrm{RH}_q}$. Combining (4.116) and (4.117), we then obtain

$$\begin{aligned} (\mathcal{M}_\Sigma(w^s \mathbf{1}_{\Delta(x,r)}))^{1/s} &\leq \frac{Cr^{-(n-1)/s'} w(\Delta(x,r))}{(2^j r)^{(n-1)/s}} \\ &= \frac{Cw(\Delta(x,r))2^{-j(n-1)/s}}{r^{n-1}} \quad \text{on } \Delta_j, \end{aligned} \tag{4.118}$$

since $1/s + 1/s' = 1$. Now, making use of our dyadic decomposition, we find that

$$\begin{aligned} &\int_{\Sigma \setminus \Delta(x,2r)} f \cdot (\mathcal{M}_\Sigma(w^s \mathbf{1}_{\Delta(x,r)}))^{1/s} \, d\sigma \\ &\quad \leq C \sum_{j=1}^{\infty} \frac{w(\Delta(x,r))2^{-j(n-1)/s}}{r^{n-1}} \int_{\Delta_j} f \, d\sigma. \end{aligned} \tag{4.119}$$

We can then use (2.49) to estimate for each $j \in \mathbb{N}$,

$$\begin{aligned} \int_{\Delta_j} f \, d\sigma &\leq \int_{\Delta(x, 2^{j+1}r)} fww^{-1} \, d\sigma \\ &\leq [w]_{A_1} \Big(\fint_{\Delta(x,2^{j+1}r)} w \, d\sigma \Big)^{-1} \int_{\Delta(x,2^{j+1}r)} fw \, d\sigma \\ &\leq C \frac{[w]_{A_1} \cdot (2^j r)^{n-1}}{w(\Delta(x, r^{j+1}r))}, \end{aligned} \tag{4.120}$$

where we made use of the Ahlfors regularity of Σ to obtain the last inequality. Inserting the above bound into (4.119) (and remembering $1/s + 1/s' = 1$), we obtain

$$\begin{aligned}
&\int_{\Sigma\setminus\Delta(x,2r)} f\cdot\left(\mathcal{M}_\Sigma(w^s\mathbf{1}_{\Delta(x,r)})\right)^{1/s}\,d\sigma \\
&\quad\le C\sum_{j=1}^{\infty}\frac{w(\Delta(x,r))}{w(\Delta(x,2^jr))}2^{j(n-1)/s'}\int_{\Delta(x,2^{j+1}r)} fw\,d\sigma \\
&\quad\le C\sum_{j=1}^{\infty}\left(\frac{\sigma(\Delta(x,r))}{\sigma(\Delta(x,2^jr))}\right)^{1/q'}2^{j(n-1)/s'}\int_{\Delta(x,2^{j+1}r)} fw\,d\sigma \\
&\quad\le C\sum_{j=1}^{\infty}2^{j(n-1)(1/s'-1/q')}\int_{\Delta(x,2^{j+1}r)} fw\,d\sigma,
\end{aligned}\tag{4.121}$$

where $C = C(\Sigma, n, [w]_{A_1}, [w]_{\mathrm{RH}_q}) \in (0,\infty)$. The second inequality above is implied by (2.68) in the case that $\Delta := \Delta(x, 2^jr)$ and $E := \Delta(x,r)$, and the final inequality in (4.121) follows from the Ahlfors regularity of Σ. At this point, we again separate the case that $x = x_0$ from $0 < r \le |x - x_0|/4$.

If $x = x_0$, then by Lemma 4.13, there exists $C = C(\alpha,\lambda) \in (0,\infty)$ such that

$$\begin{aligned}
&\sum_{j=1}^{\infty}2^{j(n-1)(1/s'-1/q')}\int_{\Delta(x_0,2^{j+1}r)} fw\,d\sigma \\
&\quad\le C\left\{\sum_{j=1}^{\infty}2^{j[\lambda-\alpha+(n-1)/s'-(n-1)/q']}\right\}r^{\lambda-\alpha}\|f\|_{M^{1,\lambda}(\Sigma,w_\alpha)}.
\end{aligned}\tag{4.122}$$

The above geometric series converges to a finite number when (4.65) is satisfied and when s is sufficiently close to 1, since such conditions ensure $(n-1)/s'$ is small enough to in turn force

$$\lambda-\alpha+\frac{n-1}{s'}-\frac{n-1}{q'}<0.\tag{4.123}$$

Therefore, (4.122), (4.121), and (4.113) imply

$$\mathrm{II}\le Cr^{\lambda-\alpha}\|f\|_{M^{1,\lambda}(\Sigma,w_\alpha)}\quad\text{if } x=x_0,\tag{4.124}$$

for some $C = C(\Sigma, n, q, \lambda, \alpha, [w]_{A_1}, [w]_{\mathrm{RH}_q}) \in (0,\infty)$.

If instead we have $0 < r < |x - x_0|/4$, then there exists $N \in \mathbb{N}$, with $N \ge 2$, such that

$$2^Nr<|x-x_0|\le 2^{N+1}r.\tag{4.125}$$

We distinguish the two case $j \le N - 2$ and $j \ge N - 1$. When $j \le N - 2$, we proceed as in (4.95)–(4.97) to obtain

$$\int_{\Delta(x,2^{j+1}r)} fw\,d\sigma \leq 2^{\alpha}|x - x_0|^{-\alpha} \int_{\Delta(x,2^{j+1}r)} f(y)|y - x_0|^{\alpha} w(y)\,d\sigma(y) \tag{4.126}$$
$$\leq 2^{\alpha+\lambda}|x - x_0|^{-\alpha}(2^j r)^{\lambda}\|f\|_{M^{1,\lambda}(\Sigma,w_\alpha)}.$$

Alternatively, when $j \geq N - 1$, we have the inclusion

$$\Delta(x, 2^{j+1}r) \subseteq \Delta(x_0, 2^{j+3}r) \tag{4.127}$$

(cf. (4.98)), so by Lemma 4.13 there exists a constant $C = C(\alpha, \lambda) \in (0, \infty)$ such that

$$\int_{\Delta(x,2^{j+1}r)} fw\,d\sigma \leq \int_{\Delta(x_0,2^{j+3}r)} fw\,d\sigma \tag{4.128}$$
$$\leq C(2^j r)^{\lambda-\alpha}\|f\|_{M^{1,\lambda}(\Sigma,w_\alpha)}.$$

Thus, using (4.126) and (4.128) together with (4.121) we obtain

$$\int_{\Sigma\setminus\Delta(x,2r)} f\mathcal{M}_\Sigma(w^s\mathbf{1}_{\Delta(x,r)})^{\frac{1}{s}}\,d\sigma \tag{4.129}$$
$$\leq C\sum_{j=1}^{N-2} |x - x_0|^{-\alpha} r^{\lambda} 2^{j[\lambda+(n-1)/s'-(n-1)/q']}\|f\|_{M^{1,\lambda}(\Sigma,w_\alpha)}$$
$$+ C\sum_{j=N-1}^{\infty} r^{\lambda-\alpha}2^{j[\lambda-\alpha+(n-1)/s'-(n-1)/q']}\|f\|_{M^{1,\lambda}(\Sigma,w_\alpha)}$$

for some $C = C(\Sigma, n, \alpha, \lambda, [w]_{A_1}, [w]_{RH_q}) \in (0, \infty)$. When (4.65) is satisfied and when we choose s sufficiently close to 1, we find that the exponent of 2 in the first sum in (4.129) is negative, and hence the first sum is bounded independently of N by comparison to a geometric series. Likewise, for said choice of s and assuming (4.65), we find that the exponent of 2 in the second sum in (4.129) is the tail end of a convergent geometric series with

$$\sum_{j=N-1}^{\infty} r^{\lambda-\alpha}2^{j[\lambda-\alpha+(n-1)/s'-(n-1)/q']}\|f\|_{M^{1,\lambda}(\Sigma,w_\alpha)} \tag{4.130}$$
$$\leq r^{\lambda-\alpha}2^{N[\lambda-\alpha+(n-1)/s'-(n-1)/q']}\|f\|_{M^{1,\lambda}(\Sigma,w_\alpha)}$$
$$\leq C|x - x_0|^{-\alpha}r^{\lambda}\|f\|_{M^{1,\lambda}(\Sigma,w_\alpha)},$$

where $C \in (0, \infty)$ depends only on λ and α. Indeed, recall (4.125) which yields that $r|x - x_0|^{-1} \geq 2^{-N-1}$. Thus, since $\alpha \geq 0$,

$$r^{\alpha}|x - x_0|^{-\alpha} \geq 2^{-\alpha} \cdot 2^{-N\alpha}. \tag{4.131}$$

Furthermore, our particular choice of s ensures $\lambda + \frac{n-1}{s'} = \frac{n-1}{q'} < 0$, and hence we have $-\alpha > \lambda - \alpha + \frac{n-1}{s'} - \frac{n-1}{q'}$, so that it's now apparent how (4.130) follows from (4.131). Together, (4.130), (4.129), and (4.113) imply that

$$\mathrm{II} \leq C|x - x_0|^{-\alpha} r^{\lambda} \|f\|_{M^{p,\lambda}(\Sigma, w_\alpha)} \quad \text{if } 0 < r \leq |x - x_0|, \tag{4.132}$$

where $C \in (0, \infty)$ depends exclusively on Σ, n, q, λ, α, $[w]_{A_1}$, and $[w]_{\mathrm{RH}_q}$.

Pressing forward to the final stage in the proof of Subcase 1B, recall (4.112) and (4.113), and combine the bounds in (4.114), (4.124), and (4.132) to obtain

$$\begin{aligned} \|g\|_{L^1(\Delta(x,r),w)} &\leq C_*(\mathrm{I} + \mathrm{II}) \\ &\leq \begin{cases} KC_* r^{\lambda-\alpha} \|f\|_{M^{1,\lambda}(\Sigma, w_\alpha)} & \text{if } x = x_0, \\ KC_* |x - x_0|^{-\alpha} r^{\lambda} \|f\|_{M^{1,\lambda}(\Sigma, w_\alpha)} & \text{if } 0 < r \leq |x - x_0|/4, \end{cases} \end{aligned} \tag{4.133}$$

where $K = K(\Sigma, n, q, \lambda, \alpha, [w]_{A_1}, [w]_{\mathrm{RH}_q}) \in (0, \infty)$ and C_* is as in (4.111). Consequently, thanks to the bounds in (4.69) (with $p = 1$), we see that

$$\frac{1}{r^{\lambda}} \int_{\Delta(x,r)} g(y)|y - x_0|^{\alpha} w(y)\, d\sigma(y) \leq KC_* \|f\|_{M^{1,\lambda}(\Sigma, w_\alpha)}, \tag{4.134}$$

in either the case that $x = x_0$ or $0 < r \leq |x - x_0|/4$. Finally, as seen in (4.108), having established the estimate in (4.134) for these two cases is enough to conclude the desired result upon taking a supremum on both sides of (4.134) over all $x \in \Sigma$, and $r \in (0, 2\,\mathrm{diam}(\Sigma))$.

Case 2: *Assume $p_0 \in (1, \infty)$*. For each $\omega \in A_{p_0}(\Sigma, \sigma)$, there exists a constant $C_\omega \in (0, \infty)$ whose nature is specified in the statement of the theorem such that

$$\|g\|_{L^{p_0}(\Sigma, \omega)} \leq C_\omega \|f\|_{L^{p_0}(\Sigma, \omega)}. \tag{4.135}$$

By the classical Rubio de Francia extrapolation theorem (cf. Proposition 2.41), we have that for all integrability exponents $s \in (1, \infty)$ and every Muckenhoupt weight $w \in A_s(\Sigma, \sigma)$, there exist two constants $K, W \in (0, \infty)$ (whose natures are detailed in the statement of Proposition 2.41) such that

$$\|g\|_{L^s(\Sigma, w)} \leq K \cdot \Big(\sup_{\substack{\omega \in A_{p_0}(\Sigma, \sigma) \\ [\omega]_{A_{p_0}} \leq W}} C_\omega \Big) \|f\|_{L^s(\Sigma, w)}. \tag{4.136}$$

Assuming $p \in (1, \infty)$ and $w \in A_p(\Sigma, \sigma)$, by item (1) of Proposition 2.29, we can choose $s \in (1, p)$ such that $w \in A_{p/s}(\Sigma, \sigma)$. Then note that (4.136) implies

$$\|g^s\|_{L^1(\Sigma, \omega)} \leq C_\omega \|f^s\|_{L^1(\Sigma, \omega)} \tag{4.137}$$

for all $\omega \in A_s(\Sigma, \sigma)$, and specifically for all $\omega \in A_1(\Sigma, \sigma)$, since it happens that $A_1(\Sigma, \sigma) \subseteq A_s(\Sigma, \sigma)$ (cf. (2.51)). If we work in the case that (4.65) holds, we may then extrapolate using our result from Subcase 1A to obtain

$$\|g^s\|_{M^{p/s,\lambda}(\Sigma,\omega_a)} \leq K \cdot \Big(\sup_{\substack{\omega \in A_1 \\ [\omega]_{A_1} \leq W}} C_\omega \Big) \cdot \|f^s\|_{M^{p/s,\lambda}(\Sigma,\omega_a)}, \tag{4.138}$$

for some constants $K, W \in (0, \infty)$ whose precise natures are specified in the statement of the theorem, and for all $\omega \in A_{p/s}(\Sigma, \sigma)$, where $\omega_a := |\cdot - x_0|^a \omega$ is a weight on Σ. In particular, (4.138) holds for w_a, because $s \in (1, p)$ was chosen so as to ensure $w \in A_{p/s}(\Sigma, \sigma)$. By an earlier discussion in (4.3), we conclude upon taking sth roots on both sides of (4.138) that

$$\|g\|_{M^{p,\lambda}(\Sigma,w_a)} \leq C_w \|f\|_{M^{p,\lambda}(\Sigma,w_a)}, \tag{4.139}$$

for $C_w \in (0, \infty)$ as in (4.66). This finally completes the proof of all claims in Theorem 4.19. □

As we will see in the next chapter, having this extrapolation result will allow us to develop a Calderón–Zygmund theory for Muckenhoupt-weighted Morrey spaces based off of work previously done in the case of Muckenhoupt-weighted Lebesgue spaces. This will eventually allow us to use layer potentials to prove solvability results for various boundary value problems with boundary datum in a Muckenhoupt-weighted Morrey space.

5 Singular integral operators on weighted Morrey spaces

In this section, we endeavor to develop a Calderón–Zygmund theory for Muckenhoupt-weighted Morrey spaces. Much progress concerning singular integral operators in the Euclidean setting for various types of Morrey spaces (e. g., weighted, generalized, variable exponent) has been made; cf. [29, 30, 31, 48, 49, 50, 70, 129, 133] and the references therein. The Calderón–Zygmund theory we develop here will be in the optimal geometric setting of uniformly rectifiable sets.

The first result we present shows that the Hardy–Littlewood maximal operator is bounded on weighted Morrey spaces. Here, we already have an instance in how useful the extrapolation result of Theorem 4.19 is for going beyond what is already understood about Muckenhoupt-weighted Lebesgue spaces.

Proposition 5.1. *Let $\Sigma \subseteq \mathbb{R}^n$ be a closed* ADR *set and write $\sigma := \mathcal{H}^{n-1}\lfloor\Sigma$. Fix two integrability exponents $p, q \in (1,\infty)$, choose a Muckenhoupt weight $w \in A_p(\Sigma,\sigma) \cap \mathrm{RH}_q(\Sigma,\sigma)$, and select some threshold $\lambda \in (0,(n-1)/q')$, where $q' := q/(q-1)$. Finally, choose some $a \in [0,\lambda)$, fix a reference point $x_0 \in \Sigma$, and abbreviate $w_a := w|\cdot - x_0|^a$, a weight on Σ.*

Then the Hardy–Littlewood maximal operator induces a mapping

$$\mathcal{M}_\Sigma : M^{p,\lambda}(\Sigma, w_a) \to M^{p,\lambda}(\Sigma, w_a) \tag{5.1}$$

which is well-defined, sublinear, and bounded. In particular, there exists a constant $C \in (0,\infty)$, depending on the ADR *character of Σ, n, p, q, λ, a, $[w]_{\mathrm{RH}_q}$, and $[w]_{A_p}$ such that*

$$\|\mathcal{M}_\Sigma f\|_{M^{p,\lambda}(\Sigma,w_a)} \leq C\|f\|_{M^{p,\lambda}(\Sigma,w_a)} \tag{5.2}$$

for every $f \in M^{p,\lambda}(\Sigma, w_a)$.

Proof. Define the collection of pairs of σ-measurable functions

$$\mathcal{F} := \left\{(h, \mathcal{M}_\Sigma h) : h \in L^1\left(\Sigma, \frac{\sigma(x)}{1+|x|^{n-1}}\right)\right\}. \tag{5.3}$$

Fix $f \in M^{p,\lambda}(\Sigma, w_a)$, then $(f, \mathcal{M}_\Sigma f) \in \mathcal{F}$, due to the embedding result of Proposition 4.15. We will apply Theorem 4.19 (the Morrey space extrapolation) with $\mathcal{M}_\Sigma f$ playing the role of g. Fix $p_0 \in (1,\infty)$ and $\omega \in A_{p_0}(\Sigma,\sigma)$. In this context, notice that without loss of generality, we may further assume $f \in L^{p_0}(\Sigma,\omega)$. Otherwise, we find (4.62) is trivially satisfied. From results in harmonic analysis in doubling metric spaces (see, e. g., [60]), there exists a constant $C \in (0,\infty)$, depending on the ADR character of Σ, n, and p_0 such that

$$\|\mathcal{M}_\Sigma f\|_{L^{p_0}(\Sigma,\omega)} \leq C[\omega]_{A_{p_0}}^{1/(p_0-1)}\|f\|_{L^{p_0}(\Sigma,\omega)}. \tag{5.4}$$

https://doi.org/10.1515/9783111458274-005

Given the nature of the multiplicative constant in the right-hand side of (5.4), the hypotheses of Theorem 4.19 are satisfied. Hence, there exists a constant $K \in (0, \infty)$, depending on the Ahlfors regularity constant of Σ, n, p, q λ, α, $[w]_{A_p}$, and $[w]_{\mathrm{RH}_q}$, such that

$$\|\mathcal{M}_\Sigma f\|_{M^{p,\lambda}(\Sigma, w_\alpha)} \leq K\|f\|_{M^{p,\lambda}(\Sigma, w_\alpha)}. \tag{5.5}$$

In view of the arbitrary nature of $f \in M^{p,\lambda}(\Sigma, w_\alpha)$, we see that (5.1) is a well-defined and bounded operator, as wanted. □

For weighted Lebesgue spaces, the boundedness of the Hardy–Littlewood maximal operator characterizes the Muckenhoupt class (cf. (2.43)). The following proposition categorizes the kinds of weights for which the Hardy–Littlewood maximal operator is bounded on a weighted Morrey space.

Proposition 5.2. *Let $\Sigma \subseteq \mathbb{R}^n$ be a closed ADR set, and write $\sigma := \mathcal{H}^{n-1}\lfloor\Sigma$. Fix an integrability exponent $p \in (1, \infty)$ and a threshold $\lambda \in [0, n-1)$. Also, choose an arbitrary weight function $w : \Sigma \to [0, \infty]$. If the Hardy–Littlewood maximal operator*

$$\mathcal{M}_\Sigma : M^{p,\lambda}(\Sigma, w) \to M^{p,\lambda}(\Sigma, w) \tag{5.6}$$

is bounded, then w belongs to the Muckenhoupt class $A_{p+\frac{\lambda}{n-1}}(\Sigma, \sigma)$.

Proof. Start by defining $p_* := p + \frac{\lambda}{n-1} \in (1, \infty)$ along with its Hölder conjugate $p'_* \in (1, \infty)$. Fix an arbitrary surface ball $\Delta_* \subseteq \Sigma$ with radius $r_* \in (0, 2\,\mathrm{diam}(\Sigma))$. Also, choose some $x_* \in \Delta_*$. We can estimate

$$\begin{aligned}\mathcal{M}_\Sigma(w^{1-p'_*} \cdot \mathbf{1}_{\Delta_*})(x_*) &\approx \sup_{\Delta \ni x} \fint_\Delta w^{1-p'_*} \cdot \mathbf{1}_{\Delta_*}\, d\sigma \\ &\geq \fint_{\Delta_*} w^{1-p'_*}\, d\sigma.\end{aligned} \tag{5.7}$$

Upon raising both sides to the pth power, multiplying by w, and integrating over Δ_*, we obtain

$$\int_{\Delta_*} |\mathcal{M}_\Sigma(w^{1-p'_*} \cdot \mathbf{1}_{\Delta_*})|^p w\, d\sigma \geq C\Big(\int_{\Delta_*} w\, d\sigma\Big)\Big(\fint_\Delta w^{1-p'_*}\, d\sigma\Big)^p. \tag{5.8}$$

On the other hand, if we define $s := \frac{p(n-1)}{n-1-\lambda} \in (1, \infty)$, then we can use the boundedness assumption on (5.6) and Proposition 4.17 to estimate

$$\begin{aligned}\int_{\Delta_*} |\mathcal{M}_\Sigma(w^{1-p'_*} \cdot \mathbf{1}_{\Delta_*})|^p w\, d\sigma &\leq r_*^\lambda \|\mathcal{M}_\Sigma(w^{1-p'_*} \cdot \mathbf{1}_{\Delta_*})\|^p_{M^{p,\lambda}(\Sigma, w)} \\ &\leq C r_*^\lambda \|w^{1-p'_*} \cdot \mathbf{1}_{\Delta_*}\|^p_{M^{p,\lambda}(\Sigma, w)}\end{aligned} \tag{5.9}$$

$$\leq Cr_*^{\lambda}\|w^{1-p'_*}\cdot \mathbf{1}_{\Delta_*}\|^p_{L^s(\Sigma,w^{s/p})}$$
$$= Cr_*^{\lambda}\Big(\int_{\Delta_*} w^{(1-p'_*)s}\cdot w^{s/p}\,d\sigma\Big)^{\frac{n-1-\lambda}{n-1}}.$$

Notice that since $1-p'_* = 1-\frac{p_*}{p_*-1} = \frac{1}{1-p_*}$, we actually have (bearing in mind the definitions of s and p_*)

$$\begin{aligned}(1-p'_*)s+\frac{s}{p} &= (1-p'_*)\cdot\Big[\frac{p(n-1)}{n-1-\lambda}+\frac{(1-p_*)(n-1)}{n-1-\lambda}\Big]\\ &= (1-p'_*)\cdot\frac{(p+1-p_*)(n-1)}{n-1-\lambda}\\ &= (1-p'_*)\cdot\frac{(1-\frac{\lambda}{n-1})(n-1)}{n-1-\lambda}\\ &= 1-p'_*.\end{aligned} \tag{5.10}$$

Using this fact in (5.9), we can further estimate via the Ahlfors regularity of Σ,

$$\begin{aligned}\int_{\Delta_*}|\mathcal{M}_\Sigma(w^{1-p'_*}\cdot\mathbf{1}_{\Delta_*})|^p w\,d\sigma &\leq Cr_*^{\lambda}\Big(\int_{\Delta_*} w^{1-p'_*}\,d\sigma\Big)^{\frac{n-1-\lambda}{n-1}}\\ &= Cr_*^{\lambda}\sigma(\Delta_*)^{\frac{n-1-\lambda}{n-1}}\Big(⨍_{\Delta_*} w^{1-p'_*}\,d\sigma\Big)^{\frac{n-1-\lambda}{n-1}}\\ &\leq Cr_*^{\lambda}r_*^{n-1-\lambda}\Big(⨍_{\Delta_*} w^{1-p'_*}\,d\sigma\Big)^{\frac{n-1-\lambda}{n-1}}\\ &\leq C\sigma(\Delta_*)\Big(⨍_{\Delta_*} w^{1-p'_*}\,d\sigma\Big)^{\frac{n-1-\lambda}{n-1}}.\end{aligned} \tag{5.11}$$

If we combine (5.8) with (5.11), then we obtain

$$\Big(\int_{\Delta_*} w\,d\sigma\Big)\Big(⨍_{\Delta} w^{1-p'_*}\,d\sigma\Big)^p \leq C\sigma(\Delta_*)\Big(⨍_{\Delta_*} w^{1-p'_*}\,d\sigma\Big)^{\frac{n-1-\lambda}{n-1}}, \tag{5.12}$$

or equivalently,

$$\Big(⨍_{\Delta_*} w\,d\sigma\Big)\Big(⨍_{\Delta} w^{1-p'_*}\,d\sigma\Big)^{p-\frac{n-1-\lambda}{n-1}} \leq C. \tag{5.13}$$

Notice that $p - \frac{n-1-\lambda}{n-1} = p - 1 + \frac{\lambda}{n-1} = p_* - 1$. The constant $C \in (0, \infty)$ in (5.13) depends only on Σ, n, p, λ, and $\|\mathcal{M}_\Sigma\|_{M^{p,\lambda}(\Sigma,w)\to M^{p,\lambda}(\Sigma,w)}$. Thus, taking a supremum over all surface balls $\Delta_* \subseteq \Sigma$ (bearing in mind (2.47)) shows that $w \in A_{p_*}(\Sigma, \sigma)$, which is the desired result. □

Remark 5.3. It is worth commenting for $p \in (1, \infty)$ that in the endpoint case $\lambda = 0$, since we have $M^{p,\lambda}(\Sigma, w) = L^p(\Sigma, w)$, we actually recover the well-known fact that if the Hardy–Littlewood maximal operator is bounded on the weighted Lebesgue space, $L^p(\Sigma, w)$, then w necessarily belongs to $A_p(\Sigma, \sigma)$.

The Hardy–Littlewood maximal operator is a valuable tool in harmonic analysis. Having established several results concerning its behavior in the context of weighted Morrey spaces, our next order of business is developing a Calderón–Zygmund theory for the brand of weighted Morrey spaces under our consideration.

Remark 5.4. Let $\Sigma \subseteq \mathbb{R}^n$ be closed, and σ any Borel measure on Σ. To understand why the particular weighted Lebesgue space $L^1(\Sigma, \frac{\sigma(x)}{1+|x|^{n-1}})$ is a natural environment in which to speak of Calderón–Zygmund operators, consider a complex-valued integral kernel $k \in \mathscr{C}^0(\mathbb{R}^n \setminus \{0\})$, where k is odd and positive homogeneous of degree $1-n$. Then given any $f \in L^1(\Sigma, \frac{\sigma(x)}{1+|x|^{n-1}})$, for all $\varepsilon > 0$ and $x \in \Sigma$ fixed, we find that

$$\int_{\substack{y\in\Sigma \\ |x-y|>\varepsilon}} |k(x-y)||f(y)|\, d\sigma(y) < +\infty. \tag{5.14}$$

Indeed, we start with the following preliminary estimate for k. Fix any $x \in \mathbb{R}^n \setminus \{0\}$, then

$$k(x) = k\left(|x|\frac{x}{|x|}\right) = |x|^{1-n} k\left(\frac{x}{|x|}\right). \tag{5.15}$$

Note that $\frac{x}{|x|} \in S^{n-1}$, which is a compact subset of $\mathbb{R}^n \setminus \{0\}$. Therefore, k is bounded on S^{n-1}, so we see from (5.15) that

$$|k(x)| \le \frac{C_0}{|x|^{n-1}} \quad \text{for all } x \in \mathbb{R}^n \setminus \{0\}, \tag{5.16}$$

where $C_0 := \sup_{x\in S^{n-1}} |k(x)| \in [0, \infty)$. Using this first estimate, we can show that for all $\varepsilon > 0$ fixed and $x \in \Sigma$ fixed,

$$|k(x-y)| \le \frac{C}{1+|y|^{n-1}} \quad \text{for all } y \in \Sigma \setminus \overline{B(x,\varepsilon)}, \tag{5.17}$$

with $C = C(k, n, x, \varepsilon) \in [0, \infty)$. To see this, let $y \in \Sigma \setminus \overline{B(x,\varepsilon)}$. Then $|x-y| > \varepsilon$ and

$$|y| \le |x-y| + |x| = |x-y| + \frac{|x|}{\varepsilon}\varepsilon < \left(1 + \frac{|x|}{\varepsilon}\right)|x-y|, \tag{5.18}$$

so that

$$1 + |y| \leq C_{x,\varepsilon}|x - y|. \tag{5.19}$$

Note also by Lemma 2.18 that $1 + |y|^{n-1} \approx (1 + |y|)^{n-1}$, whose implicit constants depend only on n. Hence, $1 + |y|^{n-1} \leq C_{n,x,\varepsilon}|x - y|^{n-1}$. Subsequently, the bound in (5.16) yields

$$|k(x - y)| \leq \frac{C_0}{|x - y|^{n-1}} \leq \frac{C}{1 + |y|^{n-1}}, \tag{5.20}$$

as in (5.17). We then see for any $f \in L^1(\Sigma, \frac{\sigma(x)}{1+|x|^{n-1}})$ that

$$\int\limits_{\substack{y\in\Sigma \\ |x-y|>\varepsilon}} |k(x - y)||f(y)|\, d\sigma(y) \leq C \int\limits_{\Sigma} \frac{|f(y)|}{1 + |y|^{n-1}}\, d\sigma(y) < +\infty, \tag{5.21}$$

as claimed.

We now record the following useful continuous embedding of weighted Lebesgue spaces [88]. Of particular note, this next lemma and Remark 5.4 show that singular integral operators of Calderón–Zygmund type acting on functions in weighted Lebesgue spaces are well-defined objects.

Lemma 5.5. *Let $\Sigma \subseteq \mathbb{R}^n$ be a closed* ADR *set and write $\sigma := \mathcal{H}^{n-1}\lfloor\Sigma$. For all $p \in (1, \infty)$ and Muckenhoupt weights $w \in A_p(\Sigma, \sigma)$, there exists $p_0 \in (1, p]$ such that*

$$L^p(\Sigma, w) \hookrightarrow L^q\left(\Sigma, \frac{\sigma(x)}{1 + |x|^{n-1}}\right) \quad \textit{continuously} \tag{5.22}$$

for each fixed $q \in (0, p_0)$. In particular,

$$L^p(\Sigma, w) \hookrightarrow L^1\left(\Sigma, \frac{\sigma(x)}{1 + |x|^{n-1}}\right) \quad \textit{continuously.} \tag{5.23}$$

Our next proposition, debuted in [79, Proposition 3.10], summarizes a Calderón–Zygmund theory for Muckenhoupt-weighted Lebesgue spaces. It is included as a reference not only because we will establish a parallel result at the level of weighted Morrey spaces, but also because this proposition serves as the starting point in the proof of Theorem 5.7, via extrapolation.

Proposition 5.6. *Let $\Sigma \subseteq \mathbb{R}^n$ denote a closed uniformly rectifiable set and abbreviate $\sigma := \mathcal{H}^{n-1}\lfloor\Sigma$. Assume $N = N(n) \in \mathbb{N}$ is a sufficiently large integer and consider a complex-valued function $k \in \mathscr{C}^N(\mathbb{R}^n \setminus \{0\})$, which is odd and positive homogeneous of degree $1 - n$. Also, fix an integrability exponent $p \in (1, \infty)$, along with a Muckenhoupt weight $w \in A_p(\Sigma, \sigma)$. Now, for each truncation parameter $\varepsilon > 0$ and each $f \in L^1(\Sigma, \frac{\sigma(x)}{1+|x|^{n-1}})$*

define[1]

$$T_\varepsilon f(x) := \int_{\substack{y\in\Sigma \\ |x-y|>\varepsilon}} k(x-y)f(y)\,d\sigma(y) \quad \textit{for each } x \in \Sigma, \tag{5.24}$$

$$T_{\max} f(x) := \sup_{\varepsilon>0} |T_\varepsilon f(x)| \quad \textit{for each } x \in \Sigma, \tag{5.25}$$

$$Tf(x) := \lim_{\varepsilon\to 0^+} T_\varepsilon f(x) \quad \textit{for } \sigma\textit{-a. e. } x \in \Sigma. \tag{5.26}$$

Then $T_{\max} f$ *is meaningfully defined in* (5.25) *as a* σ*-measurable function on* Σ *for all functions* $f \in L^1(\Sigma, \frac{\sigma(x)}{1+|x|^{n-1}})$*. In addition, there exists a constant* $C \in (0,\infty)$*, depending exclusively on* n*,* p*,* $[w]_{A_p}$*, and the UR constants of* Σ*, which stays bounded if* $[w]_{A_p}$ *stays bounded with the additional property that for each* $f \in L^p(\Sigma, w)$ *one has*

$$\|T_{\max} f\|_{L^p(\Sigma,w)} \le C\Big(\sum_{|\alpha|\le N} \sup_{S^{n-1}} |\partial^\alpha k|\Big)\|f\|_{L^p(\Sigma,w)}. \tag{5.27}$$

In particular,

> *the truncated integral operators* $T_\varepsilon : L^p(\Sigma,w) \to L^p(\Sigma,w)$ *are well-defined, linear, and bounded in a uniform fashion with respect to the truncation parameter* $\varepsilon > 0$. (5.28)

Moreover, for each $f \in L^1(\Sigma, \frac{\sigma(x)}{1+|x|^{n-1}})$ *the limit defining* $Tf(x)$ *in* (5.26) *exists at* σ*-a. e. point* $x \in \Sigma$*, the function* Tf *thus defined is* σ*-measurable on* Σ*, and the operator*

$$T : L^p(\Sigma,w) \longrightarrow L^p(\Sigma,w) \tag{5.29}$$

is well-defined, linear, and bounded. Let $p' \in (1,\infty)$ *denote the Hölder conjugate exponent of* p *and, with* $w' := w^{1-p'} \in A_{p'}(\Sigma,\sigma)$*, consider the natural identification*

$$\big(L^p(\Sigma,w)\big)^* = L^{p'}(\Sigma,w'). \tag{5.30}$$

Then, under the canonical integral pairing $(f,g) \mapsto \int_\Sigma fg\,d\sigma$*, it follows that*

> *the (real) transpose of the operator* (5.29) *is the operator* $-T : L^{p'}(\Sigma,w') \to L^{p'}(\Sigma,w')$. (5.31)

Next, let $\Omega \subseteq \mathbb{R}^n$ *be an open set such that* $\partial\Omega$ *is uniformly rectifiable, and abbreviate* $\sigma := \mathcal{H}^{n-1}\lfloor\partial\Omega$*. Fix some* $p \in (1,\infty)$ *along with a Muckenhoupt weight* $w \in A_p(\partial\Omega,\sigma)$*, and pick an aperture parameter* $\kappa > 0$*. With the integral kernel* k *as before, for each*

1 As noted in Remark 5.4, $(T_\varepsilon f)(x)$ is well-defined via an absolutely convergent integral for each function $f \in L^1(\Sigma, \frac{\sigma(x)}{1+|x|^{n-1}})$ and each $x \in \Sigma$.

$f \in L^p(\partial\Omega, w)$ define

$$\mathcal{T}f(x) := \int_{\partial\Omega} k(x-y)f(y)d\sigma(y) \quad \text{for each } x \in \Omega. \tag{5.32}$$

Then there exists a constant $C \in (0,\infty)$ which depends exclusively on n, p, $[w]_{A_p}$, and the UR *character of $\partial\Omega$ with the property that for each function $f \in L^p(\partial\Omega, w)$, one has*

$$\|\mathcal{N}_\kappa(\mathcal{T}f)\|_{L^p(\partial\Omega,w)} \leq C\Big(\sum_{|\alpha|\leq N} \sup_{S^{n-1}}|\partial^\alpha k|\Big)\|f\|_{L^p(\partial\Omega,w)}. \tag{5.33}$$

Finally, for each function $f \in L^1(\partial\Omega, \frac{\sigma(x)}{1+|x|^{n-1}})$ one has the jump-formula

$$(\mathcal{T}f\big|_{\partial\Omega}^{\kappa-\text{n.t.}})(x) = \frac{1}{2\sqrt{-1}}\widehat{k}(\nu(x))f(x) + (Tf)(x) \tag{5.34}$$

at σ-a. e. point $x \in \partial_*\Omega$, where $\widehat{k}$ denotes the Fourier transform of k and $\nu(x)$ is the geometric measure theoretic outward unit normal to Ω at the point x. In particular, the jump-formula (5.34) *is valid for each function $f \in L^p(\partial\Omega, w)$.*

The strategy now is to combine the estimates in Theorem 5.6 with our extrapolation result, Theorem 4.19, to obtain the following Calderón–Zygmund theory at the level of weighted Morrey spaces. We also note the importance of the embedding result, Proposition 4.15, which ensures the validity of a jump relation for functions belonging to a weighted Morrey space.

Theorem 5.7. *Let $\Sigma \subseteq \mathbb{R}^n$ denote a closed uniformly rectifiable set and abbreviate $\sigma := \mathcal{H}^{n-1}\lfloor\Sigma$. Assume $N = N(n) \in \mathbb{N}$ is a sufficiently large integer and consider some complex-valued function $k \in \mathscr{C}^N(\mathbb{R}^n \setminus \{0\})$, which is odd and positive homogeneous of degree $1-n$. Also, fix two integrability exponents $p, q \in (1,\infty)$, a Muckenhoupt weight $w \in A_p(\Sigma,\sigma) \cap \mathrm{RH}_q(\Sigma,\sigma)$, and some threshold $\lambda \in (0,(n-1)/q')$, where $q' := q/(q-1)$. Lastly, fix a reference point $x_0 \in \Sigma$, and abbreviate $w_a := |\cdot - x_0|^a w$, a weight on Σ.*

Then the operator $T_{\max}$ as in (5.25)*, induces a well-defined, linear, and bounded mapping on $M^{p,\lambda}(\Sigma, w_a)$ with*

$$\|T_{\max}f\|_{M^{p,\lambda}(\Sigma,w_a)} \leq C\Big(\sum_{|\gamma|\leq N} \sup_{S^{n-1}}|\partial^\gamma k|\Big)\|f\|_{M^{p,\lambda}(\Sigma,w_a)}, \tag{5.35}$$

where the constant $C \in (0,\infty)$ depends only on the UR *constants of Σ, n, p, q, λ, a, x_0, $[w]_{\mathrm{RH}_q}$, and on $[w]_{A_p}$ in such that a way that C stays bounded whenever $[w]_{A_p}$ stays bounded.*

Furthermore, the operator T as defined in (5.26)*, induces a mapping*

$$T : M^{p,\lambda}(\Sigma, w_a) \to M^{p,\lambda}(\Sigma, w_a) \tag{5.36}$$

which is well-defined, linear, and bounded.

Next, let $\Omega \subseteq \mathbb{R}^n$ be an open set such that $\partial\Omega$ is a uniformly rectifiable set and abbreviate $\sigma := \mathcal{H}^{n-1}\lfloor\partial\Omega$. Pick an aperture parameter $\kappa > 0$. With integral kernel k as before, define for each $f \in M^{p,\lambda}(\partial\Omega, w_a)$ the operator

$$\mathcal{T}f(x) := \int_{\partial\Omega} k(x-y)f(y)\,d\sigma(y) \quad \text{for each } x \in \Omega. \tag{5.37}$$

Then there exists a constant $C \in (0,\infty)$, which depends only on the UR *character of Σ, n, p, q, λ, a, x_0, $[w]_{\mathrm{RH}_q}$, and on $[w]_{A_p}$ in such a way that C stays bounded whenever $[w]_{A_p}$ stays bounded, with the property that*

$$\|\mathcal{N}_\kappa(\mathcal{T}f)\|_{M^{p,\lambda}(\partial\Omega,w_a)} \le C\Big(\sum_{|\gamma|\le N}\sup_{S^{n-1}}|\partial^\gamma k|\Big)\|f\|_{M^{p,\lambda}(\partial\Omega,w_a)}, \tag{5.38}$$

for each $f \in M^{p,\lambda}(\partial\Omega, w_a)$. Additionally, for each $f \in M^{p,\lambda}(\partial\Omega, w_a)$ one has the jump formula

$$\big(\mathcal{T}f\big|^{\kappa-\mathrm{n.t.}}_{\partial\Omega}\big)(x) = \frac{1}{2\sqrt{-1}}\widehat{k}(\nu(x))f(x) + (Tf)(x) \tag{5.39}$$

at σ-a. e. point $x \in \partial_*\Omega$, where $\widehat{k}$ denotes the Fourier transform of k and $\nu(x)$ denotes the geometric measure theoretic outward unit normal to Ω at the point x.

Proof. We first show that (5.35) holds. For that purpose, consider the family of pairs of σ-measurable functions

$$\mathcal{F}_1 := \Big\{(h, T_{\max}h) : h \in L^1\Big(\Sigma, \frac{\sigma(x)}{1+|x|^{n-1}}\Big)\Big\}. \tag{5.40}$$

Note that the σ-measurability of $T_{\max}h$ is guaranteed by Proposition 5.6. Fix some arbitrary $f \in M^{p,\lambda}(\Sigma, w_a)$. The embedding result of Proposition 4.15 allows us to conclude that $(f, T_{\max}f) \in \mathcal{F}_1$. We turn to Theorem 4.19 with $T_{\max}f$ playing the role of g. Fix $p_0 \in (1,\infty)$ and $\omega \in A_{p_0}(\Sigma,\sigma)$. In this context, we may assume without loss of generality that $f \in L^{p_0}(\Sigma,\omega)$. Otherwise, hypothesis (4.62) is trivially true. By Proposition 5.6, we find that

$$\|T_{\max}f\|_{L^{p_0}(\Sigma,\omega)} \le C_\omega \cdot \Big(\sum_{|\gamma|\le N}\sup_{S^{n-1}}|\partial^\gamma k|\Big)\|f\|_{L^{p_0}(\Sigma,\omega)}, \tag{5.41}$$

where $C_\omega \in (0,\infty)$ is such that C_ω stays bounded whenever $[\omega]_{A_{p_0}}$ is bounded. Thus, the hypotheses of Theorem 4.19 are satisfied. Consequently, there exist two finite, positive constants, K and W, such that

$$\begin{aligned}&\|T_{\max}f\|_{M^{p,\lambda}(\Sigma,w_a)}\\ &\quad\le K\cdot\Big(\sup_{\substack{\omega\in A_{p_0}(\Sigma,\sigma)\\ [\omega]_{A_{p_0}}<W}} C_\omega\Big)\cdot\Big(\sum_{|\gamma|\le N}\sup_{S^{n-1}}|\partial^\gamma k|\Big)\|f\|_{M^{p,\lambda}(\Sigma,w_a)}.\end{aligned} \tag{5.42}$$

This is the desired inequality (5.35).

Next, we treat the principal value operator T, defined in (5.36). The fact that T is bounded follows from the first part of this proof, because given any $f \in M^{p,\lambda}(\Sigma, w_a)$, (5.35) shows that

$$\begin{aligned}\|Tf\|_{M^{p\lambda}(\Sigma,w_a)} &\leq \|T_{\max} f\|_{M^{p,\lambda}(\Sigma,w_a)} \\ &\leq C\Big(\sum_{|\gamma|\leq N} \sup_{S^{n-1}} |\partial^\gamma k|\Big)\|f\|_{M^{p,\lambda}(\Sigma,w_a)},\end{aligned} \tag{5.43}$$

which thanks to the lattice property of Morrey spaces (cf. Proposition 4.7), shows that (5.36) is well-defined, and bounded as wanted. Note that the σ-measurability of Tf follows from Proposition 4.15 and Proposition 5.6.

Next, we seek nontangential maximal operator estimates. We proceed using a strategy involving extrapolation. Consider the family of pairs σ-measurable functions

$$\mathcal{F}_2 := \Big\{(h, \mathcal{N}_\kappa(\mathcal{T}h)) : h \in L^1\Big(\partial\Omega, \frac{\sigma(x)}{1+|x|^{n-1}}\Big)\Big\}. \tag{5.44}$$

First, observe that $\mathcal{F}_2$ is a collection of pairs of σ-measurable functions, since $\mathcal{N}_\kappa(\mathcal{T}h)$ is σ-measurable as mentioned in Remark 2.13. Next, suppose $f \in M^{p,\lambda}(\partial\Omega, w_a)$. By Proposition 4.15, we have the membership $(f, \mathcal{N}_\kappa(\mathcal{T}f)) \in \mathcal{F}_2$. We again utilize Theorem 4.19 with $\mathcal{N}_\kappa(\mathcal{T}f)$ now playing the role of g. For that purpose, fix $p_0 \in (1,\infty)$ and let $\omega \in A_{p_0}(\partial\Omega, \sigma)$. We may assume without loss of generality that $f \in L^{p_0}(\partial\Omega, \omega)$, so that hypothesis (4.62) is not otherwise trivially satisfied. By Proposition 5.6, specifically (5.33), there exists a constant $C_\omega \in (0,\infty)$, whose nature is specified in the statement of Proposition 5.6, such that

$$\|\mathcal{N}_\kappa(\mathcal{T}f)\|_{L^{p_0}(\partial\Omega,\omega)} \leq C_\omega\Big(\sum_{|\gamma|\leq N} \sup_{S^{n-1}} |\partial^\gamma k|\Big)\|f\|_{L^{p_0}(\partial\Omega,\omega)}. \tag{5.45}$$

Given the nature of the constant C_w, the hypotheses of Theorem 4.19 are witnessed to be satisfied, and hence there exist constants $K, W \in (0,\infty)$ such that

$$\begin{aligned}&\|\mathcal{N}_\kappa(\mathcal{T}f)\|_{M^{p,\lambda}(\partial\Omega,w_a)} \\ &\quad\leq K\cdot\Big(\sup_{\substack{\omega\in A_{p_0}(\partial\Omega,\sigma)\\ [\omega]_{A_p}<W}} C_\omega\Big)\Big(\sum_{|\gamma|\leq N} \sup_{S^{n-1}} |\partial^\gamma k|\Big)\|f\|_{M^{p,\lambda}(\partial\Omega,w_a)}.\end{aligned} \tag{5.46}$$

This is the desired inequality in (5.38).

Finally, the jump relation in (5.39) is an immediate consequence of the embedding result in Proposition 4.15 and the jump relation in (5.34) of Proposition 5.6. This completes the proof of Theorem 5.7. □

We also have the following interesting way of measuring the size of boundary-to-domain singular integral operators of convolution type acting on spaces of Muckenhoupt-weighted Morrey functions.

Proposition 5.8. *Let $\Omega \subseteq \mathbb{R}^n$ be an AR domain, write $\sigma := \mathcal{H}^{n-1}\lfloor\partial\Omega$, and denote $\delta_{\partial\Omega}(x) := \operatorname{dist}(x, \partial\Omega)$ for each $x \in \mathbb{R}^n$, i. e., the distance function to $\partial\Omega$. Next, fix $N = N(n) \in \mathbb{N}$ sufficiently large and some complex-valued function $k \in \mathscr{C}^N(\mathbb{R}^n \setminus \{0\})$ which is odd, and positive homogeneous of degree $1-n$. For each $f \in L^1(\partial\Omega, \frac{\sigma(x)}{1+|x|^{n-1}})$ define*

$$(\mathcal{T}f)(x) := \int_{\partial\Omega} k(x-y)f(y)\,d\sigma(y) \quad \text{at each } x \in \Omega. \tag{5.47}$$

Finally, fix two integrability exponents $p, q \in (1, \infty)$, a Muckenhoupt weight $w \in A_p(\partial\Omega, \sigma) \cap \mathrm{RH}_q(\partial\Omega, \sigma)$, and a threshold $\lambda \in (0, (n-1)/q')$, where q' is the Hölder conjugate to q.

Then for each $\alpha \in \mathbb{N}_0^n$ with $|\alpha| \le N$ there exists $C_\alpha \in (0, \infty)$, depending on α, k, the ADR character of $\partial\Omega$, n, p, q, λ, $[w]_{A_p}$, and $[w]_{\mathrm{RH}_q}$ such that

$$\begin{aligned}&\sup_{x\in\Omega}\{\delta_{\partial\Omega}^{|\alpha|-\lambda/p}(x)\cdot w(\Delta(x_*, 2\delta_{\partial\Omega}(x)))^{1/p}\cdot|\partial^\alpha(\mathcal{T}f)(x)|\}\\&\quad\le C_\alpha\|f\|_{M^{p,\lambda}(\partial\Omega,w)}\end{aligned} \tag{5.48}$$

for all $f \in M^{p,\lambda}(\partial\Omega, w)$, where $x_* \in \partial\Omega$ is chosen such that $|x-x_*| = \delta_{\partial\Omega}(x)$.

Proof. Because $k \in \mathscr{C}^N(\mathbb{R}^n \setminus \{0\})$ is odd and positive homogeneous of degree $1-n$, for each $x \in \Omega$ we can estimate

$$\begin{aligned}&\delta_{\partial\Omega}^{|\alpha|-\lambda/p}(x)|\partial^\alpha(\mathcal{T}f)(x)|\\&\quad\le \delta_{\partial\Omega}^{|\alpha|-\lambda/p}(x)\int_{\partial\Omega}|(\partial^\alpha k)(x-y)||f(y)|\,d\sigma(y)\\&\quad\le \delta_{\partial\Omega}^{|\alpha|-\lambda/p}(x)\cdot\Big(\sum_{|\alpha|\le N}\sup_{S^{n-1}}|\partial^\alpha k|\Big)\cdot\int_{\partial\Omega}\frac{|f(y)|}{|x-y|^{n-1+|\alpha|}}\,d\sigma(y).\end{aligned} \tag{5.49}$$

Next, choose $x_* \in \partial\Omega$ such that $|x-x_*| = \delta_{\partial\Omega}(x)$. Define the dyadic annuli

$$\Delta_0 := \Delta(x_*, 2\delta_{\partial\Omega}(x)) \quad \text{and} \quad \Delta_j := \Delta(x_*, 2^{j+1}\delta_{\partial\Omega}(x)) \setminus \Delta(x_*, 2^j\delta_{\partial\Omega}(x)) \tag{5.50}$$

for each $j \in \mathbb{N}$. Then by the Ahlfors regularity of $\partial\Omega$, and since $\delta_{\partial\Omega}(x) \le |x-y|$ for all $y \in \partial\Omega$, we can estimate

$$\delta_{\partial\Omega}^{|\alpha|-\lambda/p}(x)\int_{\partial\Omega}\frac{|f(y)|}{|x-y|^{n-1+|\alpha|}}\,d\sigma(y) \tag{5.51}$$

$$
\begin{aligned}
&= \delta_{\partial\Omega}^{|\alpha|-\lambda/p}(x) \cdot \sum_{j=0}^{\infty} \int_{\Delta_j} \frac{|f(y)|}{|x-y|^{n-1+|\alpha|}} \, d\sigma(y) \\
&\leq C\delta_{\partial\Omega}^{|\alpha|-\lambda/p}(x) \cdot \sum_{j=0}^{\infty} (2^j \delta_{\partial\Omega}(x))^{-|\alpha|} \fint_{\Delta(x_*, 2^{j+1}\delta_{\partial\Omega}(x))} |f(y)| \, d\sigma(y) \\
&\leq C[w]_{A_p}^{1/p} \cdot \delta_{\partial\Omega}^{-\lambda/p}(x) \cdot \sum_{j=0}^{\infty} 2^{-j|\alpha|} \Big(\fint_{\Delta(x_*, 2^{j+1}\delta_{\partial\Omega}(x))} |f|^p \, dw \Big)^{1/p},
\end{aligned}
$$

where the last inequality makes use of (2.76) in Lemma 2.34, and $C \in (0, \infty)$ depends only on the ADR character of $\partial\Omega$. Now, since $w \in \mathrm{RH}_q(\partial\Omega, \sigma)$, we can use item (3) in Proposition 2.33 and the Ahlfors regularity of $\partial\Omega$ to estimate

$$
\begin{aligned}
\Big(\frac{w(\Delta(x_*, 2\delta_{\partial\Omega}(x)))}{w(\Delta(x_*, 2^{j+1}\delta_{\partial\Omega}(x)))} \Big)^{1/p} &\leq [w]_{\mathrm{RH}_q}^{1/p} \Big(\frac{\sigma(\Delta(x_*, 2\delta_{\partial\Omega}(x)))}{\sigma(\Delta(x_*, 2^{j+1}\delta_{\partial\Omega}(x)))} \Big)^{1/(pq')} \\
&\leq C 2^{-j(n-1)/(pq')},
\end{aligned} \tag{5.52}
$$

for some $C \in (0, \infty)$ depending on the ADR character of $\partial\Omega$, p, and $[w]_{\mathrm{RH}_q}$. We can unpack the integral average in the last line of (5.51), and use (5.52) in combination with (5.49) to obtain

$$
\begin{aligned}
&\delta_{\partial\Omega}^{|\alpha|-\lambda/p}(x) |\partial^\alpha (\mathcal{T} f)(x)| \\
&\quad \leq C\delta_{\partial\Omega}^{-\lambda/p}(x) \cdot \sum_{j=0}^{\infty} 2^{-j[(n-1)/(pq')+|\alpha|]} \times \\
&\quad\quad \times w(\Delta(x_*, 2\delta_{\partial\Omega}(x)))^{-1/p} \cdot \Big(\int_{\Delta(x_*, 2^{j+1}\delta_{\partial\Omega}(x))} |f|^p \, dw \Big)^{1/p} \\
&\quad \leq C \sum_{j=0}^{\infty} 2^{-j[(n-1)/(pq')+|\alpha|-\lambda/p]} \times \\
&\quad\quad \times w(\Delta(x_*, 2\delta_{\partial\Omega}(x)))^{-1/p} \cdot \|f\|_{M^{p,\lambda}(\partial\Omega, w)} \\
&\quad = C w(\Delta(x_*, 2\delta_{\partial\Omega}(x)))^{-1/p} \cdot \|f\|_{M^{p,\lambda}(\partial\Omega, w)}.
\end{aligned} \tag{5.53}
$$

Note that the second inequality follows from multiplying each term in the sum above and below by $2^{(j+1)(\lambda/p)}$, and (4.2). The final equality in (5.53) follows from the fact that

$$
\frac{n-1}{pq'} + |\alpha| - \frac{\lambda}{p} > 0 \iff \lambda - \frac{n-1}{q'} - |\alpha| < 0, \tag{5.54}
$$

which happens even in the case that $|\alpha| = 0$, since $\lambda \in (0, (n-1)/q')$ by assumption. The desired estimate in (5.48) then follows from (5.53) upon multiplication by $w(\Delta(x_*, 2\delta_{\partial\Omega}(x)))^{-1/p}$ and taking a supremum over all $x \in \Omega$. □

Before stating the next few results, we introduce the following notation which will be of use for upcoming estimates.

Definition 5.9. For each $m \in \mathbb{N}_0$ and $t \in [0, \infty)$, define

$$t^{\langle 0 \rangle} := 1 \tag{5.55}$$

and, if $m \geq 1$,

$$t^{\langle m \rangle} := \begin{cases} 0 & \text{if } t = 0, \\ t \cdot \underbrace{\ln(\cdots \ln(\ln(1/t))\cdots)}_{m \text{ natural logarithms}} & \text{if } 0 < t \leq ({}^{m}e)^{-1}, \\ ({}^{m}e)^{-1} & \text{if } t > ({}^{m}e)^{-1}, \end{cases} \tag{5.56}$$

where ${}^{m}e$ is the mth tetration of e (involving m copies of e, combined by exponentiation), i. e.,

$${}^{m}e := \underbrace{e^{e^{\cdot^{\cdot^{\cdot^{e}}}}}}_{m \text{ copies of } e}, \quad \text{the } m\text{th fold exponentiation of } e. \tag{5.57}$$

We also agree to set ${}^{0}e := 1$.

Remark 5.10. Note that for any $m \in \mathbb{N}$, we have that $\lim_{t \to 0^+} t^{\langle m \rangle} = 0$.

The following proposition is originally stated and proved in [79, Theorem 4.4, p. 215]. While it details a similar type of bound as in (5.27) of Proposition 5.6, we find that at the cost of more algebraic structure on the integral kernel, i. e., with chord-dot-normal structure, we can achieve a more refined estimate on the operator norm in terms of the BMO seminorm of the outward unit normal.

Proposition 5.11. *Let $\Omega \subseteq \mathbb{R}^n$ be a UR domain, abbreviate $\sigma := \mathcal{H}^{n-1}\lfloor\partial\Omega$, and denote by ν the geometric measure theoretic outward unit normal to Ω. Fix an integrability exponent $p \in (1, \infty)$ and a Muckenhoupt weight $w \in A_p(\partial\Omega, \sigma)$. Let $N = N(n) \in \mathbb{N}$ be sufficiently large and consider a complex-valued function $k \in \mathscr{C}^N(\mathbb{R}^n \setminus \{0\})$ which is even, and positive homogeneous of degree $-n$. For each $\varepsilon > 0$, let T_ε act on each $f \in L^1(\partial\Omega, \frac{\sigma(x)}{1+|x|^{n-1}})$ according to*

$$(T_\varepsilon f)(x) := \int\limits_{\substack{y \in \partial\Omega \\ |x-y|>\varepsilon}} \langle x - y, \nu(y)\rangle k(x - y) f(y)\, d\sigma(y) \quad \textit{for each } x \in \partial\Omega. \tag{5.58}$$

Define also the associated maximal operator

$$(T_{\max} f)(x) := \sup_{\varepsilon>0} \big|(T_\varepsilon f)(x)\big| \quad \textit{for each } x \in \partial\Omega. \tag{5.59}$$

Then for each $m \in \mathbb{N}$ there exists a constant $C_m \in (0, \infty)$ depending only on m, the UR *character of $\partial\Omega$, n, p, and $[w]_{A_p}$ such that*

$$\|T_{\max}\|_{L^p(\partial\Omega,w)\to L^p(\partial\Omega,w)} \leq C_m\Big(\sum_{|\alpha|\leq N}\sup_{S^{n-1}}|\partial^\alpha k|\Big)\|\nu\|^{\langle m\rangle}_{[\mathrm{BMO}(\partial\Omega,\sigma)]^n}. \tag{5.60}$$

As before, our strategy will be to use the estimates from Proposition 5.11 with our extrapolation result in Theorem 4.19 to obtain small norm estimates for Calderón–Zygmund singular integral operators with a chord-dot-normal structure in the integral kernel in terms of the BMO seminorm of the outward unit normal, now at the level of weighted Morrey spaces.

Theorem 5.12. *Let $\Omega \subseteq \mathbb{R}^n$ be a* UR *domain, abbreviate $\sigma := \mathcal{H}^{n-1}\lfloor\partial\Omega$, and denote by ν the geometric measure theoretic outward unit normal to Ω. Let $N = N(n) \in \mathbb{N}$ be sufficiently large and consider a complex-valued function $k \in \mathscr{C}^N(\mathbb{R}^n \setminus \{0\})$, which is even, and positive homogeneous of degree $-n$. Let $T_{\max}$ be the maximal operator whose action on $f \in L^1(\partial\Omega, \frac{\sigma(x)}{1+|x|^{n-1}})$ is as in (5.59). Now, fix two integrability exponents $p, q \in (1, \infty)$, a Muckenhoupt weight $w \in A_p(\partial\Omega, \sigma) \cap \mathrm{RH}_q(\partial\Omega, \sigma)$, and some threshold $\lambda \in (0, (n-1)(q-1)/q)$, where $q' := q/(q-1)$. Lastly, fix $a \in [0, \lambda)$, a reference point $x_0 \in \partial\Omega$, and abbreviate $w_a := |\cdot - x_0|^a w$, a weight on $\partial\Omega$.*

Then for each $m \in \mathbb{N}$ there exists a constant $C_m \in (0, \infty)$, depending only on m, the UR *character of $\partial\Omega$, n, p, q, λ, a, $[w]_{\mathrm{RH}_q}$ and $[w]_{A_p}$, such that*

$$\begin{aligned}&\|T_{\max}\|_{M^{p,\lambda}(\partial\Omega,w_a)\to M^{p,\lambda}(\partial\Omega,w_a)}\\&\quad\leq C_m\Big(\sum_{|\alpha|\leq N}\sup_{S^{n-1}}|\partial^\alpha k|\Big)\|\nu\|^{\langle m\rangle}_{[\mathrm{BMO}(\partial\Omega,\sigma)]^n}.\end{aligned} \tag{5.61}$$

Proof. Consider the family of pairs

$$\mathcal{F} := \Big\{(h, T_{\max}h) : h \in L^1\Big(\partial\Omega, \frac{\sigma(x)}{1+|x|^{n-1}}\Big)\Big\}. \tag{5.62}$$

Proposition 5.6 shows that $T_{\max}f$ is σ-measurable, so we see that $\mathcal{F}$ is a collection of pairs of σ-measurable functions. Let $f \in M^{p,\lambda}(\partial\Omega, w)$, so that $(f, T_{\max}f) \in \mathcal{F}$ thanks to the embedding in Proposition 4.15. Now, fix an integrability exponent $p_0 \in (1, \infty)$, and any Muckenhoupt weight $\omega \in A_{p_0}(\partial\Omega, \sigma)$. Since we plan to apply Theorem 4.19 with $T_{\max}f$ playing the role of g, we may assume without loss of generality that f belongs to $L^{p_0}(\partial\Omega, \omega)$, which ensures that the hypothesis in (4.62) is not otherwise trivially true. By (5.61) of Theorem 5.11, the following inequality is satisfied:

$$\begin{aligned}&\|T_{\max}f\|_{L^{p_0}(\partial\Omega,\omega)}\\&\quad\leq C_{m,\omega}\cdot\Big(\sum_{|\alpha|\leq N}\sup_{S^{n-1}}|\partial^\alpha k|\Big)\cdot\|\nu\|^{\langle m\rangle}_{[\mathrm{BMO}(\partial\Omega,\sigma)]^n}\|f\|_{L^{p_0}(\partial\Omega,\omega)},\end{aligned} \tag{5.63}$$

where $C_{m,\omega} \in (0,\infty)$ is such that $C_{m,\omega}$ stays bounded whenever $[\omega]_{A_{p_0}}$ is bounded. Since $\omega \in A_{p_0}(\partial\Omega,\sigma)$ is arbitrary, it follows that the hypotheses of Theorem 4.19 are satisfied, and consequently, since (4.65) holds, there exist constants $K, W \in (0,\infty)$ such that

$$\|T_{\max} f\|_{M^{p,\lambda}(\partial\Omega,w_a)} \le K \cdot \Big(\sup_{\substack{\omega\in A_{p_0}(\Sigma,\sigma)\\ [\omega]_{A_{p_0}}<W}} C_{m,\omega} \Big) \cdot \Big(\sum_{|\alpha|\le N} \sup_{S^{n-1}} |\partial^\alpha k| \Big) \times \\ \times \|\nu\|^{\langle m\rangle}_{[\mathrm{BMO}(\partial\Omega,\sigma)]^n} \|f\|_{M^{p,\lambda}(\partial\Omega,w_a)}, \tag{5.64}$$

which is equivalent to (5.61), given the arbitrary nature of $f \in M^{p,\lambda}(\partial\Omega, w_a)$. □

The following result may be regarded as a corollary of Proposition 5.11. See [79, Corollary 4.2, pp. 292–293] for a proof. It gives bounds on the operator norms at the level of Muckenhoupt-weighted Lebesgue spaces of principal value singular integral operators of chord-dot-normal type in terms of the BMO seminorm of the outward unit normal.

Corollary 5.13. *Let $\Omega \subseteq \mathbb{R}^n$ be a* UR *domain and abbreviate $\sigma := \mathcal{H}^{n-1}\lfloor\partial\Omega$. Let ν denote the geometric measure theoretic outward unit normal to Ω. Fix an integrability exponent $p \in (1,\infty)$ and a Muckenhoupt weight $w \in A_p(\partial\Omega,\sigma)$. Let $k \in \mathscr{C}^N(\mathbb{R}^n \setminus \{0\})$, with $N = N(n) \in \mathbb{N}$ sufficiently large, such that k is even and positive homogeneous of degree $-n$. Consider the following principal value singular integral operators who act on each $f \in L^p(\partial\Omega, \frac{\sigma(x)}{1+|x|^{n-1}})$ according to*

$$(Tf)(x) := \lim_{\varepsilon\to 0^+} \int\limits_{\substack{y\in\partial\Omega\\ |x-y|>\varepsilon}} \langle x-y, \nu(y)\rangle k(x-y) f(y)\, d\sigma(y) \tag{5.65}$$

and

$$(T^{\#}f)(x) := \lim_{\varepsilon\to 0^+} \int\limits_{\substack{y\in\partial\Omega\\ |x-y|>\varepsilon}} \langle y-x, \nu(x)\rangle k(x-y) f(y)\, d\sigma(y), \tag{5.66}$$

for σ-a. e. $x \in \partial\Omega$.

Then for each $m \in \mathbb{N}$ there exists a constant $C_m \in (0,\infty)$, depending only on the UR *character of $\partial\Omega$, n, p, and $[w]_{A_p}$, such that*

$$\|T\|_{L^p(\partial\Omega,w)\to L^p(\partial\Omega,w)} \le C_m \Big(\sum_{|\alpha|\le N} \sup_{S^{n-1}} |\partial^\alpha k| \Big) \|\nu\|^{\langle m\rangle}_{[\mathrm{BMO}(\partial\Omega,\sigma)]^n} \tag{5.67}$$

and

$$\|T^{\#}\|_{L^p(\partial\Omega,w)\to L^p(\partial\Omega,w)} \le C_m \Big(\sum_{|\alpha|\le N} \sup_{S^{n-1}} |\partial^\alpha k| \Big) \|\nu\|^{\langle m\rangle}_{[\mathrm{BMO}(\partial\Omega,\sigma)]^n}. \tag{5.68}$$

Remark 5.14. In the context of Corollary 5.13, note that for σ-a. e. point $x \in \partial\Omega$ we can write

$$(Tf)(x) = \lim_{\varepsilon \to 0^+} \int_{\substack{y \in \partial\Omega \\ |x-y|>\varepsilon}} \langle x - y, \nu(y) \rangle k(x-y) f(y) \, d\sigma(y) = \sum_{j=1}^{n} (T_j f_j)(x), \tag{5.69}$$

where T_j is defined as in (5.26), but with the integral kernel $k_j(x) := x_j k(x)$ for all $x \in \mathbb{R}^n \setminus \{0\}$, and $f_j(y) := \nu_j(y) f(y)$ for σ-a. e. point $y \in \partial\Omega$. For each $j \in \{1, \dots, n\}$, because k is even and positive homogeneous of degree $-n$, we see that the integral kernel k_j must be odd and positive homogenous of degree $1 - n$. Since $\nu_j \in L^\infty(\partial\Omega, \sigma)$, we know that $f_j \in L^1(\partial\Omega, \frac{\sigma(x)}{1+|x|^{n-1}})$ for each $j \in \{1, \dots, n\}$, and as such, Tf is a sum of operators of the type covered generally by Proposition 5.6, and hence $Tf : \partial\Omega \to \mathbb{C}$ is well-defined as a σ-measurable function.

We have a similar decomposition for $T^\#$. For σ-a. e. point $x \in \partial\Omega$,

$$(T^\# f)(x) = \lim_{\varepsilon \to 0^+} \int_{\substack{y \in \partial\Omega \\ |x-y|>\varepsilon}} \langle y - x, \nu(x) \rangle k(x-y) f(y) \, d\sigma(y) = -\sum_{j=1}^{n} (T_j^\# f_j)(x), \tag{5.70}$$

where for all $j \in \{1, \dots, n\}$ each $T_j^\#$ is defined as in (5.26), but with the integral kernel $k_j(x) := x_j k(x)$ for all $x \in \mathbb{R}^n \setminus \{0\}$, and where $f_j(y) := \nu_j(x) f(x)$ for σ-a. e. point $y \in \partial\Omega$. As in the case of T, we can easily see that $T^\#$ is also a sum of operators of the type covered generally by Proposition 5.6, likewise making the mapping $T^\# f : \partial\Omega \to \mathbb{C}$ well-defined as a σ-measurable function.

At this point, we once again look toward Theorem 4.19 to transform estimates as in (5.67)–(5.68) into estimates at the level of Muckenhoupt-weighted Morrey spaces.

Theorem 5.15. *Let $\Omega \subseteq \mathbb{R}^n$ be a UR domain, abbreviate $\sigma := \mathcal{H}^{n-1} \lfloor \partial\Omega$, and denote by ν the geometric measure theoretic outward unit normal to Ω. Let $N = N(n) \in \mathbb{N}$ be sufficiently large, and fix some complex-valued function $k \in \mathscr{C}^N(\mathbb{R}^n \setminus \{0\})$, which is even and positive homogeneous of degree $-n$. Let T and $T^\#$ act on each $f \in L^1(\partial\Omega, \frac{\sigma(x)}{1+|x|^{n-1}})$ according to (5.65) and (5.66), respectively. Next, fix any two integrability exponents $p, q \in (1, \infty)$, a Muckenhoupt weight $w \in A_p(\partial\Omega, \sigma) \cap \mathrm{RH}_q(\partial\Omega, \sigma)$, and some threshold $\lambda \in (0, (n-1)/q')$, where $q' := q/(q-1)$. Lastly, fix $\alpha \in [0, \lambda)$, a reference point $x_0 \in \partial\Omega$, and abbreviate $w_\alpha := |\cdot - x_0|^\alpha w$, a weight function on $\partial\Omega$.*

Then for each $m \in \mathbb{N}$ there exists a constant $C_m \in (0, \infty)$, depending only on m, the UR character of $\partial\Omega$, n, p, q, λ, α, $[w]_{\mathrm{RH}_q}$, and $[w]_{A_p}$ such that

$$\|T\|_{M^{p,\lambda}(\partial\Omega,w_a)\to M^{p,\lambda}(\partial\Omega,w_a)} \leq C_m\Big(\sum_{|\gamma|\leq N}\sup_{S^{n-1}}|\partial^\gamma k|\Big)\|\nu\|^{\langle m\rangle}_{[\mathrm{BMO}(\partial\Omega,\sigma)]^n} \tag{5.71}$$

and

$$\|T^\#\|_{M^{p,\lambda}(\partial\Omega,w_a)\to M^{p,\lambda}(\partial\Omega,w_a)} \leq C_m\Big(\sum_{|\gamma|\leq N}\sup_{S^{n-1}}|\partial^\gamma k|\Big)\|\nu\|^{\langle m\rangle}_{[\mathrm{BMO}(\partial\Omega,\sigma)]^n}. \tag{5.72}$$

Proof. Define the two families of pairs

$$\mathcal{F}_1 := \Big\{(h, Th) : h \in L^1\Big(\partial\Omega, \frac{\sigma(x)}{1+|x|^{n-1}}\Big)\Big\}, \tag{5.73}$$

and

$$\mathcal{F}_2 := \Big\{(h, T^\# h) : h \in L^1\Big(\partial\Omega, \frac{\sigma(x)}{1+|x|^{n-1}}\Big)\Big\}. \tag{5.74}$$

Proposition 5.6 and Remark 5.14 guarantee that Th and $T^\# h$ are σ-measurable, so $\mathcal{F}_1$ and $\mathcal{F}_2$ are collections of pairs of σ-measurable functions. Let $f \in M^{p,\lambda}(\partial\Omega, w_a)$, so that by the embedding in Proposition 4.15, we have $(f, Tf) \in \mathcal{F}_1$ and $(f, T^\# f) \in \mathcal{F}_2$. Fix an integrability exponent $p_0 \in (1,\infty)$ and a Muckenhoupt weight $\omega \in A_{p_0}(\partial\Omega,\sigma)$. Since our plan is to use Theorem 4.19, we may assume without loss of generality that $f \in L^{p_0}(\partial\Omega,\omega)$. Otherwise, (4.62) is trivially satisfied. With this assumption in hand, Corollary 5.13 implies that there exists a constant $C_{m,\omega} \in (0,\infty)$ such that

$$\|Tf\|_{L^{p_0}(\partial\Omega,w)} \leq C_{m,\omega}\Big(\sum_{|\beta|\leq N}\sup_{S^{n-1}}|\partial^\alpha k|\Big)\|\nu\|^{\langle m\rangle}_{[\mathrm{BMO}(\partial\Omega,\sigma)]^n}\|f\|_{L^{p_0}(\partial\Omega,w)} \tag{5.75}$$

and

$$\|T^\# f\|_{L^{p_0}(\partial\Omega,w)} \leq C_{m,\omega}\Big(\sum_{|\beta|\leq N}\sup_{S^{n-1}}|\partial^\alpha k|\Big)\|\nu\|^{\langle m\rangle}_{[\mathrm{BMO}(\partial\Omega,\sigma)]^n}\|f\|_{L^{p_0}(\partial\Omega,w)}. \tag{5.76}$$

Note that $C_{m,\omega}$ stays bounded whenever $[\omega]_{A_{p_0}}$ stays bounded. Therefore, the hypotheses of Theorem 4.19 are satisfied, and hence there exist two constants $K, W \in (0,\infty)$ for which

$$\|Tf\|_{M^{p,\lambda}(\partial\Omega,w_a)} \leq K\Big(\sup_{\substack{\omega\in A_{p_0}(\partial\Omega,\sigma)\\ [\omega]_{A_{p_0}}<W}} C_{m,\omega}\Big)\Big(\sum_{|\beta|\leq N}\sup_{S^{n-1}}|\partial^\alpha k|\Big)\times$$
$$\times\|\nu\|^{\langle m\rangle}_{[\mathrm{BMO}(\partial\Omega,\sigma)]^n}\|f\|_{M^{p,\lambda}(\partial\Omega,w_a)}, \tag{5.77}$$

and

$$\|T^{\#}f\|_{M^{p,\lambda}(\partial\Omega,w_a)} \leq K\Big(\sup_{\substack{\omega\in A_{p_0}(\partial\Omega,\sigma)\\ [\omega]_{A_{p_0}}<W}} C_{m,\omega}\Big)\Big(\sum_{|\beta|\leq N}\sup_{S^{n-1}}|\partial^\alpha k|\Big)\times$$
$$\times\|\nu\|^{\langle m\rangle}_{[\mathrm{BMO}(\partial\Omega,\sigma)]^n}\|f\|_{M^{p,\lambda}(\partial\Omega,w_a)}. \tag{5.78}$$

Thus, given the arbitrary nature of f, we retrieve the operator norm estimates in (5.71) and (5.72) as desired. □

The following proposition may be found in [79, Theorem 4.3, pp. 294–295]. It is not only a sharper Rubio de Francia extrapolation result than what was initially introduced in Proposition 2.41, but it also provides commutator estimates for bounded linear operators on weighted Lebesgue spaces with the operation of pointwise multiplication by an essentially bounded function.

Proposition 5.16. *Let $\Sigma \subseteq \mathbb{R}^n$ be a closed* ADR *set, and write $\sigma := \mathcal{H}^{n-1}\lfloor\Sigma$. Fix an arbitrary integrability exponent $p_0 \in (1,\infty)$ along with a nondecreasing function $\Phi : (0,\infty) \to (0,\infty)$. Finally, let T be a linear operator, which is bounded on $L^{p_0}(\Sigma, w)$ for every $w \in A_{p_0}(\Sigma,\sigma)$, with operator norm $\leq \Phi([w]_{A_{p_0}})$.*

Then for each integrability exponent $p \in (1,\infty)$ there exist constants $C_1, C_2 \in (0,\infty)$, which depend exclusively on the dimension n, the exponents p_0, p, and the ADR *character of Σ, with the property that for any Muckenhoupt weight $w \in A_p(\Sigma,\sigma)$ the operator*

$$T : L^p(\Sigma, w) \longrightarrow L^p(\Sigma, w) \tag{5.79}$$

is well-defined, linear, and bounded, with operator norm

$$\|T\|_{L^p(\Sigma,w)\to L^p(\Sigma,w)} \leq C_1\cdot\Phi\big(C_2\cdot[w]_{A_p}^{\max\{1,(p_0-1)/(p-1)\}}\big). \tag{5.80}$$

In addition, given any $p \in (1,\infty)$ along with some $w \in A_p(\Sigma,\sigma)$, there exists a constant $C = C(\Sigma, n, p_0, p, [w]_{A_p}) \in (0,\infty)$ with the property that for every complex-valued function $b \in L^\infty(\Sigma,\sigma)$ one has (with C_1 as before)

$$\big\|[M_b, T]\big\|_{L^p(\Sigma,w)\to L^p(\Sigma,w)} \leq C_1\cdot\Phi(C)\|b\|_{\mathrm{BMO}(\Sigma,\sigma)}, \tag{5.81}$$

where

$$[M_b, T]f := b(Tf) - T(bf) \quad \text{for each } f \in L^p(\Sigma, w), \tag{5.82}$$

i. e., the commutator of T considered as in (5.79) *and the operator* M_b *of pointwise multiplication on* $L^p(\Sigma, w)$ *by the function b.*

We will eventually be able to use our extrapolation result in Theorem 4.19 to obtain a version of Proposition 5.16 but for Muckenhoupt-weighted Morrey spaces. In particular, we will obtain commutator estimates. In order to do that, we first we need the following set inclusion of Muckenhoupt-weighted Morrey spaces into a union of Muckenhoupt-weighted Lebesgue spaces. The proof we employ relies on later results and developments about Muckenhoupt-weighted Block spaces, a brand of spaces introduced in Section 6, which act as a pre-dual to Muckenhoupt-weighted Morrey spaces (cf. Theorem 6.29).

Lemma 5.17. *Let* $\Sigma \subseteq \mathbb{R}^n$ *be a closed ADR set, and write* $\sigma := \mathcal{H}^{n-1}\lfloor\Sigma$. *For each integrability exponent* $p_0 \in (1,\infty)$, *one has the inclusion*

$$\bigcup_{\substack{p,q\in(1,\infty)\\ w\in A_p(\Sigma,\sigma)\cap \mathrm{RH}_q(\Sigma,\sigma)\\ 0\le\lambda<q(n-1)/(q-1)}} M^{p,\lambda}(\Sigma,w) \subseteq \bigcup_{\omega\in A_{p_0}(\Sigma,\sigma)} L^{p_0}(\Sigma,\omega). \tag{5.83}$$

In particular, for any two fixed integrability exponents $p, q \in (1,\infty)$, *an arbitrary Muckenhoupt weight* $w \in A_p(\Sigma,\sigma) \cap \mathrm{RH}_q(\Sigma,\sigma)$, *and some threshold* $\lambda \in (0,(n-1)/q')$, *where* $q' := q/(q-1)$, *one has*

$$M^{p,\lambda}(\Sigma,w) \subseteq \bigcup_{\omega\in A_{p_0}(\Sigma,\sigma)} L^{p_0}(\Sigma,\omega). \tag{5.84}$$

Proof. Fix $p_0 \in (1,\infty)$ arbitrary. Pick $p,q \in (1,\infty)$ along with $w \in A_p(\Sigma,\sigma)$ and some $\lambda \in (0,(n-1)/q')$. Also, fix $f \in M^{p,\lambda}(\Sigma,w)$ arbitrary. Our goal is to construct some Muckenhoupt weight $\omega \in A_{p_0}(\Sigma,\sigma)$ for which the function f belongs to $L^{p_0}(\Sigma,\omega)$. Observe that if $f = 0$ on Σ, then the conclusion is trivially true, so we now assume f is not identically zero.

For all $\phi \in M^{p,\lambda}(\Sigma,w)$ define

$$\mathcal{R}\phi := \sum_{k=0}^{\infty} \frac{\mathcal{M}_\Sigma^k \phi}{2^k \|\mathcal{M}_\Sigma\|^k_{M^{p,\lambda}(\Sigma,w)\to M^{p,\lambda}(\Sigma,w)}}, \tag{5.85}$$

where $\mathcal{M}_\Sigma$ denotes the Hardy–Littlewood maximal function on Σ (cf. (2.26)) with the convention that for all $\phi \in M^{p,\lambda}(\Sigma,w)$ we denote $\mathcal{M}_\Sigma^0 := |\phi|$ and

$$\mathcal{M}_\Sigma^k \phi := \underbrace{\mathcal{M}_\Sigma(\mathcal{M}_\Sigma(\cdots(\mathcal{M}_\Sigma}_{k\text{ times}}\phi)\cdots)). \tag{5.86}$$

First, note that $\mathcal{R} : M^{p,\lambda}(\Sigma,w) \to M^{p,\lambda}(\Sigma,w)$ is well-defined, since Proposition 5.1 says that $\mathcal{M}_\Sigma : M^{p,\lambda}(\Sigma,w) \to M^{p,\lambda}(\Sigma,w)$ is a bounded operator. Moreover, $\mathcal{R}$ enjoys the following three properties:

(1) $\|\mathcal{R}\phi\|_{M^{p,\lambda}(\Sigma,w)} \le 2\|\phi\|_{M^{p,\lambda}(\Sigma,w)}$ for each $\phi \in M^{p,\lambda}(\Sigma, w)$,
(2) $|\phi| \le \mathcal{R}\phi$ pointwise σ-a. e. on Σ for all $\phi \in M^{p,\lambda}(\Sigma, w)$,
(3) For all $\phi \in M^{p,\lambda}(\Sigma, w)$ not identically zero σ-a. e. on Σ, the function $\mathcal{R}\phi$ belongs to $A_1(\Sigma,\sigma)$ and

$$[\mathcal{R}\phi]_{A_1} \le C_\Sigma \|\mathcal{M}_\Sigma\|_{M^{p,\lambda}(\Sigma,w)\to M^{p,\lambda}(\Sigma,w)}, \tag{5.87}$$

where the constant $C_\Sigma \in (0,\infty)$ depends only on Σ.

To justify (1), note that for each $k \in \mathbb{N}_0$, as a consequence of Proposition 5.1, we have $\mathcal{M}_\Sigma^k\phi \in M^{p,\lambda}(\Sigma, w)$ with

$$\|\mathcal{M}_\Sigma^k\phi\|_{M^{p,\lambda}(\Sigma,w)} \le \|\mathcal{M}_\Sigma\|^k_{M^{p,\lambda}(\Sigma,w)\to M^{p,\lambda}(\Sigma,w)} \cdot \|\phi\|_{M^{p,\lambda}(\Sigma,w)}. \tag{5.88}$$

Hence, by (5.85) we can conclude

$$\|\mathcal{R}\phi\|_{M^{p,\lambda}(\Sigma,w)} \le \left(\sum_{k=0}^{\infty}\frac{1}{2^k}\right)\|\phi\|_{M^{p,\lambda}(\Sigma,w)} = 2\|\phi\|_{M^{p,\lambda}(\Sigma,w)}. \tag{5.89}$$

To justify (2), fix $\phi \in M^{p,\lambda}(\Sigma, w)$ and observe that the first term in the series defining $\mathcal{R}\phi$ is just $|\phi|$ and the rest of the terms are nonnegative by the definition of $\mathcal{M}_\Sigma$ (2.26).

Finally, to justify (3) consider $\phi \in M^{p,\lambda}(\Sigma, w)$ not identically zero σ-a. e. on Σ. Then $0 < \mathcal{R}\phi < +\infty$ at σ-a. e. point on Σ by (1) and Remark 4.3. Also, $\mathcal{R}\phi$ is a σ-measurable function which is not identically zero σ-a. e. on Σ by (2). By Lemma 2.24 and the definition of $\mathcal{R}\phi$ (5.85), we have

$$\begin{aligned}\mathcal{M}_\Sigma(\mathcal{R}\phi) &\le \sum_{k=0}^{\infty}\frac{\mathcal{M}^{k+1}\phi}{2^k\|\mathcal{M}_\Sigma\|^k_{M^{p,\lambda}(\Sigma,w)\to M^{p,\lambda}(\Sigma,w)}} \\ &= 2\|\mathcal{M}_\Sigma\|_{M^{p,\lambda}(\Sigma,w)\to M^{p,\lambda}(\Sigma,w)}\sum_{k=0}^{\infty}\frac{\mathcal{M}^{k+1}\phi}{2^{k+1}\|\mathcal{M}_\Sigma\|^{k+1}_{M^{p,\lambda}(\Sigma,w)\to M^{p,\lambda}(\Sigma,w)}} \\ &\le 2\|\mathcal{M}_\Sigma\|_{M^{p,\lambda}(\Sigma,w)\to M^{p,\lambda}(\Sigma,w)} \cdot \mathcal{R}\phi.\end{aligned} \tag{5.90}$$

Hence, by Lemma 2.30 with $C := 2\|\mathcal{M}_\Sigma\|_{M^{p,\lambda}(\Sigma,w)\to M^{p,\lambda}(\Sigma,w)}$ and $w := \mathcal{R}\phi$, we obtain the desired result.

Moving on, let $p' := p/(p-1)$ denote the Hölder conjugate exponent to p and let $w' := w^{-p'/p} \in A_{p'}(\Sigma,\sigma)$ denote the conjugate weight to w. Bring in the Block space $\mathcal{B}^{p',\lambda}(\Sigma, w')$, which is defined in relation to p', w', $\varepsilon := (p'-1)/q' \in (0, p'-1)$, and λ, as prescribed by Definition 6.7. Note that $w' \in A_{p'-\varepsilon}(\Sigma,\sigma)$ as a consequence of Proposition 6.35. Additionally, due to this choice of ε, we have

$$(0, \varepsilon(n-1)/(p'-1)) = (0, (n-1)/q'), \tag{5.91}$$

the latter interval being the natural range for λ in the context of Block spaces (cf. Definition 6.7). Next, for all $\psi \in \mathcal{B}^{p',\lambda}(\Sigma, w')$ define

$$\mathcal{Q}\psi := \sum_{k=0}^{\infty} \frac{\mathcal{M}_{\Sigma}^{k}\psi}{2^k \|\mathcal{M}_{\Sigma}\|^{k}_{\mathcal{B}^{p',\lambda}(\Sigma,w')\to\mathcal{B}^{p',\lambda}(\Sigma,w')}}. \tag{5.92}$$

We now require a later result, Proposition 7.1,[2] which says that the mapping given by $\mathcal{M}_{\Sigma} : \mathcal{B}^{p',\lambda}(\Sigma, w') \to \mathcal{B}^{p',\lambda}(\Sigma, w')$ is bounded. Then in a similar way as with $\mathcal{R}$, we have that $\mathcal{Q} : \mathcal{B}^{p',\lambda}(\Sigma, w') \to \mathcal{B}^{p',\lambda}(\Sigma, w')$ is well-defined, and enjoys the following properties:

(1') $\|\mathcal{Q}\psi\|_{\mathcal{B}^{p',\lambda}(\Sigma,w')} \le 2\|\psi\|_{\mathcal{B}^{p',\lambda}(\Sigma,w')}$ for all $\psi \in \mathcal{B}^{p',\lambda}(\Sigma, w')$,

(2') $|\psi| \le \mathcal{Q}\psi$ pointwise on Σ for all $\psi \in \mathcal{B}^{p',\lambda}(\Sigma, w')$,

(3') For all $\psi \in \mathcal{B}^{p',\lambda}(\Sigma, w')$ not identically zero σ-a. e. on Σ, the function $\mathcal{Q}\psi$ belongs to $A_1(\Sigma, \sigma)$ and

$$[\mathcal{Q}\psi]_{A_1} \le C_{\Sigma}\|\mathcal{M}_{\Sigma}\|_{\mathcal{B}^{p',\lambda}(\Sigma,w')\to\mathcal{B}^{p',\lambda}(\Sigma,w')}, \tag{5.93}$$

where the constant $C_{\Sigma} \in (0, \infty)$ depends only on Σ.

The proofs of (1')–(3') are analogous to those of (1)–(3), so are omitted. Now, define $\widetilde{f} := \frac{|f|}{\|f\|_{M^{p,\lambda}(\Sigma,w)}}$, which is a well-defined function on Σ, given our present assumption that f is not identically zero. Moreover, it is clear that $\widetilde{f}$ is nonnegative with $\widetilde{f} \in M^{p,\lambda}(\Sigma, w)$, since $f \in M^{p,\lambda}(\Sigma, w)$. By (2'), we have that $\widetilde{f} \le \mathcal{R}\widetilde{f}$. Unpacking the definition of $\widetilde{f}$ and raising both sides to the $1 - p_0$ power obtain

$$(\mathcal{R}\widetilde{f})^{1-p_0} \le \|f\|^{p_0-1}_{M^{p,\lambda}(\Sigma,w)}|f|^{1-p_0}. \tag{5.94}$$

Note that by (3'), we know $\mathcal{R}\widetilde{f} \in A_1(\Sigma, \sigma)$, and hence $0 < \mathcal{R}\widetilde{f} < \infty$ at σ-a. e. point on Σ. In particular, $(\mathcal{R}\widetilde{f})^{1-p_0}$ is a well-defined number in $(0, \infty)$ at σ-a. e. point in Σ.

Now, choose some $g \in \mathcal{B}^{p',\lambda}(\Sigma, w')$ with $\|g\|_{\mathcal{B}^{p',\lambda}(\Sigma,w')} = 1$ and define the weight

$$\omega := (\mathcal{R}\widetilde{f})^{1-p_0}(\mathcal{Q}g). \tag{5.95}$$

By item (5) of Proposition 2.29, bearing in mind (3) and (3'), we see that $\omega \in A_{p_0}(\Sigma, \sigma)$ with

$$\begin{aligned}[\omega]_{A_{p_0}} \le C_{\Sigma}\|\mathcal{M}_{\Sigma}\|^{p_0-1}_{M^{p,\lambda}(\Sigma,w)\to M^{p,\lambda}(\Sigma,w)} \times \\ \times \|\mathcal{M}_{\Sigma}\|_{\mathcal{B}^{p',\lambda}(\Sigma,w')\to\mathcal{B}^{p',\lambda}(\Sigma,w')}.\end{aligned} \tag{5.96}$$

As such,

2 This result is proved independently of Lemma 5.17.

$$\begin{aligned}\|f\|_{L^{p_0}(\Sigma,\omega)} &= \Big(\int_\Sigma |f|^{p_0}(\mathcal{R}\widetilde{f})^{1-p_0}(\mathcal{Q}g)\,d\sigma\Big)^{1/p_0} \\ &\le \|f\|^{1-1/p_0}_{M^{p,\lambda}(\Sigma,w)}\Big(\int_\Sigma |f|(\mathcal{Q}g)\,d\sigma\Big)^{1/p_0} \\ &\le \|f\|^{1-1/p_0}_{M^{p,\lambda}(\Sigma,w)}\|f\|^{1/p_0}_{M^{p,\lambda}(\Sigma,w)}\|\mathcal{Q}g\|^{1/p_0}_{\mathcal{B}^{p',\lambda}(\Sigma,w')} \\ &= 2\|f\|_{M^{p,\lambda}(\Sigma,w)} < \infty.\end{aligned} \tag{5.97}$$

The first line uses (5.95), the second line follows from (5.94), the third line follows from Theorem 6.28, and the last line follows from (1) and the fact that we have chosen g so that $\|g\|_{\mathcal{B}^{p',\lambda}(\Sigma,w')} = 1$. Therefore, the estimate in (5.97) shows that $f \in L^{p_0}(\Sigma,\omega)$ as wanted, subsequently establishing the desired inclusion in (5.83). □

We now present an extrapolation result, which serves two functions. First, it takes us from Muckenhoupt-weighted Lebesgue spaces to Muckenhoupt-weighted Morrey spaces, and second, it establishes commutator estimates on the latter scale of spaces.

Theorem 5.18. *Let $\Sigma \subseteq \mathbb{R}^n$ be a closed ADR set, and write $\sigma := \mathcal{H}^{n-1}\lfloor\Sigma$. Fix $p_0 \in (1,\infty)$ with some nondecreasing function $\Phi : (0,\infty) \to (0,\infty)$ and let T be a linear operator, which is bounded on $L^{p_0}(\Sigma,w)$ for every weight $w \in A_{p_0}(\Sigma,\sigma)$, with operator norm $\le \Phi([w]_{A_{p_0}})$. Let $q \in (1,\infty)$ with Hölder conjugate exponent q'. Lastly, fix a threshold $\lambda \in [0,(n-1)/q')$.*

Then for every integrability exponent $p \in (1,\infty)$ there exist constants $C_1, C_2 \in (0,\infty)$, depending only on Σ, n, p_0, and there exist constants $K, W \in (0,\infty)$, depending only on the ADR character of Σ, n, p, q, λ, $[w]_{\mathrm{RH}_q}$, and $[w]_{A_p}$, such that for every Muckenhoupt weight $w \in A_p(\Sigma,\sigma) \cap \mathrm{RH}_q(\Sigma,\sigma)$, the operator

$$T : M^{p,\lambda}(\Sigma,w) \to M^p(\Sigma,w) \tag{5.98}$$

is well-defined, linear, and bounded, with operator norm

$$\|T\|_{M^{p,\lambda}(\Sigma,w)\to M^{p,\lambda}(\Sigma,w)} \le K\cdot C_1\cdot\Big(\sup_{\substack{\omega\in A_{p_0}(\Sigma,\sigma)\\ [\omega]_{A_{p_0}}\le W}} \Phi(C_2\cdot[\omega]_{A_{p_0}})\Big). \tag{5.99}$$

Moreover, there exists a constant $C \in (0,\infty)$, depending on Σ, n, p_0, and $[\omega]_{A_{p_0}}$, and constants $K', W' \in (0,\infty)$, depending on Σ, n, p, q, λ, $[w]_{\mathrm{RH}_q}$, and $[w]_{A_p}$ with the property that for each $b \in L^\infty(\Sigma,\sigma)$, one has

$$\begin{aligned}&\|[M_b,T]\|_{M^{p,\lambda}(\Sigma,w)\to M^{p,\lambda}(\Sigma,w)} \\ &\quad\le K'\cdot C_1\cdot\Big(\sup_{\substack{\omega\in A_{p_0}\\ [\omega]_{A_{p_0}}\le W'}}\Phi(C)\Big)\|b\|_{\mathrm{BMO}(\Sigma,\sigma)},\end{aligned} \tag{5.100}$$

where $[M_b, T]$ is the commutator of T and the operator M_b of pointwise multiplication by the function b, whose action is defined via

$$[M_b, T]f := b(Tf) - T(bf) \quad \textit{for each } f \in M^{p,\lambda}(\Sigma, w). \tag{5.101}$$

Proof. Consider the family of pairs

$$\mathcal{F} := \Big\{(f, Tf) : f \in \bigcup_{\omega \in A_{p_0}(\Sigma,\sigma)} L^{p_0}(\Sigma, \omega)\Big\}. \tag{5.102}$$

By assumption, for all $\omega \in A_{p_0}(\Sigma, \sigma)$, we have

$$\int_\Sigma |\psi|^{p_0}\, d\omega \le \Phi([\omega]_{A_{p_0}}) \int_\Sigma |\phi|^{p_0}\, d\omega \quad \text{for all } (\phi, \psi) \in \mathcal{F}. \tag{5.103}$$

Now, fix $p \in (1, \infty)$ and $q \in (1, \infty)$ along with some $w \in A_p(\Sigma, \sigma) \cap \mathrm{RH}_q(\Sigma, \sigma)$. Also, fix some threshold $\lambda \in [0, (n-1)/q')$, where q' denotes the Hölder conjugate to q. The extrapolation result in Theorem 4.19 implies there exist constants $K, W \in (0, \infty)$ such that

$$\|\psi\|_{M^{p,\lambda}(\Sigma,w)} \le C\|\phi\|_{M^{p,\lambda}(\Sigma,w)} \quad \text{for all } (\phi, \psi) \in \mathcal{F}, \tag{5.104}$$

where the precise nature of the constant $C \in (0, \infty)$ is described in the statement of the theorem. Hence,

$$\|Tf\|_{M^{p,\lambda}(\Sigma,w)} \le C\|f\|_{M^{p,\lambda}(\Sigma,w)} \quad \text{for all } f \in \bigcup_{\omega \in A_{p_0}(\Sigma,\sigma)} L^{p_0}(\Sigma, \omega). \tag{5.105}$$

Lemma 5.17 ensures we have the inclusion in (5.84), and so (5.105) and Proposition 4.7 imply that the mapping $T : M^{p,\lambda}(\Sigma, w) \to M^{p,\lambda}(\Sigma, w)$ is well-defined, linear, and bounded.

Define now the family of pairs

$$\mathcal{G} := \Big\{(f, [M_b, T]f) : f \in \bigcup_{\omega \in A_{p_0}(\Sigma,\sigma)} L^{p_0}(\Sigma, \omega)\Big\}. \tag{5.106}$$

By Proposition 5.16, there exist constants C and C_1, depending on Σ, n, p_0, and $[\omega]_{A_{p_0}}$, such that

$$\|\psi\|_{L^{p_0}(\Sigma,\omega)} \le C_1 \cdot \Phi(C)\|b\|_{\mathrm{BMO}(\Sigma,\sigma)}\|\phi\|_{L^{p_0}(\Sigma,\omega)} \tag{5.107}$$

for all pairs $(\phi, \psi) \in \mathcal{G}$. Now, fix two integrability exponents $p, q \in (1, \infty)$, a weight $w \in A_p(\Sigma, \sigma) \cup \mathrm{RH}_q(\Sigma, \sigma)$, and some threshold $\lambda \in [0, (n-1)/q')$. Note that (5.107) shows that the hypotheses of Theorem 4.19 are satisfied. In particular, note that Proposition 5.16 implies that $C_1\Phi(C)$ stays bounded as $[w]_{A_{p_0}}$ stays bounded. Thus, there exist constants $K, W \in (0, \infty)$ such that

$$\|\psi\|_{M^{p,\lambda}(\Sigma,w)} \leq \Big(K \cdot \sup_{\substack{\omega\in A_{p_0}(\Sigma,\sigma)\\ [\omega]_{A_{p_0}}\leq W}} \Phi(C)\Big) \cdot \|b\|_{\mathrm{BMO}(\Sigma,\sigma)} \|\phi\|_{M^{p,\lambda}(\Sigma,w)} \tag{5.108}$$

for all $(\phi, \psi) \in \mathcal{G}$. Thanks to (5.106), this means that

$$\begin{aligned} &\|[M_b, T]f\|_{M^{p,\lambda}(\Sigma,w)} \leq C\|b\|_{\mathrm{BMO}(\Sigma,\sigma)}\|f\|_{M^{p,\lambda}(\Sigma,w)} \\ &\text{for all } f \in \bigcup_{\omega\in A_{p_0}(\Sigma,\sigma)} L^{p_0}(\Sigma, \omega). \end{aligned} \tag{5.109}$$

According to Lemma 5.17, the inclusion in (5.84) holds, so in fact (5.109) implies that the operator $[M_b, T] : M^{p,\lambda}(\Sigma, w) \to M^{p,\lambda}(\Sigma, w)$ is well-defined and bounded with estimates as in (5.99). This completes the proof of Theorem 5.18. □

The next set of results show that the principal value singular integral operators introduced in Proposition 5.6, among other related operators like chord-dot-normal operators, and commutators are continuous on local Lebesgue spaces.

The following result can be traced back to [86].

Proposition 5.19. *Let $\Sigma \subseteq \mathbb{R}^n$ be a UR set and write $\sigma := \mathcal{H}^{n-1}\lfloor\Sigma$. Let $N = N(n) \in \mathbb{N}$ be sufficiently large and fix $k \in \mathscr{C}^N(\mathbb{R}^n \setminus \{0\})$, odd, positive homogeneous of degree $1-n$. For each $f \in L^1(\Sigma, \frac{\sigma(x)}{1+|x|^{n-1}})$ define*

$$(Tf)(x) := \lim_{\varepsilon\to 0^+} \int_{\substack{y\in\Sigma\\ |x-y|>\varepsilon}} k(x-y)f(y)\, d\sigma(y) \quad \textit{for } \sigma\textit{-a. e. point } x \in \Sigma. \tag{5.110}$$

Then Tf is well-defined as a σ-measurable function on Σ, and for all $p \in (1, \infty)$, the operator T induces a mapping

$$T : L^1\Big(\Sigma, \frac{\sigma(x)}{1+|x|^{n-1}}\Big) \cap L^p_{\mathrm{loc}}(\Sigma, \sigma) \to L^p_{\mathrm{loc}}(\Sigma, \sigma), \tag{5.111}$$

which is linear and continuous.

An immediate corollary of the above result is the continuity on local Lebesgue spaces of commutators between principal value singular integral operators and the operation of pointwise multiplication by essentially bounded functions.

Corollary 5.20. *Let $\Sigma \subseteq \mathbb{R}^n$ be a UR set and write $\sigma := \mathcal{H}^{n-1}\lfloor\Sigma$. Choose $N = N(n) \in \mathbb{N}$ a sufficiently large integer and fix $k \in \mathscr{C}^N(\mathbb{R}^n \setminus \{0\})$, assumed to be odd and positive homogeneous of degree $1-n$. For each $f \in L^1(\Sigma, \frac{\sigma(x)}{1+|x|^{n-1}})$, define Tf as in (5.110). Select any $b \in L^\infty(\Sigma, \sigma)$, and denote by M_b the operation of pointwise multiplication by b. Then*

$$[M_b, T] : L^1\Big(\Sigma, \frac{\sigma(x)}{1+|x|^{n-1}}\Big) \cap L^p_{\mathrm{loc}}(\Sigma, \sigma) \to L^p_{\mathrm{loc}}(\Sigma, \sigma) \tag{5.112}$$

is well-defined, linear, and continuous, where $[M_b, T]$ is the commutator of M_b and T, defined according to $[M_b, T]f := b \cdot (Tf) - T(bf)$ for all functions $f \in L^1(\Sigma, \frac{\sigma(x)}{1+|x|^{n-1}})$.

Proof. This follows from Proposition 5.19 the fact that the operation of pointwise multiplication by an essentially bounded function is continuous on both $L^1(\Sigma, \frac{\sigma(x)}{1+|x|^{n-1}})$ and $L^p_{\mathrm{loc}}(\Sigma, \sigma)$. □

This final corollary to Proposition 5.19 that we present concerns the continuity of chord-dot-normal singular integral operators on $L^p_{\mathrm{loc}}(\Sigma, \sigma)$.

Corollary 5.21. *Let $\Omega \subseteq \mathbb{R}^n$ be a UR domain, abbreviate $\sigma := \mathcal{H}^{n-1}\lfloor\partial\Omega$, and denote by ν the geometric measure theoretic outward unit normal to Ω. Let $N = N(n) \in \mathbb{N}$ be sufficiently large and fix $k \in \mathscr{C}^N(\mathbb{R}^n \setminus \{0\})$, which is even, positive homogeneous of degree $-n$. For each $f \in L^1(\partial\Omega, \frac{\sigma(x)}{1+|x|^{n-1}})$, define*

$$(Tf)(x) := \lim_{\varepsilon \to 0^+} \int_{\substack{y \in \partial\Omega \\ |x-y|>\varepsilon}} \langle \nu(y), x - y\rangle k(x-y) f(y)\, d\sigma(y) \tag{5.113}$$

and

$$(T^{\#}f)(x) := \lim_{\varepsilon \to 0^+} \int_{\substack{y \in \partial\Omega \\ |x-y|>\varepsilon}} \langle \nu(x), x - y\rangle k(x-y) f(y)\, d\sigma(y) \tag{5.114}$$

at σ-a. e. point $x \in \partial\Omega$. Then the operators

$$T, T^{\#} : L^1\Big(\partial\Omega, \frac{\sigma(x)}{1+|x|^{n-1}}\Big) \cap L^p_{\mathrm{loc}}(\partial\Omega, \sigma) \to L^p_{\mathrm{loc}}(\partial\Omega, \sigma) \tag{5.115}$$

are well-defined, linear, and continuous.

Proof. Notice that T can be written as sum of operators of the type contained in Proposition 5.19. Indeed, for each $j \in \{1, \dots, n\}$ if we define the kernel $k_j(z) := z_j k(z)$ for all $z = (z_1, \dots, z_n) \in \mathbb{R}^n \setminus \{0\}$, then k_j is odd, and positive homogeneous of degree $1 - n$. Furthermore, for each $f \in L^1(\partial\Omega, \frac{\sigma(x)}{1+|x|^{n-1}})$ and σ-a. e. point $x \in \partial\Omega$ we have

$$\begin{aligned}(Tf)(x) &= \lim_{\varepsilon \to 0^+} \sum_{j=1}^{n} \int_{\substack{y \in \partial\Omega \\ |x-y|>\varepsilon}} \nu_j(y)(x_j - y_j) k(x-y) f(y)\, d\sigma(y) \\ &= \sum_{j=1}^{n} \lim_{\varepsilon \to 0^+} \int_{\substack{y \in \partial\Omega \\ |x-y|>\varepsilon}} k_j(x-y)\nu_j(y) f(y)\, d\sigma(y) \\ &= \sum_{j=1}^{n} T_j(\nu_j f)(x),\end{aligned} \tag{5.116}$$

where T_j is as in (5.110), but with k_j now playing the role of k. Since $\nu \in [L^\infty(\partial\Omega, \sigma)]^n$ and the operation of pointwise multiplication by essentially bounded functions is continuous on both $L^1(\partial\Omega, \frac{\sigma(x)}{1+|x|^{n-1}})$ and $L^p_{\mathrm{loc}}(\partial\Omega, \sigma)$, we see that Proposition 5.19 implies the desired result. The case of $T^\#$ is handled analogously. □

As a final proposition for this section, we note that the maximal operator, $T_{\max}$ is also continuous on local Lebesgue spaces; cf. [88].

Proposition 5.22. *Let $\Sigma \subseteq \mathbb{R}^n$ be a closed UR set and write $\sigma := \mathcal{H}^{n-1}\lfloor\Sigma$. Let $N = N(n) \in \mathbb{N}$ be sufficiently large and consider an integral kernel $k \in \mathscr{C}^N(\mathbb{R}^n \setminus \{0\})$, which is odd and positive homogeneous of degree $1 - n$. For each function $f \in L^1(\Sigma, \frac{\sigma(x)}{1+|x|^{n-1}})$, define the maximal operator*

$$T_{\max} f(x) := \sup_{\varepsilon>0} \int\limits_{\substack{y\in\Sigma \\ |x-y|>\varepsilon}} k(x-y)f(y)\,d\sigma(y) \quad \text{for each } x \in \Sigma. \tag{5.117}$$

Then for each $p \in (1, \infty)$ the operator

$$T_{\max} : L^1\Big(\Sigma, \frac{\sigma(x)}{1+|x|^{n-1}}\Big) \cap L^p_{\mathrm{loc}}(\Sigma, \sigma) \to L^p_{\mathrm{loc}}(\Sigma, \sigma) \tag{5.118}$$

is well-defined and continuous.

6 Weighted Block spaces

Our original goal was to establish a well-posedness result for boundary value problems on Muckenhoupt-weighted Morrey spaces. While solvability will follow largely thanks to the work contained in Section 9.1, the problem of uniqueness will lead us to consider functions that pair well with Morrey functions in an integral sense; cf. Proposition 11.4 as it applies to the proof of uniqueness in Theorem 11.9. Because of this, a solid duality theory for Muckenhoupt-weighted Morrey spaces is needed, and this section details a functional analytic treatment of a brand of spaces that act as a pre-dual to Muckenhoupt-weighted Morrey spaces, the so-called Muckenhoupt-weighted Block spaces.

6.1 Definitions and basic properties

As the name suggests, Block spaces are constructed from pieces of smaller functions, which we call blocks. We will begin by introducing the notion of a block and eventually work up to the notion of a Muckenhoupt-weighted Block space.

Definition 6.1. Let $\Sigma \subseteq \mathbb{R}^n$ be a closed Ahlfors regular set and abbreviate $\sigma := \mathcal{H}^{n-1}\lfloor\Sigma$. Fix an integrability exponent $q \in (1,\infty)$, a Muckenhoupt weight $\omega \in A_q(\Sigma,\sigma)$, and some threshold $\lambda \in [0,\infty)$. A function $b : \Sigma \to \mathbb{C}$ is called a (q,λ,ω)-block if $b \in L^q(\Sigma,\omega)$ and there exist a location $x_0 \in \Sigma$ and a scale $R \in (0, 2\operatorname{diam}(\Sigma))$ (both of which may depend on b) such that the following localization and normalization condition hold:

$$\begin{cases} \operatorname{supp} b \subseteq \Delta(x_0,R), \\ \|b\|_{L^q(\Sigma,\omega)} \leq R^{\lambda(1/q-1)}. \end{cases} \tag{6.1}$$

The reader is alerted to the fact that there is no multiplicative constant in the last line above. These functions will be used as building blocks to create Muckenhoupt-weighted Block spaces.

Remark 6.2. As a quick example of a (q,λ,ω)-block, consider the function $b := c\mathbf{1}_{\Delta(x_0,R)}$, where $x_0 \in \Sigma$ and $R \in (0,2\operatorname{diam}(\Sigma))$ are arbitrary, and $c := R^{\lambda(1/q-1)}\omega(\Delta(x_0,R))^{-1/q}$. Indeed, $\operatorname{supp} b \subseteq \Delta(x_0,R)$ and

$$\|b\|_{L^q(\Sigma,\omega)} = c\|\mathbf{1}_{\Delta(x_0,R)}\|_{L^q(\Sigma,\omega)} = c \cdot \omega\big(\Delta(x_0,R)\big)^{1/q} = R^{\lambda(1/q-1)}. \tag{6.2}$$

Before proceeding, we record a useful lemma that gives an estimate involving the L^1-norm of a (q,λ,ω)-block and dyadic annuli. This type of estimate will also show up frequently in Chapter 7 throughout our development of a Calderón–Zygmund theory for weighted Block spaces.

https://doi.org/10.1515/9783111458274-006

Lemma 6.3. *Let $\Sigma \subseteq \mathbb{R}^n$ be a closed* ADR *set, and abbreviate $\sigma := \mathcal{H}^{n-1}\lfloor\Sigma$. Fix an integrability exponent $q \in (1,\infty)$, along with a Muckenhoupt weight $\omega \in A_q(\Sigma,\sigma)$. Let $\varepsilon \in (0, q-1)$ be such that $\omega \in A_{q-\varepsilon}(\Sigma,\sigma)$, and fix a threshold $\lambda \in (0, \varepsilon(n-1)/(q-1))$. Suppose b is a (q,λ,ω)-block on Σ. If $x_0 \in \Sigma$ and $R \in (0, 2\operatorname{diam}(\Sigma))$ are such that*

$$\operatorname{supp} b \subseteq \Delta(x_0,R) \quad \textit{and} \quad \|b\|_{L^q(\Sigma,\omega)} \le R^{\lambda(1/q-1)}, \tag{6.3}$$

then there exists a constant $C \in (0,\infty)$, depending on the ADR *character of Σ, n, q, $[\omega]_{A_{q-\varepsilon}}$, and $[\omega]_{A_q}$ such that*

$$\frac{\omega(\Delta(x_0,2^{j+1}R)\setminus\Delta(x_0,2^jR))^{1/q}}{(2^jR)^{n-1}} \int\limits_{\Delta(x_0,R)} |b|\,d\sigma \le C2^{-j(n-1)(\varepsilon/q)} \cdot R^{\lambda(1/q-1)} \tag{6.4}$$

for each $j \in \mathbb{N}$.

Remark 6.4. Retain the context of Lemma 6.3. It should be noted that such a ε as in the statement of Lemma 6.3 always exists, thanks to the openness property of the Muckenhoupt class; cf. item (1) of Proposition 2.29. In this case, ε will necessarily depend on Σ, n, the choice of q, and $[\omega]_{A_q}$. However, it should be noted that the ε, which appears in the statement of the lemma need not be the one guaranteed by the theory of Muckenhoupt weights. For example, in the case that $\omega \in A_q(\Sigma,\sigma)$ is actually constant, then any $\varepsilon \in (0,q-1)$ would suffice to ensure the membership of ω to $A_{q-\varepsilon}(\Sigma,\sigma)$ as well.

Proof of Lemma 6.3. To make notation easier, write

$$\Delta_j := \Delta(x_0,2^{j+1}R)\setminus\Delta(x_0,2^jR) \quad \text{for each } j \in \mathbb{N}. \tag{6.5}$$

Then we may estimate

$$\begin{aligned}
\frac{\omega(\Delta_j)^{1/q}}{(2^jR)^{n-1}} \int\limits_{\Delta(x_0,R)} |b|\,d\sigma &\le \frac{C\sigma(\Delta(x_0,R))}{\sigma(\Delta(x_0,2^jR))}\omega(\Delta(x_0,2^{j+1}R))^{1/q} \fint\limits_{\Delta(x_0,R)} |b|\,d\sigma \\
&\le \frac{C\sigma(\Delta(x_0,R))}{\sigma(\Delta(x_0,2^jR))}\omega(\Delta(x_0,2^{j+1}R))^{1/q}\Big(\fint\limits_{\Delta(x_0,R)} |b|^q\,d\omega\Big)^{1/q} \\
&\le \frac{C\sigma(\Delta(x_0,R))}{\sigma(\Delta(x_0,2^jR))}\Big(\frac{\omega(\Delta(x_0,2^{j+1}R))}{\omega(\Delta(x_0,R))}\Big)^{1/q} R^{\lambda(1/q-1)}.
\end{aligned} \tag{6.6}$$

The first inequality follows using the Ahlfors regularity of Σ and the fact that $\Delta_j \subseteq \Delta(x_0,2^{j+1}R)$. The second inequality is a consequence of Lemma 2.34. The last inequality follows from unpacking the integral average and (6.3). Now, since $\omega \in A_{q-\varepsilon}(\Sigma,\sigma)$, item (4) of Proposition 2.29 and the Ahlfors regularity of Σ allow us to estimate

$$\left(\frac{\omega(\Delta(x_0,2^{j+1}R))}{\omega(\Delta(x_0,R))}\right)^{1/q} \leq [\omega]_{A_{q-\varepsilon}}^{1/q}\left(\frac{\sigma(\Delta(x_0,2^{j+1}R))}{\sigma(\Delta(x_0,R))}\right)^{(q-\varepsilon)/q} \leq C[\omega]_{A_{q-\varepsilon}}^{1/q}\left(\frac{(2^{j+1}R)^{n-1}}{R^{n-1}}\right)^{(q-\varepsilon)/q} \leq C[\omega]_{A_{q-\varepsilon}}^{1/q}\, 2^{j(n-1)(1-\varepsilon/q)}. \tag{6.7}$$

It follows from (6.6) and (6.7), and Ahlfors regularity that

$$\frac{\omega(\Delta_j)^{1/q}}{(2^jR)^{n-1}} \int_{\Delta(x_0,R)} |b|\,d\sigma \leq C[\omega]_{A_{q-\varepsilon}}^{1/q}\frac{R^{n-1}}{(2^jR)^{n-1}}\cdot 2^{j(n-1)(1-\varepsilon/q)}\cdot R^{\lambda(1/q-1)} \leq C[\omega]_{A_{q-\varepsilon}}^{1/q}\cdot 2^{-j(n-1)(\varepsilon/q)}\cdot R^{\lambda(1/q-1)}. \tag{6.8}$$

This is the desired inequality advertised in (6.4). □

Before introducing Block spaces, we have the following initial estimate for blocks, which places each block in a certain weighted Lebesgue space. It is worth mentioning that the bounds we obtain are independent of the choice of block. This is a crucial result that will help us when working with Block spaces, whose members are functions made up of a sequence of blocks which have been "glued" together in a ℓ^1 sense.

Lemma 6.5. *Let $\Sigma \subseteq \mathbb{R}^n$ be a closed ADR set, and abbreviate $\sigma := \mathcal{H}^{n-1}\lfloor\Sigma$. Fix an integrability exponent $q \in (1,\infty)$ along with a Muckenhoupt weight $\omega \in A_q(\Sigma,\sigma)$. Let $\varepsilon \in (0,q-1)$ be such that $\omega \in A_{q-\varepsilon}(\Sigma,\sigma)$, and select an arbitrary threshold $\lambda \in (0,\varepsilon(n-1)/(q-1))$. There exists $q_* \in (1,q)$ with the following significance. For all $q_0 \in [1,q_*)$, there exists a constant $C_0 \in (0,\infty)$, depending only on Σ, n, q, ε, λ, $[\omega]_{A_q-\varepsilon}$, and $[\omega]_{A_q}$ with the property that*

$$\|b\|_{L^{q_0}(\Sigma,\frac{\sigma(x)}{1+|x|^{n-1}})} \leq C_0 \tag{6.9}$$

for every (q,λ,ω)-block b. In particular,

$$\|b\|_{L^{1}(\Sigma,\frac{\sigma(x)}{1+|x|^{n-1}})} \leq C_0 \tag{6.10}$$

for all (q,λ,ω)-blocks, b.

Proof. Fix some (q,λ,ω)-block b on Σ. Then $b \in L^q(\Sigma,\omega)$ and according to (6.1) there exists some $x_0 \in \Sigma$ and some $R \in (0,2\operatorname{diam}(\Sigma))$ such that

$$\operatorname{supp} b \subseteq \Delta(x_0,R) \quad \text{and} \quad \|b\|_{L^q(\Sigma,\omega)} \leq R^{\lambda(1/q-1)}. \tag{6.11}$$

Fix a reference point $x_* \in \Sigma$. The geometry of the problem leads us to split the proof into three cases, each of which will produce a bound on the L^{q_0} norm of the block b, which holds for some small range of q_0's. Taking the largest bound and the smallest range in each of the three cases will complete the proof.

Case 1: ***Assume*** $R \in (0,1)$ ***and*** $|x_0 - x_*| \le 2$**.** To start,

$$\text{choose } q_0 \in [1, q-1) \text{ sufficiently close to 1 such that } \omega \text{ belongs to } A_{q/q_0}(\Sigma,\sigma). \tag{6.12}$$

That this is possible is ensured by item (1) of Proposition 2.29. Then

$$\begin{aligned}\int_\Sigma \frac{|b(x)|^{q_0}}{1+|x|^{n-1}}\,d\sigma(x) &\le CR^{n-1} \fint_{\Delta(x_0,R)} |b|^{q_0}\,d\sigma \\ &\le [\omega]_{A_{q/q_0}}^{q_0/q}\Big(\fint_{\Delta(x_0,R)} |b|^q\,d\omega\Big)^{q_0/q} \\ &\le \frac{CR^{n-1+q_0\lambda(1/q-1)}}{\omega(\Delta(x_0,R))^{q/q_0}}.\end{aligned} \tag{6.13}$$

The first inequality follows from the support condition on b in (6.11), the Ahlfors regularity of Σ, and the fact that $1+|x|^{n-1} \ge 1$ for all $x \in \Sigma$. The second inequality follows from Lemma 2.34 (whose applicability in the present case is ensured by (6.12)), and the third inequality follows from the normalization condition on b in (6.11). Next, observe that the inclusion $\Delta(x_0,R) \subseteq \Delta(x_*,3)$ holds. Indeed, for each $x \in \Delta(0,R)$,

$$|x-x_*| \le |x-x_0| + |x_0-x_*| < R+2 < 3, \tag{6.14}$$

thanks to our current hypothesis that $|x_0-x_*| < 2$ and $R < 1$. Using item (4) of Proposition 2.29, we have

$$\left(\frac{\sigma(\Delta(x_0,R))}{\sigma(\Delta(x_*,3))}\right)^{q-\varepsilon} \le [\omega]_{A_{q-\varepsilon}} \frac{\omega(\Delta(x_0,R))}{\omega(\Delta(x_*,3))}, \tag{6.15}$$

which together with the Ahlfors regularity of Σ, and after raising both sides of (6.15) to the q_0/q power implies

$$\frac{1}{\omega(\Delta(x_0,R))^{q_0/q}} \le C\left(\frac{[\omega]_{A_{q-\varepsilon}}}{\omega(\Delta(x_*,3))}\right)^{q_0/q}\left(\frac{3^{n-1}}{R^{n-1}}\right)^{(q_0/q)(q-\varepsilon)}. \tag{6.16}$$

Combining this with (6.13), we obtain

$$\int_\Sigma \frac{|b(x)|^{q_0}}{1+|x|^{n-1}}\,d\sigma(x) \le C_1 R^{(n-1)[1-(q_0/q)(q-\varepsilon)]+q_0\lambda(1/q-1)}, \tag{6.17}$$

for some constant $C_1 \in (0,\infty)$ independent of R and x_0. We want to bound the right-hand side of (6.17) independently of R. Recalling the present hypothesis that $R \in (0,1)$, if

$$(n-1)[1-(q_0/q)(q-\varepsilon)] + q_0\lambda(1/q-1) > 0, \tag{6.18}$$

then

$$R^{(n-1)[1-(q_0/q)(q-\varepsilon)]+q_0\lambda(1/q-1)} \le 1. \tag{6.19}$$

By assumption, $\varepsilon(n-1)/(q-1) > \lambda$, and a moment's reflection shows that this is equivalent to (6.18) when $q_0 := 1$. Since the dependence on q_0 is continuous, we can conclude

$$\begin{aligned}&\text{there exists } q_1 \in (1, q-1), \text{ depending only on } n, q, \varepsilon, \text{ and } \lambda\\ &\text{such that (6.18) holds for every } q_0 \in [1, q_1).\end{aligned} \tag{6.20}$$

From (6.17)–(6.20), we see for any $q_0 \in [1, q_1)$ that

$$\|b\|^{q_0}_{L^{q_0}(\Sigma, \frac{\sigma(x)}{1+|x|^{n-1}})} \le C_1 R^{(n-1)[1-(q_0/q)(q-\varepsilon)]+q_0\lambda(1/q-1)} \le C_1, \tag{6.21}$$

for some $C_1 \in (0, \infty)$, independent of the block b. Note that the monotonicity of the Muckenhoupt class (cf. (2.51)) ensures that the computations leading up to (6.17) remain valid for all $q_0 \in [1, q_1)$.

Case 2: *Assume $R \in (0,1)$ and $|x_0 - x_*| > 2$.* First, let q_0 be as in (6.12). Now, we remark by item (3) of Lemma 2.18 that $1 + |x|^{n-1} \approx |x_0 - x_*|^{n-1}$ uniformly for points $x \in \Delta(x_0, R)$, with implicit constants that depend only on n, and x_*. We may then write

$$\begin{aligned}\int_\Sigma \frac{|b(x)|^{q_0}}{1+|x|^{n-1}}\, d\sigma(x) &\le \frac{C}{|x_0 - x_*|^{n-1}} \int_{\Delta(x_0,R)} |b|^{q_0}\, d\sigma\\ &\le \frac{CR^{n-1}}{|x_0 - x_*|^{n-1}} [\omega]^{q_0/q}_{A_{q/q_0}} \Big(⨍_{\Delta(x_0,R)} |b|^q\, d\omega\Big)^{q_0/q}\\ &\le \frac{CR^{n-1+q_0\lambda(1/q-1)}}{|x_0 - x_*|^{n-1}} \frac{1}{\omega(\Delta(x_0,R))^{q_0/q}}.\end{aligned} \tag{6.22}$$

The second inequality above follows from creating an integral average, the Ahlfors regularity of Σ, and Lemma 2.34 with $f := |b|^{q_0}$, $w := \omega$, and $p := q/q_0 \in (1, \infty)$, bearing in mind (6.12). The third inequality in (6.22) follows the normalization condition on b in (6.11). Next, note that

$$\Delta(x_0, R) \subseteq \Delta(x_*, (3/2)|x_0 - x_*|). \tag{6.23}$$

Indeed, for any $x \in \Delta(x_0, R)$ we have

$$\begin{aligned}|x - x_*| &\le |x - x_0| + |x_0 - x_*| < R + |x_0 - x_*|\\ &< 1 + |x_0 - x_*| < (3/2)|x_0 - x_*|,\end{aligned} \tag{6.24}$$

as we are currently working under the assumptions $R < 1$ and $|x_0 - x_*| > 2$. This justifies the inclusion in (6.23). In turn, since $\omega \in A_{q-\varepsilon}(\Sigma, \sigma)$, we find that (6.23) and item (4) of

Proposition 2.29 allow us to estimate

$$\left(\frac{\sigma(\Delta(x_0,R))}{\sigma(\Delta(x_*,(3/2)|x_0-x_*|))}\right)^{q-\varepsilon} \le [\omega]_{A_{q-\varepsilon}} \frac{\omega(\Delta(x_0,R))}{\omega(\Delta(x_*,(3/2)|x_0-x_*|))}. \tag{6.25}$$

Using this, the fact that we presently have $|x_0 - x_*| > 2$, and the Ahlfors regularity of Σ, we obtain (after rearranging (6.25) and raising both sides to the q_0/q power)

$$\frac{1}{\omega(\Delta(x_0,R))^{q_0/q}} \le C\left(\frac{[\omega]_{A_{q-\varepsilon}}}{\omega(\Delta(x_*,3))}\right)^{q_0/q}\left(\frac{|x_0-x_*|^{n-1}}{R^{n-1}}\right)^{(q_0/q)(q-\varepsilon)}. \tag{6.26}$$

Combining this result with (6.22) yields

$$\int_\Sigma \frac{|b(x)|^{q_0}}{1+|x|^{n-1}}\,d\sigma(x) \le C\frac{R^{(n-1)[1-(q_0/q)(q-\varepsilon)]+q_0\lambda(1/q-1)}}{|x_0-x_*|^{(n-1)[1-(q_0/q)(q-\varepsilon)]}}, \tag{6.27}$$

for some constant $C \in (0,\infty)$ depending only on the Ahlfors regularity constants of Σ, n, x_*, ω, q, and q_0.

As in Case 1, we seek to bound the left-hand side of (6.27) independently of R and x_0, and hence independently of the block b. Because we are presently working under the assumption that $R < 1$, we have

$$R^{(n-1)[1-(q_0/q)(q-\varepsilon)]+q_0\lambda(1/q-1)} < 1 \tag{6.28}$$

provided (6.18) holds as before. With q_1 as in (6.20), we find that (6.17) holds for all $q_0 \in [1,q_1)$. Since we are also assuming that $|x_0 - x_*| > 2$, if $(n-1)(1-(q_0/q)(q-\varepsilon)) > 0$, which is equivalent to the condition that $q_0 < q/(q-\varepsilon)$, then

$$\frac{1}{|x_0-x_*|^{(n-1)[1-(q_0/q)(q-\varepsilon)]}} \le \frac{1}{2^{(n-1)[1-(q_0/q)(q-\varepsilon)]}}, \tag{6.29}$$

and the above bound is clearly independent of x_0. Therefore, choosing

$$q_2 \in \left(1, \min\left\{q_1, \frac{q}{q-\varepsilon}\right\}\right) \tag{6.30}$$

to begin with makes (6.29) and (6.28) true for all $q_0 \in [1,q_2)$.

The monotonicity of the Muckenhoupt class (cf. (2.51)) ensures that for all $q_0 \in [1,q_2)$ we still have $\omega \in A_{q/q_0}(\Sigma,\sigma)$, and hence the computations leading up to the bound in (6.27) remain valid for any $q_0 \in [1,q_2)$. It follows from (6.27)–(6.29) that

$$\|b\|^{q_0}_{L^{q_0}(\Sigma,\frac{\sigma(x)}{1+|x|^{n-1}})} \le \frac{C}{2^{(n-1)[1-(q_0/q)(q-\varepsilon)]}} =: C_2, \tag{6.31}$$

and $C_2 \in (0,\infty)$, witnessed to be independent of the block b.

Case 3: *Assume R* ≥ 1. By Lemma 5.5, there exists $q_3 \in (1, q]$ such that

$$L^q(\Sigma, \omega) \hookrightarrow L^{q_0}\left(\Sigma, \frac{\sigma(x)}{1 + |x|^{n-1}}\right) \tag{6.32}$$

continuously for each $q_0 \in [1, q_3)$. As a result, for each such q_0, there exists some constant $C_3 \in (0, \infty)$, independent of b, such that

$$\|b\|_{L^{q_0}(\Sigma, \frac{\sigma(x)}{1+|x|^{n-1}})} \leq C_0 \|b\|_{L^q(\Sigma,\omega)} \leq C_3 R^{\lambda(1/q-1)} \leq C_3. \tag{6.33}$$

The penultimate inequality comes from (6.11), while the last inequality follows from the fact that $\lambda(1/q - 1) < 0$, which forces $R^{\lambda(1/q-1)} \leq 1$, given that we are presently assuming $R \geq 1$.

To complete the proof, recall (6.21), (6.31), and (6.33). As such, choosing $q_* := \min\{q_2, q_3\}$ and $C_0 := \max\{C_1^{1/q_0}, C_2^{1/q_0}, C_3\}$ establishes the desired result. □

Because of the bound in (6.10) and bearing in mind Remark 5.4, we find that singular integral operators of Calderón–Zygmund type are well-defined objects when acting on (q, λ, ω)-blocks.

By definition, (q, λ, ω)-blocks belong to $L^q(\Sigma, \omega)$, but as the following corollary shows, (q, λ, ω)-blocks belong to the unweighted Lebesgue space $L^{q_0}(\Sigma, \sigma)$ for a small range of q_0, and in particular for $q_0 = 1$. A more general result of this flavor is obtained later in Proposition 6.17.

Corollary 6.6. *Let $\Sigma \subseteq \mathbb{R}^n$ be a closed ADR set, and write $\sigma := \mathcal{H}^{n-1}\lfloor\Sigma$. Fix an integrability exponent $q \in (1, \infty)$ along with a Muckenhoupt weight $\omega \in A_q(\Sigma, \sigma)$. Let $\varepsilon \in (0, q - 1)$ be such that $\omega \in A_{q-\varepsilon}(\Sigma, \sigma)$ and fix a threshold $\lambda \in (0, \varepsilon(n - 1)/(q - 1))$. Then there exists $q_* \in (1, q)$ with the following significance. For all $q_0 \in [1, q_*)$, every (q, λ, ω)-block on Σ belongs to $L^1(\Sigma, \frac{\sigma(x)}{1+|x|^{n-1}}) \cap L^{q_0}(\Sigma, \sigma)$.*

Proof. Fix a (q, λ, ω)-block b on Σ. Then there exist a location $x_0 \in \Sigma$ and a scale $R \in (0, 2\,\mathrm{diam}(\Sigma))$ such that $\mathrm{supp}\, b \subseteq \Delta(x_0, R)$. Define the quantity $C_0 := \inf_{x \in \Delta(x_0,R)} \frac{1}{1+|x|^{n-1}}$. Then $0 < C_0 \leq \frac{1}{1+|x|^{n-1}}$ for all $x \in \mathrm{supp}\, b$. Lemma 6.5 implies that there exists $q_* \in (1, q)$ such that $b \in L^{q_0}(\Sigma, \frac{\sigma(x)}{1+|x|^{n-1}})$ for all $q_0 \in [1, q_*)$. Subsequently,

$$\begin{aligned} \|b\|_{L^{q_0}(\Sigma,\sigma)} &= \left(\int_{\Delta(x_0,R)} |b|^{q_0}\, d\sigma \right)^{1/q_0} \\ &\leq C_0^{-1} \|b\|_{L^{q_0}(\Sigma, \frac{\sigma(x)}{1+|x|^{n-1}})} < +\infty. \end{aligned} \tag{6.34}$$

Hence, $b \in L^{q_0}(\Sigma, \sigma)$ as well. □

We are now ready to introduce the definition of a Muckenhoupt-weighted Block space.

Definition 6.7. Let $\Sigma \subseteq \mathbb{R}^n$ be a closed ADR set, and write $\sigma := \mathcal{H}^{n-1}\lfloor\Sigma$. Fix an integrability exponent $q \in (1,\infty)$ along with a Muckenhoupt weight $\omega \in A_q(\Sigma,\sigma)$. Let $\varepsilon \in (0, q-1)$ be such that $\omega \in A_{q-\varepsilon}(\Sigma,\sigma)$.[1] Also, fix an arbitrary threshold $\lambda \in [0, \varepsilon(n-1)/(q-1))$. Suppose

$$g = \sum_{j=1}^{\infty} \lambda_j b_j \quad \text{in } L^1\Big(\Sigma, \frac{\sigma(x)}{1+|x|^{n-1}}\Big) \tag{6.35}$$

for some $\{\lambda_j\}_{j\in\mathbb{N}} \in \ell^1$ and $\{b_j\}_{j\in\mathbb{N}}$ a sequence of (q,λ,ω)-blocks. Call such a series a `block-representation` of g. Define the `Muckenhoupt-weighted Block space` as the set

$$\mathcal{B}^{q,\lambda}(\Sigma,\omega) := \Big\{g \in L^1\Big(\Sigma, \frac{\sigma(x)}{1+|x|^{n-1}}\Big) : \text{there exists a block-representation of } g\Big\}. \tag{6.36}$$

A note on terminology: when we say "block function" we mean a function f on Σ such that f belongs to a Block space, as opposed to simply a "block," which will mean a function on Σ in the sense of Definition 6.1.

There are a couple of observations to make in light of Definition 6.7. The first of which is that any sequence of blocks, together with a sequence in ℓ^1 may be suitably "glued" together to create a block function.

Lemma 6.8. *Let $\Sigma \subseteq \mathbb{R}^n$ be a closed ADR set, and write $\sigma := \mathcal{H}^{n-1}\lfloor\Sigma$. Fix an integrability exponent $q \in (1,\infty)$ along with a Muckenhoupt weight $\omega \in A_q(\Sigma,\sigma)$. Next, select any $\varepsilon \in (0,q-1)$ for which $\omega \in A_{q-\varepsilon}(\Sigma,\sigma)$, and fix an arbitrary threshold $\lambda \in (0,\varepsilon(n-1)/(q-1))$.*

Then for all $\{\lambda_j\}_{j\in\mathbb{N}} \in \ell^1$ and all families $\{b_j\}_{j\in\mathbb{N}}$ of (q,λ,ω)-blocks on Σ, the series

$$\sum_{j=1}^{\infty} \lambda_j b_j \quad \textit{converges in } L^1\Big(\Sigma, \frac{\sigma(x)}{1+|x|^{n-1}}\Big). \tag{6.37}$$

Proof. Let $\{\lambda_j\}_{j\in\mathbb{N}} \in \ell^1$ and let $\{b_j\}_{j\in\mathbb{N}}$ be a sequence of (q,λ,ω)-blocks on Σ. Looking at the tail end of the series, we find that

$$\begin{aligned}\lim_{N\to\infty}\Big\|\sum_{j=N}^{\infty}\lambda_j b_j\Big\|_{L^1(\Sigma,\frac{\sigma(x)}{1+|x|^{n-1}})} &\le \lim_{N\to\infty}\sum_{j=N}^{\infty}|\lambda_j|\|b_j\|_{L^1(\Sigma,\frac{\sigma(x)}{1+|x|^{n-1}})}\\ &\le C\lim_{N\to\infty}\sum_{j=N}^{\infty}|\lambda_j| = 0,\end{aligned} \tag{6.38}$$

where the last inequality is a consequence of Lemma 6.5 in the case that $q_0 = 1$. The desired conclusion follows. □

1 Note that the existence of such an ε is guaranteed by item (1) of Proposition 2.29, though this ε need not be the same as the one guaranteed by the theory; cf. Remark 6.4 for a similar discussion in this regard.

Our next observation is that Muckenhoupt-weighted Block spaces are not just sets of functions, but come with a vector space structure.

Proposition 6.9. *Let $\Sigma \subseteq \mathbb{R}^n$ be a closed ADR set and write $\sigma := \mathcal{H}^{n-1}\lfloor\Sigma$. Fix an integrability exponent $q \in (1,\infty)$ along with a Muckenhoupt weight $\omega \in A_q(\Sigma,\sigma)$. Finally, choose $\varepsilon \in (0, q-1)$ such that $\omega \in A_{q-\varepsilon}(\Sigma,\sigma)$, and select an arbitrary threshold $\lambda \in (0, \varepsilon(n-1)/(q-1))$.*

Then the space $\mathcal{B}^{q,\lambda}(\Sigma,\omega)$, defined in (6.36), is a vector space over $\mathbb{C}$.

Proof. Let $f_1, f_2 \in \mathcal{B}^{q,\lambda}(\Sigma,\omega)$, and fix two constants $\alpha^{(1)}, \alpha^{(2)} \in \mathbb{C}$. Then there exist block representations $\sum_{j=1}^{\infty} \lambda_j^{(1)} b_j^{(1)}$ and $\sum_{j=1}^{\infty} \lambda_j^{(2)} b_j^{(2)}$ of f_1 and f_2, respectively. Subsequently, we find that the series $\sum_{k=1}^{2}\sum_{j=1}^{\infty} \alpha^k \lambda_j^{(k)} b_j^{(k)}$ converges to $\alpha^{(1)} f_1 + \alpha^{(2)} f_2$ in $L^1(\Sigma, \frac{\sigma(x)}{1+|x|^{n-1}})$. Also, note that

$$\{\alpha^{(k)}\lambda_j^{(k)}\}_{(k,j)\in\{1,2\}\times\mathbb{N}} \in \ell^1 \tag{6.39}$$

as $\{\lambda_j^k\}_{j\in\mathbb{N}} \in \ell^1$ for each fixed $k \in \{1,2\}$. Indeed,

$$\sum_{k=1}^{2}\sum_{j=1}^{\infty} |\alpha^{(k)}\lambda_j^k| \leq |\alpha^{(1)}| \cdot \|\{\lambda_j^{(1)}\}\|_{\ell^1} + |\alpha^{(2)}| \cdot \|\{\lambda_j^{(2)}\}\|_{\ell^1} < +\infty. \tag{6.40}$$

Hence, $\sum_{k=1}^{2}\sum_{j=1}^{\infty} \alpha^k \lambda_j^{(k)} b_j^{(k)}$ is a block representation of $\alpha^{(1)} f_1 + \alpha^{(2)} f_2$, which by definition of $\mathcal{B}^{q,\lambda}(\Sigma,\omega)$ implies that $\alpha^{(1)} f_1 + \alpha^{(2)} f_2 \in \mathcal{B}^{q,\lambda}(\Sigma,\omega)$, as desired. □

The Block space $\mathcal{B}^{q,\lambda}(\Sigma,\omega)$ comes equipped with a norm made explicit below.

Proposition 6.10. *Let $\Sigma \subseteq \mathbb{R}^n$ be a closed ADR set and write $\sigma := \mathcal{H}^{n-1}\lfloor\Sigma$. Fix $q \in (1,\infty)$ and $\omega \in A_q(\Sigma,\sigma)$. Next, choose $\varepsilon \in (0, q-1)$ such that $\omega \in A_{q-\varepsilon}(\Sigma,\sigma)$ and fix a threshold $\lambda \in (0, \varepsilon(n-1)/(q-1))$. Then the function $\|\cdot\|_{\mathcal{B}^{q,\lambda}(\Sigma,\omega)} : \mathcal{B}^{q,\lambda}(\Sigma,\omega) \to [0,\infty)$ defined by*

$$\|g\|_{\mathcal{B}^{q,\lambda}(\Sigma,\omega)} := \inf\left\{\sum_{j=1}^{\infty} |\lambda_j| : \sum_{j=1}^{\infty} \lambda_j b_j \text{ is a block representation of } g\right\} \tag{6.41}$$

for each $g \in \mathcal{B}^{q,\lambda}(\Sigma,\omega)$ is a norm on $\mathcal{B}^{q,\lambda}(\Sigma,\omega)$.

Proof. Let $g \in \mathcal{B}^{q,\lambda}(\Sigma,\sigma)$. The fact that the set over which the infimum is being taken in (6.41) is nonempty and bounded below by 0 implies that the infimum exists as a well-defined number in $[0,\infty)$. Taking an infimum over all block representations of g ensures that the norm is independent of the choice of representation. To see that this is indeed a norm, first note that positive definiteness follows from the definition. Next, we make the following claim:

$$\|\alpha g\|_{\mathcal{B}^{q,\lambda}(\Sigma,\omega)} = |\alpha|\|g\|_{\mathcal{B}^{q,\lambda}(\Sigma,\omega)} \quad \text{for all } \alpha \in \mathbb{C}. \tag{6.42}$$

Indeed, if $a = 0$, the claim is trivial. If $a \in \mathbb{C} \setminus \{0\}$, then we see that

$$\begin{aligned}
\|ag\|_{\mathcal{B}^{q,\lambda}(\Sigma,\omega)} &= \inf\left\{\sum_{j=1}^{\infty}|\lambda_j| : \sum_{j=1}^{\infty}\lambda_j b_j \text{ is a block representation of } ag\right\} \\
&= \inf\left\{|a|\sum_{j=1}^{\infty}\left|\frac{\lambda_j}{a}\right| : \sum_{j=1}^{\infty}\frac{\lambda_j}{a}b_j \text{ is a block representation of } g\right\} \\
&= |a|\|g\|_{\mathcal{B}^{q,\lambda}(\Sigma,\omega)},
\end{aligned} \tag{6.43}$$

establishing the claim.

Finally, to see that the triangle inequality holds, let $g_1, g_2 \in \mathcal{B}^{q,\lambda}(\Sigma,\omega)$. Note that the set of all block representations of g_1+g_2 contains the set of sums of block-representations of g_1 with block representations of g_2. Consequently,

$$\begin{aligned}
\|g_1 + g_2\|_{\mathcal{B}^{p,\lambda}(\Sigma,\omega)} &= \inf\left\{\sum_{j=1}^{\infty}|\lambda_j| : \sum_{j=1}^{\infty}\lambda_j b_j \text{ is a block representation of } g_1 + g_2\right\} \qquad (6.44) \\
&\leq \inf\left\{\sum_{j=1}^{\infty}|\lambda_j^{(1)}| + |\lambda_j^{(2)}| : \sum_{j=1}^{\infty}\lambda_j^{(1)}b_j^{(1)} \text{ and } \sum_{j=1}^{\infty}\lambda_j^{(2)}b_j^{(2)}\right. \\
&\qquad \left.\text{are block representations of } g_1 \text{ and } g_2, \text{ respectively}\right\} \\
&= \|g_1\|_{\mathcal{B}^{p,\lambda}(\Sigma,\omega)} + \|g_2\|_{\mathcal{B}^{p,\lambda}(\Sigma,\omega)},
\end{aligned}$$

as wanted □

The Block space $\mathcal{B}^{q,\lambda}(\Sigma,\omega)$ is not only a normed vector space, but also a complete vector space with respect to the norm defined in (6.41).

Theorem 6.11. *Let $\Sigma \subseteq \mathbb{R}^n$ be a closed* ADR *set and write $\sigma := \mathcal{H}^{n-1}\lfloor\Sigma$. Fix $q \in (1,\infty)$ and $\omega \in A_q(\Sigma,\sigma)$. Let $\varepsilon \in (0, q-1)$ be such that ω belongs to $A_{q-\varepsilon}(\Sigma,\sigma)$ and select some threshold $\lambda \in (0, \varepsilon(n-1)/(q-1))$.*

Then the space $\mathcal{B}^{q,\lambda}(\Sigma,\omega)$, equipped with the norm defined in (6.41) *is a Banach space.*

Proof. We have seen that $(\mathcal{B}^{q,\lambda}(\Sigma,\omega), \|\cdot\|_{\mathcal{B}^{q,\lambda}(\Sigma,\omega)})$ is a normed vector space (Proposition 6.9 and Proposition 6.10), so it remains only to show that this space is complete. To that end, let $\{f_j\}_{j\in\mathbb{N}}$ be a Cauchy sequence in $\mathcal{B}^{q,\lambda}(\Sigma,\omega)$. Then for every $k \in \mathbb{N}$, there exists $j_k \in \mathbb{N}$ such that

$$\|f_j - f_\ell\|_{\mathcal{B}^{q,\lambda}(\Sigma,\omega)} < 2^{-k} \quad \text{whenever } j, \ell \geq j_k. \tag{6.45}$$

Without loss of generality, we may assume $1 \leq j_1 < j_2 < \cdots$. In particular, we have

$$\|f_{j_k} - f_{j_{k+1}}\|_{\mathcal{B}^{q,\lambda}(\Sigma,\omega)} < 2^{-k} \quad \text{for all } k \in \mathbb{N}. \tag{6.46}$$

This and (6.41) further imply that for each $k \in \mathbb{N}$ there exist a sequence $\{\lambda_i^k\}_{i\in\mathbb{N}} \in \ell^1$ and a sequence of (q,λ,ω)-blocks $\{b_i^k\}_{i\in\mathbb{N}}$ on Σ such that $f_{j_k} - f_{j_{k+1}} = \sum_{i=1}^{\infty} \lambda_i^k b_i^k$, with convergence of the series in $L^1(\Sigma, \frac{\sigma(x)}{1+|x|^{n-1}})$ and

$$\sum_{i=1}^{\infty} |\lambda_i^k| < 2^{-k}. \tag{6.47}$$

Now, consider

$$f := f_{j_1} - \sum_{k=1}^{\infty}\sum_{i=1}^{\infty} \lambda_i^k b_i^k. \tag{6.48}$$

Note by Remark 6.8 that the above series converges in $L^1(\Sigma, \frac{\sigma(x)}{1+|x|^{n-1}})$ because $\{\lambda_i^k\}_{(i,k)\in\mathbb{N}^2}$ is a sequence in ℓ^1. Indeed, by (6.47) we have

$$\sum_{k=1}^{\infty}\sum_{i=1}^{\infty} |\lambda_i^k| < 2. \tag{6.49}$$

Thus, f is meaningfully defined as a function in $L^1(\Sigma, \frac{\sigma(x)}{1+|x|^{n-1}})$, and by definition of $\mathcal{B}^{q,\lambda}(\Sigma,\omega)$ and its vector space structure, we can see that f is in $\mathcal{B}^{q,\lambda}(\Sigma,\omega)$. We will now show that there exists a subsequence of the Cauchy sequence $\{f_j\}_{j\in\mathbb{N}}$ that converges to f in $\mathcal{B}^{q,\lambda}(\Sigma,\omega)$. This will imply that the entire sequence converges in $\mathcal{B}^{q,\lambda}(\Sigma,\omega)$. To that end, note that for all $N \in \mathbb{N}$, we may write

$$f_{j_{N+1}} = f_{j_1} + \sum_{k=1}^{N} (f_{j_{k+1}} - f_{j_k}) = f_{j_1} - \sum_{k=1}^{N}\sum_{i=1}^{\infty} \lambda_i^k b_i^k. \tag{6.50}$$

This implies

$$f - f_{j_{N+1}} = - \sum_{k=N+1}^{\infty}\sum_{i=1}^{\infty} \lambda_i^k b_i^k. \tag{6.51}$$

Hence,

$$\begin{aligned} \limsup_{N\to\infty} \|f - f_{j_{N+1}}\|_{\mathcal{B}^{q,\lambda}(\Sigma,\omega)} &\le \limsup_{N\to\infty} \sum_{k=N+1}^{\infty}\sum_{i=1}^{\infty} |\lambda_i^k| \\ &\le \lim_{N\to\infty} \sum_{k=N+1}^{\infty} 2^{-k} \\ &= \lim_{N\to\infty} 2^{-N} = 0, \end{aligned} \tag{6.52}$$

completing the proof. □

It is then straightforward to see that given a block representation for a block function that the partial sums converge in norm to said block function.

Proposition 6.12. *Let $\Sigma \subseteq \mathbb{R}^n$ be a closed ADR set, and write $\sigma := \mathcal{H}^{n-1}\lfloor\Sigma$. Fix an integrability exponent $q \in (1,\infty)$ along with a Muckenhoupt weight $\omega \in A_q(\Sigma,\sigma)$. Finally, choose $\varepsilon \in (0, q-1)$ such that $\omega \in A_{q-\varepsilon}(\Sigma,\sigma)$, and fix some arbitrary threshold $\lambda \in (0, \varepsilon(n-1)/(q-1))$.*

If $g \in \mathcal{B}^{q,\lambda}(\Sigma,\omega)$ and has a block representation $g = \sum_{j=1}^{\infty}\lambda_j b_j$, then the partial sums, defined for each $m \in \mathbb{N}$ as $g_m := \sum_{j=1}^{m}\lambda_j b_j$, converge to g in $\mathcal{B}^{q,\lambda}(\Sigma,\omega)$ as $m \to \infty$.

Proof. Let $g \in \mathcal{B}^{q,\lambda}(\Sigma,\sigma)$ with block representation $g = \sum_{j=1}^{\infty}\lambda_j b_j$. In particular, $\{\lambda_j\}_{j\in\mathbb{N}} \in \ell^1$. By the vector space property of weighted Block spaces (see Remark 6.9), it follows that $g - g_m \in \mathcal{B}^{q,\lambda}(\Sigma,\omega)$ for each $m \in \mathbb{N}$. Moreover,

$$g - g_m = \sum_{j=m+1}^{\infty}\lambda_j b_j \in L^1\Big(\Sigma, \frac{\sigma(x)}{1+|x|^{n-1}}\Big) \tag{6.53}$$

by Definition 6.7, and hence using (6.41), we arrive at

$$\|g - g_m\|_{\mathcal{B}^{q,\lambda}(\Sigma,\omega)} \le \sum_{j=m+1}^{\infty}|\lambda_j| \to 0 \quad \text{as } m \to \infty, \tag{6.54}$$

since $\{\lambda_j\}_{j\in\mathbb{N}} \in \ell^1$ by assumption. □

We now present several properties that Block spaces enjoy. The first property states that pointwise multiplication by an essentially bounded function is a well-defined, linear, and bounded operation on a Muckenhoupt-weighted Block space.

Proposition 6.13. *Let $\Sigma \subseteq \mathbb{R}^n$ be a closed ADR set and write $\sigma := \mathcal{H}^{n-1}\lfloor\Sigma$. Fix an integrability exponent $q \in (1,\infty)$ along with a Muckenhoupt weight $\omega \in A_q(\Sigma,\sigma)$. Pick any $\varepsilon \in (0, q-1)$ with the property that $\omega \in A_{q-\varepsilon}(\Sigma,\sigma)$, and fix an arbitrary threshold $\lambda \in (0, \varepsilon(n-1)/(q-1))$. Let $g \in L^\infty(\Sigma,\sigma)$ and let M_g denote the operation of pointwise multiplication by g. That is, for any σ-measurable function $f : \Sigma \to \mathbb{C}$,*

$$(M_g f)(x) := g(x)f(x) \quad \text{for all } x \in \Sigma. \tag{6.55}$$

Then the operator M_g is well-defined, linear, and bounded on $\mathcal{B}^{q,\lambda}(\Sigma,\omega)$ with

$$\|M_g\|_{\mathcal{B}^{q,\lambda}(\Sigma,\omega)\to\mathcal{B}^{q,\lambda}(\Sigma,\omega)} \le \|g\|_{L^\infty(\Sigma,\sigma)}. \tag{6.56}$$

Proof. It is clear from definition that M_g is linear. As far as boundedness of M_g is concerned, this is trivially true if $g = 0$. Henceforth, assume $\|g\|_{L^\infty(\Sigma,\sigma)} > 0$. Now, let $h \in \mathcal{B}^{q,\lambda}(\Sigma,\omega)$. Then by Definition 6.7 there exist a collection $\{b_j\}_{j\in\mathbb{N}}$ of (q,λ,ω)-blocks, and a sequence $\{\lambda_j\}_{j\in\mathbb{N}} \in \ell^1$ such that $h = \sum_{j=1}^{\infty}\lambda_j b_j$ in $L^1(\Sigma, \frac{\sigma(x)}{1+|x|^{n-1}})$. We can then write

$$M_g h = gh = \sum_{j=1}^{\infty} \widetilde{\lambda_j}\widetilde{b_j}, \tag{6.57}$$

where $\widetilde{\lambda_j} := \|g\|_{L^\infty(\Sigma,\sigma)}\lambda_j$ and $\widetilde{b_j} := \|g\|^{-1}_{L^\infty(\Sigma,\sigma)} g b_j$. Since each b_j is a (q,λ,ω)-block, for each $j \in \mathbb{N}$ there exist $x_j \in \Sigma$ and $R_j \in (0, 2\,\mathrm{diam}(\Sigma))$ such that

$$\begin{aligned} &\operatorname{supp} b_j \subseteq \Delta(x_j, R_j), \quad \text{and} \\ &\|b_j\|_{L^q(\Sigma,\omega)} \le R_j^{\lambda(1/q-1)}. \end{aligned} \tag{6.58}$$

It follows that $\widetilde{b_j}$ is also a (q,λ,ω)-block for each $j \in \mathbb{N}$, since

$$\begin{aligned} &\operatorname{supp} \widetilde{b_j} \subseteq \operatorname{supp} b_j \subseteq \Delta(x_j, R_j), \quad \text{and} \\ &\|\widetilde{b_j}\|_{L^q(\Sigma,\omega)} = \|g\|^{-1}_{L^\infty(\Sigma,\sigma)}\|g b_j\|_{L^q(\Sigma,\omega)} \le \|b_j\|_{L^q(\Sigma,\omega)} \le R_j^{\lambda(1/q-1)}. \end{aligned} \tag{6.59}$$

Note also that $\{\widetilde{\lambda_j}\}_{j\in\mathbb{N}} \in \ell^1$, because

$$\sum_{j=1}^{\infty} |\widetilde{\lambda_j}| = \|g\|_{L^\infty(\Sigma,\sigma)} \sum_{j=1}^{\infty} |\lambda_j| < +\infty. \tag{6.60}$$

By Lemma 6.8, $M_g h$ has a block representation, and hence $M_g h \in \mathcal{B}^{q,\lambda}(\Sigma,\omega)$ with

$$\|M_g h\|_{\mathcal{B}^{q,\lambda}(\Sigma,\omega)} \le \|g\|_{L^\infty(\Sigma,\sigma)} \sum_{j=1}^{\infty} |\lambda_j|. \tag{6.61}$$

Taking an infimum over all block representations of h in the above inequality yields

$$\|M_g h\|_{\mathcal{B}^{q,\lambda}(\Sigma,\omega)} \le \|g\|_{L^\infty(\Sigma,\sigma)} \|h\|_{\mathcal{B}^{q,\lambda}(\Sigma,\omega)}. \tag{6.62}$$

The arbitrary nature of h implies that (6.56) holds. □

The second property we present states that a Block space is a lattice of functions.

Proposition 6.14. *Let $\Sigma \subseteq \mathbb{R}^n$ be a closed* ADR *set and write $\sigma := \mathcal{H}^{n-1}\lfloor\Sigma$. Fix an integrability exponent $q \in (1,\infty)$ along with a Muckenhoupt weight $\omega \in A_q(\Sigma,\sigma)$. Choose some $\varepsilon \in (0, q-1)$ for which $\omega \in A_{q-\varepsilon}(\Sigma,\sigma)$, and select an arbitrary threshold $\lambda \in (0, \varepsilon(n-1)/(q-1))$.*

Suppose f is a σ-measurable function and $g \in \mathcal{B}^{q,\lambda}(\Sigma,\sigma)$ with $|f| \le g$ at σ-a. e. point on Σ. Then $f \in \mathcal{B}^{q,\lambda}(\Sigma,\omega)$ with $\|f\|_{\mathcal{B}^{q,\lambda}(\Sigma,\omega)} \le \|g\|_{\mathcal{B}^{q,\lambda}(\Sigma,\omega)}$.

Proof. Let f and g be as above. Define $h : \Sigma \to \mathbb{R}$ by setting for each $x \in \Sigma$,

$$h(x) := \begin{cases} f(x)/g(x) & \text{if } f(x) \neq 0, \\ 0 & \text{if } g(x) = 0. \end{cases} \tag{6.63}$$

Then h is σ-measurable and $|h| \leq 1$ for σ-a. e. point on Σ. Moreover, from (6.63) and the fact that f vanishes whenever g vanishes, we have that $f = hg$ at σ-a. e. point on Σ. As a result of this and Proposition 6.13,

$$\|f\|_{\mathcal{B}^{q,\lambda}(\Sigma,\omega)} \leq \|h\|_{L^\infty(\Sigma,\sigma)}\|g\|_{\mathcal{B}^{q,\lambda}(\Sigma,\omega)} \leq \|g\|_{\mathcal{B}^{q,\lambda}(\Sigma,\omega)}, \tag{6.64}$$

which is the desired result. □

This third property of Block spaces that we present entails that the block norms of a function and its absolute value are equal. This is something true for both Morrey spaces and Lebesgue spaces, which is immediate from the definitions of their respective norms.

Corollary 6.15. *Let $\Sigma \subseteq \mathbb{R}^n$ be a closed ADR set, and write $\sigma := \mathcal{H}^{n-1}\lfloor\Sigma$. Fix an integrability exponent $q \in (1,\infty)$ along with a Muckenhoupt weight $\omega \in A_q(\Sigma,\sigma)$. Choose $\varepsilon \in (0, q-1)$ such that $\omega \in A_{q-\varepsilon}(\Sigma,\sigma)$, and fix $\lambda \in (0, \varepsilon(n-1)/(q-1))$. If $f \in \mathcal{B}^{q,\lambda}(\Sigma,\omega)$, then*

$$\|f\|_{\mathcal{B}^{q,\lambda}(\Sigma,\omega)} = \||f|\|_{\mathcal{B}^{q,\lambda}(\Sigma,\omega)}. \tag{6.65}$$

Proof. Let $f \in \mathcal{B}^{q,\lambda}(\Sigma,\omega)$. Then $||f|| \leq |f|$, and applying Proposition 6.14, we have the membership $|f| \in \mathcal{B}^{q,\lambda}(\Sigma,\omega)$ with $\||f|\|_{\mathcal{B}^{q,\lambda}(\Sigma,\omega)} \leq \|f\|_{\mathcal{B}^{q,\lambda}(\Sigma,\omega)}$. On the other hand, since $|f| \in \mathcal{B}^{q,\lambda}(\Sigma,\omega)$ as we have just seen, and because it holds that $|f| \leq ||f||$, we obtain via Proposition 6.14 the reverse inequality $\|f\|_{\mathcal{B}^{q,\lambda}(\Sigma,\omega)} \leq \||f|\|_{\mathcal{B}^{q,\lambda}(\Sigma,\omega)}$, establishing the claim. □

Recall that for Muckenhoupt-weighted Morrey spaces, when the threshold λ is equal to zero, we obtain a Lebesgue space, showing that Lebesgue spaces are an endpoint on the scale of Morrey spaces; see (4.4). Interestingly, Lebesgue spaces are also an endpoint on the scale of Block spaces, though it is not as obvious to see from the block norm (6.41) as was the case for Morrey spaces, as witnessed below.

Proposition 6.16. *Suppose $\Sigma \subseteq \mathbb{R}^n$ is a closed Ahlfors regular set, and abbreviate $\sigma := \mathcal{H}^{n-1}\lfloor\Sigma$. Fix an integrability exponent $q \in (1,\infty)$, and a Muckenhoupt weight $\omega \in A_q(\Sigma,\sigma)$. Then*

$$\mathcal{B}^{q,0}(\Sigma,\omega) = L^q(\Sigma,w) \quad \textit{with coincidence of norms.} \tag{6.66}$$

Proof. Suppose $g \in \mathcal{B}^{q,0}(\Sigma,\omega)$. Then for any $\varepsilon > 0$ there exists $\{\lambda_j\}_{j\in\mathbb{N}} \in \ell^1$ and $\{b_j\}_{j\in\mathbb{N}}$, a sequence of $(q,0,\omega)$-blocks on Σ, for which

$$g = \sum_{j=1}^{\infty}\lambda_j b_j \quad \text{in } L^1\Big(\Sigma, \frac{\sigma(x)}{1+|x|^{n-1}}\Big) \text{ with } \sum_{j=1}^{\infty}|\lambda_j| < (1+\varepsilon)\|g\|_{\mathcal{B}^{q,0}(\Sigma,\omega)}. \tag{6.67}$$

For each $m \in \mathbb{N}$, define $g_m := \sum_{j=1}^{m}\lambda_j b_j$. From (6.1) with $\lambda = 0$, we see that the property of being a $(q,0,\omega)$-block on Σ entails $\|b_j\|_{L^q(\Sigma,\omega)} \leq 1$ for each $j \in \mathbb{N}$. Subsequently,

$$\begin{aligned}\|g_m\|_{L^q(\Sigma,\omega)} &\le \sum_{j=1}^{m} |\lambda_j| \|b_j\|_{L^q(\Sigma,\omega)} \\ &\le \sum_{j=1}^{m} |\lambda_j| \le \sum_{j=1}^{\infty} |\lambda_j| \quad \text{for each } m \in \mathbb{N}.\end{aligned} \tag{6.68}$$

A similar estimate as above shows that $\{g_m\}_{m\in\mathbb{N}}$ is Cauchy in $L^q(\Sigma,\omega)$, which implies that there exists $\widetilde{g} \in L^q(\Sigma,\omega)$ such that

$$g_m \to \widetilde{g} \quad \text{in } L^q(\Sigma,\omega) \text{ as } m \to \infty. \tag{6.69}$$

Now, (6.67), implies that there exists a subsequence $\{m_k\}_{k\in\mathbb{N}} \subseteq \mathbb{N}$ such that $g_{m_k} \to g$ pointwise σ-a. e. on Σ as $k \to \infty$. Additionally, (6.69) implies there exists a sub-subsequence $\{m_{k_\ell}\}_{\ell\in\mathbb{N}} \subseteq \mathbb{N}$ such that $g_{m_{k_\ell}} \to \widetilde{g}$ pointwise σ-a. e. on Σ as $\ell \to \infty$. Therefore, we see that $\widetilde{g} = g$ pointwise σ-a. e. on Σ. This, together with (6.67) and (6.68) imply

$$\begin{aligned}\|g\|_{L^q(\Sigma,\omega)} &= \|\widetilde{g}\|_{L^q(\Sigma,\omega)} = \lim_{m\to\infty} \|g_m\|_{L^q(\Sigma,\omega)} \\ &\le \sum_{j=1}^{\infty} |\lambda_j| \le (1+\varepsilon)\|g\|_{\mathcal{B}^{q,0}(\Sigma,\omega)}.\end{aligned} \tag{6.70}$$

In view of the arbitrary nature of $\varepsilon > 0$, we ultimately obtain

$$\|g\|_{L^q(\Sigma,\omega)} \le \|g\|_{\mathcal{B}^{q,0}(\Sigma,\omega)}. \tag{6.71}$$

For the reverse inequality, fix a reference point $x_0 \in \Sigma$, let $\varepsilon > 0$, and assume that $g \in L^q(\Sigma,\omega)$. Then $\|g\|_{L^q(\Sigma,\omega)} < +\infty$, so there exists an increasing sequence of radii $\{R_j\}_{j\in\mathbb{N}} \subseteq (0,\infty)$ such that

$$\|g \cdot \mathbf{1}_{\Delta(x_0,R_{j+1})\setminus\Delta(x_0,R_j)}\|_{L^q(\Sigma,\omega)} < \frac{\varepsilon}{2^j}. \tag{6.72}$$

At this point, we may also assume that g is not identically zero, for in that case, membership of g to $\mathcal{B}^{q,0}(\Sigma,\omega)$ is immediate. Now, decompose in a pointwise fashion

$$g = g \cdot \mathbf{1}_{\Delta(x_0,R_1)} + \sum_{j=1}^{\infty} g \cdot \mathbf{1}_{\Delta(x_0,R_{j+1})\setminus\Delta(x_0,R_j)} \quad \text{on } \Sigma. \tag{6.73}$$

Call $\lambda_0 := \|g\|_{L^q(\Sigma,\omega)}$ and $b_0 := g \cdot \mathbf{1}_{\Delta(x_0,R_1)}\|g\|^{-1}_{L^q(\Sigma,\omega)}$. Also, for each $j \in \mathbb{N}$ set $\lambda_j := \frac{\varepsilon}{2^j}$ and $b_j := \frac{2^j}{\varepsilon} g \cdot \mathbf{1}_{\Delta(x_0,R_{j+1})\setminus\Delta(x_0,R_j)}$. Then by construction, for each $j \in \mathbb{N}_0$ we find that $\|b_j\|_{L^q(\Sigma,\omega)} \le 1$, making b_j a $(q,0,\omega)$-block on Σ in light of (6.1) (with $\lambda := 0$ and $R := R_{j+1}$). Moreover, our choice of λ_j and b_j ensures that $g = \sum_{j=0}^{\infty} \lambda_j b_j$ is a block representation for g, since $\{\lambda_j\}_{j\in\mathbb{N}} \in \ell^1$, placing g in $\mathcal{B}^{q,0}(\Sigma,\omega)$ (cf. Definition 6.7). Recalling (6.41), we also find that

$$\|g\|_{\mathcal{B}^{q,0}(\Sigma,\omega)} \leq \sum_{j=0}^{\infty} |\lambda_j| = \|g\|_{L^q(\Sigma,\omega)} + \sum_{j=1}^{\infty} \frac{\varepsilon}{2^j} \tag{6.74}$$
$$= \|g\|_{L^q(\Sigma,\omega)} + \varepsilon.$$

Since $\varepsilon > 0$ was arbitrary, it must be the case that

$$\|g\|_{\mathcal{B}^{q,0}(\Sigma,\omega)} \leq \|g\|_{L^q(\Sigma,\omega)}, \tag{6.75}$$

which together with (6.70) establishes the desired result. □

Remarkably, as (4.4) and Proposition 6.16 detail, Lebesgue spaces find themselves as endpoints on the scales of both Morrey and Block spaces. While by no means a justification, this is some evidence toward the duality of Morrey and Block spaces. We will expand on these ideas later in this chapter.

One very important property for Muckenhoupt-weighted Block spaces, especially in the context of developing a Calderón–Zygmund theory on this scale of spaces, is the following embedding result, reminiscent of the result involving weighted Morrey spaces in Proposition 4.15. Namely, we show that Muckenhoupt-weighted Block spaces continuously embed into a certain weighted Lebesgue space.

Proposition 6.17. *Let $\Sigma \subseteq \mathbb{R}^n$ be a closed ADR set, and write $\sigma := \mathcal{H}^{n-1}\lfloor\Sigma$. Fix an integrability exponent $q \in (1,\infty)$ along with a Muckenhoupt weight $\omega \in A_q(\Sigma,\sigma)$. Let $\varepsilon \in (0,q-1)$ be such that $\omega \in A_{q-\varepsilon}(\Sigma,\sigma)$ and fix a threshold $\lambda \in (0,\varepsilon(n-1)/(q-1))$.*

Then there exists $q_* \in (1,q)$ such that for all $q_0 \in [1,q_*)$ one has

$$\mathcal{B}^{q,\lambda}(\Sigma,\omega) \hookrightarrow L^{q_0}\left(\Sigma, \frac{\sigma(x)}{1+|x|^{n-1}}\right) \quad \textit{continuously.} \tag{6.76}$$

As a corollary, for all $q_0 \in [1,q_*)$,

$$\mathcal{B}^{q,\lambda}(\Sigma,\omega) \hookrightarrow L^1\left(\Sigma, \frac{\sigma(x)}{1+|x|^{n-1}}\right) \cap L^{q_0}_{\mathrm{loc}}(\Sigma,\sigma) \quad \textit{continuously.} \tag{6.77}$$

Proof. Let $f \in \mathcal{B}^{q,\lambda}(\Sigma,\omega)$. Then there exist $\{\lambda_j\}_{j\in\mathbb{N}} \in \ell^1$ and $\{b_j\}_{j\in\mathbb{N}}$, a sequence of (q,λ,ω)-blocks on Σ, such that $f = \sum_{j=1}^{\infty} \lambda_j b_j$ with convergence in $L^1(\Sigma, \frac{\sigma(x)}{1+|x|^{n-1}})$. As a result, there exists an increasing sequence $\{N_k\}_{k\in\mathbb{N}} \subseteq \mathbb{N}$ such that $\sum_{j=1}^{N_k} \lambda_j b_j$ converges pointwise σ-a. e. on Σ to f as $k \to \infty$. By Lemma 6.5,

> there exists an integrability exponent $q_* \in (1,q)$ such that for each $q_0 \in [1,q_*)$, every (q,λ,ω)-block b belongs to $L^{q_0}(\Sigma, \frac{\sigma(x)}{1+|x|^{n-1}})$, with $\|b\|_{L^{q_0}(\Sigma, \frac{\sigma(x)}{1+|x|^{n-1}})} \leq C_0$ for some constant $C_0 \in (0,\infty)$ independent of b. (6.78)

Subsequently, for each $q_0 \in [1,q_*)$, we may write

$$\|f\|_{L^{q_0}(\Sigma,\frac{\sigma(x)}{1+|x|^{n-1}})} = \left(\int_\Sigma \liminf_{k\to\infty}\left|\sum_{j=1}^{N_k}\lambda_j b_j(x)\right|^{q_0}\frac{d\sigma(x)}{1+|x|^{n-1}}\right)^{1/q_0} \tag{6.79}$$
$$\le \liminf_{k\to\infty}\left(\int_\Sigma\left|\sum_{j=1}^{N_k}\lambda_j b_j(x)\right|^{q_0}\frac{d\sigma(x)}{1+|x|^{n-1}}\right)^{1/q_0}$$
$$\le \liminf_{k\to\infty}\sum_{j=1}^{N_k}|\lambda_j|\left(\int_\Sigma |b_j(x)|^{q_0}\frac{d\sigma(x)}{1+|x|^{n-1}}\right)^{1/q_0}$$
$$\le C_0\sum_{j=1}^{\infty}|\lambda_j|.$$

The first inequality is a consequence of Fatou's lemma, the second inequality follows from Minkowski's inequality, and the final inequality is a consequence of (6.78). Taking an infimum over all block representations of f yields

$$\|f\|_{L^{q_0}(\Sigma,\frac{\sigma(x)}{1+|x|^{n-1}})} \le C_0\|f\|_{\mathcal{B}^{q,\lambda}(\Sigma,\omega)}, \tag{6.80}$$

from which the desired result follows, given the arbitrary nature of f.

To prove the corollary, fix $q_0 \in [1, q_*)$ and note that the first part of the proposition shows that

$$\mathcal{B}^{q,\lambda}(\Sigma,\omega) \hookrightarrow L^1\left(\Sigma, \frac{\sigma(x)}{1+|x|^{n-1}}\right) \cap L^{q_0}\left(\Sigma, \frac{\sigma(x)}{1+|x|^{n-1}}\right) \tag{6.81}$$

continuously. Showing that $L^{q_0}(\Sigma, \frac{\sigma(x)}{1+|x|^{n-1}}) \hookrightarrow L^{q_0}_{\rm loc}(\Sigma,\sigma)$ continuously will establish the result. For that purpose, let $f \in L^{q_0}(\Sigma, \frac{\sigma(x)}{1+|x|^{n-1}})$. Then for any compact set $K \subseteq \Sigma$, one has $1 \approx \frac{1}{1+|x|^{n-1}}$ uniformly in $x \in K$ by Lemma 2.18, item (3). Therefore, we have

$$\left(\int_K |f(x)|^{q_0}\,d\sigma(x)\right)^{1/q_0} \le C\left(\int_K \frac{|f(x)|}{1+|x|^{n-1}}\,d\sigma(x)\right)^{1/q_0} \tag{6.82}$$
$$\le C\|f\|_{L^{q_0}(\Sigma,\frac{\sigma(x)}{1+|x|^{n-1}})},$$

and the proof of the corollary is complete. □

Remark 6.18. It is clear from Proposition 6.17 that if $f \in \mathcal{B}^{q,\lambda}(\Sigma,\omega)$, then $f \in L^\infty(\Sigma,\sigma)$.

Remark 6.19. Recall Remark 5.4. Suppose $\Sigma \subseteq \mathbb{R}^n$ is a closed set and σ is any Borel measure on Σ. If $k \in \mathscr{C}^0(\mathbb{R}^n\setminus\{0\})$ is positive homogeneous of degree $1-n$, then Proposition 6.17 shows for each $\varepsilon > 0$ and each f belonging to a Block space that

$$\int_{\substack{y\in\Sigma\\ |x-y|>\varepsilon}} k(x-y)f(y)\,d\sigma(y) \tag{6.83}$$

is well-defined via an absolutely convergent integral. That is, we can freely consider integral operators of Calderón–Zygmund type as they act on Block spaces.

The defining property of a Muckenhoupt-weighted Block space is the existence of a block representation. However, if we know a function belongs to a Muckenhoupt-weighted Block space a priori, then there is a way to choose a block representation, which has some structure involving a dyadic grid.

Lemma 6.20. *Let $\Sigma \subseteq \mathbb{R}^n$ be a closed ADR set, and write $\sigma := \mathcal{H}^{n-1}\lfloor\Sigma$. Fix an integrability exponent $q \in (1,\infty)$ along with a Muckenhoupt weight $\omega \in A_q(\Sigma,\sigma)$. Choose a parameter $\varepsilon \in (0,q-1)$ such that $\omega \in A_{q-\varepsilon}(\Sigma,\sigma)$, and select an arbitrary threshold $\lambda \in (0,\varepsilon(n-1)/(q-1))$. Finally, bring in a dyadic grid structure $\mathbb{D}(\Sigma)$ on Σ as in Lemma 2.42.*

Then given any $f \in \mathcal{B}^{q,\lambda}(\Sigma,\omega)$ with $\|f\|_{\mathcal{B}^{q,\lambda}(\Sigma,\omega)} \leq 1$, one has

$$f = \sum_{Q\in\mathbb{D}(\Sigma)} \lambda(Q)b(Q), \tag{6.84}$$

where $\{\lambda(Q)\}_{Q\in\mathbb{D}(\Sigma)}$ is a sequence of nonnegative numbers, which satisfy $\sum_{Q\in\mathbb{D}(\Sigma)}\lambda(Q) \leq C_*$ for some $C_* \in (0,\infty)$, depending only on Σ, n, q, and λ, and for each $Q \in \mathbb{D}(\Sigma)$ the function $b(Q)$ is a (q,λ,ω)-block on Σ.

Moreover, there exists a purely dimensional constant $c_\Sigma \in (0,\infty)$ such that if $k \in \mathbb{Z}$ and $Q \in \mathbb{D}_k(\Sigma)$ with center $x_Q \in \Sigma$ and side length $\ell(Q) := 2^{-k}$, then $\operatorname{supp} b(Q) \subseteq \Delta(x_Q, c_\Sigma\ell(Q))$ with $\|b(Q)\|_{L^q(\Sigma,\omega)} \leq (c_\Sigma\ell(Q))^{\lambda(1/q-1)}$.

Proof. If $f \in \mathcal{B}^{q,\lambda}(\Sigma,\omega)$ with $\|f\|_{\mathcal{B}^{q,\lambda}(\Sigma,\omega)} \leq 1$, then by quasioptimality there exist a sequence of (q,λ,ω)-blocks $\{b_j\}_{j\in\mathbb{N}}$ and a sequence $\{\lambda_j\}_{j\in\mathbb{N}} \in \ell^1$ such that

$$f = \sum_{j\in\mathbb{N}} \lambda_j b_j \quad \text{and} \quad \sum_{j\in\mathbb{N}} |\lambda_j| < 2. \tag{6.85}$$

We are free to assume without loss of generality that $\lambda_j \neq 0$ for all $j \in \mathbb{N}$. Note that for each $j \in \mathbb{N}$ the (q,λ,ω)-block b_j satisfies $\operatorname{supp} b_j \subseteq \Delta_j$, where $\Delta_j \subseteq \Sigma$ is a surface ball of radius $r_j \in (0, 2\operatorname{diam}(\Sigma))$. Then for each $j \in \mathbb{N}$ there exists $k \in \mathbb{Z}$ with $k > \kappa_\Sigma + 2$ (where κ_Σ is defined in relation to Σ as in (2.92)), which satisfies $2^{-k} \leq r_j \leq 2^{-k+1}$. Also, since $\bigcup_{Q\in\mathbb{D}_k(\Sigma)} Q$ is dense in Σ (cf. item (7) of Lemma 2.42) and Δ_j is relatively open in Σ, there exists $Q \in \mathbb{D}_k(\Sigma)$ (with center $x_Q \in \Sigma$ and side length $\ell(Q) := 2^{-k}$) such that $Q \cap \Delta_j \neq \varnothing$. Subsequently,

$$\text{there exists a purely-dimensional constant } c_\Sigma \in (0,\infty) \text{ such that } \Delta_j \subseteq \Delta(x_Q, c_\Sigma\ell(Q)). \tag{6.86}$$

Define in relation to Q the surface ball $\widetilde{Q} := \Delta(x_Q, c_\Sigma\ell(Q))$, and note that the assignment $\mathbb{D}(\Sigma) \ni Q \mapsto \widetilde{Q}$ is unambiguous.

Next, for each $j \in \mathbb{N}$ select one and only one of the (conceivably many) dyadic cubes, say $Q_j \in \mathbb{D}_k(\Sigma)$, which intersects Δ_j. It follows that the function $F : \mathbb{N} \to \mathbb{D}(\Sigma)$, defined according to $F(j) := Q_j$ for each $j \in \mathbb{N}$ is well-defined. Subsequently, we may decompose f as

$$f = \sum_{j\in\mathbb{N}} \lambda_j b_j = \sum_{Q\in\mathrm{Range}(F)} \sum_{j\in F^{-1}(\{Q_j\})} \lambda_j b_j, \tag{6.87}$$

where the second equality is justified because

$$\sum_{j\in\mathbb{N}} |\lambda_j||b_j| \in \mathcal{B}^{q,\lambda}(\Sigma,\omega) \subseteq L^1\Big(\Sigma, \frac{\sigma(x)}{1+|x|^{n-1}}\Big) \tag{6.88}$$

(cf. Proposition 6.17), which shows that for σ-a. e. point $x \in \Sigma$, the first series appearing in (6.87) is absolutely convergent.

Moving on, for each $Q \in \mathbb{D}(\Sigma)$, define

$$\widetilde{\lambda}(Q) := \begin{cases} \sum_{j\in F^{-1}(\{Q\})} |\lambda_j| & \text{if } Q \in \mathrm{Range}(F), \\ 0 & \text{if } Q \notin \mathrm{Range}(F), \end{cases} \tag{6.89}$$

and

$$\widetilde{b}(Q) := \begin{cases} (\sum_{j\in F^{-1}(\{Q\})} |\lambda_j|)^{-1}(\sum_{j\in F^{-1}(\{Q\})} \lambda_j b_j) & \text{if } Q \in \mathrm{Range}(F), \\ 0 & \text{if } Q \notin \mathrm{Range}(F). \end{cases} \tag{6.90}$$

Note that $\widetilde{b}(Q)$ is well-defined for each $Q \in \mathbb{D}(\Sigma)$, since we are presently assuming $\lambda_j \neq 0$ for all $j \in \mathbb{N}$. Because of (6.89) and (6.90), we see that

$$\begin{aligned} \sum_{Q\in\mathrm{Range}(F)} \sum_{j\in F^{-1}(\{Q\})} \lambda_j b_j &= \sum_{Q\in\mathrm{Range}(F)} \widetilde{\lambda}(Q)\widetilde{b}(Q) \\ &= \sum_{Q\in\mathbb{D}(\Sigma)} \widetilde{\lambda}(Q)\widetilde{b}(Q). \end{aligned} \tag{6.91}$$

Note also that $\widetilde{\lambda}(Q) \in [0,\infty)$ for each $Q \in \mathbb{D}(\Sigma)$, and hence, using (6.89),

$$\begin{aligned} \sum_{Q\in\mathbb{D}(\Sigma)} |\widetilde{\lambda}(Q)| &= \sum_{Q\in\mathrm{Range}(F)} \widetilde{\lambda}(Q) = \sum_{Q\in\mathrm{Range}(F)} \sum_{j\in F^{-1}(\{Q\})} |\lambda_j| \\ &= \sum_{j\in\mathbb{N}} |\lambda_j| \leq 2, \end{aligned} \tag{6.92}$$

the last inequality being a consequence of (6.85). This shows that the sequence $\{\widetilde{\lambda}(Q)\}_{Q\in\mathbb{D}(\Sigma)}$ belongs to ℓ^1. Concerning $\widetilde{b}(Q)$, if $Q \notin \mathrm{Range}(F)$, then $\widetilde{b}(Q) = 0$, which is trivially a (q,λ,ω)-block on Σ. Supposing that $Q \in \mathrm{Range}(F)$, we find

$$
\begin{aligned}
\|\widetilde{b}(Q)\|_{L^q(\Sigma,\omega)} &= \Big(\sum_{j\in F^{-1}(\{Q\})}|\lambda_j|\Big)^{-1}\Big\|\sum_{j\in F^{-1}(\{Q\})}\lambda_j b_j\Big\|_{L^q(\Sigma,\omega)} \\
&\leq \Big(\sum_{j\in F^{-1}(\{Q\})}|\lambda_j|\Big)^{-1}\sum_{j\in F^{-1}(\{Q\})}|\lambda_j|\|b_j\|_{L^q(\Sigma,\omega)} \\
&\leq \Big(\sum_{j\in F^{-1}(\{Q\})}|\lambda_j|\Big)^{-1}\sum_{j\in F^{-1}(\{Q\})}|\lambda_j|r_j^{\lambda(1/q-1)} \\
&\leq \Big(\sum_{j\in F^{-1}(\{Q\})}|\lambda_j|\Big)^{-1}\Big(\sum_{j\in F^{-1}(\{Q\})}|\lambda_j|\Big)\ell(Q)^{\lambda(1/q-1)} \\
&= C_*\big(c_\Sigma\ell(Q)\big)^{\lambda(1/q-1)},
\end{aligned}
\tag{6.93}
$$

where $C_* := c_\Sigma^{-\lambda(1/q-1)} \in (0,\infty)$ (which depends only on Σ, n, q, and λ), and c_Σ is as in (6.86). The first line above follows from (6.90), the second line uses the triangle inequality, the third line follows since b_j is a (q,λ,ω)-block, supported in the surface ball $\Delta_j \subseteq \Sigma$ of radius $r_j \in (0,\infty)$, and the fourth line is a consequence of our definition of F and having $\ell(Q) \leq r_j$ and $\lambda(1/q-1) < 0$. Recall our definition $\widetilde{Q} := \Delta(x_Q, c_\Sigma\ell(Q))$, which is defined unambiguously in relation to $Q \in \mathbb{D}(\Sigma)$, and observe that $\operatorname{supp}\widetilde{b}(Q) \subseteq \widetilde{Q}$. Indeed, for all $j \in F^{-1}(\{Q\})$, we have that $F(j) = Q$, which given the definition of F means that $\Delta_j \cap Q \neq \varnothing$, and hence $\Delta_j \subseteq \widetilde{Q}$, bearing in mind (6.86). Given that $\operatorname{supp} b_j \subseteq \Delta_j$ for each $j \in \mathbb{N}$, we find that $\operatorname{supp}(\sum_{j\in F^{-1}(\{Q\})}\lambda_j b_j) \subseteq \widetilde{Q}$ as well. Thus, recalling (6.90) and (6.93), we can conclude that $\widetilde{b}(Q)$ is a constant multiple of a (q,λ,ω)-block on Σ, supported in $\widetilde{Q}$.

Entering the final stage of this proof, for each cube $Q \in \mathbb{D}(\Sigma)$ denote $\lambda(Q) := C_*\widetilde{\lambda}(Q)$ and $b(Q) := C_*^{-1}\widetilde{b}(Q)$ with C_* as in (6.93). Therefore, in view of (6.87) and (6.91), we have

$$
f = \sum_{Q\in\mathbb{D}(\Sigma)}\lambda(Q)b(Q), \tag{6.94}
$$

which is in fact a block representation for f. Indeed, given the above discussion regarding $\widetilde{b}(Q)$, we conclude that $b(Q)$ is a genuine (q,λ,ω)-block on Σ with

$$
\begin{aligned}
&\operatorname{supp} b(Q) \subseteq \widetilde{Q} := \Delta\big(x_Q, c_\Sigma\ell(Q)\big) \quad \text{and} \\
&\|b(Q)\|_{L^q(\Sigma,\omega)} \leq \big(c_\Sigma\ell(Q)\big)^{\lambda(1/q-1)}.
\end{aligned}
\tag{6.95}
$$

Moreover, (6.92) shows that

$$
\sum_{Q\in\mathbb{D}(\Sigma)}\lambda(Q) \leq 2C_*. \tag{6.96}
$$

At this point, we have established all desired conclusions. □

As a final property of Block spaces, one interesting feature they exhibit that fails to be true for Morrey spaces is that Block spaces are separable; i. e., Block spaces contain a countable dense subset; cf. Proposition 6.22 below. Separability is an important feature of Block spaces from the perspective of the Sequential Banach–Alaoglu theorem (cf. [121, Theorem 3.17, p. 70]), which asserts that

given a separable topological vector space X, if the sequence $\{\Lambda_j\}_{j\in\mathbb{N}} \subseteq X^*$ has the property that there exists a neighborhood V of the origin in X for which $\sup_{x\in V}\sup_{j\in\mathbb{N}}|\Lambda_j x| < +\infty$ then there exists a subsequence $\{\Lambda_{j_k}\}_{k\in\mathbb{N}}$ of $\{\Lambda_j\}_{j\in\mathbb{N}}$ which is weak-$*$ convergent to some $\Lambda \in X^*$, i. e., $\Lambda_{j_k}x \to \Lambda x$ in $\mathbb{C}$ as $k \to \infty$ for each fixed $x \in X$. (6.97)

We shall soon discuss the duality paring of Block spaces and Morrey spaces in Theorem 6.29, and with this duality in hand, the separability of Block spaces is necessary to apply the sequential Banach-Alaoglu theorem to conclude that

any bounded sequence in $M^{p,\lambda}(\Sigma, w)$ has a sub-sequence which is weak-$*$ convergent. (6.98)

Before proving the separability of Block spaces, it will be important to note the following result from [86] regarding the separability of Lebesgue spaces with Borel-regular measures.

Proposition 6.21. *Let (X,τ) be a second countable topological space and consider a sigma-finite, Borel-regular measure $\mu : \mathfrak{M} \to [0,+\infty]$ on X. Then the measure μ is separable,*[2] *and for each integrability exponent $p \in (0,\infty)$ the Lebesgue space $L^p(X,\mu)$ is separable.*

In [79], it is also noted that for any weight function $w \in L^1_{\mathrm{loc}}(\Sigma,\sigma)$ it follows that

the measure $dw := w\,d\sigma$ is complete, locally finite (hence also sigma-finite), separable, and Borel-regular on Σ. (6.99)

In particular, (6.99) applies to the case that the weight w is a Muckenhoupt weight (cf. Remark 2.28), and hence we can conclude from Proposition 6.21 that

the Lebesgue space $L^p(\Sigma, w)$ is separable in the case that $\Sigma \subseteq \mathbb{R}^n$ is an Ahlfors regular set, $p \in (1,\infty)$, and $w \in A_\infty(\Sigma,\sigma)$, where $\sigma := \mathcal{H}^{n-1}\lfloor\Sigma$. (6.100)

We now present the proof of separability of Muckenhoupt-weighted Block spaces.

---

2 Given a measure space $(X,\mathfrak{M},\mu)$, consider the equivalence relation $A \sim B \iff \mu(A \triangle B) = 0$, where $\triangle$ denotes the set theoretic symmetric difference. Consider the space of equivalence relations $\{[A] : A \in \mathfrak{M}, \mu(A) < +\infty\}$, equipped with the metric $([A],[B]) \mapsto \mu(A \triangle B)$. The measure μ is called separable provided this metric space is separable.

Proposition 6.22. *Let $\Sigma \subseteq \mathbb{R}^n$ be a closed ADR set and write $\sigma := \mathcal{H}^{n-1}\lfloor\Sigma$. Fix an integrability exponent $q \in (1,\infty)$ along with a Muckenhoupt weight $\omega \in A_q(\Sigma,\sigma)$. Let $\varepsilon \in (0,q-1)$ be such that $\omega \in A_{q-\varepsilon}(\Sigma,\sigma)$, and fix a threshold $\lambda \in (0,\varepsilon(n-1)/(q-1))$.*

Then the Muckenhoupt-weighted Block space $\mathcal{B}^{q,\lambda}(\Sigma,\omega)$ is separable.

Proof. Begin by fixing $x_0 \in \Sigma$. For each $j \in \mathbb{N}$, let $\Delta_j := \Delta(x_0, j)$ denote the surface ball centered at x_0 with radius j. From (6.100), we know that $L^q(\Delta_j,\omega)$ is separable, and hence there exists a countable dense subset $\mathcal{F}_j \subseteq L^q(\Sigma,\omega)$. Now, extend by zero each function in $\mathcal{F}_j$ and call the resulting collection of functions $\widetilde{\mathcal{F}}_j$. Observe by Remark 6.2 that each function belonging to $\widetilde{\mathcal{F}}_j$ is a constant multiple of some (q,λ,ω)-block on Σ. As such, $\widetilde{\mathcal{F}}_j \subseteq \mathcal{B}^{q,\lambda}(\Sigma,\omega)$ for each $j \in \mathbb{N}$. Now, let $\mathcal{F}$ denote the collection of all finite sums of functions belonging to $\bigcup_{j\in\mathbb{N}} \widetilde{\mathcal{F}}_j$, so that

$$\mathcal{F} \text{ is a countable subset of } \mathcal{B}^{q,\lambda}(\Sigma,\omega). \tag{6.101}$$

We now claim that $\mathcal{F}$ is dense in $\mathcal{B}^{q,\lambda}(\Sigma,\omega)$. To facilitate the proof of this claim, we first show that

$$\begin{gathered}\text{for any } \varepsilon > 0 \text{ and any choice of } \eta \in \mathbb{C} \text{ along with a } (q,\lambda,\omega)\text{-block,}\\ b, \text{ there exists } f \in \mathcal{F} \text{ such that } \|\eta b - f\|_{\mathcal{B}^{q,\lambda}(\Sigma,\omega)} \le \varepsilon.\end{gathered} \tag{6.102}$$

As such, fix $\varepsilon > 0$ and choose $\eta \in \mathbb{C}$ along with some (q,λ,ω)-block, b. By (6.1), there exist $x_1 \in \Sigma$ and $R \in (0,\infty)$ with the property that

$$\operatorname{supp} b \subseteq \Delta(x_1,R) \quad \text{and} \quad \|b\|_{L^q(\Sigma,\omega)} \le R^{\lambda(1/q-1)}. \tag{6.103}$$

It follows that there exists some $j_0 \in \mathbb{N}$ for which

$$\operatorname{supp}(\eta b) \subseteq \Delta_{j_0} \quad \text{and} \quad (\eta b)|_{\Delta_{j_0}} \in L^q(\Delta_{j_0},\omega). \tag{6.104}$$

By the density of $\mathcal{F}_{j_0}$ in $L^q(\Delta_{j_0},\omega)$ there exists $f \in \mathcal{F}_{j_0}$ such that

$$\|\eta b - f\|_{L^q(\Delta_{j_0},\omega)} < R^{\lambda(1/q-1)}\varepsilon. \tag{6.105}$$

Now, if $\widetilde{f}$ denotes the extension by zero of f, we have that $\widetilde{f} \in \widetilde{\mathcal{F}}_{j_0} \subseteq \mathcal{F}$, and if we define $h := \eta b - \widetilde{f}$ on Σ, we see that h satisfies

$$\operatorname{supp} h \subseteq \Delta_{j_0} \quad \text{and} \quad \|h\|_{L^q(\Sigma,\omega)} \le R^{\lambda(1/q-1)}\varepsilon. \tag{6.106}$$

This implies that $\varepsilon^{-1}h$ is a (q,λ,ω)-block on Σ, so that by Remark 6.23, we find that $\|\varepsilon^{-1}h\|_{\mathcal{B}^{q,\lambda}(\Sigma,\omega)} \le 1$, which entails

$$\|\eta b - f\|_{\mathcal{B}^{q,\lambda}(\Sigma,\omega)} \le \varepsilon \tag{6.107}$$

as wanted.

Let us now return to the original claim that $\mathcal{F}$ is dense in $\mathcal{B}^{q,\lambda}(\Sigma,\omega)$. Pick some $g \in \mathcal{B}^{q,\lambda}(\Sigma,\omega)$ and note that for any $\varepsilon > 0$, by Proposition 6.24 there exists $N \in \mathbb{N}$ such that the collection of (q,λ,ω)-blocks $\{b_j\}_{j=1}^N$ and the collection of numbers $\{\lambda_j\}_{j=1}^N \subseteq \mathbb{C}$ satisfy

$$\left\| g - \sum_{j=1}^{N} \lambda_j b_j \right\|_{\mathcal{B}^{q,\lambda}(\Sigma,\omega)} < \frac{\varepsilon}{2}. \tag{6.108}$$

Using the result from (6.102), for each $j \in \{1,\dots,N\}$ there exists some $f_j \in \mathcal{F}$ such that

$$\|\lambda_j b_j - f_j\|_{\mathcal{B}^{q,\lambda}(\Sigma,\omega)} < \frac{\varepsilon}{2N}. \tag{6.109}$$

Because of (6.108) and (6.109), it follows via the triangle inequality that

$$\begin{aligned}\left\| f - \sum_{j=1}^{N} f_j \right\|_{\mathcal{B}^{q,\lambda}(\Sigma,\omega)} &\le \left\| f - \sum_{j=1}^{N} \lambda_j b_j \right\|_{\mathcal{B}^{q,\lambda}(\Sigma,\omega)} + \sum_{j=1}^{N} \|\lambda_j b_j - f_j\|_{\mathcal{B}^{q,\lambda}(\Sigma,\omega)} \\ &< \frac{\varepsilon}{2} + \frac{\varepsilon}{2} = \varepsilon. \end{aligned} \tag{6.110}$$

Given the arbitrary nature of ε, (6.101), and the fact that $\sum_{j=1}^{N} f_j \in \mathcal{F}$, this establishes the claim that $\mathcal{F}$ is dense in $\mathcal{B}^{q,\lambda}(\Sigma,\omega)$, and hence shows that $\mathcal{B}^{q,\lambda}(\Sigma,\omega)$ is separable as advertised. □

6.2 Basic examples

We now would like to demonstrate that Blocks spaces are rich function spaces by showing several nontrivial examples of block functions.

As a preliminary comment to this section that we remark that blocks not only belong to the Block space, but are bounded above in norm by 1.

Remark 6.23. Let $\Sigma \subseteq \mathbb{R}^n$ be a closed Ahlfors regular set. Note that we can write any (q,λ,ω)-block b on Σ as $b = \sum_{j=1}^{\infty} \lambda_j b_j$ with

$$\lambda_j := \begin{cases} 1 & \text{if } j = 1, \\ 0 & \text{if } j \neq 1, \end{cases} \quad \text{and} \quad b_j := \begin{cases} b & \text{if } j = 1, \\ 0 & \text{if } j \neq 1, \end{cases} \tag{6.111}$$

for each $j \in \mathbb{N}$. Thus, by the definition of $\|\cdot\|_{\mathcal{B}^{q,\lambda}(\Sigma,\omega)}$ we find that

$$b \in \mathcal{B}^{q,\lambda}(\Sigma,\omega) \quad \text{and} \quad \|b\|_{\mathcal{B}^{q,\lambda}(\Sigma,\omega)} \le 1 \tag{6.112}$$

for all (q,λ,ω)-blocks b.

Our next observation, stated below, is that the set of all finite linear combinations of blocks is dense in the Block space.

Proposition 6.24. *Let $\Sigma \subseteq \mathbb{R}^n$ be a closed* ADR *set, and write $\sigma := \mathcal{H}^{n-1}\lfloor\Sigma$. Fix an integrability exponent $q \in (1,\infty)$ along with a Muckenhoupt weight $\omega \in A_q(\Sigma,\sigma)$. Finally, choose $\varepsilon \in (0,q-1)$ such that $\omega \in A_{q-\varepsilon}(\Sigma,\sigma)$, and fix an arbitrary threshold $\lambda \in (0,\varepsilon(n-1)/(q-1))$.*

Then the set of all finite linear combinations of (q,λ,ω)-blocks on Σ is a dense subset of $\mathcal{B}^{q,\lambda}(\Sigma,\omega)$.

Proof. Clearly, such functions belong to $\mathcal{B}^{q,\lambda}(\Sigma,\omega)$ by Remark 6.23 and the fact that $\mathcal{B}^{q,\lambda}(\Sigma,\omega)$ is a vector space (see Proposition 6.9). Now, suppose $f \in \mathcal{B}^{q,\lambda}(\Sigma,\omega)$. Then there exist a sequence $\{\lambda_j\}_{j\in\mathbb{N}} \in \ell^1$ and a family $\{b_j\}_{j\in\mathbb{N}}$ of (q,λ,ω)-blocks on Σ such that

$$f = \sum_{j=1}^{\infty} \lambda_j b_j \quad \text{with convergence in } L^1\Big(\Sigma, \frac{\sigma(x)}{1+|x|^{n-1}}\Big). \tag{6.113}$$

Define the partial sums $f_N := \sum_{n=1}^{N} \lambda_j b_j$ for each $N \in \mathbb{N}$. Then the sequence of partial sums $\{f_N\}_{N\in\mathbb{N}}$ belongs to the space of all finite linear combinations of (q,λ,ω)-blocks. We then see that

$$\|f - f_N\|_{\mathcal{B}^{q,\lambda}(\Sigma,\omega)} \le \sum_{j=N+1}^{\infty} |\lambda_j| \to 0 \quad \text{as } N \to \infty, \tag{6.114}$$

thanks to (6.41) and the fact that $\{\lambda_j\}_{j\in\mathbb{N}} \in \ell^1$. □

One wealth of block functions comes from the space of essentially bounded functions with compact support.

Proposition 6.25. *Let $\Sigma \subseteq \mathbb{R}^n$ be a closed* ADR *set and write $\sigma := \mathcal{H}^{n-1}\lfloor\Sigma$. Fix an integrability exponent $q \in (1,\infty)$ along with a Muckenhoupt weight $\omega \in A_q(\Sigma,\sigma)$. Let $\varepsilon \in (0,q-1)$ be such that $\omega \in A_{q-\varepsilon}(\Sigma,\sigma)$ and fix $\lambda \in (0,\varepsilon(n-1)/(q-1))$. Finally, let $x_0 \in \Sigma$ and $R \in (0, 2\operatorname{diam}(\Sigma))$ be arbitrary.*

Then $\mathbf{1}_{\Delta(x_0,R)} \in \mathcal{B}^{q,\lambda}(\Sigma,\omega)$ and in fact

$$L^{\infty}_{\text{comp}}(\Sigma,\sigma) \subseteq \mathcal{B}^{q,\lambda}(\Sigma,\omega). \tag{6.115}$$

Proof. Remark 6.2 together with the vector space structure of $\mathcal{B}^{q,\lambda}(\Sigma,\omega)$ imply that $\mathbf{1}_{\Delta(x_0,R)} \in \mathcal{B}^{q,\lambda}(\Sigma,\omega)$. Now, let $f \in L^{\infty}_{\text{comp}}(\Sigma,\sigma)$. Then there exists $C \in (0,\infty)$ such that $|f| \le C$ at σ-a. e. point on Σ. Also, given a reference point $x_0 \in \Sigma$, there exists $R \in (0, 2\operatorname{diam}(\Sigma))$ such that $\operatorname{supp} f$ is contained in $\Delta(x_0,R)$. As such,

$$|f| \le C\mathbf{1}_{\Delta(x_0,R)} \quad \text{at } \sigma\text{-a. e. point on } \Sigma. \tag{6.116}$$

Consequently, by the lattice property of $\mathcal{B}^{q,\lambda}(\Sigma,\omega)$ (cf. Proposition 6.14) and the fact that $\mathbf{1}_{\Delta}(x_0,R) \in \mathcal{B}^{q,\lambda}(\Sigma,\omega)$, we see that $f \in \mathcal{B}^{q,\lambda}(\Sigma,\omega)$ as well. □

We can also show that functions with the right type of decay enjoy membership to Block spaces, as witnessed below.

Lemma 6.26. *Let $\Sigma \subseteq \mathbb{R}^n$ be a closed ADR set and write $\sigma := \mathcal{H}^{n-1}\lfloor\Sigma$. Fix an integrability exponent $q \in (1,\infty)$ along with a Muckenhoupt weight $\omega \in A_q(\Sigma,\sigma)$. Fix also some $\varepsilon \in (0, q-1)$ for which $\omega \in A_{q-\varepsilon}(\Sigma,\sigma)$, and finally set a threshold $\lambda \in (0, \varepsilon(n-1)/(q-1))$.*

Then $f : \Sigma \to \mathbb{R}$, given by

$$f(x) := \frac{1}{1+|x|^{n-1}}, \quad \textit{for all } x \in \Sigma, \tag{6.117}$$

belongs to $\mathcal{B}^{q,\lambda}(\Sigma,\omega)$.

As a corollary of this and the lattice property for Block spaces detailed in Proposition 6.14, any σ-measurable function $f : \Sigma \to \mathbb{R}$ with the property that there exists $C \in (0,\infty)$ such that $|f| \leq C(1+|\cdot|)^{n-1}$ at σ-a. e. point on Σ belongs to $\mathcal{B}^{q,\lambda}(\Sigma,\omega)$.

Proof. Fix an arbitrary reference point $x_* \in \Sigma$. Then

$$\frac{1}{1+|x|^{n-1}} \approx \frac{1}{1+|x-x_*|^{n-1}} \quad \text{uniformly for all } x \in \Sigma \tag{6.118}$$

by item (2) of Lemma 2.18. Using the lattice property of $\mathcal{B}^{q,\lambda}(\Sigma,\omega)$ (cf. Proposition 6.14), we find that the membership of f follows if we can show that the function

$$g(x) := \frac{1}{1+|x-x_*|^{n-1}} \quad \text{for all } x \in \Sigma \tag{6.119}$$

belongs to $\mathcal{B}^{q,\lambda}(\Sigma,\omega)$. For that purpose, define $\Delta_0 := \Delta(x_*,2)$ and for each $j \in \mathbb{N}$ define $\Delta_j := \Delta(x_*,2^{j+1}) \setminus \Delta(x_*,2^j)$, so that

$$g = \sum_{j=0}^{\infty} g\mathbf{1}_{A_j} \quad \text{pointwise on } \Sigma. \tag{6.120}$$

Define $b_j := c_j g\mathbf{1}_{A_j}$ for all $j \in \mathbb{N}_0$, where

$$c_j := \begin{cases} \frac{(2^{j+1})^{\lambda(1/q-1)}}{\|g\mathbf{1}_{A_j}\|_{L^q(\Sigma,\omega)}} & \text{if } \|g\mathbf{1}_{A_j}\|_{L^q(\Sigma,\omega)} \neq 0, \\ 0 & \text{if } \|g\mathbf{1}_{A_j}\|_{L^q(\Sigma,\omega)} = 0. \end{cases} \tag{6.121}$$

By construction, each b_j is a (q,λ,ω)-block on Σ with $\operatorname{supp} b_j \subseteq \Delta(x_*,2^{j+1})$ as well as $\|b_j\|_{L^q(\Sigma,\omega)} \leq 2^{(j+1)\lambda(1/q-1)}$. Now, for each $j \in \mathbb{N}$ define $\lambda_j := c_j^{-1}$ provided $c_j \neq 0$, and set $\lambda_j := 0$ otherwise.

Fix $x \in A_0$, so that $|x-x_0| < 2$. Then

$$\frac{1}{1+|x-x_*|^{n-1}} \leq \frac{1}{1+2^{n-1}} < 1, \tag{6.122}$$

and hence recalling (6.119),

$$\|g\mathbf{1}_{A_0}\|_{L^q(\Sigma,\omega)} \le \omega(A_0)^{1/q} < +\infty. \tag{6.123}$$

Next, fix $j \in \mathbb{N}$ and let $x \in A_j$. Then $|x - x_*| > 2^j$, which implies that

$$\frac{1}{1+|x-x_*|^{n-1}} < \frac{1}{1+2^j} < \frac{1}{2^j}. \tag{6.124}$$

Consequently,

$$\begin{aligned} \|g\mathbf{1}_{A_j}\|_{L^q(\Sigma,\omega)} &\le 2^{-j(n-1)}\omega(A_j)^{1/q} \\ &\le \omega(\Delta(x_*,2))^{1/q}2^{-j(n-1)}\left(\frac{\omega(\Delta(x_*,2^{j+1}))}{\omega(\Delta(x_*,2))}\right)^{1/q} \\ &\le C\cdot 2^{-j(n-1)}\cdot 2^{j(n-1)(1-\varepsilon/q)} \\ &= C\cdot 2^{-j(n-1)(\varepsilon/q)}. \end{aligned} \tag{6.125}$$

The first inequality uses (6.119) and (6.124). The third inequality is a consequence of the following estimate:

$$\begin{aligned} \left(\frac{\omega(\Delta(x_*,2^{j+1}))}{\omega(\Delta(x_*,2))}\right)^{1/q} &\le [\omega]_{q-\varepsilon}^{1/q}\left(\frac{\sigma(\Delta(x_*,2^{j+1}))}{\sigma(\Delta(x_*,2))}\right)^{(q-\varepsilon)/q} \\ &\le C[\omega]_{q-\varepsilon}^{1/q}\left(\frac{2^{(j+1)(n-1)}}{2^{n-1}}\right)^{(q-\varepsilon)/q} \\ &= C[\omega]_{q-\varepsilon}^{1/q}2^{j(n-1)(1-\varepsilon/q)}, \end{aligned} \tag{6.126}$$

where the first inequality above follows from item (4) of Proposition 2.29, and the second inequality follows from the Ahlfors regularity of Σ. Therefore, given (6.121), the definition of λ_j, (6.123), and (6.125) we obtain

$$\sum_{j=0}^{\infty}|\lambda_j| \le \sum_{j=0}^{\infty}2^{-j\lambda(1/q-1)}\cdot 2^{-j(n-1)(\varepsilon/q)} \tag{6.127}$$

$$= C\sum_{j=0}^{\infty}2^{-j(1-1/q)[\varepsilon(n-1)/(q-1)-\lambda]} < +\infty. \tag{6.128}$$

Convergence of the geometric series follows from the initial hypothesis that $\lambda \in (\varepsilon(n-1)/(q-1))$. It follows by Lemma 6.8 that $g = \sum_{j=0}^{\infty}\lambda_j b_j$ is a block-representation for g, so we see that $g \in \mathcal{B}^{q,\lambda}(\Sigma,\omega)$ and ultimately that the original function f belongs to $\mathcal{B}^{q,\lambda}(\Sigma,\omega)$ as claimed. □

Remark 6.27. Note that we can prove Lemma 6.26 using an alternative method. Namely, as in the proof of Lemma 6.26 we need only show the function $g : \Sigma \to (0,\infty)$, defined in (6.119) belongs to $\mathcal{B}^{q,\lambda}(\Sigma,\omega)$. By Lemma 2.25, for each $x \in \Sigma$ we have

$$g(x) = \frac{1}{1+|x-x_*|^{n-1}} \le C\mathcal{M}_\Sigma(\mathbf{1}_{\Delta(x_*,1)})(x), \tag{6.129}$$

for some constant $C \in (0,\infty)$ that depends only on Σ and n. As such, the lattice property of Block spaces (Proposition 6.14) yields

$$\left\| \frac{1}{1+|\cdot - x_*|^{n-1}} \right\|_{\mathcal{B}^{q,\lambda}(\Sigma,\omega)} \le C\|\mathcal{M}(\mathbf{1}_{\Delta(x_*,1)})\|_{\mathcal{B}^{q,\lambda}(\Sigma,\omega)} \tag{6.130}$$

$$\le C\|\mathbf{1}_{\Delta(x_*,1)}\|_{\mathcal{B}^{q,\lambda}(\Sigma,\omega)}, \tag{6.131}$$

where the last inequality follows from Proposition 7.1, which states that the operator $\mathcal{M}_\Sigma : \mathcal{B}^{q,\lambda}(\Sigma,\omega) \to \mathcal{B}^{q,\lambda}(\Sigma,\omega)$ is bounded.[3] Moving on, observe that Remark 6.2 implies that $\|\mathbf{1}_{\Delta(x_*,1)}\|_{\mathcal{B}^{q,\lambda}(\Sigma,\omega)} < +\infty$, so in fact (6.130) shows

$$\|g\|_{\mathcal{B}^{q,\lambda}(\Sigma,\omega)} = \left\| \frac{1}{1+|\cdot - x_*|^{n-1}} \right\|_{\mathcal{B}^{q,\lambda}(\Sigma,\omega)} < +\infty, \tag{6.132}$$

and we have the desired membership of g to $\mathcal{B}^{q,\lambda}(\Sigma,\omega)$.

6.3 Duality theory

We have the following Hölder-type inequality for Morrey functions and functions in a Block space. This is a crucial step in ultimately proving that a Muckenhoupt-weighted Block space is the pre-dual of a Muckenhoupt-weighted Morrey space for suitable integrability exponents and choice of λ.

Theorem 6.28. *Let $\Sigma \subseteq \mathbb{R}^n$ be a closed ADR set and write $\sigma := \mathcal{H}^{n-1}\lfloor\Sigma$, and let $p, p' \in (1,\infty)$ with $1/p + 1/p' = 1$. Fix $w \in A_p(\Sigma,\sigma)$ and denote the conjugate weight $w' := w^{-p'/p} \in A_{p'}(\Sigma,\sigma)$. Let $\varepsilon \in (0, p'-1)$ be such that $w' \in A_{p'-\varepsilon}(\Sigma,\sigma)$ and assume $\lambda \in (0, \varepsilon(n-1)/(p'-1))$. Bring in the weighted Morrey space $M^{p,\lambda}(\Sigma,w)$ and also the weighted Block space $\mathcal{B}^{p',\lambda}(\Sigma,w')$ associated with p', w', ε, and λ in the sense of Definition 6.7.*

Then for every $f \in M^{p,\lambda}(\Sigma,w)$ and $g \in \mathcal{B}^{p',\lambda}(\Sigma,w')$, one has

$$\int_\Sigma |fg|\,d\sigma \le \|f\|_{M^{p,\lambda}(\Sigma,w)}\|g\|_{\mathcal{B}^{p',\lambda}(\Sigma,w')}. \tag{6.133}$$

Proof. We begin with the simplest case when g is a (p',λ,w')-block.

3 Although Proposition 7.1 appears later in this work, the proof is independent of this result.

Claim: If b is a (p',λ,w')-block and $f \in M^{p,\lambda}(\Sigma,w)$, then

$$\int_\Sigma |f||b|\,d\sigma \le \|f\|_{M^{p,\lambda}(\Sigma,w)}. \tag{6.134}$$

To see this, let b be a (p',λ,w')-block. Then there exist a location $x_0 \in \Sigma$ and a scale $R \in (0, 2\operatorname{diam}(\Sigma))$ such that $\operatorname{supp} b \subseteq \Delta(x_0,R)$. Observe that if $f \in M^{p,\lambda}(\Sigma,w)$, then by Hölder's inequality,

$$\begin{aligned}\int_\Sigma |f||b|\,d\sigma &= \int_{\Delta(x_0,R)} |f||b|w^{1/p}w^{-1/p}\,d\sigma \\ &\le \Big(\int_{\Delta(x_0,R)} |f|^p\,dw\Big)^{1/p}\Big(\int_{\Delta(x_0,R)} |b|^{p'}\,dw'\Big)^{1/p'}.\end{aligned} \tag{6.135}$$

Because b is a (p',λ,w')-block, we know that $\|b\|_{L^{p'}(\Sigma,w')} \le R^{\lambda(1/p'-1)}$ (cf. Definition 6.1). Thus, we may further estimate

$$\begin{aligned}\int_\Sigma |f||b|\,d\sigma &\le \Big(\int_{\Delta(x_0,R)} |f|^p\,dw\Big)^{1/p} R^{\lambda(1/p'-1)} \\ &= \Big(\frac{1}{R^\lambda}\int_{\Delta(x_0,R)} |f|^p\,dw\Big)^{1/p} \\ &\le \|f\|_{M^{p,\lambda}(\Sigma,w)},\end{aligned} \tag{6.136}$$

whereby the last inequality follows from the definition of the Morrey norm (4.2). This proves the claim.

Generalizing to an arbitrary block function, let $g \in \mathcal{B}^{p',\lambda}(\Sigma,w')$ and consider a block representation $g = \sum_{j=1}^\infty \lambda_j b_j$, where convergence is in $L^1(\Sigma, \frac{\sigma(x)}{1+|x|^{n-1}})$. Define the partial sums $g_N := \sum_{j=1}^N \lambda_j b_j$ for each $N \in \mathbb{N}$. Then there exists a subsequence $\{g_{N_k}\}_{k\in\mathbb{N}}$ such that $g_{N_k} \to g$ pointwise σ-a. e. on Σ as $k \to \infty$. Consequently,

$$\begin{aligned}\int_\Sigma |f||g|\,d\sigma &= \int_\Sigma \liminf_{k\to\infty} |f||g_{N_k}|\,d\sigma \\ &\le \liminf_{k\to\infty}\int_\Sigma |f||g_{N_k}|\,d\sigma \\ &\le \liminf_{k\to\infty}\int_\Sigma |f|\Big(\sum_{j=1}^{N_k}|\lambda_j||b_j|\Big)\,d\sigma \\ &\le \liminf_{k\to\infty}\sum_{j=1}^{N_k}|\lambda_j|\|f\|_{M^{p,\lambda}(\Sigma,w)}\end{aligned} \tag{6.137}$$

$$\leq \|f\|_{M^{p,\lambda}(\Sigma,w)} \sum_{j=1}^{\infty} |\lambda_j|.$$

The first inequality follows from Fatou's lemma, the second inequality from the triangle inequality, and the third inequality follows from claim (6.134) applied to each b_j. Recalling the definition of the block norm (6.41), we see that taking an infimum in (6.137) over all block representations of g establishes the desired result. □

We now present a crucial result that will eventually lead to uniqueness of solutions for a particular boundary value problem on Morrey spaces. That is, in a suitable context, the Muckenhoupt-weighted Block space $\mathcal{B}^{q,\lambda}(\Sigma,\omega)$ is the predual of the Muckenhoupt-weighted Morrey space $M^{p,\lambda}(\Sigma,w)$, when q is the Hölder conjugate of p, and ω is the conjugate weight to w.

Theorem 6.29. *Let $\Sigma \subseteq \mathbb{R}^n$ be a closed* ADR *set and write $\sigma := \mathcal{H}^{n-1}\lfloor\Sigma$. Fix integrability exponents $p, p' \in (1,\infty)$ with $1/p + 1/p' = 1$, and fix a Muckenhoupt weight $w \in A_p(\Sigma,\sigma)$. Abbreviate $w' := w^{-p'/p} \in A_{p'}(\Sigma,\sigma)$, the conjugate weight to w, and let $\varepsilon \in (0, p'-1)$ be such that $w' \in A_{p'-\varepsilon}(\Sigma,\sigma)$, and fix $\lambda \in (0, \varepsilon(n-1)/(p'-1))$. Bring in the weighted Morrey space $M^{p,\lambda}(\Sigma,w)$ and also the weighted Block space $\mathcal{B}^{p',\lambda}(\Sigma,w')$, associated with p', w', ε, and λ in the sense of Definition* 6.7.

Then the space $\mathcal{B}^{p',\lambda}(\Sigma,w')$ is the predual of $M^{p,\lambda}(\Sigma,w)$, i. e.,

$$\left(\mathcal{B}^{p',\lambda}(\Sigma,w')\right)^* = M^{p,\lambda}(\Sigma,w). \tag{6.138}$$

More precisely, the mapping

$$M^{p,\lambda}(\Sigma,w) \ni f \mapsto \Lambda_f \in \left(\mathcal{B}^{p',\lambda}(\Sigma,w')\right)^*, \tag{6.139}$$

with

$$\Lambda_f(g) := \int_\Sigma fg\,d\sigma \quad \text{for each } g \in \mathcal{B}^{p',\lambda}(\Sigma,w'), \tag{6.140}$$

is a well-defined, linear, and continuous isometric isomorphism with bounded inverse.

Proof. Fix $f \in M^{p,\lambda}(\Sigma,w)$. We begin by observing that the linear (by definition) functional $\Lambda_f : \mathcal{B}^{p',\lambda}(\Sigma,w') \to \mathbb{C}$ is well-defined via an absolutely convergent integral thanks to Theorem 6.28. This ensures that the assignment in (6.139) is also well-defined and linear. In addition, Theorem 6.28 gives

$$\begin{aligned} |\Lambda_f(g)| &\leq \int_\Sigma |f||g|\,d\sigma \\ &\leq \|f\|_{M^{p,\lambda}(\Sigma,w)}\|g\|_{\mathcal{B}^{p',\lambda}(\Sigma,w')} \quad \text{for all } g \in \mathcal{B}^{p',\lambda}(\Sigma,w'), \end{aligned} \tag{6.141}$$

which implies that Λ_f is continuous with the operator norm estimate

$$\|\Lambda_f\|_{(\mathcal{B}^{p',\lambda}(\Sigma,w'))^*} \leq \|f\|_{M^{p,\lambda}(\Sigma,w)}. \tag{6.142}$$

In summary, we have shown that the assignment in (6.139) is well-defined, linear, and continuous.

It remains to show injectivity and surjectivity of the mapping in (6.139). Granted this, the boundedness of the inverse of this mapping is a consequence of the open mapping theorem, whose present applicability is ensured by the fact that $M^{p,\lambda}(\Sigma,w)$ and $\mathcal{B}^{p',\lambda}(\Sigma,w')$ are Banach spaces (cf. Propositions 4.5 and 6.11, respectively).

Injectivity: We will show that if $\Lambda_f(g) = 0$ for every $g \in \mathcal{B}^{p',\lambda}(\Sigma,w')$, then $f = 0$. Suppose $f \in M^{p,\lambda}(\Sigma,w)$ is such that $\int_\Sigma fg\,d\sigma = 0$ for every $g \in \mathcal{B}^{p',\lambda}(\Sigma,w')$. Choose $g := \mathbf{1}_{\Delta_R}$, where $\Delta_R \subseteq \Sigma$ is an arbitrary surface ball of radius $R \in (0, 2\operatorname{diam}(\Sigma))$. Note that such a choice of g belongs to $\mathcal{B}^{p',\lambda}(\Sigma,w')$ in view of Proposition 6.25. Ultimately, bearing in mind that $f \in M^{p,\lambda}(\Sigma,w) \subseteq L^1_{\mathrm{loc}}(\Sigma,\sigma)$ thanks to Lemma 4.14, we see that Corollary 2.23 ensures that $f = 0$ at σ-a. e. point on Σ, as wanted.

Surjectivity: Given $\Lambda \in (\mathcal{B}^{p',\lambda}(\Sigma,w'))^*$, we need to find some function $f \in M^{p,\lambda}(\Sigma,w)$ such that $\Lambda_f = \Lambda$. For that purpose fix $\Delta_R \subseteq \Sigma$, an arbitrary surface ball of radius $R \in (0, 2\operatorname{diam}(\Sigma))$. Define

$$\iota_{\Delta_R} : L^{p'}(\Delta_R, w') \to \mathcal{B}^{p',\lambda}(\Sigma,w') \tag{6.143}$$

according to

$$\iota_{\Delta_R}(g) := \begin{cases} g & \text{on } \Delta_R \\ 0 & \text{on } \Sigma \setminus \Delta_R, \end{cases} \qquad \text{for all } g \in L^{p'}(\Delta_R, w'). \tag{6.144}$$

This is well-defined, since for any $g \in L^{p'}(\Sigma,w')$, not identically zero, the function $R^{\lambda(1/p'-1)}\|g\|^{-1}_{L^{p'}(\Sigma,w')}\iota_{\Delta_R}(g)$ is by construction a (p',λ,w')-block on Σ (cf. (6.1)). As such, if $g \in L^{p'}(\Delta_R,w')$, then the vector space property of $\mathcal{B}^{p',\lambda}(\Sigma,w')$ (cf. Proposition (6.9)) guarantees membership of $\iota_{\Delta_R}(g)$ to the Block space. That is,

$$\begin{aligned} &\iota_{\Delta_R}(g) \quad \text{belongs to } \mathcal{B}^{p',\lambda}(\Sigma,w') \text{ with} \\ &\|\iota_{\Delta_R}(g)\|_{\mathcal{B}^{p',\lambda}(\Sigma,w')} \leq R^{-\lambda(1/p'-1)}\|g\|_{L^{p'}(\Delta_R,w')}, \end{aligned} \tag{6.145}$$

for all $g \in \mathcal{B}^{p',\lambda}(\Sigma,w')$, where the norm estimate above is a consequence of Remark 6.23. Equivalently, we have the operator norm estimate

$$\|\iota_{\Delta_R}\|_{L^{p'}(\Sigma,w')\to\mathcal{B}^{p',\lambda}(\Sigma,w')} \leq R^{-\lambda(1/p'-1)}. \tag{6.146}$$

As a result, the composition $\Lambda \circ \iota_{\Delta_R}$ is a bounded linear functional on $L^{p'}(\Delta_R,w')$. That is, $\Lambda \circ \iota_{\Delta_R} \in (L^{p'}(\Delta_R,w'))^*$ with

$$\|\Lambda \circ \iota_{\Delta_R}\|_{(L^{p'}(\Delta_R,w'))^*} \leq \|\Lambda\|_{(\mathcal{B}^{p',\lambda}(\Sigma,w'))^*} R^{-\lambda(1/p'-1)}. \tag{6.147}$$

By the standard Riesz representation theorem for Lebesgue spaces [42, Theorem 6.5, p. 190]

$$\begin{aligned} &\text{there exists a unique } f_{\Delta_R} \in L^p(\Delta_R, w') \text{ such that} \\ &(\Lambda \circ \iota_{\Delta_R})(h) = \int_{\Delta_R} h f_{\Delta_R}\, dw' \quad \text{for all } h \in L^{p'}(\Delta_R, w'), \end{aligned} \tag{6.148}$$

and, moreover (in view of (6.147)),

$$\|f_{\Delta_R}\|_{L^p(\Delta_R,w')} \leq R^{-\lambda(1/p'-1)}\|\Lambda\|_{(\mathcal{B}^{p',\lambda}(\Sigma,w'))^*}. \tag{6.149}$$

At this point, we need to construct a function f that is independent of the choice of surface ball Δ_R. For that purpose, suppose that $x_1, x_2 \in \Sigma$ and $R_1, R_2 \in (0, 2\,\mathrm{diam}(\Sigma))$ are such that $\Delta(x_1, R_1) \subseteq \Delta(x_2, R_2)$. Choose some arbitrary $g \in L^{p'}(\Delta_{R_1}, w')$ and define the extension

$$\widetilde{g} := \begin{cases} g & \text{in } \Delta(x_1, R_1) \\ 0 & \text{in } \Delta(x_2, R_2) \setminus \Delta(x_1, R_1). \end{cases} \tag{6.150}$$

Because this is an extension by zero, $\widetilde{g} \in L^{p'}(\Delta(x_2, R_2), w')$. Note also that we have $\iota_{\Delta(x_1,R_1)}(g) = \iota_{\Delta(x_2,R_2)}(\widetilde{g})$ with both functions belonging to $\mathcal{B}^{p',\lambda}(\Sigma, w')$ thanks to (6.146). As a consequence, with $f_{\Delta(x_j,R_j)}$ ($i \in \{1,2\}$) denoting the functions guaranteed by the Riesz representation theorem as in (6.148), we obtain

$$\begin{aligned} \int_{\Delta(x_1,R_1)} f_{\Delta(x_1,R_1)} g\, dw' &= (\Lambda \circ \iota_{\Delta(x_2,R)})(\widetilde{g}) \\ &= \int_{\Delta(x_2,R_2)} f_{\Delta(x_2,R_2)} \widetilde{g}\, dw' \\ &= \int_{\Delta(x_1,R_1)} (f_{\Delta(x_2,R_2)}|_{\Delta(x_1,R_1)}) g\, dw'. \end{aligned} \tag{6.151}$$

By uniqueness in the Riesz representation, $f_{\Delta(x_2,R_2)}|_{\Delta(x_1,R_1)} = f_{\Delta(x_1,R_1)}$, which implies that there exists a unique function $F \in L^p_{\mathrm{loc}}(\Sigma, w')$ such that $F|_{\Delta_R} = f_{\Delta_R}$ for all surface balls $\Delta_R \subseteq \Sigma$.

Define $f := Fw'$ on Σ. Then having fixed $x_0 \in \Sigma$ and $j \in \mathbb{N}$, we have

$$f(x) = F(x)w'(x) = f_{\Delta(x_0,j)}(x)w'(x) \quad \text{for all } x \in \Delta(x_0, j), \tag{6.152}$$

and we remark that f is well-defined per the above discussion. Consider an arbitrary surface ball $\Delta_R \subseteq \Sigma$ with $R \in (0, 2\,\mathrm{diam}(\Sigma))$, and choose a radius $R_0 \in \mathbb{N}$ large enough so that $\Delta_R \subseteq \Delta(x_0, R_0)$. Then (6.152) and the compatibility property shown in (6.151) imply

$$f = f_{\Delta(x_0,R_0)} w' = f_{\Delta_R} w' \quad \text{on } \Delta_R. \tag{6.153}$$

With (6.153) in mind and the fact that $w' = w^{-p'/p} = w^{1-p'}$, we observe that for every surface ball $\Delta_R \subseteq \Sigma$ with radius $R \in (0, 2\,\mathrm{diam}(\Sigma))$,

$$\begin{aligned}\Big(\frac{1}{R^\lambda}\int_{\Delta_R} |f|^p\,dw\Big)^{1/p} &= R^{-\lambda/p}\Big(\int_{\Delta_R} |f_{\Delta_R}|^p w^{1-p'}\,d\sigma\Big)^{1/p} \qquad (6.154)\\ &= R^{-\lambda/p}\|f_{\Delta_R}\|_{L^p(\Delta_R,w')}\\ &\le R^{-\lambda/p}R^{-\lambda(1/p'-1)}\|\Lambda\|_{(\mathcal{B}^{p',\lambda}(\Sigma,w'))^*}\\ &= \|\Lambda\|_{(\mathcal{B}^{p',\lambda}(\Sigma,w'))^*}, \qquad (6.155)\end{aligned}$$

where we used (6.149) to obtain the inequality. Since $R^{-\lambda/p}R^{-\lambda(1/p'-1)} = 1$, the above bound is independent of Δ_R. Recalling (4.2), we see that (6.154) implies

$$f \in M^{p,\lambda}(\Sigma, w) \quad \text{and} \quad \|f\|_{M^{p,\lambda}(\Sigma,w)} \le \|\Lambda\|_{(\mathcal{B}^{p',\lambda}(\Sigma,w'))^*}. \tag{6.156}$$

Next, because of (6.140) and (6.153), for any given (p', λ, w')-block b on Σ, say with $\operatorname{supp} b \subseteq \Delta_R$, we have

$$\Lambda_f(b) = \int_\Sigma fb\,d\sigma = \int_\Sigma bf_{\Delta_R} w'\,d\sigma. \tag{6.157}$$

Additionally, since $b|_{\Delta_R} \in L^{p'}(\Delta_R, w)$ with $\operatorname{supp} b \subseteq \Delta_R$, using (6.144), (6.148), and (6.157), we find that

$$\begin{aligned}\Lambda(b) = (\Lambda \circ \iota_{\Delta_R})(b|_{\Delta_R}) &= \int_{\Delta_R} (b|_{\Delta_R}) f_{\Delta_R}\,dw' \qquad (6.158)\\ &= \int_{\Delta_R} bf_{\Delta_R} w'\,d\sigma = \Lambda_f(b),\end{aligned}$$

so that Λ_f and Λ agree on individual (p', λ, w')-blocks. Hence, Λ_f and Λ agree on finite linear combinations of (p', λ, w')-blocks on Σ, which constitutes a dense subset of $\mathcal{B}^{p',\lambda}(\Sigma, w')$ (cf. Proposition 6.24). Since Λ_f and Λ are both continuous, it follows that the actions of Λ_f and Λ agree on the whole space $\mathcal{B}^{p',\lambda}(\Sigma, w')$. Simply put, we have found a function $f \in M^{p,\lambda}(\Sigma, w)$ for which $\Lambda_f = \Lambda$ in $(\mathcal{B}^{p',\lambda}(\Sigma, w'))^*$, proving surjectivity of the mapping in (6.139). Moreover, (6.142) and (6.156) imply $\|f\|_{M^{p,\lambda}(\Sigma,w)} = \|\Lambda_f\|_{(\mathcal{B}^{p',\lambda}(\Sigma,w'))^*}$. This finally completes the proof of Theorem 6.29. □

Remark 6.30. The comment in (4.4) and Proposition 6.16 hint at the fact that the duality result in Theorem 6.29 can be extended to accommodate the endpoint case $\lambda = 0$. However, in the context of Theorem 6.29, it should be noted that the duality pairing between $L^p(\Sigma, w)$ and $L^{p'}(\Sigma, w')$ that one would obtain from Theorem 6.29 when $\lambda = 0$ is not the canonical duality pairing given by the Riesz representation theorem for Lebesgue spaces. We differ to Remark 10.40 for a more detailed discussion and justification of this alternate duality pairing for weighted Lebesgue spaces.

The duality paring in Theorem 6.29 allows us to give an alternate characterization of the block norm in terms of an integral pairing.

Corollary 6.31. *Let $\Sigma \subseteq \mathbb{R}^n$ be a closed* ADR *set and write $\sigma := \mathcal{H}^{n-1}\lfloor\Sigma$. Fix integrability exponents $p, p' \in (1, \infty)$ with $1/p + 1/p' = 1$, and fix a Muckenhoupt weight $w \in A_p(\Sigma, \sigma)$. Abbreviate $w' := w^{-p'/p} \in A_{p'}(\Sigma, \sigma)$, the conjugate weight to w, and let $\varepsilon \in (0, p' - 1)$ be such that $w' \in A_{p'-\varepsilon}(\Sigma, \sigma)$. Finally, fix $\lambda \in (0, \varepsilon(n-1)/(p'-1))$, and bring in the weighted Block space $\mathcal{B}^{p',\lambda}(\Sigma, w')$ associated to p', w', ε, and λ in the sense of Definition* 6.7.

Then for each $g \in \mathcal{B}^{p',\lambda}(\Sigma, w')$, one has

$$\|g\|_{\mathcal{B}^{p',\lambda}(\Sigma,w')} = \sup_{\substack{f \in M^{p,\lambda}(\Sigma,w) \\ \|f\|_{M^{p,\lambda}(\Sigma,w)} \leq 1}} \int_\Sigma |fg| \, d\sigma = \sup_{\substack{f \in M^{p,\lambda}(\Sigma,w) \\ \|f\|_{M^{p,\lambda}(\Sigma,w)} \leq 1}} \left| \int_\Sigma fg \, d\sigma \right|. \tag{6.159}$$

Proof. Given $g \in \mathcal{B}^{p',\lambda}(\Sigma, w')$, a well-known corollary of the Hahn–Banach theorem guarantees the existence of some functional $\Lambda \in (\mathcal{B}^{p',\lambda}(\Sigma, w'))^*$ with $\|\Lambda\|_{(\mathcal{B}^{p',\lambda}(\Sigma,w'))^*} = 1$ and $\|g\|_{\mathcal{B}^{p',\lambda}(\Sigma,w')} = \Lambda(g)$ (note Proposition 6.11 ensures $\mathcal{B}^{p',\lambda}(\Sigma, \omega)$ is Banach). According to Theorem 6.29, we have $\Lambda = \Lambda_f$ for some $f \in M^{p,\lambda}(\Sigma, w)$ with

$$\|f\|_{M^{p,\lambda}(\Sigma,w)} = \|\Lambda\|_{(\mathcal{B}^{p',\lambda}(\Sigma,w'))^*} = 1, \tag{6.160}$$

where Λ_f is as in (6.140). Subsequently,

$$\begin{aligned} \|g\|_{\mathcal{B}^{p',\lambda}(\Sigma,w')} &= \Lambda_f(g) = \int_\Sigma fg \, d\sigma \leq \left| \int_\Sigma fg \, d\sigma \right| \\ &\leq \sup_{\substack{f \in M^{p,\lambda}(\Sigma,w) \\ \|f\|_{M^{p,\lambda}(\Sigma,w)} \leq 1}} \left| \int_\Sigma fg \, d\sigma \right| \\ &\leq \sup_{\substack{f \in M^{p,\lambda}(\Sigma,w) \\ \|f\|_{M^{p,\lambda}(\Sigma,w)} \leq 1}} \int_\Sigma |fg| \, d\sigma. \end{aligned} \tag{6.161}$$

The opposite inequality in (6.159) readily follows from Theorem 6.28. □

As a consequence of the duality pairing between Muckenhoupt-weighted Block spaces and Muckenhoupt-weighted Morrey spaces, we obtain the following corollary, which is a weak-$*$ version of Lebesgue's dominated convergence theorem at the level of Muckenhoupt-weighted Morrey spaces.

Corollary 6.32. *Let $\Sigma \subseteq \mathbb{R}^n$ be a closed ADR set and write $\sigma := \mathcal{H}^{n-1}\lfloor\Sigma$. Fix two integrability exponents $p, p' \in (1,\infty)$ with $1/p + 1/p' = 1$, and a Muckenhoupt weight $w \in A_p(\Sigma,\sigma)$. Abbreviate $w' := w^{-p'/p} \in A_{p'}(\Sigma,\sigma)$, the conjugate weight to w. Choose some $\varepsilon \in (0, p'-1)$ for which $w' \in A_{p'-\varepsilon}(\Sigma,\sigma)$, and fix a threshold $\lambda \in (0, \varepsilon(n-1)/(p'-1))$. Bring in the weighted Block space $\mathcal{B}^{p',\lambda}(\Sigma, w')$ associated to p', w', ε, and λ in the sense of Definition 6.7. Finally, suppose $\{f_j\}_{j\in\mathbb{N}} \subseteq M^{p,\lambda}(\Sigma, w)$ is a sequence of functions with the property that*

$$\begin{aligned}&f(x) := \lim_{j\to\infty} f_j(x) \text{ exists for } \sigma\text{-a. e. point } x \in \Sigma, \text{ and there}\\ &\text{exists some } g \in M^{p,\lambda}(\Sigma, w) \text{ such that for each } j \in \mathbb{N} \text{ one}\\ &\text{has } |f_j(x)| \le |g(x)| \text{ for } \sigma\text{-a. e. point } x \in \Sigma.\end{aligned} \tag{6.162}$$

Then $f \in M^{p,\lambda}(\Sigma, w)$ and $f_j \to f$ as $j \to \infty$ weak-$*$ in $M^{p,\lambda}(\Sigma, w)$, i. e.,

$$\lim_{j\to\infty} \int_\Sigma f_j h \, d\sigma = \int_\Sigma fh \, d\sigma \quad \text{for each } h \in \mathcal{B}^{p',\lambda}(\Sigma, w'). \tag{6.163}$$

Proof. Note that f is a pointwise limit of σ-measurable functions, so f is σ-measurable. Also, from (6.162), we see that $|f(x)| \le |g(x)|$ for σ-a. e. point $x \in \Sigma$. The fact that $f \in M^{p,\lambda}(\Sigma, w)$ then follows because $M^{p,\lambda}(\Sigma, w)$ is a function lattice; cf. Proposition 4.7. To demonstrate weak-$*$ convergence, let $h \in \mathcal{B}^{p',\lambda}(\Sigma, w')$. Then $gh \in L^1(\Sigma,\sigma)$ as a consequence of Theorem 6.28, and we see that

$$\begin{aligned}&\lim_{j\to\infty}(f_j(x) - f(x)) = 0 \text{ for } \sigma\text{-a. e. point } x \in \Sigma, \text{ and for each } j \in \mathbb{N} \text{ we}\\ &\text{have } |(f_j(x) - f(x))h(x)| \le 2|g(x)h(x)| \text{ at } \sigma\text{-a. e. point } x \in \Sigma.\end{aligned} \tag{6.164}$$

Subsequently, we find via Lebesgue's dominated convergence theorem that

$$\lim_{j\to\infty} \int_\Sigma |(f_j - f)h| \, d\sigma = 0, \tag{6.165}$$

establishing (6.163) as wanted. □

Remark 6.33. Fix now $q \in (1,\infty)$, $\omega \in A_q(\Sigma,\sigma)$, and let $\varepsilon \in (0, q-1)$ be such that $\omega \in A_{q-\varepsilon}(\Sigma,\sigma)$, and $\lambda \in (0, \varepsilon(n-1)/(q-1))$. A similar formulation of Corollary 6.32 holds with $\mathcal{B}^{q,\lambda}(\Sigma,\omega)$ in place of $M^{p,\lambda}(\Sigma, w)$, where $\mathcal{B}^{q,\lambda}(\Sigma,\omega)$ is defined in relation to q, ω, ε, and λ in the sense of Definition 6.7. In this case, weak-$*$ convergence of $\{f_j\}_{j\in\mathbb{N}} \subseteq \mathcal{B}^{q,\lambda}(\Sigma,\omega)$ to $f \in \mathcal{B}^{q,\lambda}(\Sigma,\omega)$ should be understood as $\lim_{j\to\infty} \int_\Sigma f_j g \, d\sigma = \int_\Sigma fg \, d\sigma$ for each $g \in M^{p,\lambda}(\Sigma, w)$. The proof remains essentially unchanged, since $\mathcal{B}^{q,\lambda}(\Sigma,\omega)$ is also a function lattice; cf. Proposition 6.14.

Remark 6.34. Fix $q \in (1,\infty)$ with $q' := q/(q-1)$ along with a Muckenhoupt weight $\omega \in A_q(\Sigma,\sigma)$ satisfying $\omega' := \omega^{-q'/q} \in A_{q'}(\Sigma,\sigma) \cap \mathrm{RH}_p(\Sigma,\sigma)$ for some $p \in (1,\infty)$. Denote the Hölder conjugate $p' := p/(p-1)$ and define the parameter $\varepsilon := (q-1)/p' \in (0,q-1)$ so that $\omega \in A_{q-\varepsilon}(\Sigma,\sigma)$ as a consequence of Proposition 6.35, and select a threshold $\lambda \in (0,\varepsilon(n-1)/(q-1))$. Associate to these choices of q, ω, ε, and λ the weighted Block space $\mathcal{B}^{q,\lambda}(\Sigma,\omega)$ as in Definition 6.7. One can then obtain a version of Corollary 6.32 with $\mathcal{B}^{q,\lambda}(\Sigma,\omega)$ in place of $M^{p,\lambda}(\Sigma,w)$, which is a genuine weak-$*$ convergence result. This naturally involves the Muckenhoupt-weighted *vanishing* Morrey space $\mathring{M}^{q',\lambda}(\Sigma,\omega')$ (cf. Definition 10.20), the pre-dual to $\mathcal{B}^{q,\lambda}(\Sigma,\omega)$ (cf. Theorem 10.39).

The Block space $\mathcal{B}^{p',\lambda}(\Sigma,w')$, the pre-dual of $M^{p,\lambda}(\Sigma,w)$, is constructed in such a way that requires the choice of some $\varepsilon \in (0,p'-1)$ for which $w' \in A_{p'-\varepsilon}(\Sigma,\sigma)$. The existence of such an ε is always guaranteed by the self-improving property of Muckenhoupt weights (see item (1) of Proposition 2.29). While guaranteed, the actual value of ε is far from obvious. However, the following result provides an explicit choice for the value of ε, provided the weight w also belongs to a reverse Hölder class. Note that the following proposition also reconciles the two different upper bounds for λ found in our theory of Morrey spaces in Theorem 4.19 and our theory of Block spaces in Theorem 6.29.

Proposition 6.35. *Let $\Sigma \subseteq \mathbb{R}^n$ be a closed ADR set and write $\sigma := \mathcal{H}^{n-1}\lfloor\Sigma$. Fix four integrability exponents $p,p',q,q' \in (1,\infty)$ with the property that $1/p + 1/p' = 1$ and $1/q + 1/q' = 1$. Also, choose any Muckenhoupt weight $w \in A_p(\Sigma,\sigma) \cap \mathrm{RH}_q(\Sigma,\sigma)$ and bring in $w' := w^{-p'/p} \in A_{p'}(\Sigma,\sigma)$, its conjugate weight.*

Then $w' \in A_{p'-\varepsilon}(\Sigma,\sigma)$ when $\varepsilon := (p'-1)/q'$.

It is worth commenting that $\varepsilon := (p'-1)/q' \in (0,p'-1)$, and hence the above conclusion involves a meaningfully formulated Muckenhoupt class.

Proof of Proposition 6.35. With ε as above, begin by defining $s := p'-\varepsilon$ and observe that

$$s = p' - \varepsilon = p' - (p'-1)/q' = \frac{p'(q'-1)+1}{q'} > \frac{q'-1+1}{q'} = 1, \tag{6.166}$$

which ensures $s \in (1,\infty)$. Note that if

$$s' := \frac{p'(q'-1)+1}{(p'-1)(q'-1)}, \tag{6.167}$$

then $1/s + 1/s' = 1$. Since $1/q + 1/q' = 1$, we see that $q = q'/(q'-1)$, and subsequently

$$\begin{aligned} 1 - s' &= \frac{(p'-1)(q'-1) - p'(q'-1) - 1}{(p'-1)(q'-1)} \\ &= \frac{-q'}{(p'-1)(q'-1)} = \frac{-q}{p'-1}, \end{aligned} \tag{6.168}$$

and

$$s-1=\frac{p'(q'-1)+1-q'}{q'}=\frac{(p'-1)(q'-1)}{q'}=\frac{p'-1}{q}. \tag{6.169}$$

Fix $\Delta \subseteq \Sigma$, an arbitrary surface ball. Recall $1/p+1/p'=1$, which implies $p=p'/(p'-1)$. Consequently, we have

$$\Big(\fint_{\Delta} w'\,d\sigma\Big)\Big(\fint_{\Delta}(w')^{1-s'}\,d\sigma\Big)^{s-1}=\Big(\fint_{\Delta} w^{-p'/p}\,d\sigma\Big)\Big(\fint_{\Delta} w^{q}\,d\sigma\Big)^{(p'-1)/q}. \tag{6.170}$$

Because $w\in \mathrm{RH}_q(\Sigma,\sigma)$, there exists a constant $C\in(0,\infty)$, such that

$$\Big(\fint_{\Delta} w^{q}\,d\sigma\Big)^{1/q}\le C\fint_{\Delta} w\,d\sigma. \tag{6.171}$$

Noting that $-p'/p=1-p'$, from (6.170) and (6.171) we obtain

$$\begin{aligned}\Big(\fint_{\Delta} w'\,d\sigma\Big)\Big(\fint_{\Delta}(w')^{1-s'}\,d\sigma\Big)^{s-1}&\le C^{p'-1}\Big(\fint_{\Delta} w\,d\sigma\Big)^{p'-1}\Big(\fint_{\Delta} w^{1-p'}\,d\sigma\Big)\\&\le C^{p'-1}[w]_{A_p}^{p'-1},\end{aligned} \tag{6.172}$$

for all surface balls $\Delta\subseteq\Sigma$, where the last inequality follows from (2.47). Because $C\in(0,\infty)$ is independent of Δ, this ultimately implies that

$$[w']_{A_{p'-\varepsilon}}=[w']_{A_s}\le C[w]_{A_p}^{1/(p'-1)}<+\infty \tag{6.173}$$

thanks to (2.47). Thus, $w'\in A_{p'-\varepsilon}(\Sigma,\sigma)$ as claimed. □

Having established a solid theory for Block spaces from a functional analytic perspective, we are now in a position to establish a Calderón–Zygmund theory. While the information contained throughout this past chapter is interesting in its own right, our motivation for the study of Muckenhoupt-weighted Block spaces is out of necessity. More concretely, our journey to provide uniqueness results for boundary value problems with boundary data in a Muckenhoupt-weighted Morrey space has required us to study a space of functions that pair well in an integral sense with functions belonging to a Muckenhoupt-weighted Morrey space.

7 Singular integral operators on weighted Block spaces

In this chapter, we develop a Calderón–Zygmund theory for Muckenhoupt-weighted Block spaces. In doing so, we find that the extrapolation result presented in Chapter 4 is no longer of use when dealing with Block spaces. As such, to prove that various operators of interest are bounded on Block spaces without extrapolation is delicate and requires a significant amount of work. The strategy is to first establish estimates for an operator as it acts on individual blocks. This requires a dyadic decomposition along with hard estimates, which eventually show that these operators map individual blocks to block functions. We then use various embedding results concerning Block spaces and continuity results about our operators of interest to "glue" together estimates for individual blocks to general block functions.

Just as the Hardy–Littlewood maximal operator is bounded on classes of Muckenhoupt-weighted Morrey spaces (cf. Proposition 5.1), we can also demonstrate that the Hardy–Littlewood maximal operator is bounded on Muckenhoupt-weighted Block spaces. This makes the object of our first result in this chapter.

Proposition 7.1. *Let $\Sigma \subseteq \mathbb{R}^n$ be a closed* ADR *set and write $\sigma := \mathcal{H}^{n-1}\lfloor\Sigma$. Fix an integrability exponent $q \in (1,\infty)$ along with a Muckenhoupt weight $\omega \in A_q(\Sigma,\sigma)$. Choose $\varepsilon \in (0, q-1)$ such that $\omega \in A_{q-\varepsilon}(\Sigma,\sigma)$, and fix a threshold $\lambda \in (0, \varepsilon(n-1)/(q-1))$.*

Then the Hardy–Littlewood maximal operator $\mathcal{M}_\Sigma$ induces a sublinear mapping

$$\mathcal{M}_\Sigma : \mathcal{B}^{q,\lambda}(\Sigma,\omega) \to \mathcal{B}^{q,\lambda}(\Sigma,\omega), \tag{7.1}$$

which is well-defined and bounded. More specifically, there exists a constant $C \in (0,\infty)$, depending only on the ADR *character of Σ, n, q, λ, ε, $[\omega]_{A_{q-\varepsilon}}$, and $[\omega]_{A_q}$ such that*

$$\|\mathcal{M}_\Sigma g\|_{\mathcal{B}^{q,\lambda}(\Sigma,\omega)} \le C\|g\|_{\mathcal{B}^{q,\lambda}(\Sigma,\omega)}, \tag{7.2}$$

for all $g \in \mathcal{B}^{q,\lambda}(\Sigma,\omega)$.

Proof. We begin by showing that $\mathcal{M}_\Sigma b \in \mathcal{B}^{q,\lambda}(\Sigma,\omega)$ for a given (q,λ,ω)-block, b. For that purpose, fix such a (q,λ,ω)-block, b. Then there exist $x_0 \in \Sigma$ and $R \in (0, 2\,\mathrm{diam}(\Sigma))$ such that

$$\mathrm{supp}\, b \subseteq \Delta(x_0,R) \quad \text{and} \quad \|b\|_{L^q(\Sigma,\omega)} \le R^{\lambda(1/q-1)}. \tag{7.3}$$

Define the dyadic annuli

$$B_0 := \Delta(x_0,2R) \quad \text{and} \quad B_j := \Delta(x_0,2^{j+1}R) \setminus \Delta(x_0,2^jR) \quad \text{for all } j \in \mathbb{N}. \tag{7.4}$$

For $j \in \mathbb{N}_0$ set $\widetilde{b}_j := (\mathcal{M}_\Sigma b)\mathbf{1}_{B_j}$. Then

https://doi.org/10.1515/9783111458274-007

$$\mathcal{M}_\Sigma b = \sum_{j=0}^{\infty} \widetilde{b}_j \quad \text{pointwise on } \Sigma. \tag{7.5}$$

Consider first the case when $j = 0$. Immediately from the definition of $\widetilde{b}_0$, we have

$$\operatorname{supp} \widetilde{b}_0 \subseteq \Delta(x_0, 2R). \tag{7.6}$$

Additionally,

$$\|\widetilde{b}_0\|_{L^q(\Sigma,\omega)} \le \|\mathcal{M}_\Sigma b\|_{L^q(\Sigma,\omega)} \le C\|b\|_{L^q(\Sigma,\omega)} \le CR^{\lambda(1/q-1)}, \tag{7.7}$$

where the second inequality follows from the boundedness of $\mathcal{M}_\Sigma$ on $L^q(\Sigma, \omega)$ (cf. (2.43)), and the constant $C \in (0, \infty)$ is independent of b. With C as above, define

$$\lambda_0 := C \quad \text{and} \quad b_0 := \lambda^{-1}\widetilde{b}_0. \tag{7.8}$$

Because of (7.6) and (7.7), it is clear from (6.1) that b_0 is a (q, λ, ω)-block on Σ.

Consider now $j \in \mathbb{N}$ arbitrary. First, observe that

$$\operatorname{supp} \widetilde{b}_j \subseteq B_j \subseteq \Delta(x_0, 2^{j+1}R). \tag{7.9}$$

Now, fix $x \in B_j$. For a given radius $r \in (0, 2\operatorname{diam}(\Sigma))$, assume that we have $\Delta(x, r) \cap \Delta(x_0, R) \neq \varnothing$. Otherwise, the support condition on b in (7.3) implies

$$⨍_{\Delta(x,r)} |b|\, d\sigma = 0. \tag{7.10}$$

Consequently, this forces $r > (2^j - 1)R \ge \frac{1}{2}2^j R$. By Ahlfors regularity and the fact that $\Delta(x, r) \cap \operatorname{supp} b \subseteq \Delta(x_0, R)$, we have

$$\begin{aligned} ⨍_{\Delta(x,r)} |b|\, d\sigma &\le \frac{C}{r^{n-1}} \int_{\Delta(x_0,R)} |b|\, d\sigma \\ &\le \frac{C}{(2^j R)^{n-1}} \int_{\Delta(x_0,R)} |b|\, d\sigma. \end{aligned} \tag{7.11}$$

Bearing in mind (7.10)–(7.11), we conclude that there exists some constant $C \in (0, \infty)$, depending only on n and the ADR character of Σ, such that

$$(\mathcal{M}_\Sigma b)(x) \le \frac{C}{(2^j R)^{n-1}} \int_{\Delta(x_0,R)} |b|\, d\sigma \quad \text{for all } x \in B_j. \tag{7.12}$$

This allows us via Lemma 6.3 to estimate

$$\begin{aligned}\|\widetilde{b}_j\|_{L^q(\Sigma,\omega)} = \|\mathcal{M}_\Sigma b\|_{L^q(B_j,\omega)} &\leq \frac{C\omega(B_j)^{1/q}}{(2^jR)^{n-1}}\int_{\Delta(x_0,R)}|b|\,d\sigma \\ &\leq C2^{-j(n-1)(\varepsilon/q)}\cdot R^{\lambda(1/q-1)} \\ &= C\frac{2^{-j(n-1)(\varepsilon/q)}}{(2^{j+1})^{\lambda(1/q-1)}}(2^{j+1}R)^{\lambda(1/q-1)},\end{aligned} \tag{7.13}$$

for some $C \in (0,\infty)$, depending only on Σ, n, q, $[\omega]_{A_{q-\varepsilon}}$, and $[\omega]_{A_q}$. Then with C as above, define for each $j \in \mathbb{N}$,

$$\lambda_j := C\frac{2^{-j(n-1)(\varepsilon/q)}}{(2^{j+1})^{\lambda(1/q-1)}} \quad \text{and} \quad b_j := \lambda_j^{-1}\widetilde{b}_j. \tag{7.14}$$

Keeping in mind (7.9), the definition of b_j in (7.14), and the estimate in (7.13), we can conclude that

$$\operatorname{supp} b_j \subseteq \Delta(x_0, 2^{j+1}R) \quad \text{and} \quad \|b_j\|_{L^q(\partial\Omega,\omega)} \leq (2^{j+1}R)^{\lambda(1/q-1)}, \tag{7.15}$$

which demonstrates that b_j is a (q,λ,ω)-block. Since $\widetilde{b}_j = \lambda_j b_j$, from (7.5) we see that

$$\mathcal{M}_\Sigma b = \sum_{j=0}^{\infty}\lambda_j b_j \quad \text{pointwise on } \Sigma \tag{7.16}$$

and, moreover,

$$\sum_{j=1}^{\infty}|\lambda_j| \leq C\sum_{j=1}^{\infty}(2^{-j})^{(n-1)(\varepsilon/q)+\lambda(1/q-1)} \tag{7.17}$$

$$= C\sum_{j=1}^{\infty}2^{-j(1-1/q)[\varepsilon(n-1)/(q-1)-\lambda]} < +\infty, \tag{7.18}$$

thanks to (7.14). The above geometric series converges thanks to the hypothesis that $\lambda \in (0, \varepsilon(n-1)/(q-1))$, which ensures that

$$(1-1/q)[\varepsilon(n-1)/(q-1)-\lambda] > 0. \tag{7.19}$$

Thus, by Remark 6.8 the series $\sum_{j=0}^{\infty}\lambda_j b_j$ is a block representation for $\mathcal{M}_\Sigma b$. Hence,

for any (q,λ,ω)-block, b, one has $\mathcal{M}_\Sigma b \in \mathcal{B}^{q,\lambda}(\Sigma,\omega)$ with $\|\mathcal{M}_\Sigma b\|_{\mathcal{B}^{q,\lambda}(\partial\Omega,\omega)} \leq C_*$ for some constant $C_* \in (0,\infty)$, which is independent of the choice of block, b. (7.20)

Moving on to general block functions, fix $g \in \mathcal{B}^{q,\lambda}(\Sigma,\omega)$. There exists a quasi-optimal block representation

$$g = \sum_{j=1}^{\infty} \lambda_j b_j \quad \text{with} \quad \sum_{j=1}^{\infty} |\lambda_j| \le 2\|g\|_{\mathcal{B}^{q,\lambda}(\Sigma,\omega)}. \tag{7.21}$$

Since by design the convergence of the series is in $L^1(\Sigma, \frac{\sigma(x)}{1+|x|^{n-1}})$, we see (passing to a subsequence if necessary) that the series also converges pointwise σ-a. e. on Σ. Therefore,

$$|g| \le \sum_{j=1}^{\infty} |\lambda_j||b_j| \quad \text{pointwise } \sigma\text{-a. e. on } \Sigma, \tag{7.22}$$

which along with Lemma 2.24 further implies that

$$0 \le \mathcal{M}_\Sigma g \le \sum_{j=1}^{\infty} |\lambda_j|(\mathcal{M}_\Sigma b_j) \quad \text{pointwise everywhere on } \Sigma. \tag{7.23}$$

Claim: *The series $\sum_{j=1}^{\infty} |\lambda_j|(\mathcal{M}_\Sigma b_j)$ converges in $\mathcal{B}^{q,\lambda}(\Sigma,\omega)$.*

Let $M, N \in \mathbb{N}$, and without loss of generality take $N \le M$. Then

$$\begin{aligned}\Big\|\sum_{j=N}^{M} |\lambda_j|(\mathcal{M}_\Sigma b_j)\Big\|_{\mathcal{B}^{q,\lambda}(\Sigma,\omega)} &\le \sum_{j=N}^{M} |\lambda_j|\|\mathcal{M}_\Sigma b_j\|_{\mathcal{B}^{q,\lambda}(\Sigma,\omega)} \\ &\le C_* \sum_{j=N}^{M} |\lambda_j|,\end{aligned} \tag{7.24}$$

where the last inequality follows from (7.20). Since $\{\lambda_j\}_{j\in\mathbb{N}} \in \ell^1$, the above inequality shows that the sequence $\{\sum_{j=1}^{N} |\lambda_j|(\mathcal{M}_\Sigma b_j)\}_{N\in\mathbb{N}}$ is Cauchy in $\mathcal{B}^{q,\lambda}(\Sigma,\omega)$. By Theorem 6.11, we know $\mathcal{B}^{q,\lambda}(\Sigma,\omega)$ is a Banach space, so there exists some $g_* \in \mathcal{B}^{q,\lambda}(\Sigma,\omega)$ such that $\sum_{j=1}^{N} |\lambda_j|(\mathcal{M}_\Sigma b_j)$ converges to g_* in $\mathcal{B}^{q,\lambda}(\Sigma,\omega)$ as $N \to \infty$. This establishes the claim.

Pressing forward, by the continuous embedding

$$\mathcal{B}^{q,\lambda}(\Sigma,\omega) \hookrightarrow L^1\Big(\Sigma, \frac{\sigma(x)}{1+|x|^{n-1}}\Big), \tag{7.25}$$

established in Proposition 6.17, we find that $\sum_{j=1}^{N} |\lambda_j|(\mathcal{M}_\Sigma b_j)$ also converges to g_* pointwise σ-a. e. on Σ as $N \to \infty$. Knowing this and using (7.23), we find that

$$0 \le \mathcal{M}_\Sigma g \le g_* \in \mathcal{B}^{q,\lambda}(\Sigma,\omega). \tag{7.26}$$

Note by Proposition 6.17 that $g \in L^1_{\mathrm{loc}}(\Sigma,\sigma)$, and hence the function $\mathcal{M}_\Sigma g$ is σ-measurable. Therefore, the lattice property of $\mathcal{B}^{q,\lambda}(\Sigma,\omega)$ (cf. Proposition 6.14) and (7.26) show that $\mathcal{M}_\Sigma g$ belongs to $\mathcal{B}^{q,\lambda}(\Sigma,\omega)$, establishing well definiteness of the operator in (7.1). Moreover, by the continuity of $\|\cdot\|_{\mathcal{B}^{q,\lambda}(\Sigma,\omega)}$, we can estimate

$$\|g_*\|_{\mathcal{B}^{q,\lambda}(\Sigma,\omega)} = \lim_{N\to\infty} \Big\|\sum_{j=1}^{N} |\lambda_j|(\mathcal{M}_\Sigma b_j)\Big\|_{\mathcal{B}^{q,\lambda}(\Sigma,\omega)} \tag{7.27}$$

$$\begin{aligned}
&\leq \limsup_{N\to\infty} \sum_{j=1}^{N} |\lambda_j| \|\mathcal{M}_\Sigma b_j\|_{\mathcal{B}^{q,\lambda}(\Sigma,\omega)} \\
&\leq C_* \sum_{j=1}^{\infty} |\lambda_j| \\
&\leq C\|g\|_{\mathcal{B}^{q,\lambda}(\Sigma,\omega)}.
\end{aligned}$$

The second inequality follows from (7.20), and the last inequality follows from quasioptimality (7.21). As such, (7.26), the lattice property of $\mathcal{B}^{q,\lambda}(\Sigma,\omega)$ (Proposition 6.14) and (7.27) imply

$$\|\mathcal{M}_\Sigma g\|_{\mathcal{B}^{q,\lambda}(\Sigma,\omega)} \leq \|g_*\|_{\mathcal{B}^{q,\lambda}(\Sigma,\omega)} \leq C\|g\|_{\mathcal{B}^{q,\lambda}(\Sigma,\omega)} < +\infty \tag{7.28}$$

for some constant $C \in (0,\infty)$, independent of g. Thus, given the arbitrary nature of $g \in \mathcal{B}^{q,\lambda}(\Sigma,\omega)$, we find that $\mathcal{M}_\Sigma$ is bounded on $\mathcal{B}^{q,\lambda}(\Sigma,\omega)$ with proper quantitative control of the norm. □

This next result is a counterpart to the boundedness result of the convolution-type principal value singular integral operator in a Muckenhoupt-weighted Morrey space, which appears in Theorem 5.7, but now with boundedness at the level of Muckenhoupt-weighted Block spaces.

Theorem 7.2. *Let $\Sigma \subseteq \mathbb{R}^n$ be a closed UR set and write $\sigma := \mathcal{H}^{n-1}\lfloor\Sigma$. Let $N = N(n) \in \mathbb{N}$ be sufficiently large, and let $k \in \mathscr{C}^N(\mathbb{R}^n \setminus \{0\})$ be a complex-valued integral kernel which is odd, and positive homogeneous of degree $1-n$. Define the principal value singular integral operator T, which acts on each $f \in L^1(\Sigma, \frac{\sigma(x)}{1+|x|^{n-1}})$ according to*

$$(Tf)(x) := \lim_{\varepsilon\to 0^+} \int_{\substack{y\in\Sigma \\ |x-y|>\varepsilon}} k(x-y)f(y)\,d\sigma(y) \quad \text{for } \sigma\text{-a. e. point } x \in \Sigma. \tag{7.29}$$

Now, fix an integrability exponent $q \in (1,\infty)$, along with an arbitrary Muckenhoupt weight $\omega \in A_q(\Sigma,\sigma)$. Let $\varepsilon \in (0, q-1)$ be such that $\omega \in A_{q-\varepsilon}(\Sigma,\sigma)$, and fix a threshold $\lambda \in (0, \varepsilon(n-1)/(q-1))$.

Then the operator T induces a mapping

$$T : \mathcal{B}^{q,\lambda}(\Sigma,\omega) \to \mathcal{B}^{q,\lambda}(\Sigma,\omega) \tag{7.30}$$

which is well-defined, linear, and bounded. Moreover, there exists a constant $C \in (0,\infty)$, depending only on the UR character of Σ, n, q, ε, λ, $[\omega]_{A_{q-\varepsilon}}$, and $[\omega]_{A_q}$ such that

$$\|Tg\|_{\mathcal{B}^{q,\lambda}(\Sigma,\omega)} \leq C\Big(\sum_{|\alpha|\leq N} \sup_{S^{n-1}} |\partial^\alpha k|\Big)\|g\|_{\mathcal{B}^{q,\lambda}(\Sigma,\omega)}, \tag{7.31}$$

for all $g \in \mathcal{B}^{q,\lambda}(\Sigma,\omega)$.

Remark 7.3. It is pertinent to remark at this stage that we have the following results: (1) The Hardy–Littlewood maximal operator is bounded on Muckenhoupt-weighted Morrey spaces, cf. Proposition 5.1, (2) The Hardy–Littlewood maximal operator is bounded on Muckenhoupt-weighted Block spaces; cf. Proposition 7.1. Looking ahead, we will obtain (3) Muckenhoupt-weighted Morrey and Block spaces are mutually associated with one another in the context of generalized Banach function spaces (a. k. a. Kothë duality); cf. Theorem 10.46. This allows us to prove an extrapolation theorem for Muckenhoupt-weighted Block spaces, which transforms estimates at the level of Muckenhoupt-weighted Lebesgue spaces into estimates at the level of Muckenhoupt-weighted Block spaces (see Theorem 10.55 for details). It is tempting to use this extrapolation to transfer the results in Proposition 5.6 into a Calderón–Zygmund theory for Muckenhoupt-weighted Block spaces (see, e. g., how extrapolation at the level of weighted Morrey spaces is a useful tool in proving Theorem 5.7). While this approach does yield results in the spirit of, e. g., Theorem 7.2, it is nonetheless more restrictive than the direct proof we present here, which accommodates a more inclusive range of indices, e. g., a choice of the parameter ε, a quantity implicit in the definition of a weighted Block space (cf. Definition 6.7), which in turn affects the range of λ. This is because the extrapolation result in Theorem 10.55 requires the background presence of the weighted Morrey space dual to the Muckenhoupt-weighted Block space in question. As such, matters must be arranged so that the range for λ in the Block space setting coincides with the range for λ in the Morrey space setting. The direct proofs presented throughout Chapter 7, e. g., the proof of Theorem 7.2, make no reference to the weighted Morrey space, which is dual to the Muckenhoupt-weighted Block space under consideration, absolving the results of any additional restrictions on the parameters involved.

Proof of Theorem 7.2. As a preliminary case, observe that when k is identically zero, T is the zero operator, and hence (7.31) holds. Also, without loss of generality, we may assume that the integral kernel k is such that

$$M := \sum_{|\alpha|\le N} \sup_{S^{n-1}} |\partial^\alpha k| = 1 \tag{7.32}$$

Otherwise, simply replace k by k/M.

Now, begin by considering the case $g = b$, where b is a (q,λ,ω)-block on Σ. Then there exists $x_0 \in \Sigma$ and $R \in (0, 2\operatorname{diam}(\Sigma))$ such that

$$\operatorname{supp} b \subseteq \Delta(x_0,R) \quad \text{and} \quad \|b\|_{L^q(\Sigma,\omega)} \le R^{\lambda(1/q-1)}. \tag{7.33}$$

Define

$$B_0 := \Delta(x_0,2R) \quad \text{and} \quad B_j := \Delta(x_0,2^{j+1}R) \setminus \Delta(x_0,2^jR) \quad \text{for all } j \in \mathbb{N}. \tag{7.34}$$

For $j \in \mathbb{N}_0$, set $\widetilde{b}_j := (Tb)\mathbf{1}_{B_j}$. Then

$$Tb = \sum_{j=0}^{\infty} \widetilde{b}_j \quad \text{pointwise on } \Sigma. \tag{7.35}$$

Start by considering the case in which $j = 0$. Observe that

$$\operatorname{supp} \widetilde{b}_0 \subseteq \Delta(x_0, 2R). \tag{7.36}$$

We can also estimate

$$\|\widetilde{b}_0\|_{L^q(\Sigma,\omega)} \leq \|Tb\|_{L^q(\Sigma,\omega)} \leq C\|b\|_{L^q(\Sigma,\omega)} \leq CR^{\lambda(1/q-1)}, \tag{7.37}$$

where $C \in (0, \infty)$ is independent of b. The second inequality is a consequence of Proposition 5.6, and the last inequality follows from (7.33). With C as above, define

$$\lambda_0 := 2^{-\lambda(1/q-1)} C \quad \text{and} \quad b_0 := \lambda_0^{-1} \widetilde{b}_0. \tag{7.38}$$

Then the definition of b_0, (7.36), and (7.37) imply that b_0 is a (q, λ, ω)-block on Σ.

We now turn our attention to the case $j \in \mathbb{N}$. Observe first that

$$\operatorname{supp} \widetilde{b}_j \subseteq B_j \subseteq \Delta(x_0, 2^{j+1} R). \tag{7.39}$$

Next, for each $x \in B_j$, we find that

$$(Tb)(x) = \int_{\Delta(x_0,R)} k(x-y) b(y)\, d\sigma(y), \tag{7.40}$$

where we can omit the limit in the definition of T, because b is supported in $\Delta(x_0, R)$ and $\operatorname{dist}(\Delta(x_0, R), B_j) \geq R > 0$. By properties of the integral kernel, k (see Remark 5.4), for each $x \in B_j$ we may estimate

$$|(Tb)(x)| \leq C \int_{\Delta(x_0,R)} \frac{|b(y)|}{|x-y|^{n-1}}\, d\sigma(y). \tag{7.41}$$

To continue our estimate above, we make the following geometric observation. Namely,

$$|x_0 - x| \approx |x - y| \text{ uniformly for } x \in B_j \text{ and } y \in \Delta(x_0, R) \tag{7.42}$$

with implicit constants independent of j. Indeed,

$$|x_0 - x| \leq |x_0 - y| + |y - x| < R + |y - x| < 2^{-j}|x - x_0| + |y - x|, \tag{7.43}$$

where the second inequality is due to $x \in B_j$ implying $|x - x_0| \geq 2^j R$. Then (7.43) forces

$$(1 - 2^{-j})|x_0 - x| \leq |x - y|, \quad \text{and hence } \frac{1}{2}|x_0 - x| \leq |x - y|. \tag{7.44}$$

On the other hand,

$$\begin{aligned}|x-y| &\le |x-x_0|+|x_0-y| < |x-x_0|+R \\ &\le |x-x_0|+2^{-j}|x-x_0| = (1+2^{-j})|x-x_0| \le 2|x-x_0|.\end{aligned} \tag{7.45}$$

This establishes (7.42).

Since $x \in B_j$, we have that $2^j R \le |x-x_0|$, so in fact (7.42) implies that $2^j R \le C|x-y|$ for all $y \in \Delta(x_0,R)$ for some constant $C \in (0,\infty)$, independent of j. Combining this with (7.41) then yields

$$|(Tb)(x)| \le \frac{C}{(2^j R)^{n-1}} \int_{\Delta(x_0,R)} |b|\,d\sigma \quad \text{for all } x \in B_j. \tag{7.46}$$

As a consequence of Lemma 6.3, we see that

$$\begin{aligned}\|\widetilde{b}_j\|_{L^q(\Sigma,\omega)} = \|Tb\|_{L^q(B_j,\omega)} &\le \frac{C\omega(B_j)^{1/q}}{(2^j R)^{n-1}} \int_{\Delta(x_0,R)} |b|\,d\sigma \\ &\le C2^{-j(n-1)(\varepsilon/q)} \cdot R^{\lambda(1/q-1)} \\ &= C\frac{2^{-j(n-1)(\varepsilon/q)}}{(2^{j+1})^{\lambda(1/q-1)}}(2^{j+1}R)^{\lambda(1/q-1)},\end{aligned} \tag{7.47}$$

for some $C \in (0,\infty)$, depending only on Σ, n, q, $[\omega]_{A_{q-\varepsilon}}$, and $[\omega]_{A_q}$. With C as above, define for each $j \in \mathbb{N}$,

$$\lambda_j := C\frac{2^{-j(n-1)(\varepsilon/q)}}{(2^{j+1})^{\lambda(1/q-1)}} \quad \text{and} \quad b_j := \lambda_j^{-1}\widetilde{b}_j. \tag{7.48}$$

At this stage, it is apparent from the definition of $\widetilde{b}_j$, (7.48), and the estimate in (7.47) that

$$\operatorname{supp} b_j \subseteq \Delta(x_0, 2^{j+1}R) \quad \text{with } \|b_j\|_{L^q(\partial\Omega,\omega)} \le (2^{j+1}R)^{\lambda(1/q-1)}, \tag{7.49}$$

which shows that b_j is a (q,λ,ω)-block. Since $\widetilde{b_j} = \lambda_j b_j$, we have pointwise on Σ,

$$Tb = \sum_{j=0}^{\infty} \lambda_j b_j, \tag{7.50}$$

and in addition (thanks to (7.48)),

$$\begin{aligned}\sum_{j=1}^{\infty} |\lambda_j| &\le C\sum_{j=1}^{\infty} (2^{-j})^{(n-1)(\varepsilon/q)+\lambda(1/q-1)} \\ &= C\sum_{j=1}^{\infty} 2^{-j(1-1/q)[\varepsilon(n-1)/(q-1)-\lambda]} < +\infty.\end{aligned} \tag{7.51}$$

The geometric series converges by the hypothesis that λ belongs to the interval $(0, \varepsilon(n-1)/(q-1))$, which ensures that

$$(1 - 1/q)[\varepsilon(n-1)/(q-1) - \lambda] > 0. \tag{7.52}$$

Thus, by Lemma 6.8, the series $\sum_{j=0}^{\infty} \lambda_j b_j$ is a block representation for Tb, and hence

for any (q, λ, ω)-block, b, one has $Tb \in \mathcal{B}^{q,\lambda}(\Sigma, \omega)$ with $\|Tb\|_{\mathcal{B}^{q,\lambda}(\partial\Omega,\omega)} \leq C_*$ for some constant $C_* \in (0, \infty)$, which is independent of the choice of block, b. (7.53)

Now, consider some general block function $g \in \mathcal{B}^{q,\lambda}(\Sigma, \omega)$. Then there exists a quasioptimal block representation

$$g = \sum_{j=1}^{\infty} \lambda_j b_j \quad \text{with} \sum_{j=1}^{\infty} |\lambda_j| \leq 2\|g\|_{\mathcal{B}^{q,\lambda}(\Sigma,\omega)}. \tag{7.54}$$

Define for each $M \in \mathbb{N}$ the partial sums $g_M := \sum_{j=1}^{M} \lambda_j b_j$. Then by Proposition 6.17 and Proposition 5.19, we find that $Tg_M \to Tg$ as $M \to \infty$ in $L^{q_0}_{\text{loc}}(\Sigma, \sigma)$ for some $q_0 \in (1, \infty)$. Hence, there exists a sequence $\{M_k\}_{k \in \mathbb{N}} \subseteq \mathbb{N}$ such that

$$Tg = \lim_{k\to\infty} \sum_{j=1}^{M_k} \lambda_j T b_j \quad \text{at } \sigma\text{-a. e. point on } \Sigma. \tag{7.55}$$

Now, (7.53) shows that $Tb_j \in \mathcal{B}^{q,\lambda}(\Sigma, \omega)$ for each $j \in \mathbb{N}$. Moreover, the estimate in (7.53) allows us to conclude

$$\left\| \sum_{j=N}^{M} \lambda_j T b_j \right\|_{\mathcal{B}^{q,\lambda}(\Sigma,\omega)} \leq C_* \sum_{j=N}^{M} |\lambda_j| \tag{7.56}$$

for all $N, M \in \mathbb{N}$ with $N < M$. Because $\{\lambda_j\}_{j\in\mathbb{N}} \in \ell^1$, the right-hand side of the above estimate goes to 0 as $N, M \to \infty$, which shows that $\{\sum_{j=1}^{M} \lambda_j T b_j\}_{M\in\mathbb{N}}$ is a Cauchy sequence in $\mathcal{B}^{q,\lambda}(\Sigma, \omega)$. Since $\mathcal{B}^{q,\lambda}(\Sigma, \omega)$ is a Banach space (Theorem 6.11), there exists some $G \in \mathcal{B}^{q,\lambda}(\Sigma, \omega)$ for which

$$G = \lim_{M\to\infty} \sum_{j=1}^{M} \lambda_j T b_j \quad \text{in } \mathcal{B}^{q,\lambda}(\Sigma, \omega), \tag{7.57}$$

and as such

$$\begin{aligned} \|G\|_{\mathcal{B}^{q,\lambda}(\Sigma,\omega)} &= \lim_{M\to\infty} \left\| \sum_{j=1}^{M} \lambda_j T b_j \right\|_{\mathcal{B}^{q,\lambda}(\Sigma,\omega)} \\ &\leq C_* \sum_{j=1}^{\infty} |\lambda_j| \leq C\|g\|_{\mathcal{B}^{q,\lambda}(\Sigma,\omega)}, \end{aligned} \tag{7.58}$$

where the first inequality uses (7.53) and the last inequality above follows from quasioptimality (7.54). Combining (7.57) with the continuous embedding

$$\mathcal{B}^{q,\lambda}(\Sigma,\omega) \hookrightarrow L^1\Big(\Sigma, \frac{\sigma(x)}{1+|x|^{n-1}}\Big), \tag{7.59}$$

given in Proposition 6.17, we find that $G = \lim_{k\to\infty} \sum_{j=1}^{M_k} \lambda_j Tb_j$ with convergence of the limit in $L^1(\Sigma, \frac{\sigma(x)}{1+|x|^{n-1}})$. Consequently, there exists a subsequence $\{M_{k_l}\}_{l\in\mathbb{N}}$ for which

$$G = \lim_{l\to\infty} \sum_{j=1}^{M_{k_l}} \lambda_j Tb_j \quad \text{at } \sigma\text{-a. e. point on } \Sigma. \tag{7.60}$$

In concert, (7.55) and (7.60) guarantee $G = Tg$ at σ-a. e. point on Σ. As a consequence of this and the estimate in (7.58), we ultimately see that $Tg \in \mathcal{B}^{q,\lambda}(\Sigma,\omega)$ with

$$\|Tg\|_{\mathcal{B}^{q,\lambda}(\Sigma,\omega)} = \|G\|_{\mathcal{B}^{q,\lambda}(\Sigma,\omega)} \leq C\|g\|_{\mathcal{B}^{q,\lambda}(\Sigma,\omega)}$$

for some $C \in (0,\infty)$, independent of g. This is the desired result. □

The following Lemma (cf. [88, Proposition 2.2.5, pp. 276–278] for a more general version) is referred to as Cotlar's inequality. We will use it to demonstrate the boundedness of the maximal operator $T_{\max}$ on Muckenhoupt-weighted Block spaces. Cotlar's inequality provides a way to control the maximal operator in terms of the principal value singular integral operator and the Hardy–Littlewood maximal operator.

Lemma 7.4 (Cotlar's inequality). *Let $\Sigma \subseteq \mathbb{R}^n$ be a closed UR set and abbreviate $\sigma := \mathcal{H}^{n-1}\lfloor\Sigma$. Let $N = N(n) \in \mathbb{N}$ be sufficiently large, and let $k \in \mathscr{C}^N(\mathbb{R}^n \setminus \{0\})$ be a complex-valued integral kernel, which is odd, and positive homogeneous of degree $1-n$. For each $\varepsilon > 0$, define the truncated singular integral operator T_ε, which acts on each $f \in L^1(\Sigma, \frac{\sigma(x)}{1+|x|^{n-1}})$ according to*

$$(T_\varepsilon f)(x) := \int_{\substack{y\in\Sigma \\ |x-y|>\varepsilon}} k(x-y)f(y)\,d\sigma(y) \quad \textit{for each } x \in \Sigma. \tag{7.61}$$

Define also T and $T_{\max}$, which both act on each $f \in L^1(\Sigma, \frac{\sigma(x)}{1+|x|^{n-1}})$ according to[1]

$$(Tf)(x) := \lim_{\varepsilon\to 0^+} (T_\varepsilon f)(x) \quad \textit{for } \sigma\textit{-a. e. } x \in \Sigma, \tag{7.62}$$

and

1 The current geometric assumptions guarantee the well definiteness of T in (7.62); cf. [88].

$$(T_{\max}f)(x) := \sup_{\varepsilon>0} |(T_\varepsilon f)(x)| \quad \textit{for each } x \in \Sigma. \tag{7.63}$$

Then for each $f \in L^1(\Sigma, \frac{\sigma(x)}{1+|x|^{n-1}})$*, there exists a constant* $C \in (0, \infty)$*, depending only on* n*, and the Ahlfors regularity constants of* Σ *such that*

$$(T_{\max}f)(x) \leq C \cdot \mathcal{M}_\Sigma(Tf)(x) + C \cdot (\mathcal{M}_\Sigma f)(x) \quad \textit{for all } x \in \Sigma, \tag{7.64}$$

where $\mathcal{M}_\Sigma$ *denotes the Hardy–Littlewood maximal operator on* Σ*.*

We are now in a position to present the result that the maximal operator $T_{\max}$ is bounded on Muckenhoupt-weighted Block spaces. Compare with Theorem 5.7, which concerns the boundedness of the maximal operator on Muckenhoupt-weighted Morrey spaces. Interestingly, we are able to prove the boundedness of the Hardy–Littlewood maximal operator and the principal value operator, T, on Muckenhoupt-weighted Block spaces first (cf. Proposition 7.1 and Theorem 7.2, respectively) and then use Cotlar's inequality to extend those boundedness results to the larger (and hence more difficult to accommodate) maximal operator, $T_{\max}$. In relation to Proposition 7.5, similar comments as in Remark 7.3 apply.

Proposition 7.5. *Let* $\Sigma \subseteq \mathbb{R}^n$ *be a closed* UR *set and write* $\sigma := \mathcal{H}^{n-1}\lfloor\Sigma$*. Let* $N = N(n) \in \mathbb{N}$ *be sufficiently large and consider a complex-valued integral kernel,* $k \in \mathscr{C}^N(\mathbb{R}^n \setminus \{0\})$*, which is odd, and positive homogeneous of degree* $1-n$*. Define the maximal operator* $T_{\max}$*, which acts on each* $g \in L^1(\Sigma, \frac{\sigma(x)}{1+|x|^{n-1}})$ *according to*

$$(T_{\max}g)(x) := \sup_{\varepsilon>0} \Bigg| \int_{\substack{y\in\Sigma \\ |x-y|>\varepsilon}} k(x-y)g(y)\, d\sigma(y) \Bigg| \quad \textit{for each } x \in \Sigma. \tag{7.65}$$

Fix an integrability exponent $q \in (1, \infty)$*, along with a Muckenhoupt weight* $\omega \in A_q(\Sigma, \sigma)$*. Lastly, let* $\varepsilon \in (0, q-1)$ *be such that* $\omega \in A_{q-\varepsilon}(\Sigma, \sigma)$*, and fix* $\lambda \in (0, \varepsilon(n-1)/(q-1))$*.*

Then $T_{\max}$ *induces an operator*

$$T_{\max} : \mathcal{B}^{q,\lambda}(\Sigma, \omega) \to \mathcal{B}^{q,\lambda}(\Sigma, \omega), \tag{7.66}$$

which is well-defined, sublinear, and bounded. Moreover, there exists a constant $C \in (0, \infty)$*, depending only on* Σ*,* n*,* q*,* ε*,* λ*,* $[\omega]_{A_{q-\varepsilon}}$*, and* $[\omega]_{A_q}$*, with the property that*

$$\|T_{\max}g\|_{\mathcal{B}^{q,\lambda}(\Sigma,\omega)} \leq C\Big(\sum_{|\alpha|\leq N} \sup_{S^{n-1}} |\partial^\alpha k|\Big) \|g\|_{\mathcal{B}^{q,\lambda}(\Sigma,\omega)} \tag{7.67}$$

for all $g \in \mathcal{B}^{q,\lambda}(\Sigma, \omega)$*.*

Proof. In a preliminary stage, consider the case in which k is identically 0. Then clearly, $T_{\max}$ is the zero operator and (7.67) holds. Also, without loss of generality we may assume

$$M := \sum_{|\alpha|\le N} \sup_{S^{n-1}} |\partial^\alpha k| = 1. \tag{7.68}$$

Since otherwise, we may just replace k by k/M.

First, note that for each (q,λ,ω)-block, b, the function $T_{\max} b$ on Σ is σ-measurable, because $b \in L^q(\Sigma,\omega)$ and $T_{\max} : L^q(\Sigma,\omega) \to L^q(\Sigma,\omega)$ is a well-defined operator by Proposition 5.6. Also, $b \in L^1(\Sigma, \frac{\sigma(x)}{1+|x|^{n-1}})$ by Lemma 6.5. As such, Cotlar's inequality (Lemma 7.4) applies, and we find that there exists some constant $C \in (0,\infty)$, independent of b, such that

$$(T_{\max} b)(x) \le C \cdot \mathcal{M}_\Sigma(Tb)(x) + C \cdot (\mathcal{M}_\Sigma b)(x) \quad \text{for each } x \in \Sigma. \tag{7.69}$$

By Theorem 7.2, the principal value singular integral operator T is well-defined, linear, and bounded on $\mathcal{B}^{q,\lambda}(\Sigma,\omega)$, hence $Tb \in \mathcal{B}^{q,\lambda}(\Sigma,\omega)$ with

$$\|Tb\|_{\mathcal{B}^{q,\lambda}(\Sigma,\omega)} \le C\|b\|_{\mathcal{B}^{q,\lambda}(\Sigma,\omega)} \tag{7.70}$$

for some constant $C \in (0,\infty)$, independent of b. Also, by Proposition 7.1, we know that $\mathcal{M}_\Sigma$ is bounded on the Block space $\mathcal{B}^{q,\lambda}(\Sigma,\omega)$, and thus (7.69), the σ-measurability of $T_{\max} b$, and the lattice property of $\mathcal{B}^{q,\lambda}(\Sigma,\omega)$ (cf. Proposition 6.14) imply $T_{\max} b \in \mathcal{B}^{q,\lambda}(\Sigma,\omega)$ with

$$\begin{aligned} \|T_{\max} b\|_{\mathcal{B}^{q,\lambda}(\Sigma,\omega)} &\le C_* \|Tb\|_{\mathcal{B}^{q,\lambda}(\Sigma,\omega)} + C_* \|b\|_{\mathcal{B}^{q,\lambda}(\Sigma,\omega)} \\ &\le C_* \|b\|_{\mathcal{B}^{q,\lambda}(\Sigma,\omega)} \end{aligned} \tag{7.71}$$

for some constant $C_* \in (0,\infty)$, independent of b. Note the second inequality above follows from (7.70). Recalling Remark 6.23, we know that $\|b\|_{\mathcal{B}^{q,\lambda}(\Sigma,\omega)} \le 1$, and hence (7.71) implies that

$$\begin{aligned} &\text{for any } (q,\lambda,\omega)\text{-block, } b, \text{ one has } T_{\max} b \in \mathcal{B}^{q,\lambda}(\Sigma,\omega) \text{ with} \\ &\|T_{\max} b\|_{\mathcal{B}^{q,\lambda}(\Sigma,\omega)} \le C_* \text{ for some positive, finite constant } C_*, \\ &\text{which is independent of the choice of block, } b. \end{aligned} \tag{7.72}$$

Now, fix some $g \in \mathcal{B}^{q,\lambda}(\Sigma,\omega)$ along with a quasioptimal block representation

$$g = \sum_{j=1}^{\infty} \lambda_j b_j \quad \text{with } \sum_{j=1}^{\infty} |\lambda_j| \le 2\|g\|_{\mathcal{B}^{q,\lambda}(\Sigma,\omega)}. \tag{7.73}$$

Define for each $M \in \mathbb{N}$ the partial sums

$$g_M := \sum_{j=1}^{M} \lambda_j b_j \in L^1\Big(\Sigma, \frac{\sigma(x)}{1+|x|^{n-1}}\Big) \cap L^{q_0}_{\mathrm{loc}}(\Sigma,\sigma), \tag{7.74}$$

where the above membership is due to Corollary 6.6 and holds for some $q_0 \in (1, q-1)$. By Proposition 6.17 and Proposition 5.22, we see that $T_{\max} g_M$ converges to $T_{\max} g$ as $M \to \infty$ in $L^{q_0}_{\rm loc}(\Sigma, \sigma)$. As such, there exists a sequence $\{M_k\}_{k\in\mathbb{N}} \subseteq \mathbb{N}$ for which

$$T_{\max} g = \lim_{k\to\infty} T_{\max} g_{M_k} \quad \text{at } \sigma\text{-a. e. point on } \Sigma. \tag{7.75}$$

Claim: *Let $M, N \in \mathbb{N}$ with $M > N$. Then*

$$\|T_{\max} g_M - T_{\max} g_N\|_{\mathcal{B}^{q,\lambda}(\Sigma,\omega)} \le \|T_{\max}(g_M - g_N)\|_{\mathcal{B}^{q,\lambda}(\Sigma,\omega)}. \tag{7.76}$$

To see this, note that $T_{\max}$ is sublinear by definition. Choose two functions f, g belonging to $L^1(\Sigma, \frac{\sigma(x)}{1+|x|^{n-1}}) \cap L^{q_0}_{\rm loc}(\Sigma, \sigma)$, where q_0 is as in (7.74). Recall Proposition 5.22, which ensures that $T_{\max} f$ and $T_{\max} g$ belong to $L^{q_0}_{\rm loc}(\Sigma, \sigma)$. In particular, said functions are finite σ-a. e. on Σ. It follows that we have

$$T_{\max}(f + g) \le T_{\max} f + T_{\max} g \quad \text{pointwise on } \Sigma. \tag{7.77}$$

Replacing f by $f - g$ above implies

$$T_{\max} f - T_{\max} g \le T_{\max}(f - g) \quad \sigma\text{-a. e. on } \Sigma. \tag{7.78}$$

Similarly, replacing g by $g - f$ in (7.77) implies

$$T_{\max} g - T_{\max} f \le T_{\max}(g - f) \quad \sigma\text{-a. e. on } \Sigma. \tag{7.79}$$

Since $T_{\max}(f - g)$ is symmetric in f and g, (7.78) and (7.79) readily imply

$$|T_{\max} f - T_{\max} g| \le T_{\max}(f - g) \quad \sigma\text{-a. e. on } \Sigma. \tag{7.80}$$

In particular (in view of (7.74)), the above inequality shows that

$$|T_{\max} g_M - T_{\max} g_N| \le T_{\max}(g_M - g_N). \tag{7.81}$$

But $g_M - g_N = \sum_{j=N+1}^{M} \lambda_j b_j$ so by sublinearity of $T_{\max}$,

$$T_{\max}(g_M - g_N) \le \sum_{j=N+1}^{M} |\lambda_j| T_{\max} b_j \in \mathcal{B}^{q,\lambda}(\Sigma, \omega). \tag{7.82}$$

Since $T_{\max} g_M - T_{\max} g_N$ is σ-measurable, (7.81), (7.82), the lattice property of $\mathcal{B}^{q,\lambda}(\Sigma, \omega)$ (cf. Proposition 6.14), and Remark 6.15 establish the claim.

With the claim established, we may use (7.82) and the lattice property of Block spaces in Proposition 6.14 to estimate

$$\begin{aligned} \|T_{\max}g_M - T_{\max}g_N\|_{\mathcal{B}^{q,\lambda}(\Sigma,\omega)} &\leq \|T_{\max}(g_M - g_N)\|_{\mathcal{B}^{q,\lambda}(\Sigma,\omega)} \\ &\leq \sum_{j=N+1}^{M} |\lambda_j| \|T_{\max}b_j\|_{\mathcal{B}^{q,\lambda}(\Sigma,\omega)} \\ &\leq C_* \sum_{j=N+1}^{M} |\lambda_j|, \end{aligned} \tag{7.83}$$

whereby the last inequality follows from (7.72). Since $\{\lambda_j\}_{j\in\mathbb{N}} \in \ell^1$, we see that $\{T_{\max}g_M\}_{M\in\mathbb{N}}$ is a Cauchy sequence in $\mathcal{B}^{q,\lambda}(\Sigma,\omega)$. Subsequently, the fact that $\mathcal{B}^{q,\lambda}(\Sigma,\omega)$ is a Banach space (cf. Theorem 6.11) means there exists some $G \in \mathcal{B}^{q,\lambda}(\Sigma,\omega)$ for which

$$G = \lim_{M\to\infty} T_{\max}g_M \quad \text{in } \mathcal{B}^{q,\lambda}(\Sigma,\omega), \tag{7.84}$$

and as such,

$$\begin{aligned} \|G\|_{\mathcal{B}^{q,\lambda}(\Sigma,\omega)} &= \lim_{M\to\infty} \|T_{\max}g_M\|_{\mathcal{B}^{q,\lambda}(\Sigma,\omega)} \\ &\leq \lim_{M\to\infty} \sum_{j=1}^{M} |\lambda_j| \|T_{\max}b_j\|_{\mathcal{B}^{q,\lambda}(\Sigma,\omega)} \\ &\leq C_* \sum_{j=1}^{\infty} |\lambda_j| \\ &\leq C\|g\|_{\mathcal{B}^{q,\lambda}(\Sigma,\omega)}. \end{aligned} \tag{7.85}$$

The first inequality follows by the sublinearity of $T_{\max}$, the second inequality follows from (7.72), and the last inequality follows from quasioptimality in (7.73). Because of the continuous embedding $\mathcal{B}^{q,\lambda}(\Sigma,\omega) \hookrightarrow L^1(\Sigma, \frac{\sigma(x)}{1+|x|^{n-1}})$, established in Proposition 6.17, there exists a subsequence $\{M_{k_l}\}_{l\in\mathbb{N}}$ for which

$$G = \lim_{l\to\infty} T_{\max}g_{M_{k_l}} \quad \text{at } \sigma\text{-a. e. point on } \Sigma. \tag{7.86}$$

In concert, (7.75) and (7.86) guarantee $G = T_{\max}g$ at σ-a. e. point on Σ. Hence, the lattice property of Block spaces (cf. Proposition 6.14), and the estimate in (7.85) show that

$$\|T_{\max}g\|_{\mathcal{B}^{q,\lambda}(\Sigma,\omega)} = \|G\|_{\mathcal{B}^{q,\lambda}(\Sigma,\omega)} \leq C\|g\|_{\mathcal{B}^{q,\lambda}(\Sigma,\omega)} \tag{7.87}$$

for some constant $C \in (0,\infty)$, independent of g. This concludes the proof. □

This next result provides norm estimates including the nontangential maximal operator in the context of Muckenhoupt-weighted Block spaces. See (5.38) of Theorem 5.7 for a similar statement in the context of Muckenhoupt-weighted Morrey spaces. Of particular note is that a jump relation continues to hold for functions belonging to a weighted Block space. Note that comments as in Remark 7.3 continue to be relevant in the context of Theorem 7.6.

Theorem 7.6. *Let $\Omega \subseteq \mathbb{R}^n$ be an open set with uniformly rectifiable boundary, $\partial\Omega$, and abbreviate $\sigma := \mathcal{H}^{n-1}\lfloor\partial\Omega$. Let $N = N(n) \in \mathbb{N}$ be sufficiently large, and consider a complex-valued function $k \in \mathscr{C}^N(\mathbb{R}^n \setminus \{0\})$, which is odd and positive homogeneous of degree $1-n$. For each function $g \in L^1(\partial\Omega, \frac{\sigma(x)}{1+|x|^{n-1}})$ define*

$$\mathcal{T}g(x) := \int_{\partial\Omega} k(x-y)g(y)\,d\sigma(y), \quad \text{for each } x \in \Omega. \tag{7.88}$$

Fix an arbitrary aperture parameter $\kappa \in (0,\infty)$, an integrability exponent $q \in (1,\infty)$, a Muckenhoupt weight $\omega \in A_q(\Sigma,\sigma)$, choose $\varepsilon \in (0, q-1)$ such that $\omega \in A_{q-\varepsilon}(\Sigma,\sigma)$, and fix a threshold $\lambda \in (0, \varepsilon(n-1)/(q-1))$.

Then for each $g \in \mathcal{B}^{q,\lambda}(\partial\Omega,\omega)$ one has $\mathcal{N}_\kappa(\mathcal{T}g) \in \mathcal{B}^{q,\lambda}(\partial\Omega,\omega)$. Moreover, there exists a constant $C \in (0,\infty)$, depending only on $\partial\Omega$, n, q, $[\omega]_{A_q}$, λ, k, and κ, so that

$$\|\mathcal{N}_\kappa(\mathcal{T}g)\|_{\mathcal{B}^{q,\lambda}(\partial\Omega,\omega)} \leq C\|g\|_{\mathcal{B}^{q,\lambda}(\partial\Omega,\omega)}, \tag{7.89}$$

for all $g \in \mathcal{B}^{q,\lambda}(\partial\Omega,\omega)$.

Additionally, for each $g \in \mathcal{B}^{q,\lambda}(\partial\Omega,\omega)$, one has the jump formula

$$\big(\mathcal{T}g\big|^{\kappa-n.t.}_{\partial\Omega}\big)(x) = \frac{1}{2\sqrt{-1}}\widehat{k}(\nu(x))g(x) + (Tg)(x) \quad \text{at } \sigma\text{-a. e. } x \in \partial_*\Omega, \tag{7.90}$$

where $\nu(x)$ denotes the geometric measure theoretic outward unit normal to Ω at x, the function $\widehat{k}$ denotes the Fourier transform of k, and T is the principal value singular integral operator defined in (7.29).

Proof. Concerning the jump formula in (7.90), this is an immediate consequence of the embedding result established in Proposition 6.17 along with the jump formula provided in (5.34) of Proposition 5.6.

To establish (7.89), consider first the case when $g = b$, an arbitrary (q,λ,ω)-block on $\partial\Omega$. Then there exists $x_0 \in \partial\Omega$ and $R \in (0, \operatorname{diam}(\partial\Omega))$ such that

$$\operatorname{supp} b \subseteq \Delta(x_0,R) \quad \text{and} \quad \|b\|_{L^q(\partial\Omega,\omega)} \leq R^{\lambda(1/q-1)}. \tag{7.91}$$

Define the dyadic annuli

$$B_0 := \Delta(x_0,2R) \quad \text{and} \quad B_j := \Delta(x_0,2^{j+1}R) \setminus \Delta(x_0,2^jR) \quad \text{for all } j \in \mathbb{N}. \tag{7.92}$$

Then write $\widetilde{b}_j := (\mathcal{N}_\kappa(\mathcal{T}b))\mathbf{1}_{B_j}$ for each $j \in \mathbb{N}_0$, so that we have

$$\mathcal{N}_\kappa(\mathcal{T}b) = \sum_{j=0}^{\infty} \widetilde{b}_j \quad \text{pointwise on } \partial\Omega. \tag{7.93}$$

Focus first on $\widetilde{b}_0$. We see that

$$\operatorname{supp}\widetilde{b}_0 \subseteq \Delta(x_0, 2R), \tag{7.94}$$

and

$$\|\widetilde{b}_0\|_{L^q(\partial\Omega,\omega)} \le \|\mathcal{N}_\kappa(\mathcal{T}b)\|_{L^q(\partial\Omega,\omega)} \le C\|b\|_{L^q(\partial\Omega,\omega)} \le CR^{\lambda(1/q-1)} \tag{7.95}$$

for some constant $C \in (0,\infty)$, where the second inequality follows by (5.33) of Proposition 5.6 and (7.91). With C as above, define

$$\lambda_0 := C \quad \text{and} \quad b_0 := \lambda^{-1}\widetilde{b}_0. \tag{7.96}$$

From this definition, we can see by (7.94)–(7.95) that b_0 is a (q,λ,ω)-block on $\partial\Omega$.

We now turn our attention to $\widetilde{b}_j$ for $j \in \mathbb{N}$ arbitrary. Note first that

$$\operatorname{supp}\widetilde{b}_j \subseteq B_j \subseteq \Delta(x_0, 2^{j+1}R). \tag{7.97}$$

Having fixed $j \in \mathbb{N}$, pick $x \in B_j$. We then claim that there exists a constant $C \in (0,\infty)$, independent of x, such that

$$|x - x_0| \le C|y - z| \quad \text{for all } z \in \Gamma_\kappa(x) \text{ and } y \in \Delta(x_0, R). \tag{7.98}$$

To see this, observe that for $z \in \Gamma_\kappa(x)$ and $y \in \Delta(x_0, R)$, we may estimate

$$\begin{aligned} |x - x_0| &\le |x - z| + |z - y| + |y - x_0| \\ &< |x - z| + |z - y| + R \\ &\le (1+\kappa)|y - z| + |y - z| + 2^{-j}|x - x_0|. \end{aligned} \tag{7.99}$$

The last inequality uses that $|x - x_0| \ge 2^j$ (since $x \in B_j$), the definition of $\Gamma_\kappa(x)$, and the fact that $y \in \partial\Omega$. After absorbing the very last term into the leftmost side, this establishes the claim with $C := (1 - 2^{-1})^{-1}(2+\kappa)$. As a result of this estimate and properties of the integral kernel k (cf. (5.16) in Remark 5.4), and since $x \in B_j$ entails $2^jR \le |x - x_0|$, for each $z \in \Gamma_\kappa(x)$ we have

$$\begin{aligned} |(\mathcal{T}b)(z)| &\le C \int_{\Delta(x_0,R)} \frac{1}{|z-y|^{n-1}} |b(y)|\, d\sigma(y) \\ &\le \frac{C}{(2^jR)^{n-1}} \int_{\Delta(x_0,R)} |b|\, d\sigma. \end{aligned} \tag{7.100}$$

As $z \in \Gamma_\kappa(x)$ was arbitrary, Definition 2.12 and the above estimate implies

$$\mathcal{N}_\kappa(\mathcal{T}b)(x) \le \frac{C}{(2^j R)^{n-1}} \int_{\Delta(x_0,R)} |b|\, d\sigma \quad \text{for each } x \in B_j. \tag{7.101}$$

Consequently, Lemma 6.3 yields that

$$\begin{aligned} \|\widetilde{b}_j\|_{L^q(\partial\Omega,\omega)} = \|\mathcal{N}_\kappa(\mathcal{T}b)\|_{L^q(B_j,\omega)} &\le \frac{C\omega(B_j)^{1/q}}{(2^j R)^{n-1}} \int_{\Delta(x_0,R)} |b|\, d\sigma \\ &\le C2^{-j(n-1)(\varepsilon/q)} \cdot R^{\lambda(1/q-1)} \\ &= C\frac{2^{-j(n-1)(\varepsilon/q)}}{(2^{j+1})^{\lambda(1/q-1)}}(2^{j+1}R)^{\lambda(1/q-1)}. \end{aligned} \tag{7.102}$$

With C as above, define

$$\lambda_j := C\frac{2^{-j(n-1)(\varepsilon/q)}}{(2^{j+1})^{\lambda(1/q-1)}} \quad \text{and} \quad b_j := \lambda_j^{-1}\widetilde{b}_j. \tag{7.103}$$

Then (7.97) and (7.102) imply that

$$\operatorname{supp} b_j \subseteq \Delta(x_0, 2^{j+1}R) \quad \text{and} \quad \|b_j\|_{L^q(\partial\Omega,\omega)} \le (2^{j+1}R)^{\lambda(1/q-1)}, \tag{7.104}$$

showing that b_j is a (q,λ,ω)-block. Since $\widetilde{b}_j = \lambda_j b_j$ for all $j \in \mathbb{N}_0$, we have that

$$\mathcal{N}_\kappa(\mathcal{T}b) = \sum_{j=0}^{\infty} \lambda_j b_j \quad \text{pointwise on } \partial\Omega. \tag{7.105}$$

In fact, from the definition of λ_j in (7.103), we see

$$\begin{aligned} \sum_{j=1}^{\infty} |\lambda_j| &\le C\sum_{j=1}^{\infty} 2^{-j[(n-1)(\varepsilon/q)+\lambda(1/q-1)]} \\ &= C\sum_{j=1}^{\infty} 2^{-j(1-1/q)[\varepsilon(n-1)/(q-1)-\varepsilon]} =: C_* < +\infty. \end{aligned} \tag{7.106}$$

The above geometric series converges thanks to the present hypothesis that $\lambda \in (0, \varepsilon(n-1)/(q-1))$, which ensures that

$$(1-1/q)[\varepsilon(n-1)/(q-1)-\varepsilon] > 0. \tag{7.107}$$

By Remark 6.8, we can conclude that $\sum_{j=0}^{\infty} \lambda_j b_j$ is a block representation for $\mathcal{N}_\kappa(\mathcal{T}b)$. Hence,

> for any (q,λ,ω)-block, b, one has $\mathcal{N}_\kappa(\mathcal{T}b) \in \mathcal{B}^{q,\lambda}(\Sigma,\omega)$ with the estimate $\|\mathcal{N}_\kappa(\mathcal{T}b)\|_{\mathcal{B}^{q,\lambda}(\partial\Omega,\omega)} \le C_*$ for some constant $C_* \in (0,\infty)$, which is independent of b. (7.108)

The next stage in this proof addresses general block functions. Fix an arbitrary $g \in \mathcal{B}^{q,\lambda}(\Sigma, \omega)$. Then there exists a quasioptimal block representation

$$g = \sum_{j=1}^{\infty} \lambda_j b_j \quad \text{with} \sum_{j=1}^{\infty} |\lambda_j| \leq 2\|g\|_{\mathcal{B}^{q,\lambda}(\partial\Omega,\omega)}. \tag{7.109}$$

Define the partial sums $g_m := \sum_{j=1}^{m} \lambda_j b_j$, and set

$$u := \mathcal{T}g \quad \text{and} \quad u_m := \mathcal{T}g_m \quad \text{for all } m \in \mathbb{N}. \tag{7.110}$$

Claim 1: $u_m \to u$ *as* $m \to \infty$ *pointwise in* Ω.

First, observe that $g_m \to g$ as $m \to \infty$ in $\mathcal{B}^{q,\lambda}(\partial\Omega, \omega)$. Proposition 6.17 shows that $\mathcal{B}^{q,\lambda}(\partial\Omega, \omega)$ continuously embeds into $L^1(\Sigma, \frac{\sigma(x)}{1+|x|^{n-1}})$, and hence $g_m \to g$ as $m \to \infty$ in $L^1(\Sigma, \frac{\sigma(x)}{1+|x|^{n-1}})$ as well. Fix $x \in \Omega$. Then

$$\begin{aligned} |u_m(x) - u(x)| &= |(\mathcal{T}g_m)(x) - (\mathcal{T}g)(x)| \\ &\leq \int_{\partial\Omega} |k(x-y)||g_m(y) - g(y)|\, d\sigma(y) \\ &\leq C_{x,k}\|g_m - g\|_{L^1(\partial\Omega, \frac{\sigma(x)}{1+|x|^{n-1}})} \to 0 \quad \text{as } m \to \infty. \end{aligned} \tag{7.111}$$

The last inequality above follows from properties of the integral kernel k (cf. (5.17)). This establishes the claim.

Claim 2: $\mathcal{N}_\kappa u \leq \limsup_{m\to\infty} \mathcal{N}_\kappa u_m$ *pointwise everywhere on* $\partial\Omega$.

Fix $x \in \partial\Omega$ and choose $y \in \Gamma_\kappa(x)$ arbitrary. By Claim 1, $u_m(y) \to u(y)$ as $m \to \infty$, and hence

$$|u(y)| = \lim_{m\to\infty} |u_m(y)| \leq \lim_{m\to\infty} \Big(\sup_{z\in\Gamma_\kappa(x)} |u_m(z)| \Big) = \limsup_{m\to\infty} (\mathcal{N}_\kappa u_m)(x), \tag{7.112}$$

by the continuity of u_m. As $y \in \Gamma_\kappa(x)$ was arbitrary, taking a supremum over all $y \in \Gamma_\kappa(x)$ in the above inequality establishes the claim.

Pressing forward, for each $m \in \mathbb{N}$, the sublinearity of $\mathcal{N}_\kappa$ implies

$$\mathcal{N}_\kappa u_m \leq \mathcal{N}_\kappa \Big(\sum_{j=1}^{m} \lambda_j \mathcal{T} b_j \Big) \leq \sum_{j=1}^{m} |\lambda_j| \mathcal{N}_\kappa(\mathcal{T} b_j). \tag{7.113}$$

By (7.108), we know $\mathcal{N}_\kappa(\mathcal{T}b_j) \in \mathcal{B}^{q,\lambda}(\partial\Omega, \omega)$ for each $j \in \mathbb{N}$. Since $\mathcal{N}_\kappa u_m$ is σ-measurable for each $m \in \mathbb{N}$ (cf. Remark 2.13), by the lattice property of $\mathcal{B}^{q,\lambda}(\partial\Omega, \omega)$ (cf. Proposition 6.14) we also find that $\mathcal{N}_\kappa u_m \in \mathcal{B}^{q,\lambda}(\partial\Omega, \omega)$ for every $m \in \mathbb{N}$. Moreover, for each $m \in \mathbb{N}$ we may estimate using the sublinearity of $\mathcal{N}_\kappa$,

$$\begin{aligned}\|\mathcal{N}_\kappa u_m\|_{\mathcal{B}^{q,\lambda}(\partial\Omega,\omega)} &\le \Big\|\sum_{j=1}^{m}|\lambda_j|\mathcal{N}_\kappa(\mathcal{T}b_j)\Big\|_{\mathcal{B}^{q,\lambda}(\partial\Omega,\omega)} \\ &\le \sum_{j=1}^{m}|\lambda_j|\|\mathcal{N}_\kappa(\mathcal{T}b_j)\|_{\mathcal{B}^{q,\lambda}(\partial\Omega,\omega)} \\ &\le C_*\sum_{j=1}^{m}|\lambda_j|,\end{aligned} \tag{7.114}$$

where the last estimate follows from (7.108). Let $m, m' \in \mathbb{N}$. Without loss of generality, take $m < m'$. By similar reasoning as above,

$$\begin{aligned}\|\mathcal{N}_\kappa u_m - \mathcal{N}_\kappa u_{m'}\|_{\mathcal{B}^{q,\lambda}(\partial\Omega,\omega)} &\le \|\mathcal{N}_\kappa(u_m - u_{m'})\|_{\mathcal{B}^{q,\lambda}(\partial\Omega,\omega)} \\ &\le C_*\sum_{j=m}^{m'}|\lambda_j|,\end{aligned} \tag{7.115}$$

with the first inequality guaranteed by Lemma 2.16 and the lattice property of $\mathcal{B}^{q,\lambda}(\Sigma,\omega)$ (cf. Proposition 6.14). In view of the fact that $\{\lambda_j\}_{j\in\mathbb{N}} \in \ell^1$, this implies $\{\mathcal{N}_\kappa u_m\}_{m\in\mathbb{N}}$ is a Cauchy sequence in $\mathcal{B}^{q,\lambda}(\Sigma,\omega)$. By Proposition 6.11, we know $\mathcal{B}^{q,\lambda}(\Sigma,\omega)$ is a Banach space, so $\{\mathcal{N}_\kappa u_m\}_{m\in\mathbb{N}}$ converges to some $G \in \mathcal{B}^{q,\lambda}(\Sigma,\omega)$ as $m \to \infty$. Without loss of generality, after relabeling we may assume $\mathcal{N}_\kappa u_m \to G$ in $\mathcal{B}^{q,\lambda}(\Sigma,\omega)$ and σ-a. e. on $\partial\Omega$. Since $\|\cdot\|_{\mathcal{B}^{q,\lambda}(\partial\Omega,\omega)}$ is continuous,

$$\|G\|_{\mathcal{B}^{q,\lambda}(\partial\Omega,\omega)} = \lim_{m\to\infty}\|\mathcal{N}_\kappa u_m\|_{\mathcal{B}^{q,\lambda}(\partial\Omega,\omega)} \le C\|g\|_{\mathcal{B}^{q,\lambda}(\partial\Omega,\omega)}, \tag{7.116}$$

where the last inequality follows by (7.114) and quasioptimality in (7.109). By Claim 2 and σ-a. e. convergence of $\{N_\kappa u_m\}_{m\in\mathbb{N}}$, we find $\mathcal{N}_\kappa u \le G$, so by the lattice property of Block spaces (cf. Proposition 6.14), we find that $\mathcal{N}_\kappa u \in \mathcal{B}^{q,\lambda}(\partial\Omega,\omega)$ with

$$\|\mathcal{N}_\kappa(\mathcal{T}g)\|_{\mathcal{B}^{q,\lambda}(\partial\Omega,\omega)} \le \|G\|_{\mathcal{B}^{q,\lambda}(\partial\Omega,\omega)} \le C\|g\|_{\mathcal{B}^{q,\lambda}(\partial\Omega,\omega)}, \tag{7.117}$$

by (7.110) and (7.116). This completes the proof of Theorem 7.6. □

So far, we have consistently favored the nontangential maximal operator as a means of measuring the size of a function defined in an open set. As the following proposition demonstrates, it is possible to employ with equal effectiveness, the so-called tangential maximal function from Definition 2.17.

Proposition 7.7. *Let Ω an open, nonempty, proper subset of $\mathbb{R}^n$ with $\partial\Omega$ an Ahlfors regular set and abbreviate $\sigma := \mathcal{H}^{n-1}\lfloor\partial\Omega$. Fix any power $M \ge n-1$, and any aperture parameter $\kappa > 0$. Then the following results are true:*

(i) *Select any two integrability exponents $p, q \in (1,\infty)$, a Muckenhoupt weight $w \in A_p(\partial\Omega, w) \cap \mathrm{RH}_q(\partial\Omega,\sigma)$, and a threshold $\lambda \in (0,(n-1)/q')$, where $q' := q/(q-1)$.*

Then

$$\|u_M^{\max}\|_{M^{p,\lambda}(\partial\Omega,w)} \approx \|\mathcal{N}_\kappa u\|_{M^{p,\lambda}(\partial\Omega,w)} \tag{7.118}$$

in a uniform fashion for Lebesgue measurable functions $u : \Omega \to \mathbb{C}$.

(ii) *Select an integrability exponent* $q \in (1,\infty)$ *along with a Muckenhoupt weight* $\omega \in A_q(\partial\Omega,\sigma)$. *Choose a parameter* $\varepsilon \in (0, q-1)$ *for which* $\omega \in A_{q-\varepsilon}(\partial\Omega,\sigma)$, *and select any threshold* $\lambda \in (0, \varepsilon(n-1)/(q-1))$. *Then*

$$\|u_M^{\max}\|_{\mathcal{B}^{q,\lambda}(\partial\Omega,\omega)} \approx \|\mathcal{N}_\kappa u\|_{\mathcal{B}^{q,\lambda}(\partial\Omega,\omega)} \tag{7.119}$$

in a uniform fashion for Lebesgue measurable functions $u : \Omega \to \mathbb{C}$.

Proof. We may argue much as in [86, Corollary 8.5.6, pp. 729–730], which we are permitted to do given the Hardy–Littlewood maximal operator is bounded on the present scales of Muckenhoupt-weighted Morrey and Block spaces (cf. Propositions 5.1 and 7.1, respectively), as well as the lattice property that both scales of spaces enjoy; cf. Proposition 4.7 and 6.14. □

As a consequence of our duality theory encapsulated in Proposition 6.28 and Theorem 6.29, the (real) transpose of the principal value singular integral operator T acting on a Block space is given by $T^\top = -T$, which acts on the Morrey space dual to the original Block space. See (5.31) in Theorem 5.6 for an analogous statement at the level of Lebesgue spaces.

Theorem 7.8. *Let* $\Sigma \subseteq \mathbb{R}^n$ *be a closed uniformly rectifiable set and write* $\sigma := \mathcal{H}^{n-1}\lfloor\Sigma$. *Let* $N = N(n) \in \mathbb{N}$ *be sufficiently large and consider a complex-valued integral kernel* $k \in \mathscr{C}^N(\mathbb{R}^n \setminus \{0\})$, *which is odd and positive homogeneous of degree* $1-n$. *Define the principal value operator* T, *which acts on each* $g \in L^1(\Sigma, \frac{\sigma(x)}{1+|x|^{n-1}})$ *according to*

$$(Tg)(x) := \lim_{\varepsilon\to 0^+} \int_{\substack{y\in\Sigma\\ |x-y|>\varepsilon}} k(x-y)g(y)\,d\sigma(y) \quad \textit{for } \sigma\textit{-a. e. } x \in \Sigma. \tag{7.120}$$

Fix two integrability exponents $p, q \in (1,\infty)$, *and define* $p' := p/(p-1)$ *and* $q' := q/(q-1)$. *Fix some Muckenhoupt weight* $w \in A_p(\Sigma,\sigma) \cap \mathrm{RH}_q(\Sigma,\sigma)$, *and denote the conjugate weight to* w *by* $w' := w^{-p'/p} \in A_{p'}(\Sigma,\sigma)$. *Define* $\varepsilon := (p'-1)/q' \in (0, p'-1)$, *so that* w' *belongs to* $A_{p'-\varepsilon}(\Sigma,\sigma)$ *thanks to Proposition* 6.35, *and fix some arbitrary threshold*

$$\lambda \in (0, \varepsilon(n-1)/(p'-1)) = (0, (n-1)/q'). \tag{7.121}$$

Lastly, bring in the weighted Block space $\mathcal{B}^{p',\lambda}(\Sigma, w')$ *associated with* p', w', ε, *and* λ *in the sense of Definition* 6.7.

Then the operator $T : \mathcal{B}^{p',\lambda}(\Sigma, w') \to \mathcal{B}^{p',\lambda}(\Sigma, w')$ *induces a well-defined, linear, and bounded operator* $-T : M^{p,\lambda}(\Sigma, w) \to M^{p,\lambda}(\Sigma, w)$ *with*

$$\int_{\Sigma} (Tg)f \, d\sigma = -\int_{\Sigma} g(Tf) \, d\sigma \tag{7.122}$$

for all $f \in M^{p,\lambda}(\Sigma, w)$ *and all* $g \in \mathcal{B}^{p',\lambda}(\Sigma, w')$. *Concisely,* $T^{\top} = -T$.

Proof. Consider first for each $\varepsilon > 0$ the truncated operators T_ε, acting on each function $f \in L^1(\Sigma, \frac{\sigma(x)}{1+|x|^{n-1}})$ according to

$$(T_\varepsilon f)(x) := \int_{\substack{y\in\Sigma \\ |x-y|>\varepsilon}} k(x-y)f(y) \, d\sigma(y) \quad \text{at each } x \in \Sigma. \tag{7.123}$$

Then $Tf(x) = \lim_{\varepsilon\to 0^+}(T_\varepsilon f)(x)$ for σ-a. e. point $x \in \Sigma$, and we define as usual the maximal operator $(T_{\max} f)(x) := \sup_{\varepsilon>0} |(T_\varepsilon f)(x)|$ for each $x \in \Sigma$. Let $f \in M^{p,\lambda}(\Sigma, w)$ and $g \in \mathcal{B}^{p',\lambda}(\Sigma, w')$. Note that since $T_\varepsilon g$ is σ-measurable (cf. Proposition 6.17 and Theorem 5.6), and since $|T_\varepsilon g| \leq T_{\max} g$ for all $\varepsilon > 0$, the lattice property of $\mathcal{B}^{p',\lambda}(\Sigma, w')$ (cf. Proposition 6.14) implies that

$$\begin{aligned} \|T_\varepsilon g\|_{\mathcal{B}^{p',\lambda}(\Sigma,w')} &\leq \|T_{\max} g\|_{\mathcal{B}^{p',\lambda}(\Sigma,w')} \\ &\leq C\|g\|_{\mathcal{B}^{p',\lambda}(\Sigma,w')} < +\infty, \end{aligned} \tag{7.124}$$

where the second inequality follows from Proposition 7.5, and $C \in (0, \infty)$ is independent of g. Thus, the operator $T_\varepsilon : \mathcal{B}^{p',\lambda}(\Sigma, w') \to \mathcal{B}^{p',\lambda}(\Sigma, w')$ is well-defined, linear, and bounded.

If $g \in \mathcal{B}^{p',\lambda}(\Sigma, w')$, then there exist $\{\lambda_j\}_{j\in\mathbb{N}} \in \ell^1$ and $\{b_j\}_{j\in\mathbb{N}}$, a sequence of (p', λ, w')-blocks so that $g = \sum_{j=1}^{\infty} \lambda_j b_j$. If we define for each $M \in \mathbb{N}$, the partial sums $g_M := \sum_{j=1}^{M} \lambda_j b_j$, then because each $b_j \in L^1(\Sigma, \sigma)$ (cf. Corollary 6.6), we find that $g_M \in L^1(\Sigma, \sigma)$ for each $M \in \mathbb{N}$. Consequently,

$$\begin{aligned} &\int_{\Sigma}\int_{\Sigma} |k(x-y)|\mathbf{1}_{\{|x-y|>\varepsilon\}}(x,y)|g_M(y)||f(x)| \, d\sigma(y) \, d\sigma(x) \\ &\quad \leq C_\varepsilon \int_{\Sigma}\int_{\Sigma} \frac{|f(x)|}{1+|x|^{n-1}} |g_M(y)| \, d\sigma(x) \, d\sigma(y) \\ &\quad \leq C_\varepsilon \int_{\Sigma} |g_M(x)| \, d\sigma(x) < +\infty. \end{aligned} \tag{7.125}$$

The first estimate follows from estimates on the integral kernel k, established in (5.4). The second inequality follows from the continuous embedding in Proposition 4.15. Due to this, Fubini's theorem applies, and we may write

$$\begin{aligned}
\int_\Sigma (T_\varepsilon g_M) f \, d\sigma &= \int_\Sigma \Big[\int_{\substack{y\in\Sigma \\ |x-y|>\varepsilon}} k(x-y) g_M(y) \, d\sigma(y) \Big] f(x) \, d\sigma(x) \qquad (7.126) \\
&= \int_\Sigma \Big[\int_\Sigma k(x-y) \mathbf{1}_{\{|x-y|>\varepsilon\}}(x,y) g_M(y) \, d\sigma(y) \Big] f(x) \, d\sigma(x) \\
&= \int_\Sigma \Big[\int_\Sigma k(x-y) \mathbf{1}_{\{|x-y|>\varepsilon\}}(x,y) f(x) \, d\sigma(x) \Big] g_M(y) \, d\sigma(y) \\
&= -\int_\Sigma \Big[\int_{\substack{x\in\Sigma \\ |y-x|>\varepsilon}} k(y-x) f(x) \, d\sigma(x) \Big] g_M(y) \, d\sigma(y) \\
&= -\int_\Sigma g_M (T_\varepsilon f) \, d\sigma.
\end{aligned}$$

Note that the fourth equality above follows by the odd parity of the integral kernel k.

We now use Lebesgue's dominated convergence theorem to make an analogous statement about T. Observe that $|(T_\varepsilon g_M) f| \le (T_{\max} g_M)|f|$ for all $\varepsilon > 0$. Moreover, $(T_{\max} g_M)|f| \in L^1(\Sigma, \sigma)$ by (6.133) of Proposition 6.28, whose application here is justified since $f \in M^{p,\lambda}(\Sigma, w)$ and $T_{\max} g_M$ belongs to $\mathcal{B}^{p',\lambda}(\Sigma, w')$ (cf. Proposition 7.5). Therefore,

$$\lim_{\varepsilon \to 0^+} \int_\Sigma (T_\varepsilon g_M) f \, d\sigma = \int_\Sigma (T g_M) f \, d\sigma. \qquad (7.127)$$

On the other hand, $|g_M (T_\varepsilon f)| \le |g_M| (T_{\max} f)$ for all $\varepsilon > 0$, and $|g_M|(T_{\max} f)$ is in $L^1(\Sigma, \sigma)$ again by (6.133) of Proposition 6.28, which indeed applies since $g_M \in \mathcal{B}^{p',\lambda}(\Sigma, w')$ and $T_{\max} f \in M^{p,\lambda}(\Sigma, w)$ (cf. Theorem 5.7). Therefore,

$$\lim_{\varepsilon \to 0^+} \int_\Sigma g_M (T_\varepsilon f) \, d\sigma = \int_\Sigma g_M (T f) \, d\sigma. \qquad (7.128)$$

Thus, (7.127) and (7.128) together with (7.126) imply

$$\int_\Sigma (T g_M) f \, d\sigma = -\int_\Sigma g_M (T f) \, d\sigma. \qquad (7.129)$$

We now pass to the limit as $M \to \infty$. To do this, note that $g_M \to g$ in $\mathcal{B}^{p',\lambda}(\Sigma, w')$ as $M \to \infty$. We may then estimate the difference

$$\begin{aligned}
\Big| \int_\Sigma (T g_M) f \, d\sigma - \int_\Sigma (T g) f \, d\sigma \Big| &\le \int_\Sigma |T(g - g_M)| |f| \, d\sigma \qquad (7.130) \\
&\le \|T(g_M - g)\|_{\mathcal{B}^{p',\lambda}(\Sigma, w')} \|f\|_{M^{p,\lambda}(\Sigma, w)} \\
&\le C \|g_M - g\|_{\mathcal{B}^{p',\lambda}(\Sigma, w')} \|f\|_{M^{p,\lambda}(\Sigma, w)} \to 0
\end{aligned}$$

as $M \to \infty$, where the second inequality is a consequence of Theorem 6.28, and the third inequality follows from the boundedness of T on $\mathcal{B}^{p',\lambda}(\Sigma, w')$, established in Proposition 7.2. It is thus clear that

$$\lim_{M\to\infty} \int_{\Sigma} (Tg_M)f \, d\sigma = \int_{\Sigma} (Tg)f \, d\sigma. \tag{7.131}$$

Similarly, we have that

$$\begin{aligned} \left| \int_{\Sigma} g_M(Tf) \, d\sigma - \int_{\Sigma} g(Tf) \, d\sigma \right| &\le \int_{\Sigma} |g_M - g||Tf| \, d\sigma \\ &\le \|g_M - g\|_{\mathcal{B}^{p',\lambda}(\Sigma,w')} \|Tf\|_{M^{p,\lambda}(\Sigma,w)} \\ &\le C\|g_M - g\|_{\mathcal{B}^{p',\lambda}(\Sigma,w')} \|f\|_{M^{p,\lambda}(\Sigma,w)} \to 0 \end{aligned} \tag{7.132}$$

as $M \to \infty$, where the second inequality is again a consequence of Theorem 6.28, and the third inequality a consequence of the boundedness of T on $M^{p,\lambda}(\Sigma, w)$, established in Proposition 5.7. We thus have that

$$\lim_{N\to\infty} \int_{\Sigma} g_M(Tf) \, d\sigma = \int_{\Sigma} g(Tf) \, d\sigma. \tag{7.133}$$

Then (7.129), together with (7.131) and (7.133) imply (7.122) as desired. □

Remark 7.9. It is worth mentioning that the above theorem can be proved alternatively in the following way. Namely, consider the partial sums $g_N := \sum_{j=1}^{N} \lambda_j b_j$ as before. Then there exists some surface ball $\Delta_{g_N} \subseteq \Sigma$ such that $\operatorname{supp} g_N \subseteq \Delta_{g_N}$. We may then write $f = f\mathbf{1}_{2\Delta_{g_N}} + f\mathbf{1}_{\Sigma\setminus\Delta_{g_N}}$. Since $f\mathbf{1}_{2\Delta_{g_N}} \in L^q(\Sigma, \omega)$ by Remark 4.2, Proposition 5.6 applies, and we find that

$$\int_{\Sigma} (Tg_N)f\mathbf{1}_{\Delta_{g_N}} \, d\sigma = -\int_{\Sigma} g_N(T(f\mathbf{1}_{\Delta_{g_N}})) \, d\sigma \tag{7.134}$$

Now, if $y \in \operatorname{supp} g_N$ and $x \in \Sigma \setminus \Delta_{g_N}$, then the limit defining T is superfluous, and if $R \in (0, \infty)$ denotes the radius of Δ_{g_N} we find arguing as in (7.125) that

$$\int_{\Sigma}\int_{\Sigma} |k(x-y)|\mathbf{1}_{|x-y|>R}|g_N(y)||f(x)| \, d\sigma(y) \, d\sigma(x) < +\infty, \tag{7.135}$$

so that by Fubini's theorem, and an argument similar to that of (7.126), we obtain

$$\int_{\Sigma} (Tg_N)f\mathbf{1}_{\Sigma\setminus\Delta_{g_N}} \, d\sigma = -\int_{\Sigma} g_N(T(f\mathbf{1}_{\Sigma\setminus\Delta_{g_N}})) \, d\sigma. \tag{7.136}$$

By the linearity of the integral and the operator T, (7.134) and (7.136) show that

$$\int_{\Sigma}(Tg_N)f\,d\sigma = -\int_{\Sigma} g_N(Tf)\,df\sigma. \tag{7.137}$$

To transfer this from a statement involving g_N to a statement involving g, we argue much as in (7.130)–(7.132). This establishes the result.

Corollary 7.10. *Fix any two integrability exponents $p, q \in (1,\infty)$ and define $p' := p/(p-1)$ and $q' := q/(q-1)$. Select an arbitrary Muckenhoupt weight $w \in A_p(\mathbb{R}, \mathcal{L}^1) \cap \mathrm{RH}_q(\mathbb{R}, \mathcal{L}^1)$ with $w' := w^{-p'/p} \in A_{p'}(\mathbb{R}, \mathcal{L}^1)$ denoting its conjugate weight. Also, define $\varepsilon := (p'-1)/q'$, so that $w' \in A_{p'-\varepsilon}(\mathbb{R}, \mathcal{L}^1)$ in view of Proposition 6.35, and fix a threshold*

$$\lambda \in (0, \varepsilon(n-1)/(p'-1)) = (0, (n-1)/q'). \tag{7.138}$$

Finally, suppose $f_j \to f$ weak-$*$ in $M^{p,\lambda}(\mathbb{R}, w)$.

Then $Hf_j \to Hf$ weak-$*$ in $M^{p,\lambda}(\mathbb{R}, w)$, where H denotes the Hilbert transform on $\mathbb{R}$, which acts on each $f \in L^1(\mathbb{R}, \frac{dx}{1+|x|})$ according to

$$(Hf)(x) := \lim_{\varepsilon\to 0^+} \frac{1}{\pi} \int_{\substack{y\in\mathbb{R}\\ |x-y|>\varepsilon}} \frac{f(y)}{x-y}\,dy \quad \text{at } \mathcal{L}^1\text{-a. e. point } x \in \mathbb{R}. \tag{7.139}$$

Proof. First, observe that the Hilbert transform on $\mathbb{R}$ (considered as a subset of $\mathbb{R}^2$) is a singular integral operator of convolution type with integral kernel $k(z) := \frac{1}{\pi z}$ for all $z \in \mathbb{R} \setminus \{0\}$, which is positive homogeneous of degree -1. As such, Theorem 7.8 applies, so given any $g \in \mathcal{B}^{p',\lambda}(\mathbb{R}, \mathcal{L}^1)$,

$$\int_{\mathbb{R}} (Hf_j)g\,d\mathcal{L}^1 = -\int_{\mathbb{R}} f_j(Hg)\,d\mathcal{L}^1 \xrightarrow{j\to\infty} -\int_{\mathbb{R}} f(Hg)\,d\mathcal{L}^1 = \int_{\mathbb{R}} (Hf)g\,d\mathcal{L}^1. \tag{7.140}$$

This demonstrates the desired result. Note that we can apply the weak-$*$ convergence of the sequence $\{f_j\}_{j\in\mathbb{N}}$ because Theorem 7.2 guarantees Hg is in $\mathcal{B}^{p',\lambda}(\mathbb{R}, \mathcal{L}^1)$. □

Remark 7.11. Now, let $q \in (1,\infty)$ with $q' := q/(q-1)$. Choose any $\omega \in A_q(\mathbb{R}, \mathcal{L}^1)$ and let $p \in (1,\infty)$ be such that

$$\omega' := \omega^{-q'/q} \in A_{q'}(\mathbb{R}, \mathcal{L}^1) \cap \mathrm{RH}_p(\mathbb{R}, \mathcal{L}^1). \tag{7.141}$$

Additionally, take $\varepsilon := (q-1)/p' \in (0, q-1)$, where p' is the Hölder conjugate of p, and assume

$$\lambda \in (0, \varepsilon(n-1)/(q-1)) = (0, (n-1)/p'). \tag{7.142}$$

Note that $\mathcal{B}^{q,\lambda}(\mathbb{R}, w)$ is a well-defined weighted Block space in the sense of Definition 6.7, since Proposition 6.35 ensures $\omega \in A_{q-\varepsilon}(\mathbb{R}, \mathcal{L}^1)$ for this choice of ε. Then

Corollary 7.10 continues to hold with $\mathcal{B}^{q,\lambda}(\mathbb{R},\omega)$ in place of $M^{p,\lambda}(\mathbb{R},w)$. In this context, weak-$*$ convergence of $\{f_j\}_{j\in\mathbb{N}}$ to some f in $\mathcal{B}^{q,\lambda}(\mathbb{R},\omega)$ should be understood as $\lim_{j\to\infty}\int_{\mathbb{R}} f_j g\,d\mathcal{L}^1 = \int_{\mathbb{R}} fg\,d\mathcal{L}^1$ for each $g \in M^{q',\lambda}(\mathbb{R},\omega')$. However, we can further replace $M^{q',\lambda}(\mathbb{R},\omega')$ by $\mathring{M}^{q',\lambda}(\mathbb{R},\omega')$ if we wish. The proof remains essentially unchanged, but instead of Theorem 7.2, we appeal to Theorem 5.7 to get the membership of Hg to $M^{q',\lambda}(\mathbb{R},\omega')$, whereas, working in terms of genuine weak-$*$ convergence, which involves weighted vanishing Morrey spaces, we use Theorem 10.30 instead to conclude that $Hg \in \mathring{M}^{q',\lambda}(\mathbb{R},\omega')$. See Chapter 10 for a detailed discussion of Muckenhoupt-weighted vanishing Morrey spaces and a precise definition of $\mathring{M}^{q',\lambda}(\mathbb{R},\omega')$.

The main progression of this work has been to use a previously developed Calderón–Zygmund theory for Muckenhoupt weighted Lebesgue spaces along with a powerful extrapolation result to develop a similar theory at the level of Muckenhoupt-weighted Morrey spaces. In contrast, we have developed a Calderón–Zygmund theory from scratch for Muckenhoupt-weighted Block spaces. However, Remark 7.9 as well as the duality theory of Theorem 6.29 and the relevance of said theory to Theorem 7.8 opens the door to the possibility of developing a Calderón–Zygmund theory for Muckenhoupt-weighted Morrey spaces based off of the newly established Calderón–Zygmund theory for Muckenhoupt-weighted Block spaces.

This next theorem is reminiscent of Theorem 5.15, but at the level of Muckenhoupt-weighted Block spaces. It provides a bound on operator norms of principal value singular integral operators of chord-dot-normal type in terms of the BMO seminorm of the outward unit normal. A similar statement as in Remark 7.3 continues to hold in the context of Theorem 7.12.

Theorem 7.12. *Let $\Omega \subseteq \mathbb{R}^n$ be a UR domain, abbreviate $\sigma := \mathcal{H}^{n-1}\lfloor\partial\Omega$, and denote ν as the geometric measure theoretic outward unit normal to Ω. Let $N = N(n) \in \mathbb{N}$ be sufficiently large and fix a complex-valued function $k \in \mathscr{C}^N(\mathbb{R}^n \setminus \{0\})$, which is even and positive homogeneous of degree $-n$. For all $f \in L^1(\partial\Omega, \frac{\sigma(x)}{1+|x|^{n-1}})$, define*

$$T^{\#}f(x) := \lim_{\varepsilon\to 0^+} \int\limits_{\substack{y\in\partial\Omega\\ |x-y|>\varepsilon}} \langle \nu(x), x-y\rangle k(x-y) f(y)\,d\sigma(y) \tag{7.143}$$

for each $x \in \partial\Omega$. Fix an integrability exponent $q \in (1,\infty)$, a Muckenhoupt weight $\omega \in A_q(\partial\Omega,\sigma)$, choose $\varepsilon \in (0, q-1)$ such that $\omega \in A_{q-\varepsilon}(\partial\Omega,\sigma)$, and fix a threshold $\lambda \in (0, \varepsilon(n-1)/(q-1))$.

Then for all $m \in \mathbb{N}$ there exists a constant $C_m \in (0,\infty)$, depending only on m, $\partial\Omega$, n, q, ε, λ, $[\omega]_{A_{q-\varepsilon}}$, and $[\omega]_{A_q}$ such that

$$\|T^{\#}\|_{\mathcal{B}^{q,\lambda}(\Sigma,\omega)\to\mathcal{B}^{q,\lambda}(\Sigma,\omega)} \le C_m \cdot \Big(\sum_{|\alpha|\le N} \sup_{S^{n-1}} |\partial^\alpha k|\Big) \cdot \|\nu\|^{\langle m\rangle}_{[\mathrm{BMO}(\partial\Omega,\sigma)]^n}. \tag{7.144}$$

Proof. To begin, consider the simple case in which $\|\nu\|_{[\mathrm{BMO}(\partial\Omega,\sigma)]^n} = 0$. If we have that $\|\nu\|_{[\mathrm{BMO}(\partial\Omega,\sigma)]^n} = 0$, then ν is constant by definition; cf. 2.81. It is a geometric measure theoretic fact that any Ahlfors regular domain whose geometric measure theoretic outward unit normal is constant is necessarily a half-space [86, Proposition 5.10.12, p. 470]. In particular, for $x, y \in \partial\Omega$, we have $\langle \nu(x), x - y\rangle = 0$, so that $T^{\#}$ is the zero operator, and (7.144) holds. Also, in the case that k is identically 0, we see by definition that $T^{\#}$ is the 0 operator, and $\sum_{|\alpha|\le N} \sup_{S^{n-1}} |\partial^\alpha k| = 0$, so that (7.144) again holds. Without loss of generality, we may assume

$$M := \sum_{|\alpha|\le N} \sup_{S^{n-1}} |\partial^\alpha k| = 1. \tag{7.145}$$

Otherwise, we work with k/M instead of k.

Now, consider the case $g = b$, where b is an arbitrary (q, λ, ω)-block on $\partial\Omega$. There exist $x_0 \in \partial\Omega$ and $R \in (0, \operatorname{diam}(\partial\Omega))$ such that

$$\operatorname{supp} b \subseteq \Delta(x_0, R) \quad \text{and} \quad \|b\|_{L^q(\partial\Omega,\omega)} \le R^{\lambda(1/q-1)}. \tag{7.146}$$

Define the dyadic annuli

$$B_0 := \Delta(x_0, 2R) \quad \text{and} \quad B_j := \Delta(x_0, 2^{j+1}R) \setminus \Delta(x_0, 2^j R) \quad \text{for all } j \in \mathbb{N} \tag{7.147}$$

and write $\widetilde{b}_j := (T^{\#} b)\mathbf{1}_{B_j}$ for each $j \in \mathbb{N}_0$, so that

$$T^{\#} b = \sum_{j=0}^{\infty} \widetilde{b}_j \quad \text{pointwise on } \partial\Omega. \tag{7.148}$$

Focus first on $\widetilde{b}_0$. We see that

$$\operatorname{supp} \widetilde{b}_0 \subseteq \Delta(x_0, 2R). \tag{7.149}$$

Additionally, by Corollary 5.13, we can estimate

$$\begin{aligned} \|\widetilde{b}_0\|_{L^q(\partial\Omega,\omega)} &\le \|T^{\#} b\|_{L^q(\partial\Omega,\omega)} \\ &\le C_m \|\nu\|^{\langle m\rangle}_{[\mathrm{BMO}(\partial\Omega,\sigma)]^n} \|b\|_{L^q(\partial\Omega,\omega)} \\ &\le C_m \|\nu\|^{\langle m\rangle}_{[\mathrm{BMO}(\partial\Omega,\sigma)]^n} R^{\lambda(1/q-1)}, \end{aligned} \tag{7.150}$$

for some constant $C_m \in (0, \infty)$, independent of b. With C_m as above, define

$$\lambda_0 := C_m \|\nu\|^{\langle m\rangle}_{[\mathrm{BMO}(\partial\Omega,\sigma)]^n} 2^{-\lambda(1/q-1)} \quad \text{and} \quad b_0 := \lambda_0^{-1} \widetilde{b}_0. \tag{7.151}$$

Then by construction, (7.149) and (7.150) imply that b_0 is a (q, λ, ω)-block on $\partial\Omega$.

Focus now on $\widetilde{b}_j$ with $j \in \mathbb{N}$ arbitrary. It is clear that

$$\operatorname{supp}\widetilde{b}_j \subseteq B_j \subseteq \Delta(x_0, 2^{j+1}R). \tag{7.152}$$

For each $x \in B_j$, since b is supported in $\Delta(x_0, R)$, we have

$$T^{\#}b(x) = \int\limits_{y\in\Delta(x_0,R)} \langle \nu(x), x-y\rangle k(x-y)b(y)\,d\sigma(y), \tag{7.153}$$

where we are permitted to omit the limit in the definition of $T^{\#}$ because of the fact that $\operatorname{dist}(\Delta(x_0,R), B_j) \geq R > 0$. Note that

$$|x-x_0| \approx |x-y| \text{ uniformly for } x \in B_j \text{ and } y \in \Delta(x_0,R). \tag{7.154}$$

Indeed, given $x \in B_j$ and $y \in \Delta(x_0,R)$, since $R \leq 2^{-j}|x-x_0|$, we have

$$\begin{aligned} |x\quad y| &\leq |x-x_0| + |x_0-y| < |x-x_0| + R \\ &\leq |x-x_0| + 2^{-j}|x-x_0| \leq 2|x-x_0|. \end{aligned} \tag{7.155}$$

On the other hand,

$$|x-x_0| \leq |x-y| + |y-x_0| < |x-y| + R \leq |x-y| + 2^{-j}|x-x_0|, \tag{7.156}$$

which implies

$$|x-x_0| \leq (1-2^{-j})^{-1}|x-y| \leq 2|x-y|. \tag{7.157}$$

This establishes (7.154).

Define the quantities

$$\nu^{\omega}_{\Delta(x_0,R)} := \fint\limits_{\Delta(x_0,R)} \nu\,d\omega, \tag{7.158}$$

and

$$M_{\Delta}(x) := \int\limits_{y\in\Delta(x_0,R)} \langle \nu^{\omega}_{\Delta(x_0,R)}, x-y\rangle k(x-y)b(y)\,d\sigma(y) \tag{7.159}$$

for all $x \in B_j$. We may then estimate, using the Cauchy–Schwarz inequality, properties of the integral kernel $|z|k(z)$ (cf. Remark 5.4) and (7.154),

$$\begin{aligned} |T^{\#}b(x) - M_{\Delta}(x)| &\leq \int\limits_{\Lambda(x_0,R)} |\nu(x) - \nu^{\omega}_{\Delta(x_0,R)}| \frac{C}{|x-y|^{n-1}} |b(y)|\,d\sigma(y) \\ &\leq \frac{C|\nu(x) - \nu^{\omega}_{\Delta(x_0,R)}|}{|x-x_0|^{n-1}} \int\limits_{\Delta(x_0,R)} |b|\,d\sigma, \quad \text{for all } x \in B_j. \end{aligned} \tag{7.160}$$

Recall that $|x - x_0| \geq 2^j R$ since $x \in B_j$, so that the above estimate yields

$$\begin{aligned}
&\Big(\int_{B_j} |T^{\#} b(x) - M_\Delta(x)|^q \, d\omega(x)\Big)^{1/q} \\
&\leq \frac{C}{(2^j R)^{n-1}} \Big(\int_{\Delta(x_0,R)} |b| \, d\sigma\Big)\Big(\int_{B_j} |\nu(x) - \nu^{\omega}_{\Delta(x_0,R)}|^q \, d\omega(x)\Big)^{1/q} \\
&\leq \frac{CR^{n-1}}{(2^j R)^{n-1}} \Big(\fint_{\Delta(x_0,R)} |b| \, d\sigma\Big)\Big(\int_{\Delta(x_0,2^{j+1}R)} |\nu(x) - \nu^{\omega}_{\Delta(x_0,R)}|^q \, d\omega(x)\Big)^{1/q} \\
&\leq \frac{C[\omega]^{1/q}_{A_q}}{2^{j(n-1)}} \Big(\fint_{\Delta(x_0,R)} |b|^q \, d\omega\Big)^{1/q}\Big(\int_{\Delta(x_0,2^{j+1}R)} |\nu(x) - \nu^{\omega}_{\Delta(x_0,R)}|^q \, d\omega(x)\Big)^{1/q} \\
&\leq \frac{CR^{\lambda(1/q-1)}}{2^{j(n-1)} \cdot \omega(\Delta(x_0,R))^{1/q}} \Big(\int_{\Delta(x_0,2^{j+1}R)} |\nu(x) - \nu^{\omega}_{\Delta(x_0,R)}|^q \, d\omega(x)\Big)^{1/q}.
\end{aligned} \tag{7.161}$$

The second inequality follows from Ahlfors regularity after creating the integral average, the third inequality uses Lemma 2.34, and the last inequality uses (7.146). Turning our attention to the remaining integral term, we estimate

$$\begin{aligned}
&\Big(\int_{\Delta(x_0,2^{j+1}R)} |\nu(x) - \nu^{\omega}_{\Delta(x_0,R)}|^q \, d\omega(x)\Big)^{1/q} \\
&= \omega(\Delta(x_0, 2^{j+1}R))^{1/q}\Big(\fint_{\Delta(x_0,2^{j+1}R)} |\nu(x) - \nu^{\omega}_{\Delta(x_0,R)}|^q \, d\omega(x)\Big)^{1/q} \\
&\leq \omega(\Delta(x_0, 2^{j+1}R))^{1/q}\Big\{\Big(\fint_{\Delta(x_0,2^{j+1}R)} |\nu(x) - \nu^{\omega}_{\Delta(x_0,2^{j+1}R)}|^q \, d\omega(x)\Big)^{1/q} \\
&\quad + \sum_{k=0}^{j}\Big(\fint_{\Delta(x_0,2^{j+1}R)} |\nu_{\Delta(x_0,2^{k+1}R)} - \nu^{\omega}_{\Delta(x_0,2^k R)}|^q \, d\omega\Big)^{1/q}\Big\}. \\
&\leq \omega(\Delta(x_0, 2^{j+1}R))^{1/q}\Big\{C\|\nu\|_{[\mathrm{BMO}(\partial\Omega,\sigma)]^n} \\
&\quad + \sum_{k=0}^{j}\Big(\fint_{\Delta(x_0,2^{j+1}R)} |\nu_{\Delta(x_0,2^{k+1}R)} - \nu^{\omega}_{\Delta(x_0,2^k R)}|^q \, d\omega\Big)^{1/q}\Big\}.
\end{aligned} \tag{7.162}$$

The last inequality follows from Lemmas 2.38 and 2.39. For $k \in \{0, 1, \dots, j\}$, we have

$$
\begin{aligned}
|v^{\omega}_{\Delta(x_0,2^{k+1}R)} - v^{\omega}_{\Delta(x_0,2^kR)}| &= \Big| \fint_{\Delta(x_0,2^kR)} (v - v^{\omega}_{\Delta(x_0,2^{k+1}R)})\, d\omega \Big| \\
&\leq \frac{1}{\omega(\Delta(x_0,2^kR))} \int_{\Delta(x_0,2^kR)} |v - v^{\omega}_{\Delta(x_0,2^{k+1}R)}|\, d\omega \\
&\leq \frac{\omega(\Delta(x_0,2^{k+1}R))}{\omega(\Delta(x_0,2^kR))} \fint_{\Delta(x_0,2^{k+1}R)} |v - v^{\omega}_{\Delta(x_0,2^{k+1}R)}|\, d\omega \\
&\leq \frac{\omega(\Delta(x_0,2^{k+1}R))}{\omega(\Delta(x_0,2^kR))} \Big(\fint_{\Delta(x_0,2^{k+1}R)} |v - v^{\omega}_{\Delta(x_0,2^{k+1}R)}|^q\, d\omega \Big)^{1/q} \\
&\leq C[\omega]_{A_q} \Big(\frac{\sigma(\Delta(x_0,2^{k+1}R))}{\sigma(\Delta(x_0,2^kR))} \Big)^q \|v\|_{[\mathrm{BMO}(\partial\Omega,\omega)]^n} \\
&\leq C\|v\|_{[\mathrm{BMO}(\partial\Omega,\sigma)]^n}.
\end{aligned} \tag{7.163}
$$

The third inequality follows from Hölder's inequality, the fourth inequality follows from doubling (item (4) of Proposition 2.29) and Lemma 2.38, and the last inequality follows from the Ahlfors regularity of $\partial\Omega$ and Lemma 2.39. Hence, (7.163) and (7.162) imply

$$
\begin{aligned}
&\Big(\int_{\Delta(x_0,2^{j+1}R)} |v(x) - v^{\omega}_{\Delta(x_0,R)}|^q\, d\omega(x) \Big)^{1/q} \\
&\quad \leq Cj\omega(\Delta(x_0,2^{j+1}R))^{1/q} \|v\|_{[\mathrm{BMO}(\partial\Omega,\sigma)]^n}.
\end{aligned} \tag{7.164}
$$

Inserting this estimate into (7.161) yields

$$
\begin{aligned}
&\Big(\int_{B_j} |T^{\#}b(x) - M_{\Delta}(x)|^q\, d\omega(x) \Big)^{1/q} \\
&\quad \leq \Big(\frac{CjR^{\lambda(1/q-1)}}{2^{j(n-1)}} \Big) \Big(\frac{\omega(\Delta(x_0,2^{j+1}R))}{\omega(\Delta(x_0,R))} \Big)^{1/q} \|v\|_{[\mathrm{BMO}(\partial\Omega,\sigma)]^n}.
\end{aligned} \tag{7.165}
$$

Fix $x \in B_j$. We now focus on estimating $M_{\Delta}(x)$. We have by properties of the integral kernel k (cf. Remark 5.4) and (7.154),

$$
\begin{aligned}
|M_{\Delta}(x)| &\leq \int_{\Delta(x_0,R)} |\langle v^{\omega}_{\Delta(x_0,R)}, x - y\rangle| |k(x-y)| |b(y)|\, d\sigma(y) \\
&\leq \frac{C}{(2^jR)^n} \int_{\Delta(x_0,R)} |\langle v^{\omega}_{\Delta(x_0,R)}, x - y\rangle| |b(y)|\, d\sigma(y).
\end{aligned} \tag{7.166}
$$

Subsequently, applying Lemma 2.40 with $\lambda := 1$, $R := 2^{j+1}R$, and $z := x_0$, we obtain via the Cauchy–Schwarz inequality,

$$\begin{aligned}
|\langle \nu^{\omega}_{\Delta(x_0,R)}, x-y\rangle| &\le |\langle \nu_{\Delta(x_0,2^{j+1}R)}, x-y\rangle| \\
&\quad + |\nu^{\omega}_{\Delta(x_0,R)} - \nu_{\Delta(x_0,2^{j+1}R)}||x-y| \\
&\le C2^{j+1}R\|\nu\|_{[\mathrm{BMO}(\partial\Omega,\sigma)]^n} \\
&\quad + |\nu^{\omega}_{\Delta(x_0,R)} - \nu^{\omega}_{\Delta(x_0,2^{j+1}R)}||x-y| \\
&\quad + |\nu^{\omega}_{\Delta(x_0,2^{j+1}R)} - \nu_{\Delta(x_0,2^{j+1}R)}||x-y| \\
&\le C2^{j+1}R\|\nu\|_{[\mathrm{BMO}(\partial\Omega,\sigma)]^n} + C\|\nu\|_{[\mathrm{BMO}(\partial\Omega,\sigma)]^n} \\
&\quad + C2^{j}R\|\nu\|_{[\mathrm{BMO}(\partial\Omega,\sigma)]^n} \\
&\le C2^{j}Rj\|\nu\|_{[\mathrm{BMO}(\partial\Omega,\sigma)]^n}
\end{aligned} \tag{7.167}$$

for each $y \in \Delta(x_0, R)$. The third inequality above follows from j applications of (7.163) after adding and subtracting $\nu^{\omega}_{\Delta(x_0,2^kR)}$ for each $k \in \{1,\ldots,j\}$; and also the fact that for any $R > 0$ (in particular $R = 2^{j+1}R$),

$$\begin{aligned}
|\nu_{\Delta(x_0,R)} - \nu^{\omega}_{\Delta(x_0,R)}| &\le \fint_{\Delta(x_0,R)} |\nu(x) - \nu^{\omega}_{\Delta(x_0,R)}|\,d\sigma(x) \\
&\le [\omega]^{1/q}_{A_p}\Big(\fint_{\Delta(x_0,R)} |\nu - \nu^{\omega}_{\Delta(x_0,R)}|^q\,d\omega\Big)^{1/q} \\
&\le C\|\nu\|_{[\mathrm{BMO}_q(\partial\Omega,\omega)]^n} \\
&\le C\|\nu\|_{[\mathrm{BMO}(\partial\Omega,\sigma)]^n}.
\end{aligned} \tag{7.168}$$

Note the use of Lemmas 2.38 and 2.39 to obtain the final inequality above. Ultimately, the estimates in (7.166) and (7.167) imply

$$\begin{aligned}
|M_\Delta(x)| &\le \frac{Cj}{(2^jR)^{n-1}}\|\nu\|_{[\mathrm{BMO}(\partial\Omega,\sigma)]^n}\int_{\Delta(x_0,R)} |b|\,d\sigma \\
&\le \frac{CjR^{n-1}}{(2^jR)^{n-1}}\|\nu\|_{[\mathrm{BMO}(\partial\Omega,\sigma)]^n}\fint_{\Delta(x_0,R)} |b|\,d\sigma, \\
&\le \frac{Cj}{2^{j(n-1)}}\cdot\frac{[\omega]^{1/q}_{A_q}}{\omega(\Delta(x_0,R))^{1/q}}\|\nu\|_{[\mathrm{BMO}(\partial\Omega,\sigma)]^n}\Big(\int_{\Delta(x_0,R)} |b|^q\,d\omega\Big)^{1/q} \\
&\le \frac{CjR^{\lambda(1/q-1)}}{2^{j(n-1)}\omega(\Delta(x_0,R))^{1/q}}\|\nu\|_{[\mathrm{BMO}(\partial\Omega,\sigma)]^n}.
\end{aligned} \tag{7.169}$$

The second inequality uses Ahlfors regularity, the third uses Lemma 2.34, and the last inequality follows from (7.146). Consequently, we find that

$$\left(\int_{B_j} |M_\Delta|^q \, d\omega\right)^{1/q} \leq \left(\frac{CjR^{\lambda(1/q-1)}}{2^{j(n-1)}}\right)\left(\frac{\omega(\Delta(x_0, 2^{j+1}R))}{\omega(\Delta(x_0, R))}\right)^{1/q} \|\nu\|_{[\mathrm{BMO}(\partial\Omega,\sigma)]^n}. \tag{7.170}$$

Combining this result with (7.165), we obtain

$$\begin{aligned} \|\widetilde{b}_j\|_{L^q(\partial\Omega,\omega)} &= \|T^{\#}b\|_{L^q(B_j,\omega)} \\ &\leq \|T^{\#}b - M_\Delta\|_{L^q(B_j,\omega)} + \|M_\Delta\|_{L^q(B_j,\omega)} \\ &\leq \left(\frac{CjR^{\lambda(1/q-1)}}{2^{j(n-1)}}\right)\left(\frac{\omega(\Delta(x_0, 2^{j+1}R))}{\omega(\Delta(x_0, R))}\right)^{1/q} \|\nu\|_{[\mathrm{BMO}(\partial\Omega,\sigma)]^n} \\ &\leq \frac{CjR^{\lambda(1/q-1)}2^{j(n-1)(q-\varepsilon)/q}}{2^{j(n-1)}} \|\nu\|_{[\mathrm{BMO}(\partial\Omega,\sigma)]^n} \\ &= CjR^{\lambda(1/q-1)}2^{-j(n-1)(\varepsilon/q)} \|\nu\|_{[\mathrm{BMO}(\partial\Omega,\sigma)]^n}, \end{aligned} \tag{7.171}$$

where the last inequality follows from the fact that $\omega \in A_{q-\varepsilon}(\partial\Omega, \sigma)$, doubling (item (4) of Proposition 2.29), and the Ahlfors regularity of $\partial\Omega$.

Now, with C as in (7.171), define

$$\lambda_j := Cj\|\nu\|_{[\mathrm{BMO}(\partial\Omega,\sigma)]^n} 2^{-\lambda(j+1)(1/q-1)-j(n-1)(\varepsilon/q)} \quad \text{and} \quad b_j := \lambda_j^{-1}\widetilde{b}_j \tag{7.172}$$

From this definition, we see because of (7.152) and (7.171) that in fact, b_j is a (q, λ, ω)-block on $\partial\Omega$ for each $j \in \mathbb{N}$. Since $\widetilde{b}_j = \lambda_j b_j$ for all $j \in \mathbb{N}_0$, we have that

$$T^{\#}b = \sum_{j=0}^{\infty} \lambda_j b_j \quad \text{pointwise on } \partial\Omega. \tag{7.173}$$

Moreover, from (7.172), we see

$$\begin{aligned} \sum_{j=1}^{\infty} |\lambda_j| &\leq C\|\nu\|_{[\mathrm{BMO}(\partial\Omega,\sigma)]^n} \sum_{j=1}^{\infty} j2^{-j[(n-1)(\varepsilon/q)-\lambda(1-1/q)]} \\ &= C\|\nu\|_{[\mathrm{BMO}(\partial\Omega,\sigma)]^n} \sum_{j=1}^{\infty} j2^{-j(1-1/q)[\varepsilon(n-1)/(q-1)-\lambda]}, \\ &\leq C_* \|\nu\|_{[\mathrm{BMO}(\partial\Omega,\sigma)]^n} \end{aligned} \tag{7.174}$$

for some constant $C_* \in (0, \infty)$. The above series converges by the hypothesis that $\lambda \in (0, \varepsilon(n-1)/(q-1))$, which ensures that

$$(1 - 1/q)[\varepsilon(n-1)/(q-1) - \lambda] > 0. \tag{7.175}$$

Lemma 6.8 shows that $\sum_{j=0}^{\infty} \lambda_j b_j$ is a block representation for $T^{\#}b$. Note also, since $t \leq t^{\langle m \rangle}$ for all $t \in [0,1]$, so the estimate in (7.174) and the definition of λ_0 in (7.151) imply that

$$\begin{aligned}\|T^{\#}b\|_{\mathcal{B}^{q,\lambda}(\partial\Omega,\omega)} &\leq \sum_{j=0}^{\infty}|\lambda_j| \\ &\leq C_m \cdot \|\nu\|^{\langle m\rangle}_{[\mathrm{BMO}(\partial\Omega,\sigma)]^n}.\end{aligned} \tag{7.176}$$

All together, we can conclude for any $m \in \mathbb{N}$ that

> for any (q, λ, ω)-block, b, one has $T^{\#}b \in \mathcal{B}^{q,\lambda}(\partial\Omega, \omega)$ with the estimate $\|T^{\#}b\|_{\mathcal{B}^{q,\lambda}(\partial\Omega,\omega)} \leq C_m\|\nu\|^{\langle m\rangle}_{[\mathrm{BMO}(\partial\Omega,\sigma)]^n}$ for some constant $C_m \in (0, \infty)$, independent of b. (7.177)

Generalizing now from individual blocks to generic block functions, fix some arbitrary $g \in \mathcal{B}^{q,\lambda}(\partial\Omega, \omega)$ with a quasioptimal block representation

$$g = \sum_{j=1}^{\infty}\lambda_j b_j \quad \text{with} \sum_{j=1}^{\infty}|\lambda_j| \leq 2\|g\|_{\mathcal{B}^{q,\lambda}(\partial\Omega,\omega)}. \tag{7.178}$$

For each $M \in \mathbb{N}$, define the partial sums $g_M := \sum_{j=1}^{M}\lambda_j b_j$. By Proposition 6.17, there exists some $q_0 \in (1, q-1)$ such that

$$\mathcal{B}^{q,\lambda}(\partial\Omega, \omega) \hookrightarrow L^1\Big(\partial\Omega, \frac{\sigma(x)}{1+|x|^{n-1}}\Big) \cap L^{q_0}_{\mathrm{loc}}(\partial\Omega, \sigma). \tag{7.179}$$

Since $g_M \to g$ as $M \to \infty$ in $\mathcal{B}^{q,\lambda}(\partial\Omega, \omega)$ thanks to Proposition 6.12, it follows that $g_M \to g$ in $L^1(\partial\Omega, \frac{\sigma(x)}{1+|x|^{n-1}})$ as $M \to \infty$. As a result of this and Corollary 5.21, we have that $T^{\#}g_M \to T^{\#}g$ as $M \to \infty$ in $L^{q_0}_{\mathrm{loc}}(\partial\Omega, \sigma)$. Thus, there exists a sequence $\{M_k\}_{k\in\mathbb{N}} \subseteq \mathbb{N}$ such that $T^{\#}g_{M_k} \to T^{\#}g$ at σ-a. e. point on $\partial\Omega$ as $k \to \infty$. Therefore, by the linearity of $T^{\#}$,

$$T^{\#}g = \lim_{k\to\infty}\sum_{j=1}^{M_k}\lambda_j T^{\#}b_j \quad \text{at } \sigma\text{-a. e. point on } \partial\Omega. \tag{7.180}$$

By (7.177), we have that $T^{\#}b_j \in \mathcal{B}^{q,\lambda}(\partial\Omega, \omega)$ for each $j \in \mathbb{N}$, along with the estimate

$$\|T^{\#}b_j\|_{\mathcal{B}^{q,\lambda}(\partial\Omega,\omega)} \leq C_m\|\nu\|^{\langle m\rangle}_{[\mathrm{BMO}(\partial\Omega,\sigma)]^n}. \tag{7.181}$$

Now, $\{\lambda_j\}_{j\in\mathbb{N}} \in \ell^1$, so that

$$\Big\|\sum_{j=M}^{M'}\lambda_j T^{\#}b_j\Big\|_{\mathcal{B}^{q,\lambda}(\partial\Omega,\omega)} \leq \Big(\sum_{j=M}^{M'}|\lambda_j|\Big)\cdot C_m \cdot \|\nu\|^{\langle m\rangle}_{[\mathrm{BMO}(\partial\Omega,\sigma)]^n} \to 0 \tag{7.182}$$

as $M, M' \to \infty$, showing that $\{\sum_{j=1}^{M}\lambda_j T^{\#}b_j\}_{M\in\mathbb{N}}$ is a Cauchy sequence in $\mathcal{B}^{q,\lambda}(\partial\Omega, \omega)$. By Theorem 6.11, the space $\mathcal{B}^{q,\lambda}(\partial\Omega, \omega)$ is Banach, so there exists some $G \in \mathcal{B}^{q,\lambda}(\partial\Omega, \omega)$ for which

$$G = \lim_{M\to\infty} \sum_{j=1}^{M} \lambda_j T^{\#} b_j \quad \text{in } \mathcal{B}^{q,\lambda}(\partial\Omega, \omega), \tag{7.183}$$

and hence

$$\begin{aligned} \|G\|_{\mathcal{B}^{q,\lambda}(\partial\Omega,\omega)} &= \lim_{M\to\infty} \left\| \sum_{j=1}^{M} \lambda_j T^{\#} b_j \right\|_{\mathcal{B}^{q,\lambda}(\partial\Omega,\omega)} \\ &\le C_m \|\vec{v}\|^{\langle m\rangle}_{[\mathrm{BMO}(\partial\Omega,\sigma)]^n} \|g\|_{\mathcal{B}^{q,\lambda}(\partial\Omega,\omega)}, \end{aligned} \tag{7.184}$$

where the last inequality follows from the inequality in (7.182) and quasioptimality in (7.178). Note that

$$G = \lim_{k\to\infty} \sum_{j=1}^{M_k} \lambda_j T^{\#} b_j \quad \text{in } L^1\Big(\partial\Omega, \frac{\sigma(x)}{1+|x|^{n-1}}\Big) \tag{7.185}$$

by (7.183) and the embedding in (7.179). Hence, there exists a subsequence $\{M_{k_l}\}_{l\in\mathbb{N}}$ such that

$$G(x) = \lim_{l\to\infty} \sum_{j=1}^{M_{k_l}} \lambda_j T^{\#} b_j(x) \quad \text{for } \sigma\text{-a. e. point } x \in \partial\Omega. \tag{7.186}$$

In concert, (7.180) and (7.186) guarantee $G = T^{\#} g$ at σ-a.e point on $\partial\Omega$. As a consequence of this, the lattice property of $\mathcal{B}^{q,\lambda}(\Sigma,\omega)$ (cf. Proposition 6.14) and (7.184), we see $T^{\#} g$ belongs to $\mathcal{B}^{q,\lambda}(\partial\Omega,\omega)$ with

$$\begin{aligned} \|T^{\#} g\|_{\mathcal{B}^{q,\lambda}(\partial\Omega,\omega)} &= \|G\|_{\mathcal{B}^{q,\lambda}(\partial\Omega,\omega)} \\ &\le C_m \|\vec{v}\|^{\langle m\rangle}_{[\mathrm{BMO}(\partial\Omega,\sigma)]^n} \|g\|_{\mathcal{B}^{q,\lambda}(\partial\Omega,\omega)}. \end{aligned} \tag{7.187}$$

This establishes the desired result given that $g \in \mathcal{B}^{q,\lambda}(\Sigma,\omega)$ was arbitrarily chosen. □

We now present a bound for the commutator of a principal value singular integral operator of convolution type with the operation of pointwise multiplication by an essentially bounded function. Moreover, we can bound the operator norm of the commutator in terms of the BMO seminorm of said essentially bounded function. Note that such an observation as in Remark 7.3 continues to be relevant in the context of Theorem 7.13.

Theorem 7.13. *Let $\Sigma \subseteq \mathbb{R}^n$ be a UR set and abbreviate $\sigma := \mathcal{H}^{n-1}\lfloor\Sigma$. Let $N = N(n) \in \mathbb{N}$ be sufficiently large and fix a complex-valued function $k \in \mathscr{C}^N(\mathbb{R}^n \setminus \{0\})$, such that k is odd and positive homogeneous of degree $1-n$. Fix $h \in L^\infty(\Sigma,\sigma)$, and for each $f \in L^1(\Sigma, \frac{\sigma(x)}{1+|x|^{n-1}})$ define*

$$\mathfrak{C}f(x) := \lim_{\varepsilon\to 0^+} \int\limits_{\substack{y\in\Sigma \\ |x-y|>\varepsilon}} (h(x)-h(y))k(x-y)f(y)\,d\sigma(y) \tag{7.188}$$

for σ-a. e. point $x \in \Sigma$. Select an integrability exponent $q \in (1,\infty)$, a Muckenhoupt weight $\omega \in A_q(\Sigma,\sigma)$, and choose $\varepsilon \in (0,q-1)$ such that ω belongs to $A_{q-\varepsilon}(\Sigma,\sigma)$. Finally, let $\lambda \in (0,\varepsilon(n-1)/(q-1))$ be arbitrary. Then $\mathfrak{C}$ induced a well-defined, linear, and bounded mapping

$$\mathfrak{C} : \mathcal{B}^{q,\lambda}(\Sigma,\omega) \to \mathcal{B}^{q,\lambda}(\Sigma,\omega). \tag{7.189}$$

Moreover, there exists a constant $C \in (0,\infty)$, depending only on the ADR *character of Σ, n, q, ε, λ, $[\omega]_{A_{q-\varepsilon}}$, and $[\omega]_{A_q}$, such that*

$$\|\mathfrak{C}\|_{\mathcal{B}^{q,\lambda}(\Sigma,\omega)\to\mathcal{B}^{q,\lambda}(\Sigma,\omega)} \leq C \cdot \Big(\sum_{|\alpha|\leq N} \sup_{S^{n-1}} |\partial^\alpha k|\Big) \cdot \|h\|_{\mathrm{BMO}(\Sigma,\sigma)}. \tag{7.190}$$

Remark 7.14. Note that if T is defined as in (7.29), then $\mathfrak{C}f = [T,h]f$, where formally, $[T,h]f := h\cdot(Tf) - T(hf)$.

Proof of Theorem 7.13. As a preliminary case, consider what happens in the event that $\|h\|_{\mathrm{BMO}(\Sigma,\sigma)} = 0$. Then h must be constant a. e., and as a result one has $h - h(x) = 0$ on Σ for σ-a. e. point $x \in \Sigma$. Consequently, $\mathfrak{C}$ is the zero operator and (7.190) holds. Note also that $\mathfrak{C}$ is the zero operator when k is identically 0 and (7.190) again holds. Therefore, assume k is nonzero and without loss of generality assume

$$M := \sum_{|\alpha|\leq N} \sup_{S^{n-1}} |\partial^\alpha k| = 1. \tag{7.191}$$

In the event that $M \neq 1$, we may replace k by k/M.

We now analyze the action of $\mathfrak{C}$ on an individual (q,λ,ω)-block on Σ. Fix such a block, b. Then there exist $x_0 \in \Sigma$ and $R \in (0, 2\,\mathrm{diam}(\Sigma))$ such that

$$\operatorname{supp} b \subseteq \Delta(x_0,R) \quad \text{and} \quad \|b\|_{L^q(\Sigma,\omega)} \leq R^{\lambda(1/q-1)}. \tag{7.192}$$

Define the dyadic annuli

$$B_0 := \Delta(x_0,2R) \quad \text{and} \quad B_j := \Delta(x_0,2^{j+1}R) \setminus \Delta(x_0,2^jR) \quad \text{for all } j \in \mathbb{N}, \tag{7.193}$$

and write $\widetilde{b}_j := (\mathfrak{C}b)\mathbf{1}_{B_j}$ for each $j \in \mathbb{N}_0$, so that

$$\mathfrak{C}b = \sum_{j=0}^{\infty} \widetilde{b}_j \quad \text{pointwise on } \Sigma. \tag{7.194}$$

We draw our attention first to $\widetilde{b}_0$. It is clear from definitions that

$$\operatorname{supp} \widetilde{b}_0 \subseteq \Delta(x_0,2R). \tag{7.195}$$

Additionally, keeping in mind Remark 7.14, since $b \in L^q(\Sigma, \omega)$, we find by Theorem 5.16 that there exists some constant $C \in (0, \infty)$, independent of b, such that

$$\begin{aligned} \|\widetilde{b}_0\|_{L^q(\Sigma,\omega)} \le \|\mathfrak{C}b\|_{L^q(\Sigma,\omega)} &\le C\|h\|_{\mathrm{BMO}(\Sigma,\sigma)}\|b\|_{L^q(\Sigma,\omega)} \\ &\le C\|h\|_{\mathrm{BMO}(\Sigma,\sigma)}R^{\lambda(1/q-1)}. \end{aligned} \tag{7.196}$$

Note the last inequality follows from (7.192). With C as above, define

$$\lambda_0 := C\|h\|_{\mathrm{BMO}(\Sigma,\sigma)}2^{-\lambda(1/q-1)} \quad \text{and} \quad b_0 := \lambda_0^{-1}\widetilde{b}_0, \tag{7.197}$$

then (7.195) and (7.196) imply that b_0 is a (q, λ, ω)-block on Σ.

We now draw our attention to $\widetilde{b}_j$ with $j \in \mathbb{N}$ arbitrary. Then

$$\operatorname{supp} \widetilde{b}_j \subseteq B_j \subseteq \Delta(x_0, 2^{j+1}R). \tag{7.198}$$

We also have the following preliminary estimate:

$$\begin{aligned} &\Big(\int_{B_j} |(\mathfrak{C}b)(x)|^q \, d\omega(x)\Big)^{1/q} \\ &\le \Big(\int_{x\in B_j}\Big|\int_{y\in\Delta(x_0,R)} |h(x) - h(y)||k(x-y)||b(y)|\, d\sigma(y)\Big|^q \, d\omega(x)\Big)^{1/q} \\ &\le \frac{C}{(2^jR)^{n-1}}\Big(\int_{x\in\Delta(x_0,2^{j+1}R)}\Big|\int_{y\in\Delta(x_0,R)} |h(x) - h(y)||b(y)|\, d\sigma(y)\Big|^q \, d\omega(x)\Big)^{1/q} \\ &\le \frac{C}{(2^jR)^{n-1}}\int_{y\in\Delta(x_0,R)} |b(y)|\Big(\int_{x\in\Delta(x_0,2^{j+1}R)} |h(x) - h(y)|^q \, d\omega(x)\Big)^{1/q} d\sigma(y). \end{aligned} \tag{7.199}$$

Note that in the first inequality the limit in the definition of $\mathfrak{C}$ is omitted because x and y are separated as witnessed below:

$$\operatorname{dist}(x, y) \ge \operatorname{dist}(B_j, \Delta(x_0, R)) \ge R > 0. \tag{7.200}$$

The second inequality in (7.199) follows from properties of the integral kernel k (cf. (5.16)) and the fact that

$$|x - y|^{1-n} \approx |x - x_0|^{1-n} \approx 2^jR^{1-n} \tag{7.201}$$

uniformly for $x \in B_j$ and $y \in \Delta(x_0, R)$ (cf. the argument justifying (7.154)). The last inequality in (7.199) follows from Minkowski's inequality.

Now, for each fixed $y \in \Delta(x_0, R)$, we focus on the innermost integral in (7.199). For $z \in \Sigma$ and $r \in (0, 2\operatorname{diam}(\Sigma))$ define $h^{\omega}_{\Delta(z,r)} := \fint_{\Delta(z,r)} h \, d\omega$. Then by the triangle inequality

$$\begin{aligned}
&\Big(\int_{x\in\Delta(x_0,2^{j+1}R)} |h(x)-h(y)|^q\,d\omega(x)\Big)^{1/q} && (7.202)\\
&\le \Big(\int_{x\in\Delta(x_0,2^{j+1}R)} |h(x)-h^{\omega}_{\Delta(x_0,R)}|^q\,d\omega(x)\Big)^{1/q}\\
&\quad+\omega(\Delta(x_0,2^{j+1}R))^{1/q}|h(y)-h^{\omega}_{\Delta(x_0,R)}|\\
&\le \Big(\int_{\Delta(x_0,2^{j+1}R)} |h(x)-h^{\omega}_{\Delta(x_0,2^{j+1}R)}|^q\,d\omega(x)\Big)^{1/q}\\
&\quad+\omega(\Delta(x_0,2^{j+1}R))^{1/q}\sum_{i=0}^{j}|h^{\omega}_{\Delta(x_0,2^{i+1}R)}-h^{\omega}_{\Delta(x_0,2^{i}R)}|\\
&\quad+\omega(\Delta(x_0,2^{j+1}R))^{1/q}|h(y)-h^{\omega}_{\Delta(x_0,R)}|.\\
&\le C(j+2)\omega(\Delta(x_0,2^{j+1}R))^{1/q}\|h\|_{\mathrm{BMO}(\Sigma,\sigma)}\\
&\quad+\omega(\Delta(x_0,2^{j+1}R))^{1/q}|h(y)-h^{\omega}_{\Delta(x_0,R)}|.
\end{aligned}$$

In the first inequality we added and subtracted $h^{\omega}_{\Delta(x_0,R)}$. For the second inequality, we added and subtracted $h^{\omega}_{\Delta(x_0,2^{i+1}R)}$ for each $i\in\{1,2,\dots,j\}$. The last inequality follows from creating an integral average, 2.81, the equivalence of BMO seminorms in Lemmas 2.38 and 2.39, and the following estimate, which holds for each $i\in\{0,1,\dots,j\}$:

$$\begin{aligned}
&|h^{\omega}_{\Delta(x_0,2^{i+1}R)}-h^{\omega}_{\Delta(x_0,2^{i}R)}| && (7.203)\\
&=\Big|\fint_{\Delta(x_0,2^{i}R)}(h-h^{\omega}_{\Delta(x_0,2^{i+1}R)})\,d\omega\Big|\\
&\le\frac{\omega(\Delta(x_0,2^{i+1}R))}{\omega(\Delta(x_0,2^{i}R))}\fint_{\Delta(x_0,2^{j+1}R)}|h-h^{\omega}_{\Delta(x_0,2^{i+1}R)}|\,d\omega\\
&\le C\|h\|_{\mathrm{BMO}(\Sigma,\sigma)}.
\end{aligned}$$

Here, the constant $C\in(0,\infty)$ depends only on Σ, n, q, and $[\omega]_{A_q}$, since the last inequality follows from doubling property of ω (cf. Proposition 2.29) and the fact that $\|\cdot\|_{\mathrm{BMO}(\Sigma,\omega)}\approx\|\cdot\|_{\mathrm{BMO}(\Sigma,\sigma)}$ by Lemma 2.39. Ultimately, we find that

$$\begin{aligned}
&\Big(\int_{x\in\Delta(x_0,2^{j+1}R)} |h(x)-h(y)|^q\,d\omega(x)\Big)^{1/q} && (7.204)\\
&\le Cj\omega(\Delta(x_0,2^{j+1}R))^{1/q}+\omega(\Delta(x_0,2^{j+1}R))^{1/q}|h(y)-h^{\omega}_{\Delta(x_0,R)}|.
\end{aligned}$$

Plugging this result back into (7.199), we have

$$\left(\int_{B_j} |(\mathfrak{C}b)(x)|^q \, d\omega(x)\right)^{1/q} \leq \mathrm{I} + \mathrm{II}, \tag{7.205}$$

where

$$\mathrm{I} := \frac{Cj\omega(\Delta(x_0, 2^{j+1}R))^{1/q}}{(2^j R)^{n-1}} \|h\|_{\mathrm{BMO}(\Sigma,\sigma)} \int_{\Delta(x_0,R)} |b| \, d\sigma, \tag{7.206}$$

and

$$\mathrm{II} := \frac{C\omega(\Delta(x_0, 2^{j+1}R))^{1/q}}{(2^j R)^{n-1}} \int_{\Delta(x_0,R)} |b||h - h^{\omega}_{\Delta(x_0,R)}| \, d\sigma. \tag{7.207}$$

Treating I first, we find

$$\begin{aligned} \mathrm{I} &\leq \frac{Cj\omega(\Delta(x_0, 2^{j+1}R))^{1/q}}{2^{j(n-1)}} \|h\|_{\mathrm{BMO}(\Sigma,\sigma)} \fint_{\Delta(x_0,R)} |b| \, d\sigma \\ &\leq \frac{Cj\omega(\Delta(x_0, 2^{j+1}R))^{1/q}}{2^{j(n-1)}} \|h\|_{\mathrm{BMO}(\Sigma,\sigma)} [\omega]_{A_q}^{1/q} \left(\fint_{\Delta(x_0,R)} |b|^q \, d\omega\right)^{1/q} \\ &\leq \frac{CjR^{\lambda(1/q-1)}}{2^{j(n-1)}} \left(\frac{\omega(\Delta(x_0, 2^{j+1}R))}{\omega(\Delta(x_0, R))}\right)^{1/q} \|h\|_{\mathrm{BMO}(\Sigma,\sigma)}. \end{aligned} \tag{7.208}$$

The first inequality follows by creating the integral average in (7.206) and Ahlfors regularity, the second inequality follows from Lemma 2.34, and the last inequality follows from unpacking an integral average and (7.192).

Turning our attention to the treatment of II, let $q' \in (1, \infty)$ denote the Hölder conjugate of q (i. e., $1/q + 1/q' = 1$) and introduce the conjugate weight $\omega' := \omega^{-q'/q} \in A_{q'}(\Sigma, \sigma)$. Fix some $s \in (1, \infty)$ to be specified later. Then

$$\begin{aligned} &\int_{\Delta(x_0,R)} |b||h - h^{\omega}_{\Delta(x_0,R)}| \, d\sigma \\ &\leq \left(\int_{\Delta(x_0,R)} |b|^q \, d\omega\right)^{1/q} \left(\int_{\Delta(x_0,R)} |h - h^{\omega}_{\Delta(x_0,R)}|^{q'} \, d\omega'\right)^{1/q'} \\ &\leq R^{\lambda(1/q-1)} \left(\int_{\Delta(x_0,R)} |h - h^{\omega}_{\Delta(x_0,R)}|^{sq'} \, d\omega\right)^{1/sq'} \left(\int_{\Delta(x_0,R)} \omega^{(-1/s-q'/q)s'} \, d\sigma\right)^{1/(s'q')} \\ &\leq CR^{\lambda(1/q-1)} \omega(\Delta(x_0, R))^{1/(sq')} \|h\|_{\mathrm{BMO}(\Sigma,\sigma)} \left(\int_{\Delta(x_0,R)} \omega^{(-1/s-q'/q)s'} \, d\sigma\right)^{1/(s'q')}. \end{aligned} \tag{7.209}$$

The first inequality follows from multiplying by $\omega^{1/q}\omega^{-1/q}$ and using Hölder's inequality. The second inequality follows from another application of Hölder's inequality, but with exponent s and decomposing the measure of integration according to $d\omega' = \omega' \, d\sigma = \omega^{1/s}\omega^{-1/s-q'/q} \, d\sigma$. For the second inequality, we also use (7.192). The last inequality follows from Lemmas 2.38 and 2.39, which ensure $\|\cdot\|_{\mathrm{BMO}_{sq'}(\Sigma,\omega)} \approx \|\cdot\|_{\mathrm{BMO}(\Sigma,\sigma)}$. Now, define $c \in (1,\infty)$ to be such that $1/(s'q') + 1/c = 1$. Then if c' denotes the Hölder conjugate of c, we see that $c' = s'q'$ and

$$\begin{aligned}\Big(-\frac{1}{s}-\frac{q'}{q}\Big)s' &= \Big(-1+\frac{1}{s'}-\frac{q'}{q}\Big)s' = -s'+1-\frac{c'}{q} \\ &= -s'+1-c'\Big(1-\frac{1}{q'}\Big) = -s'+1-c'+\frac{c'}{q'} \\ &= -s'+1-c'+s' = 1-c'.\end{aligned} \tag{7.210}$$

As a result,

$$\begin{aligned}&\omega(\Delta(x_0,R))^{1/(sq')}\Big(\int_{\Delta(x_0,R)} \omega^{(-1/s-q'/q)s'} \, d\sigma\Big)^{1/(s'q')} \\ &= \omega(\Delta(x_0,R))^{1/(sq')}\sigma(\Delta(x_0,R))^{1/c'}\Big(\fint_{\Delta(x_0,R)} \omega^{1-c'} \, d\sigma\Big)^{1/c'}.\end{aligned} \tag{7.211}$$

We now choose $s := \frac{(q-1)(q-\varepsilon)}{\varepsilon}$, which belongs to the interval $(1,\infty)$, since $\varepsilon < q-1$. It follows from this judicious choice of s that $c = q-\varepsilon$. Indeed, recalling that $1/c + 1/c' = 1$ and $c' = s'q'$, as well as that $1/s + 1/s' = 1$, we find

$$\begin{aligned}c &= \frac{s'q'}{s'q'-1} = \frac{sq}{sq-(s-1)(q-1)} = \frac{sq}{s+q-1} \\ &= \frac{(q-1)(q-\varepsilon)q}{(q-1)(q-\varepsilon)+\varepsilon(q-1)} = \frac{(q-1)(q-\varepsilon)q}{q(q-1)} = q-\varepsilon.\end{aligned} \tag{7.212}$$

Thus, we have $\omega \in A_c(\Sigma,\sigma)$ by assumption. We then have by (2.47) that

$$\Big(\fint_{\Delta(x_0,R)} \omega \, d\sigma\Big)\Big(\fint_{\Delta(x_0,R)} \omega^{1-c'} \, d\sigma\Big)^{c-1} \le [\omega]_{A_c} < +\infty, \tag{7.213}$$

which implies that

$$\fint_{\Delta(x_0,R)} \omega^{1-c'} \, d\sigma \le [\omega]_{A_c}^{1/(c-1)}\Big(\frac{\sigma(\Delta(x_0,R))}{\omega(\Delta(x_0,R))}\Big)^{1/(c-1)}. \tag{7.214}$$

Inserting the above estimate into (7.211), we obtain

$$\begin{aligned}
\omega(\Delta(x_0,R))^{1/(sq')}\Big(\int_{\Delta(x_0,R)} \omega^{(-1/s-q'/q)s'}\,d\sigma\Big)^{1/(s'q')} \\
&\leq [\omega]_{A_c}^{1/(c-1)} \frac{\sigma(\Delta(x_0,R))^{1/c'+1/[c'(c-1)]}}{\omega(\Delta(x_0,R))^{1/[c'(c-1)]-1/(sq')}}. \\
&= [\omega]_{A_c}^{1/(c-1)} \frac{\sigma(x_0,R)}{\omega(\Delta(x_0,R))^{1/q}} \\
&\leq C[\omega]_{A_c}^{1/(c-1)} \frac{R^{n-1}}{\omega(\Delta(x_0,R))^{1/q}}.
\end{aligned} \tag{7.215}$$

The last inequality above follows from the Ahlfors regularity of Σ. The equality in (7.215) is obtained by observing that $c' = c/(c-1)$, from which it follows that

$$\frac{1}{c'} + \frac{1}{c'(c-1)} = \frac{1}{c'}\Big(1+\frac{1}{c-1}\Big) = \frac{1}{c'}\Big(\frac{c}{c-1}\Big) = 1, \tag{7.216}$$

and

$$\begin{aligned}
\frac{1}{c'(c-1)} - \frac{1}{sq'} &= \frac{1}{c} - \frac{1}{sq'} = \frac{1}{c} - \frac{1}{q'}\Big(1-\frac{1}{s'}\Big) \\
&= \frac{1}{c} - \frac{1}{q'} + \frac{1}{c'} = 1 - \frac{1}{q'} = \frac{1}{q}.
\end{aligned} \tag{7.217}$$

Therefore, using the estimate in (7.215) with that of (7.209), we obtain from (7.207) the estimate

$$\mathrm{II} \leq C\frac{R^{\lambda(1/q-1)}}{2^{j(n-1)}}\Big(\frac{\omega(\Delta(x_0,2^{j+1}R))}{\omega(\Delta(x_0,R))}\Big)^{1/q}\|h\|_{\mathrm{BMO}(\Sigma,\sigma)}, \tag{7.218}$$

where $C \in (0,\infty)$ depends on Σ, n, q, ε, and $[\omega]_{A_{q-\varepsilon}}$ (recalling that $c = q-\varepsilon$ in view of our choice of s after (7.211)). Note the similarity in the estimate for I, obtained in (7.208). Taking the worse estimate, we find thanks to (7.205)–(7.208) and (7.218) that

$$\|\mathfrak{C}b\|_{L^q(B_j,\omega)} \leq \frac{CjR^{\lambda(1/q-1)}}{2^{j(n-1)}}\Big(\frac{\omega(\Delta(x_0,2^{j+1}R))}{\omega(\Delta(x_0,R))}\Big)^{1/q}\|h\|_{\mathrm{BMO}(\Sigma,\sigma)}. \tag{7.219}$$

Since $\omega \in A_{q-\varepsilon}(\Sigma,\sigma)$, we can use item (4) of Proposition 2.29, under the identifications $p := q-\varepsilon$, $\Delta := \Delta(x_0,2^{j+1}R)$, and $E := \Delta(x_0,R)$, to estimate

$$\begin{aligned}
\Big(\frac{\omega(\Delta(x_0,2^{j+1}R))}{\omega(\Delta(x_0,R))}\Big)^{1/q} &\leq C\Big(\frac{\sigma(\Delta(x_0,2^{j+1}R))}{\sigma(\Delta(x_0,R))}\Big)^{(q-\varepsilon)/q} \\
&\leq C2^{j(n-1)(q-\varepsilon)/q},
\end{aligned} \tag{7.220}$$

where the last inequality is a consequence of the Ahlfors regularity of Σ. Subsequently, (7.219)–(7.220) imply

$$\|\widetilde{b}_j\|_{L^q(\Sigma,\omega)} = \|\mathfrak{C}b\|_{L^q(B_j,\omega)} \le \frac{CjR^{\lambda(1/q-1)}}{2^{j(n-1)(\varepsilon/q)}}\|h\|_{\mathrm{BMO}(\Sigma,\sigma)}, \tag{7.221}$$

for some constant $C \in (0,\infty)$, depending only on Σ, n, q, ε, $[\omega]_{A_{q-\varepsilon}}$, and $[\omega]_{A_q}$.

With C as in (7.221), for each $j \in \mathbb{N}$ define

$$\lambda_j := Cj\|h\|_{\mathrm{BMO}(\Sigma,\sigma)}2^{-\lambda(j+1)(1/q-1)-j(n-1)(\varepsilon/q)} \quad \text{and} \quad b_j := \lambda_j^{-1}\widetilde{b}_j. \tag{7.222}$$

With these definitions, we see from (7.198) and (7.221) that the function b_j is a (q,λ,ω)-block on Σ, centered at x_0 with radius $2^{j+1}R$. Since $\widetilde{b}_j = \lambda_j b_j$ for each $j \in \mathbb{N}_0$, we see from (7.194) that

$$\mathfrak{C}b = \sum_{j=0}^{\infty}\lambda_j b_j \quad \text{pointwise on } \Sigma. \tag{7.223}$$

Moreover, from (7.222), we find that

$$\begin{aligned}\sum_{j=1}^{\infty}|\lambda_j| &= C\|h\|_{\mathrm{BMO}(\Sigma,\sigma)}\sum_{j=1}^{\infty}j2^{-j[(n-1)(\varepsilon/q)-\lambda(1-1/q)]} \\ &= C\|h\|_{\mathrm{BMO}(\Sigma,\sigma)}\sum_{j=1}^{\infty}j2^{-j(1-1/q)[\varepsilon(n-1)/(q-1)-\lambda]} < +\infty.\end{aligned} \tag{7.224}$$

The above series converges because the hypothesis that λ belongs to the interval $(0,\varepsilon(n-1)/(q-1))$ ensures that

$$(1-1/q)\big[\varepsilon(n-1)/(q-1)-\lambda\big] > 0. \tag{7.225}$$

From Lemma 6.8, we see that (7.223) is a block representation for b. Also, from (7.224) and the definition of λ_0 in (7.197), we find that

$$\|\mathfrak{C}b\|_{\mathcal{B}^{q,\lambda}(\Sigma,\omega)} \le \sum_{j=0}^{\infty}|\lambda_j| \le C_*\|h\|_{\mathrm{BMO}(\Sigma,\sigma)} \tag{7.226}$$

for some constant $C_* \in (0,\infty)$, depending only on Σ, n, q, ε, λ, $[\omega]_{A_{q-\varepsilon}}$, and $[\omega]_{A_q}$. In summary,

for any (q,λ,ω)-block, b, one has $\mathfrak{C}b \in \mathcal{B}^{q,\lambda}(\Sigma,\omega)$ with $\|\mathfrak{C}b\|_{\mathcal{B}^{q,\lambda}(\Sigma,\omega)} \le C_*$ for some constant $C_* \in (0,\infty)$, which is independent of the choice of block, b. (7.227)

We now turn our attention to general block functions. Let $g \in \mathcal{B}^{q,\lambda}(\Sigma,\sigma)$ be arbitrary with a quasioptimal block representation

$$g = \sum_{j=1}^{\infty}\lambda_j b_j \quad \text{with} \quad \sum_{j=1}^{\infty}|\lambda_j| \le 2\|g\|_{\mathcal{B}^{q,\lambda}(\Sigma,\omega)}. \tag{7.228}$$

For each $M \in \mathbb{M}$ define the partial sums $g_M := \sum_{j=1}^{M} \lambda_j b_j$. By Proposition 6.17, there exists some $q_0 \in (1, q-1)$ such that

$$\mathcal{B}^{q,\lambda}(\Sigma, \omega) \hookrightarrow L^1\Big(\Sigma, \frac{\sigma(x)}{1+|x|^{n-1}}\Big) \cap L^{q_0}_{\mathrm{loc}}(\Sigma, \sigma) \tag{7.229}$$

continuously. Thus, (in view of Proposition 6.12), we find that $g_M \to g$ in $L^1(\Sigma, \frac{\sigma(x)}{1+|x|^{n-1}})$ as $M \to \infty$. Additionally, by Corollary 5.20, we find that $\mathfrak{C} g_M \to \mathfrak{C} g$ as $M \to \infty$ in $L^{q_0}_{\mathrm{loc}}(\Sigma, \sigma)$ as well. In particular, there exists a sequence $\{M_k\}_{k \in \mathbb{N}} \subseteq \mathbb{N}$ such that

$$\mathfrak{C} g(x) = \lim_{k \to \infty} \sum_{j=1}^{M_k} \lambda_j \mathfrak{C} b_j(x) \quad \text{at } \sigma\text{-a. e. point } x \in \Sigma. \tag{7.230}$$

Note that we have used the linearity of $\mathfrak{C}$ as well. Let $M, M' \in \mathbb{N}$, and without loss of generality take $M < M'$. Then

$$\begin{aligned} \|\mathfrak{C} g_M - \mathfrak{C} g_{M'}\|_{\mathcal{B}^{q,\lambda}(\Sigma,\omega)} &\leq \sum_{j=M+1}^{M'} |\lambda_j| \|\mathfrak{C} b_j\|_{\mathcal{B}^{q,\lambda}(\Sigma,\omega)} \\ &\leq C \sum_{j=M+1}^{M'} |\lambda_j|. \end{aligned} \tag{7.231}$$

Note that the last inequality follows from (7.227) as it applies to b_j for each $j \in \mathbb{N}$. Because $\{\lambda_j\}_{j \in \mathbb{N}} \in \ell^1$, the above estimate implies $\{\mathfrak{C} g_M\}_{M \in \mathbb{N}}$ is a Cauchy sequence in $\mathcal{B}^{q,\lambda}(\Sigma, \omega)$. Since $\mathcal{B}^{q,\lambda}(\Sigma, \omega)$ is a Banach space (cf. Theorem 6.11), there exists some $G \in \mathcal{B}^{q,\lambda}(\Sigma, \omega)$ for which

$$G = \lim_{M \to \infty} \sum_{j=1}^{M} \lambda_j \mathfrak{C} b_j \quad \text{in } \mathcal{B}^{q,\lambda}(\Sigma, \omega). \tag{7.232}$$

Consequently, thanks to (7.227),

$$\begin{aligned} \|G\|_{\mathcal{B}^{q,\lambda}(\Sigma,\omega)} &= \lim_{m \to \infty} \Big\| \sum_{j=1}^{m} \lambda_j \mathfrak{C} b_j \Big\|_{\mathcal{B}^{q,\lambda}(\Sigma,\omega)} \\ &\leq \lim_{m \to \infty} |\lambda_j| \|g_m\|_{\mathcal{B}^{q,\lambda}(\Sigma,\omega)} \\ &\leq C_* \|h\|_{\mathrm{BMO}(\Sigma,\sigma)} \sum_{j=1}^{\infty} |\lambda_j| \\ &\leq C_* \|h\|_{\mathrm{BMO}(\Sigma,\sigma)} \|g\|_{\mathcal{B}^{q,\lambda}(\Sigma,\omega)}, \end{aligned} \tag{7.233}$$

where the last inequality follows by quasioptimality, (7.228). Note that by (7.232) and the embedding in Proposition 6.17, we find that

$$G = \lim_{k\to\infty} \sum_{j=1}^{M_k} \lambda_j \mathfrak{C} b_j \quad \text{in } L^1\Big(\Sigma, \frac{\sigma(x)}{1+|x|^{n-1}}\Big), \tag{7.234}$$

as well. Thus, there exists a subsequence $\{M_{k_\ell}\}_{\ell\in\mathbb{N}}$ such that

$$G = \lim_{\ell\to\infty} \sum_{j=1}^{M_{k_\ell}} \lambda_j \mathfrak{C} b_j \quad \text{at } \sigma\text{-a. e. point on } \Sigma. \tag{7.235}$$

In concert, (7.230) and (7.235) guarantee $G = \mathfrak{C}g$ at σ-a. e. point on Σ. Subsequently, by the lattice property of $\mathcal{B}^{q,\lambda}(\Sigma,\sigma)$ (cf. Proposition 6.14) and (7.233) we can conclude that $\mathfrak{C}g \in \mathcal{B}^{q,\lambda}(\Sigma,\omega)$ with

$$\|\mathfrak{C}g\|_{\mathcal{B}^{q,\lambda}(\Sigma,\omega)} = \|G\|_{\mathcal{B}^{q,\lambda}(\Sigma,\omega)} \le C\|h\|_{\mathrm{BMO}(\Sigma,\sigma)}\|g\|_{\mathcal{B}^{q,\lambda}(\Sigma,\omega)}, \tag{7.236}$$

which is the desired result given the arbitrary nature of $g \in \mathcal{B}^{q,\lambda}(\Sigma,\sigma)$. □

The following theorem provides a similar result to that of Theorem 7.12, with $\nu(x)$ replaced by $\nu(y)$ in (7.143). This theorem may also be regarded as a corollary to Theorem 7.12 and Theorem 7.13. It is worth noting that comments similar to those of Remark 7.3 are valid in the context of Theorem 7.15.

Theorem 7.15. *Let $\Omega \subseteq \mathbb{R}^n$ be a UR domain, abbreviate $\sigma := \mathcal{H}^{n-1}\lfloor\partial\Omega$, and denote ν as the geometric measure theoretic outward unit normal to Ω. Let $N = N(n) \in \mathbb{N}$ be sufficiently large and fix a complex-valued function $k \in \mathscr{C}^N(\mathbb{R}^n \setminus \{0\})$, which is even and positive homogeneous of degree $-n$. For all $f \in L^1(\partial\Omega, \frac{\sigma(x)}{1+|x|^{n-1}})$, define*

$$Tf(x) := \lim_{\varepsilon\to 0^+} \int_{\substack{y\in\partial\Omega\\ |x-y|>\varepsilon}} \langle \nu(y), x-y\rangle k(x-y) f(y)\, d\sigma(y) \tag{7.237}$$

for σ-a. e. point $x \in \partial\Omega$. Fix an integrability exponent $q \in (1,\infty)$, a Muckenhoupt weight $\omega \in A_q(\partial\Omega,\sigma)$, choose $\varepsilon \in (0, q-1)$ such that $\omega \in A_{q-\varepsilon}(\partial\Omega,\sigma)$, and fix a threshold $\lambda \in (0, \varepsilon(n-1)/(q-1))$. Then for all $m \in \mathbb{N}$ there exists a constant $C_m \in (0,\infty)$, depending only on m, Σ, n, q, ε, λ, $[\omega]_{A_{q-\varepsilon}}$, and $[\omega]_{A_q}$ such that

$$\|T\|_{\mathcal{B}^{q,\lambda}(\partial\Omega,\omega)\to\mathcal{B}^{q,\lambda}(\partial\Omega,\omega)} \le C_m \cdot \Big(\sum_{|\alpha|\le N} \sup_{S^{n-1}} |\partial^\alpha k|\Big) \cdot \|\nu\|^{\langle m\rangle}_{[\mathrm{BMO}(\partial\Omega,\sigma)]^n}. \tag{7.238}$$

Proof. We begin by writing $T = T^\# + (T - T^\#)$ and observing that $T - T^\#$ is of commutator type. Indeed, if for each $j \in \{1,\dots,n\}$ we define the integral kernel $k_j(z) := z_j k(z)$ for all $z \in \mathbb{R}^n \setminus \{0\}$ and

$$T_{k_j} f(x) := \lim_{\varepsilon\to 0^+} \int_{\substack{y\in\partial\Omega\\ |x-y|>\varepsilon}} k_j(x-y) f(y)\, d\sigma(y) \tag{7.239}$$

for σ-a. e. $x \in \partial\Omega$, then we see that

$$
\begin{aligned}
(T - T^{\#})f(x) &= \lim_{\varepsilon\to 0^+} \int\limits_{\substack{y\in\partial\Omega\\ |x-y|>\varepsilon}} \langle v(y) - v(x), x - y\rangle k(x-y)f(y)\,d\sigma(y) \qquad (7.240)\\
&= \sum_{j=1}^{n} \lim_{\varepsilon\to 0^+} \int\limits_{\substack{y\in\partial\Omega\\ |x-y|>\varepsilon}} (v_j(x) - v_j(y))(x_j - y_j)k(x-y)f(y)\,d\sigma(y)\\
&= \sum_{j=1}^{n} v(x)(T_{k_j}f)(x) - T_{k_j}(v_jf)(x)\\
&= \sum_{j=1}^{n} [M_{v_j}, T_{k_j}]f(x).
\end{aligned}
$$

By Theorem 7.12 and Theorem 7.13, we obtain the estimate

$$
\begin{aligned}
\|T\|_{\mathcal{B}^{q,\lambda}(\partial\Omega,\omega)\to\mathcal{B}^{q,\lambda}(\partial\Omega,\omega)} &\le \|T^{\#}\|_{\mathcal{B}^{q,\lambda}(\partial\Omega,\omega)\to\mathcal{B}^{q,\lambda}(\partial\Omega,\omega)} \qquad (7.241)\\
&\quad + \sum_{j=1}^{n} \|[M_{v_j}, T_{k_j}]\|_{\mathcal{B}^{q,\lambda}(\partial\Omega,\omega)\to\mathcal{B}^{q,\lambda}(\partial\Omega,\omega)}\\
&\le C_m\Big(\sum_{|\alpha|\le N} \sup_{S^{n-1}} |\partial^\alpha k|\Big) \cdot \|v\|^{\langle m\rangle}_{[\mathrm{BMO}(\partial\Omega)]^n}\\
&\quad + C\Big(\sum_{|\alpha|\le N} \sup_{S^{n-1}} |\partial^\alpha k|\Big) \sum_{j=1}^{n} \|v_j\|_{\mathrm{BMO}(\partial\Omega,\sigma)}\\
&\le C_m\Big(\sum_{|\alpha|\le N} \sup_{S^{n-1}} |\partial^\alpha k|\Big) \cdot \|v\|^{\langle m\rangle}_{[\mathrm{BMO}(\partial\Omega,\sigma)]^n},
\end{aligned}
$$

where we have simply taken the worse estimate involving $\|v\|_{[\mathrm{BMO}(\partial\Omega,\sigma)]^n}$ to obtain the last inequality. This establishes the desired result. □

8 Weighted Morrey- and Block-based Sobolev spaces

In Chapters 4 and 6, we built up a functional analytic framework for Muckenhoupt-weighted Morrey and Block spaces, respectively. To consider different types of boundary value problems like the regularity problem and the transmission problem, we need to develop Muckenhoupt-weighted Morrey-based and Block-based Sobolev spaces. It is also of interest to develop and study homogeneous versions of these spaces. However, to even begin we will need a suitable notion of weak derivatives on closed Ahlfors regular sets.

Let $\Omega \subseteq \mathbb{R}^n$ be an AR domain and abbreviate $\sigma := \mathcal{H}^{n-1}\lfloor\partial\Omega$. As usual, let $\nu = (\nu_1, \dots, \nu_n) \in \mathbb{R}^n$ denote the geometric measure theoretic outward unit normal to Ω. We take Ω to be an AR domain, rather than considering some arbitrary open set with Ahlfors regular boundary, because the added condition that $\mathcal{H}^{n-1}(\partial\Omega \setminus \partial_*\Omega) = 0$ in the definition of an AR domain (cf. Definition 2.5) ensures that ν is defined σ-a. e. on $\partial\Omega$. The family of first-order tangential derivative operators, $\partial_{\tau_{jk}}$, with $j, k \in \{1, \dots, n\}$, acts on each function $\varphi \in \mathscr{C}_0^\infty(\mathbb{R}^n)$ according to

$$\partial_{\tau_{jk}}\varphi := \nu_j(\partial_k\varphi)\big|_{\partial\Omega} - \nu_k(\partial_j\varphi)\big|_{\partial\Omega}. \tag{8.1}$$

Given any $\varphi, \psi \in \mathscr{C}_0^\infty(\mathbb{R}^n)$, the following integration by parts formula holds on the boundary:

$$\int_{\partial\Omega} (\partial_{\tau_{jk}}\varphi)\psi \, d\sigma = -\int_{\partial\Omega} \varphi(\partial_{\tau_{jk}}\psi) \, d\sigma \quad \text{for all } j, k \in \{1, \dots, n\}. \tag{8.2}$$

To see this, define the vector field $\vec{F} := \partial_k(\varphi\psi)e_j - \partial_j(\varphi\psi)e_k$ (where $\{e_i\}_{1\le i\le n}$ is the standard orthonormal basis for $\mathbb{R}^n$). Then $\vec{F}$ is smooth and compactly supported, so the divergence theorem (cf. [36, Theorem 5.16, p. 236]) applies, and hence

$$\int_{\Omega} \operatorname{div}\vec{F} \, d\mathcal{L}^n = \int_{\partial\Omega} \nu \cdot (\vec{F}|_{\partial\Omega}) \, d\sigma. \tag{8.3}$$

But $\operatorname{div}\vec{F} = \partial_j\partial_k(\varphi\psi) - \partial_k\partial_j(\varphi\psi) = 0$, because the smoothness of φ and ψ ensures that the second-order mixed partial derivatives of $\varphi\psi$ agree. On the other hand,

$$\begin{aligned} \nu \cdot (\vec{F}|_{\partial\Omega}) &= \nu_j\partial_k(\varphi\psi)\big|_{\partial\Omega} - \nu_k\partial_j(\varphi\psi)\big|_{\partial\Omega} \\ &= \nu_j\big(\varphi(\partial_k\psi) + (\partial_k\varphi)\psi\big)\big|_{\partial\Omega} - \nu_k\big(\varphi(\partial_j\psi) + (\partial_j\varphi)\psi\big)\big|_{\partial\Omega} \\ &= \partial_{\tau_{jk}}\varphi + \partial_{\tau_{jk}}\psi, \end{aligned} \tag{8.4}$$

so (8.2) follows from (8.3) and (8.4).

The integration by parts formula in (8.2) motivates the definition of a weak (tangential) derivative on $\partial\Omega$.

https://doi.org/10.1515/9783111458274-008

Definition 8.1. Let $\Omega \subseteq \mathbb{R}^n$ be an AR domain and write $\sigma := \mathcal{H}^{n-1}\lfloor\partial\Omega$. Given $f \in L^1_{\mathrm{loc}}(\partial\Omega, \sigma)$, we say that f has a weak tangential derivative in $L^1_{\mathrm{loc}}(\partial\Omega, \sigma)$ if for all indices $j, k \in \{1, \ldots, n\}$ there exists a function f_{jk} belonging to $L^1_{\mathrm{loc}}(\partial\Omega, \sigma)$ such that

$$\int_{\partial\Omega} f \partial_{\tau_{jk}} \varphi \, d\sigma = - \int_{\partial\Omega} f_{jk} \varphi \, d\sigma \quad \text{for all } \varphi \in \mathscr{C}_0^\infty(\mathbb{R}^n). \tag{8.5}$$

Note that such a function f_{jk} is unique σ-a. e. on $\partial\Omega$, and hence well-defined. Indeed, see [86, Proposition 3.7.2, p. 281], which allows us to conclude that

$$f = 0 \quad \text{at } \sigma\text{-a. e. point on } \partial\Omega \quad \Longleftrightarrow \quad \int_{\partial\Omega} f\varphi \, d\sigma = 0 \quad \text{for all } \varphi \in \mathscr{C}_0^\infty(\mathbb{R}^n). \tag{8.6}$$

From here on out, we will frequently identify the function f_{jk} with $\partial_{\tau_{jk}} f$. Also, for $p \in [1, \infty]$, we shall denote by $L^p_{1,\mathrm{loc}}(\partial\Omega, \sigma)$ the collection of all functions $f \in L^p_{\mathrm{loc}}(\partial\Omega, \sigma)$ with $\partial_{\tau_{jk}} f \in L^p_{\mathrm{loc}}(\partial\Omega, \sigma)$ for all $j, k \in \{1, \ldots, n\}$.

Definition 8.2. Given $f \in L^1_{1,\mathrm{loc}}(\partial\Omega, \sigma)$, define the tangential gradient of f as the vector-valued function

$$\nabla_{\tan} f := (\nu_j \partial_{\tau_{jk}} f)_{1 \le k \le n}, \tag{8.7}$$

where we remind the reader that we are using the convention of implied summation over repeated indices.

It has been shown in [87, Proposition 11.4.2, p. 729] that if $\Omega \subseteq \mathbb{R}^n$ is actually a UR domain, and $p \in (1, \infty)$, then for every $f \in L^p_{1,\mathrm{loc}}(\partial\Omega, \sigma)$ and for each pair of indices $j, k \in \{1, \ldots, n\}$, one has

$$\partial_{\tau_{jk}} f = \nu_j (\nabla_{\tan} f)_k - \nu_k (\nabla_{\tan} f)_j. \tag{8.8}$$

Note that the above formula follows immediately from definitions (cf. (8.1) and (8.7)) in the case that f is smooth, but that in general (8.8) is highly nontrivial, and is heavily dependent on the underlying geometry.

Remark 8.3. If $f, g \in L^1_{1,\mathrm{loc}}(\partial\Omega, \sigma)$, then for any $\alpha, \beta \in \mathbb{C}$ the function $\alpha f + \beta g$ belongs to $L^1_{1,\mathrm{loc}}(\partial\Omega, \sigma)$ and

$$\partial_{\tau_{jk}}(\alpha f + \beta g) = \alpha(\partial_{\tau_{jk}} f) + \beta(\partial_{\tau_{jk}} g) \quad \text{for all } j, k \in \{1, \ldots, n\}. \tag{8.9}$$

To see this, fix $j, k \in \{1, \ldots, n\}$ and $\varphi \in \mathscr{C}_0^\infty(\mathbb{R}^n)$. Then

$$\begin{aligned}\int_{\partial\Omega}(\alpha f+\beta g)\partial_{\tau_{jk}}\varphi\,d\sigma &= \alpha\int_{\partial\Omega} f\partial_{\tau_{jk}}\varphi\,d\sigma+\beta\int_{\partial\Omega} g\partial_{\tau_{jk}}\varphi\,d\sigma \\ &= -\alpha\int_{\partial\Omega}(\partial_{\tau_{jk}}f)\varphi\,d\sigma-\beta\int_{\partial\Omega}(\partial_{\tau_{jk}}g)\varphi\,d\sigma \\ &= -\int_{\partial\Omega}[\alpha(\partial_{\tau_{jk}}f)+\beta(\partial_{\tau_{jk}}g)]\varphi\,d\sigma.\end{aligned} \tag{8.10}$$

Since $\alpha(\partial_{\tau_{jk}}f)+\beta(\partial_{\tau_{jk}}g)\in L^1_{\mathrm{loc}}(\partial\Omega,\sigma)$ and using the uniqueness of weak derivatives, we find that (8.9) holds. Ultimately, this shows that $L^1_{1,\mathrm{loc}}(\partial\Omega,\sigma)$ is a vector space over $\mathbb{C}$.

We now define Muckenhoupt-weighted Block-based Sobolev spaces.

Definition 8.4. To set the stage, let $\Omega\subseteq\mathbb{R}^n$ denote an AR domain and abbreviate $\sigma:=\mathcal{H}^{n-1}\lfloor\partial\Omega$. Fix an integrability exponent $q\in(1,\infty)$ along with a Muckenhoupt weight $\omega\in A_q(\partial\Omega,\sigma)$. Choose $\varepsilon\in(0,q-1)$ such that $\omega\in A_{q-\varepsilon}(\partial\Omega,\sigma)$ (cf. item (i) of Proposition 2.29), and fix any threshold $\lambda\in(0,\varepsilon(n-1)/(q-1))$. Define the `Muckenhoupt-weighted Block-based Sobolev space` according to

$$\mathcal{B}^{q,\lambda}_1(\partial\Omega,\omega):=\{f\in\mathcal{B}^{q,\lambda}(\partial\Omega,\omega):\partial_{\tau_{jk}}f\text{ exists in a weak sense and }\partial_{\tau_{jk}}f\in\mathcal{B}^{q,\lambda}(\partial\Omega,\omega)\text{ for all }j,k\in\{1,\dots,n\}\}. \tag{8.11}$$

We can also endow $\mathcal{B}^{q,\lambda}_1(\partial\Omega,\omega)$ with a natural norm as follows. For each function $f\in\mathcal{B}^{q,\lambda}_1(\partial\Omega,\omega)$, define

$$\|f\|_{\mathcal{B}^{q,\lambda}_1(\partial\Omega,\omega)}:=\|f\|_{\mathcal{B}^{q,\lambda}(\partial\Omega,\omega)}+\sum_{j,k=1}^n\|\partial_{\tau_{jk}}f\|_{\mathcal{B}^{q,\lambda}(\partial\Omega,\omega)}. \tag{8.12}$$

Remark 8.5. Retaining the context of Definition 8.4, Proposition 6.17 provides the continuous embedding $\mathcal{B}^{q,\lambda}(\partial\Omega,\omega)\hookrightarrow L^1_{\mathrm{loc}}(\partial\Omega,\sigma)$, so that it makes sense to talk about weak tangential derivatives (in the sense of Definition 8.1) of functions belonging to $\mathcal{B}^{q,\lambda}(\partial\Omega,\omega)$.

Remark 8.6. It follows from Remark 8.3 that the Muckenhoupt-weighted Block-based Sobolev space $\mathcal{B}^{q,\lambda}_1(\partial\Omega,\omega)$ is a vector subspace of $\mathcal{B}^{q,\lambda}(\partial\Omega,\omega)$. Additionally, the norm properties of $\|\cdot\|_{\mathcal{B}^{q,\lambda}_1(\partial\Omega,\omega)}$ are inherited from those of $\|\cdot\|_{\mathcal{B}^{q,\lambda}(\partial\Omega,\omega)}$ and Remark 8.3.

In fact, as witnessed by the following proposition, the norm described in (8.12) makes $\mathcal{B}^{q,\lambda}_1(\partial\Omega,\omega)$ a Banach space.

Proposition 8.7. *Let $\Omega\subseteq\mathbb{R}^n$ be an AR domain and write $\sigma:=\mathcal{H}^{n-1}\lfloor\partial\Omega$. Fix an integrability exponent $q\in(1,\infty)$ with $q':=q/(q-1)$ along with a Muckenhoupt weight $\omega\in A_q(\partial\Omega,\sigma)$. Let $\varepsilon\in(0,q-1)$ be such that ω belongs to $A_{q-\varepsilon}(\partial\Omega,\sigma)$. Lastly, fix any $\lambda\in(0,\varepsilon(n-1)/(q-1))$.*

Then the Muckenhoupt-weighted Block-based Sobolev space $\mathcal{B}_1^{q,\lambda}(\partial\Omega,\omega)$, when equipped with the norm $\|\cdot\|_{\mathcal{B}_1^{q,\lambda}(\partial\Omega,\omega)}$ is a Banach space.

Proof. We begin by observing that $(\mathcal{B}_1^{q,\lambda}(\partial\Omega,\omega),\|\cdot\|_{\mathcal{B}_1^{q,\lambda}(\partial\Omega,\omega)})$ is a normed vector space, as mentioned in Remark 8.6. It remains to show $\mathcal{B}_1^{q,\lambda}(\partial\Omega,\omega)$ is complete with respect to its norm. For that purpose, let $\{f_m\}_{m\in\mathbb{N}}$ be a Cauchy sequence in $\mathcal{B}_1^{q,\lambda}(\partial\Omega,\omega)$. Because

$$\|f_{m_1}-f_{m_2}\|_{\mathcal{B}^{q,\lambda}(\partial\Omega,\omega)} \leq \|f_{m_1}-f_{m_2}\|_{\mathcal{B}_1^{q,\lambda}(\partial\Omega,\omega)} \tag{8.13}$$

for all $m_1,m_2\in\mathbb{N}$, we see that $\{f_m\}_{m\in\mathbb{N}}$ is also Cauchy in $\mathcal{B}^{q,\lambda}(\partial\Omega,\omega)$. Since the space $\mathcal{B}^{q,\lambda}(\partial\Omega,\omega)$ is Banach by Theorem 6.11, there exists a function $f\in\mathcal{B}^{q,\lambda}(\partial\Omega,\omega)$ for which

$$\lim_{m\to\infty}\|f_m-f\|_{\mathcal{B}^{q,\lambda}(\partial\Omega,\omega)}=0. \tag{8.14}$$

Similarly, for each $j,k\in\{1,\dots,n\}$, from (8.12) we find that $\{\partial_{\tau_{jk}}f_m\}_{m\in\mathbb{N}}$ is Cauchy in $\mathcal{B}^{q,\lambda}(\partial\Omega,\omega)$. Hence, for each $j,k\in\{1,\dots,n\}$, there exists a function $f_{jk}\in\mathcal{B}^{q,\lambda}(\partial\Omega,\omega)$ with the property that

$$\lim_{m\to\infty}\|\partial_{\tau_{jk}}f_m-f_{jk}\|_{\mathcal{B}^{q,\lambda}(\partial\Omega,\omega)}=0. \tag{8.15}$$

At this stage, fix an arbitrary $\varphi\in\mathscr{C}_0^\infty(\mathbb{R}^n)$. Then for each $j,k\in\{1,\dots,n\}$ fixed

$$\begin{aligned}\int_{\partial\Omega} f(\partial_{\tau_{jk}}\varphi)\,d\sigma &= \lim_{m\to\infty}\int_{\partial\Omega} f_m(\partial_{\tau_{jk}}\varphi)\,d\sigma\\ &= -\lim_{m\to\infty}\int_{\partial\Omega}(\partial_{\tau_{jk}}f_m)\varphi\,d\sigma\\ &= -\int_{\partial\Omega} f_{jk}\varphi\,d\sigma.\end{aligned} \tag{8.16}$$

Note that the above integrals are well-defined objects in view of Remark 8.5. The first and last equalities in (8.16) make use of the fact that φ and $\partial_{\tau_{jk}}\varphi$ have compact support, as well as the continuous embedding $\mathcal{B}^{q,\lambda}(\partial\Omega,\omega)\hookrightarrow L^1_{\mathrm{loc}}(\partial\Omega,\sigma)$ (established in Proposition 6.17) together with (8.14) and (8.15). The second equality in (8.16) uses the assumption that $f_m\in\mathcal{B}_1^{q,\lambda}(\partial\Omega,\omega)$ and Definition 8.1. This allows us to conclude that $\partial_{\tau_{jk}}f=f_{jk}\in\mathcal{B}^{q,\lambda}(\partial\Omega,\omega)$, and hence f belongs to $\mathcal{B}_1^{q,\lambda}(\partial\Omega,\omega)$ demonstrating the completeness of $\mathcal{B}_1^{q,\lambda}(\partial\Omega,\omega)$. □

We can also define Muckenhoupt-weighted Morrey-based Sobolev spaces.

Definition 8.8. Let $\Omega\subseteq\mathbb{R}^n$ be an AR domain and write $\sigma:=\mathcal{H}^{n-1}\lfloor\partial\Omega$. Fix two integrability exponents $p,q\in(1,\infty)$ along with a Muckenhoupt weight $w\in A_p(\partial\Omega,\sigma)\cap\mathrm{RH}_q(\partial\Omega,\sigma)$.

Select a threshold $\lambda \in (0, (n-1)/q')$, where $q' := q/(q-1)$. Define the Muckenhoupt-weighted Morrey-based Sobolev space according to

$$M_1^{p,\lambda}(\partial\Omega, w) := \{f \in M^{p,\lambda}(\partial\Omega, w) : \partial_{\tau_{jk}} f \text{ exists in a weak sense} \\ \text{and } \partial_{\tau_{jk}} f \in M^{p,\lambda}(\partial\Omega, w) \text{ for all } j,k \in \{1,\dots,n\}\}. \tag{8.17}$$

We can also equip $M_1^{p,\lambda}(\partial\Omega, w)$ with a natural norm, defined for each function $f \in M_1^{p,\lambda}(\partial\Omega, w)$ according to

$$\|f\|_{M_1^{p,\lambda}(\partial\Omega,w)} := \|f\|_{M^{p\lambda}(\partial\Omega,w)} + \sum_{j,k=1}^{n} \|\partial_{\tau_{jk}} f\|_{M^{p,\lambda}(\partial\Omega,w)}. \tag{8.18}$$

Remark 8.9. Retaining the context of Definition 8.8, note that from the continuous embedding $M^{p,\lambda}(\partial\Omega, w) \hookrightarrow L^1_{\mathrm{loc}}(\partial\Omega, \sigma)$ (cf. Proposition 4.16), we see that it makes sense to speak of weak tangential derivatives (see Definition 8.1) in the context of functions belonging to $M^{p\lambda}(\partial\Omega, w)$.

Remark 8.10. As a consequence of Remark 8.3, we see that the Morrey-based Sobolev space $M_1^{p,\lambda}(\partial\Omega, w)$ is a vector subspace of $M^{p,\lambda}(\partial\Omega, w)$. Moreover, the norm properties of $\|\cdot\|_{M_1^{p,\lambda}(\partial\Omega,w)}$ are inherited from $\|\cdot\|_{M^{p,\lambda}(\partial\Omega,w)}$ and Remark 8.3.

Our next order of business is to show that the norm in (8.18) turns $M_1^{p,\lambda}(\partial\Omega, w)$ into a Banach space.

Proposition 8.11. *Let $\Omega \subseteq \mathbb{R}^n$ be an AR domain and set $\sigma := \mathcal{H}^{n-1}\lfloor\partial\Omega$. Fix two integrability exponents $p, q \in (1,\infty)$ along with a Muckenhoupt weight $w \in A_p(\partial\Omega, \sigma) \cap \mathrm{RH}_q(\partial\Omega, \sigma)$, and a threshold $\lambda \in (0, (n-1)/q')$, where $q' := q/(q-1)$. The Muckenhoupt-weighted Morrey-based Sobolev space $M_1^{p,\lambda}(\partial\Omega, w)$ when equipped with the norm $\|\cdot\|_{M_1^{p,\lambda}(\partial\Omega,w)}$ is complete, and hence Banach.*

Proof. The proof parallels that of Proposition 8.7 with $M^{p,\lambda}(\partial\Omega, w)$ and $M_1^{p,\lambda}(\partial\Omega, w)$ in place of $\mathcal{B}^{q,\lambda}(\partial\Omega, \omega)$ and $\mathcal{B}_1^{q,\lambda}(\partial\Omega, \omega)$, respectively. This time we use Proposition 4.5 and Proposition 4.16 in place of Proposition 6.11 and Proposition 6.17, respectively. □

The Sobolev spaces discussed so far are *inhomogeneous* due to the presence of both the function itself and its tangential derivatives in the definition of the norm. Our next task is to define a *homogeneous* version of $M_1^{p,\lambda}(\partial\Omega, w)$. The definition we present is inspired by work done in [79, Section 8.1].

Definition 8.12. Let $\Omega \subseteq \mathbb{R}^n$ be an AR domain and set $\sigma := \mathcal{H}^{n-1}\lfloor\partial\Omega$. Fix two integrability exponents $p, q \in (1,\infty)$, along with a Muckenhoupt weight $w \in A_p(\partial\Omega, \sigma) \cap \mathrm{RH}_q(\partial\Omega, \sigma)$, and a threshold $\lambda \in (0, (n-1)/q')$, where $q' := q/(q-1)$. Define the Muckenhoupt-weighted Morrey-based homogeneous Sobolev space via

$$\dot{M}_1^{p,\lambda}(\partial\Omega, w) := \Big\{ f \in L^1\Big(\partial\Omega, \frac{\sigma(x)}{1+|x|^n}\Big) \cap M_{\mathrm{loc}}^{p,\lambda}(\partial\Omega, w) : \\ \partial_{\tau_{jk}} f \in M^{p,\lambda}(\partial\Omega, w) \text{ for each } j,k \in \{1,\dots,n\} \Big\}, \tag{8.19}$$

where

$$M_{\mathrm{loc}}^{p,\lambda}(\partial\Omega, w) := \{ f : \partial\Omega \to \mathbb{C} : f\mathbf{1}_\Delta \in M^{p,\lambda}(\partial\Omega, w) \\ \text{for each surface ball } \Delta \subseteq \partial\Omega \}. \tag{8.20}$$

A simple, yet significant observation is that constants belong to $\dot{M}_1^{p,\lambda}(\partial\Omega, w)$. Indeed, any constant belongs to $L^1(\partial\Omega, \frac{\sigma(x)}{1+|x|^n})$, and thanks to Lemma 4.9, constants belong to $M_{\mathrm{loc}}^{p,\lambda}(\partial\Omega, w)$ as well.

One can equip $\dot{M}_1^{p,\lambda}(\partial\Omega, w)$ with the following seminorm, defined for each function $f \in \dot{M}_1^{p,\lambda}(\partial\Omega, w)$ according to

$$\|f\|_{\dot{M}_1^{p,\lambda}(\partial\Omega,w)} := \sum_{j,k=1}^{n} \|\partial_{\tau_{jk}} f\|_{M^{p,\lambda}(\partial\Omega,w)}. \tag{8.21}$$

Notice that $\|f\|_{\dot{M}_1^{p,\lambda}(\partial\Omega,w)} = 0$ whenever $f \in \dot{M}_1^{p,\lambda}(\partial\Omega, w)$ is constant. Therefore, we will find it useful to work with the space $\dot{M}_1^{p,\lambda}(\partial\Omega, w)/\sim$, the quotient space of equivalence classes $[\cdot]$ modulo constant functions in $\dot{M}_1^{p,\lambda}(\partial\Omega, w)$. Let us also introduce

$$\|[f]\|_{\dot{M}_1^{p,\lambda}(\partial\Omega,w)/\sim} := \sum_{j,k=1}^{n} \|\partial_{\tau_{jk}} f\|_{M^{p,\lambda}(\partial\Omega,w)} \tag{8.22}$$

for all $f \in \dot{M}_1^{p,\lambda}(\partial\Omega, w)$. In fact, as we shall see in Proposition 8.15, stated later, (8.22) defines a genuine norm on the quotient space $\dot{M}_1^{p,\lambda}(\partial\Omega, w)/\sim$ and, moreover, $\dot{M}_1^{p,\lambda}(\partial\Omega, w)/\sim$ is complete with respect to this norm when Ω has suitable regularity. Before we state this result, a lemma providing several useful estimates is needed, and we also introduce the following useful notation. Let $\phi \in \mathscr{C}_0^\infty(\mathbb{R}^n)$ with $\phi \equiv 1$ on $B(0,1)$ and $\operatorname{supp}\phi \subseteq B(0,2)$. Fix a reference point $x_0 \in \partial\Omega$, and for each $r \in (0,\infty)$ define

$$\phi_r(x) := \phi\Big(\frac{x-x_0}{r}\Big) \quad \text{for each } x \in \mathbb{R}^n. \tag{8.23}$$

For each $r \in (0,\infty)$, denote $\Delta_r := B(x_0, r) \cap \partial\Omega$ and given any $f \in L^1_{\mathrm{loc}}(\partial\Omega, \sigma)$ define

$$f_r := (\phi_r|_{\partial\Omega}) \cdot (f - f_{\Delta_{2r}}), \quad \text{where } f_{\Delta_{2r}} := \fint_{\Delta_{2r}} f \, d\sigma. \tag{8.24}$$

We now present the lemma alluded to earlier.

Lemma 8.13. *Let $\Omega \subseteq \mathbb{R}^n$ be a two-sided* NTA *domain whose topological boundary $\partial\Omega$ is an unbounded Ahlfors regular set, and write $\sigma := \mathcal{H}^{n-1}\lfloor\partial\Omega$. Fix an arbitrary reference point $x_0 \in \partial\Omega$, any two integrability exponents $p, q \in (1,\infty)$, a Muckenhoupt weight $w \in A_p(\partial\Omega,\sigma) \cap \mathrm{RH}_q(\partial\Omega,\sigma)$, and a threshold $\lambda \in (0,(n-1)/q')$, where $q' := q/(q-1)$. Finally, choose some function $f \in L^1(\partial\Omega, \frac{\sigma(x)}{1+|x|^n}) \cap L^1_{1,\mathrm{loc}}(\partial\Omega, w)$, and for each $r \in (0,\infty)$ define the surface ball $\Delta_r := B(x_0,r) \cap \partial\Omega$ and $f_{\Delta_r} := \fint_{\Delta_r} f\, d\sigma$. The following are true:*

(1) *There exists a constant $C \in (0,\infty)$, depending on Ω, n, p, q, λ, $[w]_{A_p}$, $[w]_{\mathrm{RH}_q}$, and x_0 such that*

$$\sup_{r>0} \big\| |f - f_{\Delta_r}| \cdot \mathbf{1}_{\Delta_r} \big\|_{M^{p,\lambda}(\partial\Omega,w)} \le C \sum_{j,k=1}^{n} \|\partial_{\tau_{jk}} f\|_{M^{p,\lambda}(\partial\Omega,w)}. \tag{8.25}$$

(2) *For each scale $r \in (0,\infty)$, there exists a constant $C_r \in (0,\infty)$ depending on r, $\partial\Omega$, n, p, q, λ, $[w]_{A_p}$, $[w]_{\mathrm{RH}_q}$, and x_0 such that*

$$\int_{\partial\Omega} \frac{|f(x) - f_{\Delta_r}|}{1+|x|^n}\, d\sigma(x) \le \frac{C_r}{\|\mathbf{1}_{\Delta_r}\|_{M^{p,\lambda}(\partial\Omega,w)}} \sum_{j,k=1}^{n} \|\partial_{\tau_{jk}} f\|_{M^{p,\lambda}(\partial\Omega,w)}. \tag{8.26}$$

(3) *There exists a constant $C \in (0,\infty)$, depending on $\partial\Omega$, n, p, q, λ, $[w]_{A_p}$, $[w]_{\mathrm{RH}_q}$, and x_0 such that*

$$\sup_{r>0} \|\nabla_{\tan} f_r\|_{[M^{p,\lambda}(\partial\Omega,w)]^n} \le C \|\nabla_{\tan} f\|_{[M^{p,\lambda}(\partial\Omega,w)]^n}, \tag{8.27}$$

where the notation f_r is as in (8.24).

(4) *If $f \in \dot{M}_1^{p,\lambda}(\partial\Omega,w)$, then $(f - f_{\Delta_r})\mathbf{1}_{\Delta_r} \in M^{p,\lambda}(\partial\Omega,w)$ for each scale $r \in (0,\infty)$.*

Proof. The proof of items (1)–(3) of this lemma follow in the exact same way as in the proof of [79, Lemma 8.1, p. 515], although we now appeal to our extrapolation result, Theorem 4.19. Note that Lemma 8.13 continues to hold when the space $M^{p,\lambda}(\partial\Omega,w)$ is replaced by the Muckenhoupt-weighted Lebesgue space, $L^p(\partial\Omega,w)$ (cf. [79, Proposition 2.25, pp. 154–155]). The conclusion of item (4) follows from item (1) and (8.19). □

Remark 8.14. Assume that $\Omega \subseteq \mathbb{R}^n$ is a two-sided NTA domain with unbounded boundary $\partial\Omega$, and set $\sigma := \mathcal{H}^{n-1}\lfloor\partial\Omega$. Then the homogeneous Muckenhoupt-weighted Morrey-based Sobolev space $\dot{M}_1^{p,\lambda}(\partial\Omega,w)$, defined in (8.19), may be equivalently characterized via

$$\dot{M}_1^{p,\lambda}(\partial\Omega,w) = \Big\{ f \in L^1\Big(\partial\Omega, \frac{\sigma(x)}{1+|x|^n}\Big) : \partial_{\tau_{jk}} f \in M^{p,\lambda}(\partial\Omega,w) \text{ for each } j,k \in \{1,\dots,n\} \Big\}. \tag{8.28}$$

This follows from the fact that if $f \in L^1(\partial\Omega, \frac{\sigma(x)}{1+|x|^n})$ is such that $\partial_{\tau_{jk}} f$ is in $M^{p,\lambda}(\partial\Omega, w)$ for each $j, k \in \{1, \dots, n\}$, then necessarily f also belongs to $L^1(\partial\Omega, \frac{\sigma(x)}{1+|x|^n}) \cap M^{p,\lambda}_{\text{loc}}(\partial\Omega, w)$. To see this, fix such an f and notice that $f \in L^1_{1,\text{loc}}(\partial\Omega, \sigma)$. Subsequently, given our present geometric setting, (8.25) of Lemma 8.13 applies. As such, for any $x_0 \in \partial\Omega$ and every $r > 0$ we have

$$\begin{aligned} \|f \mathbf{1}_{\Delta_r}\|_{M^{p,\lambda}(\partial\Omega,w)} &\le \big\| |f - f_{\Delta_r}| \cdot \mathbf{1}_{\Delta_r} \big\|_{M^{p,\lambda}(\partial\Omega,w)} \\ &\quad + |f_{\Delta_r}| \cdot \|\mathbf{1}_{\Delta_r}\|_{M^{p,\lambda}(\partial\Omega,w)} \\ &\le C \sum_{j,k=1}^{n} \|\partial_{\tau_{jk}} f\|_{M^{p,\lambda}(\partial\Omega,w)} \\ &\quad + |f_{\Delta_r}| \cdot \|\mathbf{1}_{\Delta_r}\|_{M^{p,\lambda}(\partial\Omega,w)} < \infty, \end{aligned} \tag{8.29}$$

where $\Delta_r := \Delta(x_0, r)$ and $f_{\Delta_r} := ⨍_{\Delta_r} f \, d\sigma$. Note that finiteness above is thanks to our assumption on f and Lemma 4.9. Thus, according to (8.20), we have $f \in M^{p,\lambda}_{\text{loc}}(\partial\Omega, w)$, and hence $f \in L^1(\partial\Omega, \frac{\sigma(x)}{1+|x|^n}) \cap M^{p,\lambda}_{\text{loc}}(\partial\Omega, w)$ as wanted.

Here, we record the fact that $\dot{M}^{p,\lambda}_1(\partial\Omega, w)/\sim$ in an appropriate context is complete with respect to its norm.

Proposition 8.15. *Let $\Omega \subseteq \mathbb{R}^n$ be a two-sided* NTA *domain whose topological boundary $\partial\Omega$ is an unbounded Ahlfors regular set, and write $\sigma := \mathcal{H}^{n-1}\lfloor\partial\Omega$. Fix two integrability exponents $p, q \in (1, \infty)$, along with a Muckenhoupt weight $w \in A_p(\partial\Omega, \sigma) \cap \mathrm{RH}_q(\partial\Omega, \sigma)$, and a threshold $\lambda \in (0, (n-1)/q')$, where $q' := q/(q-1)$. Then (8.22) is a genuine norm, and the quotient space $\dot{M}^{p,\lambda}_1(\partial\Omega, w)/\sim$ is a Banach space when equipped with said norm.*

Proof. With Lemma 8.13 in hand, the proof follows analogously to that of the proof of [79, Proposition 8.4, p. 517]. It is worth mentioning that we require Proposition 4.15 in adapting the proof. □

We would also like a suitable notion of a Muckenhoupt-weighted Block-based homogeneous Sobolev space. We again define this according to the model set in [79, Section 8.1].

Definition 8.16. Let $\Omega \subseteq \mathbb{R}^n$ be an AR domain and write $\sigma := \mathcal{H}^{n-1}\lfloor\partial\Omega$. Fix an integrability exponent $q \in (1, \infty)$ along with any Muckenhoupt weight $\omega \in A_q(\partial\Omega, \sigma)$. Choose some $\varepsilon \in (0, q-1)$ for which $\omega \in A_{q-\varepsilon}(\partial\Omega, \sigma)$ and select any threshold $\lambda \in (0, \varepsilon(n-1)/(q-1))$. The `Muckenhoupt-weighted Block-based homogeneous Sobolev space` is defined according to

$$\begin{aligned} \dot{\mathcal{B}}^{q,\lambda}_1(\partial\Omega, \omega) := \Big\{ f \in L^1\Big(\partial\Omega, \frac{\sigma(x)}{1+|x|^n}\Big) \cap \mathcal{B}^{q,\lambda}_{\text{loc}}(\partial\Omega, \omega) :& \\ \partial_{\tau_{jk}} f \in \mathcal{B}^{q,\lambda}(\partial\Omega, \omega) \text{ for each } j, k \in \{1, \dots, n\} \Big\},& \end{aligned} \tag{8.30}$$

where the space $\mathcal{B}^{q,\lambda}_{\mathrm{loc}}(\partial\Omega,\omega)$ is defined as

$$\mathcal{B}^{q,\lambda}_{\mathrm{loc}}(\partial\Omega,\omega) := \{f : \partial\Omega \to \mathbb{C} : f\mathbf{1}_\Delta \in \mathcal{B}^{q,\lambda}(\partial\Omega,\omega) \text{ for each surface ball } \Delta \subseteq \partial\Omega\}. \tag{8.31}$$

It is of relevance to note as is apparent from the above definition and Proposition 6.25 that constants belong to $\dot{\mathcal{B}}^{q,\lambda}_1(\partial\Omega,\omega)$.

At this point, equip the space $\dot{\mathcal{B}}^{q,\lambda}_1(\partial\Omega,\omega)$ with the following seminorm:

$$\|f\|_{\dot{\mathcal{B}}^{q,\lambda}_1(\partial\Omega,\omega)} := \sum_{j,k=1}^{n} \|\partial_{\tau_{jk}} f\|_{\mathcal{B}^{q,\lambda}(\partial\Omega,\omega)}, \tag{8.32}$$

for all $f \in \dot{\mathcal{B}}^{q,\lambda}_1(\partial\Omega,\omega)$. This is only a seminorm, since $\|f\|_{\dot{\mathcal{B}}^{q,\lambda}_1(\partial\Omega,\omega)} = 0$ whenever f is constant. We therefore bring in the space $\dot{\mathcal{B}}^{q,\lambda}_1(\partial\Omega,\omega)/\sim$, the quotient space of equivalence classes $[\,\cdot\,]$ modulo constant functions in $\dot{\mathcal{B}}^{q,\lambda}_1(\partial\Omega,\omega)$. We can define the following genuine norm on this space via

$$\|[f]\|_{\dot{\mathcal{B}}^{q,\lambda}_1(\partial\Omega,\omega)/\sim} := \sum_{j,k=1}^{n} \|\partial_{\tau_{jk}} f\|_{\mathcal{B}^{q,\lambda}_1(\partial\Omega,\omega)} \tag{8.33}$$

for each $f \in \dot{\mathcal{B}}^{q,\lambda}_1(\partial\Omega,\omega)$. We shall shortly see that the space $\dot{\mathcal{B}}^{q,\lambda}_1(\partial\Omega,\omega)/\sim$ is complete with respect to this norm. First, we have an analogous result to Lemma 8.13 at the level of Block spaces.

Lemma 8.17. *Let $\Omega \subseteq \mathbb{R}^n$ be a two-sided* NTA *domain whose topological boundary $\partial\Omega$ is an unbounded Ahlfors regular set, and write $\sigma := \mathcal{H}^{n-1}\lfloor\partial\Omega$. Fix an integrability exponent $q \in (1,\infty)$ along with a Muckenhoupt weight $\omega \in A_q(\partial\Omega,\sigma)$, which also satisfies $\omega' := \omega^{-q'/q} \in A_{q'}(\partial\Omega,\sigma) \cap \mathrm{RH}_p(\partial\Omega,\sigma)$ for some $p \in (1,\infty)$. Define $\varepsilon := (q-1)/p'$, where $p' := p/(p-1)$, so that ω belongs to $A_{q-\varepsilon}(\partial\Omega,\sigma)$ by Proposition* 6.35. *Lastly, select any threshold $\lambda \in (0, \varepsilon(n-1)/(q-1))$. Then similar statements as in items (1)–(4) of Lemma* 8.13 *continue to hold when the space $M^{p,\lambda}(\partial\Omega,w)$ is replaced by $\mathcal{B}^{q,\lambda}(\partial\Omega,\omega)$, defined as in Definition* 6.7 *in relation to these choices of q, ω, ε, and λ.*

Proof. Since we also have an extrapolation result for Block spaces in Theorem 10.55, we can argue analogously as in the proof of Lemma 8.13, which we recall closely follows [79, Lemma 8.1, p. 515], and makes use of [79, Proposition 2.25, pp. 154–155]. □

Remark 8.18. Thanks to Lemma 8.17, if $\Omega \subseteq \mathbb{R}^n$ is a two-sided NTA domain with unbounded boundary, then we can offer an equivalent characterization of $\dot{\mathcal{B}}^{q,\lambda}_1(\partial\Omega,\omega)$, much as we did for $\dot{M}^{p,\lambda}_1(\partial\Omega,w)$ in Remark 8.14. Namely, in this geometric setting, we can write

$$\dot{\mathcal{B}}_1^{q,\lambda}(\partial\Omega, \omega) := \left\{ f \in L^1\left(\partial\Omega, \frac{\sigma(x)}{1+|x|^n}\right) : \partial_{\tau_{jk}} f \in \mathcal{B}^{q,\lambda}(\partial\Omega, \omega) \right. \\ \left. \text{for each } j, k \in \{1, \ldots, n\} \right\}. \tag{8.34}$$

We now show that $\dot{\mathcal{B}}_1^{q,\lambda}(\partial\Omega, \omega)/\sim$ is Banach in a suitable context.

Proposition 8.19. *Let $\Omega \subseteq \mathbb{R}^n$ be a two-sided* NTA *domain whose topological boundary $\partial\Omega$ is an unbounded Ahlfors regular set, and write $\sigma := \mathcal{H}^{n-1}\lfloor\partial\Omega$. Fix an integrability exponent $q \in (1, \infty)$ along with a Muckenhoupt weight $\omega \in A_q(\partial\Omega, \sigma)$, which satisfies $\omega' := \omega^{-q'/q} \in A_{q'}(\partial\Omega, \sigma) \cap \mathrm{RH}_p(\partial\Omega, \sigma)$ for some $p \in (1, \infty)$. Define $\varepsilon := (q-1)/p' \in (0, q-1)$ so that $\omega \in A_{q-\varepsilon}(\partial\Omega, \sigma)$ by Proposition* 6.35 *and let $\lambda \in (0, \varepsilon(n-1)/(q-1))$.*

Then (8.33) *defines a genuine norm, and the space $\dot{\mathcal{B}}_1^{q,\lambda}(\partial\Omega, \omega)/\sim$ is Banach when equipped with said norm.*

Proof. Thanks to Lemma 8.17, the proof follows closely that of the proof of [79, Proposition 8.4, p. 517], and we note the necessity of Proposition 6.17 in adapting the proof to this case. □

9 Layer potentials in weighted Morrey and Block spaces

The Calderón–Zygmund theory developed in Chapters 5 and 7 equip us with the necessary tools to produce various boundedness results for the different layer potentials introduced throughout Chapter 3. The results in this chapter will serve as catalysts to solve the various boundary value problems appearing later in Chapter 11.

9.1 Layer potentials in weighted Morrey spaces

As mentioned in Chapter 3, the double layer potential is a mechanism for producing many null solutions to a given weakly elliptic, second-order, homogeneous, constant complex coefficient, $M \times M$ system in $\mathbb{R}^n$. Our first result here shows that property of the double layer potential is not lost when considering its action on functions belonging to a Muckenhoupt-weighted Morrey space. As we saw in Theorem 5.7, we also have a jump relation between general boundary-to-domain singular integral operators and their corresponding principal value singular integral counterparts as they act on Muckenhoupt-weighted Morrey functions. Naturally, we continue to have a jump relation between the more specific double layer potentials as they apply to functions in a weighted Morrey space.

Theorem 9.1. *Let $\Omega \subseteq \mathbb{R}^n$ be a* UR *domain and abbreviate $\sigma := \mathcal{H}^{n-1}\lfloor\partial\Omega$. Let L be a weakly elliptic, second-order, homogeneous, constant complex coefficient $M \times M$ system in $\mathbb{R}^n$. Now, fix two integrability exponents $p, q \in (1, \infty)$ along with a Muckenhoupt weight $w \in A_p(\partial\Omega, \sigma) \cap \mathrm{RH}_q(\partial\Omega, \sigma)$. Fix also a threshold $\lambda \in (0, (n-1)/q')$, where $q' := q/(q-1)$ is the Hölder conjugate of q. Finally, fix a reference point $x_0 \in \partial\Omega$ and a power $a \in [0, \lambda)$, and introduce the weight function $w_a := w|\cdot - x_0|^a$ on $\partial\Omega$.*

Then for each $f \in [M^{p,\lambda}(\partial\Omega, w_a)]^M$ and any $A \in \mathfrak{A}_L$, one has

$$\mathcal{D}_A f \in [\mathscr{C}^\infty(\Omega)]^M \quad \text{and} \quad L(\mathcal{D}_A f) = 0 \quad \text{in } \Omega, \tag{9.1}$$

where $\mathcal{D}_A$ is as in (3.18). *Moreover, for each aperture parameter $\kappa \in (0, \infty)$, the following jump relation holds:*

$$(\mathcal{D}_A f)\Big|_{\partial\Omega}^{\kappa-n.t.} = \Big(\frac{1}{2}I + K_A\Big)f \quad \text{at } \sigma\text{-a. e. point on } \partial\Omega, \tag{9.2}$$

where K_A is the boundary-to-boundary double layer potential, introduced in (3.19).

Proof. The continuous embedding $M^{p,\lambda}(\partial\Omega, w_a) \hookrightarrow L^1(\partial\Omega, \frac{\sigma(x)}{1+|x|^{n-1}})$ of Proposition 4.15 paired with (3.21) establishes (9.1). Proposition 4.15 along with the jump relation for the double layer, recorded in (3.22), establish (9.2). □

https://doi.org/10.1515/9783111458274-009

As seen in Chapter 3, the modified single layer potential is also a mechanism for producing many null solutions to a given weakly elliptic, homogeneous, $M \times M$ system of order two with constant complex coefficients. It turns out as witnessed by the proposition below that this continues to be true in the context of weighted Morrey spaces. Also, the modified single layer potential acting on functions from a weighted Morrey space satisfies a jump relation involving the conormal derivative, which brings in the so-called "transpose" operator $K_A^{\#}$; cf. (3.20).

Theorem 9.2. *Let $\Omega \subseteq \mathbb{R}^n$ be a UR domain and abbreviate $\sigma := \mathcal{H}^{n-1}\lfloor\partial\Omega$. Let L be a weakly elliptic, second-order, homogeneous, constant complex coefficient, $M \times M$ system in $\mathbb{R}^n$. Fix two integrability exponents $p, q \in (1, \infty)$ along with a Muckenhoupt weight $w \in A_p(\partial\Omega, \sigma) \cap \mathrm{RH}_q(\partial\Omega, \sigma)$, and a threshold $\lambda \in (0, (n-1)/q')$, where $q' := q/(q-1)$. Finally, fix a reference point $x_0 \in \partial\Omega$ and a power $a \in [0, \lambda)$, and introduce the weight function $w_a := w|\cdot - x_0|^a$ on $\partial\Omega$.*

Then for each $f \in [M^{p,\lambda}(\partial\Omega, w_a)]^M$, one has

$$\mathscr{S}_{\mathrm{mod}} f \in \left[\mathscr{C}^\infty(\Omega)\right]^M \quad \text{and} \quad L(\mathscr{S}_{\mathrm{mod}} f) = 0 \quad \text{in } \Omega. \tag{9.3}$$

Moreover, for any aperture parameter $\kappa \in (0, \infty)$ and any $A \in \mathfrak{A}_L$, the following jump relation holds:

$$\partial_\nu^A(\mathscr{S}_{\mathrm{mod}} f) = \left(-\frac{1}{2}I + K_{A^\top}^{\#}\right)f \quad \text{at } \sigma\text{-a. e. point on } \partial\Omega, \tag{9.4}$$

where $K_{A^\top}^{\#}$ is as in (3.20) *and ∂_ν^A is as in Definition* 3.18.

Proof. The continuous embedding in Proposition 4.15 and (3.73) justify (9.3). Proposition 4.15 along with the jump relation for the single layer given in (3.75) justify (9.4). □

The Calderón–Zygmund theory described in Theorem 5.7 allows us to obtain estimates for the nontangential maximal operator as it acts on the double layer at the level of weighted Morrey spaces.

Proposition 9.3. *Let $\Omega \subseteq \mathbb{R}^n$ be a UR domain and write $\sigma := \mathcal{H}^{n-1}\lfloor\partial\Omega$. Let L be a weakly elliptic, second-order, homogeneous, constant complex coefficient, $M \times M$ system in $\mathbb{R}^n$. Fix two integrability exponents $p, q \in (1, \infty)$ along with a Muckenhoupt weight $w \in A_p(\partial\Omega, \sigma) \cap \mathrm{RH}_q(\partial\Omega, \sigma)$, and a threshold $\lambda \in (0, (n-1)/q')$, where $q' := q/(q-1)$. Finally, fix a reference point $x_0 \in \partial\Omega$ and a power $a \in [0, \lambda)$, and introduce the weight function $w_a := w|\cdot - x_0|^a$ on $\partial\Omega$.*

Then for any aperture parameter $\kappa \in (0, \infty)$ and any $A \in \mathfrak{A}_L$ there exists a constant $C \in (0, \infty)$, depending on the UR *character of $\partial\Omega$, n, κ, p, q, λ, a, $[w]_{\mathrm{RH}_q}$, $[w]_{A_p}$, and A such that*

$$\|\mathcal{N}_\kappa(\mathcal{D}_A f)\|_{M^{p,\lambda}(\partial\Omega, w_a)} \leq C\|f\|_{[M^{p,\lambda}(\partial\Omega, w_a)]^M} \tag{9.5}$$

for all $f \in [L^1(\partial\Omega, \frac{\sigma(x)}{1+|x|^{n-1}})]^M$.

Proof. Given some function $f \in [L^1(\partial\Omega, \frac{\sigma(x)}{1+|x|^{n-1}})]^M$, and a matrix-valued integral kernel $k \in [\mathscr{C}^\infty(\mathbb{R}^n \setminus \{0\})]^{M\times M}$, define the boundary-to-domain operator

$$\mathcal{T}_k f(x) := \int_{\partial\Omega} k(x-y)f(y)\,d\sigma(y) \quad \text{for each } x \in \Omega, \tag{9.6}$$

which emphasizes the dependence on the kernel k. Next, fix any choice of $A = (a_{jk}^{\alpha\beta})_{\substack{1\le\alpha,\beta\le M\\ 1\le j,k\le n}} \in \mathfrak{A}_L$. By the embedding result in Proposition 4.15 and the definition of $\mathcal{D}_A$ in (3.18), for each $f = (f_\alpha)_{1\le\alpha\le M} \in [M^{p,\lambda}(\partial\Omega, w_a)]^M$ we may write (keeping in mind the summation convention over repeated indices)

$$(\mathcal{D}_A f)(x) = \big(-a_{jk}^{\beta\alpha}\big(\mathcal{T}_{\partial_j E_{\gamma\beta}}(M_{\nu_k} f_\alpha)\big)(x)\big)_{1\le\gamma\le M} \quad \text{for all } x \in \Omega, \tag{9.7}$$

where M_{ν_k} denotes the operation of pointwise multiplication by ν_k, the kth component of the geometric measure theoretic outward unit normal, and $E := (E_{\alpha\beta})_{1\le\alpha,\beta\le M}$ is the fundamental solution associated with L as in Proposition 3.4. Hence, for any $\kappa \in (0,\infty)$ and for all $x \in \partial\Omega$,

$$(\mathcal{N}_\kappa(\mathcal{D}_A f))(x) \le \max_{1\le\gamma\le M} \{|a_{jk}^{\beta\alpha}|\mathcal{N}_\kappa\big(\mathcal{T}_{\partial_j E_{\gamma\beta}}(M_{\nu_k} f_\alpha)\big)(x)\}. \tag{9.8}$$

The lattice property of Morrey spaces (cf. Proposition 4.7) along with Theorem 5.7, specifically (5.38), Remark 4.4, and the convention in (2.1) allow us to conclude from (9.8) that for all $a \in [0,\lambda)$,

$$\|\mathcal{N}_\kappa(\mathcal{D}_A f)\|_{M^{p,\lambda}(\partial\Omega,w_a)} \le C\|f\|_{[M^{p,\lambda}(\partial\Omega,w_a)]^M}, \tag{9.9}$$

where the constant $C \in (0,\infty)$ is now seen to depend on the coefficient tensor A in addition to $\partial\Omega, n, p, q, \lambda, a, [w]_{\mathrm{RH}_q}$, and $[w]_{A_p}$ as Theorem 5.7 implies. Note that the present applicability of Theorem 5.7 is ensured by Proposition 3.4, which gives that the integral kernel $\partial_j E_{\gamma\beta} \in \mathscr{C}^\infty(\mathbb{R}^n \setminus \{0\})$ is odd, and positive homogeneous of degree $1-n$. □

For functions belonging to a Morrey-based Sobolev space, we are also able to establish estimates for the nontangential maximal operator as it acts on the gradient of the double layer. We can also conclude that the nontangential restriction to the boundary of the gradient of the double layer exists as a σ-measurable function.

Proposition 9.4. *Let $\Omega \subseteq \mathbb{R}^n$ be a* UR *domain and write $\sigma := \mathcal{H}^{n-1}\lfloor\partial\Omega$. Let L be a weakly elliptic, second-order, homogeneous constant complex coefficient, $M \times M$ system in $\mathbb{R}^n$. Now, fix two integrability exponents $p, q \in (1,\infty)$ along with a Muckenhoupt weight $w \in A_p(\partial\Omega,\sigma) \cap \mathrm{RH}_q(\partial\Omega,\sigma)$, and a threshold $\lambda \in (0,(n-1)/q')$, where $q' := q/(q-1)$. Finally, fix a reference point $x_0 \in \partial\Omega$ and a power $a \in [0,\lambda)$, and introduce the weight $w_a := w|\cdot - x_0|^a$ on $\partial\Omega$.*

Then for any given aperture parameter $\kappa \in (0,\infty)$ and any $A \in \mathfrak{A}_L$ there exists a constant $C \in (0,\infty)$, depending on the UR *character of $\partial\Omega$, n, κ, p, q, λ, a, $[w]_{\mathrm{RH}_q}$, $[w]_{A_p}$, and A such that*

$$\|\mathcal{N}_\kappa(\nabla \mathcal{D}_A f)\|_{M^{p,\lambda}(\partial\Omega,w_a)} \leq C\|f\|_{[M_1^{p,\lambda}(\partial\Omega,w_a)]^M} \tag{9.10}$$

for all $f \in [M_1^{p,\lambda}(\partial\Omega,w_a)]^M$. Moreover, $(\nabla(\mathcal{D}_A f))\big|_{\partial\Omega}^{\kappa-\mathrm{n.t.}}$ exists σ-a. e. on $\partial\Omega$.

Proof. Let $f \in [M_1^{p,\lambda}(\partial\Omega,w_a)]^M$. Proposition 4.15 and the definition of $M_1^{p,\lambda}(\partial\Omega,w_a)$ (cf. Definition (8.8)) ensures that $f \in [L^1(\partial\Omega,\frac{\sigma(x)}{1+|x|^{n-1}})]^M$ and $\partial_{\tau_{jk}} f_a \in L^1(\partial\Omega,\frac{\sigma(x)}{1+|x|^{n-1}})$ for each $j,k \in \{1,\dots,n\}$ and $a \in \{1,\dots,M\}$. Therefore, by Proposition 3.7 for each $\ell \in \{1,\dots,n\}$ and each $x \in \Omega$ (with the summation convention over repeated indices in mind),

$$\partial_\ell(\mathcal{D}_A f)(x) = \big(a_{jk}^{\beta a}(\mathcal{T}_{\partial_j E_{\gamma\beta}}(\partial_{\tau_{ek}} f_a))(x)\big)_{1\leq\gamma\leq M}. \tag{9.11}$$

Here, $E = (E_{\alpha\beta})_{1\leq\alpha,\beta\leq M}$ is the fundamental solution associated to L as in Proposition 3.4. By the sublinearity of $\mathcal{N}_\kappa$, we may estimate for each $x \in \Omega$,

$$(\mathcal{N}_\kappa(\nabla\mathcal{D}_A f))(x) \leq \max_{\substack{1\leq\gamma\leq M\\ 1\leq\ell\leq n}} \{|a_{jk}^{\beta a}|\mathcal{N}_\kappa(\mathcal{T}_{\partial_j E_{\gamma\beta}}(\partial_{\tau_{ek}} f_a))(x)\}. \tag{9.12}$$

The integral kernel $\partial_j E_{\gamma\beta}$ is odd and positive homogeneous of degree $1-n$ for each $j \in \{1,\dots,n\}$ and $\gamma,\beta \in \{1,\dots M\}$ as a consequence of Proposition 3.4. Also, since $\partial_{\tau_{ek}} f_a \in M^{p,\lambda}(\partial\Omega,w_a)$ for each $\ell,k \in \{1,\dots,n\}$ and $a \in \{1,\dots,M\}$, we may appeal to Proposition 5.7, specifically (5.38), and the lattice property of weighted Morrey spaces (cf. Proposition 4.7) to conclude from (9.12) that there exists a constant $C \in (0,\infty)$, depending only on the UR character of $\partial\Omega$, n, κ, p, q, λ, a, $[w]_{\mathrm{RH}_q}$, $[w]_{A_p}$, and A such that

$$\begin{aligned}\|\mathcal{N}_\kappa(\nabla\mathcal{D}_A f)\|_{M^{p,\lambda}(\partial\Omega,w_a)} &\leq C \max_{\substack{1\leq a\leq M\\ 1\leq\ell,k\leq n}} \|\partial_{\tau_{ek}} f_a\|_{M^{p,\lambda}(\partial\Omega,w_a)} \\ &\leq C\|f\|_{[M_1^{p,\lambda}(\partial\Omega,w_a)]^M},\end{aligned} \tag{9.13}$$

which is the desired control of the norm.

The last claim that $(\nabla(\mathcal{D}_A f))\big|_{\partial\Omega}^{\kappa-\mathrm{n.t.}}$ exists σ-a. e. on $\partial\Omega$ is a direct consequence of Proposition 3.7 and the embedding in Proposition 4.15. □

The structure of the modified single layer allows us to obtain estimates on the non-tangential maximal operator acting on the gradient of the modified single layer.

Proposition 9.5. *Let $\Omega \subseteq \mathbb{R}^n$ be a* UR *domain and write $\sigma := \mathcal{H}^{n-1}\lfloor\partial\Omega$. Let L be a weakly elliptic, second-order, homogeneous, constant complex coefficient, $M\times M$ system in $\mathbb{R}^n$. Now, fix any two integrability exponents $p,q \in (1,\infty)$, a Muckenhoupt weight $w \in A_p(\partial\Omega,\sigma)\cap \mathrm{RH}_q(\partial\Omega,\sigma)$, and a threshold $\lambda \in (0,(n-1)/q')$, where $q' := q/(q-1)$. Finally, fix $x_0 \in \partial\Omega$ and $a \in [0,\lambda)$ and denote the weight function $w_a := |\cdot - x_0|^a w$ on $\partial\Omega$.*

Then for any given aperture parameter $\kappa \in (0,\infty)$ there exists a constant $C \in (0,\infty)$ depending only on the UR *character of Ω, n, κ, p, q, λ, a, $[w]_{\mathrm{RH}_q}$, $[w]_{A_p}$, and L such that*

$$\|\mathcal{N}_\kappa(\nabla \mathscr{S}_{\mathrm{mod}} f)\|_{M^{p,\lambda}(\partial\Omega,w_a)} \leq C\|f\|_{[M^{p,\lambda}(\partial\Omega,w_a)]^M} \tag{9.14}$$

for all $f \in [M^{p,\lambda}(\partial\Omega, w_a)]^M$.

Proof. For any $f \in [M^{p,\lambda}(\partial\Omega, w_a)]^M$, observe that the embedding in Proposition 4.15 and the formula in (3.72) show that for each index $\ell \in \{1,\dots,n\}$ that $\partial_\ell(\mathscr{S}_{\mathrm{mod}} f)$ can be written as

$$\partial_\ell(\mathscr{S}_{\mathrm{mod}} f)(x) = \Big(\int_{\partial\Omega} (\partial_\ell E_{\alpha\beta})(x-y) f_\beta(y)\, d\sigma(y)\Big)_{1\leq\alpha\leq M} \tag{9.15}$$

for each $x \in \Omega$, where $E = (E_{\alpha\beta})_{1\leq\alpha,\beta\leq M}$ denotes the fundamental solution of L as in Proposition 3.4 and we are using the summation convention over repeated indices. Thus, each component of $\partial_\ell(\mathscr{S}_{\mathrm{mod}} f)$ is a sum of operators of the type described in (5.37) of Theorem 5.7. Indeed, for each $\ell \in \{1,\dots,n\}$ and $\alpha,\beta \in \{1,\dots,M\}$ the integral kernel $\partial_\ell E_{\alpha\beta}$ is smooth, odd, and positive homogeneous of degree $1-n$ by Proposition 3.4. Therefore, Theorem 5.7, specifically (5.38), applies and we find by the lattice property of Morrey spaces (cf. Proposition 4.7) that for each $\ell \in \{1,\dots,n\}$ the following estimate holds:

$$\|\mathcal{N}_\kappa(\partial_\ell \mathscr{S}_{\mathrm{mod}} f)\|_{M^{p,\lambda}(\partial\Omega,w_a)} \leq C \max_{1\leq\alpha\leq M} \|f_\alpha\|_{M^{p,\lambda}(\partial\Omega,_a)}, \tag{9.16}$$

where $C \in (0,\infty)$ is some constant independent of f. By the sublinearity of $\mathcal{N}_\kappa$ and the lattice property of Morrey spaces (cf. Proposition 4.7), we see

$$\begin{aligned}\|\mathcal{N}_\kappa(\nabla \mathscr{S}_{\mathrm{mod}} f)\|_{M^{p,\lambda}(\partial\Omega,w_a)} &\leq \max_{1\leq\ell\leq n}\|\mathcal{N}_\kappa(\partial_\ell \mathscr{S}_{\mathrm{mod}} f)\|_{M^{p,\lambda}(\partial\Omega,w_a)} \\ &\leq C\|f\|_{[M^{p,\lambda}(\partial\Omega,w_a)]^M},\end{aligned} \tag{9.17}$$

where the last inequality follows from (9.16) and the convention in (2.1). This is the desired result. ☐

We now would like to show that the boundary-to-boundary double layer and its “transpose” are bounded operators on Morrey spaces.

Proposition 9.6. *Let $\Omega \subseteq \mathbb{R}^n$ be a* UR *domain and write $\sigma := \mathcal{H}^{n-1}\lfloor\partial\Omega$. Let L be a weakly elliptic, second-order, homogeneous, constant complex coefficient, $M \times M$ system in $\mathbb{R}^n$. Now, fix any two integrability exponents $p, q \in (1,\infty)$ along with a Muckenhoupt weight $w \in A_p(\partial\Omega,\sigma) \cap \mathrm{RH}_q(\partial\Omega,\sigma)$, and a threshold $\lambda \in (0,(n-1)/q')$, where q' is the Hölder conjugate of q. Finally, fix a reference point $x_0 \in \partial\Omega$ and a power $a \in [0,\lambda)$, and introduce the weight $w_a := w|\cdot - x_0|^a$ on $\partial\Omega$.*

Then the boundary-to-boundary double layer potential and its transpose induce mappings

$$K_A : [M^{p,\lambda}(\partial\Omega, w_a)]^M \to [M^{p,\lambda}(\partial\Omega, w_a)]^M \tag{9.18}$$

and

$$K_A^\# : [M^{p,\lambda}(\partial\Omega, w_a)]^M \to [M^{p,\lambda}(\partial\Omega, w_a)]^M, \tag{9.19}$$

which are well-defined, linear, and bounded. Moreover, for each function $f \in [M^{p,\lambda}(\partial\Omega, w_a)]^M$ *the following estimates hold:*

$$\|K_A f\|_{[M^{p,\lambda}(\partial\Omega,w_a)]^M} \le C\|f\|_{[M^{p,\lambda}(\partial\Omega,w_a)]^M} \tag{9.20}$$

and

$$\|K_A^\# f\|_{[M^{p,\lambda}(\partial\Omega,w_a)]^M} \le C\|f\|_{[M^{p,\lambda}(\partial\Omega,w_a)]^M}, \tag{9.21}$$

where the constant $C \in (0,\infty)$ *depends on the* UR *character of* $\partial\Omega$, n, p, q, λ, a, $[w]_{\mathrm{RH}_q}$, $[w]_{A_p}$, *and* A.

Proof. Given some function $f \in [L^1(\partial\Omega, \frac{\sigma(x)}{1+|x|^{n-1}})]^M$, and in integral kernel $k \in \mathscr{C}^\infty(\mathbb{R}^n \setminus \{0\})$, define the principal value singular integral operator

$$T_k f(x) := \lim_{\varepsilon\to 0^+} \int_{\substack{y\in\partial\Omega \\ |x-y|>\varepsilon}} k(x-y) f(y)\, d\sigma(y) \quad \text{for } \sigma\text{-a. e. } x \in \Omega, \tag{9.22}$$

which emphasizes the dependence on the kernel k. Then per the definition of K_A, given in (3.19), we find that for any function $f = (f_a)_{1\le a\le M}$ belonging to $[M^{p,\lambda}(\partial\Omega, w_a)]^M$,

$$(K_A f)(x) = (-a_{jk}^{\beta a}(T_{\partial_j E_{\gamma\beta}}(M_{\nu_k} f_a))(x))_{1\le\gamma\le M} \tag{9.23}$$

for σ-a. e. $x \in \partial\Omega$. Here, $E := (E_{\alpha\beta})_{1\le\alpha,\beta\le M}$ is the fundamental solution associated with L as in Proposition 3.4, and M_{ν_j} denotes the operation of pointwise multiplication by ν_j, the jth component of the geometric measure theoretic outward unit normal to Ω. Notice that $\partial_j E_{\gamma\beta} \in \mathscr{C}^\infty(\mathbb{R}^n \setminus \{0\})$ is positive homogeneous of degree $1-n$ by Proposition 3.4 for each $j \in \{1,\dots,n\}$ and $\gamma, \beta \in \{1,\dots,M\}$. Therefore, Theorem 5.7 applies, specifically (5.36), and we may conclude, bearing in mind the convention in (2.1), that the operator in (9.18) is well-defined, linear, and bounded. Note that (5.35) of Theorem 5.7 also gives the desired control of the operator norm in terms of a constant depending on $\partial\Omega$, n, p, q, λ, a, $[w]_{\mathrm{RH}_q}$, and $[w]_{A_p}$, but that (9.23) shows that the constant appearing in (9.20) necessarily depends on the choice of coefficient tensor A as well.

The claim pertaining to $K_A^\#$ is established in a similar manner, but now using (3.20) instead of (3.19). □

We have seen in Proposition 9.6 that K_A and $K_A^\#$ are bounded operators on weighted Morrey spaces. In particular, K_A and $K_A^\#$ have finite operator norms. In the case that

$A \in \mathfrak{A}_L^{\mathrm{dis}}$, we can further refine this conclusion by providing an upper bound on their respective operator norms, which now takes into account the BMO seminorm of the outward unit normal to the underlying domain. This next result is a crucial estimate for purposes pertaining to the solvability of boundary value problems. The reader is reminded that the symbol $t^{\langle m\rangle}$ has been defined in Definition 5.9.

Theorem 9.7. *Let $\Omega \subseteq \mathbb{R}^n$ be a UR domain and abbreviate $\sigma := \mathcal{H}^{n-1}\lfloor\partial\Omega$. Let L be a weakly elliptic, second-order, homogeneous, constant complex coefficient, $M \times M$ system in $\mathbb{R}^n$, which has a distinguished coefficient tensor. Fix two integrability exponents $p, q \in (1,\infty)$ along with a Muckenhoupt weight $w \in A_p(\partial\Omega,\sigma) \cap \mathrm{RH}_q(\partial\Omega,\sigma)$. Fix also a threshold $\lambda \in [0,(n-1)/q')$, where $q' := q/(q-1)$ is the Hölder conjugate of q. Finally, fix a reference point $x_0 \in \partial\Omega$ and a power $a \in [0,\lambda)$, and introduce the weight function $w_a := w|\cdot - x_0|^a$ on $\partial\Omega$.*

Then for each $m \in \mathbb{N}$, assuming $A \in \mathfrak{A}_L^{\mathrm{dis}}$, we have

$$\|K_A\|_{[M^{p,\lambda}(\partial\Omega,w_a)]^M \to [M^{p,\lambda}(\partial\Omega,w_a)]^M} \le C_m \cdot \|\nu\|^{\langle m\rangle}_{[\mathrm{BMO}(\partial\Omega,\sigma)]^n} \tag{9.24}$$

and

$$\|K_A^{\#}\|_{[M^{p,\lambda}(\partial\Omega,w_a)]^M \to [M^{p,\lambda}(\partial\Omega,w_a)]^M} \le C_m \cdot \|\nu\|^{\langle m\rangle}_{[\mathrm{BMO}(\partial\Omega,\sigma)]^n}, \tag{9.25}$$

where ν denotes the geometric measure theoretic outward unit normal to Ω and the constant $C_m \in (0,\infty)$ depends on m, the UR character of $\partial\Omega$, n, p, q, λ, a, $[w]_{\mathrm{RH}_q}$, $[w]_{A_p}$, and A.

Remark 9.8. It is worth pointing out that an estimate like in (9.24) or (9.25) necessarily forces the coefficient tensor to be distinguished. Indeed, whenever $\Omega \subseteq \mathbb{R}^n$ is a half-space, ν is constant (cf. [86, Proposition 5.10.12, p. 470]), and hence $\|\nu\|_{[\mathrm{BMO}(\partial\Omega,\sigma)]^n} = 0$. This implies K_A and $K_A^{\#}$ are the zero operators, which according to Proposition 3.16 is equivalent to having $A \in \mathfrak{A}_L^{\mathrm{dis}}$.

Proof of Theorem 9.7. Proposition 3.16 says that the fact that the system L has a distinguished coefficient tensor $A \in \mathfrak{A}_L^{\mathrm{dis}}$ implies that there exist two matrix-valued functions $k, k^{\#} \in [\mathscr{C}^\infty(\mathbb{R}^n \setminus \{0\})]^{M\times M}$ which are both even, and positive homogeneous of degree $-n$ such that for each function $f \in [L^1(\partial\Omega, \frac{\sigma(x)}{1+|x|^{n-1}})]^M$,

$$(K_A f)(x) = \lim_{\varepsilon\to 0^+} \int_{\substack{y\in\partial\Omega\\ |x-y|>\varepsilon}} \langle \nu(y), x-y\rangle k(x-y) f(y)\, d\sigma(y) \tag{9.26}$$

and

$$(K_A^{\#} f)(x) = \lim_{\varepsilon\to 0^+} \int_{\substack{y\in\partial\Omega\\ |x-y|>\varepsilon}} \langle \nu(x), y-x\rangle k^{\#}(x-y) f(y)\, d\sigma(y) \tag{9.27}$$

at σ-a. e. point $x \in \partial\Omega$. In particular, (9.26) and (9.27) hold for each function $f \in [M^{p,\lambda}(\partial\Omega, w_a)]^M$ due to the embedding in Proposition 4.15. Consequently, the structure of the integral kernels permits us to invoke Theorem 5.15, specifically (5.71) and (5.72), to conclude for each $m \in \mathbb{N}$ there exists a constant $C_m \in (0, \infty)$, depending on m, $\partial\Omega$, n, p, q, λ, a, $[w]_{\mathrm{RH}_q}$, and $[w]_{A_p}$, such that

$$\|K_A f\|_{[M^{p,\lambda}(\partial\Omega,w_a)]^M} \leq C_m \|\nu\|^{\langle m\rangle}_{[\mathrm{BMO}(\partial\Omega,\sigma)]^n} \cdot \|f\|_{[M^{p,\lambda}(\partial\Omega,w_a)]^M} \tag{9.28}$$

and

$$\|K_A^{\#} f\|_{[M^{p,\lambda}(\partial\Omega,w_a)]^M} \leq C_m \|\nu\|^{\langle m\rangle}_{[\mathrm{BMO}(\partial\Omega,\sigma)]^n} \cdot \|f\|_{[M^{p,\lambda}(\partial\Omega,w_a)]^M}, \tag{9.29}$$

for all $f \in [M^{p,\lambda}(\partial\Omega, w_a)]^M$, which implies (9.24) and (9.25) since f was arbitrary. □

We also have that K_A is bounded on Muckenhoupt-weighted Morrey-based Sobolev spaces for any suitable coefficient tensor A.

Proposition 9.9. *Let $\Omega \subseteq \mathbb{R}^n$ be a* UR *domain and write $\sigma := \mathcal{H}^{n-1}\lfloor\partial\Omega$. Suppose L is a weakly elliptic, second-order, homogeneous, constant complex coefficient, $M \times M$ system in $\mathbb{R}^n$. Fix any two integrability exponents $p, q \in (1, \infty)$, a Muckenhoupt weight $w \in A_p(\partial\Omega, \sigma) \cap \mathrm{RH}_q(\partial\Omega, \sigma)$, and some threshold $\lambda \in [0, (n-1)/q')$, where $q' := q/(q-1)$.*

Then for any $A \in \mathfrak{A}_L$, the boundary-to-boundary double layer potential associated to A (cf. (3.19)*), induces a mapping*

$$K_A : [M_1^{p,\lambda}(\partial\Omega, w)]^M \to [M_1^{p,\lambda}(\partial\Omega, w)]^M \tag{9.30}$$

which is well-defined, linear, and bounded. Moreover, there exists a constant $C \in (0, \infty)$, depending only on the UR *character of $\partial\Omega$, n, p, q, λ, $[w]_{\mathrm{RH}_q}$, $[w]_{A_p}$, and A such that*

$$\|K_A f\|_{[M_1^{p,\lambda}(\partial\Omega,w)]^M} \leq C\|f\|_{[M_1^{p,\lambda}(\partial\Omega,w)]^M} \tag{9.31}$$

for all $f \in [M_1^{p,\lambda}(\partial\Omega, w)]^M$.

Proof. Let $f \in M_1^{p,\lambda}(\partial\Omega, w)$, so that by Proposition 9.6,

$$\|K_A f\|_{M^{p,\lambda}(\partial\Omega,w)} \leq C\|f\|_{M^{p,\lambda}(\partial\Omega,w)}. \tag{9.32}$$

Let $j, k \in \{1, \dots, n\}$. It remains to show that $\partial_{\tau_{jk}}(K_A f)$ is similarly bounded. We turn to Proposition 3.8, whose present applicability is ensured by the embedding in Proposition 4.16. Consequently, we have the identity

$$\partial_{\tau_{jk}}(K_A f) = K_A(\partial_{\tau_{jk}} f) + U_{jk}(\nabla_{\tan} f), \tag{9.33}$$

which holds at σ-a. e. point on $\partial\Omega$. Note that we have retained the notation for U_{jk}, defined in (3.30). Since $\partial_{\tau_{jk}} f \in [M^{p,\lambda}(\partial\Omega, w)]^M$, we may appeal to Proposition 9.6 to say that

$$\|K_A(\partial_{\tau_{jk}}f)\|_{[M^{p,\lambda}(\partial\Omega,w)]^M} \leq C\|\partial_{\tau_{jk}}f\|_{[M^{p,\lambda}(\partial\Omega,w)]^M} \tag{9.34}$$

for some constant $C \in (0,\infty)$, independent of f. Now, notice for every $\gamma \in \{1,\dots,M\}$ that each piece of $(U_{jk}(\nabla_{\tan}f))_\gamma$ is an operator of the type described in Theorem 5.18. That is, a commutator between the operation of pointwise multiplication by a component of ν and $T_{\partial_r E_{\gamma\beta}}$, defined for each $f \in [M^{p,\lambda}(\partial\Omega,w)]^M$ according to

$$(T_{\partial_r E_{\gamma\beta}}f)(x) := \lim_{\varepsilon\to 0^+} \int\limits_{\substack{y\in\partial\Omega \\ |x-y|>\varepsilon}} (\partial_r E_{\gamma\beta})(x-y)f(y)\,d\sigma(y) \tag{9.35}$$

at σ-a. e. point $x \in \partial\Omega$. For each $r \in \{1,\dots,n\}$ and $\beta,\gamma \in \{1,\dots,M\}$, the integral kernel $\partial_r E_{\gamma\beta}$ is odd and positive homogeneous of degree $1-n$ thanks to Proposition 3.4. Recall that $F_{\alpha s} := (\nabla_{\tan}f_\alpha)_s$ (i. e., the sth component of $\nabla_{\tan}f_\alpha$) in the definition of U_{jk}. Thus, as $F_{\alpha s}$ belongs to $M^{p,\lambda}(\partial\Omega,w)$, we may appeal to Theorem 5.18 to obtain the estimate

$$\|U_{jk}(\nabla_{\tan}f)\|_{[M^{p,\lambda}(\partial\Omega,w)]^M} \leq C\|\nu\|_{[\mathrm{BMO}(\partial\Omega,\sigma)]^n} \cdot \|\partial_{\tau_{jk}}f\|_{[M^{p,\lambda}(\partial\Omega,w)]^M} \tag{9.36}$$

for some $C \in (0,\infty)$, independent of f. Thus, (9.33) along with the estimates in (9.34) and (9.36) yield

$$\|\partial_{\tau_{jk}}(K_A f)\|_{[M^{p,\lambda}(\partial\Omega,w)]^M} \leq C\|\partial_{\tau_{jk}}f\|_{[M^{p,\lambda}(\partial\Omega,w)]^M}. \tag{9.37}$$

This, together with (9.32), ultimately implies (9.31), and subsequently, given the arbitrary nature of f, we see that the operator in (9.30) is well-defined and bounded. □

If the coefficient tensor A in Proposition 9.9 is distinguished, we likewise obtain more refined small norm estimates in the spirit of (9.24) but now with an extra degree of regularity involving a Muckenhoupt-weighted Morrey-based Sobolev space.

Theorem 9.10. *Let $\Omega \subseteq \mathbb{R}^n$ be a* UR *domain and abbreviate $\sigma := \mathcal{H}^{n-1}\lfloor\partial\Omega$. Suppose L is a weakly elliptic, second-order, homogeneous, constant complex coefficient, $M \times M$ system in $\mathbb{R}^n$, which has a distinguished coefficient tensor. Fix two integrability exponents $p,q \in (1,\infty)$, an arbitrary Muckenhoupt weight $w \in A_p(\partial\Omega,\sigma) \cap \mathrm{RH}_q(\partial\Omega,\sigma)$, and some threshold $\lambda \in [0,(n-1)/q')$, where $q' := q/(q-1)$.*

Then for each $m \in \mathbb{N}$ and $A \in \mathfrak{A}_L^{\mathrm{dis}}$, there exists a constant $C_m \in (0,\infty)$, depending only on m, the UR *character of $\partial\Omega$, n, p, q, λ, $[w]_{\mathrm{RH}_q}$, $[w]_{A_p}$, and A such that*

$$\|K_A\|_{[M_1^{p,\lambda}(\partial\Omega,w)]^M\to[M_1^{p,\lambda}(\partial\Omega,w)]^M} \leq C_m \cdot \|\nu\|^{\langle m\rangle}_{[\mathrm{BMO}(\partial\Omega,\sigma)]^n}, \tag{9.38}$$

where ν denotes the geometric measure theoretic outward unit normal to Ω.

Proof. Let $f \in [M_1^{p,\lambda}(\partial\Omega,w)]^M$. Since A is a distinguished coefficient tensor, we may appeal to Theorem 9.7 to obtain for each $m \in \mathbb{N}$ some $C_m \in (0,\infty)$ such that

$$\|K_A f\|_{[M^{p,\lambda}(\partial\Omega,w)]^M} \leq C_m \cdot \|\nu\|^{\langle m\rangle}_{[\mathrm{BMO}(\partial\Omega,\sigma)]^n} \cdot \|f\|_{[M^{p,\lambda}(\partial\Omega,w)]^M}. \tag{9.39}$$

Note that for each $j,k \in \{1,\dots,n\}$ that (9.33) continues to hold as does the estimate in (9.36). However, instead of (9.34), the fact that A is distinguished permits us to again appeal to Theorem 9.7, and we see for each $m \in \mathbb{N}$ there exists a constant $C_m \in (0,\infty)$ for which

$$\|K_A(\partial_{\tau_{jk}} f)\|_{[M^{p,\lambda}(\partial\Omega,w)]^M} \leq C_m \cdot \|\nu\|^{\langle m\rangle}_{[\mathrm{BMO}(\partial\Omega,\sigma)]^n} \cdot \|\partial_{\tau_{jk}} f\|_{[M^{p,\lambda}(\partial\Omega,w)]^M}. \tag{9.40}$$

Therefore, (9.33), (9.36), and (9.40) allow us to conclude that

$$\|\partial_{\tau_{jk}}(K_A f)\|_{[M^{p,\lambda}(\partial\Omega,w)]^M} \leq C_m \cdot \|\nu\|^{\langle m\rangle}_{[\mathrm{BMO}(\partial\Omega,\sigma)]^n} \cdot \|\partial_{\tau_{jk}} f\|_{[M^{p,\lambda}(\partial\Omega,w)]^M}. \tag{9.41}$$

Hence, recalling (9.39), the above inequality implies the for each $m \in \mathbb{N}$ there exists some constant $C_m \in (0,\infty)$, depending only on m, the UR character of $\partial\Omega$, n, p, q, λ, $[w]_{A_p}$, and Λ such that

$$\|K_A f\|_{[M_1^{p,\lambda}(\partial\Omega,w)]^M} \leq C_m \cdot \|\nu\|^{\langle m\rangle}_{[\mathrm{BMO}(\partial\Omega,\sigma)]^n} \cdot \|f\|_{[M_1^{p,\lambda}(\partial\Omega,w)]^M}, \tag{9.42}$$

which yields (9.38) given the arbitrary nature of $f \in [M_1^{p,\lambda}(\partial\Omega,w)]^M$. □

This next result provides an invertibility result for certain operators involving the boundary-to-boundary double layer potential on Muckenhoupt-weighted Morrey-based Sobolev spaces provided a suitable underlying geometry and the existence of a distinguished coefficient tensor. Moreover, this invertibility comes with quantitative control on the tangential gradient.

Proposition 9.11. *Let $\Omega \subseteq \mathbb{R}^n$ be a UR domain and write $\sigma := \mathcal{H}^{n-1}\lfloor\partial\Omega$. Suppose L is a weakly elliptic, second-order, homogeneous, constant complex coefficient, $M \times M$ system in $\mathbb{R}^n$, which has a distinguished coefficient tensor. Fix two integrability exponents $p,q \in (1,\infty)$, an arbitrary Muckenhoupt weight $w \in A_p(\partial\Omega,\sigma) \cap \mathrm{RH}_q(\partial\Omega,\sigma)$, and some threshold $\lambda \in [0,(n-1)/q')$, where $q' := q/(q-1)$.*

Then for each $z \in \mathbb{C}$ there exists $\delta \in (0,1)$ with the following significance. When Ω is a δ-AR domain and $A \in \mathfrak{A}_L^{\mathrm{dis}}$, the operator

$$zI + K_A : [M_1^{p,\lambda}(\partial\Omega,w)]^M \to [M_1^{p,\lambda}(\partial\Omega,w)]^M \tag{9.43}$$

is invertible and there exists a constant $C \in (0,\infty)$ with the property that

$$\|\nabla_{\tan} g\|_{[M^{p,\lambda}(\partial\Omega,w)]^{n\cdot M}} \leq C\|\nabla_{\tan} f\|_{[M_1^{p,\lambda}(\partial\Omega,w)]^M}, \tag{9.44}$$

for all $f \in [M_1^{p,\lambda}(\partial\Omega,w)]^M$, where $g := (zI + K_A)^{-1} f \in [M_1^{p,\lambda}(\partial\Omega,w)]^M$.

Proof. To begin, the fact that g is a well-defined function belonging to $[M_1^{p,\lambda}(\partial\Omega, w)]^M$ when Ω is a δ-AR domain for sufficiently small δ is a consequence of the estimate in (9.38), which implies for $A \in \mathfrak{A}_L^{\text{dis}}$ and $\delta \in (0,1)$ small enough that

$$\|K_A\|_{[M_1^{p,\lambda}(\partial\Omega,w)]^M \to [M_1^{p,\lambda}(\partial\Omega,w)]^M} \leq |z|, \tag{9.45}$$

so that the operator (9.43) is invertible via a Neumann series (cf. Lemma 11.2 and note we are using the fact that $[M_1^{p,\lambda}(\partial\Omega, w)]^M$ is a Banach space, a fact established in Proposition 8.11).

Now, recall (3.32), which allows us to write for each $j, k \in \{1, \dots, n\}$ (bearing in mind (8.2))

$$\begin{aligned} \partial_{\tau_{jk}} f &= (zI + K_A)(\partial_{\tau_{jk}} g) + U_{jk}(\nabla_{\tan} g) \\ &= (zI + K_A)(\partial_{\tau_{jk}} g) + U_{jk}\big((\nu_r \partial_{\tau_{rs}} g_\alpha)_{\substack{1\leq\alpha\leq M \\ 1\leq s\leq n}}\big) \end{aligned} \tag{9.46}$$

at σ-a. e. point on $\partial\Omega$, where $\nu = (\nu_r)_{1\leq r\leq n}$ denotes the geometric measure theoretic outward unit normal to Ω. Let us agree to abbreviate

$$\nabla_\tau f := (\partial_{\tau_{jk}} f)_{1\leq j,k\leq n} \quad \text{and} \quad \nabla_\tau g := (\partial_{\tau_{jk}} g)_{1\leq j,k\leq n}. \tag{9.47}$$

Additionally, let M_{ν_r} denote the operation of pointwise multiplication by the rth component of ν, and let π_{rs}^α be the "coordinate projection" operator, defined according to $\pi_{rs}^\alpha(X) = X_{rs}^\alpha$ for each $X = (X_{rs}^\alpha)_{\substack{1\leq M\leq\alpha \\ 1\leq r,s\leq n}} \in \mathbb{C}^{M\cdot n^2}$. Lastly, define the operator

$$R : [M^{p,\lambda}(\partial\Omega, w)]^{M\cdot n^2} \to [M^{p,\lambda}(\partial\Omega, w)]^{M\cdot n^2} \tag{9.48}$$

via

$$R := K_A + \big(U_{jk} \circ (M_{\nu_r} \circ \pi_{rs}^\alpha)_{\substack{1\leq\alpha\leq M \\ 1\leq s\leq n}}\big)_{1\leq j,k\leq n}, \tag{9.49}$$

where we understand the action of K_A on some function $F = (F_{rs}^\alpha)_{\substack{1\leq\alpha\leq M \\ 1\leq r,s\leq n}}$ belonging to $[M^{p,\lambda}(\partial\Omega, w)]^{M\cdot n^2}$ as

$$K_A F = \big(K_A(F_{rs}^\alpha)_{1\leq\alpha\leq M}\big)_{1\leq r,s\leq n}. \tag{9.50}$$

With this notation in hand, (9.46) may be recast as

$$\nabla_\tau f = (zI + R)(\nabla_\tau g). \tag{9.51}$$

Then using (9.48), (9.24) of Theorem 9.7, (3.30), Theorem 5.18 (recalling that for each $j, k \in \{1, \dots, n\}$ we have that U_{jk} is a sum of commutators between components of the outward unit normal and boundary-to-domain integral operators with integral kernels that are first-order derivatives of entries of the fundamental solution to L in the sense of Proposition 3.4; cf. (3.30)), we can conclude for each $m \in \mathbb{N}$ that

$$\|R\|_{[M^{p,\lambda}(\partial\Omega,w)]^{M\cdot n^2}\to[M^{p,\lambda}(\partial\Omega,w)]^{M\cdot n^2}} \leq C_m\|\nu\|^{\langle m\rangle}_{[\mathrm{BMO}(\partial\Omega,\sigma)]^n} \tag{9.52}$$

for some constant $C_m \in (0,\infty)$, whose nature is specified in the statement of Theorem 9.7. Note the use of Proposition 5.6 (which applies thanks to Proposition 3.4) to ensure that the integral operators that make up the commutators appearing in the definition of U_{jk} satisfy the hypotheses of Theorem 5.18.

Consequently, when $\delta \in (0,1)$ is small enough to begin with, since $[M^{p,\lambda}(\partial\Omega,w)]^{M\cdot n^2}$ is a Banach space (cf. Proposition 4.5), we may conclude that $zI + R$ is invertible on $[M^{p,\lambda}(\partial\Omega,w)]^{M\cdot n^2}$ via a Neumann series (cf. Lemma 11.2) with appropriate operator norm estimates on the inverse. As such, (9.51) and (9.47) allow us to conclude that (9.44) holds. □

We now demonstrate that in the context of Muckenhoupt-weighted Morrey spaces, the modified double layer potential, introduced in (3.34), is also a good mechanism for producing null solutions to the kinds of weakly elliptic systems we have in mind.

Theorem 9.12. *Let $\Omega \subseteq \mathbb{R}^n$ be a UR domain and abbreviate $\sigma := \mathcal{H}^{n-1}\lfloor\partial\Omega$. Let L be a weakly elliptic, second-order, homogeneous, constant complex coefficient, $M \times M$ system in $\mathbb{R}^n$. Fix two integrability exponents $p,q \in (1,\infty)$, a Muckenhoupt weight $w \in A_p(\partial\Omega,\sigma) \cap \mathrm{RH}_q(\partial\Omega,\sigma)$, and any threshold $\lambda \in (0,(n-1)/q')$, where $q' := q/(q-1)$. Finally, choose some $A \in \mathfrak{A}_L$. Then for each $f \in [\dot{M}^{p,\lambda}_1(\partial\Omega,w)]^M$ one has*

$$\mathcal{D}_{A,\mathrm{mod}}f \in [\mathscr{C}^\infty(\Omega)]^M \quad \text{and} \quad L(\mathcal{D}_{A,\mathrm{mod}}f) = 0 \quad \text{in } \Omega. \tag{9.53}$$

Additionally, both $(\mathcal{D}_{A,\mathrm{mod}}f)\big|^{\kappa-\mathrm{n.t.}}_{\partial\Omega}$ and $(\nabla(\mathcal{D}_{A,\mathrm{mod}}f))\big|^{\kappa-\mathrm{n.t.}}_{\partial\Omega}$ exist σ-a. e. on $\partial\Omega$ and in fact,

$$(\mathcal{D}_{A,\mathrm{mod}}f)\big|^{\kappa-\mathrm{n.t.}}_{\partial\Omega} = \Big(\frac{1}{2}I + K_{A,\mathrm{mod}}\Big)f \quad \text{at } \sigma\text{-a. e. point on } \partial\Omega, \tag{9.54}$$

and there exists a constant $C \in (0,\infty)$, depending only on Ω, n, A, κ, p, q, λ, $[w]_{\mathrm{RH}_q}$, and $[w]_{A_p}$ such that $\mathcal{N}_\kappa(\nabla(\mathcal{D}_{A,\mathrm{mod}}f)) \in M^{p,\lambda}(\partial\Omega,w)$ with

$$\|\mathcal{N}_\kappa(\nabla(\mathcal{D}_{A,\mathrm{mod}}f))\|_{M^{p,\lambda}(\partial\Omega,w)} \leq C\|f\|_{[\dot{M}^{p,\lambda}_1(\partial\Omega,w)]^M}. \tag{9.55}$$

Proof. To begin, (8.19) and (3.36) establish the claims in (9.53). Using (8.19), but now paired with (3.50), allows us to conclude (9.54). Next, observe that (8.19), Proposition 4.15, and (3.43) ensure the σ-a. e. existence of $(\nabla(\mathcal{D}_{A,\mathrm{mod}}f))\big|^{\kappa-\mathrm{n.t.}}_{\partial\Omega}$ on $\partial\Omega$. Finally, (3.42) allows us to argue much as in (9.11)–(9.13) in the proof of Proposition 9.4 to obtain the estimate in (9.55) as wanted. □

As witnessed by the following proposition, the conormal derivative of the modified double layer potential induces a well-defined mapping from a homogeneous Muckenhoupt-weighted Morrey-based Sobolev space into a Muckenhoupt-weighted Morrey space.

Proposition 9.13. *Let $\Omega \subseteq \mathbb{R}^n$ be a UR domain, abbreviate $\sigma := \mathcal{H}^{n-1}\lfloor\partial\Omega$, and let ν denote the geometric measure theoretic outward unit normal to Ω. Let L be a weakly elliptic, second-order, homogeneous, constant complex coefficient, $M \times M$ system in $\mathbb{R}^n$. Next, fix any two integrability exponents $p, q \in (1,\infty)$, an arbitrary Muckenhoupt weight $w \in A_p(\partial\Omega,\sigma) \cap \mathrm{RH}_q(\partial\Omega,\sigma)$, and some threshold $\lambda \in (0,(n-1)/q')$, where $q' := q/(q-1)$. Finally, choose some $A \in \mathfrak{A}_L$. Then the operator*

$$\partial_\nu^A \mathcal{D}_{A,\mathrm{mod}} : \big[\dot{M}_1^{p,\lambda}(\partial\Omega,w)\big]^M \to \big[M^{p,\lambda}(\partial\Omega,w)\big]^M \tag{9.56}$$

is well-defined, linear, and bounded, where ∂_ν^A denotes the conormal derivative associated to A as in Definition 3.18.

Proof. Fix $f \in [\dot{M}_1^{p,\lambda}(\partial\Omega,w)]^M$ arbitrary, and write $u := \mathcal{D}_{A,\mathrm{mod}}f$ for brevity. Given (3.18), it follows that there exists a constant $C \in (0,\infty)$ such that

$$|\partial_\nu^A u| \le C\big|(\nabla u)\big|_{\partial\Omega}^{\kappa-\mathrm{n.t.}}\big| \le C\mathcal{N}_\kappa(\nabla u) \quad \text{at } \sigma\text{-a.e. point on } \partial\Omega. \tag{9.57}$$

Because $f \in [\dot{M}_1^{p,\lambda}(\partial\Omega,w)]^M$, from Theorem 9.12 we find that $\mathcal{N}_\kappa(\nabla u)$ belongs to $M^{p,\lambda}(\partial\Omega,w)$, and hence the lattice property of $M^{p,\lambda}(\partial\Omega,w)$ (cf. Proposition 4.7) entails the membership of $\partial_\nu^A u$ to $[M^{p,\lambda}(\partial\Omega,w)]^M$ with

$$\|\partial_\nu^A u\|_{[M^{p,\lambda}(\partial\Omega,w)]^M} \le C\|f\|_{[\dot{M}_1^{p,\lambda}(\partial\Omega,w)]^M}. \tag{9.58}$$

This implies the desired result given the arbitrary nature of f. □

This next result shows that the modified boundary-to-boundary double layer is a well-defined operator on homogeneous Muckenhoupt-weighted Morrey-based Sobolev spaces.

Proposition 9.14. *Let $\Omega \subseteq \mathbb{R}^n$ be a UR domain and write $\sigma := \mathcal{H}^{n-1}\lfloor\partial\Omega$. Let L be a weakly elliptic, second-order, homogeneous, constant complex coefficient, $M \times M$ system in $\mathbb{R}^n$. Next, fix two integrability exponents $p, q \in (1,\infty)$, a Muckenhoupt weight $w \in A_p(\partial\Omega,\sigma) \cap \mathrm{RH}_q(\partial\Omega,\sigma)$, and any threshold $\lambda \in (0,(n-1)/q')$, where q' is the Hölder conjugate to q. Finally, choose $A \in \mathfrak{A}_L$. Then the operator*

$$K_{A,\mathrm{mod}} : \big[\dot{M}_1^{p,\lambda}(\partial\Omega,w)\big]^M \to \big[\dot{M}_1^{p,\lambda}(\partial\Omega,w)\big]^M \tag{9.59}$$

is well-defined, linear, and bounded. Consequently, (bearing in mind (3.51) *which says that $K_{A,\mathrm{mod}}$ maps constants to constants), the operator*

$$[K_{A,\mathrm{mod}}] : \big[\dot{M}_1^{p,\lambda}(\partial\Omega, w)/\!\sim\big]^M \to \big[\dot{M}_1^{p,\lambda}(\partial\Omega, w)/\!\sim\big]^M \quad \textit{defined as} \tag{9.60}$$
$$[K_{A,\mathrm{mod}}][f] := [K_{A,\mathrm{mod}} f] \in \big[\dot{M}_1^{p,\lambda}(\partial\Omega, w)/\!\sim\big]^M,$$

for all functions $f \in [\dot{M}_1^{p,\lambda}(\partial\Omega, w)/\!\sim]^M$ *is also well-defined and bounded, where* $\dot{M}_1^{p,\lambda}(\partial\Omega, w)/\!\sim$ *is the quotient space of equivalence classes* $[\,\cdot\,]$ *modulo constant functions in* $\dot{M}_1^{p,\lambda}(\partial\Omega, w)$.

Proof. Let $f \in [\dot{M}_1^{p,\lambda}(\partial\Omega, w)]^M$. From Theorem 9.12, we find that the boundary traces $(\mathcal{D}_{A,\mathrm{mod}} f)\big|_{\partial\Omega}^{\kappa-\mathrm{n.t.}}$ and $(\partial_j(\mathcal{D}_{A,\mathrm{mod}} f))\big|_{\partial\Omega}^{\kappa-\mathrm{n.t.}}$ exist σ-a. e. on $\partial\Omega$ for each $j \in \{1, \dots, n\}$. We also know from (8.19) and (3.45), whose present applicability is ensured by Corollary 2.35, that $\mathcal{N}_\kappa(\mathcal{D}_{A,\mathrm{mod}} f)$ belongs to $L^1_{\mathrm{loc}}(\partial\Omega, \sigma)$. Additionally, Proposition 4.16 and (9.55) ensure that $\mathcal{N}_\kappa(\nabla(\mathcal{D}_{A,\mathrm{mod}} f))$ belongs to $L^1_{\mathrm{loc}}(\partial\Omega, \sigma)$ as well. Let $j, k \in \{1, \dots, n\}$. The preceding facts and the geometry of Ω allow us to appeal to [79, Proposition 2.22, p. 148] to conclude (bearing in mind (9.54)) that

$$\begin{aligned}\left|\partial_{\tau_{jk}}\left(\tfrac{1}{2}I + K_{A,\mathrm{mod}}\right)f\right| &= \left|\partial_{\tau_{jk}}\big((\mathcal{D}_{A,\mathrm{mod}} f)\big|_{\partial\Omega}^{\kappa-\mathrm{n.t.}}\big)\right| \\ &\le 2\mathcal{N}_\kappa(\nabla(\mathcal{D}_{A,\mathrm{mod}} f))\end{aligned} \tag{9.61}$$

at σ-a. e. point on $\partial\Omega$. We then appeal to (9.55) and Proposition 4.7 to conclude that $\partial_{\tau_{jk}}(K_{A,\mathrm{mod}} f)$ belongs to $[M^{p,\lambda}(\partial\Omega, w)]^M$. As such, (8.21) yields that in fact $K_{A,\mathrm{mod}} f \in [\dot{M}_1^{p,\lambda}(\partial\Omega, w)]^M$. Moreover, (9.61), (9.55), Proposition 4.7, and (8.21) imply

$$\|K_{A,\mathrm{mod}} f\|_{[\dot{M}_1^{p,\lambda}(\partial\Omega, w)]^M} \le C\|f\|_{[\dot{M}_1^{p,\lambda}(\partial\Omega, w)]^M} \tag{9.62}$$

for some constant $C \in (0, \infty)$ independent of f. This demonstrates the desired well definiteness and boundedness of the operator in (9.59). The analogous result concerning the operator in (9.60) then follows from the well definiteness and boundedness of the operator in (9.59) and definitions. □

We are also able to obtain small-norm geometric estimates for the modified boundary-to-boundary double layer potential operator provided the writing of said layer potential is done with a distinguished coefficient tensor.

Theorem 9.15. *Let* $\Omega \subseteq \mathbb{R}^n$ *be a two-sided* NTA *domain such that* $\partial\Omega$ *is an unbounded Ahlfors regular set, abbreviate* $\sigma := \mathcal{H}^{n-1}\lfloor\partial\Omega$, *and denote by* ν *the geometric measure theoretic outward unit normal to* Ω. *Let* L *be a weakly elliptic, second-order, homogeneous, constant complex coefficient,* $M\times M$ *system in* $\mathbb{R}^n$ *with the property that* $\mathfrak{A}_L^{\mathrm{dis}} \neq \varnothing$. *Next, fix two integrability exponents* $p, q \in (1, \infty)$, *a Muckenhoupt weight* $w \in A_p(\partial\Omega, \sigma) \cap \mathrm{RH}_q(\partial\Omega, \sigma)$, *and any threshold* $\lambda \in (0, (n-1)/q')$, *where* $q' = q/(q-1)$.

Then assuming $A \in \mathfrak{A}_L^{\mathrm{dis}}$, *for each* $m \in \mathbb{N}$, *there exists* $C_m \in (0, \infty)$, *depending only on* m, *the Ahlfors regularity constants of* $\partial\Omega$, n, L, p, q, λ, $[w]_{\mathrm{RH}_q}$, *and* $[w]_{A_p}$ *such that (recalling* (5.9))

$$\|[K_{A,\mathrm{mod}}]\|_{[\dot{M}_1^{p,\lambda}(\partial\Omega,w)/\sim]^M\to[\dot{M}_1^{p,\lambda}(\partial\Omega,w)/\sim]^M} \leq C_m\|\nu\|^{\langle m\rangle}_{[\mathrm{BMO}(\partial\Omega,\sigma)]^n}. \tag{9.63}$$

Proof. Notice that under our present assumptions that Ω is a UR domain (cf. [79, (2.48), p. 38]). Using Remark 3.11, whose present applicability in this setting is ensured by (8.19) and Corollary 2.35, we can argue as in the proof of Theorem 9.10 which yielded (9.41) to instead obtain for $f \in [\dot{M}_1^{p,\lambda}(\partial\Omega, w)]^M$ arbitrary and any $j, k \in \{1, \dots, n\}$,

$$\|\partial_{\tau_{jk}}(K_{A,\mathrm{mod}}f)\|_{[M^{p,\lambda}(\partial\Omega,w)]^M} \leq C_m\|\nu\|^{\langle m\rangle}_{[\mathrm{BMO}(\partial\Omega,\sigma)]^n}\cdot\|\partial_{\tau_{jk}}f\|_{[M^{p,\lambda}(\partial\Omega,w)]^M} \tag{9.64}$$

which then implies the desired estimate in (9.63) in view of (8.22). □

Having established a solid layer potential theory on weighted Morrey spaces, we will almost immediately be able to tackle the problem of solvability for boundary value problems with boundary data belonging to weighted Morrey spaces. However, we cannot directly apply this layer potential theory on weighted Morrey spaces to the problem of uniqueness. In order to obtain a full well-posedness result, we need to consider a brand of spaces which act as a pre-dual to weighted Morrey spaces. The significance of this relationship will be further explained in Chapter 11.

9.2 Layer potentials in weighted Block spaces

At this point, we are in a position to develop a layer potential theory for Muckenhoupt-weighted Block spaces. While the developments here will almost immediately allow us to obtain solvability results for boundary value problems with data in a Muckenhoupt-weighted Block space, the relationship that Muckenhoupt-weighted Block spaces have with Muckenhoupt-weighted Morrey spaces via duality will eventually allow us to treat the problem of uniqueness for boundary value problems with data in a Muckenhoupt-weighted Morrey space thanks to the results in this section.

Our first result in this section details the behavior of the double layer potential acting on functions belonging to a Muckenhoupt-weighted Block space. Specifically, the double layer continues to produce smooth null solutions to weakly elliptic, second-order, homogeneous, constant complex coefficient systems when acting on any block function, and we understand the nontangential boundary trace via a jump formula.

Theorem 9.16. *Let $\Omega \subseteq \mathbb{R}^n$ be a UR domain and abbreviate $\sigma := \mathcal{H}^{n-1}\lfloor\partial\Omega$. Let L be a weakly elliptic, second-order, homogeneous, constant complex coefficient, $M \times M$ system in $\mathbb{R}^n$. Fix an integrability exponent $q \in (1, \infty)$ along with a Muckenhoupt weight $\omega \in A_q(\partial\Omega, \sigma)$. Finally, choose some $\varepsilon \in (0, q-1)$ for which $\omega \in A_{q-\varepsilon}(\partial\Omega, \sigma)$ and select a threshold $\lambda \in (0, \varepsilon(n-1)/(q-1))$.*

Then for each $g \in [\mathcal{B}^{q,\lambda}(\partial\Omega, \omega)]^M$ and any $A \in \mathfrak{A}_L$ one has

$$\mathcal{D}_A g \in [\mathscr{C}^\infty(\Omega)]^M \quad \text{and} \quad L(\mathcal{D}_A g) = 0 \quad \text{in } \Omega, \tag{9.65}$$

where $\mathcal{D}_A$ is the boundary-to-domain double layer potential; cf. (3.18). *Moreover, the following jump relation holds:*

$$(\mathcal{D}_A g)\Big|_{\partial\Omega}^{\kappa-\text{n.t.}} = \Big(\frac{1}{2}I + K_A\Big)g \quad \text{at } \sigma\text{-a. e. point on } \partial\Omega. \tag{9.66}$$

where K_A is the boundary-to-boundary double layer; cf. (3.19).

Proof. The continuous embedding $\mathcal{B}^{q,\lambda}(\partial\Omega,\omega) \hookrightarrow L^1(\partial\Omega, \frac{\sigma(x)}{1+|x|^{n-1}})$ of Proposition 6.17 paired with (3.21) demonstrates (9.65). Proposition 6.17 and the jump relation for the double layer potential given in (3.22) ensure (9.66) holds. □

An alternative way to produce null solutions to weakly-elliptic, second-order, homogeneous, constant complex coefficient systems is using the modified single layer potential introduced in Definition 3.17. Our next result shows that Muckenhoupt-weighted Block spaces are a good functional analytic environment from the perspective of the modified single layer potential.

Theorem 9.17. *Let $\Omega \subseteq \mathbb{R}^n$ be a UR domain and abbreviate $\sigma := \mathcal{H}^{n-1}\lfloor\partial\Omega$. Let L be a weakly elliptic, second-order, homogeneous, constant complex coefficient, $M \times M$ system in $\mathbb{R}^n$. Fix an integrability exponent $q \in (1,\infty)$ along with a Muckenhoupt weight $\omega \in A_q(\partial\Omega,\sigma)$. Choose some $\varepsilon \in (0, q-1)$ for which $\omega \in A_{q-\varepsilon}(\partial\Omega,\sigma)$ and select a threshold $\lambda \in (0, \varepsilon(n-1)/(q-1))$.*

Then for each $g \in [\mathcal{B}^{q,\lambda}(\partial\Omega,\omega)]^M$ and any $A \in \mathcal{A}_L$, one has

$$\mathscr{S}_{\text{mod}} g \in \big[\mathscr{C}^\infty(\Omega)\big]^M \quad \text{and} \quad L(\mathscr{S}_{\text{mod}} g) = 0 \quad \text{in } \Omega, \tag{9.67}$$

where S_{mod} is as in (3.71). *Moreover, for any aperture parameter $\kappa \in (0,\infty)$, the following jump formula holds for the conormal derivative* (*cf.* (3.74)):

$$\partial_\nu^A(\mathscr{S}_{\text{mod}} g) = \Big(-\frac{1}{2}I + K^\#_{A^\top}\Big)g \quad \text{at } \sigma\text{-a. e. point on } \partial\Omega, \tag{9.68}$$

where $K^\#_{A^\top}$ denotes the transpose boundary-to-boundary double layer potential associated with $A^\top$ as in (3.20).

Proof. The continuous embedding in (6.77) from Proposition 6.17 and (3.73) lets us conclude (9.67). Similarly, (9.68) follows from Proposition 6.17 along with the jump relation for the single layer given in (3.75). □

An important aspect of boundary value problems is some sort of size condition on solutions. We will measure the size of solutions in terms of the nontangential maximal operator. This next proposition shows that we have the proper control of the nontangential maximal operator applied to the double layer potential as it acts on a function belonging to a Muckenhoupt-weighted Block space.

Proposition 9.18. *Let $\Omega \subseteq \mathbb{R}^n$ be a UR domain and write $\sigma := \mathcal{H}^{n-1}\lfloor\partial\Omega$. Let L be a weakly elliptic, second-order, homogeneous, constant complex coefficient, $M \times M$ system in $\mathbb{R}^n$. Fix an integrability exponent $q \in (1,\infty)$ along with a Muckenhoupt weight $\omega \in A_q(\partial\Omega,\sigma)$. Finally, choose some $\varepsilon \in (0, q-1)$ for which $\omega \in A_{q-\varepsilon}(\partial\Omega,\sigma)$ and select a threshold $\lambda \in (0, \varepsilon(n-1)/(q-1))$.*

Then for every aperture parameter $\kappa \in (0,\infty)$ and $A \in \mathfrak{A}_L$ there exists a constant $C \in (0,\infty)$, depending on the UR character of $\partial\Omega$, n, q, ε, λ, $[\omega]_{A_{q-\varepsilon}}$, $[\omega]_{A_q}$, κ, and A such that

$$\|\mathcal{N}_\kappa(\mathcal{D}_A g)\|_{\mathcal{B}^{q,\lambda}(\partial\Omega,\omega)} \le C\|g\|_{[\mathcal{B}^{q,\lambda}(\partial\Omega,\omega)]^M} \tag{9.69}$$

for all $g \in [\mathcal{B}^{q,\lambda}(\partial\Omega,\omega)]^M$.

Proof. Given some function $g \in L^1(\partial\Omega, \frac{\sigma(x)}{1+|x|^{n-1}})$, and an integral kernel $k \in \mathscr{C}^\infty(\mathbb{R}^n \setminus \{0\})$, define the boundary-to-domain operator

$$\mathcal{T}_k g(x) := \int_{\partial\Omega} k(x-y)g(y)\,d\sigma(y) \quad \text{for each } x \in \Omega, \tag{9.70}$$

which emphasizes the dependence on the kernel k. Then by the embedding result in Proposition 6.17 and the definition of $\mathcal{D}_A$ in (3.18), we may write for any function $g = (g_\alpha)_{1\le\alpha\le M} \in [\mathcal{B}^{q,\lambda}(\partial\Omega,\omega)]^M$,

$$(\mathcal{D}_A g)(x) = \big(-a_{jk}^{\beta\alpha}(\mathcal{T}_{\partial_j E_{\gamma\beta}}(M_{\nu_k} g_\alpha))(x)\big)_{1\le\gamma\le M} \quad \text{for all } x \in \partial\Omega, \tag{9.71}$$

where M_{ν_k} denotes the operation of pointwise multiplication by ν_k, the kth component of the geometric measure theoretic outward unit normal, and $E := (E_{\alpha\beta})_{1\le\alpha,\beta\le M}$ is the fundamental solution associated to L as in Proposition 3.4. Hence, for all $x \in \partial\Omega$,

$$(\mathcal{N}_\kappa(\mathcal{D}_A g))(x) \le \max_{1\le\gamma\le M}\{|a_{jk}^{\beta\alpha}|\mathcal{N}_\kappa(\mathcal{T}_{\partial_j E_{\gamma\beta}}(M_{\nu_k} g_\alpha))(x)\}. \tag{9.72}$$

The lattice property of Block spaces (cf. Proposition 6.14) along with Theorem 7.6, specifically (7.89), Remark 6.13, and the convention in (2.1) allow us to conclude from (9.72) that

$$\|\mathcal{N}_\kappa(\mathcal{D}_A g)\|_{\mathcal{B}^{q,\lambda}(\partial\Omega,\omega)} \le C\|g\|_{[\mathcal{B}^{q,\lambda}(\partial\Omega,\omega)]^M}, \tag{9.73}$$

where the constant $C \in (0,\infty)$ is now seen to depend on the coefficient tensor A in addition to $\partial\Omega$, n, κ, q, ε, λ, $[\omega]_{A_{q-\varepsilon}}$, and $[\omega]_{A_q}$, as Theorem 7.6 implies. Note that the present applicability of Theorem 7.6 is ensured by Proposition 3.4, which gives that the integral kernel $\partial_j E_{\gamma\beta} \in \mathscr{C}^\infty(\mathbb{R}^n \setminus \{0\})$ is odd, and positive homogeneous of degree $1-n$ for each $j \in \{1,\dots,n\}$ and $\gamma, \beta \in \{1,\dots,M\}$. □

Our next result shows that even with one extra degree of smoothness, i. e., in the context of Muckenhoupt-weighted Block-based Sobolev spaces of order one, the double layer potential still has the proper size estimates involving the nontangential maximal operator.

Proposition 9.19. *Let $\Omega \subseteq \mathbb{R}^n$ be a UR domain and write $\sigma := \mathcal{H}^{n-1}\lfloor\partial\Omega$. Let L be a weakly elliptic, second-order, homogeneous, constant complex coefficient, $M \times M$ system in $\mathbb{R}^n$. Fix an integrability exponent $q \in (1,\infty)$ along with a Muckenhoupt weight $\omega \in A_q(\partial\Omega,\sigma)$. Finally, choose some $\varepsilon \in (0, q-1)$ for which $\omega \in A_{q-\varepsilon}(\partial\Omega,\sigma)$ and select a threshold $\lambda \in (0, \varepsilon(n-1)/(q-1))$.*

Then for every aperture parameter $\kappa \in (0,\infty)$ and any $A \in \mathfrak{A}_L$, there exists a constant $C \in (0,\infty)$, depending on the UR character of $\partial\Omega$, n, q, ε, λ, $[\omega]_{A_{q-\varepsilon}}$, $[\omega]_{A_q}$, κ, and A such that

$$\|\mathcal{N}_\kappa(\nabla\mathcal{D}_A g)\|_{\mathcal{B}^{q,\lambda}(\partial\Omega,\omega)} \leq C\|g\|_{[\mathcal{B}_1^{q,\lambda}(\partial\Omega,\omega)]^M} \tag{9.74}$$

for all $g \in [\mathcal{B}_1^{q,\lambda}(\partial\Omega,\omega)]^M$. Moreover, $(\nabla(\mathcal{D}_A g))\big|_{\partial\Omega}^{\kappa-\mathrm{n.t.}}$ exists at σ-a. e. point on $\partial\Omega$.

Proof. Retain the notation for boundary-to-boundary domain operators emphasizing the dependence on the integral kernel, as in (9.70). Thanks to Proposition 6.17 and Definition 8.11, we may use Proposition 3.7 to write for each $\ell \in \{1,\dots,n\}$ and every $g \in \mathcal{B}_1^{q,\lambda}(\partial\Omega,\omega)$,

$$\partial_\ell(\mathcal{D}_A g)(x) = \big(a_{jk}^{\beta\alpha}(\mathcal{T}_{\partial_j E_{\gamma\beta}}(\partial_{\tau_{\ell k}} g_\alpha))(x)\big)_{1\leq\gamma\leq M} \quad \text{for each } x \in \Omega, \tag{9.75}$$

where $E := (E_{\alpha\beta})_{1\leq\alpha,\beta\leq M}$ is the fundamental solution associated to L as in Proposition 3.4. By the sublinearity of the nontangential maximal operator,

$$(\mathcal{N}_\kappa(\nabla\mathcal{D}_A g))(x) \leq \max_{\substack{1\leq\gamma\leq M\\ 1\leq\ell\leq n}}\{|a_{jk}^{\beta\alpha}|\mathcal{N}_\kappa(\mathcal{T}_{\partial_j E_{\gamma\beta}}(\partial_{\tau_{\ell k}} g_\alpha))(x)\} \tag{9.76}$$

for each $x \in \Omega$. Note for each $j \in \{1,\dots,n\}$ and $\gamma,\beta \in \{1,\dots,M\}$ the integral kernel $\partial_j E_{\gamma\beta} \in \mathscr{C}^\infty(\mathbb{R}^n \setminus \{0\})$ is odd, and positive homogeneous of degree $1-n$ by Proposition 3.4. Subsequently, since each $\partial_{\tau_{\ell k}} g \in \mathcal{B}^{q,\lambda}(\partial\Omega,\omega)$ by assumption, we may appeal to Theorem 7.6, specifically (7.89), and the lattice property of Block spaces (cf. Proposition 6.14) to conclude from (9.76) that there exists some constant $C \in (0,\infty)$, whose nature is specified in the statement of the proposition such that

$$\begin{aligned}\|\mathcal{N}_\kappa(\nabla\mathcal{D}_A g)\|_{\mathcal{B}^{q,\lambda}(\partial\Omega,\omega)} &\leq C \max_{\substack{1\leq\alpha\leq M\\ 1\leq\ell,k\leq n}} \|\partial_{\tau_{\ell k}} g_\alpha\|_{\mathcal{B}^{q,\lambda}(\partial\Omega,\omega)} \\ &\leq C\|g\|_{[\mathcal{B}_1^{q,\lambda}(\partial\Omega,\omega)]^M},\end{aligned} \tag{9.77}$$

which is the desired control of the norm.

The last claim about the existence of $(\nabla(\mathcal{D}_A g))\big|^{\kappa-\mathrm{n.t.}}_{\partial\Omega}$ at σ-a. e. point on $\partial\Omega$ follows directly from Proposition 3.7 thanks to the embedding in Proposition 6.17. □

We are also able to obtain proper control of the nontangential maximal operator applied to the gradient of the modified single layer potential acting on functions from a Muckenhoupt-weighted Block space.

Proposition 9.20. *Let $\Omega \subseteq \mathbb{R}^n$ be a UR domain and write $\sigma := \mathcal{H}^{n-1}\lfloor\partial\Omega$. Let L be a second-order, weakly elliptic, homogeneous, constant complex coefficient, $M \times M$ system in $\mathbb{R}^n$, with coefficient tensor $A \in \mathfrak{A}_L$. Fix an integrability exponent $q \in (1,\infty)$ along with a Muckenhoupt weight $\omega \in A_q(\partial\Omega,\sigma)$. Finally, choose some $\varepsilon \in (0,q-1)$ for which $\omega \in A_{q-\varepsilon}(\partial\Omega,\sigma)$ and set a threshold $\lambda \in (0,\varepsilon(n-1)/(q-1))$.*

Then for every aperture parameter $\kappa \in (0,\infty)$ there exists a constant $C \in (0,\infty)$, depending on the UR character of $\partial\Omega$, n, κ, q, ε, λ, $[\omega]_{A_q}$, and A such that

$$\big\|\mathcal{N}_\kappa(\nabla\mathscr{S}_{\mathrm{mod}}g)\big\|_{\mathcal{B}^{q,\lambda}(\partial\Omega,\omega)} \leq C\|g\|_{[\mathcal{B}^{q,\lambda}(\partial\Omega,\omega)]^M} \tag{9.78}$$

for all $g \in [\mathcal{B}^{q,\lambda}(\partial\Omega,\omega)]^M$.

Proof. Fix any $g = (g_\beta)_{1\leq\beta\leq M} \in [\mathcal{B}^{q,\lambda}(\partial\Omega,\omega)]^M$. We may use the embedding in Proposition 6.17 along with (3.72) to write

$$\partial_\ell(\mathscr{S}_{\mathrm{mod}}g)(x) = \bigg(\int_{\partial\Omega}(\partial_\ell E_{\alpha\beta})(x-y)g_\beta(y)\,d\sigma(y)\bigg)_{1\leq\alpha\leq M} \tag{9.79}$$

for each $x \in \Omega$ and each $\ell \in \{1,\dots,n\}$, where $E = (E_{\alpha\beta})_{1\leq\alpha,\beta\leq M}$ denotes the fundamental solution to L as in Proposition 3.4. Note the use of summation over repeated indices in (9.79), so that thanks to Proposition 3.4 each component of $\partial_\ell(\mathscr{S}_{\mathrm{mod}}g)$ is a sum of operators of the type concerned in Theorem 7.6. As such, we are guaranteed from (7.89) the existence of some constant $C \in (0,\infty)$, depending only on $\partial\Omega$, n, q, ε, λ, $[\omega]_{A_{q-\varepsilon}}$, $[\omega]_{A_q}$, and κ such that

$$\big\|\mathcal{N}_\kappa(\partial_\ell(\mathscr{S}_{\mathrm{mod}}g))\big\|_{\mathcal{B}^{q,\lambda}(\partial\Omega,\omega)} \leq C\max_{1\leq\alpha\leq M}\|g_\alpha\|_{\mathcal{B}^{q,\lambda}(\partial\Omega,\omega)}. \tag{9.80}$$

Using the sublinearity of $\mathcal{N}_\kappa$ and the lattice property of Block spaces (cf. Proposition 6.14), we my estimate

$$\begin{aligned}\big\|\mathcal{N}_\kappa(\nabla\mathscr{S}_{\mathrm{mod}}g)\big\|_{\mathcal{B}^{q,\lambda}(\partial\Omega,\omega)} &\leq C\max_{1\leq\ell\leq n}\big\|\mathcal{N}_\kappa(\partial_\ell(\mathscr{S}_{\mathrm{mod}}g))\big\|_{\mathcal{B}^{q,\lambda}(\partial\Omega,\omega)} &(9.81)\\ &\leq C\|g\|_{\mathcal{B}^{q,\lambda}(\partial\Omega,\omega)}, &(9.82)\end{aligned}$$

where we have used (9.80) and the convention in (2.1). This completes the proof. □

This next result shows that the boundary-to-boundary double layer potential and the transpose boundary-to-boundary double layer potential both induce bounded operators on Muckenhoupt-weighted Block spaces.

Proposition 9.21. *Let $\Omega \subseteq \mathbb{R}^n$ be a UR domain and write $\sigma := \mathcal{H}^{n-1}\lfloor\partial\Omega$. Let L be a weakly elliptic, second-order, homogeneous, constant complex coefficient, $M \times M$ system in $\mathbb{R}^n$. Fix an integrability exponent $q \in (1,\infty)$ along with a Muckenhoupt weight $\omega \in A_q(\partial\Omega,\sigma)$. Lastly, choose some $\varepsilon \in (0, q-1)$ for which $\omega \in A_{q-\varepsilon}(\partial\Omega,\sigma)$ and select a threshold $\lambda \in (0, \varepsilon(n-1)/(q-1))$.*

Then for any $A \in \mathfrak{A}_L$ the boundary-to-boundary double layer potential associated to A (cf. (3.19)) and its transpose (cf. (3.20)) induce mappings

$$K_A : [\mathcal{B}^{q,\lambda}(\partial\Omega,\omega)]^M \to [\mathcal{B}^{q,\lambda}(\partial\Omega,\omega)]^M \tag{9.83}$$

and

$$K_A^\# : [\mathcal{B}^{q,\lambda}(\partial\Omega,\omega)]^M \to [\mathcal{B}^{q,\lambda}(\partial\Omega,\omega)]^M, \tag{9.84}$$

which are well-defined, linear, and bounded.

Proof. Given some function $g \in L^1(\partial\Omega, \frac{\sigma(x)}{1+|x|^{n-1}})$, and an integral kernel $k \in \mathscr{C}^\infty(\mathbb{R}^n \setminus \{0\})$, define the principal value singular integral operator

$$T_k g(x) := \lim_{\varepsilon\to 0^+} \int\limits_{\substack{y\in\partial\Omega \\ |x-y|>\varepsilon}} k(x-y)g(y)\,d\sigma(y) \quad \text{for } \sigma\text{-a. e. } x \in \Omega, \tag{9.85}$$

which highlights the dependence on the kernel k. Then per the definition of K_A, given in (3.19), and thanks to the embedding in Proposition 6.17, we find that for any function $g = (g_\alpha)_{1\le\alpha\le M} \in [\mathcal{B}^{q,\lambda}(\partial\Omega,\omega)]^M$,

$$(K_A g)(x) = \big(-a_{jk}^{\beta\alpha}(T_{\partial_j E_{\gamma\beta}}(M_{\nu_k} g_\alpha))(x)\big)_{1\le\gamma\le M}. \tag{9.86}$$

for σ-a. e. $x \in \partial\Omega$, where $E := (E_{\alpha\beta})_{1\le\alpha,\beta\le M}$ is the fundamental solution associated to L as in Proposition 3.4, and M_{ν_k} denotes multiplication by the kth component of ν, the geometric measure theoretic outward unit normal to Ω. Notice that the integral kernel $\partial_j E_{\gamma\beta} \in \mathscr{C}^\infty(\mathbb{R}^n \setminus \{0\})$ is positive homogeneous of degree $1-n$ by Proposition 3.4. Therefore, Theorem 7.2 applies and we may conclude that the operator in (9.83) is well-defined, linear, and bounded.

The case for $K_A^\#$ is similarly proved but using the writing of $K_A^\#$ in (3.20) as opposed to that of K_A in (3.19). □

The following is a quick corollary involving the boundedness of the Hilbert transform on Muckenhoupt-weighted block spaces.

Corollary 9.22. *Fix an integrability exponent $q \in (1,\infty)$ and a Muckenhoupt weight $\omega \in A_q(\mathbb{R},\mathcal{L}^1)$. Choose $\varepsilon \in (0, q-1)$ such that $\omega \in A_{q-\varepsilon}(\mathbb{R},\mathcal{L}^1)$ and select any threshold $\lambda \in (0, \varepsilon/(q-1))$. Bring in the Hilbert transform H, which acts on any $f \in L^1(\mathbb{R}, \frac{1}{1+|x|})$ according to*

$$Hf(x) := \lim_{\varepsilon\to 0^+} \frac{1}{\pi} \int\limits_{\substack{y\in\mathbb{R}\\ |x-y|>\varepsilon}} \frac{f(y)}{x-y}\, dy \quad \text{at } \mathcal{L}^1\text{-a.e. point } x \in \mathbb{R}. \tag{9.87}$$

Then H induces a mapping

$$H : \mathcal{B}^{q,\lambda}(\mathbb{R}, \omega) \to \mathcal{B}^{q,\lambda}(\mathbb{R}, \omega) \tag{9.88}$$

which is well-defined, linear, and bounded.

Proof. This is an immediate consequence of Proposition 9.21 given the fact that H is a scalar multiple of the boundary-to-boundary Cauchy operator on $\mathbb{R}$, which (cf. [89, Section 1.4]) can be realized as a boundary-to-boundary double layer potential associated to the Laplacian for a specific choice of coefficient tensor. □

If a weakly elliptic, second-order, homogeneous, constant complex coefficient system has a distinguished coefficient tensor, then not only are the associated principal value and transpose double layers bounded on Muckenhoupt-weighted Block spaces, but they have geometric estimates in terms of the BMO seminorm of the outward unit normal. We will see that such an estimate is crucial for the solvability of boundary value problems with boundary datum belonging to a Muckenhoupt-weighted Block space.

Theorem 9.23. *Let $\Omega \subseteq \mathbb{R}^n$ be a* UR *domain and abbreviate $\sigma := \mathcal{H}^{n-1}\lfloor\partial\Omega$. Let L be a weakly elliptic, second-order, homogeneous, constant complex coefficient, $M \times M$ system in $\mathbb{R}^n$, which has a distinguished coefficient tensor. Now, fix an integrability exponent $q \in (1,\infty)$ along with a Muckenhoupt weight $\omega \in A_q(\partial\Omega,\sigma)$. Finally, choose some $\varepsilon \in (0, q-1)$ for which $\omega \in A_{q-\varepsilon}(\partial\Omega,\sigma)$ and select a threshold $\lambda \in (0, \varepsilon(n-1)/(q-1))$.*

Then for each $m \in \mathbb{N}$ and $A \in \mathfrak{A}_L^{\mathrm{dis}}$ we have (cf. (3.19), (3.20), *and Definition* 5.9*)*

$$\|K_A\|_{[\mathcal{B}^{q,\lambda}(\partial\Omega,\omega)]^M\to[\mathcal{B}^{q,\lambda}(\partial\Omega,\omega)]^M} \le C_m \cdot \|\nu\|^{\langle m\rangle}_{[\mathrm{BMO}(\partial\Omega,\omega)]^n} \tag{9.89}$$

and

$$\|K_A^{\#}\|_{[\mathcal{B}^{q,\lambda}(\partial\Omega,\omega)]^M\to[\mathcal{B}^{q,\lambda}(\partial\Omega,\omega)]^M} \le C_m \cdot \|\nu\|^{\langle m\rangle}_{[\mathrm{BMO}(\partial\Omega,\omega)]^n} \tag{9.90}$$

where ν denotes the geometric measure theoretic outward unit normal to Ω, and the constant $C_m \in (0,\infty)$ depends on m, the UR *character of $\partial\Omega$, n, q, ε, λ, $[\omega]_{A_{q-\varepsilon}}$, and $[\omega]_{A_q}$.*

Proof. By Proposition 3.16, the fact that the system L has a distinguished coefficient tensor A also means that there exist matrix-valued functions $k, k^{\#} \in [\mathscr{C}^\infty(\mathbb{R}^n \setminus \{0\})]^{M\times M}$, which are even, and positive homogeneous of degree $-n$ such that for each function $g \in [L^1(\partial\Omega, \frac{\sigma(x)}{1+|x|^{n-1}})]^M$ we have

$$(K_A g)(x) = \lim_{\varepsilon\to 0^+} \int\limits_{\substack{y\in\partial\Omega\\ |x-y|>\varepsilon}} \langle \nu(y), x-y\rangle k(x-y) g(y)\, d\sigma(y) \tag{9.91}$$

and

$$(K_A^{\#}g)(x) = \lim_{\varepsilon\to 0^+} \int\limits_{\substack{y\in\partial\Omega \\ |x-y|>\varepsilon}} \langle \nu(x), y-x\rangle k^{\#}(x-y)g(y)\,d\sigma(y) \tag{9.92}$$

for σ-a. e. $x \in \partial\Omega$. In particular, (9.91) and (9.92) hold for each function $g \in [\mathcal{B}^{q,\lambda}(\partial\Omega, \omega)]^M$ due to the embedding in Proposition 6.17. The structures of the integral kernels in (9.91) and (9.92) then permit us to invoke Theorem 7.15 and Theorem 7.12 to conclude that for each $m \in \mathbb{N}$ there exists a constant $C_m \in (0, \infty)$ depending on m, $\partial\Omega$, n, q, λ, $[\omega]_{A_{q-\varepsilon}}$, and $[\omega]_{A_q}$ such that

$$\|K_A g\|_{[\mathcal{B}^{q,\lambda}(\partial\Omega,\omega)]^M} \leq C_m \|\nu\|^{\langle m\rangle}_{[\mathrm{BMO}(\partial\Omega,\sigma)]^n} \cdot \|g\|_{[\mathcal{B}^{q,\lambda}(\partial\Omega,\omega)]^M}, \tag{9.93}$$

and

$$\|K_A^{\#} g\|_{[\mathcal{B}^{q,\lambda}(\partial\Omega,\omega)]^M} \leq C_m \|\nu\|^{\langle m\rangle}_{[\mathrm{BMO}(\partial\Omega,\sigma)]^n} \cdot \|g\|_{[\mathcal{B}^{q,\lambda}(\partial\Omega,\omega)]^M}, \tag{9.94}$$

for all $g \in [\mathcal{B}^{q,\lambda}(\partial\Omega, \omega)]^M$, which imply (9.89) and (9.90) given the arbitrary nature of g. □

We can also consider Muckenhoupt-weighted Block-based Sobolev spaces of order one and continue to get a boundedness result for the boundary-to-boundary double layer potential. This will be useful in the context of solving the regularity problem on Muckenhoupt-weighted Block spaces.

Proposition 9.24. *Let $\Omega \subseteq \mathbb{R}^n$ be a UR domain and write $\sigma := \mathcal{H}^{n-1}\lfloor\partial\Omega$. Let L be a weakly elliptic, second-order, homogeneous, constant complex coefficient, $M \times M$ system in $\mathbb{R}^n$. Fix an integrability exponent $q \in (1, \infty)$ along with a Muckenhoupt weight $\omega \in A_q(\partial\Omega, \sigma)$. Lastly, choose some $\varepsilon \in (0, q-1)$ for which $\omega \in A_{q-\varepsilon}(\partial\Omega, \sigma)$ and select a threshold $\lambda \in (0, \varepsilon(n-1)/(q-1))$.*

Then for any $A \in \mathfrak{A}_L$ the boundary-to-boundary double layer potential K_A (cf. (3.19)) induces a mapping

$$K_A : [\mathcal{B}_1^{q,\lambda}(\partial\Omega, \omega)]^M \to [\mathcal{B}_1^{q,\lambda}(\partial\Omega, \omega)]^M \tag{9.95}$$

which is well-defined, linear, and bounded.

Proof. The embedding in Proposition 6.17 demonstrates that (3.31) of Proposition 3.8 is satisfied, allowing us to conclude that for any function $g \in \mathcal{B}_1^{q,\lambda}(\partial\Omega, \omega)$ and any $j, k \in \{1, \ldots, n\}$,

$$\partial_{\tau_{jk}}(K_A g) = K_A(\partial_{\tau_{jk}} g) + U_{jk}(\nabla_{\tan} g) \quad \text{at } \sigma\text{-a. e. point in } \partial\Omega, \tag{9.96}$$

where we have retained the notation for U_{jk}, defined in (3.30). By Proposition 9.21, there exists some constant $C \in (0,\infty)$ such that

$$\|K_A(\partial_{\tau_{jk}} g)\|_{[\mathcal{B}^{q,\lambda}(\partial\Omega,\omega)]^M} \leq C\|\partial_{\tau_{jk}} g\|_{[\mathcal{B}^{q,\lambda}(\partial\Omega,\omega)]^M}. \tag{9.97}$$

for each $j,k \in \{1,\dots,n\}$. Now, notice for each $\gamma \in \{1,\dots,M\}$ that $(U_{jk}(\nabla_{\tan})g)_\gamma$ is a sum of operators of the type discussed in Proposition 7.13. The fact that the integral kernel $E_{\gamma\beta}$ is odd, positive homogeneous of degree $1-n$ for each $\gamma,\beta \in \{1,\dots,M\}$ is ensured by Proposition 3.4. Therefore, noting in this context that $F_{\alpha s} = (\nabla_{\tan} g_\alpha)_s$ in the definition of U_{jk} (cf. (3.30)), Proposition 7.13 ensures the estimate

$$\|U_{jk}(\nabla_{\tan} g)\|_{[\mathcal{B}^{q,\lambda}(\partial\Omega,\omega)]^M} \leq C\|\nu\|_{[\mathrm{BMO}(\partial\Omega,\sigma)]^n} \cdot \|\partial_{\tau_{jk}} g\|_{[\mathcal{B}^{q,\lambda}(\partial\Omega,\omega)]^M} \tag{9.98}$$

for some $C \in (0,\infty)$ independent of g, where ν is the geometric measure theoretic outward unit normal to Ω. It follows from (9.96) and the estimates in (9.97) and (9.98) that

$$\|\partial_{\tau_{jk}}(K_A g)\|_{[\mathcal{B}^{q,\lambda}(\partial\Omega,\omega)]^M} \leq C\|\partial_{\tau_{jk}} g\|_{[\mathcal{B}^{q,\lambda}(\partial\Omega,\omega)]^M} \tag{9.99}$$

Since Proposition 9.21 already tells us that K_A is bounded on the Block space $[\mathcal{B}^{q,\lambda}(\partial\Omega,\omega)]^M$, from (9.99) and the arbitrary nature of $g \in [\mathcal{B}^{q,\lambda}(\partial\Omega,\omega)]^M$ we ultimately find (bearing in mind (8.12)) that the operator in (9.95) is well-defined, linear, and bounded. □

Even in the context of Muckenhoupt-weighted Block-based Sobolev spaces (of order one), if a weakly elliptic, second-order, homogeneous, constant complex coefficient system has a distinguished coefficient tensor, then this leads to geometric estimates on the operator norm of the boundary-to-boundary double layer potential in terms of the BMO seminorm of the outward unit normal.

Theorem 9.25. *Let $\Omega \subseteq \mathbb{R}^n$ be a* UR *domain and abbreviate $\sigma := \mathcal{H}^{n-1}\lfloor\partial\Omega$. Let L be a weakly elliptic, second-order, homogeneous, $M \times M$ system in $\mathbb{R}^n$, which has a distinguished coefficient tensor. Now, fix an integrability exponent $q \in (1,\infty)$ along with a Muckenhoupt weight $\omega \in A_q(\partial\Omega,\sigma)$. Finally, choose some $\varepsilon \in (0,q-1)$ for which $\omega \in A_{q-\varepsilon}(\partial\Omega,\sigma)$ and select a threshold $\lambda \in (0,\varepsilon(n-1)/(q-1))$.*

Then for each $m \in \mathbb{N}$ and $A \in \mathfrak{A}_L^{\mathrm{dis}}$ one has (cf. (3.19) *and Definition* 5.9*)*

$$\|K_A\|_{[\mathcal{B}_1^{q,\lambda}(\partial\Omega,\omega)]^M \to [\mathcal{B}_1^{q,\lambda}(\partial\Omega,\omega)]^M} \leq C_m \cdot \|\nu\|^{\langle m\rangle}_{[\mathrm{BMO}(\partial\Omega,\omega)]^n}, \tag{9.100}$$

where ν denotes the geometric measure theoretic outward unit normal to Ω and the constant $C_m \in (0,\infty)$ depends on m, the UR *character of $\partial\Omega$, n, q, ε, λ, $[\omega]_{A_{q-\varepsilon}}$, and $[\omega]_{A_q}$.*

Proof. We may follow a similar approach as in the proof of Proposition 9.24. In fact, we have the same identity as in (9.96). However, as a consequence of Proposition 3.16, the existence of a distinguished coefficient tensor means that the boundary-to-boundary double layer potential K_A can be written with an integral kernel of chord-dot-normal type. Therefore, instead of relying on (9.97), we may appeal to Theorem 7.15 to obtain the more refined estimate

$$\|K_A(\partial_{\tau_{jk}}g)\|_{[\mathcal{B}^{q,\lambda}(\partial\Omega,\omega)]^M} \leq C_m \cdot \|\nu\|^{\langle m\rangle}_{[\mathrm{BMO}(\partial\Omega,\sigma)]^n} \cdot \|\partial_{\tau_{jk}}g\|_{[\mathcal{B}^{q,\lambda}(\partial\Omega,\omega)]^M}, \tag{9.101}$$

which holds for each $g \in [\mathcal{B}^{q,\lambda}_1(\partial\Omega,\omega)]^M$, each $j,k \in \{1,\dots,n\}$ and each $m \in \mathbb{N}$. The estimate in (9.98) continues to hold in this context and, therefore, allows us to conclude that

$$\|\partial_{\tau_{jk}}(K_A g)\|_{[\mathcal{B}^{q,\lambda}(\partial\Omega,\omega)]^M} \leq C_m\|\nu\|^{\langle m\rangle}_{[\mathrm{BMO}(\partial\Omega,\sigma)]^n} \cdot \|\partial_{\tau_{jk}}g\|_{[\mathcal{B}^{q,\lambda}(\partial\Omega,\omega)]^M}. \tag{9.102}$$

Since we also have

$$\|K_A g\|_{[\mathcal{B}^{q,\lambda}(\partial\Omega,\omega)]^M} \leq C_m\|\nu\|^{\langle m\rangle}_{[\mathrm{BMO}(\partial\Omega,\sigma)]^n} \cdot \|g\|_{[\mathcal{B}^{q,\lambda}(\partial\Omega,\omega)]^M}, \tag{9.103}$$

which itself is a result of Theorem 9.23, we ultimately find (bearing in mind (8.12)) that

$$\|K_A g\|_{[\mathcal{B}^{q,\lambda}_1(\partial\Omega,\omega)]^M} \leq C_m\|\nu\|^{\langle m\rangle}_{[\mathrm{BMO}(\partial\Omega,\sigma)]^n} \cdot \|g\|_{[\mathcal{B}^{q,\lambda}_1(\partial\Omega,\omega)]^M}, \tag{9.104}$$

which implies (9.100) given the arbitrary nature of $g \in [\mathcal{B}^{q,\lambda}_1(\partial\Omega,\omega)]^M$. □

We now present an invertibility result for an operator that involves the boundary-to-boundary double layer potential on Muckenhoupt-weighted Block-based Sobolev spaces. Such an invertibility result requires a nice enough underlying geometry and the existence of a distinguished coefficient tensor, but given that, this invertibility result even comes with quantitative control on the tangential gradient.

Proposition 9.26. *Let $\Omega \subseteq \mathbb{R}^n$ be a* UR *domain and write $\sigma := \mathcal{H}^{n-1}\lfloor\partial\Omega$. Suppose L is a weakly elliptic, second-order, homogeneous, constant complex coefficient, $M \times M$ system in $\mathbb{R}^n$, which has a distinguished coefficient tensor. Fix an integrability exponent $q \in (1,\infty)$ along with a Muckenhoupt weight $\omega \in A_q(\partial\Omega,\sigma)$. Lastly, choose some $\varepsilon \in (0,q-1)$ for which $\omega \in A_{q-\varepsilon}(\partial\Omega,\sigma)$, and select a threshold $\lambda \in (0,\varepsilon(n-1)/(q-1))$.*

*Then for any $z \in \mathbb{C}$ there exists $\delta \in (0,1)$ with the following significance. When Ω is a δ-*AR *domain and $A \in \mathfrak{A}^{\mathrm{dis}}_L$, the operator*

$$zI + K_A : [\mathcal{B}^{q,\lambda}(\partial\Omega,\omega)]^M \to [\mathcal{B}^{q,\lambda}(\partial\Omega,\omega)]^M \tag{9.105}$$

is invertible and there exists a constant $C \in (0,\infty)$, which satisfies

$$\|\nabla_{\tan}g\|_{[\mathcal{B}^{q,\lambda}(\partial\Omega,\omega)]^{n\cdot M}} \leq C\|\nabla_{\tan}f\|_{[\mathcal{B}^{q,\lambda}_1(\partial\Omega,\omega)]^M}. \tag{9.106}$$

for any $f \in [\mathcal{B}^{q,\lambda}_1(\partial\Omega,\omega)]^M$, where $g := (zI+K_A)^{-1}f \in [\mathcal{B}^{q,\lambda}_1(\partial\Omega,\omega)]^M$.

Proof. This is proved analogously to that of Proposition 9.11. Most notably, we now rely on Theorem 9.25, Theorem 9.23, Theorem 7.13, and Theorem 7.2, as well as the Banach space property of Block spaces, recorded in Proposition 6.11. □

We can also work with the modified double layer potential operator, introduced in (3.34), which is also a way to produce smooth null solutions for the kinds of weakly elliptic systems we are interested in. Specifically, this next result shows that this continues to be the case in the context of Muckenhoupt-weighted Block spaces, and that we even understand the nontangential trace of this object in terms of the modified boundary-to-boundary double layer; cf. (3.47).

Theorem 9.27. *Let $\Omega \subseteq \mathbb{R}^n$ be a UR domain and abbreviate $\sigma := \mathcal{H}^{n-1}\lfloor\partial\Omega$. Let L be a weakly elliptic, second-order, homogeneous, constant complex coefficient, $M \times M$ system in $\mathbb{R}^n$. Next, fix an integrability exponent $q \in (1, \infty)$ along with a Muckenhoupt weight $\omega \in A_q(\partial\Omega, \sigma)$. Finally, choose some parameter $\varepsilon \in (0, q-1)$ such that $\omega \in A_{q-\varepsilon}(\partial\Omega, \sigma)$ and pick any threshold $\lambda \in (0, \varepsilon(n-1)/(q-1))$.*

Then for each $g \in [\dot{\mathcal{B}}_1^{q,\lambda}(\partial\Omega, \omega)]^M$ and any $A \in \mathfrak{A}_L$ one has (cf. (3.34))

$$\mathcal{D}_{A,\mathrm{mod}}g \in [\mathscr{C}^\infty(\Omega)]^M \quad \text{and} \quad L(\mathcal{D}_{A,\mathrm{mod}}g) = 0 \quad \text{in } \Omega. \tag{9.107}$$

Additionally, both $(\mathcal{D}_{A,\mathrm{mod}}g)\big|_{\partial\Omega}^{\kappa-\mathrm{n.t.}}$ and $(\nabla(\mathcal{D}_{A,\mathrm{mod}}g))\big|_{\partial\Omega}^{\kappa-\mathrm{n.t.}}$ exist σ-a. e. on $\partial\Omega$ and in fact,

$$(\mathcal{D}_{A,\mathrm{mod}}g)\big|_{\partial\Omega}^{\kappa-\mathrm{n.t.}} = \Big(\frac{1}{2}I + K_{A,\mathrm{mod}}\Big)g \quad \text{at } \sigma\text{-a. e. point on } \partial\Omega \tag{9.108}$$

(cf. (3.47)). Moreover, there exists a constant $C \in (0, \infty)$, depending only on Ω, n, q, λ, $[\omega]_{A_{q-\varepsilon}}$, $[\omega]_{A_q}$, κ, and A such that $\mathcal{N}_\kappa(\nabla(\mathcal{D}_{A,\mathrm{mod}}g)) \in \mathcal{B}^{q,\lambda}(\partial\Omega, \omega)$ with

$$\|\mathcal{N}_\kappa(\nabla(\mathcal{D}_{A,\mathrm{mod}}g))\|_{\mathcal{B}^{q,\lambda}(\partial\Omega,\omega)} \le C\|g\|_{[\dot{\mathcal{B}}_1^{q,\lambda}(\partial\Omega,\omega)]^M} \tag{9.109}$$

for each $g \in [\dot{\mathcal{B}}_1^{q,\lambda}(\partial\Omega, \omega)]^M$.

Proof. We can use (8.30) together with (3.36) to conclude (9.107). Next, (8.30) and (3.50) imply the jump formula in (9.108). We also observe that (8.30), Proposition 6.17, and (3.43) yield the σ-a. e. existence of $(\nabla(\mathcal{D}_{A,\mathrm{mod}}g))\big|_{\partial\Omega}^{\kappa-\mathrm{n.t.}}$ on $\partial\Omega$. Lastly, because of (3.42), we can argue similarly as in Proposition 9.19, specifically (9.75)–(9.77) to obtain (9.109) as wanted. □

The conormal derivative of the modified boundary-to-domain double layer potential is an operator, which behaves well in the context of Muckenhoupt-weighted Block spaces as this next proposition shows.

Proposition 9.28. *Let $\Omega \subseteq \mathbb{R}^n$ be a UR domain, abbreviate $\sigma := \mathcal{H}^{n-1}\lfloor\partial\Omega$, and let ν denote the outward unit normal to Ω. Let L be a weakly elliptic, second-order, homogeneous,* *constant complex coefficient, $M \times M$ system in $\mathbb{R}^n$. Next, fix an integrability exponent $q \in (1, \infty)$ along with a Muckenhoupt weight $\omega \in A_q(\partial\Omega, \sigma)$. Finally, choose some parameter $\varepsilon \in (0, q-1)$ so that $\omega \in A_{q-\varepsilon}(\partial\Omega, \sigma)$ and select any threshold $\lambda \in (0, \varepsilon(n-1)/(q-1))$.*

Then for any $A \in \mathfrak{A}_L$ the operator

$$\partial_\nu^A \mathcal{D}_{A,\mathrm{mod}} : [\dot{\mathcal{B}}_1^{q,\lambda}(\partial\Omega, \omega)]^M \to [\mathcal{B}^{q,\lambda}(\partial\Omega, \omega)]^M \tag{9.110}$$

is well-defined, linear, and bounded, where ∂_ν^A denotes the conormal derivative associated to A as in (3.18).

Proof. Because we now have Theorem 9.27, we can argue analogously as in the proof of Proposition 9.13, also bearing in mind Proposition 6.14, i. e., the lattice property of Block spaces. □

We are also able to work with the modified boundary-to-boundary double layer potential in the context of homogeneous Muckenhoupt-weighted Block-based Sobolev spaces.

Proposition 9.29. *Let $\Omega \subseteq \mathbb{R}^n$ be a UR domain and write $\sigma := \mathcal{H}^{n-1}\lfloor\partial\Omega$. Let L be a weakly elliptic, second-order, homogeneous, constant complex coefficient, $M \times M$ system in $\mathbb{R}^n$. Next, fix an integrability exponent $q \in (1, \infty)$ along with a Muckenhoupt weight $\omega \in A_q(\partial\Omega, \sigma)$. Now, choose some parameter $\varepsilon \in (0, q-1)$ such that $\omega \in A_{q-\varepsilon}(\partial\Omega, \sigma)$ and pick any threshold $\lambda \in (0, \varepsilon(n-1)/(q-1))$. Finally, choose $A \in \mathfrak{A}_L$. Then the operator*

$$K_{A,\mathrm{mod}} : [\dot{\mathcal{B}}_1^{q,\lambda}(\partial\Omega, \omega)]^M \to [\dot{\mathcal{B}}_1^{q,\lambda}(\partial\Omega, \omega)]^M \tag{9.111}$$

is well-defined, linear, and bounded. Consequently, (bearing in mind (3.51) *which says that says $K_{A,\mathrm{mod}}$ maps constants to constants), the operator*

$$\begin{aligned} [K_{A,\mathrm{mod}}] : [\dot{\mathcal{B}}_1^{q,\lambda}(\partial\Omega, \omega)/\sim]^M \to [\dot{\mathcal{B}}_1^{q,\lambda}(\partial\Omega, \omega)/\sim]^M \quad \textit{defined as} \\ [K_{A,\mathrm{mod}}][f] := [K_{A,\mathrm{mod}} f] \in [\dot{\mathcal{B}}_1^{q,\lambda}(\partial\Omega, \omega)/\sim]^M, \end{aligned} \tag{9.112}$$

for all functions $f \in [\dot{\mathcal{B}}_1^{q,\lambda}(\partial\Omega, \omega)/\sim]^M$ is also well-defined, linear, and bounded, where $\dot{\mathcal{B}}_1^{q,\lambda}(\partial\Omega, \omega)/\sim$ is the quotient space of equivalence classes $[\cdot]$ modulo constant functions in $\dot{\mathcal{B}}_1^{q,\lambda}(\partial\Omega, \omega)$.

Proof. Start by fixing $g \in [\dot{\mathcal{B}}_1^{q,\lambda}(\partial\Omega, \omega)]^M$ arbitrary. Thanks to Theorem 9.27, we know that $(\mathcal{D}_{A,\mathrm{mod}} g)\big|_{\partial\Omega}^{\kappa-\mathrm{n.t.}}$ and $(\partial_j(\mathcal{D}_{A,\mathrm{mod}} g))\big|_{\partial\Omega}^{\kappa-\mathrm{n.t.}}$ exist σ-a. e. on $\partial\Omega$ for each $j \in \{1, \ldots, n\}$. We also know from (8.30) and (3.45), whose relevance is ensured by Corollary 2.35, that $\mathcal{N}_\kappa(\mathcal{D}_{A,\mathrm{mod}} g)$ belongs to $L^1_{\mathrm{loc}}(\partial\Omega, \sigma)$. Moreover, Proposition 6.17 and (9.109) ensure that $\mathcal{N}_\kappa(\nabla(\mathcal{D}_{A,\mathrm{mod}} g))$ belongs to $L^1_{\mathrm{loc}}(\partial\Omega, \sigma)$ as well. From this and given the geometry of Ω, it follows from [79, Proposition 2.22, p. 148] (keeping in mind (9.108)) that

$$\left|\partial_{\tau_{jk}}\left(\frac{1}{2}I+K_A\right)g\right| = \left|\partial_{\tau_{jk}}\left((\mathcal{D}_{A,\mathrm{mod}}g)\big|_{\partial\Omega}^{\kappa-\mathrm{n.t.}}\right)\right| \le 2\mathcal{N}_\kappa(\nabla(\mathcal{D}_{A,\mathrm{mod}}g)) \tag{9.113}$$

at σ-a. e. point on $\partial\Omega$. We then turn to (9.109) and Proposition 6.14 to say that $\partial_{\tau_{jk}}(K_{A,\mathrm{mod}}g)$ belongs to $[\mathcal{B}^{q,\lambda}(\partial\Omega,\omega)]^M$ for each $j,k\in\{1,\dots,n\}$. As such, (8.32) implies the membership of $K_{A,\mathrm{mod}}g$ to $[\dot{\mathcal{B}}_1^{q,\lambda}(\partial\Omega,\omega)]^M$. Moreover, (9.113), (9.109), Proposition 6.14, and (8.32) provide the estimate

$$\|K_{A,\mathrm{mod}}g\|_{[\dot{\mathcal{B}}_1^{q,\lambda}(\partial\Omega,\omega)]^M} \le C\|g\|_{[\dot{\mathcal{B}}_1^{q,\lambda}(\partial\Omega,\omega)]^M} \tag{9.114}$$

for some constant $C\in(0,\infty)$ independent of f. This implies the desired well definiteness and boundedness of the operator in (9.111) given the arbitrary nature of g. The analogous result concerning the operator in (9.112) then follows from the well definiteness and boundedness of the operator in (9.111) and definitions. □

Small norm estimates involving the BMO seminorm of the outward unit normal are also possible for the modified boundary-to-boundary double layer potential operator in the context of homogeneous Muckenhoupt-weighted Block-based Sobolev spaces under the additional assumption that the layer potential is written using a distinguished coefficient tensor.

Theorem 9.30. *Let $\Omega\subseteq\mathbb{R}^n$ be a 2-sided* NTA *domain such that $\partial\Omega$ is an unbounded Ahlfors regular set, abbreviate $\sigma:=\mathcal{H}^{n-1}\lfloor\partial\Omega$, and denote by ν the geometric measure theoretic outward unit normal to Ω. Let L be a weakly elliptic, second-order, homogeneous, constant complex coefficient, $M\times M$ system in $\mathbb{R}^n$ with the property that $\mathfrak{A}_L^{\mathrm{dis}}\neq\varnothing$. Next, fix an integrability exponent $q\in(1,\infty)$ along with a Muckenhoupt weight $\omega\in A_q(\partial\Omega,\sigma)$. Finally, choose $\varepsilon\in(0,q-1)$ such that $\omega\in A_{q-\varepsilon}(\partial\Omega,\sigma)$ and pick any threshold $\lambda\in(0,\varepsilon(n-1)/(q-1))$.*

Then for each $m\in\mathbb{N}$, with $A\in\mathfrak{A}_L^{\mathrm{dis}}$, there exists a constant $C_m\in(0,\infty)$, depending only on m, the Ahlfors regularity constants of $\partial\Omega$, n, L, q, λ, $[\omega]_{A_{q-\varepsilon}}$, and $[\omega]_{A_q}$ such that (recalling (5.9)*)*

$$\left\|[K_{A,\mathrm{mod}}]\right\|_{[\dot{\mathcal{B}}_1^{q,\lambda}(\partial\Omega,\omega)/\sim]^M\to[\dot{\mathcal{B}}_1^{q,\lambda}(\partial\Omega,\omega)/\sim]^M} \le C_m\|\nu\|_{[\mathrm{BMO}(\partial\Omega,\sigma)]^n}^{\langle m\rangle}. \tag{9.115}$$

Proof. Note that our current geometric assumptions on Ω imply that Ω is a UR domain (cf. [79, (2.48), p. 38]). Recall Remark 3.11, which we can appeal to in this setting thanks to (8.30) and Corollary 2.35. Thus, we can argue much as in the proof of Theorem 9.25 leading up to (9.102) to similarly obtain for any $j,k\in\{1,\dots,n\}$ the following estimate:

$$\|\partial_{\tau_{jk}}(K_{A,\mathrm{mod}}g)\|_{[\mathcal{B}^{q,\lambda}(\partial\Omega,\omega)]^M} \le C_m\cdot\|\nu\|_{[\mathrm{BMO}(\partial\Omega,\sigma)]^n}^{\langle m\rangle}\cdot\|\partial_{\tau_{jk}}g\|_{[\mathcal{B}^{q,\lambda}(\partial\Omega,\omega)]^M}, \tag{9.116}$$

which holds for any $g\in[\dot{\mathcal{B}}_1^{q,\lambda}(\partial\Omega,\omega)]^M$, where the constant $C_m\in(0,\infty)$ is as in the statement of Theorem 9.30. This then implies the desired estimate in (9.115) in view of (8.33) and the arbitrary nature of g. □

9.3 Invertibility of the single layer potential

We would like to study how S_{mod}, the modified boundary-to-boundary single layer, defined in (3.76), acts on Muckenhoupt-weighted Morrey and Block spaces. Namely, is such an operator well-defined in these functional analytic environments, and what is the correct range space for such an operator? Only with these preliminaries sorted will we be able to tackle the question of invertibility of this operator on Muckenhoupt-weighted Morrey and Block spaces.

Before exploring these ideas further, we need a lemma from [89], which details how the modified boundary-to-boundary single layer accommodates tangential derivatives.

Lemma 9.31. *Let $\Omega \subseteq \mathbb{R}^n$ be a UR domain, and set $\sigma := \mathcal{H}^{n-1}\lfloor\partial\Omega$. Let L be a weakly elliptic, homogeneous, constant complex coefficient, second-order, $M \times M$ system in $\mathbb{R}^n$. Denote by $\nu = (\nu_1, \ldots, \nu_n)$ the geometric measure theoretic outward unit normal to Ω, and let E denote the fundamental solution associated to L as in Proposition 3.4. Also, recall S_{mod} from (3.76). Finally, fix an integrability exponent $p \in (1, \infty)$.*

Then for any $j, k \in \{1, \ldots, n\}$ and any function f belonging to the intersection $[L^1(\partial\Omega, \frac{\sigma(x)}{1+|x|^{n-1}}) \cap L^p_{\mathrm{loc}}(\partial\Omega, \sigma)]^M$, one has

$$\partial_{\tau_{jk}}(S_{\mathrm{mod}}f)(x) = \lim_{\varepsilon\to 0^+} \int_{\substack{y\in\partial\Omega\\ |x-y|>\varepsilon}} \{\nu_j(x)(\partial_k E)(x-y) - \nu_k(x)(\partial_j E)(x-y)\} f(y)\, d\sigma(y) \tag{9.117}$$

at σ-a. e. point $x \in \partial\Omega$.

We now show that the boundary-to-boundary single layer as it maps functions from a Muckenhoupt-weighted Morrey space into the corresponding homogeneous Sobolev space is a well-defined operator.

Proposition 9.32. *Let $\Omega \subseteq \mathbb{R}^n$ be a UR domain and set $\sigma := \mathcal{H}^{n-1}\lfloor\partial\Omega$. Let L be a weakly elliptic, second-order, homogeneous, constant complex coefficient, $M \times M$ system in $\mathbb{R}^n$. Fix two integrability exponents $p, q \in (1, \infty)$ along with a Muckenhoupt weight $w \in A_p(\partial\Omega, \sigma) \cap \mathrm{RH}_q(\partial\Omega, \sigma)$, and a threshold $\lambda \in (0, (n-1)/q')$, where $q' := q/(q-1)$.*

Then the operator S_{mod}, originally introduced in (3.76), induces a mapping

$$S_{\mathrm{mod}} : [M^{p,\lambda}(\partial\Omega, w)]^M \to [\dot{M}^{p,\lambda}_1(\partial\Omega, w)]^M \tag{9.118}$$

which is well-defined, linear, and bounded when the target space is endowed with the seminorm defined in (8.21).

Proof. In [89], it is shown for each $\varepsilon > 0$ that the operator

$$S_{\mathrm{mod}} : \left[L^1\left(\partial\Omega, \frac{\sigma(x)}{1+|x|^{n-1-\varepsilon}}\right)\right]^M \to \left[L^1\left(\partial\Omega, \frac{\sigma(x)}{1+|x|^{n}}\right)\right]^M \tag{9.119}$$

is well-defined, linear, and bounded. As such, the continuous embedding in Proposition 4.15 shows that

$$S_{\mathrm{mod}} : [M^{p,\lambda}(\partial\Omega, w)]^M \to \Big[L^1\Big(\partial\Omega, \frac{\sigma(x)}{1+|x|^n}\Big)\Big]^M \tag{9.120}$$

is well-defined, linear, and bounded as well. Additionally, [86] shows that

$$S_{\mathrm{mod}} : \Big[L^1\Big(\partial\Omega, \frac{\sigma(x)}{1+|x|^{n-1}}\Big) \cap L^p_{\mathrm{loc}}(\partial\Omega, w)\Big]^M \to [L^p_{\mathrm{loc}}(\partial\Omega, w)]^M \tag{9.121}$$

is well-defined, linear, and continuous. Note that for each function f in $[M^{p,\lambda}(\partial\Omega, w)]^M$ and each $j, k \in \{1, \dots, n\}$ we have via the embedding in Proposition 4.16 and Lemma 9.31,

$$\begin{aligned}\partial_{\tau_{jk}}(S_{\mathrm{mod}} f)(x) = \lim_{\varepsilon\to 0^+} \int\limits_{\substack{y\in\partial\Omega\\|x-y|>\varepsilon}} &\{\nu_j(x)(\partial_k E)(x-y) \\ &- \nu_k(x)(\partial_j E)(x-y)\} f(y)\, d\sigma(y)\end{aligned} \tag{9.122}$$

for σ-a. e. point $x \in \partial\Omega$. Because of Proposition 3.4, the integral kernel appearing in (9.122) is odd and positive homogeneous of degree $1-n$, and hence Theorem 5.7 applies, specifically the boundedness of the operator in (5.36). Thus, recalling (8.19), while also bearing in mind Remark 4.4 and Proposition 4.7, we see that $\partial_{\tau_{jk}}(S_{\mathrm{mod}} f)$ belongs to $[M^{p,\lambda}(\partial\Omega, w)]^M$ with proper control of the norm. As such, (9.118) is a well-defined, linear, and bounded operator. □

We can also show that the modified boundary-to-boundary single layer potential maps functions belonging to a Muckenhoupt-weighted Block space into the corresponding homogeneous Sobolev space.

Proposition 9.33. *Let $\Omega \subseteq \mathbb{R}^n$ be a* UR *domain and write $\sigma := \mathcal{H}^{n-1}\lfloor\partial\Omega$. Let L be a weakly elliptic, second-order, homogeneous, constant complex coefficient, $M \times M$ system in $\mathbb{R}^n$. Fix an integrability exponent $q \in (1, \infty)$ along with a Muckenhoupt weight $\omega \in A_q(\partial\Omega, \sigma)$. Choose some parameter $\varepsilon \in (0, q-1)$ for which $\omega \in A_{q-\varepsilon}(\partial\Omega, \sigma)$, and select any threshold $\lambda \in (0, \varepsilon(n-1)/(q-1))$. Then the operator S_{mod}, originally introduced in* (3.76), *induces a mapping*

$$S_{\mathrm{mod}} : [\mathcal{B}^{q,\lambda}(\partial\Omega, \omega)]^M \to [\dot{\mathcal{B}}^{q,\lambda}_1(\partial\Omega, \omega)]^M \tag{9.123}$$

which is well-defined, linear, and bounded when the target space is endowed with the seminorm defined in (8.32).

Proof. The proof follows closely that of Proposition 9.32 with the following notable changes. First, we use the continuous embedding in (6.76) (with $q_0 := 1$) to say that

$$S_{\text{mod}} : [\mathcal{B}^{q,\lambda}(\partial\Omega,\omega)]^M \to \left[L^1\left(\partial\Omega, \frac{\sigma(x)}{1+|x|^n}\right)\right]^M \tag{9.124}$$

is a well-defined, linear, and bounded operator. Next, we use (6.77) in tandem with Lemma 9.31 to yield the equality in (9.122), but for functions f belonging to $[\mathcal{B}^{q,\lambda}(\partial\Omega,\omega)]^M$. Finally, we turn to Theorem 7.2, (8.30), Proposition 6.13, and Proposition 6.14 to obtain the desired boundedness of the operator in (9.123). □

We introduce another lemma that provides a useful identity in which S_{mod} appears.

Lemma 9.34. *Let $\Omega \subseteq \mathbb{R}^n$ be a UR domain and set $\sigma := \mathcal{H}^{n-1}\lfloor\partial\Omega$. Let L be a weakly elliptic, second-order, homogeneous, constant complex coefficient $M \times M$ system in $\mathbb{R}^n$ and suppose $A = (a_{jk}^{\alpha\beta})_{\substack{1\leq j,k\leq n\\ 1\leq\alpha,\beta\leq M}} \in \mathfrak{A}_L$. Fix two integrability exponents $p, q \in (1,\infty)$, along with a Muckenhoupt weight $w \in A_p(\partial\Omega,\sigma)\cap \mathrm{RH}_q(\partial\Omega,\sigma)$, and a threshold $\lambda \in (0,(n-1)/q')$, where $q' := q/(q-1)$. Then for any $f \in [M^{p,\lambda}(\partial\Omega,w)]^M$ one has*

$$\begin{aligned}\Big(\tfrac{1}{2}I + K^{\#}_{A^\top}\Big)\Big(\Big(-\tfrac{1}{2}I + K^{\#}_{A^\top}\Big)f\Big)(x) \\ = \Bigg(\lim_{\varepsilon\to 0^+} \int\limits_{\substack{y\in\partial\Omega\\ |x-y|>\varepsilon}} \nu_r(x) a_{rs}^{\mu\gamma} a_{jk}^{\beta\alpha} (\partial_j E_{\gamma\beta})(x-y)\partial_{\tau_{sk}}(S_{\text{mod}} f)_\alpha(y)\, d\sigma(y)\Bigg)_{1\leq\mu\leq M}\end{aligned} \tag{9.125}$$

at σ-a. e. point $x \in \partial\Omega$, where $\nu = (\nu_1,\dots,\nu_n)$ is the geometric theoretic outward unit normal to Ω, and $K^{\#}_{A^\top}$ is the transpose boundary-to-boundary double layer associated with $A^\top$ as in (3.20), and S_{mod} is the operator defined in (3.76).

Moreover, for each sequence of functions $\{f_j\}_{j\in\mathbb{N}} \subseteq [M^{p,\lambda}(\partial\Omega,w)]^M$, which is weak-$*$ convergent to some $f \in [M^{p,\lambda}(\partial\Omega,w)]^M$ and for each test function $\varphi \in [\mathrm{Lip}_c(\partial\Omega)]^M$ one has

$$\lim_{j\to\infty} \int_{\partial\Omega} \langle S_{\text{mod}} f_j, \varphi\rangle\, d\sigma = \int_{\partial\Omega} \langle S_{\text{mod}} f, \varphi\rangle\, d\sigma. \tag{9.126}$$

Finally, a similar statement holds when $M^{p,\lambda}(\partial\Omega,w)$ is replaced by $\mathcal{B}^{q,\lambda}(\partial\Omega,\omega)$, where now $q \in (1,\infty)$ with $q' := q/(q-1)$, $\omega \in A_q(\partial\Omega,\sigma)$ is such that

$$\omega' := \omega^{-q'/q} \in A_{q'}(\partial\Omega,\sigma) \cap \mathrm{RH}_p(\partial\Omega,\sigma) \quad \textit{for some } p \in (1,\infty), \tag{9.127}$$

$p' := p/(p-1)$, $\varepsilon := (q-1)/p' \in (0,q-1)$ so that $\omega \in A_{q-\varepsilon}(\partial\Omega,\sigma)$ by Proposition 6.35, and $\lambda \in (0,\varepsilon(n-1)/(q-1))$. In the latter scenario, weak-$*$ convergence of a sequence $\{f_j\}_{j\in\mathbb{N}} \subseteq [\mathcal{B}^{q,\lambda}(\partial\Omega,\omega)]^M$ to $f \in [\mathcal{B}^{q,\lambda}(\partial\Omega,\omega)]^M$ can be understood as genuine weak-$*$ convergence, which involves weighted vanishing Morrey spaces (cf. Definition 10.20), or simply that $\lim_{j\to\infty} \int_{\partial\Omega} f_j g\, d\sigma = \int_{\partial\Omega} fg\, d\sigma$ for each $g \in M^{q',\lambda}(\partial\Omega,\omega')$.

Proof. The fact that (9.125) holds for any function in $[M^{p,\lambda}(\partial\Omega, w)]^M$ follows from the inclusion found in Lemma 5.17 and [89], which shows that the identity in (9.125) holds for any $f \in [L^p(\partial\Omega, w)]^M$ with $p \in (1, \infty)$ and $w \in A_p(\partial\Omega, \sigma)$ both arbitrary.

For the treatment of (9.126), fix a sequence $\{f_j\}_{j\in\mathbb{N}} \subseteq [M^{p,\lambda}(\partial\Omega, w)]^M$, which is weak-$*$ convergent to some $f \in [M^{p,\lambda}(\partial\Omega, w)]^M$. That is to say

$$\lim_{j\to\infty} \int_{\partial\Omega} \langle f_j, g\rangle \, d\sigma = \int_{\partial\Omega} \langle f, g\rangle \, d\sigma \tag{9.128}$$

for each $g \in [\mathcal{B}^{p',\lambda}(\partial\Omega, w')]^M$, where p' is the Hölder conjugate of p and $w' := w^{-p'/p}$ is the dual weight to w. Now, choose some arbitrary test function $\varphi \in [\mathrm{Lip}_c(\Sigma)]^M$. Fix a reference point $x_0 \in \partial\Omega$ and a large enough radius $R \in (0, \infty)$ to ensure that $\operatorname{supp}\varphi \subseteq \Delta(x_0, R)$. Now, for each $j \in \mathbb{N}$ we decompose the distributional pairing between $S_{\mathrm{mod}} f_j$ and φ in the following way:

$$\langle S_{\mathrm{mod}} f_j, \varphi\rangle = \mathrm{I}_j + \mathrm{II}_j + \mathrm{III}_j, \tag{9.129}$$

where

$$\begin{aligned}
\mathrm{I}_j &:= \int_{\partial\Omega} \Big\langle \int_{\partial\Omega} \mathbf{1}_{\partial\Omega\setminus\Delta(x_0,2R)}(y)\{E(x-y) - E_*(-y)\} f_j(y)\, d\sigma(y), \varphi(x) \Big\rangle \, d\sigma(x) \qquad (9.130)\\
&= \int_{\partial\Omega} \Big\langle f_j(y), \mathbf{1}_{\partial\Omega\setminus\Delta(x_0,2R)}(y) \cdot \int_{\partial\Omega} \{E(x-y) - E_*(-y)\}^{\top} \varphi(x)\, d\sigma(x) \Big\rangle \, d\sigma(y),\\
\mathrm{II}_j &:= \int_{\partial\Omega} \Big\langle \int_{\partial\Omega} \mathbf{1}_{\partial\Omega\cap\Delta(x_0,2R)}(y)(-E_*(-y)) f_j(y)\, d\sigma(y), \varphi(x) \Big\rangle \, d\sigma(x)\\
&= -\Big\langle \int_{\partial\Omega} E_*(-y)\mathbf{1}_{\partial\Omega\cap\Delta(x_0,2R)}(y) f_j(y)\, d\sigma(y), \int_{\partial\Omega} \varphi(x)\, d\sigma(x) \Big\rangle,\\
\mathrm{III}_j &:= \int_{\partial\Omega} \Big\langle \int_{\partial\Omega} \mathbf{1}_{\partial\Omega\cap\Delta(x_0,2R)}(y) E(x-y) f_j(y)\, d\sigma(y), \varphi(x) \Big\rangle \, d\sigma(x)\\
&= \int_{\partial\Omega} \Big\langle f_j(y), \mathbf{1}_{\partial\Omega\cap\Delta(x_0,2R)}(y) \int_{\partial\Omega\cap\Delta(x_0,R)} E^{\top}(x-y)\varphi(x)\, d\sigma(x) \Big\rangle \, d\sigma(y).
\end{aligned}$$

We are now tasked with computing the limit as $j \to \infty$ of I_j, II_j, and III_j. To start, for each $y \in \partial\Omega$ define the $\mathbb{C}^M$-valued, σ-measurable function

$$g_1(y) := \mathbf{1}_{\partial\Omega\setminus\Delta(x_0,2R)}(y) \cdot \int_{\partial\Omega} \{E(x-y) - E_*(-y)\}^{\top} \varphi(x)\, d\sigma(x), \tag{9.131}$$

which may be estimated by

$$|g_1(y)| \leq \frac{C_{x_0,R,\varphi}}{1+|y|^{n-1}} \quad \text{for each } y \in \partial\Omega, \tag{9.132}$$

for some constant $C_{x_0,R,\varphi} \in (0,\infty)$. It follows from the σ-measurability of g_1, the lattice property of $\mathcal{B}^{p',\lambda}(\partial\Omega, w')$ (cf. Proposition 6.14), and Lemma 6.26 that g_1 belongs to $[\mathcal{B}^{p',\lambda}(\partial\Omega, w')]^M$, so that by (9.128), we can conclude that

$$\begin{aligned}\lim_{j\to\infty} \mathrm{I}_j &= \lim_{j\to\infty} \int_{\partial\Omega} \langle f_j(y), g_1(y)\rangle \, d\sigma(y) = \int_{\partial\Omega} \langle f(y), g_1(y)\rangle \, d\sigma(y) \\ &= \int_{\partial\Omega} \Big\langle f(y), \mathbf{1}_{\partial\Omega\setminus\Delta(x_0,2R)}(y) \int_{\partial\Omega} \{E(x-y) - E_*(-y)\}^\top \varphi(x)\, d\sigma(x) \Big\rangle \, d\sigma(y).\end{aligned} \tag{9.133}$$

Turning our attention to II_j, define the function

$$g_2(y) := -E_*(-y)\mathbf{1}_{\partial\Omega\cap\Delta(x_0,2R)}(y) \quad \text{for each point } y \in \partial\Omega. \tag{9.134}$$

We then have $g_2 \in [L^\infty_{\mathrm{comp}}(\partial\Omega,\sigma)]^{M\times M} \subseteq [\mathcal{B}^{p',\lambda}(\partial\Omega, w')]^{M\times M}$ by Proposition 6.25. As such, (9.128) allows us to compute

$$\begin{aligned}\lim_{j\to\infty} \mathrm{II}_j &= \lim_{j\to\infty} \Big\langle \int_{\partial\Omega} g_2(y) f_j(y)\, d\sigma(y), \int_{\partial\Omega} \varphi(x)\, d\sigma(x) \Big\rangle \\ &= \Big\langle \int_{\partial\Omega} g_2(y) f(y)\, d\sigma(y), \int_{\partial\Omega} \varphi(x)\, d\sigma(x) \Big\rangle \\ &= -\Big\langle \int_{\partial\Omega} E_*(-y)\mathbf{1}_{\partial\Omega\cap\Delta(x_0,2R)}(y) f(y)\, d\sigma(y), \int_{\partial\Omega} \varphi(x)\, d\sigma(x) \Big\rangle.\end{aligned} \tag{9.135}$$

Finally, to treat III_j, for each $y \in \partial\Omega$ define the $\mathbb{C}^M$-valued, σ-measurable function

$$g_3(y) := \mathbf{1}_{\partial\Omega\cap\Delta(x_0,2R)}(y) \cdot \int_{\partial\Omega\cap\Delta(x_0,R)} E(x-y)^\top \varphi(x)\, d\sigma(x). \tag{9.136}$$

The for each $x \in \operatorname{supp}\varphi \subseteq \Delta(x_0,R)$ and each $y \in \partial\Omega \cap \Delta(x_0,2R)$, we may estimate

$$|x-y| \leq |x-x_0| + |x_0-y| < R + 2R = 3R, \tag{9.137}$$

which permits us to estimate g_3 in the following way:

$$|g_3(y)| \leq C \int_{\substack{x\in\partial\Omega \\ |y-x|<3R}} \frac{|\varphi(x)|}{|y-x|^{n-2}}\, d\sigma(x) \quad \text{for each } y \in \partial\Omega, \tag{9.138}$$

where $C \in (0,\infty)$. Because of the σ-measurability of g_3, we can rely on the combination of Lemma 9.35 (cited after this proof) and Proposition 7.1 to conclude that g_3 belongs to $[\mathcal{B}^{p',\lambda}(\partial\Omega, w')]^M$. As such, using (9.128), we can write

$$\begin{aligned}\lim_{j\to\infty} \mathrm{III}_j &= \lim_{j\to\infty} \int_{\partial\Omega} \langle f_j(y), g_3(y)\rangle\, d\sigma(y) = \int_{\partial\Omega} \langle f(y), g_3(y)\rangle\, d\sigma(y) \\ &= \int_{\partial\Omega} \Big\langle f(y)\,,\, \mathbf{1}_{\partial\Omega\cap\Delta(x_0,2R)}(y) \int_{\partial\Omega\cap\Delta(x_0,R)} E(x-y)^{\top}\varphi(x)\, d\sigma(x)\Big\rangle\, d\sigma(y).\end{aligned} \tag{9.139}$$

Now, bringing everything together from (9.133), (9.135), and (9.139), and keeping in mind (9.129) and the definitions in (9.130), we can ultimately conclude that

$$\lim_{j\to\infty} \langle S_{\mathrm{mod}} f_j, \varphi\rangle = \langle S_{\mathrm{mod}} f, \varphi\rangle. \tag{9.140}$$

This is the desired result given the arbitrary nature of $\varphi \in [\mathrm{Lip}_c(\Sigma)]^M$.

To obtain similar claims in the context of $\mathcal{B}^{q,\lambda}(\partial\Omega, \omega)$, observe that (9.125) holds for any $f \in [\mathcal{B}^{q,\lambda}(\partial\Omega,\omega)]^M$ on account of the inclusion in Lemma 10.57 and [89], which says (9.125) holds for any function belonging to $L^{p_0}(\partial\Omega, w)$ where $p_0 \in (1,\infty)$ and any $w \in A_{p_0}(\partial\Omega,\sigma)$ are both arbitrary. For the analogous version of (9.126), the proof remains largely the same, but we highlight the use of the following necessary analogous results at the level of Morrey spaces. Namely, Proposition 4.7, Lemma 4.12, Corollary 4.11, and Proposition 5.1, though Proposition 10.28 is relevant when working with weighted vanishing Morrey spaces. □

Here is the lemma, mentioned in the proof of Lemma 9.34, which provides a useful bound for fractional integral operators in terms of the Hardy–Littlewood maximal operator. A more general version in the context of quasimetric measure spaces may be found in [86, Proposition 7.8.1, pp. 666–667].

Lemma 9.35. *Let $\Sigma \subseteq \mathbb{R}^n$ ($n \in \mathbb{N}$ with $n \geq 2$) be a closed set which is Ahlfors regular and abbreviate $\sigma := \mathcal{H}^{n-1}\lfloor\Sigma$. Also, fix some $\alpha \in (0,\infty)$ together with some $R \in (0,\infty)$ and, given $f \in L^1_{\mathrm{loc}}(\Sigma,\sigma)$, define*

$$F(x) := \int_{\substack{y\in\Sigma \\ |x-y|<R}} \frac{f(y)}{|x-y|^{n-1-\alpha}}\, d\sigma(y), \quad \textit{for all } x \in \Sigma. \tag{9.141}$$

Then there exists $C = C(\Sigma, n, \alpha) \in (0,\infty)$ such that, with $\mathcal{M}_\Sigma$ denoting the Hardy–Littlewood maximal operator associated with Σ (cf. (2.26)), one has

$$|F| \leq C R^{\alpha} \mathcal{M}_\Sigma f \quad \textit{on } \Sigma. \tag{9.142}$$

The following theorem is the main result of this section and details the geometric and analytic conditions under which S_{mod} acting on a Muckenhoupt-weighted Morrey space is surjective and/or injective. We are also able to obtain a similar result at the level of Muckenhoupt-weighted Block spaces.

Theorem 9.36. *Suppose $\Omega \subseteq \mathbb{R}^n$ is an AR domain and write $\sigma := \mathcal{H}^{n-1}\lfloor\partial\Omega$. Let L be a weakly elliptic, second-order, homogeneous, constant complex coefficient, $M \times M$ system in $\mathbb{R}^n$. Fix two integrability exponents $p, q \in (1,\infty)$ along with a Muckenhoupt weight $w \in A_p(\partial\Omega,\sigma) \cap \mathrm{RH}_q(\partial\Omega,\sigma)$, and a threshold $\lambda \in (0,(n-1)/q')$, where $q' := q/(q-1)$. Recall* (3.76) *and consider the operator*

$$[S_{\mathrm{mod}}] : [M^{p,\lambda}(\partial\Omega,w)]^M \to [\dot{M}_1^{p,\lambda}(\partial\Omega,w)/\sim]^M \tag{9.143}$$

defined as

$$[S_{\mathrm{mod}}]f := [S_{\mathrm{mod}}f] \in [\dot{M}_1^{p,\lambda}(\partial\Omega,w)/\sim]^M \tag{9.144}$$

for each $f \in [M^{p,\lambda}(\partial\Omega,w)]^M$. The following statements concerning $[S_{\mathrm{mod}}]$ hold:

(1) *If Ω satisfies a two-sided local John condition, then the operator* (9.143) *is well-defined, linear, and bounded.*

(2) *If $\mathfrak{A}_L^{\mathrm{dis}} \neq \varnothing$, then there exists $\delta \in (0,1)$, depending on the Ahlfors regularity constants of $\partial\Omega$, n, p, q, λ, $[w]_{A_p}$, $[w]_{\mathrm{RH}_q}$, and L such that when Ω is a δ-AR domain, it follows that the operator* (9.143) *is surjective. Moreover, this surjectivity comes with quantitative control. That is, for each $g \in [\dot{M}_1^{p,\lambda}(\partial\Omega,w)]^M$, there exists $f \in [M^{p,\lambda}(\partial\Omega,w)]^M$ and $C \in (0,\infty)$ independent of g for which $[S_{\mathrm{mod}}]f = [g]$ and*

$$\|f\|_{[M^{p,\lambda}(\partial\Omega,w)]^M} \leq C\|\nabla_{\tan}g\|_{[\dot{M}_1^{p,\lambda}(\partial\Omega,w)]^{n\cdot M}}. \tag{9.145}$$

(3) *If $\mathfrak{A}_{L^\top}^{\mathrm{dis}} \neq \varnothing$, then there exists $\delta \in (0,1)$, depending on the Ahlfors regularity constants of $\partial\Omega$, n, p, q, λ, $[w]_{A_p}$, $[w]_{\mathrm{RH}_q}$, and L such that when Ω is a δ-AR domain, it follows that the operator* (9.143) *is injective.*

(4) *Whenever both $\mathfrak{A}_L^{\mathrm{dis}} \neq \varnothing$ and $\mathfrak{A}_{L^\top}^{\mathrm{dis}} \neq \varnothing$, then there exists some $\delta \in (0,1)$, depending on the Ahlfors regularity constants of $\partial\Omega$, n, p, q, λ, $[w]_{A_p}$, $[w]_{\mathrm{RH}_q}$, and L, such that when Ω is a δ-AR domain, it follows that the operator* (9.143) *is an isomorphism.*

(5) *Analogous statements to those contained in (1)–(4) continue to hold when the spaces $M^{p,\lambda}(\partial\Omega,w)$ and $\dot{M}_1^{p,\lambda}(\partial\Omega,w)$ are replaced by $\mathcal{B}^{q,\lambda}(\partial\Omega,\omega)$ and $\dot{\mathcal{B}}_1^{q,\lambda}(\partial\Omega,\omega)$, associated as in Definition* 6.7 *to some integrability exponent $q \in (1,\infty)$ with $q' := q/(q-1)$, and a Muckenhoupt weight $\omega \in A_q(\partial\Omega,\sigma)$, which also satisfies*

$$\omega' := \omega^{-q'/q} \in A_{q'}(\partial\Omega,\sigma) \cap \mathrm{RH}_p(\partial\Omega,\sigma) \tag{9.146}$$

for some index $p \in (1,\infty)$ with $p' := p/(p-1)$, a parameter $\varepsilon := (q-1)/p'$ (which ensures $\omega \in A_{q-\varepsilon}(\partial\Omega,\sigma)$ thanks to Proposition 6.35*), and any given threshold $\lambda \in (0,\varepsilon(n-1)/(q-1))$.*

Proof. Starting with item (1) note that since Ω is an Ahlfors regular domain, the assumption that Ω also satisfies a two-sided local John condition implies that Ω is a UR domain; cf. [26, Theorem 1, p. 840] and [79, (2.87), p. 45]. Thus, Proposition 9.32 applies and guarantees that the operator (9.143) is well-defined, linear, and bounded.

To treat item (2) start by choosing $A \in \mathfrak{A}_L^{\text{dis}}$. Then matters may be arranged so that when $\|\nu\|_{[\mathrm{BMO}(\partial\Omega,\sigma)]^n} < \delta$ for small enough $\delta \in (0,1)$, then actually

$$\|K_A\|_{[M_1^{p,\lambda}(\partial\Omega,w)]^M \to [M_1^{p,\lambda}(\partial\Omega,w)]^M} < \frac{1}{2}, \tag{9.147}$$

thanks to Theorem 9.10, and

$$\|K_A\|_{[L_1^p(\partial\Omega,w)]^M \to [L_1^p(\partial\Omega,w)]^M} < \frac{1}{2}, \tag{9.148}$$

thanks to [79], so that

$$\text{the operators } \pm\tfrac{1}{2}I + K_A \text{ are invertible on } [M_1^{p,\lambda}(\partial\Omega,w)]^M \text{ and } [L_1^p(\partial\Omega,w)]^M \tag{9.149}$$

via Neumann series (cf. Lemma 11.2). Continuing on, let $\phi \in \mathscr{C}_0^\infty(\mathbb{R}^n)$ be such that $\phi \equiv 1$ on $B(0,1)$ and $\operatorname{supp}\phi \subseteq B(0,2)$. Fix a reference point $x_0 \in \partial\Omega$, and define ϕ_r as in (8.23) for each scale $r \in (0,\infty)$. We will use the same notation to denote the restriction of ϕ_r to the boundary $\partial\Omega$. Now, fix some $g \in [\dot{M}_1^{p,\lambda}(\partial\Omega,w)]^M$ and for each $r \in (0,\infty)$ define g_r as in (8.24). Notice that since we are assuming that Ω is a δ-AR domain, then Ω is necessarily a two-sided NTA domain with unbounded boundary; cf. [79, Theorem 2.5, p. 86]. As such, we can apply items (3) and (4) of Lemma 8.13 to say that

$$\|\nabla_{\tan} g_r\|_{[M^{p,\lambda}(\partial\Omega,w)]^{M\cdot n}} \le C\|\nabla_{\tan} g\|_{[M^{p,\lambda}(\partial\Omega,w)]^{M\cdot n}} \tag{9.150}$$

for some constant $C \in (0,\infty)$ independent of both g and r and that g_r is in $[M^{p,\lambda}(\partial\Omega,w)]^M$. Note by the product rule that

$$\partial_{\tau_{jk}} g_r = (\partial_{\tau_{jk}}\phi_r)\cdot(g - g_{\Delta_{2r}}) + \phi_r \cdot \partial_{\tau_{jk}} g, \tag{9.151}$$

so we see, recalling $g \in [\dot{M}_1^{p,\lambda}(\partial\Omega,w)]^M$, that $g_r \in [M_1^{p,\lambda}(\partial\Omega,w)]^M$. A similar argument, bearing in mind (8.19) and Lemma 8.13 guarantees the membership of g_r to $[L_1^p(\partial\Omega,w)]^M$ as well. Thus, we find that g_r belongs to the intersection $[M_1^{p,\lambda}(\partial\Omega,w)\cap L_1^p(\partial\Omega,w)]^M$. Now, define

$$h_r := \Big(\frac{1}{2}I + K_A\Big)^{-1}\Big(-\frac{1}{2}I + K_A\Big)^{-1} g_r. \tag{9.152}$$

Subsequently, given (9.152) and (9.149), we find that h_r is a meaningfully defined function belonging to $[M_1^{p,\lambda}(\partial\Omega,w)\cap L_1^p(\partial\Omega,w)]^M$. Additionally, the definition of h_r in (9.152), Proposition 9.11, and (9.150) imply

$$\|\nabla_{\tan} h_r\|_{[M^{p,\lambda}(\partial\Omega,w)]^{n\cdot M}} \leq C\|\nabla_{\tan} g\|_{[M^{p,\lambda}(\partial\Omega,w)]^{n\cdot M}} \tag{9.153}$$

for each $r \in (0,\infty)$, where the constant $C \in (0,\infty)$ is independent of r and g.

Now, for each $r \in (0,\infty)$ define

$$f_r := \partial_\nu^A(\mathcal{D}_A h_r) \quad \text{at } \sigma\text{-a. e. point on } \partial\Omega. \tag{9.154}$$

Then because $h_r \in [M_1^{p,\lambda}(\partial\Omega,w) \cap L_1^p(\partial\Omega,w)]^M$, we have that $\partial_\ell(\mathcal{D}_A f)\big|_{\partial\Omega}^{\kappa-\text{n.t.}}$ exists σ-a. e. on $\partial\Omega$ by Proposition 9.4. Therefore, (3.28), (3.18), (8.8), the boundedness of (5.36) from Theorem 5.7, (9.153), (9.150), and (9.149) allow us to estimate

$$\begin{aligned}\|f_r\|_{[M^{p,\lambda}(\partial\Omega,w)]^M} &\leq C\|\nabla_{\tan} h_r\|_{[M^{p,\lambda}(\partial\Omega,w)]^M} \\ &\leq C\|\nabla_{\tan} g\|_{[M^{p,\lambda}(\partial\Omega,w)]^M}.\end{aligned} \tag{9.155}$$

Also, because $h_r \in [L_1^p(\partial\Omega,w)]^M$ and Ω is connected (cf. [79, Theorem 2.5, p. 86]), we can conclude from (9.152), (9.154), and [79, Proposition 3.5, pp. 187–189] that there exists a constant $c_r \in \mathbb{C}^M$ such that

$$S_{\text{mod}} f_r = g_r + c_r \quad \text{on } \partial\Omega. \tag{9.156}$$

Let $\{r_j\}_{j\in\mathbb{N}} \subseteq (0,\infty)$ be such that $r_j \to \infty$ as $j \to \infty$. Then (9.155) implies that $\{f_{r_j}\}_{j\in\mathbb{N}}$ is a bounded sequence in $[M^{p,\lambda}(\partial\Omega,w)]^M$. As such, the sequential Banach–Alaoglu theorem (cf. (6.98)) allows us to assume without loss of generality that f_{r_j} is weak-$*$ convergent to some $f \in [M^{p,\lambda}(\partial\Omega,w)]^M$. Using (9.126) of Lemma 9.34 and (9.156), we find that for any test function ψ belonging to $[\mathrm{Lip}_c(\partial\Omega)]^M$, we have

$$\begin{aligned}\int_{\partial\Omega} \langle S_{\text{mod}} f, \psi\rangle\, d\sigma &= \lim_{j\to\infty} \int_{\partial\Omega} \langle S_{\text{mod}} f_{r_j}, \psi\rangle\, d\sigma \\ &= \lim_{j\to\infty} \int_{\partial\Omega} \langle g_{r_j} + c_{r_j}, \psi\rangle\, d\sigma \\ &= \lim_{j\to\infty} \int_{\partial\Omega} \langle \phi_{r_j} \cdot (g - g_{\Delta_{2r_j}}) + c_{r_j}, \psi\rangle\, d\sigma \\ &= \lim_{j\to\infty} \int_{\partial\Omega} \langle g - g_{\Delta_{2r_j}} + c_{r_j}, \psi\rangle\, d\sigma \\ &= \int_{\partial\Omega} \langle g, \psi\rangle\, d\sigma + \lim_{j\to\infty} \Big\langle c_{r_j} - g_{\Delta_{2r_j}}, \int_{\partial\Omega} \psi\, d\sigma \Big\rangle.\end{aligned} \tag{9.157}$$

The arbitrary nature of ψ allows us to conclude that $c_j - g_{\Delta_{2r_j}}$ converges to some constant $c \in \mathbb{C}^M$ as $j \to \infty$. Thus, (9.157) implies

$$\int_{\partial\Omega} \langle S_{\text{mod}} f, \psi\rangle \, d\sigma = \int_{\partial\Omega} \langle g + c, \psi\rangle \, d\sigma \tag{9.158}$$

for each $\psi \in [\text{Lip}_c(\partial\Omega)]^M$. Because the above equation holds for all test functions ψ, this actually shows that $S_{\text{mod}} f = g + c$ at σ-a. e. point on $\partial\Omega$ (cf. [86, Proposition 3.7.2, p. 281]). Hence, $[S_{\text{mod}}]f = [S_{\text{mod}} f] = [g]$. Given that $[g] \in [\dot{M}_1^{p,\lambda}(\partial\Omega, w)/\sim]^M$ is arbitrary, this shows that the operator in (9.143) is surjective. To demonstrate the necessary quantitative control, observe that from (9.155) we obtain

$$\begin{aligned} \|f\|_{[M^{p,\lambda}(\partial\Omega,w)]^M} &\le \limsup_{j\to\infty} \|f_{r_j}\|_{[M^{p,\lambda}(\partial\Omega,w)]^M} \\ &\le C\|\nabla_{\tan} g\|_{[M^{p,\lambda}(\partial\Omega,w)]^{n\cdot M}}. \end{aligned} \tag{9.159}$$

This is the desired estimate, completing the justification of item (2).

We now focus on (3). Suppose $f \in [M^{p,\lambda}(\partial\Omega, w)]^M$ is such that

$$[S_{\text{mod}}]f = [0] \in \left[\dot{M}_1^{p,\lambda}(\partial\Omega, w)\right]^M \tag{9.160}$$

We need to show that in fact $f = 0$. Because $[S_{\text{mod}}]f = [S_{\text{mod}} f] = [0]$, there exists $c \in \mathbb{C}^M$ such that

$$S_{\text{mod}} f = c \quad \text{at } \sigma\text{-a. e. point on } \partial\Omega. \tag{9.161}$$

Fix a coefficient tensor $A = (a_{jk}^{\alpha\beta})_{\substack{1\le j,k\le n \\ 1\le\alpha,\beta\le M}} \in \mathfrak{A}_L$. By Lemma 9.34 and (9.161), we find that

$$\left(\frac{1}{2}I + K_{A^\top}^{\#}\right)\left(\left(-\frac{1}{2}I + K_{A^\top}^{\#}\right)f\right) = 0. \tag{9.162}$$

According to Theorem 9.7, if we assume $\mathfrak{A}_{L^\top}^{\text{dis}} \neq \varnothing$ and arrange matters so that $A^\top \in \mathfrak{A}_{L^\top}^{\text{dis}}$, then there exists $\delta \in (0,1)$ such that when Ω is a δ-AR domain (i. e., $\|\nu\|_{[\text{BMO}(\partial\Omega,\sigma)]^n} < \delta$), then

$$\|K_{A^\top}^{\#}\|_{[M^{p,\lambda}(\partial\Omega,w)]^M \to [M^{p,\lambda}(\partial\Omega,w)]^M} < \frac{1}{2}, \tag{9.163}$$

and hence the operators $\pm\frac{1}{2}I + K_{A^\top}^{\#}$ are invertible on $[M^{p,\lambda}(\partial\Omega, w)]^M$ via Neumann series (cf. Lemma 11.2). Consequently, (9.162) entails $f = 0$. The linearity of the operator in (9.143) then shows that this mapping is injective.

Turning our attention to (4), note that if $\mathfrak{A}_L^{\text{dis}} \neq \varnothing$ and $\mathfrak{A}_{L^\top}^{\text{dis}} \neq \varnothing$, then items (2) and (3) guarantee that the operator (9.143) is a bijection. Given that $[\dot{M}_1^{p,\lambda}(\partial\Omega, w)/\sim]^M$ is a Banach space (cf. Proposition 8.15), it follows that the operator in (9.143) is a linear isomorphism.

For the treatment of (5), we can argue similarly as before. For item (1) in the context of Block spaces, we now must appeal to Proposition 9.33. To treat item (2) in this context, the proof remains largely unchanged, though notably we must appeal to Theorem 9.25,

Lemma 8.17, Proposition 9.26, Proposition 9.19, Theorem 7.2, and Lemma 9.34 formulated for Block spaces. To treat (3) in the context of Block spaces, we can use Lemma 9.34 formulated for Block spaces, as well as Theorem 9.23. Item (4) in the context of Block spaces then follows immediately. □

10 Generalized Banach function spaces

A Banach function space is a familiar structure in functional and harmonic analysis, and there is a significant theory that has been developed in various works, e. g., in the likes of [9] and the references therein. In fact, in [79] a theory of weighted Banach function spaces is discussed, and we also see the development of a Calderón–Zygmund theory on this brand of Banach function space. Subsequently, we see the natural application of the Calderón–Zygmund theory to various boundary value problems where the boundary datum is assumed to belong to a weighted Banach function space.

Despite being able to consider boundary value problems in such an abstract setting, function spaces of interest to us in this work like Muckenhoupt-weighted Morrey and Block spaces fall short of satisfying the necessary requirements to make them traditional Banach function spaces. However, we will define the notion of a generalized Banach function space, first introduced in [87]. Subsequently, we will see that Muckenhoupt-weighted Morrey and Block spaces nicely fit into this more inclusive functional analytic framework. Moreover, there is a solid body of work developed for generalized Banach function spaces such as an extrapolation theory, which we shall see offers an alternate proof of the result in Theorem 4.19. Not only that, but we will also show that astonishingly, we can obtain an extrapolation result at the level of Muckenhoupt-weighted Block spaces, a striking result whose direct proof has thus far been elusive.

10.1 Definitions and an extrapolation theorem

Throughout this exposition, we refer heavily to results established in [87, Section 5]. We begin with the following definition from [87, Definition 5.1.1, p. 218].

Definition 10.1. Let $(X, \mathfrak{M}, \mu)$ be an arbitrary measure space[1] and denote

$$\mathscr{M}(X,\mu) := \{f : X \to \mathbb{R} : f \text{ is } \mu\text{-measurable}\}, \tag{10.1}$$

$$\mathscr{M}_+(X,\mu) := \{f \in \mathscr{M}(X,\mu) : f \geq 0\}. \tag{10.2}$$

We call a mapping $\rho : \mathscr{M}_+(X,\mu) \to [0,\infty]$ a `function norm` provided the following conditions hold for every $f,\ g \in \mathscr{M}_+(X,\mu)$:

(P1) $\rho(f) = 0$ if and only if $f = 0$ at μ-a. e. point in X, for each $\lambda \in [0,\infty)$ one has $\rho(\lambda f) = \lambda\rho(f)$, and $\rho(f+g) \leq \rho(f) + \rho(g)$;

(P2) if $f \leq g$ at μ-a. e. point in X then $\rho(f) \leq \rho(g)$;

(P3) if $\{f_k\}_{k\in\mathbb{N}} \subset \mathscr{M}_+(X,\mu)$ is such that f_k increases to f pointwise μ-a. e. in X as $k \to \infty$, then $\rho(f_k)$ increases to $\rho(f)$ as $k \to \infty$;

1 Note that the measure μ is not required to be complete, nor even sigma-finite.

https://doi.org/10.1515/9783111458274-010

(P4) there exists a collection of sets $\{Y_j\}_{j\in\mathbb{N}} \subseteq \mathfrak{M}$ with the property that $X = \bigcup_{j=1}^{\infty} Y_j$ and $\rho(\mathbf{1}_{Y_j}) < \infty$ for all $j \in \mathbb{N}$;

(P5) there exists a collection of sets $\{Z_j\}_{j\in\mathbb{N}} \subseteq \mathfrak{M}$ with the property that $X = \bigcup_{j=1}^{\infty} Z_j$ and such that for every $j \in \mathbb{N}$ there exists some constant $c_j \in (0,\infty)$ satisfying $\int_{Z_j} f\, d\mu \le c_j \rho(f)$.

Remark 10.2. Working in the context of Definition 10.1, it is worth mentioning that (P1)–(P5) imply that the measure space $(X, \mathfrak{M}, \mu)$ is sigma-finite. Indeed, let $\{Y_j\}_{j\in\mathbb{N}}$ and $\{Z_j\}_{j\in\mathbb{N}}$ be as in (P4) and (P5), respectively. Then for all $j, k \in \mathbb{N}$, we have

$$\mu(Y_j \cap Z_k) = \int_{Z_k} \mathbf{1}_{Y_j}\, d\mu \le c_k \rho(\mathbf{1}) < +\infty, \tag{10.3}$$

where the first inequality follows from (P5) and the last from (P4). Finally, we see that

$$\bigcup_{j,k=1}^{\infty} (Y_j \cap Z_k) = \bigcup_{j=1}^{\infty}\left(Y_j \cap \left(\bigcup_{k=1}^{\infty} Z_k \right)\right) = \bigcup_{j=1}^{\infty} (Y_j \cap X) = \bigcup_{j=1}^{\infty} Y_j = X. \tag{10.4}$$

We are now ready to define a generalized Banach function space following [87, Definition 5.1.4, p. 219].

Definition 10.3. Let $(X, \mathfrak{M}, \mu)$ be an arbitrary measure space and let ρ be a function norm as in Definition 10.1. Then the set

$$\mathbb{X} := \{f \in \mathscr{M}(X,\mu) : \rho(|f|) < \infty\} \tag{10.5}$$

is called a `generalized Banach function space` on X. In this context, for each $f \in \mathscr{M}(X,\mu)$ define

$$\|f\|_{\mathbb{X}} := \rho(|f|) = \begin{cases} \rho(|f|) & \text{if } f \in \mathbb{X}, \\ +\infty & \text{if } f \notin \mathbb{X}. \end{cases} \tag{10.6}$$

In view of Definition 10.3, thanks to (10.6) we may alternatively characterize (10.5) as

$$\mathbb{X} = \{f \in \mathscr{M}(X,\mu) : \|f\|_{\mathbb{X}} < \infty\}. \tag{10.7}$$

It is immediately clear from (10.6), (P2) of Definition 10.1, and (10.7) that a generalized Banach function space satisfies a lattice property. That is,

$$\begin{gathered}\text{if } f, g : X \to \mathbb{R} \text{ are two } \mu\text{-measurable functions such that} \\ |f| \le |g| \text{ at } \mu\text{-a. e. point on } X \text{ and } g \in \mathbb{X}, \text{ then } f \text{ also belongs} \\ \text{to the space } \mathbb{X} \text{ and } \|f\|_{\mathbb{X}} \le \|g\|_{\mathbb{X}}.\end{gathered} \tag{10.8}$$

It then follows that

the operation of pointwise multiplication by an essentially bounded function $b \in L^\infty(X,\mu)$ is a bounded mapping from the space $\mathbb{X}$ into itself, with operator norm $\leq \|b\|_{L^\infty(X,\mu)}$. (10.9)

Indeed, for an arbitrary μ-measurable function f on X, we have the inequality $|bf| \leq \|b\|_{L^\infty(X,\mu)}|f|$ at μ-a. e. point on X. As such, (10.8), (10.6), and (P1) of Definition 10.1 show that $\|bf\|_{\mathbb{X}} \leq \|b\|_{L^\infty(X,\mu)}\|f\|_{\mathbb{X}}$ for all $f \in \mathscr{M}(X,\mu)$.

Remark 10.4. The requirements (P4) and (P5) are what separate what we call generalized Banach function spaces from their classically defined counterparts as is done, e. g., in [9, Definition 1.1, p. 2]. Namely, the function norm ρ associated to a *classical* Banach function space on (X,μ) satisfies (P1)–(P3) from Definition 10.1, but instead of (P4)–(P5) ρ is asked to satisfy the more restrictive requirements:

(P4') for every μ-measurable set E with $\mu(E) < +\infty$ one has $\rho(\mathbf{1}_E) < +\infty$;

(P5') for each μ-measurable set E with $\mu(E) < +\infty$ there exists a constant $C_E \in (0,\infty)$ such that $\int_X f\,d\mu \leq C_E\rho(f)$ for every $f \in \mathscr{M}_+(X,\mu)$.

Note even in the simple case of $X := \mathbb{R}$ and $\mu := \mathcal{L}^1$ that Muckenhoupt-weighted Lebesgue spaces fail to be *classical* Banach function spaces. For instance, take $p := 3$, $w := |\cdot| \in A_p(\mathbb{R},\mathcal{L}^1)$ (cf. (1.41), which guarantees this membership), and set $\rho := \|\cdot\|_{L^3(\mathbb{R},w)}$. We quickly find that w cannot possibly satisfy (P4') by taking $E := \bigcup_{j=1}^\infty [j, j+j^{-2})$. Indeed, we see that $\mathcal{L}^1(E) = \sum_{j=1}^\infty j^{-2} = \pi^2/6 < +\infty$, and yet

$$\rho(\mathbf{1}_E)^3 = \int_E |x|\,dx = \sum_{j=1}^\infty \int_j^{j+j^{-2}} x\,dx = \frac{1}{2}\sum_{j=1}^\infty (2j^{-1} + j^{-4}) > \sum_{j=1}^\infty j^{-1} = \infty. \tag{10.10}$$

Note that by similar reasoning as above (with slight modifications to the choice of weight w and the set E) it can be shown that $\|\cdot\|_{L^p(\mathbb{R},w)}$ fails to satisfy (P4') for any $p \in (1,\infty)$, preventing $L^p(\mathbb{R}^n, w)$ from being a classical Banach function space for all $p \in (1,\infty)$.

It is also shown in [87, Proposition 5.1.8, p. 222] that as the name suggests, generalized Banach function spaces are Banach spaces when equipped with the norm defined in (10.6).

Proposition 10.5. *Let $(\mathbb{X}, \|\cdot\|_{\mathbb{X}})$ be a generalized Banach function space associated with a measure space $(X,\mathfrak{M},\mu)$ as in Definition 10.3. Then the space $(\mathbb{X}, \|\cdot\|_{\mathbb{X}})$ is complete, and hence Banach.*

At this point, we introduce two subspaces of generalized Banach function spaces, stemming from the work of W. Luxemburg [78], the first of which is based on the definition below.

Definition 10.6. Let $(X, \mathfrak{M}, \mu)$ be a measure space and $\mathbb{X}$ a generalized Banach function apace on (X, μ). Denote by $\{Y_j\}_{j\in\mathbb{N}}$ the collection of subsets of X as in (P4) of Definition 10.1, and arrange matters so that this sequence of sets is increasing (cf. [87, Remark 5.1.2, p. 218]). A function $f \in \mathbb{X}$ is said to have an `absolutely continuous norm` if the following conditions are satisfied:

(a) if $\{E_j\}_{j\in\mathbb{N}} \subseteq \mathfrak{M}$ is such that there exists $j_* \in \mathbb{N}$ with the property that $E_j \subseteq Y_{j_*}$ for all $j \in \mathbb{N}$ and $\mu(E_j) \to 0$ as $j \to \infty$, then $\|f\mathbf{1}_{E_j}\|_{\mathbb{X}} \to 0$ as $j \to \infty$;[2]

(b) one has $\|f\mathbf{1}_{X\setminus Y_j}\|_{\mathbb{X}} \to 0$ as $j \to \infty$.

We are now in a position to define $\mathbb{X}_a$ as the subset of $\mathbb{X}$ containing all functions $f \in \mathbb{X}$, which have an absolutely continuous norm. In fact, [78, Theorem 1, p. 13] ensures that

$$\mathbb{X}_a \text{ is a closed subspace of } \mathbb{X}. \tag{10.11}$$

Remark 10.7. Retain the context of Definition 10.6. More in line with the definition presented in [9, Definition 3.1, p. 14], one can also define $\mathbb{X}_a$ as the collection of all functions $f \in \mathbb{X}$ with the property that $\|f\mathbf{1}_{E_j}\|_{\mathbb{X}} \to 0$ as $j \to \infty$ whenever $\{E_j\}_{j\in\mathbb{N}} \subseteq \mathfrak{M}$ is such that $\mathbf{1}_{E_j} \to 0$ μ-a. e. on X as $j \to \infty$. To see the equivalence of these characterizations of $\mathbb{X}_a$, we recall [78, Lemma 1, pp. 12–13], which says that a function $f \in \mathbb{X}$ has absolutely continuous norm if and only if the following conditions are satisfied:

(i) given $\varepsilon > 0$, there exists a number $\delta > 0$ such that $\mu(E) < \delta$ implies $\|f\mathbf{1}_E\|_{\mathbb{X}} < \varepsilon$;

(ii) if the sequence of sets $\{E_j\}_{j\in\mathbb{N}} \subseteq \mathfrak{M}$ is such that $\mathbf{1}_{E_j} \to 0$ pointwise μ-a.e. on X as $j \to \infty$, then $\|f\mathbf{1}_{E_j}\|_{\mathbb{X}} \to 0$ as $j \to \infty$.

Subsequently, we glean from the proof of [9, Lemma 3.4, p. 15] that (ii) implies (i), justifying the equivalence of definitions for an absolutely continuous norm.

The next subspace of generalized Banach function spaces we wish to consider is as follows; cf. [78, Definition 2, p. 15].

Definition 10.8. Let $(X, \mathfrak{M}, \mu)$ be a measure space and $\mathbb{X}$ a generalized Banach function space on (X, μ). Denote by $\{Y_j\}_{j\in\mathbb{N}}$ the collection of subsets of X as in (P4) of Definition 10.1, and arrange matters so that this sequence of sets is increasing (cf. [87, Remark 5.1.2, p. 218]). Denote by B the collection of all bounded, μ-measurable functions $f : \mathbb{X} \to \mathbb{C}$

2 To be more consistent with our treatment of generalized Banach function spaces, this condition differs slightly from what is presented in [78, Definition 1, p. 12], though it is not difficult to see that this reformulation of the classical definition is equivalent.

with the property that the set on which $f \neq 0$ is contained in Y_{j_*} for some $j_* \in \mathbb{N}$. Define the subspace $\mathbb{X}_b$ of $\mathbb{X}$ as the closure (in the norm topology on $\mathbb{X}$) of B in $\mathbb{X}$.

Significantly, (in the context of Definitions 10.6 and 10.8) one always has

$$\mathbb{X}_a \subseteq \mathbb{X}_b \tag{10.12}$$

as recorded in [78, Lemma 4, p. 16]. We will explore the spaces $\mathbb{X}_a$ and $\mathbb{X}_b$ in the context of Muckenhoupt-weighted Morrey spaces later in Proposition 10.52.

Generalized Banach function spaces have nice duality properties. For a standard Banach function space, these "dual" spaces are the so-called *Köthe duals*. We would like to develop an analogous notion for generalized Banach function spaces. The first step in doing so is to introduce the following function norm, which is defined in relation to a previously given function norm in the following way.

Definition 10.9. Let $(X, \mathfrak{M}, \mu)$ be a measure space and let ρ be a function norm as in Definition 10.1. Its `associated function norm` is the function $\rho' : \mathscr{M}_+(X, \mu) \to [0, \infty]$, defined according to

$$\rho'(g) := \sup\left\{\int_X fg \, d\mu : f \in \mathscr{M}_+(X, \mu),\ \rho(f) \leq 1\right\} \quad \text{for all } g \in \mathscr{M}_+(X, \mu), \tag{10.13}$$

which itself is a function norm in the context of Definition 10.1; cf. [87, Proposition 5.1.10, p. 223].

Remark 10.10. If $(X, \mathfrak{M}, \mu)$ is actually a sigma-finite measure space with a function norm ρ in the sense of Definition 10.1, then properties (P1)–(P4) for ρ imply (P5). To see this, note that [75, Proposition 3.2, p. 9][3] (whose applicability is ensured by the fact that $(X, \mathfrak{M}, \mu)$ is presently assumed sigma-finite) implies that the associated function norm ρ' from Definition 10.9 is also a function norm satisfying (P1)–(P4). Choose $\{Y_j'\}_{j\in\mathbb{N}}$ to be the family of sets in the formulation of (P4) for the associated function norm ρ'. Select some $f \in \mathscr{M}_+(X, \mu)$, and consider three cases.

Case 1: $0 < \rho(f) < \infty$. We have

$$\int_{Y_j'} f \, d\mu = \int_X f \cdot \mathbf{1}_{Y_j'} = \rho(f) \int_X \frac{f}{\rho(f)} \cdot \mathbf{1}_{Y_j'} \, d\mu \leq \rho(f)\rho'(\mathbf{1}_{Y_j'}) < +\infty, \tag{10.14}$$

so that ρ satisfies (P5) with $Z_j := Y_j'$ for each $j \in \mathbb{N}$ and any constant $c_j \in (\rho'(\mathbf{1}_{Y_j'}), \infty)$.

Case 2: $\rho(f) = \infty$. It is clear that (P5) holds for ρ with $Z_j := Y_j'$ for all $j \in \mathbb{N}$ and any choice of $c_j \in (0, \infty)$.

---

3 Instead of working with (P4), the authors assume what they call the *saturation property*. However, these notions are seen to be equivalent by [75, Proposition 2.5, p. 6] and [87, Remark 5.1.2, p. 218].

Case 3: $\rho(f) = 0$. Notice by (P1) that $f = 0$ at μ-a. e. point on X. Therefore, $\int_{Y_j'} f\, d\mu = 0$, so that any choice of $c_j \in (0, \infty)$ will ensure (P5) holds for ρ with $Z_j := Y_j'$ for all $j \in \mathbb{N}$.

With this associated function norm in hand, we can define what we call the associated space, which we will see acts as a kind of dual to a generalized Banach function space.

Definition 10.11. Let $(X, \mathfrak{M}, \mu)$ be a measure space, ρ a function norm as in Definition 10.1, and ρ' its associated function norm as in Definition 10.9. Then the generalized Banach function space on (X, μ) defined in relation to ρ' according to

$$\mathbb{X}' := \{g \in \mathscr{M}(X, \mu) : \rho'(|g|) < \infty\} \tag{10.15}$$

is called the associated space of $\mathbb{X}$ (defined in relation to ρ as in Definition 10.3).

Going back to (10.6), but implemented for ρ' in place of ρ, the formula given in (10.13) and (10.5), show that for every $g \in \mathscr{M}(X, \mu)$,

$$\begin{aligned} \|g\|_{\mathbb{X}'} = \rho'(|g|) &= \sup\left\{ \int_X f|g|\, d\mu : f \in \mathscr{M}_+(X, \mu),\ \rho(f) \le 1 \right\} \\ &= \sup\left\{ \int_X |fg|\, d\mu : f \in \mathbb{X},\ \|f\|_{\mathbb{X}} \le 1 \right\}. \end{aligned} \tag{10.16}$$

Subsequently, (10.15) may be recast as

$$\mathbb{X}' = \{g \in \mathscr{M}(X, \mu) : \|g\|_{\mathbb{X}'} < \infty\}. \tag{10.17}$$

In an analogous way to the duality pairing between two Lebesgue spaces, for a given generalized Banach function space $\mathbb{X}$ and its associated space $\mathbb{X}'$ there exists a generalized version of Hölder's inequality; cf. [87, Proposition 5.1.12, p. 226].

Proposition 10.12. *Let $\mathbb{X}$ be a generalized Banach function space on a measure space $(X, \mathfrak{M}, \mu)$ with associated space $\mathbb{X}'$. Then for every $f \in \mathbb{X}$ and every $g \in \mathbb{X}'$ the function fg is absolutely integrable on X and*

$$\int_X |fg|\, d\mu \le \|f\|_{\mathbb{X}} \|g\|_{\mathbb{X}'}. \tag{10.18}$$

We now present a version of the classical Lorentz–Luxemburg theorem (cf. [9, Theorem 2.7, p. 10]), but for generalized Banach function spaces. This is a kind of reflexivity property, which says that the associated space of an associated space is the original generalized Banach function space. This is in stark contrast to Banach spaces and their topological duals, which need not be reflexive. See [87, Proposition 5.1.14, p. 226] for a proof of the following proposition.

Proposition 10.13. *Let* $\mathbb{X}$ *be a generalized Banach function space on a measure space* $(X, \mathfrak{M}, \mu)$*. Assume* μ *is sigma-finite. Then* $\mathbb{X}'' = \mathbb{X}$ *isometrically, i. e.,* $\mathbb{X}$ *coincides with its second associated space* $\mathbb{X}'' := (\mathbb{X}')'$ *and* $\|f\|_{\mathbb{X}} = \|f\|_{\mathbb{X}''}$ *for each* $f \in \mathbb{X}$*. In fact, one has*

$$\|f\|_{\mathbb{X}} = \|f\|_{\mathbb{X}''} \quad \text{for each } f \in \mathscr{M}(X, \mu). \tag{10.19}$$

Thanks to Proposition 10.13, we obtain the following corollary, originally given in [86, Corollary 5.1.15, p. 231], which gives an alternate characterization for a norm on a given generalized Banach function space in terms of its associated space. We will see later in this section how beneficial it is to be able to use this result freely.

Corollary 10.14. *Let* $\mathbb{X}$ *be a generalized Banach function space on a measure space* $(X, \mathfrak{M}, \mu)$*. Assume that the measure* μ *is sigma-finite. Then for each* $f \in \mathscr{M}(X, \mu)$ *one has*

$$\|f\|_{\mathbb{X}} = \sup\left\{ \int_X |fg|\, d\mu : g \in \mathbb{X}',\ \|g\|_{\mathbb{X}'} \le 1 \right\}. \tag{10.20}$$

Also, [87, Proposition 5.1.16, p. 231] provides these other equivalent descriptions of the norms in a given generalized Banach function space along with its associated space. As with Corollary 10.14, we record these results now, so that we may use them later without reservation.

Proposition 10.15. *Let* $(\mathbb{X}, \|\cdot\|_{\mathbb{X}})$ *be a generalized Banach function space associated as in Definition* 10.3 *with a measure space* $(X, \mathfrak{M}, \mu)$*. Consider its associated space* $\mathbb{X}'$ *from Definition* 10.11 *with norm defined in* (10.16)*. Then*

$$\|g\|_{\mathbb{X}'} = \sup\left\{ \left| \int_X fg\, d\mu \right| : f \in \mathbb{X},\ \|f\|_{\mathbb{X}} \le 1 \right\} \quad \text{for all } g \in \mathbb{X}' \tag{10.21}$$

and if the measure μ *is sigma-finite one also has*

$$\|f\|_{\mathbb{X}} = \sup\left\{ \left| \int_X fg\, d\mu \right| : g \in \mathbb{X}',\ \|g\|_{\mathbb{X}'} \le 1 \right\} \quad \text{for all } f \in \mathbb{X}. \tag{10.22}$$

We now turn to an extrapolation theory for generalized Banach function spaces. Rubio de Francia's extrapolation theorem, a version of which we have recorded in Proposition 2.41, is a highly versatile piece of mathematical machinery. In fact, an extrapolation result at the level of Muckenhoupt-weighted Morrey spaces (cf. Theorem 4.19) was the driving force behind the development of a Calderón–Zygmund theory for this brand of Morrey space throughout Chapter 5. As it turns out, an extrapolation result at the level of generalized Banach function spaces is also possible. Loosely speaking, estimates at the level of Muckenhoupt-weighted Lebesgue spaces for a single integrability exponent and all corresponding Muckenhoupt weights imply estimates at the level of generalized Banach function spaces. In order for this to go through there is a slight caveat. Not just any generalized Banach function space will work; the Hardy–Littlewood maximal operator must be bounded on both the generalized Banach function space and its associated space. Though, this is a very reasonable assumption, given the nature of the result and the significance of the Hardy–Littlewood maximal operator in harmonic analysis and its connection to Muckenhoupt weights. Here is the extrapolation result, originally presented in [87, Theorem 5.2.1, pp. 223–224].

Theorem 10.16. *Let (X, d, μ) be a space of homogeneous type with the property that the quasi-distance $d : X \times X \to [0, \infty)$ is continuous in the product topology $\tau_d \times \tau_d$. Also, denote by $\mathcal{M}_X$ the Hardy–Littlewood maximal operator on (X, d, μ), which acts on each μ-measurable function $f : X \to \mathbb{C}$ according to*

$$(\mathcal{M}_X f)(x) := \sup_{r>0} \fint_{B_d(x,r)} |f| \, d\mu \quad \text{for each } x \in X. \tag{10.23}$$

Given a generalized Banach function space $\mathbb{X}$ on (X, μ), denote by $\mathbb{X}'$ its associated space as in Definition 10.3 *and Definition* 10.11.

Then there exists some $C_X \in (0, \infty)$, depending only on the quasidistance d and the doubling charter of μ, such that the following statements are true:

(1) *Fix some integrability exponent $p_0 \in (1, \infty)$ and, assuming*

$$\begin{aligned} &\mathcal{M}_X : \mathbb{X} \to \mathbb{X} \quad \text{and} \quad \mathcal{M}_X : \mathbb{X}' \to \mathbb{X}' \\ &\text{are well-defined bounded mappings,} \end{aligned} \tag{10.24}$$

introduce

$$\begin{aligned} W_{\mathbb{X},p_0} &:= C_X^{p_0} \cdot \|\mathcal{M}_X\|_{\mathbb{X}\to\mathbb{X}}^{p_0-1} \cdot \|\mathcal{M}_X\|_{\mathbb{X}'\to\mathbb{X}'} \\ \text{and} \quad W_{\mathbb{X}',p_0} &:= C_X^{p_0} \cdot \|\mathcal{M}_X\|_{\mathbb{X}'\to\mathbb{X}'}^{p_0-1} \cdot \|\mathcal{M}_X\|_{\mathbb{X}\to\mathbb{X}}. \end{aligned} \tag{10.25}$$

Also, suppose f, g are two μ-measurable real-valued functions defined on X, which satisfy for each Muckenhoupt weight $w \in A_{p_0}(X, d, \mu)$ with $[w]_{A_{p_0}} \le W_{\mathbb{X},p_0}$,

$$\|f\|_{L^{p_0}(X,w)} \le C_w \|g\|_{L^{p_0}(X,w)} \tag{10.26}$$

for some constant $C_w \in (0, \infty)$, which depends only on the ambient (via d, μ, p_0), and w. Then

$$\|f\|_{\mathbb{X}} \le 2^{2-1/p_0} \cdot \Big(\sup_{\substack{w \in A_{p_0}(X,d,\mu) \\ [w]_{A_{p_0}} \le W_{\mathbb{X},p_0}}} C_w \Big) \|g\|_{\mathbb{X}}. \tag{10.27}$$

Moreover, if (10.26) *holds for each Muckenhoupt weight w belonging to $A_{p_0}(X, d, \mu)$ with $[w]_{A_{p_0}} \le W_{\mathbb{X}',p_0}$, then*

$$\|f\|_{\mathbb{X}'} \le 2^{2-1/p_0} \cdot \Big(\sup_{\substack{w \in A_{p_0}(X,d,\mu) \\ [w]_{A_{p_0}} \le W_{\mathbb{X}',p_0}}} C_w \Big) \|g\|_{\mathbb{X}'}. \tag{10.28}$$

(2) *Assume*

$$\mathcal{M}_X : \mathbb{X}' \to \mathbb{X}' \quad \textit{is a well-defined bounded mapping,} \tag{10.29}$$

and define

$$W_{\mathbb{X},1} := C_X \cdot \|\mathcal{M}_X\|_{\mathbb{X}' \to \mathbb{X}'}. \tag{10.30}$$

In addition, suppose f, g are two μ-measurable real-valued functions defined on X with the property that for each Muckenhoupt weight w belonging to $A_1(X, d, \mu)$ with $[w]_{A_1} \le W_{\mathbb{X},1}$ one has

$$\|f\|_{L^1(X,w)} \le C_w \|g\|_{L^1(X,w)} \tag{10.31}$$

for some constant $C_w \in (0, \infty)$, which depends only on the ambient (via d, μ), and w. Then

$$\|f\|_{\mathbb{X}} \le 2 \cdot \Big(\sup_{\substack{w \in A_1(X,d,\mu) \\ [w]_{A_1} \le W_{\mathbb{X},1}}} C_w \Big) \|g\|_{\mathbb{X}}. \tag{10.32}$$

(3) *Assume*

$$\mathcal{M}_X : \mathbb{X} \to \mathbb{X} \quad \textit{is a well-defined bounded mapping,} \tag{10.33}$$

and define

$$W_{\mathbb{X}',1} := C_X \cdot \|\mathcal{M}_X\|_{\mathbb{X} \to \mathbb{X}}. \tag{10.34}$$

In addition, suppose f, g are two μ-measurable real-valued functions defined on X with the property that for each Muckenhoupt weight w belonging to $A_1(X, d, \mu)$ with $[w]_{A_1} \le W_{\mathbb{X}',1}$ one has

$$\|f\|_{L^1(X,w)} \le C_w \|g\|_{L^1(X,w)} \tag{10.35}$$

for some constant $C_w \in (0, \infty)$, which depends only on the ambient (via d, μ), and w. Then

$$\|f\|_{\mathbb{X}'} \le 2 \cdot \Big(\sup_{\substack{w \in A_1(X,d,\mu) \\ [w]_{A_1} \le W_{\mathbb{X}',1}}} C_w \Big) \|g\|_{\mathbb{X}'}. \tag{10.36}$$

The boundedness of the Hardy–Littlewood maximal operator on a generalized Banach function space and its associated space is crucial for the kind of extrapolation result contained in Theorem 10.16. As a consequence of the above extrapolation result, we find that such generalized Banach function spaces are closely related to the class of Muckenhoupt-weighted Lebesgue spaces via the embedding result below (cf. [87, Corollary 5.2.2, p. 240]). Interestingly, despite the very general nature of these spaces, any function belonging to a generalized Banach function space (provided it along with its associated space are stable under the action of the Hardy–Littlewood maximal operator) necessarily belongs to some Muckenhoupt-weighted Lebesgue space.

Corollary 10.17. *Assume (X, d, μ) is a space of homogeneous type with the property that the quasidistance $d : X \times X \to [0, \infty)$ is continuous in the product topology $\tau_d \times \tau_d$. Let $\mathcal{M}_X$ be the Hardy–Littlewood maximal operator on (X, d, μ) as in* (10.23). *Finally, suppose $\mathbb{X}$ is a generalized Banach function space on (X, μ) and denote by $\mathbb{X}'$ its associated space (cf. Definition* 10.3 *and Definition* 10.11*).*

Then there exists some $C_X \in (0, \infty)$, depending only on the quasidistance d and the doubling character of μ, such that

$$\mathcal{M}_X : \mathbb{X}' \to \mathbb{X}' \quad \textit{boundedly} \quad \Longrightarrow \quad \mathbb{X} \subseteq \bigcup_{\substack{w \in A_1(X,d,\mu) \\ [w]_{A_1} \le W_{\mathbb{X},1}}} L^1(X, w\mu), \tag{10.37}$$

where

$$W_{\mathbb{X},1} := C_X \cdot \|\mathcal{M}_X\|_{\mathbb{X}' \to \mathbb{X}'} \tag{10.38}$$

and

$$\mathcal{M}_X : \mathbb{X} \to \mathbb{X} \quad \textit{boundedly} \quad \Longrightarrow \quad \mathbb{X}' \subseteq \bigcup_{\substack{w \in A_1(X,d,\mu) \\ [w]_{A_1} \le W_{\mathbb{X}',1}}} L^1(X, w\mu) \tag{10.39}$$

where

$$W_{\mathbb{X}',1} := C_X \cdot \|\mathcal{M}_X\|_{\mathbb{X} \to \mathbb{X}}. \tag{10.40}$$

Finally, if

$$\begin{gathered} \mathcal{M}_X : \mathbb{X} \to \mathbb{X} \quad \textit{and} \quad \mathcal{M}_X : \mathbb{X}' \to \mathbb{X}' \\ \textit{are well-defined bounded mappings,} \end{gathered} \tag{10.41}$$

then for each $p_0 \in (1, \infty)$ one has

$$\mathbb{X} \subseteq \bigcup_{\substack{w \in A_{p_0}(X,d,\mu) \\ [w]_{A_{p_0}} \le W_{\mathbb{X},p_0}}} L^{p_0}(X, w\mu) \quad \textit{and} \quad \mathbb{X}' \subseteq \bigcup_{\substack{w \in A_{p_0}(X,d,\mu) \\ [w]_{A_{p_0}} \le W_{\mathbb{X}',p_0}}} L^{p_0}(X, w\mu) \tag{10.42}$$

where

$$\begin{aligned} W_{\mathbb{X},p_0} &:= C_X^{p_0} \cdot \|\mathcal{M}_X\|_{\mathbb{X}\to\mathbb{X}}^{p_0-1} \cdot \|\mathcal{M}_X\|_{\mathbb{X}'\to\mathbb{X}'} \\ \text{and} \quad W_{\mathbb{X}',p_0} &:= C_X^{p_0} \cdot \|\mathcal{M}_X\|_{\mathbb{X}'\to\mathbb{X}'}^{p_0-1} \cdot \|\mathcal{M}_X\|_{\mathbb{X}\to\mathbb{X}}. \end{aligned} \tag{10.43}$$

We will show how the Muckenhoupt-weighted Morrey and Block spaces developed in Chapters 4 and 6 fit into the framework of generalized Banach function spaces. In particular, we will be able to apply Theorem 10.16 to Muckenhoupt-weighted Block spaces to obtain an extrapolation result in the spirit of Theorem 4.19, but formulated for Muckenhoupt-weighted Block spaces. Similarly, we will be able to deduce a similar embedding result to that of Lemma 5.17, but formulated for weighted Block spaces by appealing to Corollary 10.17.

10.2 Weighted vanishing Morrey spaces

At this point, we now take what seems like a slight detour in our program concerning generalized Banach function spaces. Namely, we would like to develop a certain flavor of Muckenhoupt-weighted *vanishing* Morrey space in analogy with developments in [79, Section 7.1] and [87, Section 6.2] for the unweighted case. While these spaces are interesting in their own right from a functional analytic point of view, we find them particularly useful for identifying Muckenhoupt-weighted Block spaces as generalized Banach function spaces. This will be established only after characterizing our version of Muckenhoupt-weighted vanishing Morrey spaces as the pre-duals of Muckenhoupt-weighted Block spaces.

Consider a closed ADR set $\Sigma \subseteq \mathbb{R}^n$ and write $\sigma := \mathcal{H}^{n-1}\lfloor\Sigma$. Fix an integrability exponent $p \in (1,\infty)$ and a threshold $\lambda \in (0, n-1)$. The *unweighted* vanishing Morrey space $\mathring{M}^{p,\lambda}(\Sigma,\sigma)$ is a closed subspace of $M^{p,\lambda}(\Sigma,\sigma)$, defined as the closure of the *unweighted* Lebesgue space $L^s(\Sigma,\sigma)$ in $M^{p,\lambda}(\Sigma,\sigma)$, where $s := \frac{p(n-1)}{n-1-\lambda}$. The fact that $L^s(\Sigma,\sigma)$ continuously embeds into $M^{p,\lambda}(\Sigma,\sigma)$ has been noted earlier in Remark 4.18, however, a direct argument based on Hölder's inequality is as follows. For any σ-measurable function $f : \Sigma \to \mathbb{R}$, and any surface ball $\Delta \subseteq \Sigma$ with radius $r \in (0, 2\,\mathrm{diam}(\Sigma))$, we have

$$\begin{aligned} \Big(\frac{1}{r^\lambda}\int_\Delta |f|^p\,d\sigma\Big)^{1/p} &\le Cr^{(n-1-\lambda)/p}\Big(⨍_\Delta |f|^p\,d\sigma\Big)^{1/p} \\ &\le Cr^{(n-1-\lambda)/p}\Big(⨍_\Delta |f|^s\,d\sigma\Big)^{1/s} \\ &\le Cr^{(n-1-\lambda)/p}r^{-(n-1)/s}\Big(\int_\Sigma |f|^s\,d\sigma\Big)^{1/s} \\ &= C\|f\|_{L^s(\Sigma,\sigma)}. \end{aligned} \tag{10.44}$$

Note that the first and third inequality use the Ahlfors regularity of Σ, and that the constant $C \in (0,\infty)$ depends on p and the Ahlfors regularity constants of Σ. Hölder's inequality is permissible to obtain the second inequality because our choice of s and the fact that $\lambda \in (0, n-1)$ ensures $s > p$. Since Hölder's inequality in this context holds for any $s > p$, it is not until the final step where we see the significance of our choice of s as it rids any dependence on r in the final constant. Thus, taking a supremum over all surface balls on both sides of (10.44) gives

$$\|f\|_{M^{p,\lambda}(\Sigma,\sigma)} \leq C\|f\|_{L^s(\Sigma,\sigma)} \tag{10.45}$$

for some constant $C \in (0,\infty)$, depending only on Σ, n, p, and λ. See [79, Section 7.1] and [86, Section 6.2] for a solid functional analytic treatment of unweighted vanishing Morrey spaces.

Working with an Ahlfors regular set $\Sigma \subseteq \mathbb{R}^n$, abbreviating $\sigma := \mathcal{H}^{n-1}\lfloor\Sigma$ as usual, fix an integrability exponent $p \in (1,\infty)$, a Muckenhoupt weight $w \in A_p(\Sigma,\sigma) \cap \mathrm{RH}_q(\Sigma,\sigma)$, and a threshold $\lambda \in (0,(n-1)/q')$, where q' is the Hölder conjugate of q. Our goal is to establish a Muckenhoupt-weighted version of the vanishing Morrey space $\mathring{M}^{p,\lambda}(\Sigma,\sigma)$. There is one immediate difficulty with translating the definition of an unweighted vanishing Morrey space to the weighted case. Namely, unless w is a constant, the space $L^s(\Sigma,\sigma)$ does not cleanly embed into $M^{p,\lambda}(\Sigma,w)$ for any choice of $s \in (1,\infty)$. The best one can prove is a crude containment of the space of functions from $L^s(\Sigma,\sigma)$ with compact support in $M^{p,\lambda}(\Sigma,w)$ when s is suitably large, as the next result shows. Further troubling, while the boundedness of singular integral operators of Calderón–Zygmund type on the unweighted vanishing Morrey space $\mathring{M}^{p,\lambda}(\Sigma,\sigma)$ comes essentially for free based off the boundedness of such operators in Lebesgue spaces and unweighted Morrey spaces, if we define the *weighted* vanishing Morrey space as the closure of $L^s_{\mathrm{comp}}(\Sigma,\sigma)$ in $M^{p,\lambda}(\Sigma,w)$ for a suitable choice of $s \in (1,\infty)$, we cannot achieve the boundedness of such operators in as straightforward manner for the simple fact that these operators are not bounded on $L^s_{\mathrm{comp}}(\Sigma,\sigma)$. Thus, accommodating a weight in this case becomes a decidedly more challenging task.

Proposition 10.18. *Let $\Sigma \subseteq \mathbb{R}^n$ be a closed Ahlfors regular set, and write $\sigma := \mathcal{H}^{n-1}\lfloor\Sigma$. Fix two arbitrary integrability exponents $p, q \in (1,\infty)$, along with a Muckenhoupt weight $w \in A_p(\Sigma,\sigma) \cap \mathrm{RH}_q(\Sigma,\sigma)$, and select a threshold $\lambda \in (0,(n-1)/q')$, where $q' := q/(q-1)$. Lastly, define $s_0 := \frac{(n-1)pq'}{n-1-\lambda q'}$.*

Then there exists a constant $C \in (0,\infty)$, depending only on the ADR *character of Σ, n, p, q, λ, $[w]_{\mathrm{RH}_q}$, and $[w]_{A_p}$ such that*

$$\left(\frac{1}{r^\lambda}\int_{\Delta_r}|f|^p\,dw\right)^{1/p} \leq Cw(\Delta_r)^{1/p}r^{-(n-1)/(pq')}\|f\|_{L^{s_0}(\Sigma,\sigma)}, \tag{10.46}$$

for all σ-measurable functions f defined on Σ and surface balls $\Delta_r \subseteq \Sigma$ with radius $r \in (0, 2\,\mathrm{diam}(\Sigma))$.

Moreover, whenever $x_0 \in \Sigma$, $R \in (0,\infty)$, *and* f *is a* σ*-measurable function with* $\operatorname{supp} f \subseteq \Delta(x_0, R)$, *one has*

$$\|f\|_{M^{p,\lambda}(\Sigma,w)} \le Cw\big(\Delta(x_0,R)\big)^{1/p} R^{-(n-1)/(pq')} \|f\|_{L^{s_0}(\Sigma,\sigma)} \tag{10.47}$$

for some constant $C \in (0,\infty)$, *depending exclusively on the* ADR *character of* Σ, n, p, q, λ, $[w]_{\mathrm{RH}_q}$, *and* $[w]_{A_p}$. *Consequently,*

$$L^s_{\mathrm{comp}}(\Sigma,\sigma) \hookrightarrow M^{p,\lambda}(\Sigma,w) \quad \textit{continuously for all } s \in [s_0,\infty]. \tag{10.48}$$

Proof. We focus on establishing (10.46) as it implies (10.47), which itself implies (10.48). Fix $f \in L^{s_0}(\Sigma,\sigma)$ and let $\Delta_r \subseteq \Sigma$ be a surface ball with radius $r \in (0, 2\operatorname{diam}(\Sigma))$. Then

$$\begin{aligned}
\Big(\frac{1}{r^\lambda}\int_{\Delta_r} |f|^p\,dw\Big) &= w(\Delta_r)^{1/p} r^{-\lambda/p}\Big(\fint_{\Delta_r} |f|^p\,dw\Big)^{1/p} \\
&\le [w]_{\mathrm{RH}_q} w(\Delta_r)^{1/p} r^{-\lambda/p}\Big(\fint_{\Delta_r} |f|^{pq'}\,d\sigma\Big)^{1/(pq')} \\
&\le Cw(\Delta_r)^{1/p} r^{-\lambda/p}\Big(\fint_{\Delta_r} |f|^{s_0}\,d\sigma\Big)^{1/s_0} \\
&\le Cw(\Delta_r)^{1/p} r^{-\lambda/p} r^{-(n-1)/s_0}\Big(\int_{\Delta_r} |f|^{s_0}\,d\sigma\Big)^{1/s_0} \\
&\le Cw(\Delta_r)^{1/p} r^{-(n-1)/(pq')} \|f\|_{L^{s_0}(\Sigma,\sigma)}.
\end{aligned} \tag{10.49}$$

The first inequality is due to (2.64). The second inequality is a consequence of Hölder's inequality, which is applicable because $s_0 \ge pq'$. Indeed, the fact that $\lambda \in (0,(n-1)/q')$ ensures $0 < \lambda q' < n-1$, and hence $n-1-\lambda q' < n-1$ or equivalently $\frac{n-1}{n-1-\lambda q'} > 1$. The third inequality follows from the Ahlfors regularity of Σ, and the final inequality is due to our choice of s_0 as witnessed below

$$\begin{aligned}
-\frac{\lambda}{p} - \frac{n-1}{s_0} &= -\frac{\lambda}{p} - (n-1)\frac{(n-1-\lambda q')}{(n-1)pq'} \\
&= -\frac{\lambda}{p} - \frac{n-1}{pq'} + \frac{\lambda q'}{pq'} = -\frac{n-1}{pq'}.
\end{aligned} \tag{10.50}$$

The string of inequalities in (10.49) establishes (10.46).

Concerning the embedding detailed in (10.48), assume $f \in L^{s_0}_{\mathrm{comp}}(\Sigma,\sigma)$. Then there exist $x_0 \in \Sigma$ and $R \in (0,\infty)$ such that $\operatorname{supp} f \subseteq \Delta(x_0,R)$. Let $\Delta_r \subseteq \Sigma$ be a surface ball with radius $r \in (0, 2\operatorname{diam}(\Sigma))$ satisfying $\Delta_r \cap \Delta(x_0,R) \ne \varnothing$. Begin with the case that $r < R$. Then $\Delta_r \subseteq \Delta(x_0, 3R)$. Recall $w \in \mathrm{RH}_q(\Sigma,\sigma)$ and item (3) of Proposition 2.33, which allows us to write

$$\left(\frac{w(\Delta_r)}{w(\Delta(x_0,3R))}\right)^{1/p} \le \left(\frac{\sigma(\Delta_r)}{\sigma(\Delta(x_0,3R))}\right)^{1/(pq')}. \tag{10.51}$$

Using the Ahlfors regularity of Σ and the doubling nature of w (cf. item (2) of Proposition 2.29), we see that (10.51) implies

$$w(\Delta_r) \le Cw(\Delta(x_0,R))^{1/p}\left(\frac{r}{R}\right)^{(n-1)/(pq')} \tag{10.52}$$

for some constant $C \in (0,\infty)$, independent of r. Next, using (10.46) and (10.52), we obtain

$$\begin{aligned}\left(\frac{1}{r^\lambda}\int_{\Delta_r}|f|^p\,dw\right)^{1/p} &\le Cw(\Delta_r)^{1/p}r^{-(n-1)/(pq')}\|f\|_{L^{s_0}(\Sigma,\sigma)} \\ &\le Cw(\Delta(x_0,R))^{1/p}R^{-(n-1)/(pq')}\|f\|_{L^{s_0}(\Sigma,\sigma)}.\end{aligned} \tag{10.53}$$

Now, consider the case that $r \ge R$. Then using (10.46) we obtain (bearing in mind $\lambda > 0$)

$$\begin{aligned}\left(\frac{1}{r^\lambda}\int_{\Delta_r}|f|^p\,dw\right)^{1/p} &\le \left(\frac{1}{R^\lambda}\int_{\Sigma}|f|^p\,dw\right)^{1/p} \\ &= \left(\frac{1}{R^\lambda}\int_{\Delta(x_0,R)}|f|^p\,dw\right)^{1/p} \\ &\le Cw(\Delta(x_0,R))^{1/p}R^{-(n-1)/(pq')}\|f\|_{L^{s_0}(\Sigma,\sigma)}\end{aligned} \tag{10.54}$$

for some constant C, independent of r and f. Notice that both (10.53) and (10.54) show that there exists a constant $C \in (0,\infty)$, whose nature is specified in the statement of Proposition 10.18, such that

$$\left(\frac{1}{r^\lambda}\int_{\Delta_r}|f|^p\,dw\right)^{1/p} \le Cw(\Delta(x_0,R))^{1/p}R^{-(n-1)/(pq')}\|f\|_{L^{s_0}(\Sigma,\sigma)} \tag{10.55}$$

for all surface balls $\Delta_r \subseteq \Sigma$ with $r \in (0, 2\,\mathrm{diam}(\Sigma))$. Upon taking a supremum over all surface balls, we see that (10.55) implies (10.47) and subsequently the continuous embedding in (10.48) for $s := s_0$ thanks to the arbitrary nature of f. Since the Lebesgue scale of functions with compact support is nested, this completes the proof of the embedding advertised in (10.48). □

We would like to define a brand of Muckenhoupt-weighted vanishing Morrey space, which is consistent with the unweighted definition. To do this, we deduce several density results involving Muckenhoupt-weighted Morrey spaces.

Lemma 10.19. *Let $\Sigma \subseteq \mathbb{R}^n$ be a closed ADR set and write $\sigma := \mathcal{H}^{n-1}\lfloor\Sigma$. Fix any two integrability exponents $p, q \in (1,\infty)$, along with a Muckenhoupt weight $w \in A_p(\Sigma,\sigma) \cap \mathrm{RH}_q(\Sigma,\sigma)$, and a threshold $\lambda \in (0,(n-1)/q')$, where $q' := q/(q-1)$. Lastly, define $s_0 := \frac{(n-1)pq'}{n-1-\lambda q'}$.*

Then for each $f \in M^{p,\lambda}(\Sigma,w)$, the following are equivalent:

(i) *There exists a sequence $\{\phi_j\}_{j\in\mathbb{N}} \subseteq L^\infty_{\mathrm{comp}}(\Sigma,\sigma)$ such that*

$$\lim_{j\to\infty} \|f - \phi_j\|_{M^{p,\lambda}(\Sigma,w)} = 0. \tag{10.56}$$

(ii) *There exist $s \in [s_0,\infty]$ and $\{\psi_j\}_{j\in\mathbb{N}} \subseteq L^s_{\mathrm{comp}}(\Sigma,\sigma)$ such that*

$$\lim_{j\to\infty} \|f - \psi_j\|_{M^{p,\lambda}(\Sigma,w)} = 0. \tag{10.57}$$

(iii) *For all $s \in [s_0,\infty]$, there exists a sequence $\{\xi_j\}_{j\in\mathbb{N}} \subseteq L^s_{\mathrm{comp}}(\Sigma,\sigma)$ such that*

$$\lim_{j\to\infty} \|f - \xi_j\|_{M^{p,\lambda}(\Sigma,w)} = 0. \tag{10.58}$$

(iv) *There exists a sequence $\{\eta_j\}_{j\in\mathbb{N}} \subseteq \mathrm{Lip}_c(\Sigma)$ such that*

$$\lim_{j\to\infty} \|f - \eta_j\|_{M^{p,\lambda}(\Sigma,w)} = 0. \tag{10.59}$$

Proof. The implications (iv) $\Longrightarrow$ (iii) $\Longrightarrow$ (i) $\Longrightarrow$ (ii) are clear. Therefore, assume that (ii) holds with the goal to show that (iv) follows. Without loss of generality, we can assume (ii) is true for $s := s_0$. Thus, for all $\varepsilon > 0$, there exists some $j_\varepsilon \in \mathbb{N}$ for which

$$\|f - \psi_{j_\varepsilon}\|_{M^{p,\lambda}(\Sigma,w)} < \varepsilon. \tag{10.60}$$

For each $i \in \mathbb{N}$, there exists $\eta_{\varepsilon,i} \in \mathrm{Lip}_c(\Sigma)$ such that

$$\|\psi_{j_\varepsilon} - \eta_{\varepsilon,i}\|_{L^s(\Sigma,\sigma)} < \frac{1}{i}. \tag{10.61}$$

Note that matters may be arranged so that $\operatorname{supp}\eta_{\varepsilon,i} \subseteq U$, where U is a fixed open neighborhood of $\operatorname{supp}\psi_{j_\varepsilon}$. Indeed, simply multiply the original choice of $\eta_{\varepsilon,i}$ by a suitable cutoff function belonging to $\mathrm{Lip}_c(\Sigma)$, which is identically one near $\operatorname{supp}\psi_{j_\varepsilon}$. Next, given (10.60) and (10.61) along with (10.46) of Proposition 10.18, we can see that

$$\begin{aligned}\|f - \eta_{\varepsilon,i}\|_{M^{p,\lambda}} &\le \|f - \psi_{j_\varepsilon}\|_{M^{p,\lambda}} + \|\psi_{j_\varepsilon} - \eta_{\varepsilon,i}\|_{M^{p,\lambda}} \\ &\le \|f - \psi_{j_\varepsilon}\|_{M^{p,\lambda}} + C\|\psi_{j_\varepsilon} - \eta_{\varepsilon,i}\|_{L^s(\Sigma,\sigma)} \\ &< \varepsilon + \frac{C}{i},\end{aligned} \tag{10.62}$$

where the constant C depends on Σ, n, p, q, λ, $[w]_{\mathrm{RH}_q}$, and $[w]_{A_p}$. As such,

$$\limsup_{i\to\infty} \|\eta_{\varepsilon,i} - f\|_{M^{p,\lambda}(\Sigma,w)} < \varepsilon. \tag{10.63}$$

If we define $\eta_j := \eta_{\frac{1}{j},j} \in \mathrm{Lip}_c(\Sigma)$ for each $j \in \mathbb{N}$, then we can conclude from (10.63) that (iv) indeed holds. □

We are now in a position to define the following brand of Muckenhoupt-weighted vanishing Morrey space.

Definition 10.20. Let $\Sigma \subseteq \mathbb{R}^n$ be a closed ADR set, and write $\sigma := \mathcal{H}^{n-1}\lfloor\Sigma$. Fix any two integrability exponents $p, q \in (1, \infty)$ along with a Muckenhoupt weight $w \in A_p(\Sigma, \sigma) \cap \mathrm{RH}_q(\Sigma, \sigma)$, and a threshold $\lambda \in (0, (n-1)/q')$, where $q' := q/(q-1)$. The Muckenhoupt-weighted vanishing Morrey space is given by

$$\begin{aligned} &\mathring{M}^{p,\lambda}(\Sigma, w) := \text{the closure of } \mathrm{Lip}_c(\Sigma) \text{ in } M^{p,\lambda}(\Sigma, w), \\ &\text{equipped with the norm inherited from } M^{p,\lambda}(\Sigma, w). \end{aligned} \tag{10.64}$$

That this is well-defined is ensured by the fact that $\mathrm{Lip}_c(\Sigma)$ is dense in $L^{s_0}_{\mathrm{comp}}(\Sigma, \sigma)$, where $s_0 := \frac{(n-1)pq'}{n-1-\lambda q'}$, and because the latter space continuously embeds into $M^{p,\lambda}(\Sigma, w)$ as witnessed in Proposition 10.18. Note that this is consistent with the fact about unweighted vanishing Morrey spaces seen in [79, Section 7.1, p. 435], which says given the density of $\mathrm{Lip}_c(\Sigma)$ in $L^s(\Sigma, \sigma)$ where $s := \frac{p(n-1)}{n-1-\lambda}$, and the continuous embedding of the latter space into $M^{p,\lambda}(\Sigma, \sigma)$ (cf. (10.45)), one has exactly (10.64) corresponding to the case that $w \equiv 1$.

We present a result, which collects several facts concerning this brand of weighted vanishing Morrey space. In particular, the last item in Proposition 10.21 demonstrates the "vanishing" property that functions belonging to $\mathring{M}^{p,\lambda}(\Sigma, w)$ enjoy, hence the name.

Proposition 10.21. *Suppose $\Sigma \subseteq \mathbb{R}^n$ is a closed ADR set and abbreviate $\sigma := \mathcal{H}^{n-1}\lfloor\Sigma$. Fix any two integrability exponents $p, q \in (1, \infty)$, along with a Muckenhoupt weight $w \in A_p(\Sigma, \sigma) \cap \mathrm{RH}_q(\Sigma, \sigma)$, and some choice of threshold $\lambda \in (0, (n-1)/q')$, where $q' = q/(q-1)$.*

The following statements are true:

(a) *The space $(\mathring{M}^{p,\lambda}(\Sigma, w), \|\cdot\|_{M^{p,\lambda}(\Sigma,w)})$ is a Banach space.*

(b) *The space $\mathrm{Lip}_c(\Sigma)$ continuously and densely embeds into $\mathring{M}^{p,\lambda}(\Sigma, w)$.*

(c) *Define $s_0 := \frac{(n-1)pq'}{n-1-\lambda q'}$. For any $s \in [s_0, \infty]$, the space $L^s_{\mathrm{comp}}(\Sigma, \sigma)$ continuously and densely embeds into $\mathring{M}^{p,\lambda}(\Sigma, w)$.*

(d) *Each function $f \in \mathring{M}^{p,\lambda}(\Sigma, w)$ satisfies the following "vanishing" condition with respect to the scale, which is uniform in the location. Namely,*

$$\lim_{R\to 0^+} \left[\sup_{\substack{x\in\Sigma \\ r\in(0,R)}} \left(\frac{1}{r^\lambda} \int_{\Delta(x,r)} |f|^p \, dw \right)^{1/p} \right] = 0. \tag{10.65}$$

Proof. The fact that $\mathring{M}^{p,\lambda}(\Sigma, w)$ is a closed linear subspace of $M^{p,\lambda}(\Sigma, w)$ is clear from the definition. Subsequently, it is clear that (a) holds upon endowing $\mathring{M}^{p,\lambda}(\Sigma, w)$ with the inherited norm from $M^{p,\lambda}(\Sigma, w)$.

The claims in (b) and (c) follow from (10.64) and Proposition 10.18, together with Lemma 10.19.

Lastly, to prove (d), fix $f \in \mathring{M}^{p,\lambda}(\Sigma, w)$. Select $\varepsilon > 0$ arbitrary, and choose some $\phi \in L^{\infty}_{\mathrm{comp}}(\Sigma, \sigma)$ so that

$$\|f - \phi\|_{M^{p,\lambda}(\Sigma,w)} < \varepsilon \tag{10.66}$$

with $\operatorname{supp}\varphi \subseteq \Delta(x_0, \rho)$ for some $x_0 \in \Sigma$ and $\rho \in (0, 2\operatorname{diam}(\Sigma))$. That such a ϕ exists is ensured by (c). Next, choose some $R \in (0, 1)$. Then for all $x \in \Sigma$ and $r \in (0, R)$, we have

$$\begin{aligned}\left(\frac{1}{r^{\lambda}}\int_{\Delta(x,r)} |f|^p\,dw\right)^{1/p} &\le \left(\frac{1}{r^{\lambda}}\int_{\Delta(x,r)} |f-\phi|^p\,dw\right)^{1/p}\\ &\quad + \left(\frac{1}{r^{\lambda}}\int_{\Delta(x,r)} |\phi|^p\,dw\right)^{1/p}.\end{aligned} \tag{10.67}$$

Notice that the second integral on the right-hand side of (10.67) vanishes whenever $\Delta(x, r) \cap \operatorname{supp}\phi = \varnothing$. Therefore, assume $\Delta(x, r) \cap \operatorname{supp}\phi \neq \varnothing$. Appealing to (10.66) and Proposition 10.18, we can achieve the estimate

$$\left(\frac{1}{r^{\lambda}}\int_{\Delta(x,r)} |f|^p\,dw\right)^{1/p} \le \varepsilon + Cr^{-\lambda/p} w(\Delta(x,r))^{1/p}\|\phi\|_{L^{\infty}(\Sigma,\sigma)}, \tag{10.68}$$

where $C \in (0, \infty)$ depends only on $\Sigma, n, p, q, \lambda, [w]_{\mathrm{RH}_q}$, and $[w]_{A_p}$. Now, using the fact that we are assuming $\Delta(x, r) \cap \operatorname{supp}\phi \neq \varnothing$, we can conclude that $\Delta(x, r) \subseteq \Delta(x_0, 2 + \rho)$. Indeed, for $x \in \Delta(x, r) \cap \operatorname{supp}\phi$,

$$|x - x_0| \le r + \rho < R + \rho < 1 + \rho \le 2 + \rho. \tag{10.69}$$

As such, we can use item (3) of Lemma 2.33 (with $\Delta(x_0, 2 + \rho)$ and $\Delta(x, r)$ playing the roles of Δ and E, respectively) to estimate

$$\frac{w(\Delta(x,r))}{w(\Delta(x_0,2+\rho))} \le [w]_{\mathrm{RH}_q}\left(\frac{\sigma(\Delta(x,r))}{\sigma(\Delta(x_0,2+\rho))}\right)^{1/q'} \le Cr^{(n-1)/q'}, \tag{10.70}$$

where the constant $C \in (0, \infty)$ depends on Σ, q, and $[w]_{\mathrm{RH}_q}$. Notice we appeal to the Ahlfors regularity of Σ to achieve the second inequality in (10.70). Next, raise the inequality in (10.70) to the $1/p$ power, multiply across by $w(\Delta(x_0, 2 + \rho))^{1/p}$, and insert the resulting estimate into (10.68) to obtain

$$\left(\frac{1}{r^\lambda}\int_{\Delta(x,r)}|f|^p\,dw\right)^{1/p}\leq\varepsilon+Cr^{(n-1)/(pq')}r^{-\lambda/p},\tag{10.71}$$

for some constant $C\in(0,\infty)$ independent of x and r. We then have

$$\begin{aligned}\lim_{R\to0^+}\left[\sup_{\substack{x\in\Sigma\\ r\in(0,R)}}\left(\frac{1}{r^\lambda}\int_{\Delta(x,r)}|f|^p\,dw\right)^{1/p}\right]&\leq\lim_{R\to0^+}\left[\sup_{\substack{x\in\Sigma\\ r\in(0,R)}}\left(\varepsilon+Cr^{(n-1-\lambda q')/(pq')}\right)\right]\\&=\lim_{R\to0^+}\left[\varepsilon+CR^{(n-1-\lambda q')/(pq')}\right]\\&=\varepsilon,\end{aligned}\tag{10.72}$$

since $(n-1-\lambda q')/(pq')=(n-1)/s_0>0$. Thanks to the arbitrary nature of ε, this completes the justification of (d). □

A Muckenhoupt-weighted Morrey space is an example of a nonseparable Banach space, but interestingly, from items (a) and (c) of Proposition 10.21, and the fact that $L^{s_0}_{\mathrm{comp}}(\Sigma,\sigma)$ is separable (being a subspace of the separable space $L^{s_0}(\Sigma,\sigma)$, cf. [86, Lemma 3.6.4, p. 278]), we see that

$$\mathring{M}^{p,\lambda}(\Sigma,w)\text{ is a separable Banach space.}\tag{10.73}$$

The following lemma, which precipitates naturally and almost immediately from our previous results and discussions, provides some equivalent characterizations of weighted vanishing Morrey spaces.

Proposition 10.22. *Suppose $\Sigma\subseteq\mathbb{R}^n$ is a closed* ADR *set and abbreviate $\sigma:=\mathcal{H}^{n-1}\lfloor\Sigma$. Fix any two integrability exponents $p,q\in(1,\infty)$, along with a Muckenhoupt weight $w\in A_p(\Sigma,\sigma)\cap\mathrm{RH}_q(\Sigma,\sigma)$, and some choice of threshold $\lambda\in(0,(n-1)/q')$, where $q':=q/(q-1)$. Lastly, define $s_0:=\frac{(n-1)pq'}{n-1-\lambda q'}$.*

The following are equivalent:

(i) $f\in\mathring{M}^{p,\lambda}(\Sigma,w)$.

(ii) *There exists a sequence $\{\phi_j\}_{j\in\mathbb{N}}\subseteq L^\infty_{\mathrm{comp}}(\Sigma,\sigma)$ such that*

$$\lim_{j\to\infty}\|f-\phi_j\|_{M^{p,\lambda}(\Sigma,w)}=0.\tag{10.74}$$

(iii) *For any $s\in[s_0,\infty]$, there exists a sequence $\{\psi_j\}_{j\in\mathbb{N}}\subseteq L^s_{\mathrm{comp}}(\Sigma,\sigma)$ such that*

$$\lim_{j\to\infty}\|f-\psi_j\|_{M^{p,\lambda}(\Sigma,w)}=0.\tag{10.75}$$

(iv) *There exists a sequence* $\{\eta_j\}_{j\in\mathbb{N}} \subseteq \mathrm{Lip}_c(\Sigma)$ *such that*

$$\lim_{j\to\infty} \|f - \eta_j\|_{M^{p,\lambda}(\Sigma,w)} = 0. \tag{10.76}$$

Proof. The implications (i) $\Longrightarrow$ (ii) $\Longrightarrow$ (iii) follow from item (c) of Proposition 10.21. The implication (iii) $\Longrightarrow$ (iv) is a consequence of Lemma 10.19, and the implication (iv) $\Longrightarrow$ (i) follows from (10.64). □

Now that we have built a suitable functional analytic framework for weighted vanishing Morrey spaces, our goal is now to develop a Calderón–Zygmund theory on this brand of Morrey space. To begin, we start with the following general lemma from functional analysis, found in [87, Lemma 1.2.20, p. 24].

Lemma 10.23. *Let* X_0, X_1 *be a pair of quasi-Banach spaces and suppose* $T : X_0 \to X_1$ *is a bounded linear operator. Consider* $Y_j \subseteq X_j$ *linear subspaces for* $j \in \{0,1\}$ *and denote by* $\overline{Y_j}^{X_j}$ *the closure of* Y_j *in* X_j *for* $j \in \{0,1\}$.

(1) *If* $T(Y_0) \subseteq Y_1$, *then* T *induces a bounded linear mapping from* $\overline{Y_0}^{X_0}$ *into* $\overline{Y_1}^{X_1}$.
(2) *If* T *is invertible and has the property that* $T(Y_0) = Y_1$, *then* T *induces an isomorphism from* $\overline{Y_0}^{X_0}$ *onto* $\overline{Y_1}^{X_1}$.

Remark 10.24. A slight modification of the proof in [87, Lemma 1.2.20, p. 24] can be made so that Lemma 10.23 can accommodate bounded, sublinear operators.

The first consequence of Lemma 10.23 that we catalog is the fact that multiplication by essentially bounded functions leaves weighted vanishing Morrey spaces invariant.

Proposition 10.25. *Suppose* $\Sigma \subseteq \mathbb{R}^n$ *is a closed* ADR *set and abbreviate* $\sigma := \mathcal{H}^{n-1}\lfloor\Sigma$. *Fix any two integrability exponents* $p, q \in (1,\infty)$, *an arbitrary Muckenhoupt weight* $w \in A_p(\Sigma,\sigma) \cap \mathrm{RH}_q(\Sigma,\sigma)$, *and some threshold* $\lambda \in (0,(n-1)/q')$, *where* $q' := q/(q-1)$. *Lastly, fix a function* $b \in L^\infty(\Sigma,\sigma)$ *and denote by* M_b *the operation of pointwise multiplication by* b.

Then the operator

$$M_b : \mathring{M}^{p,\lambda}(\Sigma,w) \to \mathring{M}^{p,\lambda}(\Sigma,w) \tag{10.77}$$

is well-defined, linear, and bounded with

$$\|M_b f\|_{M^{p,\lambda}(\Sigma,w)} \le \|b\|_{L^\infty(\Sigma,\sigma)} \|f\|_{M^{p,\lambda}(\Sigma,w)} \tag{10.78}$$

for each $f \in \mathring{M}^{p,\lambda}(\Sigma,w)$.

Proof. Fix $b \in L^\infty(\Sigma,\sigma)$ arbitrary. From Remark 4.4, we see that the operator

$$M_b : M^{p,\lambda}(\Sigma,w) \to M^{p,\lambda}(\Sigma,w) \tag{10.79}$$

is well-defined, linear, and bounded. Additionally, since the range of M_b acting on $L^\infty_{\rm comp}(\Sigma,\sigma)$ is contained in $L^\infty_{\rm comp}(\Sigma,\sigma)$, we may appeal to (1) of Lemma 10.23 to say that M_b leaves $\overline{L^\infty_{\rm comp}(\Sigma,\sigma)}^{M^{p,\lambda}(\Sigma,\sigma)}$ invariant (note we are retaining the notation convention introduced in Lemma 10.23). But because of Proposition 10.22, we know that

$$\overline{L^\infty_{\rm comp}(\Sigma,\sigma)}^{M^{p,\lambda}(\Sigma,\sigma)} = \mathring{M}^{p,\lambda}(\Sigma,w), \tag{10.80}$$

so this is the desired result. The estimate in (10.78) now follows immediately from Remark 4.11 given that $\mathring{M}^{p,\lambda}(\Sigma,w)$ is a subspace of $M^{p,\lambda}(\Sigma,w)$. □

Now that we know that multiplying a function in a weighted-vanishing Morrey space by an essentially bounded function is still a function belonging to said weighted-vanishing Morrey space, we can show that weighted-vanishing Morrey spaces are function lattices.

Proposition 10.26. *Suppose $\Sigma \subseteq \mathbb{R}^n$ is a closed* ADR *set, and abbreviate $\sigma := \mathcal{H}^{n-1}\lfloor\Sigma$. Fix any two integrability exponents $p,q \in (1,\infty)$, an arbitrary Muckenhoupt weight $w \in A_p(\Sigma,\sigma) \cap \mathrm{RH}_q(\Sigma,\sigma)$, and some choice of threshold $\lambda \in (0,(n-1)/q')$, where $q' := q/(q-1)$.*

If $g \in \mathring{M}^{p,\lambda}(\Sigma,w)$ and f is a σ-measurable function with $|f| \le g$ at σ-a. e. point on Σ, then in fact $f \in \mathring{M}^{p,\lambda}(\Sigma,w)$ with $\|f\|_{M^{p,\lambda}(\Sigma,w)} \le \|g\|_{M^{p,\lambda}(\Sigma,w)}$.

Proof. The proof follows analogously to that of Proposition 6.14, now making use of Proposition 10.25 instead of Proposition 6.13. □

Our next order of business is the issue of whether or not the Hardy–Littlewood maximal operator is well-defined and bounded on Muckenhoupt-weighted vanishing Morrey spaces. This is the first significant step toward establishing a robust Calderón–Zygmund theory for these function spaces. However, to facilitate the proof of that result, we present a lemma concerning the Hardy–Littlewood maximal operator, which may be found in [86, Proposition 7.6.8, pp. 632–633], which expands upon the result recorded in Lemma 2.25.

Lemma 10.27. *Let $\Sigma \subseteq \mathbb{R}^n$ be a closed* ADR *set, and write $\sigma := \mathcal{H}^{n-1}\lfloor\Sigma$. Then with $\mathcal{M}_\Sigma$ denoting the Hardy–Littlewood maximal operator associated with Σ as in (2.26), one has for each fixed $x_0 \in \Sigma$ and $R \in (0,\infty)$,*

$$\big(\mathcal{M}_\Sigma(\mathbf{1}_{\Delta(x_0,R)})\big)(x) \approx \frac{R^{n-1}}{(R+|x-x_0|)^{n-1}} \tag{10.81}$$

uniformly with respect to $x \in \Sigma$, x_0, and R.

Additionally, if one defines for each $t \in [0,\infty)$ the function

$$\log_+ t := \begin{cases} 0 & \text{if } t \in [0,1] \\ \ln t & \text{if } t \in [1,\infty), \end{cases} \tag{10.82}$$

then

$$((\mathcal{M}_\Sigma \circ \mathcal{M}_\Sigma)(\mathbf{1}_{\Delta(x_0,1)}))(x) \approx \Big(\mathcal{M}_\Sigma\Big(\frac{1}{1+|\cdot - x_0|^{n-1}}\Big)\Big)(x) \approx \frac{1+\log_+|x-x_0|}{1+|x-x_0|^{n-1}} \tag{10.83}$$

uniformly with respect to $x, x_0 \in \Sigma$.

We are now ready to prove the well definiteness and boundedness of the Hardy–Littlewood maximal operator on Muckenhoupt-weighted vanishing Morrey spaces.

Proposition 10.28. *Suppose* $\Sigma \subseteq \mathbb{R}^n$ *is a closed* ADR *set and abbreviate* $\sigma := \mathcal{H}^{n-1}\lfloor\Sigma$. *Fix any two integrability exponents* $p, q \in (1,\infty)$, *along with a Muckenhoupt weight* $w \in A_p(\Sigma,\sigma) \cap \mathrm{RH}_q(\Sigma,\sigma)$, *and some choice of threshold* $\lambda \in (0,(n-1)/q')$, *where* $q' := q/(q-1)$.

The Hardy–Littlewood maximal operator

$$\mathcal{M}_\Sigma : \mathring{M}^{p,\lambda}(\Sigma,w) \to \mathring{M}^{p,\lambda}(\Sigma,w) \tag{10.84}$$

is well-defined, sublinear, and bounded.

Proof. Start by fixing $x_0 \in \Sigma$, then choose $f \in L^\infty_{\mathrm{comp}}(\Sigma,\sigma)$ arbitrary. It follows that there exists $N \in \mathbb{N}$ such that $|f| \le N \cdot \mathbf{1}_{\Delta(x_0,N)}$. As such,

$$0 \le \mathcal{M}_\Sigma f \le N \cdot \mathbf{1}_{\Delta(x_0,N)} \le C\frac{N^n}{1+|\cdot - x_0|^n} \tag{10.85}$$

pointwise on Σ. Note that the last inequality in (10.85) follows from Lemma 10.27. Since the operator

$$\mathcal{M}_\Sigma : L^\infty(\Sigma,\sigma) \to L^\infty(\Sigma,\sigma) \tag{10.86}$$

is well-defined and bounded (as seen from its definition), it follows that $\mathcal{M}_\Sigma(L^\infty_{\mathrm{comp}}(\Sigma,\sigma))$ is contained in $L^\infty(\Sigma,\sigma)$. Therefore, we can conclude that $(\mathcal{M}_\Sigma f)\cdot \mathbf{1}_{\Delta(x_0,j)} \in L^\infty(\Sigma,\sigma)$ for each $j \in \mathbb{N}$. Moreover, we may estimate freely using Proposition 4.7 together with (10.85) and Lemma 10.27,

$$\begin{aligned}
&\|\mathcal{M}_\Sigma f - (\mathcal{M}_\Sigma f)\cdot \mathbf{1}_{\Delta(x_0,j)}\|_{M^{p,\lambda}(\Sigma,w)} \\
&= \|(\mathcal{M}_\Sigma f)\cdot \mathbf{1}_{\Sigma\setminus\Delta(x_0,j)}\|_{M^{p,\lambda}(\Sigma,w)} \\
&\le C_{n,N}\Big\|\mathbf{1}_{\Sigma\setminus\Delta(x_0,j)}\cdot\frac{1}{1+|\cdot - x_0|^{n-1}}\Big\|_{M^{p,\lambda}(\Sigma,w)} \\
&= C_{n,N}\Big\|\frac{\mathbf{1}_{\Sigma\setminus\Delta(x_0,j)}}{1+\log_+|\cdot - x_0|}\cdot\frac{1+\log_+|\cdot - x_0|}{1+|\cdot - x_0|^n}\Big\|_{M^{p,\lambda}(\Sigma,w)} \\
&\le C_{n,N}\Big\|\frac{1}{\ln(j)}\cdot(\mathcal{M}_\Sigma\circ\mathcal{M}_\Sigma)(\mathbf{1}_{\Delta(x_0,1)})\Big\|_{M^{p,\lambda}(\Sigma,w)},
\end{aligned} \tag{10.87}$$

for each $j \geq 2$. Note that to obtain the last inequality in (10.87) we appeal to (10.83) as well as the fact that

$$\ln(j) \cdot \mathbf{1}_{\Sigma\setminus\Delta(x_0,j)} \leq 1 + |\cdot - x_0| \quad \text{on } \Sigma \text{ for all } j \in \mathbb{N}. \tag{10.88}$$

Indeed, when $x \in \Sigma$ is such that $|x - x_0| < j$ this is trivially true. Otherwise, for $|x - x_j| \geq j$, we have that $\mathbf{1}_{\Sigma\setminus\Delta(x_0,j)} = 1$ and $\ln(j) \leq \log_+ |x - x_0|$, bearing in mind (10.82). As such, for any $j \in \mathbb{N}$ we can estimate

$$\ln(j) \cdot \mathbf{1}_{\Sigma\setminus\Delta(x_0,j)} \leq \log_+ |x - x_0| \leq 1 + \log_+ |x - x_0| \tag{10.89}$$

for all $x \in \Sigma$, which is the desired inequality in (10.88).

Turning back to the mainstream discussion and the estimate in (10.87), observe that $(\mathcal{M}_\Sigma \circ \mathcal{M}_\Sigma)(\mathbf{1}_{\Delta(x_0,1)}) \in M^{p,\lambda}(\Sigma, w)$, due to Proposition 5.1 and the fact that $\mathbf{1}_{\Delta(x_0,1)}$ belongs to $M^{p,\lambda}(\Sigma, w)$ (cf. Lemma 4.9). Subsequently, from (10.87) we find that

$$\begin{aligned} &\lim_{j\to\infty} \left\| \mathcal{M}_\Sigma f - (\mathcal{M}_\Sigma f) \cdot \mathbf{1}_{\Delta(x_0,j)} \right\|_{M^{p,\lambda}(\Sigma,w)} \\ &\quad \leq \lim_{j\to\infty} \frac{C}{\ln(j)} \left\| (\mathcal{M}_\Sigma \circ \mathcal{M}_\Sigma)(\mathbf{1}_{\Delta(x_0,1)}) \right\|_{M^{p,\lambda}(\Sigma,w)} = 0. \end{aligned} \tag{10.90}$$

This means that the sequence $\{(\mathcal{M}_\Sigma f) \cdot \mathbf{1}_{\Delta(x_0,j)}\}_{j\in\mathbb{N}} \subseteq L^\infty_{\text{comp}}(\Sigma, \sigma)$ converges to $\mathcal{M}_\Sigma f$ in $M^{p,\lambda}(\Sigma, w)$, which according to Lemma 10.22 ensures that $\mathcal{M}_\Sigma f$ belongs to $\mathring{M}^{p,\lambda}(\Sigma, w)$. Since $f \in L^\infty_{\text{comp}}(\Sigma, \sigma)$ was arbitrary, we see that $\mathcal{M}_\Sigma(L^\infty_{\text{comp}}(\Sigma, \sigma)) \subseteq \mathring{M}^{p,\lambda}(\Sigma, w)$. Thus, a slight variant of Lemma 10.23 (cf. Remark 10.24), and making the identifications

$$\begin{aligned} T &:= \mathcal{M}_\Sigma, & X_0 = X_1 &:= M^{p,\lambda}(\Sigma, w), \\ Y_0 &:= L^\infty_{\text{comp}}(\Sigma, \sigma), & Y_1 &:= \mathring{M}^{p,\lambda}(\Sigma, w), \end{aligned} \tag{10.91}$$

we can conclude (bearing in mind (10.64)) that the operator in (10.84) is well- defined and bounded as wanted. □

We would like to present the following corollary, which demonstrates that bounded measurable functions with sufficient decay at infinity belong to Muckenhoupt-weighted-vanishing Morrey spaces. See Lemma 4.12 for a similar result involving a Muckenhoupt-weighted Morrey space.

Corollary 10.29. *Let $\Sigma \subseteq \mathbb{R}^n$ be a closed Ahlfors regular set, and write $\sigma := \mathcal{H}^{n-1}\lfloor\Sigma$. Fix any two integrability exponents $p, q \in (1, \infty)$, an arbitrary Muckenhoupt weight $w \in A_p(\Sigma, \sigma) \cap \mathrm{RH}_q(\Sigma, \sigma)$, and a threshold $\lambda \in (0, (n-1)/q')$, where $q' := q/(q-1)$. Suppose $f : \Sigma \to \mathbb{C}$ is a σ-measurable function, which for some constant $C \in (0, \infty)$, satisfies $|f(x)| \leq C(1 + |x|)^{1-n}$ at σ-a. e. point $x \in \Sigma$. Then f belongs to $\mathring{M}^{p,\lambda}(\Sigma, w)$.*

Proof. Let f be an arbitrary function as in the statement of the lemma. According to Lemma 10.27 (and Lemma 2.18), we know for a fixed choice of reference point $x_0 \in \Sigma$

that

$$f(x) \le C(1+|x|)^{1-n} \approx \mathcal{M}_\Sigma(\mathbf{1}_{\Delta(x_0,1)})(x) \tag{10.92}$$

uniformly for σ-a. e. point $x \in \Sigma$. However, since $\mathbf{1}_{\Delta(x_0,1)} \in \mathring{M}^{p,\lambda}(\Sigma, w)$ (as can be witnessed by item (b) of Proposition 10.21), Proposition 10.28 shows that $\mathcal{M}_\Sigma(\mathbf{1}_{\Delta(x_0,1)})$ belongs to $\mathring{M}^{p,\lambda}(\Sigma, w)$ as well. By the lattice property of Muckenhoupt-weighted vanishing Morrey spaces (cf. Proposition 10.26), we find that f necessarily belongs to $\mathring{M}^{p,\lambda}(\Sigma, w)$ as well, completing the proof. □

To further develop a Calderón–Zygmund theory on Muckenhoupt-weighted vanishing Morrey spaces, we turn to the task of demonstrating that principal value singular integral operators of convolution type with the right type of integral kernel are well-defined and bounded on Muckenhoupt-weighted vanishing Morrey spaces.

Theorem 10.30. *Let $\Sigma \subseteq \mathbb{R}^n$ be a UR set and write $\sigma := \mathcal{H}^{n-1}\lfloor\Sigma$. Let $N = N(n) \in \mathbb{N}$ be sufficiently large and suppose $k \in \mathscr{C}^N(\mathbb{R}^n \setminus \{0\})$ is an integral kernel, which is odd and positive homogeneous of degree $1-n$. For each $f \in L^1(\Sigma, \frac{\sigma(x)}{1+|x|^{n-1}})$, define*

$$(Tf)(x) := \lim_{\varepsilon\to 0^+} \int_{\substack{y\in\Sigma \\ |x-y|>\varepsilon}} k(x-y)f(y)\,d\sigma(y) \quad \text{at } \sigma\text{-a. e. } x \in \Sigma. \tag{10.93}$$

Finally, fix two integrability exponents $p, q \in (1,\infty)$ along with a Muckenhoupt weight $w \in A_p(\Sigma,\sigma) \cap \mathrm{RH}_q(\Sigma,\sigma)$ and a threshold $\lambda \in (0,(n-1)/q')$, where $q' := q/(q-1)$.

Then the operator in (10.93) induces a mapping

$$T : \mathring{M}^{p,\lambda}(\Sigma, w) \to \mathring{M}^{p,\lambda}(\Sigma, w), \tag{10.94}$$

which is well-defined, linear, and bounded.

Proof. Define $s_0 := \frac{(n-1)pq'}{n-1-\lambda}$. From Proposition 5.6, we know that the operator

$$T : L^{s_0}(\Sigma,\sigma) \to L^{s_0}(\Sigma,\sigma) \tag{10.95}$$

is well-defined, linear, and bounded, and so, $T(L^{s_0}_{\mathrm{comp}}(\Sigma,\sigma)) \subseteq L^{s_0}(\Sigma,\sigma)$. Fix a reference point $x_0 \in \Sigma$ and choose some $f \in L^{s_0}_{\mathrm{comp}}(\Sigma, w)$ arbitrary. Then we have that $(Tf)\cdot \mathbf{1}_{\Delta(x_0,j)} \in L^{s_0}_{\mathrm{comp}}(\Sigma,\sigma) \subseteq L^{\infty}_{\mathrm{comp}}(\Sigma,\sigma)$ for all $j \in \mathbb{N}$. If $j_0 \in \mathbb{N}$ is large enough so that $\operatorname{supp} f \subseteq \Delta(x_0, j_0)$, then for all $j \in \mathbb{N}$ with $j > j_0$, there exists a constant $C_f \in (0,\infty)$ such that

$$0 \le (|Tf|\cdot \mathbf{1}_{\Sigma\setminus\Delta(x_0,j)})(x) \le \frac{C_f \cdot \mathbf{1}_{\Sigma\setminus\Delta(x_0,j)}(x)}{1+|x-x_0|^{n-1}} \tag{10.96}$$

for all $x \in \Sigma$. Indeed, for $x \in \Delta(x_0,j)$ the inequality in (10.96) is trivially true. Otherwise, if $x \in \Sigma \setminus \Delta(x_0,j)$, then notice that

$$0 \le (|Tf| \cdot \mathbf{1}_{\Sigma\setminus\Delta(x_0,j)})(x) \le \int_{\Delta(x_0,j_0)} \frac{|f(y)|}{|y-x|^{n-1}}\,d\sigma(y) \le \frac{C_f}{1+|x-x_0|^{n-1}}. \tag{10.97}$$

To obtain the second inequality in (10.97) we appeal to (5.4), whereas the last inequality in (10.97) follows from Lemma 2.18 and the geometric observation that

$$|x-y| \approx 1+|x-x_0| \quad \text{uniformly} \quad \text{for } x \in \Sigma\setminus\Delta(x_0,j) \text{ and } y \in \Delta(x_0,j_0). \tag{10.98}$$

Indeed, notice first that for $x \in \Sigma \setminus \Delta(x_0,j)$ and $y \in \Delta(x_0,j_0)$ we can estimate

$$|x-y| \le |x-x_0| + |x_0-y| < |x-x_0| + j_0 \le j_0(1+|x-x_0|). \tag{10.99}$$

Conversely, since $j > j_0$ and assuming $x \in \Sigma \setminus \Delta(x_0,j)$ and $y \in \Delta(x_0,j_0)$, we have that $1 \le |x-y|$. Also, we may estimate

$$|x-x_0| \le |x-y| + |y-x_0| \le |x-y| + j_0 \le |x-y| + \frac{j_0}{j}|x-x_0|, \tag{10.100}$$

which implies that $|x-x_0| \le \frac{j}{j-j_0}|x-y| \le C|x-y|$ for some constant $C \in (0,\infty)$, independent of j. This, together with (10.100) shows that

$$1+|x-x_0| \le C|y-x| \quad \text{for all } x \in \Sigma\setminus\Delta(x_0,j) \text{ and } y \in \Delta(x_0,j_0), \tag{10.101}$$

which recalling the inequality in (10.99) constitutes the quantitative equivalence claimed in (10.98).

Returning from that geometric digression to the main issue at hand, note that the inequality in (10.97) allows us to write for $j > j_0$,

$$\begin{aligned} &\|Tf - (Tf)\cdot\mathbf{1}_{\Delta(x_0,j)}\|_{M^{p,\lambda}(\Sigma,w)} \\ &\quad = \||Tf|\cdot\mathbf{1}_{\Sigma\setminus\Delta(x_0,j)}\|_{M^{p,\lambda}(\Sigma,w)} \\ &\quad \le C_f\left\|\frac{\mathbf{1}_{\Sigma\setminus\Delta(x_0,j)}}{1+|\cdot-x_0|^{n-1}}\right\|_{M^{p,\lambda}(\Sigma,w)} \\ &\quad = C_f\left\|\frac{\mathbf{1}_{\Sigma\setminus\Delta(x_0,j)}}{1+\log_+|\cdot-x_0|}\cdot\frac{1+\log_+|\cdot-x_0|}{1+|\cdot-x_0|^{n-1}}\right\|_{M^{p,\lambda}(\Sigma,w)} \\ &\quad \le \frac{C_f}{\ln(j)}\|(\mathcal{M}_\Sigma\circ\mathcal{M}_\Sigma)(\mathbf{1}_{\Delta(x_0,1)})\|_{M^{p,\lambda}(\Sigma,w)}. \end{aligned} \tag{10.102}$$

Observe also that the inequalities above all rely on the lattice property of $M^{p,\lambda}(\Sigma,w)$ (cf. Proposition 4.7). Additionally, the last inequality in (10.102) makes use of (10.83), as well as a geometric argument paralleling that which led to the inequality in (10.89). Now, because of Proposition 5.1 and Lemma 4.9, we see that (10.102) implies that the sequence $\{(Tf)\cdot \mathbf{1}_{\Delta(x_0,j)}\}_{j\in\mathbb{N}} \subseteq L^{s_0}_{\mathrm{comp}}(\Sigma,\sigma)$ converges to Tf in $M^{p,\lambda}(\Sigma,w)$. Proposition 10.22 then ensures that $Tf \in \mathring{M}^{p,\lambda}(\Sigma,w)$. Given the arbitrary nature of $f \in L^{s_0}_{\mathrm{comp}}(\Sigma,\sigma)$, this shows that $T(L^{s_0}_{\mathrm{comp}}(\Sigma,\sigma)) \subseteq \mathring{M}^{p,\lambda}(\Sigma,w)$. Consequently, upon making the identifications

$$\begin{aligned} T \text{ as in (10.93)}, \quad & X_0 = X_1 := M^{p,\lambda}(\Sigma,w), \\ Y_0 = L^{s_0}_{\mathrm{comp}}(\Sigma,\sigma), \quad & Y_1 := \mathring{M}^{p,\lambda}(\Sigma,w), \end{aligned} \tag{10.103}$$

Lemma 10.23 (bearing in mind item (c) of Proposition 10.21) allows us to conclude that the operator in (10.94) is indeed well-defined, linear, and bounded. □

Having established that the Hardy–Littlewood maximal operator and the P. V. operator in (10.93) are well-defined on weighted-vanishing Morrey spaces, we can use Cotlar's inequality to obtain a well-definiteness result for the larger, so-called maximal operator, $T_{\max}$.

Theorem 10.31. *Let $\Sigma \subseteq \mathbb{R}^n$ be a* UR *set and write $\sigma := \mathcal{H}^{n-1}\lfloor\Sigma$. Let $N = N(n) \in \mathbb{N}$ be a sufficiently large integer and consider an integral kernel $k \in \mathscr{C}^N(\mathbb{R}^n \setminus \{0\})$, which is odd and positive homogeneous of degree $1-n$. For each $f \in L^1(\Sigma, \frac{\sigma(x)}{1+|x|^{n-1}})$, define*

$$(T_{\max} f)(x) := \sup_{\varepsilon>0} \Bigg| \int\limits_{\substack{y\in\Sigma\\ |x-y|>\varepsilon}} k(x-y) f(y)\, d\sigma(y) \Bigg| \quad \text{at } \sigma\text{-a. e. } x \in \Sigma. \tag{10.104}$$

Finally, fix two integrability exponents $p, q \in (1,\infty)$ along with a Muckenhoupt weight $w \in A_p(\Sigma,\sigma) \cap \mathrm{RH}_q(\Sigma,\sigma)$ and a threshold $\lambda \in (0,(n-1)/q')$, where $q' := q/(q-1)$.

Then the operator in (10.104) *induces a mapping*

$$T_{\max} : \mathring{M}^{p,\lambda}(\Sigma,w) \to \mathring{M}^{p,\lambda}(\Sigma,w), \tag{10.105}$$

which is well-defined, sublinear, and bounded.

Proof. For all $f \in \mathring{M}^{p,\lambda}(\Sigma,w)$, Cotlar's inequality (cf. Lemma 7.4) says there exists some constant $C \in (0,\infty)$, independent of f, such that

$$0 \le T_{\max} f \le C \cdot (\mathcal{M}_\Sigma(Tf)) + C \cdot (\mathcal{M}_\Sigma f) \tag{10.106}$$

pointwise σ-a. e. on Σ, where T is as in (10.93). According to Theorem 10.30, we know that $Tf \in \mathring{M}^{p,\lambda}(\Sigma,w)$, and since Proposition 10.28 tells us that $\mathcal{M}_\Sigma$ is well-defined and bounded on $\mathring{M}^{,\lambda}(\Sigma,w)$, we can conclude that both $\mathcal{M}_\Sigma(Tf)$ and $\mathcal{M}_\Sigma f$ belong to $\mathring{M}^{p,\lambda}(\Sigma,w)$. By the lattice property of $\mathring{M}^{p,\lambda}(\Sigma,w)$ (cf. Proposition 10.26), we can conclude that $T_{\max} f$ belongs to $\mathring{M}^{p,\lambda}(\Sigma,w)$ as well, and hence $T_{\max}$ is a well-defined operator in the context of (10.105). The fact that $T_{\max}$ is bounded on $\mathring{M}^{p,\lambda}(\Sigma,w)$ is a consequence of the fact that $T_{\max}$ is bounded on $M^{p,\lambda}(\Sigma, w)$ (cf. Theorem 5.7) along with the observation that $\mathring{M}^{p,\lambda}(\Sigma, w)$ is a closed linear subspace of $M^{p,\lambda}(\Sigma, w)$ with inherited norm. □

The last order of business in rounding out our Calderón–Zygmund theory for weighted-vanishing Morrey spaces is to demonstrate boundedness properties for the nontangential maximal operator applied to general boundary-to-domain singular integral operators of convolution type.

Theorem 10.32. *Let $\Omega \subseteq \mathbb{R}^n$ be an open set whose topological boundary, $\partial\Omega$, is uniformly rectifiable, and define $\sigma := \mathcal{H}^{n-1}\lfloor\partial\Omega$. Let $N = N(n) \in \mathbb{N}$ be sufficiently large and consider an integral kernel $k \in \mathscr{C}^N(\mathbb{R}^n \setminus \{0\})$, which is odd and positive homogeneous of degree $1-n$. For each $f \in L^1(\partial\Omega, \frac{\sigma(x)}{1+|x|^{n-1}})$, define*

$$(\mathcal{T}f)(x) := \int_{\partial\Omega} k(x-y)f(y)\,d\sigma(y) \quad \text{at each } x \in \Omega. \tag{10.107}$$

Finally, fix an aperture parameter $\kappa \in (0,\infty)$, and choose two integrability exponents $p, q \in (1,\infty)$, a Muckenhoupt weight $w \in A_p(\partial\Omega,\sigma) \cap \mathrm{RH}_q(\partial\Omega,\sigma)$, and some threshold $\lambda \in (0,(n-1)/q')$, where $q' := q/(q-1)$.

Then

$$\mathcal{N}_\kappa(\mathcal{T}f) \in \mathring{M}^{p,\lambda}(\partial\Omega, w) \quad \text{for each } f \in \mathring{M}^{p,\lambda}(\partial\Omega, w). \tag{10.108}$$

Additionally, there exists a constant, $C \in (0,\infty)$, depending only on the UR *character of $\partial\Omega$, n, p, q, $[w]_{\mathrm{RH}_q}$, $[w]_{A_p}$, and κ such that*

$$\|\mathcal{N}_\kappa(\mathcal{T}f)\|_{M^{p,\lambda}(\partial\Omega,w)} \le C\|f\|_{M^{p,\lambda}(\partial\Omega,w)} \tag{10.109}$$

for all $f \in \mathring{M}^{p,\lambda}(\partial\Omega, w)$.

Proof. For each function $f \in \mathring{M}^{p,\lambda}(\Sigma, w)$, we have according to [88] that there exists some constant $C \in (0,\infty)$, independent of f, such that

$$0 \le \mathcal{N}_\kappa(\mathcal{T}f) \le T_{\max}f + C\cdot(\mathcal{M}_\Sigma f) \tag{10.110}$$

pointwise on Σ, where $T_{\max}$ is as in (10.104). Then Theorem 10.31 ensures that $T_{\max}f$ belongs to $\mathring{M}^{p,\lambda}(\Sigma, w)$, and Proposition 10.28 guarantees that $\mathcal{M}_\Sigma f$ belongs to $\mathring{M}^{p,\lambda}(\Sigma, w)$. As such, (10.110) and the lattice property of $\mathring{M}^{p,\lambda}(\Sigma, w)$ (cf. Proposition 10.26) imply that $\mathcal{N}_\kappa(\mathcal{T}f) \in \mathring{M}^{p,\lambda}(\Sigma, w)$ with proper quantitative control of the norm as wanted. □

The Calderón–Zygmund theory for weighted-vanishing Morrey spaces developed so far means that not only do the estimates contained throughout Chapter 5 and Chapter 9.1 continue to hold for functions belonging to a weighted-vanishing Morrey spaces, but that operator norm estimates at the level of weighted-vanishing Morrey spaces continue to hold. Of particular note and important in relation to PDE, small norm estimates for the boundary-to-boundary double layer potential as in Theorem 9.7 continue to hold when $M^{p,\lambda}(\partial\Omega, w)$ is replaced by the smaller space $\mathring{M}^{p,\lambda}(\partial\Omega, w)$.

Theorem 10.33. *The following results continue to hold when $M^{p,\lambda}(\partial\Omega, w)$ is replaced by $\mathring{M}^{p,\lambda}(\partial\Omega, w)$: Theorem 5.7, Proposition 5.8, Theorem 5.12, Corollary 5.13, Theorem 5.15, and Theorem 5.18.*

Additionally, concerning layer potentials, analogous versions of the following results also hold when $M^{p,\lambda}(\partial\Omega, w)$ is replaced by $\mathring{M}^{p,\lambda}(\partial\Omega, w)$: Theorem 9.1, Theorem 9.2, Proposition 9.3, Proposition 9.5, Proposition 9.6, and Theorem 9.7.

Proof. This follows from Proposition 10.28, Theorem 10.30, Theorem 10.31, and Theorem 10.32. □

We are also in a position to define a Sobolev space version of Muckenhoupt-weighted-vanishing Morrey spaces.

Definition 10.34. Let $\Omega \subseteq \mathbb{R}^n$ be an AR domain and write $\sigma := \mathcal{H}^{n-1}\lfloor\partial\Omega$. Fix any two integrability exponents $p, q \in (1, \infty)$, a Muckenhoupt weight $w \in A_p(\partial\Omega, \sigma) \cap \mathrm{RH}_q(\partial\Omega, \sigma)$, and some threshold $\lambda \in (0, (n-1)/q')$, where $q' := q/(q-1)$. Define the `Muckenhoupt-weighted vanishing Morrey-based Sobolev space` via

$$\mathring{M}_1^{p,\lambda}(\partial\Omega, w) := \{f \in \mathring{M}^{p,\lambda}(\Sigma, w) : \partial_{\tau_{jk}} f \text{ exist in a weak sense} \\ \text{and } \partial_{\tau_{jk}} f \in \mathring{M}^{p,\lambda}(\partial\Omega, w) \text{ for each } j, k \in \{1, \ldots, n\}\}. \tag{10.111}$$

Since $\mathring{M}^{p,\lambda}(\partial\Omega, w) \subseteq M^{p,\lambda}(\partial\Omega, w)$, it is clear from Definition 8.8 that $\mathring{M}^{p,\lambda}(\partial\Omega, w)$ is a vector subspace of $M_1^{p,\lambda}(\partial\Omega, w)$. Subsequently, we equip $\mathring{M}_1^{p,\lambda}(\partial\Omega, w)$ with the inherited norm, defined in (8.18).

In fact, the space $\mathring{M}_1^{p,\lambda}(\partial\Omega, w)$ is a closed vector subspace of $M_1^{p,\lambda}(\partial\Omega, w)$ with respect to the norm $\|\cdot\|_{M_1^{p,\lambda}(\partial\Omega,w)}$ as this next result shows.

Proposition 10.35. *Let $\Omega \subseteq \mathbb{R}^n$ be an Ahlfors regular domain and write $\sigma := \mathcal{H}^{n-1}\lfloor\partial\Omega$. Fix any two integrability exponents $p, q \in (1, \infty)$, along with a Muckenhoupt weight $w \in A_p(\partial\Omega, \sigma) \cap \mathrm{RH}_q(\partial\Omega, \sigma)$, and some threshold $\lambda \in (0, (n-1)/q')$, where $q' := q/(q-1)$.*

Then $\mathring{M}_1^{p,\lambda}(\partial\Omega, w)$ is a Banach space when equipped with $\|\cdot\|_{M_1^{p,\lambda}(\partial\Omega,w)}$, the inherited norm introduced in (8.18).

Proof. That $\mathring{M}_1^{p,\lambda}(\partial\Omega, w)$ is Banach follows immediately from the fact that $\mathring{M}_1^{p,\lambda}(\partial\Omega, w)$ is a closed vector subspace of $M_1^{p,\lambda}(\partial\Omega, w)$ with respect to the norm $\|\cdot\|_{M_1^{p,\lambda}(\partial\Omega,w)}$. Indeed, suppose $\{f_m\}_{m\in\mathbb{N}} \subseteq \mathring{M}_1^{p,\lambda}(\partial\Omega, w)$ is a sequence that converges to f in $\mathring{M}_1^{p,\lambda}(\partial\Omega, w)$. Then $\{f_m\}_{m\in\mathbb{N}}$ converges to f in $M_1^{p,\lambda}(\partial\Omega, w)$ as well, which entails that $\{f_m\}_{m\in\mathbb{N}}$ converges to f in $M^{p,\lambda}(\partial\Omega, w)$ and $\{\partial_{\tau_{jk}} f_m\}_{m\in\mathbb{N}}$ converges to $\partial_{\tau_{jk}} f$ in $M^{p,\lambda}(\partial\Omega, w)$ for each $j, k \in \{1, \ldots, n\}$. Since $\{f_m\}_{m\in\mathbb{N}} \subseteq \mathring{M}^{p,\lambda}(\partial\Omega, w)$ and $\{\partial_{\tau_{jk}} f_m\}_{m\in\mathbb{N}}$ is a sequence in $\mathring{M}^{p,\lambda}(\partial\Omega, w)$ for each $j, k \in \{1, \ldots, n\}$, and $\mathring{M}^{p,\lambda}(\partial\Omega, w)$ is a closed subspace of $M^{p,\lambda}(\partial\Omega, w)$ (cf. (10.64)), we find that in fact $f \in \mathring{M}^{p,\lambda}(\partial\Omega, w)$ and $\partial_{\tau_{jk}} f \in \mathring{M}^{p,\lambda}(\partial\Omega, w)$ for each $j, k \in \{1, \dots, n\}$ as well, and hence $f \in \mathring{M}_1^{p,\lambda}(\partial\Omega, w)$ per (10.111) as wanted. □

We also introduce a homogeneous version of the Muckenhoupt-weighted vanishing Morrey-based Sobolev space.

Definition 10.36. Let $\Omega \subseteq \mathbb{R}^n$ be an AR domain and write $\sigma := \mathcal{H}^{n-1}\lfloor\partial\Omega$. Fix any two integrability exponents $p, q \in (1, \infty)$, along with a Muckenhoupt weight $w \in A_p(\partial\Omega, \sigma) \cap \mathrm{RH}_q(\partial\Omega, \sigma)$, and some threshold $\lambda \in (0, (n-1)/q')$, where $q' := q/(q-1)$. Define the `Muckenhoupt-weighted vanishing Morrey-based homogeneous Sobolev space` according to

$$\mathbb{M}_1^{p,\lambda}(\partial\Omega, w) := \left\{ f \in L^1\left(\partial\Omega, \frac{1}{1+|x|^n}\right) \cap M_{\mathrm{loc}}^{p,\lambda}(\partial\Omega, w) : \right. \tag{10.112}$$
$$\left. \partial_{\tau_{jk}} f \in \mathring{M}^{p,\lambda}(\partial\Omega, w) \text{ for each } j, k \in \{1, \dots, n\} \right\},$$

where $M_{\mathrm{loc}}^{p,\lambda}(\partial\Omega, w)$ is as in (8.20). We analogously equip $\mathbb{M}_1^{p,\lambda}(\partial\Omega, w)$ with the following seminorm, defined for each $f \in \mathbb{M}_1^{p,\lambda}(\partial\Omega, w)$ via

$$\|f\|_{\mathbb{M}_1^{p,\lambda}(\partial\Omega,w)} := \sum_{j,k=1}^{n} \|\partial_{\tau_{jk}} f\|_{M^{p,\lambda}(\partial\Omega,w)}. \tag{10.113}$$

Occasionally, we will work with the quotient space $\mathbb{M}_1^{p,\lambda}(\partial\Omega, w)/\sim$, i. e., the space of equivalence classes of functions belonging to $\mathbb{M}_1^{p,\lambda}(\partial\Omega, w)$ modulo constants. We equip the space $\mathbb{M}_1^{p,\lambda}(\partial\Omega, w)/\sim$ with the norm inherited from $\dot{M}_1^{p,\lambda}(\partial\Omega, w)/\sim$, i. e., (8.22). Modding constants out of $\mathbb{M}_1^{p,\lambda}(\partial\Omega, w)$ allows us to conclude that $\mathbb{M}_1^{p,\lambda}(\partial\Omega, w)/\sim$ is a Banach space as detailed in the next result.

Proposition 10.37. *Let $\Omega \subseteq \mathbb{R}^n$ be an Ahlfors regular domain and write $\sigma := \mathcal{H}^{n-1}\lfloor\partial\Omega$. Fix any two integrability exponents $p, q \in (1, \infty)$, along with a Muckenhoupt weight $w \in A_p(\partial\Omega, \sigma) \cap \mathrm{RH}_q(\partial\Omega, \sigma)$, and some threshold $\lambda \in (0, (n-1)/q')$, where $q' := q/(q-1)$.*

Then the quotient space $\mathbb{M}_1^{p,\lambda}(\partial\Omega, w)/\sim$, equipped with the inherited norm from (8.22), *is a Banach space.*

Proof. The fact that $\mathbb{M}_1^{p,\lambda}(\partial\Omega, w)/\sim$ is a Banach space follows as soon as we show that that $\mathbb{M}_1^{p,\lambda}(\partial\Omega, w)/\sim$ is a closed subspace of $\dot{M}_1^{p,\lambda}(\partial\Omega, w)/\sim$. Indeed, suppose the sequence $\{[f_m]\}_{m\in\mathbb{N}} \subseteq \mathbb{M}_1^{p,\lambda}(\partial\Omega, w)/\sim$ converges to $[f]$ for some function $f \in \dot{M}_1^{p,\lambda}(\partial\Omega, w)$. Then per (8.19), we find that

$$f \in L^1\left(\partial\Omega, \frac{\sigma(x)}{1+|x|^n}\right) \cap M_{\mathrm{loc}}^{p,\lambda}(\partial\Omega, w). \tag{10.114}$$

Additionally, thanks to (8.22), we see that for each $j, k \in \{1, \dots, n\}$ there exists some function $f_{jk} \in M^{p,\lambda}(\partial\Omega, w)$ for which

$$\lim_{m\to\infty} \|\partial_{\tau_{jk}} f_m - f_{jk}\|_{M^{p,\lambda}(\partial\Omega,w)} = 0. \tag{10.115}$$

But from (10.112), we know that $\{\partial_{\tau_{jk}} f_m\}_{m\in\mathbb{N}} \subseteq \mathring{M}^{p,\lambda}(\partial\Omega, w)$, and since $\mathring{M}^{p,\lambda}(\partial\Omega, w)$ is a closed subspace of $M^{p,\lambda}(\partial\Omega, w)$ (cf. (10.64)), it must be the case that f_{jk} belongs to $\mathring{M}^{p,\lambda}(\partial\Omega, w)$ as well for each $j, k \in \{1, \dots, n\}$. This fact and (10.114) show that $[f] \in \mathbb{M}_1^{p,\lambda}(\partial\Omega, w)/\sim$ as wanted. □

Interestingly, all of the results in Chapter 9.1, which concern layer potentials on Muckenhoupt-weighted Morrey-based Sobolev spaces, carry over if we wish to work with the smaller, Muckenhoupt-weighted vanishing Morrey-based Sobolev space.

Theorem 10.38. *The following results continue to hold when $M_1^{p,\lambda}(\partial\Omega, w)$ is replaced by $\mathring{M}_1^{p,\lambda}(\partial\Omega, w)$: Proposition 9.4, Proposition 9.9, Theorem 9.10, and Proposition 9.11.*

Additionally, the following results continue to hold when $M^{p,\lambda}(\partial\Omega, w)$ and $\dot{M}_1^{p,\lambda}(\partial\Omega, w)$ is replaced by $\mathring{M}^{p,\lambda}(\partial\Omega, w)$ and $\mathbb{M}_1^{p,\lambda}(\partial\Omega, w)$, respectively: Theorem 9.12, Proposition 9.13, Proposition 9.14, Theorem 9.15, Proposition 9.32, and Theorem 9.36.

Proof. Analogous arguments while keeping in mind Theorem 10.33 yield the desired result. □

The culmination of our results about vanishing Morrey spaces now comes to a head with the following duality result, which identifies Muckenhoupt-weighted Block spaces as the dual spaces of Muckenhoupt-weighted vanishing Morrey spaces. This is a generalization of the result in [87, Proposition 6.2.16, pp. 344–345] for the unweighted case.

Theorem 10.39. *Let $\Sigma \subseteq \mathbb{R}^n$ be a closed ADR set, and define $\sigma := \mathcal{H}^{n-1}\lfloor\Sigma$. Fix integrability exponents $p, p', q, q' \in (1,\infty)$ with $1/p + 1/p' = 1$ and $1/q + 1/q' = 1$. Fix a Muckenhoupt weight $w \in A_p(\Sigma,\sigma) \cap \mathrm{RH}_q(\Sigma,\sigma)$, and bring in the conjugate weight $w' := w^{-p'/p} \in A_{p'}(\Sigma,\sigma)$. Define*

$$\begin{aligned} &\varepsilon := (p'-1)/q' \in (0, p'-1) \quad \text{and fix a threshold} \\ &\lambda \in (0, (n-1)/q') = (0, \varepsilon(n-1)/(q-1)). \end{aligned} \tag{10.116}$$

In view of Proposition 6.35, the weight w' belongs to $A_{p'-\varepsilon}(\Sigma,\sigma)$, and hence $\mathcal{B}^{p',\lambda}(\Sigma, w')$ is a well-defined Muckenhoupt-weighted Block space in the sense of Definition 6.7 for these choices of p', w', ε, and λ.

Then the Muckenhoupt-weighted vanishing Morrey space $\mathring{M}^{p,\lambda}(\Sigma, w)$ is the pre-dual of the Muckenhoupt-weighted Block space $\mathcal{B}^{p',\lambda}(\Sigma, w')$, i. e.,

$$(\mathring{M}^{p,\lambda}(\Sigma, w))^* = \mathcal{B}^{p',\lambda}(\Sigma, w'). \tag{10.117}$$

More precisely, the mapping

$$\mathcal{B}^{p',\lambda}(\Sigma, w') \ni g \longmapsto \Lambda_g \in (\mathring{M}^{p,\lambda}(\Sigma, w))^*, \tag{10.118}$$

with

$$\Lambda_g(f) := \int_\Sigma fg\,d\sigma \quad \textit{for each } f \in \mathring{M}^{p,\lambda}(\Sigma, w), \tag{10.119}$$

is a well-defined, linear, and continuous isomorphism with bounded inverse.

A couple of comments are warranted before preceding with the proof.

Remark 10.40. In order to facilitate the proof of Theorem 10.39, we observe the following fact. Start with a closed Ahlfors regular set $\Sigma \subseteq \mathbb{R}^n$ and write $\sigma := \mathcal{H}^{n-1}\lfloor\Sigma$. Fix an integrability exponent $p \in (1, \infty)$ and a Muckenhoupt weight $w \in A_p(\Sigma, \sigma)$. The Riesz representation theorem gives the standard identification of $(L^p(\Sigma, w))^*$ with $L^{p'}(\Sigma, w)$, where $p' \in (1, \infty)$ is the Hölder conjugate of p. With $w' := w^{1-p'}$ denoting the conjugate weight to w, we introduce the alternate identification

$$\big(L^p(\Sigma, w)\big)^* = L^{p'}(\Sigma, w'), \tag{10.120}$$

which entails that the mapping

$$\begin{aligned} &L^{p'}(\Sigma, w') \ni g \longmapsto \Lambda_g \in \big(L^p(\Sigma, w)\big)^*, \quad \text{where} \\ &\Lambda_g(f) := \int_\Sigma fg\,d\sigma \quad \text{for each } f \in L^p(\Sigma, w) \end{aligned} \tag{10.121}$$

is a well-defined, linear, and continuous isomorphism with bounded inverse. We can see that the mapping in (10.121) is well-defined and bounded thanks to Hölder's inequality. We show that (10.121) defines an isomorphism in the following way. Injectivity is clear given that if $\Lambda_g(f) = 0$ for all $f \in L^p(\Sigma, w)$, then in particular $\int_\Delta g\,d\sigma = 0$ for all surface balls $\Delta \subseteq \Sigma$, which confirms $g = 0$ at σ-a. e. point on Σ on account of Corollary 2.23. For the matter of surjectivity, note that the standard identification $(L^p(\Sigma, w))^* = L^{p'}(\Sigma, w)$ from the Riesz representation means for any fixed $\Lambda \in (L^p(\Sigma, w))^*$ there exists a (σ-a. e.) unique function $g \in L^{p'}(\Sigma, w)$ for which

$$\Lambda(f) = \int_\Sigma fg\,dw \quad \text{for each } f \in L^p(\Sigma, w). \tag{10.122}$$

In order to make the identification in (10.120), define $G := gw$ on Σ, where g is as in (10.122). Then

$$\begin{aligned} \Big(\int_\Sigma |G|^{p'}\,dw'\Big)^{1/p'} &= \Big(\int_\Sigma |g|^{p'} w^{p'} w^{1-p'}\,d\sigma\Big)^{1/p'} \\ &= \Big(\int_\Sigma |g|^{p'}\,dw\Big)^{1/p'} < +\infty, \end{aligned} \tag{10.123}$$

which demonstrates that $G \in L^{p'}(\Sigma, w')$. Then

$$\int_\Sigma fG\,d\sigma = \int_\Sigma fgw\,d\sigma = \int_\Sigma fg\,dw = \Lambda(f). \tag{10.124}$$

Hence, the mapping in (10.121) is surjective. Thus, (10.121) is a linear isomorphism (with bounded inverse by the open mapping theorem), justifying the identification in (10.120).

We are now ready to present the proof of the duality result advertised in the statement of Theorem 10.39.

Proof of Theorem 10.39. To begin, notice that since $\mathring{M}^{p,\lambda}(\Sigma, w)$ is a closed linear subspace of $M^{p,\lambda}(\Sigma, w)$ (cf. (10.64)), we can use Theorem 6.28 and (10.119) to conclude that the mapping in (10.118) is well-defined, linear, and bounded. The matter of injectivity is taken care of in the following way. Namely, suppose that

$$\Lambda_g(f) = 0 \quad \text{for all } f \in \mathring{M}^{p,\lambda}(\Sigma, w). \tag{10.125}$$

Then in particular, given (10.119), we find that $\int_\Sigma g\mathbf{1}_\Delta\,d\sigma = 0$ for each surface ball $\Delta \subseteq \Sigma$. Note $\mathbf{1}_\Delta \in \mathring{M}^{p,\lambda}(\Sigma, w)$ since $L^\infty_{\mathrm{comp}}(\Sigma) \subseteq \mathring{M}^{p,\lambda}(\Sigma, w)$ (cf. Proposition 10.21). It then follows that $g = 0$ at σ-a. e. point on Σ by Corollary 2.23.

The rest of the proof is dedicated to proving the surjectivity of (10.118). As in the proof of [87, Proposition 6.2.16, pp. 344–345], we break the proof up into four steps. We will provide full details in the case when Σ is unbounded, since the situation when Σ is bounded requires only minor alterations having to do with the nature of the dyadic grid structure, described in Lemma 2.42.

Step 1: *Functional analytic preliminaries.* Consider an arbitrary sequence of Banach spaces $\mathcal{A} = \{A_j\}_{j\in\mathbb{N}}$. Define the Banach spaces

$$\ell_1(\mathcal{A}) := \left\{a = \{a_j\}_{j\in\mathbb{N}} \in \Pi_{j\in\mathbb{N}}A_j : \|a\|_{\ell_1(\mathcal{A})} := \sum_{j=1}^{\infty} \|a_j\|_{A_j} < \infty\right\}, \tag{10.126}$$

$$\ell_\infty(\mathcal{A}) := \left\{a = \{a_j\}_{j\in\mathbb{N}} \in \Pi_{j\in\mathbb{N}}A_j : \|a\|_{\ell_\infty(\mathcal{A})} := \sup_{j\in\mathbb{N}} \|a_j\|_{A_j} < \infty\right\}, \tag{10.127}$$

and write

$$c_0(\mathcal{A}) := \left\{a = \{a_j\}_{j\in\mathbb{N}} \in \ell_\infty(\mathcal{A}) : \lim_{j\to\infty} \|a_j\|_{A_j} = 0\right\}. \tag{10.128}$$

It follows that $c_0(\mathcal{A})$ is a closed subspace of $\ell_\infty(\mathcal{A})$, and hence $c_0(\mathcal{A})$ is its own Banach space when equipped with the inherited norm $\|\cdot\|_{\ell_\infty(\mathcal{A})}$.

Next, define the collection of dual spaces $\mathcal{A}^* := \{A_j^*\}_{j\in\mathbb{N}}$, where each A_j^* is the dual of A_j, and introduce the pairing

$$(a, a') := \sum_{j=1}^{\infty} (a_j, a_j') \quad \text{for each} \qquad (10.129)$$
$$a = \{a_j\}_{j \in \mathbb{N}} \in \Pi_{j \in \mathbb{N}} A_j \quad \text{and} \quad a' = \{a_j'\}_{j \in \mathbb{N}} \in \Pi_{j \in \mathbb{N}} A_j^*.$$

Then, due to [33, pp. 73–74] (cf. also [148, Lemma 1.11.1, p. 68]), we can conclude that

under the pairing (10.128), the dual of the Banach space $c_0(\mathcal{A})$ may be canonically identified with the space $\ell_1(\mathcal{A}^*)$. (10.130)

Lastly, given two Banach spaces X, Y, we agree to call a linear map $\Phi : X \to Y$ a `quasi-isometry`, provided there exists some number $\theta \in (0, 1]$ with the property that $\theta \|x\|_X \leq \|\Phi x\|_Y \leq \theta^{-1} \|x\|_X$ for each $x \in X$. As a direct consequence, in such a setting Φ maps any closed subspace of X into a closed subspace of Y isomorphically.

Step 2: *Labeling the dyadic grid on* Σ. Observe that Lemma 2.42 ensures that there exists a dyadic grid structure on Σ, denoted by $\mathbb{D}(\Sigma)$. For each dyadic cube $Q \in \mathbb{D}(\Sigma)$, denote by $\ell(Q)$ and x_Q the side length and center of Q, respectively, in the sense of Definition 2.43. Since $\mathbb{D}(\Sigma)$ is a countable set, there exists a bijective identification with $\mathbb{N}$; however, not in any canonical way. We will construct a specific identification, which suits our purposes. To do so, start by fixing a reference point $x_0 \in \Sigma$ and define

$$\operatorname{rank}(Q) := \max\{\ell(Q), \ell(Q)^{-1}, \operatorname{dist}(x_0, Q)\}, \quad \text{for all } Q \in \mathbb{D}(\Sigma). \qquad (10.131)$$

For each $j \in \mathbb{N}$, define the collection of dyadic cubes

$$\mathbb{D}_j(\Sigma) := \{Q \in \mathbb{D}(\Sigma) : j - 1 < \operatorname{rank}(Q) \leq j\}, \qquad (10.132)$$

which necessarily has finite cardinality, i. e., $\#\mathbb{D}_j(\Sigma) < +\infty$ for each $j \in \mathbb{N}$. Moreover, $\mathbb{D}(\Sigma)$ is the disjoint union of all the $\mathbb{D}_j(\Sigma)$, $j \in \mathbb{N}$. Now, label $\mathbb{D}_1(\Sigma)$ using indices from the set $\{1, 2, \ldots, \#\mathbb{D}_1(\Sigma)\}$, then label $\mathbb{D}_2(\Sigma)$ using indices from the set $\{\#\mathbb{D}_1(\Sigma) + 1, \#\mathbb{D}_1(\Sigma) + 2, \ldots, \#\mathbb{D}_1(\Sigma) + \#\mathbb{D}_2(\Sigma)\}$, and so on. Ultimately, this yields

an enumeration $Q_1, Q_2, \ldots, Q_j, Q_{j+1}, \ldots$ of $\mathbb{D}(\Sigma)$ with the property that $\operatorname{rank}(Q_j) \to \infty$ as $j \to \infty$. (10.133)

Whenever convenient, from here on out we agree to identify $\mathbb{D}(\Sigma)$ with $\mathbb{N}$ in the above manner.

Step 3: *Discretizing weighted Morrey spaces*. Start by defining the family of Banach spaces

$$\mathcal{A}_\Sigma := \{A_Q\}_{Q \in \mathbb{D}(\Sigma)} \quad \text{with } A_Q := L^p(Q, \ell(Q)^{-\lambda} w) \text{ for each } Q \in \mathbb{D}(\Sigma). \qquad (10.134)$$

Recall the framework outlined in steps 1 and 2 (in particular the identification of $\mathbb{D}(\Sigma)$ with $\mathbb{N}$ as in (10.133)), so that we have

$$\mathcal{A}_\Sigma^* = \{A_Q^*\}_{Q\in\mathbb{D}(\Sigma)} \quad \text{with } A_Q^* = L^{p'}(Q\,,\,\ell(Q)^{-\lambda}\,w') \text{ for each } Q \in \mathbb{D}(\Sigma), \tag{10.135}$$

where $w' := w^{1-p'}$ is the conjugate weight to w. Note also that we are identifying $(L^p(\Sigma, \ell(Q)^{-\lambda}w))^*$ with $L^{p'}(Q, \ell(Q)^{-\lambda}w')$ as in Remark 10.40. Also, unpacking definitions from Step 1, for each $F = \{f_Q\}_{Q\in\mathbb{D}(\Sigma)} \in \ell_\infty(\mathcal{A}_\Sigma)$ and $G = \{g_Q\}_{Q\in\mathbb{D}(\Sigma)} \in \ell_1(\mathcal{A}_\Sigma^*)$, we find that

$$\|F\|_{\ell_\infty(\mathcal{A}_\Sigma)} = \sup_{Q\in\mathbb{D}(\Sigma)} \Big(\int_Q |f_Q|^p \ell(Q)^{-\lambda}\,dw\Big)^{1/p}, \tag{10.136}$$

$$\|G\|_{\ell_1(\mathcal{A}_\Sigma^*)} = \sum_{Q\in\mathbb{D}(\Sigma)} \Big(\int_Q |g_Q|^{p'} \ell(Q)^{-\lambda}\,dw'\Big)^{1/p'}, \tag{10.137}$$

$$(F, G) = \sum_{Q\in\mathbb{D}(\Sigma)} \int_Q f_Q g_Q\,\ell(Q)^{-\lambda}\,d\sigma. \tag{10.138}$$

Next, define the linear mapping

$$M^{p,\lambda}(\Sigma, w) \ni f \overset{\Phi}{\longmapsto} F := \{f|_Q\}_{Q\in\mathbb{D}(\Sigma)} \in \ell_\infty(\mathcal{A}_\Sigma). \tag{10.139}$$

We will show that

$$\Phi \text{ is a quasiisometry, which maps } \mathrm{Lip}_c(\Sigma) \text{ into } c_0(\mathcal{A}_\Sigma). \tag{10.140}$$

To see this, first select any $f \in M^{p,\lambda}(\Sigma, w)$. Items (6) and (7) of Lemma 2.42 guarantee the existence of some number $N \in \mathbb{N}$ such that for each location $x \in \Sigma$ and scale $R \in (0, 2\,\mathrm{diam}(\Sigma))$ there are at most N dyadic cubes, say $\{Q_i\}_{i\in I} \subseteq \mathbb{D}(\Sigma)$ with $\#I \le N$, with the property that for each $i \in I$ there holds

$$\begin{aligned} &\ell(Q_i) \approx R \quad \text{with proportionality constants independent of } x, R, \\ &\text{and} \quad \sigma(\Delta(x, R) \setminus \cup_{i\in I} Q_i) = 0. \end{aligned} \tag{10.141}$$

With these properties in hand, we may estimate

$$\begin{aligned} \Big(\frac{1}{R^\lambda} \int_{\Delta(x,R)} |f|^p\,dw\Big)^{1/p} &\le C \sum_{i\in I} \Big(\ell(Q_i)^{-\lambda} \int_{Q_i} |f|^p\,dw\Big)^{1/p} \\ &\le C \sup_{Q\in\mathbb{D}(\Sigma)} \Big(\int_Q |f|^p \ell(Q)^{-\lambda}\,dw\Big)^{1/p} \\ &= C\|\Phi(f)\|_{\ell_\infty(\mathcal{A}_\Sigma)}. \end{aligned} \tag{10.142}$$

Taking a supremum over all locations $x \in \Sigma$ and all scales $R \in (0, 2\,\mathrm{diam}(\Sigma))$ (bearing in mind (4.2)) yields

$$\|f\|_{M^{p,\lambda}(\Sigma,w)} \leq C\|\Phi(f)\|_{\ell_\infty(\mathcal{A}_\Sigma)}. \tag{10.143}$$

On the other hand, (5) of Lemma 2.42 and (4.2) ensure that the inequality in the opposite direction holds, confirming that Φ is a quasiisometry.

To verify the second claim regarding Φ in (10.140), fix some arbitrary $f \in \mathrm{Lip}_c(\Sigma)$. Then

$$\|f|_Q\|_{L^p(Q,\ell(Q)^{-\lambda}w)} \leq \ell(Q)^{-\lambda/p}\|f\|_{L^\infty(\Sigma,\sigma)}w(\operatorname{supp} f)^{1/p}, \tag{10.144}$$

the right side of which goes to 0 as $\ell(Q) \to \infty$ since $\lambda > 0$. Also, since we restrict λ to the interval $(0, (n-1)/q')$, we see that $\lambda < n-1$ and so

$$\begin{aligned} \|f|_Q\|_{L^p(Q,\ell(Q)^{-\lambda}w)} &\approx \ell(Q)^{-\lambda/p}w(Q)^{1/p}\Big(\fint_Q |f|^p\,dw\Big)^{1/p} \\ &\leq \ell(Q)^{(n-1-\lambda)/p}\Big(\fint_Q |f|^p\,dw\Big)^{1/p} \\ &\leq \ell(Q)^{(n-1-\lambda)/p}\|f\|_{L^\infty(\Sigma,\sigma)} \longrightarrow 0 \quad \text{as } \ell(Q) \to 0^+. \end{aligned} \tag{10.145}$$

The first line is due to item (5) of Lemma 2.42 and second line follows from [86, (7.7.100), p. 658]. Finally, since f has compact support, we can say

$$\begin{aligned} &\text{for each fixed } \varepsilon \in (0,1), \text{ within the class of cubes } Q \in \mathbb{D}(\Sigma), \\ &\text{which satisfy } \varepsilon < \ell(Q) < \varepsilon-1, \text{ we have } \|f|_Q\|_{L^p(Q,\ell(Q)^{-\lambda}w)} \longrightarrow 0 \\ &\text{as } \operatorname{dist}(x_0, Q) \to \infty. \end{aligned} \tag{10.146}$$

Thus, recalling (10.131) and (10.133), the results in (10.144)–(10.146) show that $\Phi(f)$ actually belongs to $c_0(\mathcal{A}_\Sigma)$. This completes the justification of all properties of Φ advertised in (10.140), given the arbitrary nature of the function $f \in \mathrm{Lip}_c(\Sigma)$.

With (10.140) in hand, the continuity of Φ and (10.64) ensure that Φ is a quasiisometry, which maps $\mathring{M}^{p,\lambda}(\Sigma, w)$ into $c_0(\mathcal{A}_\Sigma)$, and hence $\Phi(\mathring{M}^{p,\lambda}(\Sigma, w))$ is a closed subspace of $c_0(\mathcal{A}_\Sigma)$ and the map

$$\Phi : \mathring{M}^{p,\lambda}(\Sigma, w) \to \Phi(\mathring{M}^{p,\lambda}(\Sigma, w)) \tag{10.147}$$

is an isomorphism, whose inverse we denote by Φ^{-1}. That is,

$$\begin{aligned} &\Phi^{-1} : \Phi(\mathring{M}^{p,\lambda}(\Sigma, w)) \to \mathring{M}^{p,\lambda}(\Sigma, w) \\ &\text{is itself an isomorphism with the property that} \\ &\Phi^{-1}(\Phi(f)) = f \quad \text{for every } f \in \mathring{M}^{p,\lambda}(\Sigma, w). \end{aligned} \tag{10.148}$$

Step 4: *Proof of the duality formula.* Fix an arbitrary functional Λ belonging to $(\mathring{M}^{p,\lambda}(\Sigma, w))^*$. Then (10.148) shows that $\Lambda \circ \Phi^{-1}$ is a continuous, linear functional on $\Phi(\mathring{M}^{p,\lambda}(\Sigma, w))$, which we recall is a closed subspace of $c_0(\mathcal{A}_\Sigma)$. By the Hahn–Banach theorem, there exists a continuous linear functional $\widetilde{\Lambda}$ on $c_0(\mathcal{A}_\Sigma)$ with the property that

$$\begin{aligned} \widetilde{\Lambda}\big|_{\Phi(\mathring{M}^{p,\lambda}(\Sigma,w))} &= \Lambda \circ \Phi^{-1} \quad \text{on } \Phi(\mathring{M}^{p,\lambda}(\Sigma, w)) \text{ and} \\ \|\widetilde{\Lambda}\|_{c_0(\mathcal{A}_\Sigma)^*} &= \|\Lambda \circ \Phi^{-1}\|_{(\Phi(\mathring{M}^{p,\lambda}(\Sigma,w)))^*} \approx \|\Lambda\|_{(\mathring{M}^{p,\lambda}(\Sigma,w))^*}. \end{aligned} \tag{10.149}$$

This then implies that $\widetilde{\Lambda}$ belongs to the dual of $\ell_\infty(\mathcal{A}_\Sigma)$, which has been identified with $\ell_1(\mathcal{A}_\Sigma^*)$. Subsequently, this entails the existence of some sequence $G = \{g_Q\}_{Q\in\mathbb{D}(\Sigma)} \in \ell_1(\mathcal{A}_\Sigma^*)$ with the property that

$$\begin{gathered} \widetilde{\Lambda}(F) = (F, G) = \sum_{Q\in\mathbb{D}(\Sigma)} \int_Q f_Q g_Q \,\ell(Q)^{-\lambda}\, d\sigma \\ \text{for each } F = \{f_Q\}_{Q\in\mathbb{D}(\Sigma)} \in c_0(\mathcal{A}_\Sigma). \end{gathered} \tag{10.150}$$

Additionally, the second line in (10.149) further yields

$$\|G\|_{\ell_1(\mathcal{A}_\Sigma^*)} \approx \|\Lambda\|_{(\mathring{M}^{p,\lambda}(\Sigma,w))^*}. \tag{10.151}$$

For each function $f \in \mathring{M}^{p,\lambda}(\Sigma, w)$, observe that $F := \{f|_Q\}_{Q\in\mathbb{D}(\Sigma)} = \Phi(f)$ belongs to the space $\Phi(\mathring{M}^{p,\lambda}(\Sigma, w)) \subseteq c_0(\mathcal{A}_\Sigma^*)$. As such, appealing to (10.149), (10.150), and (10.151), we can write

$$\begin{aligned} \Lambda(f) &= \Lambda(\Phi^{-1}(\Phi(f))) = (\Lambda \circ \Phi^{-1})(\Phi(f)) = \widetilde{\Lambda}(F) \\ &= \sum_{Q\in\mathbb{D}(\Sigma)} \int_Q (f|_Q) g_Q \,\ell(Q)^{-\lambda}\, d\sigma = \sum_{Q\in\mathbb{D}(\Sigma)} \int_Q f g_Q \,\ell(Q)^{-\lambda}\, d\sigma. \end{aligned} \tag{10.152}$$

Pressing forward, for each dyadic cube $Q \in \mathbb{D}(\Sigma)$ agree to call $\widetilde{g}_Q$ the extension by zero of $g_Q \in A_Q^* = (L^{p'}(Q, \ell(Q)^{-\lambda} w'))^*$ to all of Σ. As such, each $\widetilde{g}_Q$ belongs to $L^{p'}(\Sigma, w')$ and we have that

$$\|\widetilde{g}_Q\|_{L^{p'}(\Sigma,w')} = \|g_Q\|_{L^{p'}(Q,w')} = \ell(Q)^{\lambda/p}\|g_Q\|_{L^{p'}(Q,\ell(Q)^{-\lambda}w')}. \tag{10.153}$$

Thus, keeping in mind (10.151), we find that

$$\begin{aligned} \sum_{Q\in\mathbb{D}(\Sigma)} \ell(Q)^{-\lambda/p'} \|\widetilde{g}_Q\|_{L^{p'}(\Sigma,w')} &= \sum_{Q\in\mathbb{D}(\Sigma)} \|g_Q\|_{L^{p'}(Q,\ell(Q)^{-\lambda}w')} \\ &= \|G\|_{\ell_1(\mathcal{A}_\Sigma^*)} \approx \|\Lambda\|_{(\mathring{M}^{p,\lambda}(\Sigma,w))^*}. \end{aligned} \tag{10.154}$$

We would like to show that

$$\begin{aligned}&\text{the series } g := \textstyle\sum_{Q\in\mathbb{D}(\Sigma)} \widetilde{g}_Q \ell(Q)^{-\lambda} \text{ converges absolutely}\\ &\text{in the space } L^1(\Sigma, \tfrac{\sigma(x)}{1+|x|^{n-1}}).\end{aligned} \tag{10.155}$$

To confirm that this is the case, recall the parameter a_1 from Lemma 2.42 and define

$$\lambda_Q := a_1^{\lambda/p} \ell(Q)^{-\lambda/p'} \|\widetilde{g}_Q\|_{L^{p'}(\Sigma,w')} \in [0,\infty) \tag{10.156}$$

together with

$$b_Q := \begin{cases} \widetilde{g}_Q \ell(Q)^{-\lambda}/\lambda_Q & \text{if } \lambda_Q \neq 0,\\ 0 & \text{if } \lambda_Q = 0.\end{cases} \tag{10.157}$$

Then in fact, each b_Q is a (p',λ,w')-block on Σ in the sense of Definition 6.1 since item (5) of Lemma 2.42 yields

$$\operatorname{supp} b_Q = \operatorname{supp} \widetilde{g}_Q \subseteq Q \subseteq \Delta(x_Q, a_1 \ell(Q)) \tag{10.158}$$

(where x_Q is the center of Q as in Definition 2.43), and

$$\begin{aligned}\|b_Q\|_{L^{p'}(\Sigma,w')} &= \frac{\|\widetilde{g}_Q\|_{L^{p'}(\Sigma,w')} \ell(Q)^{-\lambda}}{a_1^{\lambda/p} \ell(Q)^{-\lambda/p'} \|\widetilde{g}_Q\|_{L^{p'}(\Sigma,w')}}\\ &= (a_1 \ell(Q))^{\lambda(1/p'-1)}.\end{aligned} \tag{10.159}$$

Moreover, (10.156) and (10.154) give that

$$\begin{aligned}&\{\lambda_Q\}_{Q\in\mathbb{D}(\Sigma)} \quad \text{is a sequence in } \ell^1 \text{ with}\\ &\sum_{Q\in\mathbb{D}(\Sigma)} |\lambda_Q| \le C \|\Lambda\|_{(\mathring{\mathcal{M}}^{p,\lambda}(\Sigma,w))^*}\end{aligned} \tag{10.160}$$

for some constant $C \in (0,\infty)$. Recalling the definition of g in (10.155) along with (10.156) and (10.157), we find that g may be rewritten as

$$g = \sum_{Q\in\mathbb{D}(\Sigma)} \lambda_Q b_Q. \tag{10.161}$$

Given that $\{b_Q\}_{Q\in\mathbb{D}(\Sigma)}$ is a family of (p',λ,w')-blocks and $\{\lambda_Q\}_{Q\in\mathbb{D}(\Sigma)} \in \ell^1$, we can conclude by Lemma 6.8 that (10.155) is true. Moreover, we can say thanks to (10.160) (bearing in mind (6.36) and (6.41)) that

$$g \in \mathcal{B}^{p',\lambda}(\Sigma, w') \quad \text{with } \|g\|_{\mathcal{B}^{p',\lambda}(\Sigma,w')} \leq C\|\Lambda\|_{(\mathring{M}^{p,\lambda}(\Sigma,w))^*}. \tag{10.162}$$

Moving into the final stage of the proof, given any $f \in L^\infty_{\text{comp}}(\Sigma, \sigma)$, we may write

$$\begin{aligned}\Lambda(f) &= \sum_{Q\in\mathbb{D}(\Sigma)} \int_Q f g_Q \ell(Q)^{-\lambda}\, d\sigma \\ &= \int_\Sigma f\Big(\sum_{Q\in\mathbb{D}(\Sigma)} \tilde{g}_Q \ell(Q)^{-\lambda}\Big)\, d\sigma \\ &= \int_\Sigma fg\, d\sigma = \Lambda_g(f).\end{aligned} \tag{10.163}$$

The first equality is implied by (10.152), keeping in mind that $L^\infty_{\text{comp}}(\Sigma, \sigma)$ is contained in $\mathring{M}^{p,\lambda}(\Sigma, w)$ (cf. Proposition 10.21). The second equality is justified by the membership of f to $L^\infty_{\text{comp}}(\Sigma, \sigma)$ and (10.155). The third equality is also a result of (10.155), and the last equality follows from (10.119). Since $L^\infty_{\text{comp}}(\Sigma, \sigma)$ is a dense subspace of $\mathring{M}^{p,\lambda}(\Sigma, w)$ (cf. Proposition 10.21), and since Λ and Λ_g are both continuous linear functionals on $\mathring{M}^{p,\lambda}(\Sigma, w)$, which coincide on $L^\infty_{\text{comp}}(\Sigma, \sigma)$, we can conclude that in fact, $\Lambda = \Lambda_g$ as functionals in $(\mathring{M}^{p,\lambda}(\Sigma, w))^*$. Given that $\Lambda \in (\mathring{M}^{p,\lambda}(\Sigma, w))^*$ was arbitrary, this completes the proof of surjectivity for the mapping in (10.118), ultimately completing the proof of Theorem 10.39. □

10.3 Morrey and Block spaces as generalized Banach function spaces

We have introduced the notion of a generalized Banach function space, and with the duality result from Theorem 10.39, we are in a position to contextualize Muckenhoupt-weighted Morrey and Block spaces within this more abstract framework. This is where we begin to see the payoff for considering vanishing Morrey spaces, and the echos of this section will be heard throughout our treatment of the various boundary value problems considered in the final chapters.

Our first result is simply to state that Muckenhoupt-weighted Morrey spaces are generalized Banach function spaces.

Proposition 10.41. *Let $\Sigma \subseteq \mathbb{R}^n$ be a closed Ahlfors regular set, and write $\sigma := \mathcal{H}^{n-1}\lfloor\Sigma$. Fix two arbitrary integrability exponents $p, q \in (1, \infty)$, along with a Muckenhoupt weight $w \in A_p(\Sigma, \sigma) \cap \mathrm{RH}_q(\Sigma, \sigma)$, and some threshold $\lambda \in (0, (n-1)/q')$, where $q' := q/(q-1)$.*

Then the mapping

$$\|\cdot\|_{M^{p,\lambda}(\Sigma,w)} : \mathscr{M}_+(\Sigma, \sigma) \to [0, \infty] \tag{10.164}$$

is a function norm in the sense of Definition 10.1, *which makes* $M^{p,\lambda}(\Sigma, w)$ *a generalized Banach function space on* Σ *in the sense of Definition* 10.3.

Proof. To start, observe that the definition of $\|\cdot\|_{M^{p,\lambda}(\Sigma,w)}$ (cf. (4.2)) immediately shows that $\|\cdot\|_{M^{p,\lambda}(\Sigma,w)}$ satisfies (P1). We also obtain (P2) as a consequence of Proposition 4.7.

Next, we show that $\|\cdot\|_{M^{p,\lambda}(\Sigma,w)}$ satisfies (P3). Suppose $\{f_j\}_{j\in\mathbb{N}}$ is a sequence in $\mathscr{M}_+(\Sigma,\sigma)$ that increases to f pointwise σ-a. e. on Σ as $j \to \infty$. Then by (P2), we can say that

$$\|f_j\|_{M^{p,\lambda}(\Sigma,w)} \leq \|f\|_{M^{p,\lambda}(\Sigma,w)} \quad \text{for all } j \in \mathbb{N}. \tag{10.165}$$

Moreover, by (P2), $\|f_j\|_{M^{p,\lambda}(\Sigma,w)} \leq \|f_{j+1}\|_{M^{p,\lambda}(\Sigma,w)}$ for all $j \in \mathbb{N}$. Taking the limit as $j \to \infty$ shows that

$$\lim_{j\to\infty} \|f_j\|_{M^{p,\lambda}(\Sigma,w)} \leq \|f\|_{M^{p,\lambda}(\Sigma,w)}. \tag{10.166}$$

To obtain the reverse inequality, fix $x_0 \in \Sigma$ and $R \in (0, 2\,\mathrm{diam}(\Sigma))$. Then bearing in mind (4.2), we find that

$$\begin{aligned} \lim_{j\to\infty} \|f_j\|_{M^{p,\lambda}(\Sigma,w)} &\geq \lim_{j\to\infty}\Big(\frac{1}{R^\lambda}\int_{\Delta(x_0,R)} f_j^p\,dw\Big)^{1/p} \\ &= \Big(\frac{1}{R^\lambda}\int_{\Delta(x_0,R)} f^p\,dw\Big)^{1/p}, \end{aligned} \tag{10.167}$$

where equality follows from Lebesgue's monotone convergence theorem. Taking a supremum over all $x_0 \in \Sigma$ and $R \in (0, 2\,\mathrm{diam}(\Sigma))$ yields

$$\lim_{j\to\infty} \|f_j\|_{M^{p,\lambda}(\Sigma,w)} \geq \|f\|_{M^{p,\lambda}(\Sigma,w)}, \tag{10.168}$$

and hence (recalling (10.166)) we see that

$$\lim_{j\to\infty} \|f_j\|_{M^{p,\lambda}(\Sigma,w)} = \|f\|_{M^{p,\lambda}(\Sigma,w)}, \tag{10.169}$$

as wanted.

To see that $\|\cdot\|_{M^{p,\lambda}(\Sigma,w)}$ satisfies (P4), fix $x_0 \in \Sigma$ and observe that we may write $\Sigma = \bigcup_{j\in\mathbb{N}} \Sigma_j$, where $\Sigma_j := \Delta(x_0, j)$ for each $j \in \mathbb{N}$. Then $\|\mathbf{1}_{\Sigma_j}\|_{M^{p,\lambda}(\Sigma,w)} < +\infty$ as a consequence of Lemma 4.9. Finally, for (P5), notice that the inequality $\int_{\Sigma_j} f\,d\sigma \leq c_j\|f\|_{M^{p,\lambda}(\Sigma,w)}$ is trivially satisfied by any choice of $c_j \in (0,\infty)$ whenever f belongs to $\mathscr{M}_+(\Sigma,\sigma)$ and $\|f\|_{M^{p,\lambda}(\Sigma,w)}$ is not finite. Therefore, suppose that $f \in \mathscr{M}_+(\Sigma,\sigma) \cap M^{p,\lambda}(\Sigma,w)$. Then by Hölder's inequality, for each $j \in \mathbb{N}$, we have

$$\begin{aligned}\int_{\Sigma_j} f\,d\sigma &= \int_{\Delta(x_0,j)} f w^{1/p} w^{-1/p}\,d\sigma \\ &\leq \Big(\int_{\Delta(x_0,j)} f^p\,dw\Big)^{1/p} w^{-p'/p}(\Delta(x_0,j)) \\ &\leq j^{\lambda/p} w^{-p'/p}(\Delta(x_0,j))\Big(\frac{1}{j^\lambda}\int_{\Delta(x_0,j)} f^p\,dw\Big)^{1/p} \\ &\leq c_j\|f\|_{M^{p,\lambda}(\Sigma,w)},\end{aligned} \tag{10.170}$$

where $c_j := j^{\lambda/p} w^{-p'/p}(\Delta(x_0,j)) < +\infty$. Alternatively, (P5) follows immediately from Proposition 4.16, which gives the continuous embedding $M^{p,\lambda}(\Sigma,w) \hookrightarrow L^1_{\mathrm{loc}}(\Sigma,\sigma)$. This completes the justification that $\|\cdot\|_{M^{p,\lambda}(\Sigma,w)}$ is a function norm, ultimately confirming that $M^{p,\lambda}(\Sigma,w)$ is a generalized Banach function space. □

We would like to show that Muckenhoupt-weighted Block spaces are also generalized Banach function spaces, however, the block norm defined in (6.41) does not lend itself to an easy proof of a monotone convergence theorem, and this is the case even when working in the unweighted setting and in the full Euclidean space. In order to work with something a little more tangible and which better suits our present purposes, we introduce the following equivalent norm on Block spaces.

Definition 10.42. Let $\Sigma \subseteq \mathbb{R}^n$ be a closed ADR set and write $\sigma := \mathcal{H}^{n-1}\lfloor\Sigma$. Fix an integrability exponent $q \in (1,\infty)$ with $q' := q/(q-1)$. Choose a Muckenhoupt weight $\omega \in A_q(\Sigma,\sigma)$ and denote by $\omega' := w^{-q'/q} \in A_{q'}(\Sigma,\sigma)$ its conjugate weight. Let $\varepsilon \in (0, q-1)$ be such that $\omega \in A_{q-\varepsilon}(\Sigma,\sigma)$, and fix a threshold $\lambda \in (0, \varepsilon(n-1)/(q-1))$. Associate as in Definition 6.7 with these choices of q, ω, ε, and λ the weighted Block space $\mathcal{B}^{q,\lambda}(\Sigma,\omega)$. Define the mapping $|||\cdot|||_{\mathcal{B}^{q,\lambda}(\Sigma,\omega)} : \mathscr{M}(\Sigma,\sigma) \to [0,\infty]$ according to

$$|||g|||_{\mathcal{B}^{q,\lambda}(\Sigma,\omega)} := \sup\Big\{\int_\Sigma |fg|\,d\sigma : f \in M^{q',\lambda}(\Sigma,\omega') \text{ and } \|f\|_{M^{q',\lambda}(\Sigma,\omega')} \leq 1\Big\} \tag{10.171}$$

for each σ-measurable function $g : \Sigma \to \mathbb{C}$.

Lemma 10.43. *Let $\Sigma \subseteq \mathbb{R}^n$ be a closed ADR set and write $\sigma := \mathcal{H}^{n-1}\lfloor\Sigma$. Fix an integrability exponent $q \in (1,\infty)$ with $q' := q/(q-1)$. Choose a Muckenhoupt weight $\omega \in A_q(\Sigma,\sigma)$ whose conjugate weight satisfies*

$$\omega' := \omega^{-q'/q} \in A_{q'}(\Sigma,\sigma) \cap \mathrm{RH}_p(\Sigma,\sigma) \tag{10.172}$$

for some $p \in (1, \infty)$. *Denote* $p' := p/(p-1)$ *and define* $\varepsilon := (q-1)/p'$, *so that* $\omega \in A_{q-\varepsilon}(\Sigma, \sigma)$ *as a consequence of Proposition* 6.35. *Lastly, fix a threshold* $\lambda \in (0, \varepsilon(n-1)/(q-1))$, *so that the weighted Block space* $\mathcal{B}^{q,\lambda}(\Sigma, \omega)$ *is well-defined in the sense of Definition* 6.7.

Then $|||\cdot|||_{\mathcal{B}^{q,\lambda}(\Sigma,\omega)}$, *defined in* (10.171), *is a norm on* $\mathcal{B}^{q,\lambda}(\Sigma, \omega)$ *that is equivalent to the standard block norm,* $\|\cdot\|_{\mathcal{B}^{q,\lambda}(\Sigma,\omega)}$, *originally introduced in* (6.41).

Proof. The fact that $|||\cdot|||_{\mathcal{B}^{q,\lambda}(\Sigma,\omega)}$ defines a norm on $\mathcal{B}^{q,\lambda}(\Sigma, \omega)$ is clear from its definition. To see that $|||\cdot|||_{\mathcal{B}^{q,\lambda}(\Sigma,\omega)}$ and $\|\cdot\|_{\mathcal{B}^{q,\lambda}(\Sigma,\omega)}$ are equivalent, notice on one hand that for any fixed function $g \in \mathcal{B}^{q,\lambda}(\Sigma, \omega)$ and any function $f \in M^{q',\lambda}(\Sigma, \omega')$ with $\|f\|_{M^{q',\lambda}(\Sigma,\omega')} \leq 1$, Theorem 6.28 says that

$$\int_\Sigma |fg|\, d\sigma \leq \|f\|_{M^{q',\lambda}(\Sigma,\omega')} \|g\|_{\mathcal{B}^{q,\lambda}(\Sigma,\omega)} \leq \|g\|_{\mathcal{B}^{q,\lambda}(\Sigma,\omega)}. \tag{10.173}$$

Taking a supremum over all $f \in M^{q',\lambda}(\Sigma, \omega')$ with $\|f\|_{M^{q',\lambda}(\Sigma,\omega')} \leq 1$ shows that

$$|||g|||_{\mathcal{B}^{q,\lambda}(\Sigma,\omega)} \leq \|g\|_{\mathcal{B}^{q,\lambda}(\Sigma,\omega)} \quad \text{for all } g \in \mathcal{B}^{q,\lambda}(\Sigma, \omega). \tag{10.174}$$

On the other hand, for any fixed $g \in \mathcal{B}^{q,\lambda}(\Sigma, \omega)$, the identification

$$\mathcal{B}^{q,\lambda}(\Sigma, \omega) = \big(\mathring{M}^{q',\lambda}(\Sigma, \omega')\big)^*, \tag{10.175}$$

given in Theorem 10.39, yields a unique functional $\Lambda_g \in (\mathring{M}^{q',\lambda}(\Sigma, \omega'))^*$ for which

$$\begin{aligned} \|g\|_{\mathcal{B}^{q,\lambda}(\Sigma,\omega)} &\approx \|\Lambda_g\|_{(\mathring{M}^{q',\lambda}(\Sigma,\omega'))^*} \\ &= \sup_{\substack{f \in \mathring{M}^{q',\lambda}(\Sigma,\omega') \\ \|f\|_{M^{q',\lambda}(\Sigma,\omega')} \leq 1}} \left| \int_\Sigma fg\, d\sigma \right| \\ &\leq |||g|||_{\mathcal{B}^{q,\lambda}(\Sigma,\omega)}. \end{aligned} \tag{10.176}$$

Since $g \in \mathcal{B}^{q,\lambda}(\Sigma, \omega)$ was arbitrary, (10.176) together with (10.174) finishes the justification of Lemma 10.43. □

Before characterizing Muckenhoupt-weighted Block spaces as generalized Banach function spaces, we present the following lemma, which provides an alternate characterization for membership to a Muckenhoupt-weighted Block space in terms of the Muckenhoupt-weighted vanishing Morrey space whose topological dual is the weighted Block space (cf. Theorem 10.39).

Lemma 10.44. *Let* $\Sigma \subseteq \mathbb{R}^n$ *be a closed ADR set, and write* $\sigma := \mathcal{H}^{n-1}\lfloor\Sigma$. *Fix an integrability exponent* $q \in (1, \infty)$ *with* $q' := q/(q-1)$. *Choose an arbitrary Muckenhoupt weight* $\omega \in A_q(\Sigma, \sigma)$ *whose conjugate weight satisfies*

$$\omega' := \omega^{-q'/q} \in A_{q'}(\Sigma,\sigma) \cap \mathrm{RH}_p(\Sigma,\sigma) \tag{10.177}$$

for some $p \in (1,\infty)$. Denote $p' := p/(p-1)$ and define $\varepsilon := (q-1)/p'$ so that $\omega \in A_{q-\varepsilon}(\Sigma,\sigma)$ by Proposition 6.35. Finally, select any threshold $\lambda \in (0, \varepsilon(n-1)/(p'-1))$ and associate to these choices of $q, \omega, \varepsilon, \lambda$ the Block space $\mathcal{B}^{q,\lambda}(\Sigma,\omega)$ as in Definition 6.7.

Then for any σ-measurable function g on Σ one has

$$g \in \mathcal{B}^{q,\lambda}(\Sigma,\omega) \quad \text{if and only if} \quad \sup_{\substack{f\in \mathring{M}^{q',\lambda}(\Sigma,\omega') \\ \|f\|_{M^{q',\lambda}(\Sigma,\omega')}\le 1}} \int_\Sigma |fg|\,d\sigma < +\infty. \tag{10.178}$$

Proof. The forward implication is straightforward given Theorem 6.28. Indeed, for any $g \in \mathcal{B}^{q,\lambda}(\Sigma,\omega)$ and any $f \in M^{q',\lambda}(\Sigma,\omega')$ (in particular for any $f \in \mathring{M}^{q',\lambda}(\Sigma,\omega')$) we have

$$\int_\Sigma |fg|\,d\sigma \le \|f\|_{M^{q',\lambda}(\Sigma,\omega')}\|g\|_{\mathcal{B}^{q,\lambda}(\Sigma,\omega)}. \tag{10.179}$$

Taking a supremum over all $f \in \mathring{M}^{q',\lambda}(\Sigma,\omega')$ with $\|f\|_{M^{q',\lambda}(\Sigma,\omega')} \le 1$ yields the desired implication.

For the reverse implication, suppose g is a σ-measurable function on Σ, which satisfies

$$C_g := \sup_{\substack{f\in \mathring{M}^{q',\lambda}(\Sigma,\omega') \\ \|f\|_{M^{q',\lambda}(\Sigma,\omega')}\le 1}} \int_\Sigma |fg|\,d\sigma < +\infty. \tag{10.180}$$

Then in particular, $\int_\Delta |g|\,d\sigma < +\infty$ for all surface balls $\Delta \subseteq \Sigma$, since $\mathbf{1}_\Delta$ belongs to $L^\infty_{\mathrm{comp}}(\Sigma,\sigma)$, a subset of $\mathring{M}^{q',\lambda}(\Sigma,\omega')$, by item (c) of Proposition 10.21. This shows that $g \in L^1_{\mathrm{loc}}(\Sigma,\sigma)$. Now, bring in the functional

$$\begin{aligned} &\Lambda_g : \mathring{M}^{q',\lambda}(\Sigma,\omega') \to \mathbb{C}, \quad \text{defined by} \\ &\Lambda_g(f) := \int_\Sigma fg\,d\sigma \quad \text{for all } f \in \mathring{M}^{q',\lambda}(\Sigma,\omega'). \end{aligned} \tag{10.181}$$

Observe that $f = 0$ at σ-a. e. point in Σ implies $\Lambda_g(f) = 0$. Otherwise, when f belongs to $\mathring{M}^{q',\lambda}(\Sigma,\omega')$, we have $\|f\|_{M^{q',\lambda}(\Sigma,\omega')} > 0$, so that the function $f/\|f\|_{M^{q',\lambda}(\Sigma,\omega')}$ is meaningfully defined and belongs to $\mathring{M}^{q',\lambda}(\Sigma,\omega')$ with $\|f/\|f\|_{M^{p,\lambda}(\Sigma,w)}\|_{M^{q',\lambda}(\Sigma,\omega')} = 1$. Subsequently,

$$\int_\Sigma \left| \frac{f}{\|f\|_{M^{q',\lambda}(\Sigma,\omega')}} \cdot g \right| d\sigma \le C_g, \tag{10.182}$$

which further implies $\int_\Sigma |fg|\,d\sigma \le C_g\|f\|_{M^{q',\lambda}(\Sigma,\omega')}$, making fg absolutely integrable on Σ. As a result, Λ_g as in (10.181) is a well-defined, linear, and bounded functional on $\mathring{M}^{q',\lambda}(\Sigma,\omega')$ with $|\Lambda_g(f)| \le C_g\|f\|_{M^{q',\lambda}(\Sigma,\omega')}$ for all $f \in \mathring{M}^{q',\lambda}(\Sigma,\omega')$. This means that $\Lambda_g \in (\mathring{M}^{q',\lambda}(\Sigma,\omega'))^*$ with $\|\Lambda_g\|_{(\mathring{M}^{q',\lambda}(\Sigma,\omega'))^*} \le C_g$. Subsequently, Theorem 10.39 ensures the existence of a unique function $\widetilde{g} \in \mathcal{B}^{q,\lambda}(\Sigma,\omega)$ with

$$\begin{aligned} \Lambda_g(f) = \int_\Sigma f\widetilde{g}\,d\sigma \quad &\text{for each } f \in \mathring{M}^{q',\lambda}(\Sigma,\omega') \\ \text{and} \quad \|\widetilde{g}\|_{\mathcal{B}^{q,\lambda}(\Sigma,\omega)} \approx \|\Lambda\|&_{(\mathring{M}^{q',\lambda}(\Sigma,\omega'))^*} \le C_g. \end{aligned} \tag{10.183}$$

Consequently, because we know that $\mathbf{1}_\Delta \in L^\infty_{\mathrm{comp}}(\Sigma,\sigma) \subseteq \mathring{M}^{q',\lambda}(\Sigma,\omega')$ for any surface ball $\Delta \subseteq \Sigma$, (10.181) and (10.183) imply that $\int_\Delta g\,d\sigma = \int_\Delta \widetilde{g}\,d\sigma$ for all surface balls $\Delta \subseteq \Sigma$. Therefore, by Corollary 2.23 (bearing in mind that $g \in L^1_{\mathrm{loc}}(\Sigma,\sigma)$), we can conclude that in fact, $g = \widetilde{g}$ at σ-a. e. point on Σ. Given the lattice property of $\mathcal{B}^{q,\lambda}(\Sigma,\omega)$ (cf. Proposition 6.14), we can confirm that $g \in \mathcal{B}^{q,\lambda}(\Sigma,\omega)$, completing the proof of the reverse implication in (10.178), and ultimately establishing the result. □

As an immediate corollary of Lemma 10.44, we find, retaining the context therein,

$$g \in \mathcal{B}^{q,\lambda}(\Sigma,\omega) \quad \text{if and only if} \quad g \in \mathscr{M}(\Sigma,\sigma) \quad \text{and} \quad |||g|||_{\mathcal{B}^{q,\lambda}(\Sigma,\omega)} < +\infty. \tag{10.184}$$

Indeed, recalling Definition 10.42, we see Theorem 6.28 demonstrates g belongs to $\mathcal{B}^{q,\lambda}(\Sigma,\omega)$ implies $g \in \mathscr{M}(\Sigma,\sigma)$ with $|||g|||_{\mathcal{B}^{q,\lambda}(\Sigma,\omega)} < +\infty$, whereas (10.178) with (10.171) and the fact that $\mathring{M}^{q',\lambda}(\Sigma,\omega') \subseteq M^{q',\lambda}(\Sigma,\omega')$ demonstrate that the converse is also true.

We are now in a position to show that Muckenhoupt-weighted Block spaces are generalized Banach function spaces.

Proposition 10.45. *Suppose $\Sigma \subseteq \mathbb{R}^n$ is a closed ADR set and abbreviate $\sigma := \mathcal{H}^{n-1}\lfloor\Sigma$. Fix an integrability exponent $q \in (1,\infty)$ with $q' := q/(q-1)$. Choose a Muckenhoupt weight $\omega \in A_q(\Sigma,\sigma)$ whose conjugate weight satisfies*

$$\omega' := \omega^{-q'/q} \in A_{q'}(\Sigma,\sigma) \cap \mathrm{RH}_p(\Sigma,\sigma) \tag{10.185}$$

for some $p \in (1,\infty)$. Denote $p' := p/(p-1)$ and define $\varepsilon := (q-1)/p'$ so that $\omega \in A_{q-\varepsilon}(\Sigma,\sigma)$ by Proposition 6.35. Lastly, fix a threshold $\lambda \in (0, \varepsilon(n-1)/(q-1))$, so that the weighted Block space $\mathcal{B}^{q,\lambda}(\Sigma,\omega)$ is well-defined in relation to q, ω, ε, and λ as per Definition 6.7.

Then the mapping (cf. Definition 10.42)

$$|||\cdot|||_{\mathcal{B}^{q,\lambda}(\Sigma,\omega)} : \mathscr{M}_+(\Sigma,\sigma) \to [0,\infty] \tag{10.186}$$

is a function norm in the sense of Definition 10.1, which makes $\mathcal{B}^{q,\lambda}(\Sigma,\omega)$, equipped with $|||\cdot|||_{\mathcal{B}^{q,\lambda}(\Sigma,\omega)}$, a generalized Banach function space on Σ in the sense of Definition 10.3.

Proof. We first observe that $|||\cdot|||_{\mathcal{B}^{q,\lambda}(\Sigma,\omega)}$ satisfies (P1) thanks to Definition 10.42, the fact that $\mathbf{1}_\Delta \in M^{q',\lambda}(\Sigma,\omega')$ for every surface ball $\Delta \subseteq \Sigma$ (cf. Lemma 4.9), and Corollary 2.23. To deduce (P2), fix two functions $g_1, g_2 \in \mathscr{M}(\Sigma, \sigma)$, which satisfy $|g_1| \leq |g_2|$ at σ-a. e. point on Σ. We must show that

$$\|g_1\|_{\mathcal{B}^{q,\lambda}(\Sigma,\omega)} \leq \|g_2\|_{\mathcal{B}^{q,\lambda}(\Sigma,\omega)} \tag{10.187}$$

as well. The case that $\|g_2\|_{\mathcal{B}^{q,\lambda}(\Sigma,\omega)} = \infty$ makes the above inequality trivially true, so suppose that $\|g_2\|_{\mathcal{B}^{q,\lambda}(\Sigma,\omega)} < +\infty$, which in turn ensures g_2 is in $L^1_{\mathrm{loc}}(\Sigma, \sigma)$, so that in particular, $|g_2| < +\infty$ at σ-a. e. point on Σ. Define the function

$$h(x) := \begin{cases} \frac{g_1(x)}{g_2(x)} & \text{if } g_2(x) \neq 0, \\ 0 & \text{if } g_2(x) = 0, \end{cases} \quad \text{at each } x \in \Sigma. \tag{10.188}$$

Then $h \in \mathcal{M}(\Sigma, \sigma)$ and we see that $|h| \leq 1$ at σ-a. e. point on Σ. For each function $f \in M^{q',\lambda}(\Sigma, \omega')$, we also find that $fg_1 = (fh)g_2$ at σ-a. e. point on Σ, and hence

$$\int_\Sigma |fg_1|\, d\sigma = \int_\Sigma |(fh)g_2|\, d\sigma \leq \int_\Sigma |fg_2|\, d\sigma. \tag{10.189}$$

Subsequently, taking a supremum on both sides over all $f \in M^{q',\lambda}(\Sigma, \omega')$ with $\|f\|_{M^{q',\lambda}(\Sigma,\omega')} \leq 1$, we obtain (10.187) as wanted.

Next, we turn to the task of showing that $\|\cdot\|_{\mathcal{B}^{q,\lambda}(\Sigma,\omega)}$ satisfies (P3). For that purpose, suppose $\{g_j\}_{j\in\mathbb{N}} \subseteq \mathscr{M}_+(\Sigma, \sigma)$ is a sequence that increases pointwise σ-a. e. to some function g as $j \to \infty$. Then for any $f \in M^{q',\lambda}(\Sigma, \omega')$, we find that $|fg_j| \leq |fg|$ for every $j \in \mathbb{N}$. Moreover, as Definition 10.42 shows, $\|g_j\|_{\mathcal{B}^{q,\lambda}(\Sigma,\omega)} \leq \|g_{j+1}\|_{\mathcal{B}^{q,\lambda}(\Sigma,\omega)}$ for all $j \in \mathbb{N}$. Thus,

$$\begin{aligned} \lim_{j\to\infty} \|g_j\|_{\mathcal{B}^{q,\lambda}(\Sigma,\omega)} &= \lim_{j\to\infty} \sup_{\substack{f\in M^{q',\lambda}(\Sigma,\omega') \\ \|f\|_{M^{q',\lambda}(\Sigma,\omega')}\leq 1}} \int_\Sigma |fg_j|\, d\sigma \\ &\leq \sup_{\substack{f\in M^{q',\lambda}(\Sigma,\omega') \\ \|f\|_{M^{q',\lambda}(\Sigma,\omega')}\leq 1}} \int_\Sigma |fg|\, d\sigma \\ &= \|g\|_{\mathcal{B}^{q,\lambda}(\Sigma,\omega)}. \end{aligned} \tag{10.190}$$

On the other hand, for an arbitrary function $f \in M^{q',\lambda}(\Sigma, \omega')$ the sequence $\{|fg_j|\}_{j\in\mathbb{N}} \subseteq \mathscr{M}_+(\Sigma, \sigma)$ increases pointwise σ-a. e. on Σ to $|fg|$ as $j \to \infty$. As a result, for any $f \in M^{q',\lambda}(\Sigma, \omega')$ with $\|f\|_{M^{q',\lambda}(\Sigma,\omega')} \leq 1$, (10.171) and Lebesgue's monotone convergence theorem allow us to write

$$\lim_{j\to\infty} \|g_j\|_{\mathcal{B}^{q,\lambda}(\Sigma,\omega)} \geq \lim_{j\to\infty} \int_\Sigma |fg_j|\, d\sigma = \int_\Sigma |fg|\, d\sigma. \tag{10.191}$$

Taking the supremum over all $f \in M^{q',\lambda}(\Sigma, \omega')$ with $\|f\|_{M^{q',\lambda}(\Sigma,\omega')} \leq 1$ permits us to conclude that

$$\lim_{j\to\infty} |||g_j|||_{\mathcal{B}^{q,\lambda}(\Sigma,\omega)} \geq |||g|||_{\mathcal{B}^{q,\lambda}(\Sigma,\omega)}. \tag{10.192}$$

Together, (10.190) and (10.192) ultimately yield

$$\lim_{j\to\infty} |||g_j|||_{\mathcal{B}^{q,\lambda}(\Sigma,\omega)} = |||g|||_{\mathcal{B}^{q,\lambda}(\Sigma,\omega)} \tag{10.193}$$

as required.

Moving on, to establish (P4), fix $x_0 \in \Sigma$ and note $\Sigma = \bigcup_{j\in\mathbb{N}} \Sigma_j$, where we define $\Sigma_j := \Delta(x_0, j)$ for each $j \in \mathbb{N}$. Then $|||\mathbf{1}_{\Sigma_j}|||_{\mathcal{B}^{q,\lambda}(\Sigma,\omega)} < +\infty$ for all $j \in \mathbb{N}$ as a consequence of (10.174) and Proposition 6.25. Finally, to show that (P5) holds for $|||\cdot|||_{\mathcal{B}^{q,\lambda}(\Sigma,\omega)}$, notice for $g \in \mathscr{M}_+(\Sigma, \sigma)$ by (10.171)

$$\begin{aligned}\int_{\Sigma_j} g\, d\sigma &= \|\mathbf{1}_{\Delta(x_0,j)}\|_{M^{q',\lambda}(\Sigma,\omega')} \int_{\Sigma} \frac{\mathbf{1}_{\Delta(x_0,j)}}{\|\mathbf{1}_{\Delta(x_0,j)}\|_{M^{q',\lambda}(\Sigma,\omega')}} \cdot g\, d\sigma \\ &\leq \|\mathbf{1}_{\Delta(x_0,j)}\|_{M^{q',\lambda}(\Sigma,\omega')} |||g|||_{\mathcal{B}^{q,\lambda}(\Sigma,\omega)} \\ &\leq c_j |||g|||_{\mathcal{B}^{q,\lambda}(\Sigma,\omega)},\end{aligned} \tag{10.194}$$

for any $c_j \in (\|\mathbf{1}_{\Delta(x_0,j)}\|_{M^{q',\lambda}(\Sigma,\omega')}, \infty)$, bearing in mind Lemma 4.9. Alternatively, (P5) follows from Proposition 6.17 (bearing in mind Lemma 10.43), which gives the continuous embedding $\mathcal{B}^{q,\lambda}(\Sigma, \omega) \hookrightarrow L^1_{\mathrm{loc}}(\Sigma, \sigma)$. This completes the justification of the fact that $|||\cdot|||_{\mathcal{B}^{q,\lambda}(\Sigma,\omega)}$ is a function norm in the sense of Definition 10.1, ultimately allowing us to confirm that $\mathcal{B}^{q,\lambda}(\Sigma, \omega)$ with function norm $|||\cdot|||_{\mathcal{B}^{q,\lambda}(\Sigma,\omega)}$ is a generalized Banach function space. □

We are now in a position to showcase the following result, the culmination of the theory developed in this chapter thus far, which identifies Muckenhoupt-weighted Morrey and Block spaces as associated spaces of each other in the sense of Definition 10.11. Here is where our development of Muckenhoupt-weighted vanishing Morrey spaces comes to fruition. Specifically, the proof of this result, which hinges on Lemma 10.44, highlights the significance of the duality pairing between weighted vanishing Morrey spaces and weighted Block spaces, outlined in Theorem 10.39. As we will see, just understanding the duality pairing between Muckenhoupt-weighted Morrey and Block spaces, detailed in Theorem 6.29, is not enough to characterize said spaces as being mutually associated with each other. It is crucial that we also understand a suitable notion of Muckenhoupt-weighted vanishing Morrey space. Interestingly, we will eventually provide an alternate proof of the following theorem, which avoids identifying Muckenhoupt-weighted vanishing Morrey spaces as the pre-duals of Muckenhoupt-weighted Block spaces, but we will see that even going this alternative route, we still cannot avoid vanishing Morrey spaces entirely.

Theorem 10.46. *Let $\Sigma \subseteq \mathbb{R}^n$ be a closed ADR set, and write $\sigma := \mathcal{H}^{n-1}\lfloor\Sigma$. Fix two integrability exponents $p, q \in (1,\infty)$ with $p' := p/(p-1)$ and $q' := q/(q-1)$. Choose a Muckenhoupt weight $w \in A_p(\Sigma,\sigma) \cap \mathrm{RH}_q(\Sigma,\sigma)$ and denote its conjugate weight by $w' := w^{-p'/p} \in A_{p'}(\Sigma,\sigma)$. Define the parameter $\varepsilon := (p'-1)/q' \in (0, p'-1)$ so that $w' \in A_{p'-\varepsilon}(\Sigma,\sigma)$ by Proposition 6.35. Finally, make any choice of threshold $\lambda \in (0,(n-1)/q') = (0, \varepsilon(n-1)/(p'-1))$. Then*[4]

$$(M^{p,\lambda}(\Sigma,w))' = \mathcal{B}^{p',\lambda}(\Sigma,w'). \tag{10.195}$$

As an immediate corollary, in view of Proposition 10.13 one also has

$$(\mathcal{B}^{p',\lambda}(\Sigma,w'))' = M^{p,\lambda}(\Sigma,w). \tag{10.196}$$

Adjusting notation, assume $q \in (1,\infty)$ with $q' := q/(q-1)$, $\omega \in A_q(\Sigma,\sigma)$ is such that its conjugate weight $\omega' := \omega^{-q'/q}$ belongs to $A_{q'}(\Sigma,\sigma) \cap \mathrm{RH}_p(\Sigma,\sigma)$ for some $p \in (1,\infty)$, $p' := p/(p-1)$. Choose $\varepsilon := (q-1)/p'$, which ensures $\omega \in A_{q-\varepsilon}(\Sigma,\sigma)$ thanks to Proposition 6.35, and fix $\lambda \in (0, \varepsilon(n-1)/(q-1))$. This entails that the weighted Block space $\mathcal{B}^{q,\lambda}(\Sigma,\omega)$, corresponding to these choices of q, ω, ε, and λ, is well-defined in the sense of Definition 6.7. Then Theorem 10.46 says that

$$(\mathcal{B}^{q,\lambda}(\Sigma,\omega))' = M^{q',\lambda}(\Sigma,\omega'). \tag{10.197}$$

Here is now the proof of Theorem 10.46.

Proof of Theorem 10.46. To begin, let $g \in \mathcal{B}^{p',\lambda}(\Sigma,w)$ be arbitrary. Then

$$\|g\|_{(M^{p,\lambda}(\Sigma,w))'} = \sup_{\substack{f\in M^{p,\lambda}(\Sigma,w)\\ \|f\|_{M^{p,\lambda}(\Sigma,w)}\le 1}} \int_\Sigma |fg|\,d\sigma = |||g|||_{\mathcal{B}^{p',\lambda}(\Sigma,w')} < +\infty,$$

where the last equality follows from (10.171). Subsequently, we find that $g \in (M^{p,\lambda}(\Sigma,w))'$, and given the arbitrary nature of $g \in \mathcal{B}^{p',\lambda}(\Sigma,w')$, we can conclude that

$$(M^{p,\lambda}(\Sigma,w))' \supseteq \mathcal{B}^{p',\lambda}(\Sigma,w'). \tag{10.198}$$

For the opposite inclusion, fix a function $g \in (M^{p,\lambda}(\Sigma,w))'$. Then in particular, since $\mathring{M}^{p,\lambda}(\Sigma,w)$ is a subspace of $M^{p,\lambda}(\Sigma,w)$, we can say that

4 with $\mathcal{B}^{p',\lambda}(\Sigma,w')$ being defined in relation to these choices of p', ε, and λ, as in Definition 6.7.

$$\sup_{\substack{f \in \mathring{M}^{p,\lambda}(\Sigma,w) \\ \|f\|_{M^{p,\lambda}(\Sigma,w)} \leq 1}} \int_{\Sigma} |fg| \, d\sigma \leq \|g\|_{(M^{p,\lambda}(\Sigma,w))'} < +\infty, \tag{10.199}$$

which according to Lemma 10.44 ensures that $g \in \mathcal{B}^{p',\lambda}(\Sigma, w')$. Since g was arbitrary, this fact allows us to conclude that

$$(M^{p,\lambda}(\Sigma, w))' \subseteq \mathcal{B}^{p',\lambda}(\Sigma, w'), \tag{10.200}$$

which recalling (10.198) completes the proof of the equality advertised in (10.195). □

Thus far, the characterization of Muckenhoupt-weighted Morrey and Block spaces as generalized Banach function spaces, presented in Theorem 10.46, has relied heavily on identifying Muckenhoupt-weighted vanishing Morrey spaces as the pre-duals of Muckenhoupt-weighted Block spaces; cf. Theorem 10.39. At this point, we would like to demonstrate that the theory may also be developed independently of that duality fact. Essentially, we will show that we can still identify Muckenhoupt-weighted Block spaces as generalized Banach function spaces by working exclusively with the space itself, avoiding any mention of their pre-duals.

Our first result in this direction is a kind of weak-$*$ compactness result, but at the level of Muckenhoupt-weighted Block spaces. That is, given a bounded sequence in a Block space, we can find a subsequence, which converges (in a weak-$*$ sense) to a limit function in the Block space.

Theorem 10.47. *Let $\Sigma \subseteq \mathbb{R}^n$ be a closed ADR set, and write $\sigma := \mathcal{H}^{n-1}\lfloor\Sigma$. Fix an integrability exponent $q \in (1,\infty)$ with $q' := q/(q-1)$ along with a Muckenhoupt weight $\omega \in A_q(\Sigma,\sigma)$ whose conjugate weight satisfies*

$$\omega' := \omega^{-q'/q} \in A_{q'}(\Sigma,\sigma) \cap \mathrm{RH}_p(\Sigma,\sigma) \tag{10.201}$$

for some $p \in (1,\infty)$. Define $\varepsilon := (q-1)/p' \in (0, q-1)$ so that $\omega \in A_{q-\varepsilon}(\Sigma,\sigma)$ by Proposition 6.35, and select an arbitrary threshold $\lambda \in (0, \varepsilon(n-1)/(q-1))$. Bring in the weighted Block space $\mathcal{B}^{q,\lambda}(\Sigma,\omega)$ defined in relation to these choices of q, ω, ε, and λ as in Definition 6.7.

Then there exists a constant $C_* \in (0,\infty)$, depending only on Σ, n, q, and λ with the following significance. For any bounded sequence of functions $\{f_j\}_{j\in\mathbb{N}} \subseteq \mathcal{B}^{q,\lambda}(\Sigma,\omega)$, there exists $f \in \mathcal{B}^{q,\lambda}(\Sigma,\omega)$ satisfying

$$\|f\|_{\mathcal{B}^{q,\lambda}(\Sigma,\omega)} \leq C_* \liminf_{j\to\infty} \|f_j\|_{\mathcal{B}^{q,\lambda}(\Sigma,\omega)}, \tag{10.202}$$

and there exists a subsequence $\{f_{j_k}\}_{k\in\mathbb{N}}$ of $\{f_j\}_{j\in\mathbb{N}}$ such that

$$\lim_{k\to\infty} \int_{\Sigma} f_{j_k} g \, d\sigma = \int_{\Sigma} fg \, d\sigma \quad \text{for each } g \in \mathring{M}^{q',\lambda}(\Sigma,\omega'). \tag{10.203}$$

Note that this is an immediate consequence of the sequential Banach–Alaoglu theorem (cf. (6.97)), noting that Theorem 10.39 gives

$$(\mathring{\mathcal{M}}^{q',\lambda}(\Sigma,\omega'))^* = \mathcal{B}^{q,\lambda}(\Sigma,\omega), \tag{10.204}$$

and that $\mathring{\mathcal{M}}^{q',\lambda}(\Sigma,\omega')$ is a separable Banach space; cf. (10.73). Alternatively, we present the following proof, which avoids the use of the duality result in Theorem 10.39.

Proof of Theorem 10.47. To begin, if $\{f_j\}_{j\in\mathbb{N}}$ is the zero sequence, then the result is trivial, so we suppose this is not the case. As such, without loss of generality, we may assume $\sup_{j\in\mathbb{N}}\|f\|_{\mathcal{B}^{q,\lambda}(\Sigma,\omega)} = 1$. Otherwise, consider the modified sequence obtained by dividing each f_j by the quantity $\sup_{j\in\mathbb{N}}\|f\|_{\mathcal{B}^{q,\lambda}(\Sigma,\omega)} \in (0,\infty)$. Starting now in earnest, note that for each $j\in\mathbb{N}$, Lemma 6.20 allows us to decompose

$$f_j = \sum_{Q\in\mathbb{D}(\Sigma)} \lambda_j(Q) b_j(Q), \tag{10.205}$$

where $\mathbb{D}(\Sigma)$ is a dyadic grid structure on Σ as in Lemma 2.42,

$$\begin{gathered}\{\lambda_j(Q)\}_{Q\in\mathbb{D}(\Sigma)} \in \ell^1 \text{ is a sequence of nonnegative numbers with}\\ \textstyle\sum_{Q\in\mathbb{D}(\Sigma)} \lambda_j(Q) \le C_* \text{ for some } C_* \in (0,\infty) \text{ that depends only on}\\ \Sigma, n, q, \text{ and } \lambda,\end{gathered} \tag{10.206}$$

and each

$$\begin{gathered}b_j(Q) \text{ is a } (q,\lambda,\omega)\text{-block on } \Sigma \text{ whose support is contained in the}\\ \text{surface ball } \widetilde{Q} := \Delta(x_Q, c_\Sigma \ell(Q)) \subseteq \Sigma, \text{ noting that } x_Q \text{ and } \ell(Q) \text{ are}\\ \text{the center and side length of } Q, \text{ respectively (cf. Definition 2.43),}\\ \text{and } c_\Sigma \in (0,\infty) \text{ is a purely geometric constant; moreover, } b_j(Q)\\ \text{satisfies } \|b_j(Q)\|_{L^q(\Sigma,\omega)} \le (c_\Sigma \ell(Q))^{\lambda(1/q-1)}.\end{gathered} \tag{10.207}$$

For each fixed $Q\in\mathbb{D}(\Sigma)$, the boundedness of $\{b_j(Q)\}_{j\in\mathbb{N}}$ in $L^q(\Sigma,\omega)$ implies there exists an increasing sequence of natural numbers $\{j_{k,Q}\}_{k\in\mathbb{N}} \subseteq \mathbb{N}$ and a function $b(Q)\in L^q(\Sigma,\omega)$ such that $b_{j_{k,Q}}(Q) \xrightarrow{\text{weak-}*} b(Q)$ as $k\to\infty$. In addition, (10.207) implies

$$\begin{aligned}&\operatorname{supp} b(Q) \subseteq \widetilde{Q},\\ &\|b(Q)\|_{L^q(\Sigma,\omega)} \le \liminf_{k\to\infty} \|b_{j_{k,Q}}(Q)\|_{L^q(\Sigma,\omega)} \le (c_\Sigma \ell(Q))^{\lambda(1/q-1)},\end{aligned} \tag{10.208}$$

showing that $b(Q)$ is a (q,λ,ω)-block on Σ. Also, from (10.206) and the Bolzano–Weierstrass theorem we see that after passing to a subsequence if necessary, there is no loss of generality in assuming that $\lambda_{j_{k,Q}}(Q) \to \lambda(Q)$ in $\mathbb{R}$ as $k\to\infty$ for some $\lambda(Q)\in[0,C_*)$, where C_* is as in (10.206).

At this point, we apply a Cantor diagonalization argument to rid any dependence of the sequence $\{j_{k,Q}\}_{k\in\mathbb{N}}$ on $Q \in \mathbb{D}(Q)$. Specifically, bring in an enumeration of the countable set $\mathbb{D}(\Sigma)$, say $\{Q_i\}_{i\in\mathbb{N}}$. Starting with Q_1, associate a sequence of indices $\{j_{k,1}\}_{k\in\mathbb{N}} \subseteq \mathbb{N}$ as above. For Q_2, select a subsequence of $\{j_{k,1}\}_{k\in\mathbb{N}}$, denoted $\{j_{k,2}\}_{k\in\mathbb{N}}$, for which $b_{j_{k,2}}(Q_2) \xrightarrow{\text{weak-}*} b(Q_2)$ in $L^q(\Sigma,\omega)$ and $\lambda_{j_{k,1}}(Q_2) \to \lambda(Q_2)$ in $\mathbb{R}$ as $k\to\infty$. Now, iterate this process. That is, for $i \in \mathbb{N}$, select a subsequence from $\{j_{k,i}\}_{k\in\mathbb{N}} \subseteq \mathbb{N}$, denoted $\{j_{k,i+1}\}_{k\in\mathbb{N}}$, for which $b_{j_{k,i+1}}(Q_{i+1}) \xrightarrow{\text{weak-}*} b(Q_{i+1})$ in $L^q(\Sigma,\omega)$ and $\lambda_{j_{k,i+1}}(Q_{i+1}) \to \lambda(Q_{i+1})$ in $\mathbb{R}$ as $k\to\infty$. Finally, take the diagonal $\{j_k\}_{k\in\mathbb{N}} := \{j_{k,k}\}_{k\in\mathbb{N}}$ and note that for all $Q \in \mathbb{D}(\Sigma)$ matters have been arranged so that $b_{j_k}(Q) \xrightarrow{\text{weak-}*} b(Q)$ in $L^q(\Sigma,\omega)$ and $\lambda_{j_k}(Q) \to \lambda(Q)$ in $\mathbb{R}$ as $k\to\infty$.

Returning to the mainstream discussion, we now define the function

$$f := \sum_{Q\in\mathbb{D}(\Sigma)} \lambda(Q)b(Q) \quad \text{on } \Sigma, \tag{10.209}$$

and observe that $f \in \mathcal{B}^{q,\lambda}(\Sigma,\omega)$, since (recalling (10.208)) the function $b(Q)$ is a (q,λ,ω)-block on Σ for each $Q \in \mathbb{D}(\Sigma)$ and $\{\lambda(Q)\}_{Q\in\mathbb{D}(\Sigma)}$ is a (nonnegative) sequence in ℓ^1. Additionally, Fatou's theorem for sequences shows

$$\sum_{Q\in\mathbb{D}(\Sigma)} \lambda(Q) \le \liminf_{k\to\infty} \sum_{Q\in\mathbb{D}(\Sigma)} \lambda_{j_k}(Q) \le C_*. \tag{10.210}$$

At this stage, our goal is to establish the weak-$*$ convergence in (10.203). To do so, start with some $g \in L^\infty_{\mathrm{comp}}(\Sigma,\sigma) \subseteq \mathring{M}^{q',\lambda}(\Sigma,\omega')$ (cf. item (c) of Proposition 10.21), and define $Q_0 \subseteq \Sigma$ to be a surface ball containing $\operatorname{supp} g$. We claim

$$\lim_{k\to\infty} \int_\Sigma f_{j_k} g\,d\sigma = \int_\Sigma fg\,d\sigma, \tag{10.211}$$

noting that $\{j_k\}_{k\in\mathbb{N}}$ is the sequence of indices coming from our Cantor diagonalization. To facilitate the proof of (10.211), we break up $\mathbb{D}(\Sigma)$ into three disjoint subsets, containing small, medium, and large cubes, respectively. Recall that given $Q \in \mathbb{D}(\Sigma)$ we denote $\widetilde{Q} := \Delta(x_Q, c_\Sigma\ell(Q))$, which is unambiguously defined in relation to Q. Now, let $c_1, c_2 \in (0,\infty)$ be constants whose natures will be elucidated later, and define

$$\begin{aligned}
\mathcal{D}_1(Q_0) &:= \{Q \in \mathbb{D}(\Sigma) : \widetilde{Q}\cap Q_0 \neq \varnothing,\ c_\Sigma\ell(Q) \le c_1\},\\
\mathcal{D}_2(Q_0) &:= \{Q \in \mathbb{D}(\Sigma) : \widetilde{Q}\cap Q_0 \neq \varnothing,\ c_\Sigma\ell(Q) \in (c_1,c_2)\},\\
\mathcal{D}_3(Q_0) &:= \{Q \in \mathbb{D}(\Sigma) : \widetilde{Q}\cap Q_0 \neq \varnothing,\ c_\Sigma\ell(Q) \ge c_2\}.
\end{aligned} \tag{10.212}$$

We concern ourselves with $\mathcal{D}_1(\Sigma)$ first. For any $k \in \mathbb{N}$ and $Q \in \mathcal{D}_1(Q_0)$, we can use the boundedness of Q_0 and Lemma 6.5 to write for some integrability exponent $q_0 \in (1,\infty)$,

$$\begin{aligned}\int_{Q_0} |b_{j_k}(Q)|\,d\sigma &\leq C_{Q_0,q_0} \int_{Q_0} \frac{|b_{j_k}(Q)(x)|}{(1+|x|^{n-1})^{q_0}}\,d\sigma(x) \\ &= C_{Q_0,q_0} \int_{\Sigma} \frac{|b_{j_k}(Q)(x)|}{(1+|x|^{n-1})^{q_0}} \mathbf{1}_{\widetilde{Q}}(x)\,d\sigma(x) \\ &\leq C_{Q_0,q_0} \|b_{j_k}\|_{L^{q_0}(\Sigma,\frac{\sigma(x)}{1+|x|^{n-1}})} \|\mathbf{1}_{\widetilde{Q}}\|_{L^{q_0'}(\Sigma,\sigma)} \\ &\leq C_{Q_0,q_0} C_0 \sigma(\widetilde{Q})^{1/q_0'} \approx C_{Q_0,q_0} C_0 \ell(Q)^{1/q_0'}.\end{aligned} \tag{10.213}$$

The first line follows from Lemma 2.18, the second line holds because $\operatorname{supp} b_{j_k}(Q) \subseteq \widetilde{Q}$, the third line is an application of Hölder's inequality (remarking that $q_0' \in (1,\infty)$ is such that $1/q_0 + 1/q_0' = 1$), the fourth line follows from Lemma 6.5, which also ensures the constant $C_0 \in (0,\infty)$ is independent of Q, and the quantitative equivalence follows from the Ahlfors regularity of Σ and item (5) of Lemma 2.42. Observe that because $b(Q)$ is also a (q,λ,ω)-block on Σ, supported in $\widetilde{Q}$, that the result in (10.213) continues to hold for $b(Q)$ in place of $b_{j_k}(Q)$. Thus, thanks to the observation we just made, (10.213), (10.206), and (10.210), there exists a constant $C \in (0,\infty)$, independent of $Q \in \mathcal{D}_1(Q_0)$, such that for any $k \in \mathbb{N}$, we may estimate

$$\begin{aligned}&\sum_{Q\in\mathcal{D}_1(Q_0)} \int_{Q_0} |\lambda_{j_k}(Q) b_{j_k}(Q) g - \lambda(Q) b(Q) g|\,d\sigma \\ &\quad\leq \|g\|_{L^\infty(\Sigma,\sigma)} \sum_{Q\in\mathcal{D}_1(Q_0)} (\lambda_{j_k}(Q) + \lambda(Q)) C_{Q_0,q_0} C_0 \ell(Q)^{1/q_0'} \\ &\quad\leq C \leq \|g\|_{L^\infty(\Sigma,\sigma)} \sup_{Q\in\mathcal{D}_1(Q_0)} \ell(Q)^{1/q_0'} \leq C\|g\|_{L^\infty(\Sigma,\sigma)} c_1^{1/q_0'},\end{aligned} \tag{10.214}$$

where $c_1 \in (0,\infty)$ is as in (10.212). From this, we see that the left-hand side of (10.214) can be made arbitrarily small provided $c_1 \in (0,\infty)$ is sufficiently small to begin with.

Next, consider the contribution coming from $\mathcal{D}_3(Q_0)$. Observe for any $k \in \mathbb{N}$ and $Q \in \mathcal{D}_3(Q_0)$, we may use the boundedness of Q_0 to estimate

$$\begin{aligned}\int_{Q_0} |b_{j_k}(Q)|\,d\sigma &= \sigma(Q_0) ⨍_{Q_0} |b_{j_k}(Q)|\,d\sigma \\ &\leq \sigma(Q_0)[\omega]_{A_q}^{1/q} \Big(⨍_{Q_0} |b_{j_k}(Q)|^q\,d\omega \Big)^{1/q} \\ &\leq \frac{\sigma(Q_0)[\omega]_{A_q}^{1/q}}{\omega(Q_0)^{1/q}} \|b_{j_k}(Q)\|_{L^q(\Sigma,\omega)} \\ &\leq C_{Q_0,\omega} (c_\Sigma \ell(Q))^{\lambda(1/q-1)},\end{aligned} \tag{10.215}$$

for some constant $C_{Q_0,\omega} \in (0,\infty)$, independent of $Q \in \mathcal{D}_3(Q_0)$. The first inequality uses Lemma 2.34, and the final inequality is a consequence of (10.207). Since $b(Q)$ is also such a (q,λ,ω)-block on Σ (cf. (10.208)), the estimate in (10.215) also holds for $b(Q)$ in place of $b_{j_k}(Q)$. Subsequently, using (10.215), (10.206), and (10.210), we find there is a constant $C \in (0,\infty)$, independent of $Q \in \mathcal{D}_3(Q_0)$ such that for each $k \in \mathbb{N}$,

$$\begin{aligned}\sum_{Q\in\mathcal{D}_3(Q_0)} \int_{Q_0} |\lambda_{j_k}(Q)b_{j_k}(Q)g - \lambda(Q)b(Q)g|\, d\sigma \\ \leq \|g\|_{L^\infty(\Sigma,\sigma)} \sum_{Q\in\mathcal{D}_1(Q_0)} (\lambda_{j_k}(Q) + \lambda(Q))C_{Q_0,\omega}(c_\Sigma \ell(Q))^{\lambda(1/q-1)} \\ \leq C\|g\|_{L^\infty(\Sigma,\sigma)} \sup_{Q\in\mathcal{D}_3(Q_0)} \ell(Q)^{\lambda(1/q-1)} \leq C\|g\|_{L^\infty(\Sigma,\sigma)} c_2^{\lambda(1/q-1)},\end{aligned} \tag{10.216}$$

for $c_2 \in (0,\infty)$ as in (10.212) and bearing in mind that $\lambda(1/q-1) < 0$. This shows that the left-hand side of (10.216) can be made arbitrarily small provided the constant $c_2 \in (0,\infty)$ is chosen sufficiently large to begin with.

We now examine the contribution coming from $\mathcal{D}_2(Q_0)$. First, (10.212) implies that $\mathcal{D}_2(Q_0)$ contains only a finite number of dyadic cubes, given that Q_0 is bounded and all dyadic cubes $Q \in \mathcal{D}_2(Q_0)$ must satisfy both $\ell(Q) \in (c_1 c_\Sigma^{-1}, c_2 c_\Sigma^{-1})$ and $Q \cap Q_0 \neq \varnothing$, meaning that there are only finitely many cubes from finitely many generations of dyadic cubes, which can intersect Q_0. Using the triangle inequality, write

$$\begin{aligned}\sum_{Q\in\mathcal{D}_2(Q_0)} \left| \int_{Q_0} \lambda_{j_k}(Q)b_{j_k}(Q)g\, d\sigma - \int_{Q_0} \lambda(Q)b(Q)g\, d\sigma \right| \\ \leq \sum_{Q\in\mathcal{D}_2(Q_0)} \left| \int_{Q_0} \lambda_{j_k}(Q)b_{j_k}(Q)g\, d\sigma - \int_{Q_0} \lambda(Q)b_{j_k}(Q)g\, d\sigma \right| \\ + \sum_{Q\in\mathcal{D}_2(Q_0)} \left| \int_{Q_0} \lambda(Q)b_{j_k}(Q)g\, d\sigma - \int_{Q_0} \lambda(Q)b(Q)g\, d\sigma \right|.\end{aligned} \tag{10.217}$$

Because $\mathcal{D}_2(Q_0)$ contains only finitely many dyadic cubes, we have $\sigma(Q) \approx 1$ uniformly for each $Q \in \mathcal{D}_2(Q_0)$, which by (10.213) shows that

$$\|b_{j_k}(Q)\|_{L^1(Q_0,\sigma)} \lesssim 1 \quad \text{and} \quad \|b(Q)\|_{L^1(Q_0,\sigma)} \lesssim 1 \tag{10.218}$$

uniformly for $k \in \mathbb{N}$ and $Q \in \mathcal{D}_2(Q_0)$. As such, there exists a constant $C \in (0,\infty)$, independent of $Q \in \mathcal{D}_2(Q_0)$ for which

$$\begin{aligned}\sum_{Q\in\mathcal{D}_2(Q_0)} \left| \int_{Q_0} \lambda_{j_k}(Q)b_{j_k}(Q)g\, d\sigma - \int_{Q_0} \lambda(Q)b_{j_k}(Q)g\, d\sigma \right| \\ \leq C\|g\|_{L^\infty(\Sigma,\sigma)} \sum_{Q\in\mathcal{D}_2(Q_0)} |\lambda_{j_k}(Q) - \lambda(Q)| \to 0 \quad \text{as } k \to \infty.\end{aligned} \tag{10.219}$$

Also, there is a constant $C \in (0,\infty)$ depending on the cardinality of $\mathcal{D}_2(Q_0)$, but independent of $Q \in \mathcal{D}_2(Q_0)$, such that

$$\sum_{Q\in\mathcal{D}_2(Q_0)} \left| \int_{Q_0} \lambda(Q) b_{j_k}(Q) g \, d\sigma - \int_{Q_0} \lambda(Q) b(Q) g \, d\sigma \right| \tag{10.220}$$
$$\leq C \|g\|_{L^\infty(\Sigma,\sigma)} \max_{Q\in\mathcal{D}_2(Q_0)} \left| \int_{Q_0} b_{j_k}(Q) \, d\sigma - \int_{Q_0} b(Q) \, d\sigma \right| \to 0$$

as $k \to \infty$, bearing in mind that $b_{j_k}(Q) \xrightarrow{\text{weak-}*} b(Q)$ in $L^q(\Sigma,\omega)$ as $k \to \infty$ and $\mathbf{1}_{Q_0} \in L^{q'}(\Sigma,\omega')$, a fact that ensures

$$\lim_{k\to\infty} \int_{Q_0} b_{j_k}(Q) \, d\sigma = \lim_{k\to\infty} \int_{\Sigma} b_{j_k}(Q) \mathbf{1}_{Q_0} \, d\sigma \tag{10.221}$$
$$= \int_{\Sigma} b(Q) \mathbf{1}_{Q_0} \, d\sigma = \int_{Q_0} b(Q) \, d\sigma.$$

Consequently, (10.219) and (10.220) show that the left-hand side of (10.217) can be made arbitrarily small provided $k \in \mathbb{N}$ is sufficiently large. In view of this fact, as well as (10.214) and (10.216), we see for suitably chosen constants $c_1, c_2 \in (0,\infty)$ and $k \in \mathbb{N}$ large enough that the quantity

$$\sum_{Q\in\mathbb{D}(\Sigma)} \left| \int_{Q_0} \lambda_{j_k}(Q) b_{j_k}(Q) g \, d\sigma - \int_{Q_0} \lambda(Q) b(Q) g \, d\sigma \right| \tag{10.222}$$
$$= \sum_{Q\in\mathcal{D}_1(Q_0)\cup\mathcal{D}_2(Q_0)\cup\mathcal{D}_3(Q_0)} \left| \int_{Q_0} \lambda_{j_k}(Q) b_{j_k}(Q) g \, d\sigma - \int_{Q_0} \lambda(Q) b(Q) g \, d\sigma \right|$$

can be made arbitrarily small. Note we are relying on the fact that the sets $\mathcal{D}_1(Q_0)$, $\mathcal{D}_2(Q_0)$, and $\mathcal{D}_3(Q_0)$ are mutually disjoint by construction; cf. (10.212). Additionally, we remark that the equality above holds in view of that fact that $b_{j_k}(Q)$ and $b(Q)$ are supported in $\widetilde{Q}$ (cf. (10.207) and (10.208)), rendering the integrals appearing on the left-hand side of (10.222) equal to zero whenever $Q \in \mathbb{D}(\Sigma)$ with $\widetilde{Q} \cap Q_0 = \varnothing$, i. e., whenever it happens that $Q \in \mathbb{D}(\Sigma) \setminus (\mathcal{D}_1(Q_0) \cup \mathcal{D}_2(Q_0) \cup \mathcal{D}_3(Q_0))$ thanks to (10.212). This justifies the validity of (10.211).

We now would like to extend the result in (10.211) from functions belonging to $L^\infty_{\mathrm{comp}}(\Sigma,\sigma)$ to functions in $\mathring{M}^{q',\lambda}(\Sigma,\omega')$. To this end, fix an arbitrary $g \in \mathring{M}^{q',\lambda}(\Sigma,\omega')$. Then according to Proposition 10.22, there exists a sequence of functions $\{g_\ell\}_{\ell\in\mathbb{N}}$ coming from $L^\infty_{\mathrm{comp}}(\Sigma,\sigma)$ for which

$$\lim_{\ell\to\infty} \|g_\ell - g\|_{M^{q',\lambda}(\Sigma,\omega')} = 0. \tag{10.223}$$

Using the triangle inequality and Theorem 6.28, we can estimate

$$\begin{aligned}
&\left|\int_\Sigma f_{j_k} g\,d\sigma - \int_\Sigma fg\,d\sigma\right| \\
&\leq \left|\int_\Sigma f_{j_k} g\,d\sigma - \int_\Sigma f_{j_k} g_\ell\,d\sigma\right| + \left|\int_\Sigma f_{j_k} g_\ell\,d\sigma - \int_\Sigma fg_\ell\,d\sigma\right| \\
&\quad + \left|\int_\Sigma fg_\ell\,d\sigma - \int_\Sigma fg\,d\sigma\right| \\
&\leq \|f_{j_k}\|_{\mathcal{B}^{q,\lambda}(\Sigma,\omega)} \|g_\ell - g\|_{\mathcal{M}^{q',\lambda}(\Sigma,\omega')} + \left|\int_\Sigma f_{j_k} g_\ell\,d\sigma - \int_\Sigma fg_\ell\,d\sigma\right| \\
&\quad + \|f\|_{\mathcal{B}^{q,\lambda}(\Sigma,\omega)} \|g_\ell - g\|_{\mathcal{M}^{q',\lambda}(\Sigma,\omega')}.
\end{aligned} \tag{10.224}$$

The boundedness of $\{f_{j_k}\}_{k\in\mathbb{N}}$ in $\mathcal{B}^{q,\lambda}(\Sigma,\omega)$, the fact that $f \in \mathcal{B}^{q,\lambda}(\Sigma,\omega)$ by construction (cf. (10.209)), (10.223), the membership of g_ℓ to $L^\infty_{\mathrm{comp}}(\Sigma,\sigma)$ for each $\ell \in \mathbb{N}$, and (10.211) are enough to glean from (10.224) that

$$\lim_{k\to\infty}\left|\int_\Sigma f_{j_k} g\,d\sigma - \int_\Sigma fg\,d\sigma\right| = 0. \tag{10.225}$$

Given the arbitrary nature of $g \in \mathring{\mathcal{M}}^{q',\lambda}(\Sigma,\omega')$, we see that (10.203) holds.

At this point, we note that our present assumption that $\sup_{j\in\mathbb{N}} \|f_j\|_{\mathcal{B}^{q,\lambda}(\Sigma,\omega)}$ equals 1 and (10.210) imply

$$\|f\|_{\mathcal{B}^{q,\lambda}(\Sigma,\omega)} \leq \sum_{Q\in\mathbb{D}(\Sigma)} \lambda(Q) \leq C_* = C_* \sup_{j\in\mathbb{N}} \|f_j\|_{\mathcal{B}^{q,\lambda}(\Sigma,\omega)}. \tag{10.226}$$

However, this self improves to (10.202) in the following way. Set

$$M := \liminf_{j\to\infty} \|f_j\|_{\mathcal{B}^{q,\lambda}(\Sigma,\omega)}, \tag{10.227}$$

which is a well-defined number in $[0,\infty)$, given the boundedness of $\{f_j\}_{j\in\mathbb{N}}$ in $\mathcal{B}^{q,\lambda}(\Sigma,\omega)$. We also remark that M is necessarily an accumulation (or cluster) point of the numerical sequence $\{\|f_j\|_{\mathcal{B}^{q,\lambda}(\Sigma,\omega)}\}_{j\in\mathbb{N}}$, so that given any $\delta > 0$ there exist infinitely many $j \in \mathbb{N}$ for which $\|f_j\|_{\mathcal{B}^{q,\lambda}(\Sigma,\omega)} \leq M + \delta$. Restricting attention from the very beginning to the subsequence labeled by these j's, and running the above argument, which has produced (10.226), we arrive at

$$\|f\|_{\mathcal{B}^{q,\lambda}(\Sigma,\omega)} \leq C_*(M+\varepsilon) = C_*\Big(\liminf_{j\to\infty} \|f_j\|_{\mathcal{B}^{q,\lambda}(\Sigma,\omega)} + \delta\Big). \tag{10.228}$$

In view of the arbitrary nature of $\delta > 0$, the above inequality establishes (10.202) as desired. This completes the proof of Theorem 10.47. $\square$

As a corollary of Theorem 10.47, we find the following integral pairing characterization of membership to a Muckenhoupt-weighted Block space.

Corollary 10.48. *Let $\Sigma \subseteq \mathbb{R}^n$ be a closed* ADR *set and write $\sigma := \mathcal{H}^{n-1}\lfloor\Sigma$. Fix an integrability exponent $q \in (1,\infty)$ with $q' := q/(q-1)$ and a Muckenhoupt weight $\omega \in A_q(\Sigma,\sigma)$ whose conjugate weight satisfies*

$$\omega' := \omega^{-q'/q} \in A_{q'}(\Sigma,\sigma) \cap \mathrm{RH}_p(\Sigma,\sigma) \tag{10.229}$$

for some $p \in (1,\infty)$. Define $\varepsilon := (q-1)/p' \in (0,q-1)$ so that $\omega \in A_{q-\varepsilon}(\Sigma,\sigma)$ by Proposition 6.35, *and select any threshold $\lambda \in (0,\varepsilon(n-1)/(q-1))$. Bring in the weighted Block space $\mathcal{B}^{q,\lambda}(\Sigma,\omega)$ defined in relation to these choices of q, ω, ε, and λ as in Definition* 6.7.

Then for any $g \in \mathscr{M}(\Sigma,\sigma)$ one has

$$\begin{aligned} &g \in \mathcal{B}^{q,\lambda}(\Sigma,\omega) \quad \textit{if and only if} \\ &\int_\Sigma |fg|\,d\sigma < +\infty \quad \textit{for all } f \in M^{q',\lambda}(\Sigma,\omega'). \end{aligned} \tag{10.230}$$

Moreover, there exists a constant $C \in (0,\infty)$, depending only on Σ, n, q, λ, and ω such that whenever (10.230) *holds, one has*

$$\|g\|_{\mathcal{B}^{q,\lambda}(\Sigma,\omega)} \le C \cdot \sup_{\substack{f\in M^{q',\lambda}(\Sigma,\omega') \\ \|f\|_{M^{q',\lambda}(\Sigma,\omega')}\le 1}} \int_\Sigma |fg|\,d\sigma. \tag{10.231}$$

Notice how the conclusion in (10.230) differs from the result already recorded in Corollary 6.31, which only provides a statement about functions belonging to $\mathcal{B}^{q,\lambda}(\Sigma,\omega)$, and fails to divulge what happens in the case that g belongs to $\mathscr{M}(\Sigma,\sigma) \setminus \mathcal{B}^{q,\lambda}(\Sigma,\omega)$.

Proof of Corollary 10.48. As mentioned above, Corollary 6.31 (or alternatively Theorem 6.28) implies the forward direction in (10.230). Therefore, suppose $g \in \mathscr{M}(\Sigma,\sigma)$ is such that $\int_\Sigma |fg|\,d\sigma < +\infty$ for all $f \in M^{q',\lambda}(\Sigma,\omega')$. Next, consider the operator $T : M^{q',\lambda}(\Sigma,\omega') \to L^1(\Sigma,\sigma)$, defined according to $Tf := fg$ for all $f \in M^{q',\lambda}(\Sigma,\omega')$. Then T is clearly linear, and its graph,

$$\begin{aligned} \Gamma(T) &:= \{(f,Tf) : f \in M^{q',\lambda}(\Sigma,\omega')\} \\ &= \{(f,fg) : f \in M^{q',\lambda}(\Sigma,\omega')\}, \end{aligned} \tag{10.232}$$

is closed in the Banach space $M^{q',\lambda}(\Sigma,\omega') \times L^1(\Sigma,\sigma)$ (cf. Proposition 4.5). Indeed, if $\{f_j\}_{j\in\mathbb{N}} \subseteq M^{q',\lambda}(\Sigma,\omega')$, $f \in M^{q',\lambda}(\Sigma,\omega')$, and $h \in L^1(\Sigma,\sigma)$ are such that $f_j \to f$ in $M^{q',\lambda}(\Sigma,\omega')$ and $f_j g \to h$ in $L^1(\Sigma,\sigma)$ as $j\to\infty$, then $\{f_j\}_{j\in\mathbb{N}}$ also converges to f in $L^1(\Sigma, \frac{\sigma(x)}{1+|x|^{n-1}}) \subseteq L^1_{\mathrm{loc}}(\Sigma,\sigma)$ (cf. Proposition 4.16), so that $f_{j_k} \to f$ as $k\to\infty$ pointwise σ-a. e. on Σ for a subsequence $\{f_{j_k}\}_{k\in\mathbb{N}}$ of $\{f_j\}_{j\in\mathbb{N}}$. Similarly, $f_{j_{k_\ell}} g \to h$ as $\ell\to\infty$ pointwise σ-a. e. on Σ for a subsequence

$\{f_{j_{k_\ell}}\}_{\ell\in\mathbb{N}}$ of $\{f_{j_k}\}_{k\in\mathbb{N}}$. Thus, $fg = h$ at σ-a. e. point on Σ and $(f,h) = (f, Tf) \in \Gamma(T)$, showing that $\Gamma(T)$ is closed. By the closed graph theorem, we can conclude that T is a bounded operator. More specifically, there exists some constant $C_0 \in (0,\infty)$ such that

$$\begin{aligned}\|fg\|_{L^1(\Sigma,\sigma)} &= \|Tf\|_{L^1(\Sigma,\sigma)} \\ &\leq C_0\|f\|_{M^{q',\lambda}(\Sigma,\omega')} \quad \text{for all } f \in M^{q',\lambda}(\Sigma,\omega').\end{aligned} \tag{10.233}$$

If we set

$$C_* := \sup_{\substack{f\in M^{q',\lambda}(\Sigma,\omega') \\ \|f\|_{M^{q',\lambda}(\Sigma,\omega')}\leq 1}} \int_\Sigma |fg|\,d\sigma, \tag{10.234}$$

then (10.233) shows that $C_* \leq C_0 < +\infty$.

Pressing forward, fix $x_0 \in \Sigma$ and define for each $j \in \mathbb{N}$ the function $g_j : \Sigma \to \mathbb{R}$ according to

$$g_j(x) := \begin{cases} \mathbf{1}_{\Delta(x_0,j)}(x)g(x) & \text{if } |g(x)| \leq j, \\ 0 & \text{if } |g(x)| > j, \end{cases} \tag{10.235}$$

for each $x \in \Sigma$. In view of this definition, $g_j \in L^\infty_{\text{comp}}(\Sigma,\sigma) \subseteq \mathcal{B}^{q,\lambda}(\Sigma,\omega)$; cf. Proposition 6.17. Thus, Corollary 6.31, (10.235), and (10.234) show that

$$\|g_j\|_{\mathcal{B}^{q,\lambda}(\Sigma,\omega)} = \sup_{\substack{f\in M^{q',\lambda}(\Sigma,\omega') \\ \|f\|_{M^{q',\lambda}(\Sigma,\omega')}}} \int_\Sigma |fg_j|\,d\sigma \leq C_* \in (0,\infty), \tag{10.236}$$

and hence $\{g_j\}_{j\in\mathbb{N}}$ is a bounded sequence in $\mathcal{B}^{q,\lambda}(\Sigma,\omega)$. Consequently, Theorem 10.47 applies, allowing us to conclude that there exists some function $g_* \in \mathcal{B}^{q,\lambda}(\Sigma,\omega) \subseteq L^1_{\text{loc}}(\Sigma,\sigma)$ (cf. Proposition 6.17) and a constant $C \in (0,\infty)$, depending only on Σ, n, q, and λ, satisfying

$$\|g_*\|_{\mathcal{B}^{q,\lambda}(\Sigma,\omega)} \leq C \liminf_{j\to\infty} \|g_j\|_{\mathcal{B}^{q,\lambda}(\Sigma,\omega)} \leq CC_*, \tag{10.237}$$

where the last inequality follows from (10.236). Additionally, Theorem 10.47 guarantees the existence of a subsequence $\{g_{j_k}\}_{k\in\mathbb{N}}$ of $\{g_j\}_{j\in\mathbb{N}}$ such that

$$\lim_{k\to\infty} \int_\Sigma g_{j_k} h\,d\sigma = \int_\Sigma g_* h\,d\sigma \quad \text{for all } h \in M^{q',\lambda}(\Sigma,\omega'). \tag{10.238}$$

Let us temporarily assume that actually g is nonnegative. In particular, (10.233) implies for each surface ball $\Delta \subseteq \Sigma$ that

$$\int_{\Delta} g\,d\sigma = \|g\mathbf{1}_\Delta\|_{L^1(\Sigma,\sigma)} < +\infty, \tag{10.239}$$

noticing that $\mathbf{1}_\Delta \in M^{q',\lambda}(\Sigma,\omega')$ (cf. Lemma 4.9), which ensures that g is in $L^1_{\mathrm{loc}}(\Sigma,\sigma)$. Then by design (see (10.235)), we know $\{g_j\}_{j\in\mathbb{N}}$ is a positive sequence of functions, which increases to g pointwise σ-a. e. on Σ as $j \to \infty$. By Lebesgue's monotone convergence theorem,

$$\lim_{j\to\infty}\int_{\Sigma} g_j h\,d\sigma = \int_{\Sigma} gh\,d\sigma \quad \text{for all } h \in \mathscr{M}_+(\Sigma,\sigma)\cap M^{q',\lambda}(\Sigma,\omega'). \tag{10.240}$$

Subsequently, choosing $h := \mathbf{1}_\Delta$ for an arbitrary surface ball $\Delta \subseteq \Sigma$ in both (10.238) and (10.240) (and keeping in mind Lemma 4.9), we obtain

$$\lim_{k\to\infty}\int_{\Delta} g_{j_k}\,d\sigma = \int_{\Delta} g_*\,d\sigma \quad\text{and}\quad \lim_{j\to\infty}\int_{\Delta} g_j\,d\sigma = \int_{\Delta} g\,d\sigma, \tag{10.241}$$

so that $g = g_*$ at σ-a. e. point on Σ by Corollary 2.23 (bearing in mind $g, g_* \in L^1_{\mathrm{loc}}(\Sigma,\sigma)$). From this fact, (10.237), and (10.234) we find that g belongs to $\mathcal{B}^{q,\lambda}(\Sigma,\omega)$ with

$$\|g\|_{\mathcal{B}^{q,\lambda}(\Sigma,\omega)} \le CC_* = C\cdot \sup_{\substack{f\in M^{q',\lambda}(\Sigma,\omega')\\ \|f\|_{M^{q',\lambda}(\Sigma,\omega')}\le 1}} \int_{\Sigma} |fg|\,d\sigma < +\infty, \tag{10.242}$$

where C depends only on Σ, n, q, and λ, which gives the desired result for $g \in \mathscr{M}_+(\Sigma,\sigma)$.

Considering now arbitrary $g \in \mathscr{M}(\Sigma,\sigma)$ (not necessarily in $\mathscr{M}_+(\Sigma,\sigma)$), which satisfies the second line of (10.230), observe that we can run the same argument as above for $|g| \in \mathscr{M}_+(\Sigma,\sigma)$ in place of g, noting that $|g|$ continues to satisfy the second line of (10.230). Thus, recalling (10.242), we obtain $|g| \in \mathcal{B}^{q,\lambda}(\Sigma,\omega)$ with

$$\big\|\,|g|\,\big\|_{\mathcal{B}^{q,\lambda}(\Sigma,\omega)} \le C\cdot \sup_{\substack{f\in M^{q',\lambda}(\Sigma,\omega')\\ \|f\|_{M^{q',\lambda}(\Sigma,\omega')}\le 1}} \int_{\Sigma} |fg|\,d\sigma < +\infty. \tag{10.243}$$

Subsequently, in view of the lattice property for Block spaces (cf. Proposition 6.14), we see that $g \in \mathcal{B}^{q,\lambda}(\Sigma,\omega)$ and that the estimate in (10.231) holds for g as desired. □

Remark 10.49. If we avail ourselves of the full force of the theory developed in this work thus far, we have the following, shorter proof of Corollary 10.48. Namely, given $g \in \mathscr{M}(\Sigma,\sigma)$, the condition in the second line of (10.230) places g in $(M^{q',\lambda}(\Sigma,\omega'))'$, which has been identified in Theorem 10.46 as the weighted Block space $\mathcal{B}^{q,\lambda}(\Sigma,\omega)$. The estimate in (10.231) holds as a consequence of Corollary 10.14.

Remark 10.50. Retaining the context of Corollary 10.48, observe that in place of the estimate in (10.231), we could also write

$$\|g\|_{\mathcal{B}^{q,\lambda}(\Sigma,\omega)} \leq C \cdot \sup_{\substack{f\in M^{q',\lambda}(\Sigma,\omega') \\ \|f\|_{M^{q',\lambda}(\Sigma,\omega')}\leq 1}} \left| \int_\Sigma fg \, d\sigma \right| \tag{10.244}$$
$$= C \cdot \sup_{\substack{f\in M^{q',\lambda}(\Sigma,\omega') \\ \|f\|_{M^{q',\lambda}(\Sigma,\omega')}=1}} \left| \int_\Sigma fg \, d\sigma \right|.$$

As the equality above is clear, we focus on verifying the inequality. Define

$$\operatorname{sgn} g(x) := \begin{cases} 1 & \text{if } g(x) > 0, \\ 0 & \text{if } g(x) = 0, \\ -1 & \text{if } g(x) < 0, \end{cases} \quad \text{for all } x \in \Sigma. \tag{10.245}$$

Notice that $|\operatorname{sgn} g| \leq 1$ on Σ and replacing f by $(\operatorname{sgn} g)f$ (which belongs to $M^{q',\lambda}(\Sigma,\omega')$ with $\|(\operatorname{sgn} g)f\|_{M^{q',\lambda}(\Sigma,\omega')} \leq 1$ whenever $f \in M^{q',\lambda}(\Sigma,\omega')$ with $\|f\|_{M^{q',\lambda}(\Sigma,\omega')} \leq 1$) in the first supremum in (10.244) gives

$$\sup_{\substack{f\in M^{q',\lambda}(\Sigma,\omega') \\ \|f\|_{M^{q',\lambda}(\Sigma,\omega')}\leq 1}} \left| \int_\Sigma fg \, d\sigma \right| \geq \sup_{\substack{f\in M^{q',\lambda}(\Sigma,\omega') \\ \|f\|_{M^{q',\lambda}(\Sigma,\omega')}\leq 1}} \left| \int_\Sigma f|g| \, d\sigma \right|. \tag{10.246}$$

Now, given any function $f \in M^{q',\lambda}(\Sigma,\omega')$ with $\|f\|_{M^{q',\lambda}(\Sigma,\omega')} \leq 1$, define the following two σ-measurable subsets of Σ:

$$A_+ := \{x \in \Sigma : f(x) \geq 0\} \quad \text{and} \quad A_- := \{x \in \Sigma : f(x) < 0\}. \tag{10.247}$$

Note that the lattice property of Morrey spaces (cf. Proposition 4.7) implies the functions $f_\pm := \pm f \mathbf{1}_{A_\pm}$ belong to $M^{q',\lambda}(\Sigma,\omega') \cap \mathscr{M}_+(\Sigma,\sigma)$ with $\|f_\pm\|_{M^{q',\lambda}(\Sigma,\omega')} \leq 1$. Additionally, $0 \leq f_\pm \leq |f|$ pointwise on Σ, and we have $f = f_+ - f_-$ and $f_+ + f_- = |f|$. Consequently,

$$\sup_{\substack{f\in M^{q',\lambda}(\Sigma,\omega') \\ \|f\|_{M^{q',\lambda}(\Sigma,\omega')}\leq 1}} \left| \int_\Sigma f|g| \, d\sigma \right| \geq \sup_{\substack{f\in M^{q',\lambda}(\Sigma,\omega') \\ \|f\|_{M^{q',\lambda}(\Sigma,\omega')}\leq 1}} \max\left\{ \int_\Sigma f_+|g| \, d\sigma, \int_\Sigma f_-|g| \, d\sigma \right\} \tag{10.248}$$
$$\geq \sup_{\substack{f\in M^{q',\lambda}(\Sigma,\omega') \\ \|f\|_{M^{q',\lambda}(\Sigma,\omega')}\leq 1}} \frac{1}{2}\left(\int_\Sigma f_+|g| \, d\sigma + \int_\Sigma f_-|g| \, d\sigma \right)$$
$$= \frac{1}{2} \sup_{\substack{f\in M^{q',\lambda}(\Sigma,\omega') \\ \|f\|_{M^{q',\lambda}(\Sigma,\omega')}\leq 1}} \int_\Sigma |f||g| \, d\sigma \, d\sigma$$
$$\geq C\|g\|_{\mathcal{B}^{q,\lambda}(\Sigma,\omega)},$$

where the last inequality follows by Corollary 10.48, and $C \in (0, \infty)$ depends only on Σ, n, q, and λ. Then (10.246) and (10.248) are enough to justify the inequality in (10.244).

Remark 10.51. Still working under the assumptions of Corollary 10.48, recall the definition of $\|\cdot\|_{\mathcal{B}^{q,\lambda}(\Sigma,\omega)}$ in (6.41). If we extend that definition to have

$$\|\cdot\|_{\mathcal{B}^{q,\lambda}(\Sigma,\omega)} : \mathscr{M}(\Sigma,\sigma) \to [0, +\infty] \tag{10.249}$$

by setting $\|g\|_{\mathcal{B}^{q,\lambda}(\Sigma,\omega)} := +\infty$ whenever $g \in \mathscr{M}(\Sigma,\sigma) \setminus \mathcal{B}^{q,\lambda}(\Sigma,\omega)$, then Corollary 10.48, Theorem 6.28, and Definition 10.42 imply

$$\|g\|_{\mathcal{B}^{q,\lambda}(\Sigma,\omega)} \approx \sup_{\substack{f \in M^{q',\lambda}(\Sigma,\omega') \\ \|f\|_{M^{q',\lambda}(\Sigma,\omega')} \leq 1}} \int_\Sigma |fg| \, d\sigma = \|\!|g|\!\|_{\mathcal{B}^{q,\lambda}(\Sigma,\omega)} \tag{10.250}$$

for all $g \in \mathscr{M}(\Sigma,\sigma)$. Thus, adopting the above convention for $\|\cdot\|_{\mathcal{B}^{q,\lambda}(\Sigma,\omega)}$, we have

$$\|g\|_{\mathcal{B}^{q,\lambda}(\Sigma,\omega)} \approx \|\!|g|\!\|_{\mathcal{B}^{q,\lambda}(\Sigma,\omega)} \quad \text{for all } g \in \mathscr{M}(\Sigma,\sigma). \tag{10.251}$$

Using this fact as well as Definitions 10.11 and 10.42, we see that

$$\big(\mathcal{B}^{q,\lambda}(\Sigma,\sigma), \|\!|\cdot|\!\|_{\mathcal{B}^{q,\lambda}(\Sigma,\sigma)}\big) = \big(M^{q',\lambda}(\Sigma,\omega')\big)', \tag{10.252}$$

or $\mathcal{B}^{q,\lambda}(\Sigma,\omega) = (M^{q',\lambda}(\Sigma,\omega))'$ for short. Since $M^{q',\lambda}(\Sigma,\omega')$ is already known to be a generalized Banach function space from Proposition 10.41 (a fact proved independently of Theorem 10.39), we know its associated space, $\mathcal{B}^{q,\lambda}(\Sigma,\omega)$ is also a generalized Banach function space (cf. Definition 10.11 and [87, Proposition 5.1.10, p. 223]). Moreover, the Lorentz–Luxemburg theorem, Proposition 10.13, yields $(\mathcal{B}^{q,\lambda}(\Sigma,\omega))' = M^{q',\lambda}(\Sigma,\omega')$. Notably,

> we have recovered the conclusions of Theorem 10.46, but independently of the duality result recorded in Theorem 10.39. (10.253)

In view of (10.253), though not immediately obvious at present, we have opened the door to an alternate proof of Theorem 10.39, which loosely says that Muckenhoupt-weighted vanishing Morrey spaces are pre-duals of Muckenhoupt-weighted Block spaces. To get there, we need the following result, which characterizes the two subspaces introduced in Definitions 10.6 and 10.8 in the context of Muckenhoupt-weighted Morrey spaces.

Proposition 10.52. *Let $\Sigma \subseteq \mathbb{R}^n$ be a closed Ahlfors regular set, and write $\sigma := \mathcal{H}^{n-1}\lfloor\Sigma$. Fix two integrability exponents $p, q \in (1,\infty)$, along with an arbitrary Muckenhoupt weight $w \in A_p(\Sigma,\sigma) \cap \mathrm{RH}_q(\Sigma,\sigma)$, and a threshold $\lambda \in (0, (n-1)/q')$, where $q' := q/(q-1)$. Then*

$$\big(M^{p,\lambda}(\Sigma,w)\big)_a = \big(M^{p,\lambda}(\Sigma,w)\big)_b = \mathring{M}^{p,\lambda}(\Sigma,w). \tag{10.254}$$

Proof. To facilitate the proof, we make the following claim:

$$\text{for any surface ball } \Delta \subseteq \Sigma, \text{ one has } \mathbf{1}_\Delta \in (M^{p,\lambda}(\Sigma, w))_a. \tag{10.255}$$

To see this, note that any function with compact support will satisfy (b) in Definition 10.6. Now, fix some $\mathbf{1}_{\Delta(x_0,R)}$, with $x_0 \in \Sigma$ and $R \in (0,\infty)$. Then for each collection of σ-measurable sets $\{E_j\}_{j\in\mathbb{N}}$ with $E_j \subseteq \Delta(x_0, R)$ for each $j \in \mathbb{N}$ and $\sigma(E_j) \to 0$ as $j \to \infty$, we see from (10.47) that

$$\|\mathbf{1}_{E_j}\|_{M^{p,\lambda}(\Sigma,w)} \le Cw(\Delta(x_0,R))^{1/p} R^{-(n-1)/(pq')} \|\mathbf{1}_{E_j}\|_{L^{s_0}(\Sigma,\sigma)} \tag{10.256}$$

for some constant $C \in (0,\infty)$, depending only on the ADR character of Σ, n, p, q, λ, $[w]_{A_p}$, and $[w]_{\mathrm{RH}_q}$, where $s_0 := \frac{(n-1)pq'}{n-1-\lambda q'}$. Since

$$\|\mathbf{1}_{E_j}\|_{L^{s_0}(\Sigma,\sigma)} = \sigma(E_j)^{1/s_0} \to 0 \quad \text{as } j \to \infty, \tag{10.257}$$

we see that (10.256) implies

$$\lim_{j\to\infty} \|\mathbf{1}_{E_j}\|_{M^{p,\lambda}(\Sigma,w)} = 0. \tag{10.258}$$

As such, item (a) in Definition 10.6 also holds for $\mathbf{1}_{\Delta(x_0,R)}$, thus verifying (10.255).

Pressing forward, Proposition 4.7 and Definition 10.6 imply that

$$L^\infty_{\mathrm{comp}}(\Sigma,\sigma) \subseteq (M^{p,\lambda}(\Sigma,w))_a. \tag{10.259}$$

Since $(M^{p,\lambda}(\Sigma,w))_a$ is closed in $M^{p,\lambda}(\Sigma,\sigma)$ (cf. (10.11)), we can then use item (c) of Proposition 10.21 to say that

$$\mathring{M}^{p,\lambda}(\Sigma,w) = \overline{L^\infty_{\mathrm{comp}}(\Sigma,\sigma)}^{M^{p,\lambda}(\Sigma,\sigma)} \subseteq (M^{p,\lambda}(\Sigma,w))_a. \tag{10.260}$$

At this point, we use item (c) of Proposition 10.21 together with Definition 10.8 to see that $(M^{p,\lambda}(\Sigma,w))_b = \mathring{M}^{p,\lambda}(\Sigma,w)$. Consequently, (10.260) and (10.12) permit us to write

$$(M^{p,\lambda}(\Sigma,w))_b = \mathring{M}^{p,\lambda}(\Sigma,w) \subseteq (M^{p,\lambda}(\Sigma,w))_a \subseteq (M^{p,\lambda}(\Sigma,w))_b, \tag{10.261}$$

which yields the desired conclusion in (10.254). □

With the identification in (10.254) in hand, we find the following result in [78, Theorem 2, p. 16] relevant. Namely,[5]

5 It is important to note that our definition of generalized Banach function space is compatible with what Luxemburg calls a Banach function space in [78], allowing us to use [78, Theorem 2, p. 16] in the context of our theory. The main difference is Luxemburg assumes the underlying measure space is sigma-finite, but this is a nonissue in view of Remark 10.2.

$$\text{for an arbitrary generalized Banach function space } \mathbb{X}, \text{ to have} \\ (\mathbb{X}_b)^* = \mathbb{X}' \text{ is necessary and sufficient to have } \mathbb{X}_a = \mathbb{X}_b. \tag{10.262}$$

Remark 10.53. As advertised before the statement of Proposition 10.52, we are now able to provide an alternate proof of Theorem 10.39. Namely, retaining the context of Theorem 10.39, we turn to (10.262) in the case that $\mathbb{X} := M^{p,\lambda}(\Sigma, w)$ (cf. Proposition 10.41), so that along with Proposition 10.52, we obtain $(\mathring{M}^{p,\lambda}(\Sigma, w))^* = (M^{p,\lambda}(\Sigma, w))'$. However, keeping Remark 10.51 in mind, specifically (10.253), we have identified the associated space $(M^{p,\lambda}(\Sigma, w))'$ with $\mathcal{B}^{p',\lambda}(\Sigma, w')$, allowing us to conclude that $(\mathring{M}^{p,\lambda}(\Sigma, w))^*$ may be identified in an isomorphic fashion with $\mathcal{B}^{p',\lambda}(\Sigma, w')$.

Remark 10.54. Establishing Proposition 10.52 also unlocks an alternative proof of Theorem 10.46 and the fact that Muckenhoupt-weighted Block spaces are generalized Banach function spaces. To see this, retain the context of Theorem 10.46 and use (10.262) with $\mathbb{X} := M^{p,\lambda}(\Sigma, w)$ (cf. Proposition 10.41). Subsequently, Proposition 10.52 and Theorem 10.39 imply

$$(M^{p,\lambda}(\Sigma, w))' = (\mathring{M}^{p,\lambda}(\Sigma, w))^* = \mathcal{B}^{p',\lambda}(\Sigma, w), \tag{10.263}$$

so that $\mathcal{B}^{p',\lambda}(\Sigma, w)$ is a generalized Banach function space, being the associated space of the generalized Banach function space $M^{p,\lambda}(\Sigma, w)$; cf. Definition 10.11 and [86, Proposition 5.1.10, p. 223].

With Muckenhoupt-weighted Morrey and Block spaces fitting neatly into the framework of generalized Banach function spaces, we are ideally positioned to apply Theorem 10.16 to obtain an extrapolation result at the level of Muckenhoupt-weighted Block spaces, which nicely parallels the extrapolation result for Muckenhoupt-weighted Morrey spaces recorded in Theorem 4.19.

Theorem 10.55. *Let $\Sigma \subseteq \mathbb{R}^n$ be a closed ADR set, and write $\sigma := \mathcal{H}^{n-1}\lfloor\Sigma$. Fix an integrability exponent $q_0 \in [1, \infty)$. Suppose that $\mathcal{F}$ is a collection of pairs of σ-measurable functions (f, g) on Σ such that*

$$\|f\|_{L^{q_0}(\Sigma,w)} \leq C_w \|g\|_{L^{q_0}(\Sigma,w)} \quad \text{for all } w \in A_{q_0}(\Sigma, \sigma), \tag{10.264}$$

where the constant $C_w \in (0, \infty)$ depends only on the ADR character of Σ, n, q_0, and $[w]_{A_{q_0}}$ in such a way that C_w stays bounded provided $[w]_{A_{q_0}}$ stays bounded. More precisely, for all $W \in (0, \infty)$,

$$\sup_{\substack{w \in A_{q_0}(\Sigma,\sigma) \\ [w]_{A_{q_0}} \leq W}} C_w < +\infty. \tag{10.265}$$

Then for any $q \in (1, \infty)$ with $q' := q/(q-1)$, any given Muckenhoupt weight $\omega \in A_q(\Sigma, w)$ whose conjugate weight satisfies

$$\omega' := \omega^{-q'/q} \in A_{q'}(\Sigma,\sigma) \cap \mathrm{RH}_p(\Sigma,\sigma) \tag{10.266}$$

for some $p \in (1,\infty)$, *and with* $\varepsilon := (q-1)/p' \in (0,q-1)$ (*which ensures that* $\omega \in A_{q-\varepsilon}(\Sigma,\sigma)$ *by Proposition* 6.35) *and* $\lambda \in (0,\varepsilon(n-1)/(q-1))$, *the following regarding the weighted Block space* $\mathcal{B}^{q,\lambda}(\Sigma,\omega)$ (*associated with these choices of* q, ω, ε, *and* λ *as in Definition* 6.7) *is true. There exist two constants* $W, K \in (0,\infty)$, *depending only on the* ADR *character of* Σ, n, q, ε, λ, $[\omega]_{A_q}$, *and* $[\omega]_{A_{q-\varepsilon}}$, *such that for all pairs* $(f,g) \in \mathcal{F}$, *one has*

$$\|f\|_{\mathcal{B}^{q,\lambda}(\Sigma,\omega)} \le C\|g\|_{\mathcal{B}^{q,\lambda}(\Sigma,\omega)}, \tag{10.267}$$

where

$$C := K \cdot \Big(\sup_{\substack{w \in A_{q_0}(\Sigma,\sigma) \\ [w]_{A_{q_0}} \le W}} C_w \Big). \tag{10.268}$$

Proof. Note that Theorem 10.46 identifies $\mathcal{B}^{q,\lambda}(\Sigma,\omega)$ and $M^{q',\lambda}(\Sigma,\omega')$ as associated spaces of one another. Since $\omega' \in A_{q'}(\Sigma,\sigma) \cap \mathrm{RH}_p(\Sigma,\sigma)$, and our choice of ε implies

$$\lambda \in (0,\varepsilon(n-1)/(q-1)) = (0,(n-1)/p'), \tag{10.269}$$

Proposition 5.1 provides the necessary boundedness of the Hardy–Littlewood maximal operator on $M^{q',\lambda}(\Sigma,\omega')$. Similarly, Proposition 7.1 ensures the Hardy–Littlewood maximal operator is bounded on $\mathcal{B}^{q,\lambda}(\Sigma,\omega)$. At this point, Theorem 10.16 applies, from which we deduce our extrapolation result for Muckenhoupt-weighted Block spaces. □

Remark 10.56. It is worth commenting on the fact that the extrapolation result stated in Theorem 10.55, when paired with the Calderón–Zygmund theory recorded in Proposition 5.6 and Proposition 5.11, can be used to build a Calderón–Zygmund theory for Muckenhoupt-weighted Block spaces, though the results found throughout Chapter 7 are not entirely recovered by going this route. Indeed, in Theorem 10.55 we demand that the weight $\omega \in A_q(\Sigma,\sigma)$ also satisfies $\omega' \in \mathrm{RH}_p(\Sigma,\sigma)$ for some $p \in (1,\infty)$. The latter membership provides us with a suitable choice of ε needed to define the weighted-Block space, namely, $\varepsilon := (q-1)/p'$. However, in a typical statement in Chapter 7 see, e. g., Theorem 7.1, we ask that $\omega \in A_q(\Sigma,\sigma)$ also belongs to $A_{q-\varepsilon}(\Sigma,\sigma)$ for *some* choice of $\varepsilon \in (0,q-1)$, and ask nothing of the conjugate weight ω' (in particular, nothing in terms of a reverse Hölder condition).

As promised earlier in this chapter, we have the following embedding result for Muckenhoupt-weighted Block spaces in the spirit of Lemma 5.17.

Lemma 10.57. *Let* $\Sigma \subseteq \mathbb{R}^n$ *be a closed* ADR *set and write* $\sigma := \mathcal{H}^{n-1}\lfloor\Sigma$. *Fix any integrability exponent* $q_0 \in (1,\infty)$. *Then for any* $q \in (1,\infty)$ *with* $q' := q/(q-1)$, *any given weight* $\omega \in A_q(\Sigma,w)$ *whose conjugate weight satisfies*

$$\omega' := \omega^{-q'/q} \in A_{q'}(\Sigma, \sigma) \cap \mathrm{RH}_p(\Sigma, \sigma) \tag{10.270}$$

for some $p \in (1, \infty)$*, and with* $\varepsilon := (q-1)/p' \in (0, q-1)$ *(which ensures* $\omega \in A_{q-\varepsilon}(\Sigma, \sigma)$ *by Proposition* 6.35*) and* $\lambda \in (0, \varepsilon(n-1)/(q-1))$*, the following inclusion involving the weighted Block space* $\mathcal{B}^{q,\lambda}(\Sigma, \omega)$ *(associated with these choices of* q*,* ω*,* ε*, and* λ *as in Definition* 6.7*) is true:*

$$\mathcal{B}^{q,\lambda}(\Sigma, \omega) \subseteq \bigcup_{w \in A_{q_0}(\Sigma,\sigma)} L^{q_0}(\Sigma, w). \tag{10.271}$$

Proof. This is a direct consequence of Corollary 10.17, given that Theorem 10.46 identifies $\mathcal{B}^{q,\lambda}(\Sigma, \omega)$ and $M^{q',\lambda}(\Sigma, \omega')$ as associated spaces of one another. Recall also Proposition 5.1 and Proposition 7.1, which demonstrate the boundedness of the Hardy–Littlewood maximal operator on $M^{q',\lambda}(\Sigma, \omega')$ and $\mathcal{B}^{q,\lambda}(\Sigma, \omega)$, respectively, so that Corollary 10.17 indeed applies. Note that $\omega' \in A_{q'}(\Sigma, \sigma) \cap \mathrm{RH}_p(\Sigma, \sigma)$ and that our choice of ε implies

$$\lambda \in (0, \varepsilon(n-1)/(q-1)) = (0, (n-1)/p'), \tag{10.272}$$

so that $M^{q',\lambda}(\Sigma, \omega')$ is the kind of weighted Morrey space under consideration in Proposition 5.1. □

11 Boundary value problems

The wealth of results and mathematical machinery we have developed thus far put us in a suitable position to deal with boundary value problems formulated in terms of Muckenhoupt weighted Morrey and Block spaces. We divide this chapter into four sections, where we consider the Dirichlet problem (Theorem 11.9), the Neumann problem (Theorem 11.22), the regularity problem, both inhomogeneous (Theorem 11.29) and homogeneous (Theorem 11.32) versions, and finally the transmission problem (Theorem 11.35).

We debut with several auxiliary results that will aid in our task of solving the aforementioned boundary value problems. As motivation for said results, we would like to first consider the Dirichlet problem on Muckenhoupt-weighted Morrey spaces and loosely describe a strategy for solving it. For that purpose, let $\Omega \subseteq \mathbb{R}^n$ be a uniformly rectifiable domain, and abbreviate $\sigma := \mathcal{H}^{n-1}\lfloor\partial\Omega$. Fix integrability exponents $p, q \in (1, \infty)$ along with a Muckenhoupt weight $w \in A_p(\partial\Omega, \sigma) \cap \mathrm{RH}_q(\partial\Omega, \sigma)$. Fix a threshold $\lambda \in (0, (n-1)/q')$, where $q' := q/(q-1)$. Let L be a weakly elliptic, second-order, homogeneous, constant complex coefficient, $M \times M$ system in $\mathbb{R}^n$. For a fixed aperture parameter $\kappa \in (0, \infty)$, consider the Dirichlet boundary value problem:

$$\begin{cases} u \in [\mathscr{C}^\infty(\Omega)]^M, \\ Lu = 0 \quad \text{in } \Omega, \\ \mathcal{N}_\kappa u \in M^{p,\lambda}(\partial\Omega, w), \\ u\big|_{\partial\Omega}^{\kappa-\text{n.t.}} = f \in [M^{p,\lambda}(\partial\Omega, w)]^M. \end{cases} \tag{11.1}$$

As a preliminary observation, we find that the boundedness of the Hardy–Littlewood maximal operator on weighted Morrey spaces can be used as a tool to show that the aperture parameter, κ, is solely present as a way to help us describe the geometry of the boundary value problem in (11.1), and has no bearing on the actual solvability of the problem. In particular, as the subsequent proposition shows, the nontangential maximal operator for any two given aperture parameters are quantitatively equivalent.

Proposition 11.1. *Let Ω be an open, nonempty, proper subset of $\mathbb{R}^n$ such that $\partial\Omega$ is an Ahlfors regular set, and write $\sigma := \mathcal{H}^{n-1}\lfloor\partial\Omega$. Fix two aperture parameters $\kappa, \widetilde{\kappa} \in (0, \infty)$. Suppose $u : \Omega \to \mathbb{C}$ is a Lebesgue measurable function. Next, fix two integrability exponents $p, q \in (1, \infty)$, a Muckenhoupt weight $w \in A_p(\partial\Omega, \sigma) \cap \mathrm{RH}_q(\partial\Omega, \sigma)$, and a threshold $\lambda \in (0, (n-1)/q')$, where $q' := q/(q-1)$. Finally, fix any $a \in [0, \lambda)$, a reference point $x_0 \in \partial\Omega$, and abbreviate $w_a := w|\cdot - x_0|^a$, a weight function on $\partial\Omega$.*

Then $\mathcal{N}_\kappa u \in M^{p,\lambda}(\partial\Omega, w_a)$ if and only if $\mathcal{N}_{\widetilde{\kappa}} u \in M^{p,\lambda}(\partial\Omega, w_a)$. More precisely, if $\kappa \le \widetilde{\kappa}$, then one has

$$\|\mathcal{N}_\kappa u\|_{M^{p,\lambda}(\partial\Omega, w_a)} \le \|\mathcal{N}_{\widetilde{\kappa}} u\|_{M^{p,\lambda}(\partial\Omega, w_a)} \le C\|\mathcal{N}_\kappa u\|_{M^{p,\lambda}(\partial\Omega, w_a)} \tag{11.2}$$

for some constant $C \in (0, \infty)$, depending only on κ, $\widetilde{\kappa}$, the ADR *character of $\partial\Omega$, n, p, q, λ, a, $[w]_{\mathrm{RH}_q}$, and $[w]_{A_p}$.*

https://doi.org/10.1515/9783111458274-011

Proof. Without loss of generality, take $\kappa \le \widetilde{\kappa}$. Suppose $\mathcal{N}_{\widetilde{\kappa}}u \in M^{p,\lambda}(\partial\Omega, w_a)$. By definitions, the σ-measurability of $\mathcal{N}_\kappa u$ (cf. Remark 2.13), and the lattice property of $M^{p,\lambda}(\partial\Omega, w_a)$ (cf. Proposition 4.7) it is clear that this implies $\mathcal{N}_\kappa u \in M^{p,\lambda}(\partial\Omega, w_a)$ with

$$\|\mathcal{N}_\kappa u\|_{M^{p,\lambda}(\partial\Omega,w)} \le \|\mathcal{N}_{\widetilde{\kappa}}u\|_{M^{p,\lambda}(\partial\Omega,w)} < +\infty. \tag{11.3}$$

Conversely, suppose $\mathcal{N}_\kappa u \in M^{p,\lambda}(\partial\Omega, w_a)$. Denote by $\mathcal{M}_{\partial\Omega}$ the Hardy–Littlewood maximal operator on $\partial\Omega$ as in (2.26). Then by [86, Theorem 8.4.6, p. 528], we may estimate

$$\mathcal{N}_\kappa u \le \mathcal{N}_{\widetilde{\kappa}}u \le C \cdot \mathcal{M}_{\partial\Omega}(\mathcal{N}_\kappa u) \quad \text{on } \partial\Omega \tag{11.4}$$

for some constant $C \in (0, \infty)$, depending only on κ, $\widetilde{\kappa}$, and the doubling character of σ. Note that the assumption that $\partial\Omega$ is Ahlfors regular implies that σ is doubling with doubling character controlled by the ADR character of Σ. By Proposition 5.1, the boundedness of $\mathcal{M}_{\partial\Omega}$ on $M^{p,\lambda}(\partial\Omega, w_a)$ shows that in fact $\mathcal{M}_{\partial\Omega}(\mathcal{N}_\kappa u) \in M^{p,\lambda}(\partial\Omega, w_a)$ with

$$\begin{aligned}\|\mathcal{N}_{\widetilde{\kappa}}u\|_{M^{p,\lambda}(\partial\Omega,w_a)} &\le C\big\|\mathcal{M}_{\partial\Omega}(\mathcal{N}_\kappa u)\big\|_{M^{p,\lambda}(\partial\Omega,w_a)} \\ &\le C\|\mathcal{N}_\kappa u\|_{M^{p,\lambda}(\partial\Omega,w_a)} < +\infty\end{aligned} \tag{11.5}$$

for some constant $C \in (0, \infty)$, independent of u. The first inequality follows from (11.4), the σ-measurability of $\mathcal{N}_{\widetilde{\kappa}}u$ (cf. Remark 2.13), and the lattice property of $M^{p,\lambda}(\partial\Omega, w_a)$ (cf. Proposition 4.7). Thus, $\mathcal{N}_{\widetilde{\kappa}}u \in M^{p,\lambda}(\partial\Omega, w_a)$ as wanted. Additionally, (11.3) and (11.5) give the desired double inequality in (11.2), establishing the result. □

It is interesting to compare Proposition 11.1 with Proposition 7.7, which provides an alternative (and stronger) way of seeing that in the context of Muckenhoupt-weighted Morrey spaces the nontangential maximal function is independent of the choice of aperture parameter.

The method of proof for existence of solutions to (11.1) we employ relies on double layer potentials. These are quite useful, because they produce null solutions to the kinds of weakly elliptic systems we consider, and we understand their nontangential boundary traces. More specifically, working in the context of the Dirichlet problem in (11.1), for any $g \in L^1(\partial\Omega, \frac{\sigma(x)}{1+|x|^{n-1}})$ and any aperture parameter $\kappa \in (0, \infty)$, the nontangential trace of the double layer potential according to (3.22) is

$$(\mathcal{D}_A g)\Big|_{\partial\Omega}^{\kappa-\text{n.t.}} = \Big(\frac{1}{2}I + K_A\Big)g \quad \text{for } \sigma\text{-a. e. point on } \partial\Omega, \tag{11.6}$$

where $\mathcal{D}_A$ and K_A are as in (3.18) and (3.19), respectively, for any coefficient tensor $A \in \mathfrak{A}_L$. We also know that $u := \mathcal{D}_A g \in [\mathscr{C}^\infty(\Omega)]^M$ will satisfy $Lu = 0$ in Ω (cf. (3.21)), but we need to find g so that $(\frac{1}{2}I + K_A)g$ equals f, the prescribed boundary datum in (11.1). This amounts to an invertibility problem for the operator $\frac{1}{2}I+K_A$ on a Muckenhoupt-weighted Morrey space. The question of invertibility of this operator is intricately linked with the geometry of the domain and the nature of the system L, but assuming both of these aspects are amenable, we ultimately achieve invertibility via a Neumann series argument. This is a standard result from functional analysis, which we state for completeness in the lemma below.

Lemma 11.2. *Let $(\mathcal{X}, \|\cdot\|_{\mathcal{X}})$ be a Banach space. Let I denote the identity operator on $\mathcal{X}$ and fix $z \in \mathbb{C} \setminus \{0\}$. If $T : \mathcal{X} \to \mathcal{X}$ is a bounded operator such that $\|T\|_{\mathcal{X}\to\mathcal{X}} < |z|$, then $zI - T$ is invertible with the inverse given via the Neumann series*

$$(zI - T)^{-1} = \sum_{j=0}^{\infty} z^{-j-1} T^j, \tag{11.7}$$

where convergence of the series is understood to be in the operator norm $\|\cdot\|_{\mathcal{X}\to\mathcal{X}}$.

With a basic strategy for existence of solutions laid out, we now turn to the question of uniqueness. This is a challenging task, and before introducing the next proposition that will be our main tool in proving uniqueness, we bring in a definition from [86, (8.9.179), p. 814].

Definition 11.3. A set Ω is `globally pathwise nontangentially accessible` provided Ω is an open, nonempty, proper subset of $\mathbb{R}^n$ for which:

> given any $\kappa > 0$ there exist $\widetilde{\kappa} \geq \kappa$ along with $c \in [1, \infty)$ such that σ-a. e. point $x \in \partial\Omega$ has the property that any $y \in \Gamma_\kappa(x)$ may be joined by a rectifiable curve $\gamma_{x,y}$ such that $\gamma_{x,y} \setminus \{x\} \subset \Gamma_{\widetilde{\kappa}}(x)$ and whose length is $\leq c|x - y|$. (11.8)

It is noted in [86, (8.9.180), p. 814] that

> any one-sided NTA domain with unbounded boundary is a globally pathwise nontangentially accessible set. (11.9)

Our strategy for demonstrating uniqueness of a solution to (11.1) is to appeal to an integral representation formula for the solution. We are granted a suitable integral representation via the following powerful result, which first appears in [88, Theorem 4.4.1, pp. 799–801]. Notice this theorem's wide range of applicability due to its "spaceless" nature. It asks the bare minimum of the functions involved so that the integral representation formula itself is a well-defined object.

Proposition 11.4. *Let Ω be an open nonempty proper subset of $\mathbb{R}^n$ (where $n \in \mathbb{N}$ with $n \geq 2$), which is globally pathwise nontangentially accessible (in the sense of (11.8)), and such that $\partial\Omega$ is unbounded and Ahlfors regular. Abbreviate $\sigma := \mathcal{H}^{n-1}\lfloor\partial\Omega$ and denote by $\nu = (\nu_1, \dots, \nu_n)$ the geometric measure theoretic outward unit normal to Ω. Next, suppose L is a weakly elliptic, second-order, homogeneous, constant complex coefficient, $M \times M$ system in $\mathbb{R}^n$. Fix an aperture parameter $\kappa \in (0, \infty)$, along with an arbitrary point $x_0 \in \Omega$, and suppose $0 < \rho < \frac{1}{4}\operatorname{dist}(x_0, \partial\Omega)$. Finally, define the compact set $K := \overline{B(x_0, \rho)}$.*

Then there exists some $\widetilde{\kappa} > 0$, which depends only on Ω and κ, with the following significance. Assume G is a matrix-valued function satisfying

$$\begin{cases} G \in [L^1_{\mathrm{loc}}(\Omega, \mathcal{L}^n)]^{M\times M}, \\ L^\top G = -\delta_{x_0} \cdot I_{M\times M} \quad \text{in } [\mathcal{D}'(\Omega)]^{M\times M} \\ (\nabla G)\big|_{\partial\Omega}^{\widetilde{\kappa}-\mathrm{n.t.}} \quad \text{exists (in } \mathbb{C}^{n\cdot M^2}) \text{ at } \sigma\text{-a. e. point on } \partial\Omega, \\ G\big|_{\partial\Omega}^{\widetilde{\kappa}-\mathrm{n.t.}} = 0 \in \mathbb{C}^{M\times M} \quad \text{at } \sigma\text{-a. e. point on } \partial\Omega, \\ \mathcal{N}_{\widetilde{\kappa}}^{\Omega\setminus K}(\nabla G) < +\infty \quad \text{at } \sigma\text{-a. e. point on } \partial\Omega, \end{cases} \tag{11.10}$$

and assume $u = (u_\beta)_{1\leq\beta\leq M}$ is a $\mathbb{C}^M$-valued function in Ω satisfying

$$\begin{cases} u \in [\mathscr{C}^\infty(\Omega)]^M, \quad Lu = 0 \quad \text{in } \Omega, \\ u\big|_{\partial\Omega}^{\kappa-\mathrm{n.t.}} \quad \text{exists at } \sigma\text{-a. e. point on } \partial\Omega, \\ \int_{\partial\Omega} \mathcal{N}_\kappa u \cdot \mathcal{N}_{\widetilde{\kappa}}^{\Omega\setminus K}(\nabla G)\, d\sigma < +\infty. \end{cases} \tag{11.11}$$

Then for any choice of a coefficient tensor $A = (a_{rs}^{\alpha\beta})_{\substack{1\leq r,s\leq n\\ 1\leq\alpha,\beta\leq M}} \in \mathfrak{A}_L$ one has the Poisson integral representation formula

$$u_\beta(x_0) = -\int_{\partial_*\Omega} \big\langle u\big|_{\partial\Omega}^{\kappa-\mathrm{n.t.}}, \partial_\nu^{A^\top} G_{\bullet\beta}\big\rangle\, d\sigma, \quad \forall \beta \in \{1,\dots,M\}, \tag{11.12}$$

where $\partial_\nu^{A^\top}$ stands for the conormal derivative associated with $A^\top$, which in the above contexts acts on the βth column of the matrix-valued function G according to

$$\partial_\nu^{A^\top} G_{\bullet\beta} := \big(\nu_r a_{sr}^{\gamma\alpha} (\partial_s G_{\gamma\beta})\big|_{\partial\Omega}^{\widetilde{\kappa}-\mathrm{n.t.}}\big)_{1\leq\alpha\leq M} \quad \text{at } \sigma\text{-a. e. point on } \partial_*\Omega, \tag{11.13}$$

for each $\beta \in \{1,\dots,M\}$.

Thus, working in the context of Proposition 11.4, if such a Green function exists for the system L at each point $x_0 \in \Omega$, and if $u\big|_{\partial\Omega}^{\kappa-\mathrm{n.t.}} = 0$ at σ-a. e. point on $\partial\Omega$, then (11.12) implies that $u(x_0) = 0$ for each $x_0 \in \Omega$, which is enough to demonstrate uniqueness of solutions to (11.1) given the linearity of the system L. Proposition 11.4 now shifts the problem of uniqueness to demonstrating the existence of a suitable Green function. An important aspect of Proposition 11.4 is that $\mathcal{N}_\kappa u$ pairs well in an integral sense with $\mathcal{N}_{\widetilde{\kappa}}^{\Omega\setminus K}(\nabla G)$ (cf. the last condition in (11.11)). The boundary value problem that we are considering in (11.1) assumes $\mathcal{N}_\kappa u$ belongs to $M^{p,\lambda}(\partial\Omega, w)$, so it is important that $\mathcal{N}_{\widetilde{\kappa}}^{\Omega\setminus K}(\nabla G)$ belongs to a space of functions, which pairs well with $M^{p,\lambda}(\partial\Omega, w)$ in an integral sense. This is our motivation for developing a functional analytic and Calderón–Zygmund theoretic framework for Muckenhoupt-weighted Block spaces in Chapters 6 and 7. Interestingly, the existence of a Green function, and by extension the uniqueness of the Dirichlet problem in (11.1), will boil down to the matter of the solvability for a regularity problem for the transpose system $L^\top$ on the Muckenhoupt-weighted Block space, which is a pre-dual to $M^{p,\lambda}(\partial\Omega, w)$. This is where the layer potential theory on Muckenhoupt-weighted Block spaces, developed in Section 9.2, plays a crucial role.

A general regularity problem for Muckenhoupt-weighted Block spaces reads as follows. Namely, let $\Omega \subseteq \mathbb{R}^n$ be an AR domain, and abbreviate $\sigma := \mathcal{H}^{n-1}\lfloor\partial\Omega$. Fix an integrability exponent $q \in (1,\infty)$, along with an arbitrary Muckenhoupt weight $\omega \in A_q(\partial\Omega,\sigma)$. Choose some $\varepsilon \in (0, q-1)$ for which $\omega \in A_{q-\varepsilon}(\partial\Omega,\sigma)$, and fix a threshold $\lambda \in (0, \varepsilon(n-1)/(q-1))$. Let L be a weakly elliptic, second-order, homogeneous, constant complex coefficient, $M \times M$ system in $\mathbb{R}^n$. For a fixed aperture parameter $\kappa \in (0,\infty)$, consider

$$\begin{cases} u \in [\mathscr{C}^\infty(\Omega)]^M, \\ Lu = 0 \quad \text{in } \Omega, \\ \mathcal{N}_\kappa u,\ \mathcal{N}_\kappa(\nabla u) \in \mathcal{B}^{q,\lambda}(\partial\Omega,\omega), \\ u\big|_{\partial\Omega}^{\kappa-\text{n.t.}} = f \in [\mathcal{B}_1^{q,\lambda}(\partial\Omega,\omega)]^M. \end{cases} \tag{11.14}$$

We treat such a regularity boundary value problem later in Theorem 11.29, which is actually a well-posedness result for (11.14). Despite appearing later in the chapter, there is no circular reasoning here, because for uniqueness of solutions to (11.1), we only require an existence result for (11.14) in our construction of a Green function, and existence of solutions to (11.14) is proved independently of the well posedness of the Dirichlet problem considered in (11.1). In fact, as we will see, existence of solutions to (11.14) follows largely from the layer potential theory in Section 9.2.

We also note with this next result that much as in the statement of (11.1), the aperture parameter κ, present in the formulation of (11.14), appears solely as a means to describe the geometry of the problem, and does not affect the solvability of (11.14).

Proposition 11.5. *Let Ω be an open nonempty proper subset of $\mathbb{R}^n$ such that $\partial\Omega$ is an Ahlfors regular set, and define $\sigma := \mathcal{H}^{n-1}\lfloor\partial\Omega$. Fix two arbitrary aperture parameters $\kappa, \widetilde{\kappa} \in (0,\infty)$. Let $u : \Omega \to \mathbb{C}$ be a Lebesgue measurable function. Next, fix an integrability exponent $q \in (1,\infty)$ along with a Muckenhoupt weight $\omega \in A_q(\partial\Omega,\sigma)$. Finally, choose $\varepsilon \in (0, q-1)$ so that $\omega \in A_{q-\varepsilon}(\partial\Omega,\sigma)$, and select a threshold $\lambda \in (0, \varepsilon(n-1)/(q'-1))$, where $q' := q/(q-1)$.*

Then $\mathcal{N}_\kappa u \in \mathcal{B}^{q,\lambda}(\partial\Omega,\omega)$ if and only if $\mathcal{N}_{\widetilde{\kappa}} u \in \mathcal{B}^{q,\lambda}(\partial\Omega,\omega)$. In fact, if $\kappa \le \widetilde{\kappa}$, then

$$\|\mathcal{N}_\kappa u\|_{\mathcal{B}^{q,\lambda}(\partial\Omega,\omega)} \le \|\mathcal{N}_{\widetilde{\kappa}} u\|_{\mathcal{B}^{q,\lambda}(\partial\Omega,\omega)} \le C\|\mathcal{N}_\kappa u\|_{\mathcal{B}^{q,\lambda}(\partial\Omega,\omega)} \tag{11.15}$$

for some constant $C \in (0,\infty)$, depending only on κ, $\widetilde{\kappa}$, the ADR *character of $\partial\Omega$, n, q, ε, λ, $[\omega]_{A_{q-\varepsilon}}$, and $[\omega]_{A_q}$.*

Proof. The same type of argument as in the proof of Proposition 11.1 applies, but now making use of Proposition 7.1 in place of Proposition 5.1, and the lattice property of $\mathcal{B}^{q,\lambda}(\partial\Omega,\omega)$, established in Proposition 6.14. □

One can also turn to Proposition 7.7 for a different and decidedly stronger way of seeing that when working with Muckenhoupt-weighted Block spaces, the nontangential maximal function is independent (in a quantitative way) of the choice of aperture parameter.

In order to construct a Green function as in Proposition 11.4 for the system in (11.1), it will be necessary to show that a difference of shifted fundamental solutions belongs to a Muckenhoupt-weighted Block-based Sobolev space. This is the content of the following lemma.

Lemma 11.6. *Let $\Omega \subseteq \mathbb{R}^n$ be an AR domain and abbreviate $\sigma := \mathcal{H}^{n-1}\lfloor\partial\Omega$. Fix an integrability exponent $q \in (1,\infty)$ along with a Muckenhoupt weight $\omega \in A_q(\partial\Omega,\sigma)$. Choose $\varepsilon \in (0, q-1)$ for which $\omega \in A_{q-\varepsilon}(\partial\Omega,\sigma)$, and select $\lambda \in (0, \varepsilon(n-1)/(q-1))$. Now, let L be a weakly elliptic, second-order, homogeneous, constant complex coefficient, $M \times M$ system in $\mathbb{R}^n$, and let the matrix-valued function $E = (E_{\alpha\beta})_{1\le\alpha,\beta\le M}$ denote the fundamental solution associated with L in the sense of Proposition 3.4. Finally, fix some $x_0 \in \Omega$ and choose some $x_1 \in \mathbb{R}^n \setminus \overline{\Omega}$.*

Then

$$(E(\cdot - x_0) - E(\cdot - x_1))\big|_{\partial\Omega} \in \big[\mathcal{B}_1^{q,\lambda}(\partial\Omega,\omega)\big]^{M\times M}. \tag{11.16}$$

Proof. For brevity, write

$$F = (F_{\alpha\beta})_{1\le\alpha,\beta\le M} := \big(E_{\alpha\beta}(\cdot - x_0) - E_{\alpha\beta}(\cdot - x_1)\big)_{1\le\alpha,\beta\le M}. \tag{11.17}$$

Given any $\alpha,\beta \in \{1,\dots,M\}$, the plan is to use Lemma 6.26 and the definition in (8.11) to conclude membership of $F_{\alpha\beta}$ to the Block-based Sobolev space $\mathcal{B}_1^{q,\lambda}(\partial\Omega,\omega)$. For that purpose, choose some point $x_* \in \partial\Omega$ such that $|x_0 - x_*| = \operatorname{dist}(x_0,\partial\Omega)$ and set $R := |x_0 - x_1|$. Proving the following two claims will establish the result.

Claim 1: $F_{\alpha\beta} \in \mathcal{B}^{q,\lambda}(\partial\Omega,\omega)$ for each $\alpha,\beta \in \{1,\dots,M\}$.

Let $x \in \partial\Omega$ and $\alpha,\beta \in \{1,\dots,M\}$. Consider the case in which $x \in \partial\Omega$ is close to x_*, say $x \in \Delta(x_*,4R)$. Then by the continuity of $F_{\alpha\beta}$ on $\mathbb{R}^n \setminus \{x_0,x_1\}$, we find that there exists a constant $C \in (0,\infty)$, which is allowed to depend on x_0 and x_*, such that

$$\begin{aligned}|F_{\alpha\beta}(x)| &\le |E_{\alpha\beta}(x - x_0)| + |E_{\alpha\beta}(x - x_1)| \\ &\le C \quad \text{for all } x \in \Delta(x_*,4R).\end{aligned} \tag{11.18}$$

The proximity of x to x_* entails (cf. (3) of Lemma 2.18) $1 \approx (1 + |x - x_*|)^{n-1}$ uniformly in $x \in \Delta(x_*,4R)$, and hence all together,

$$|F_{\alpha\beta}(x)| \le \frac{C}{1 + |x - x_*|^{n-1}} \quad \text{for all } x \in \Delta(x_*,4R). \tag{11.19}$$

Now, consider $x \in \partial\Omega \setminus \Delta(x_*,4R)$. Then since $\operatorname{dist}(x_0,\partial\Omega) < R$, we conclude $x_0, x_1 \in B(x_*,2R)$. As a consequence, $x \notin [x_0,x_1]$, the line segment from x_0 to x_1. Bearing this in mind, the mean value theorem applies and guarantees the existence of a point $z_* \in [x_0, x_1]$ such that

$$E_{\alpha\beta}(x - x_0) - E_{\alpha\beta}(x - x_1) = (\nabla E_{\alpha\beta})(x - z_*) \cdot (x_0 - x_1). \tag{11.20}$$

Consequently, there exists a constant $C \in (0, \infty)$ such that

$$\begin{aligned} |F_{\alpha\beta}(x)| &= |E_{\alpha\beta}(x - x_0) - E_{\alpha\beta}(x - x_1)| \\ &\leq \frac{C|x_0 - x_1|}{|x - z_*|^{n-1}} \\ &\leq \frac{C}{1 + |x - x_*|^{n-1}}, \end{aligned} \tag{11.21}$$

for all $x \in \partial\Omega \setminus \Delta(x_*, 4R)$. The first inequality follows from (11.20), the Cauchy–Schwarz inequality and (d) of Proposition 3.4. The last inequality in (11.21) follows from the fact that

$$1 + |x - x_*| \leq C|x - z_*| \tag{11.22}$$

for some $C \in (0, \infty)$, independent of x, along with (1) of Lemma 2.18. To see that (11.22) holds, recall that we are working in the case that $x \in \partial\Omega \setminus \Delta(x_*, 4R)$, so $4R < |x - x_*|$. Subsequently,

$$\begin{aligned} |x - x_*| &\leq |x - z_*| + |z_* - x_0| + |x_0 - x_*| \leq |x - z_*| + 2R \\ &< |x - z_*| + \frac{1}{2}|x - x_*|, \end{aligned} \tag{11.23}$$

which implies $|x - z_*| \geq \frac{1}{2}|x - x_*|$. Since

$$|x - z_*| \geq 2R \iff 1 \leq (2R)^{-1}|x - z_*|, \tag{11.24}$$

we ultimately obtain (11.22) with $C := (2R)^{-1} + 2$.

To complete the proof of Claim 1, the estimates in (11.19) and (11.21), the σ-measurability of $F_{\alpha\beta}$, the lattice property of $\mathcal{B}^{q,\lambda}(\partial\Omega, \omega)$ (cf. Proposition 6.14), Lemma 2.18, and Lemma 6.26 ensure that

$$F_{\alpha\beta} \in \mathcal{B}^{q,\lambda}(\partial\Omega, \omega) \tag{11.25}$$

as claimed.

Claim 2: $\partial_{\tau_{jk}} F_{\alpha\beta} \in \mathcal{B}^{q,\lambda}(\partial\Omega, \omega)$ for each choice of $j, k \in \{1. \ldots, n\}$ and $\alpha, \beta \in \{1, \ldots, M\}$.

Fix $\alpha, \beta \in \{1, \ldots, M\}$. First, observe that for any $j, k \in \{1, \ldots, n\}$ we have for all $x \in \partial\Omega$ that

$$\begin{aligned}(\partial_{\tau_{jk}} F_{\alpha\beta})(x) &= \partial_{\tau_{jk}}[E_{\alpha\beta}(x-x_0)] - \partial_{\tau_{jk}}[E_{\alpha\beta}(x-x_1)] \\ &= \nu_j(x)\partial_k[E_{\alpha\beta}(x-x_0)]\big|_{\partial\Omega} - \nu_k(x)\partial_j[E_{\alpha\beta}(x-x_0)]\big|_{\partial\Omega} \\ &\quad - \nu_j(x)\partial_k[E_{\alpha\beta}(x-x_1)]\big|_{\partial\Omega} + \nu_k(x)\partial_j[E_{\alpha\beta}(x-x_1)]\big|_{\partial\Omega}\end{aligned} \tag{11.26}$$

which together with item (d) of Proposition 3.4 implies that there exists some constant $C \in (0,\infty)$ for which

$$\left|\partial_{\tau_{jk}} E_{\alpha\beta}(x - x_i)\right| \leq \frac{C}{|x - x_i|^{n-1}} \quad \text{for all } x \in \partial\Omega, \text{ and } i \in \{0,1\}. \tag{11.27}$$

Because $x_0, x_1 \notin \partial\Omega$, we have for $i \in \{0,1\}$ that

$$\frac{1}{|x - x_i|^{n-1}} \approx \frac{1}{1 + |x|^{n-1}}, \tag{11.28}$$

uniformly in $x \in \partial\Omega$. As a consequence of Lemma 6.26 and the lattice property of Block spaces (cf. Proposition 6.14), we find that $\partial_{\tau_{jk}} F_{\alpha\beta}$ belongs to $\mathcal{B}^{q,\lambda}(\partial\Omega,\omega)$ for all $j,k \in \{1,\dots,n\}$. This completes the justification for Claim 2, so the proof of Lemma 11.6 is complete. □

Eventually, we would also like to solve a Dirichlet problem as in (11.1), but at the level of Muckenhoupt-weighted Block spaces. Because of our layer potential theory in Chapter 9, which takes into account both Muckenhoupt-weighted Block and Morrey spaces, such a task can be accomplished, essentially by switching the roles of Morrey and Block spaces. For example, when considering the problem of uniqueness for a Dirichlet problem on Muckenhoupt-weighted Block spaces, we will be able to reduce matters to a question of existence of solutions to a regularity problem involving the Muckenhoupt-weighted Morrey space dual to the Block space originally under consideration. As such, we will also need this next result, which demonstrates the membership of a difference of two shifted fundamental solutions to a Muckenhoupt-weighted Morrey-based Sobolev space.

Lemma 11.7. *Let $\Omega \subseteq \mathbb{R}^n$ be an AR domain and abbreviate $\sigma := \mathcal{H}^{n-1}\lfloor\partial\Omega$. Fix two integrability exponents $p, q \in (1,\infty)$, along with a Muckenhoupt weight $w \in A_p(\partial\Omega,\sigma) \cap \mathrm{RH}_q(\partial\Omega,\sigma)$, and some threshold $\lambda \in (0,(n-1)/q')$, where $q' := q/(q-1)$. Now, let L be a weakly elliptic, second-order, homogeneous, constant complex coefficient, $M \times M$ system in $\mathbb{R}^n$, and let the matrix-valued function $E = (E_{\alpha\beta})_{1\leq\alpha,\beta\leq M}$ be the fundamental solution associated with L in the sense of Proposition 3.4. Finally, fix some $x_0 \in \Omega$ and choose some $x_1 \in \mathbb{R}^n \setminus \overline{\Omega}$.*

Then

$$\left(E(\cdot - x_0) - E(\cdot - x_1)\right)\big|_{\partial\Omega} \in \left[M_1^{p,\lambda}(\partial\Omega, w)\right]^{M\times M}. \tag{11.29}$$

Proof. The proof follows closely that of Lemma 11.6, now using the lattice property of Morrey spaces (cf. Proposition 4.7) and Lemma 4.12 instead of Lemma 6.26. □

As a last preliminary result before diving into the proof of the well posedness of (11.1), we present the following lemma, appearing in [86, Lemma 8.3.7, p. 523], which will be used to show that the Green function we construct in our proof for uniqueness of solutions to (11.1) satisfies the last condition on G in (11.10). This lemma provides a pointwise estimate for the nontangential maximal operator applied to power functions with a negative exponent.

Lemma 11.8. *Let $\Omega \subseteq \mathbb{R}^n$ be an open, nonempty, proper subset of $\mathbb{R}^n$. Fix an aperture parameter $\kappa \in (0,\infty)$ and some $N \in [0,\infty)$. Then there exists some finite constant $C(N,\kappa) > 0$ such that for all $x_0 \in \Omega$ if one defines $B := B(x_0, \frac{1}{2}\delta_{\partial\Omega}(x_0))$, it follows that*

$$(\mathcal{N}_\kappa^{\Omega\setminus B}(|\cdot - x_0|^{-N}))(x) \leq C|x - x_0|^{-N} \quad \text{for all } x \in \partial\Omega. \tag{11.30}$$

11.1 The Dirichlet problem on Morrey and Block spaces

We have the following well-posedness result (in the sense of Hadamard) for the Dirichlet problem formulated in (11.1).

Theorem 11.9 (The Weighted Morrey Dirichlet BVP). *Fix $n \in \mathbb{N}$ with $n \geq 2$ along with $M \in \mathbb{N}$. Let $\Omega \subseteq \mathbb{R}^n$ be an AR domain and write $\sigma := \mathcal{H}^{n-1}\lfloor\partial\Omega$. Suppose L is a weakly elliptic, second-order, homogeneous, constant complex coefficient, $M \times M$ system in $\mathbb{R}^n$. Next, fix any two integrability exponents $p, q \in (1,\infty)$, along with a Muckenhoupt weight $w \in A_p(\partial\Omega,\sigma) \cap \mathrm{RH}_q(\partial\Omega,\sigma)$, and a threshold $\lambda \in (0,(n-1)/q')$, where $q' := q/(q-1)$. Lastly, fix an aperture parameter $\kappa \in (0,\infty)$ and consider the Dirichlet problem for L in Ω with data in a Muckenhoupt-weighted Morrey space, formulated as*

$$\begin{cases} u \in [\mathscr{C}^\infty(\Omega)]^M, \\ Lu = 0 \quad \text{in } \Omega, \\ \mathcal{N}_\kappa u \in M^{p,\lambda}(\partial\Omega, w), \\ u\big|_{\partial\Omega}^{\kappa-\text{n.t.}} = f \in [M^{p,\lambda}(\partial\Omega, w)]^M. \end{cases} \tag{11.31}$$

Then there exists $\delta \in (0,1)$, depending on the ADR character of $\partial\Omega$, n, p, q, λ, $[w]_{\mathrm{RH}_q}$, $[w]_{A_p}$, and L such that whenever

$$\|\nu\|_{[\mathrm{BMO}(\partial\Omega,\sigma)]^n} < \delta, \tag{11.32}$$

i. e., whenever Ω is a δ-AR domain, it follows that Ω is actually a UR domain and all claims below are true:

(a) [Existence, estimates, and integral representation]. *Suppose $\mathfrak{A}_L^{\text{dis}} \neq \varnothing$ and $A \in \mathfrak{A}_L^{\text{dis}}$. Then the operator*

$$\frac{1}{2}I + K_A : [M^{p,\lambda}(\partial\Omega, w)]^M \to [M^{p,\lambda}(\partial\Omega, w)]^M \tag{11.33}$$

is an isomorphism, and the function $u : \Omega \to \mathbb{C}^M$ *defined as*

$$u(x) := \Big(\mathcal{D}_A\Big[\Big(\frac{1}{2}I + K_A\Big)^{-1} f\Big]\Big)(x) \quad \text{at each } x \in \Omega \tag{11.34}$$

(with $\mathcal{D}_A$ *associated to* A *as in* (3.18)) *is a solution of* (11.31) *satisfying the estimate*

$$\|f\|_{[M^{p,\lambda}(\partial\Omega,w)]^M} \le \|\mathcal{N}_\kappa u\|_{M^{p,\lambda}(\partial\Omega,w)} \le C\|f\|_{[M^{p,\lambda}(\partial\Omega,w)]^M}, \tag{11.35}$$

for some constant $C \in (0,\infty)$, *depending only on the* ADR *character of* $\partial\Omega$, n, p, q, λ, $[w]_{\mathrm{RH}_q}$, $[w]_{A_p}$, *and* κ. *Additionally, for each* $\alpha \in \mathbb{N}_0^n$, *the solution constructed in* (11.41) *satisfies*

$$\sup_{x\in\Omega}\{\delta_{\partial\Omega}^{|\alpha|-\lambda/p}(x) w(\Delta(x_*, 2\delta_{\partial\Omega}(x)))^{1/p}|(\partial^\alpha u)(x)|\} \le C_\alpha \|f\|_{[M^{p,\lambda}(\partial\Omega,w)]^M}, \tag{11.36}$$

for some $C_\alpha \in (0,\infty)$ *independent of* f, *where* $\delta_{\partial\Omega}$ *denotes the distance function to* $\partial\Omega$, *and* $x_* \in \partial\Omega$ *is chosen such that* $|x - x_*| = \delta_{\partial\Omega}(x)$.

(b) [Uniqueness]. *If* $\mathfrak{A}_{L^\top}^{\mathrm{dis}} \neq \varnothing$, *the Dirichlet problem* (11.31) *has at most one solution.*

(c) [Well posedness]. *If* $\mathfrak{A}_L^{\mathrm{dis}} \neq \varnothing$ *and* $\mathfrak{A}_{L^\top}^{\mathrm{dis}} \neq \varnothing$, *the Dirichlet problem* (11.31) *is well posed in the sense of Hadamard. That is,* (11.31) *is uniquely solvable with naturally accompanying estimates as in* (11.35).

(d) [Sharpness]. *If* $\mathfrak{A}_L^{\mathrm{dis}} = \varnothing$, *the Dirichlet problem* (11.31) *may fail to be solvable. In fact, solvability may fail for boundary data belonging to an infinite-dimensional subspace of* $[M^{p,\lambda}(\partial\Omega, w)]^M$, *even when* $\Omega = \mathbb{R}^n_+$ *(in which scenario the* BMO*-seminorm of* v *is zero). Also, if* $\mathfrak{A}_{L^\top}^{\mathrm{dis}} = \varnothing$, *then the Dirichlet problem in* (11.31) *may have more than one solution. In fact, the vector space of null solutions may be infinite-dimensional even when* $\Omega = \mathbb{R}^n_+$.

(e) [Weighted vanishing Morrey spaces]. *Analogous results to those found within items (a)–(d) continue to hold if* $M^{p,\lambda}(\partial\Omega, w)$ *is replaced by the Muckenhoupt-weighted vanishing Morrey space* $\mathring{M}^{p,\lambda}(\partial\Omega, w)$, *i. e., for the boundary value problem below:*

$$\begin{cases} u \in [\mathscr{C}^\infty(\Omega)]^M, \\ Lu = 0 \quad \text{in } \Omega, \\ \mathcal{N}_\kappa u \in \mathring{M}^{p,\lambda}(\partial\Omega, w), \\ u\big|_{\partial\Omega}^{\kappa-\mathrm{n.t.}} = f \in [\mathring{M}^{p,\lambda}(\partial\Omega, w)]^M. \end{cases} \tag{11.37}$$

Remark 11.10. There is a delicate timing issue that warrants explanation in the formulation of Theorem 11.9. Namely, we seem to start with an AR domain, Ω, but then retroactively ask that Ω is actually a δ-AR domain for sufficiently small $\delta \in (0,1)$, so as to yield invertibility of the operator in (11.33). The constant appearing in the norm estimate that leads to the invertibility of (11.33) (cf. (11.45)) depends on the geometry of Ω through the ADR character of $\partial\Omega$. Since asking that Ω be a δ-AR domain potentially changes the geometry of our original choice of Ω, it is not clear that this change does not leave the constant in the original estimate unaffected. Put differently, it is not immediately obvious that asking that an AR domain be as flat as we like (where flatness is measured in terms of the mean oscillation of the outward unit normal) does not force the ADR character to change in an unmanageable way. However, this is not an issue if we fix some constant $C_{\rm ADR}$ at the very beginning, and subsequently consider the class of δ-AR domains with δ sufficiently small, where δ now depends on this choice of $C_{\rm ADR}$, removing the dependence of δ on Ω itself. See [79, Section 2.4] and [90, Section 3] for an in-depth discussion on the class of δ-AR domains, and examples of δ-AR domains Ω for which δ can be made arbitrarily small relative to the ADR character of $\partial\Omega$. In view of this observation, we find it enlightening to present the following version of Theorem 11.9, which offers a more precise description of the timing of the geometric and analytic quantities that are involved in our well-posedness result.

Fix $n \in \mathbb{N}$ with $n \geq 2$ and $M \in \mathbb{N}$. Suppose L is a weakly elliptic, second-order, homogeneous, constant complex coefficient, $M \times M$ system in $\mathbb{R}^n$. Fix two constants $C_{\rm ADR}, C_W \in [1, \infty)$. Next, select any two integrability exponents $p, q \in (1, \infty)$. Then for every threshold $\lambda \in (0, (n-1)/q')$, where $q' := q/(q-1)$, there exists some small

$$\delta = \delta(n, C_{\rm ADR}, C_W, L, p, q, \lambda) \in (0, 1) \tag{11.38}$$

with the following significance. Assume $\Omega \subseteq \mathbb{R}^n$ is a δ-AR domain whose topological boundary has Ahlfors regularity constants that are controlled by $C_{\rm ADR}$. Abbreviate $\sigma := \mathcal{H}^{n-1}\lfloor\partial\Omega$ and select an arbitrary Muckenhoupt weight $w \in A_p(\partial\Omega, \sigma) \cap {\rm RH}_q(\partial\Omega, \sigma)$ with $[w]_{A_p}, [w]_{{\rm RH}_q} \leq C_W$. Finally, fix an arbitrary aperture parameter $\kappa \in (0, \infty)$ and consider the Dirichlet problem below:

$$\begin{cases} u \in [\mathscr{C}^\infty(\Omega)]^M, \\ Lu = 0 \quad \text{in } \Omega, \\ \mathcal{N}_\kappa u \in M^{p,\lambda}(\partial\Omega, w), \\ u\big|_{\partial\Omega}^{\kappa-{\rm n.t.}} = f \in [M^{p,\lambda}(\partial\Omega, w)]^M. \end{cases} \tag{11.39}$$

Then Ω is actually a UR domain and the following claims are true:

(a) [Existence, estimates, and integral representation]. Suppose $\mathfrak{A}_L^{\rm dis} \neq \varnothing$ and $A \in \mathfrak{A}_L^{\rm dis}$. Then with K_A associated with A as in (3.19), the operator

$$\frac{1}{2}I + K_A : [M^{p,\lambda}(\partial\Omega, w)]^M \to [M^{p,\lambda}(\partial\Omega, w)]^M \tag{11.40}$$

is an isomorphism, and the function $u : \partial\Omega \to \mathbb{C}^M$ defined as

$$u(x) := \left(\mathcal{D}_A\left[\left(\frac{1}{2}I + K_A\right)^{-1} f\right]\right)(x) \quad \text{for all } x \in \Omega \tag{11.41}$$

(with $\mathcal{D}_A$ associated to A as in (3.18)) is a solution to the Dirichlet problem (11.39) and satisfies the estimate

$$\|f\|_{[M^{p,\lambda}(\partial\Omega,w)]^M} \le \|\mathcal{N}_\kappa u\|_{M^{p,\lambda}(\partial\Omega,w)} \le C\|f\|_{[M^{p,\lambda}(\partial\Omega,w)]^M}, \tag{11.42}$$

for some constant $C = C(n, C_{\mathrm{ADR}}, C_W, L, p, q, \lambda, \kappa) \in (0,\infty)$. Additionally, for each $\alpha \in \mathbb{N}_0^n$, the solution constructed in (11.41) satisfies

$$\sup_{x\in\Omega}\{\delta_{\partial\Omega}^{|\alpha|-\lambda/p}(x)w(\Delta(x_*, 2\delta_{\partial\Omega}(x)))^{1/p}|(\partial^\alpha u)(x)|\} \le C_\alpha\|f\|_{[M^{p,\lambda}(\partial\Omega,w)]^M}, \tag{11.43}$$

for $C_\alpha \in (0,\infty)$ independent of f, where $\delta_{\partial\Omega}$ denotes the distance function to $\partial\Omega$, and $x_* \in \partial\Omega$ is chosen such that $|x - x_*| = \delta_{\partial\Omega}(x)$.

(b) [Uniqueness]. If $\mathfrak{A}_{L^\top}^{\mathrm{dis}} \neq \varnothing$, then the Dirichlet problem (11.39) has at most one solution.

(c) [Well posedness]. If $\mathfrak{A}_L^{\mathrm{dis}} \neq \varnothing$ and $\mathfrak{A}_{L^\top}^{\mathrm{dis}} \neq \varnothing$, then the Dirichlet problem (11.39) is well posed in the sense of Hadamard. That is, (11.39) is uniquely solvable with the naturally accompanying estimates in (11.42).

It is worth mentioning that in the context of Theorem 11.9, corresponding to the case that $w \equiv 1$, we recover the result for the unweighted case, first established in [79, Theorem 7.18, pp. 481–482]. See also [86, Theorem 8.4.3, pp. 807–809]. It is also of interest to contrast our result in Theorem 11.9 with that of Z. Shen's in [137, Theorem 1.5], wherein Shen obtains a well-posedness result for the Dirichlet problem in the class of bounded Lipschitz domains (with connected boundary), whenever the system L is Legendre–Hadamard strongly elliptic, symmetric, with real coefficients, $p = 2$, $w \equiv 1$, and λ belongs to a suitable subinterval of $(0, n-1)$.

Remark 11.11. As we will see in the proof of Theorem 11.9, due to the nature of the results in Section 9.1 that we use in the proof of existence (Theorem 9.1, Proposition 9.3, and Theorem 9.7), we can also obtain a solvability result for (11.31) when the weight w appearing in the Muckenhoupt-weighted Morrey space $M^{p,\lambda}(\partial\Omega, w)$ is replaced by the weight $w_a := w|\cdot - x_0|^a$ on $\partial\Omega$, where w is as before, $x_0 \in \partial\Omega$ is an arbitrary reference point, and $a \in [0,\lambda)$.

Finally, we remark that the size condition on the solution u, appearing in (11.31) and (11.68) in terms of the nontangential maximal operator, could have been formulated using the tangential maximal operator from Definition 2.17. See Proposition 7.7 in this regard.

At last, we are ready to present the proof of Theorem 11.9.

Proof of Theorem 11.9. Starting with the claims made in item (a), the fact that Ω is a UR domain for sufficiently small δ is a consequence of [79, Theorem 2.47, p. 75]. We now turn to the task of showing existence. We will see that the estimate in (11.42) precipitates naturally from our method of demonstrating solvability via layer potentials.

We seek to show that (11.33) is an invertible operator. Note that $[M^{p,\lambda}(\partial\Omega, w)]^M$ is a Banach space by Proposition 4.5, and hence Lemma 11.2 implies that $\frac{1}{2}I + K_A$ is invertible (with bounded inverse by the open mapping theorem) on $[M^{p,\lambda}(\partial\Omega, w)]^M$ provided

$$\|K_A\|_{[M^{p,\lambda}(\partial\Omega,w)]^M \to [M^{p,\lambda}(\partial\Omega,w)]^M} < \frac{1}{2}. \tag{11.44}$$

Operating under the assumption that $\mathfrak{A}_L^{\text{dis}} \neq \varnothing$ and that A is a distinguished coefficient tensor for the system L permits us to invoke Theorem 9.7. Consequently, with ν denoting the geometric measure theoretic outward unit normal to Ω, we have for each $m \in \mathbb{N}$ (keeping in mind Definition 5.9)

$$\|K_A\|_{[M^{p,\lambda}(\partial\Omega,w)]^M \to [M^{p,\lambda}(\partial\Omega,w)]^M} \leq C_m \cdot \|\nu\|^{\langle m \rangle}_{[\text{BMO}(\partial\Omega,\sigma)]^n}, \tag{11.45}$$

where the constant $C_m \in (0, \infty)$ depends on m, the ADR character of $\partial\Omega$, n, p, q, λ, $[w]_{\text{RH}_q}$, $[w]_{A_p}$, and A. Recall (2.37) and Remark 5.10. By considering a δ-AR domain for sufficiently small $\delta \in (0, 1)$, we can force $\|\nu\|^{\langle m \rangle}_{[\text{BMO}(\partial\Omega,\sigma)]^n}$ small enough to yield the desired control on K_A in (11.44). Ultimately, for such a δ we find that (11.41) is a solution to (11.31) under the hypotheses of Theorem 11.9. Indeed, the invertibility of the operator in (11.33) implies that the function $(\frac{1}{2}I + K_A)^{-1}f$ belongs to $[M^{p,\lambda}(\partial\Omega, w)]^M$, so Theorem 9.1 ensures that u is a well-defined function in the context of (11.34), which also satisfies $u \in [\mathscr{C}^\infty(\Omega)]^M$ and $Lu = 0$ in Ω. Moreover, the nontangential trace result in Theorem 9.1 guarantees that u satisfies the boundary condition in (11.31). Lastly, Proposition 9.3 implies $\mathcal{N}_\kappa u \in M^{p,\lambda}(\partial\Omega, w)$ with the following estimate, which also uses the boundedness of $(\frac{1}{2}I + K_A)^{-1}$,

$$\begin{aligned} \|\mathcal{N}_\kappa u\|_{M^{p,\lambda}(\partial\Omega,w)} &\leq C\left\|\left(\frac{1}{2}I + K_A\right)^{-1} f\right\|_{[M^{p,\lambda}(\partial\Omega,w)]^M} \\ &\leq C\|f\|_{[M^{p,\lambda}(\partial\Omega,w)]^M}. \end{aligned} \tag{11.46}$$

This takes care of the second inequality in (11.35), whereas the first inequality in (11.35) is a consequence of (2.18). As for the estimate in (11.36), this is an immediate consequence of (3.18), item (d) of Proposition 3.4, Proposition 5.8, and the boundedness of the inverse of the operator in (11.33).

We now turn to the claim made in item (b). Establishing uniqueness for (11.31) requires that we show that if u satisfies

$$\begin{cases} u \in [\mathscr{C}^\infty(\Omega)]^M, \\ Lu = 0 \quad \text{in } \Omega, \\ \mathcal{N}_\kappa u \in M^{p,\lambda}(\partial\Omega, w), \\ u\big|^{\kappa-\text{n.t.}}_{\partial\Omega} = 0 \in \mathbb{C}^M \quad \text{at } \sigma\text{-a.e. point on } \partial\Omega, \end{cases} \tag{11.47}$$

then $u(x_0) = 0$ for each $x_0 \in \Omega$. We shall present two different proofs of uniqueness. For the first proof, we enlist the help of Proposition 11.4, and subsequently endeavor to find a suitable Green function. In order to invoke Proposition 11.4 note that thanks to [79, Theorem 2.47, p. 75] and (11.9), the geometric hypotheses on Ω are satisfied by a δ-AR domain when $\delta \in (0,1)$ is taken to be sufficiently small relative to the ADR character of $\partial\Omega$. Also, by design it is clear that u as in (11.47) satisfies all but the integrability requirement in (11.11).

To construct a Green function, we start with E, the fundamental solution to L in the sense of Proposition 3.4. Fix $x_0 \in \Omega$ and some $x_1 \in \mathbb{R}^n \setminus \overline{\Omega}$. Define the matrix-valued function $h(x) := E^\top(x - x_0) - E^\top(x - x_1)$ for each $x \in \mathbb{R}^n \setminus \{x_0, x_1\}$. Now, define for each $x \in \Omega \setminus \{x_0\}$ what will be our Green function,

$$G(x) := h(x) - v(x), \tag{11.48}$$

where the "correcting" matrix-valued function v is chosen so as to satisfy

$$\begin{cases} v \in [\mathscr{C}^\infty(\Omega)]^{M\times M}, \\ L^\top v = 0 \cdot I_{M\times M} \quad \text{in } \Omega, \\ \mathcal{N}_\kappa v,\ \mathcal{N}_\kappa(\nabla v) \in \mathcal{B}^{p',\lambda}(\partial\Omega, w'), \\ v\big|_{\partial\Omega}^{\kappa-\text{n.t.}} = h\big|_{\partial\Omega}. \end{cases} \tag{11.49}$$

Here, $p' := p/(p-1)$ and $w' := w^{-p'/p} \in A_{p'}(\partial\Omega, \sigma)$ is the conjugate weight to w. Notice for each $\beta \in \{1, \dots, M\}$ that (11.49) actually leads to a regularity boundary value problem formulated on Block spaces of the type first considered in (11.14), but now working with $L^\top$ in place of L, $v_{\bullet\beta} := (v_{\alpha\beta})_{1\le\alpha\le M}$ in place of u, and $h_{\bullet\beta} := (h_{\alpha\beta})_{1\le\alpha\le M}$ in place of f, and observing that $h_{\bullet\beta}|_{\partial\Omega} \in [\mathcal{B}_1^{p',\lambda}(\partial\Omega, w')]^M$ as a consequence of Lemma 11.6. Because we are presently assuming $\mathfrak{A}_{L^\top}^{\text{dis}} \neq \varnothing$, the existence of $v_{\bullet\beta}$ for each $\beta \in \{1, \dots, M\}$ follows from item (a) of Theorem 11.30, noting that we may need to take $\delta \in (0,1)$ smaller than previously chosen in the proof of part (a).

Pressing forward, to show that G as in (11.48) satisfies (11.10) in Proposition 11.4, first observe that by construction, $v \in [\mathscr{C}^\infty(\Omega)]^{M\times M}$ so together with (11.48) and Proposition 3.4, we can conclude that G belongs to $[L^1_{\text{loc}}(\Omega, \mathcal{L}^n)]^{M\times M}$. By item (c) of Proposition 3.4, we have

$$L^\top E^\top(\cdot - x_0) = \delta_{x_0} I_{M\times M} \quad \text{in } [\mathcal{D}'(\mathbb{R}^n)]^{M\times M} \tag{11.50}$$

and

$$L^\top E^\top(\cdot - x_1) = \delta_{x_1} I_{M\times M} \quad \text{in } [\mathcal{D}'(\mathbb{R}^n)]^{M\times M}, \tag{11.51}$$

which further implies that $L^\top E^\top(\cdot - x_1) = 0 \cdot I_{M\times M}$ in Ω. Hence,

$$\big(L^\top G_{\bullet\beta}(\cdot - x_0)\big)_\alpha = -\delta_{x_0}\delta_{\alpha\beta} \quad \text{in } [\mathcal{D}'(\Omega)]^M \tag{11.52}$$

for each $\alpha, \beta \in \{1, \ldots, M\}$. Also, the boundary trace of v in (11.49) together with (11.48) ensures that $G\big|_{\partial\Omega}^{\kappa-\text{n.t.}} = 0$. The next two observations show that $(\nabla G)\big|_{\partial\Omega}^{\kappa-\text{n.t.}}$ exists σ-a. e. on $\partial\Omega$. First, for $j \in \{0,1\}$ the smoothness of $E(\cdot - x_j)$ on $\mathbb{R}^n \setminus \{x_j\}$ guarantees that $(\nabla h)\big|_{\partial\Omega}^{\kappa-\text{n.t.}} = (\nabla h)|_{\partial\Omega}$, which exists σ-a. e. on $\partial\Omega$. Second, since we are assuming v satisfies (11.49), then for each $\alpha, \beta \in \{1, \ldots, M\}$, we have that $(\nabla v_{\alpha\beta})\big|_{\partial\Omega}^{\kappa-\text{n.t.}}$ exists σ-a. e. on $\partial\Omega$ thanks to Theorem 11.21.

Finally, suppose $0 < \rho < \frac{1}{4}\operatorname{dist}(x_0, \partial\Omega)$ and define $K := \overline{B(x_0, \rho)}$, which is a compact subset of Ω. Then not only do we have that $\mathcal{N}_\kappa^{\Omega\setminus K}(\nabla G) < +\infty$ at σ-a. e. point on $\partial\Omega$, but in fact, $\mathcal{N}_\kappa^{\Omega\setminus K}(\nabla G) \in \mathcal{B}^{p',\lambda}(\partial\Omega, w')$. Indeed, $\mathcal{N}_\kappa(\nabla v) \in \mathcal{B}^{p',\lambda}(\partial\Omega, w')$ since v satisfies (11.49). Then (d) of Proposition 3.4, Lemma 11.8, (2) of Lemma 2.18, Lemma 6.26, and Proposition 6.14 ensure that $\mathcal{N}_\kappa^{\Omega\setminus K}(\nabla h) \in \mathcal{B}^{p',\lambda}(\partial\Omega, w')$. Thus, the sublinearity of the nontangential maximal operator yields the desired membership of $\mathcal{N}_\kappa^{\Omega\setminus K}(\nabla G)$ to $\mathcal{B}^{p',\lambda}(\partial\Omega, w')$. With this in hand and the fact that $\mathcal{N}_\kappa u \in M^{p,\lambda}(\partial\Omega, w)$ by the assumption in (11.47), we see that the integrability condition in (11.11) is satisfied thanks to Theorem 6.28. All the hypotheses of Proposition 11.4 are now confirmed to hold, and as such, $u(x_0) = (u_\beta(x_0))_{1\le\beta\le M}$ has components given by

$$u_\beta(x_0) = -\int_{\partial\Omega} \langle 0, \partial_\nu^{A^\top} G_{\cdot\beta}\rangle \, d\sigma = 0, \quad \text{for each } \beta \in \{1, \ldots, M\}, \tag{11.53}$$

which shows that the solution u to (11.47) is identically zero on Ω, given that $x_0 \in \Omega$ was arbitrary. This ultimately yields the uniqueness of solutions to (11.31), in view of the linearity of the system L, concluding our first treatment of (b).

Our second proof of the fact that (11.47) implies $u \equiv 0$ in Ω relies on Lemma 5.17 with $p_0 := p$, which says there exists $\omega \in A_p(\partial\Omega, \sigma)$ such that $\mathcal{N}_\kappa u \in L^p(\partial\Omega, \omega)$, so that uniqueness then follows from [79, Theorem 6.2, pp. 367–369], a result which concerns the well posedness of an analogous formulation of the Dirichlet problem in (11.31), but with data in Muckenhoupt-weighted Lebesgue spaces.

The claim in (c) follows immediately by combining the results in (a) and (b).

Concerning item (d), we introduce the following second-order, homogeneous, real constant coefficient, 2×2 system[1]

$$L := \frac{1}{4}\begin{pmatrix} \partial_x^2 - \partial_y^2 & -2\partial_x\partial_y \\ 2\partial_x\partial_y & \partial_x^2 - \partial_y^2 \end{pmatrix}. \tag{11.54}$$

The characteristic matrix is given by

$$L(\xi) := \frac{1}{4}\begin{pmatrix} \xi_1^2 - \xi_2^2 & -2\xi_1\xi_2 \\ 2\xi_1\xi_2 & \xi_1^2 - \xi_2^2 \end{pmatrix} \quad \text{for each } \xi = (\xi_1, \xi_2) \in \mathbb{R}^2, \tag{11.55}$$

1 This system is sometimes referred to as Bitsadze's operator.

and a quick computation shows that for each $\xi \in \mathbb{R}^2 \setminus \{0\}$,

$$\det[L(\xi)] = \frac{1}{16}[(\xi_1 - \xi_2)^2 + 4\xi_1^2\xi_2^2] = \frac{1}{16}(\xi_1^2 + \xi_2^2)^2 = \frac{1}{16}|\xi|^2 \neq 0, \tag{11.56}$$

showing that L is weakly elliptic. However, [79, Example 3.5, p. 210] shows that L and $L^\top$ do not have a distinguished coefficient tensor, i. e., $\mathfrak{A}_L^{\mathrm{dis}} = \varnothing$ and $\mathfrak{A}_{L^\top}^{\mathrm{dis}} = \varnothing$. Moreover, an argument paralleling that of the proof of [79, Proposition 3.10, p. 217] shows that the space of admissible boundary data for the Dirichlet problem, formulated in terms of Muckenhoupt-weighted Morrey spaces for the system L in the upper half-plane, i. e.,

$$\begin{aligned} \{u\big|_{\partial\mathbb{R}^2_+}^{\kappa-\mathrm{n.t.}} : u \in [\mathscr{C}^2(\mathbb{R}^2_+)]^2,\ Lu = 0 \text{in } \mathbb{R}^2_+,\ \mathcal{N}_\kappa u \in M^{p,\lambda}(\mathbb{R}, w), \\ \text{and } u\big|_{\partial\mathbb{R}^2_+}^{\kappa-\mathrm{n.t.}} \text{ exists at } \mathcal{L}^1\text{-a. e. point on } \mathbb{R}\}, \end{aligned} \tag{11.57}$$

is precisely

$$\{(f, Hf) : f \in M^{p,\lambda}(\mathbb{R}, w)\}, \tag{11.58}$$

where H denotes the Hilbert transform on $\mathbb{R}$. The proof of [79, Proposition 3.10, p. 217] does make use of Lebesgue's dominated convergence theorem, but we can get around this by appealing to Corollary 6.32 and Corollary 7.10. An analogous statement concerning the space of admissible boundary data for $L^\top$ is achieved via the fact that for the system L and any open set $\Omega \subseteq \mathbb{R}^2 \equiv \mathbb{C}$, and any complex valued function $u_1 + iu_2 \in \mathscr{C}^2(\Omega)$, one has $L(u_1, u_2) = 0$ if and only if $L^\top(u_1, -u_2) = 0$ (cf. [79, (3.248), p. 212]). The second part of the statement in (d) follows from [79, Example 3.5, p. 210] with a slight modification to the argument which now uses Lemma 4.12 to place $\mathcal{N}_\kappa(\partial^\alpha u^{(m)})$ in $M^{p,\lambda}(\mathbb{R}, w)$ whenever $\alpha \in \mathbb{N}_0^n$ with $m + |\alpha| \geq 2$, and where $u^{(m)}$ is as in [79, Example 3.5, p. 210]. Concerning higher dimensions, a similar result holds for the system $L_D := \Delta - 2\nabla\,\mathrm{div}$; cf. [79, Proposition 3.13, pp. 233–234].

Finally, to address the claim made in item (e), we find that the proof of (a) is largely unchanged, thanks to Theorem 10.33. The proof of (b) is actually a consequence of (b) in the case of $M^{p,\lambda}(\partial\Omega, w)$, given that $\mathring{M}^{p,\lambda}(\partial\Omega, w)$ is a subspace of $M^{p,\lambda}(\partial\Omega, w)$, although a similar argument involving a the construction of a Green function may also be employed given the duality result contained in Theorem 10.39. Item (c) then naturally follows. Finally, item (d) is handled similarly as before, but appealing to Remark 6.34 and Remark 7.11 in place of Corollary 6.32 and Corollary 7.10. Also, we use Corollary 10.29 in place of Lemma 4.12. □

Because we have also developed a robust layer potential theory at the level of Muckenhoupt-weighted Block spaces, we are in a position to present a well-posedness result for the Dirichlet problem for weakly elliptic systems with data in a Muckenhoupt-weighted Block space.

Theorem 11.12 (The weighted Block Dirichlet BVP). *Fix $n \in \mathbb{N}$ with $n \geq 2$ along with $M \in \mathbb{N}$. Let $\Omega \subseteq \mathbb{R}^n$ be an* AR *domain and write $\sigma := \mathcal{H}^{n-1}\lfloor\partial\Omega$. Suppose L is a weakly elliptic, second-order, homogeneous, constant complex coefficient, $M \times M$ system in $\mathbb{R}^n$. Fix an integrability exponent $q \in (1,\infty)$ with Hölder conjugate $q' := q/(q-1)$, along with a Muckenhoupt weight $\omega \in A_q(\partial\Omega,\sigma)$.*

$$\text{Choose } \varepsilon \in (0, q-1) \text{ so that } \omega \in A_{q-\varepsilon}(\partial\Omega,\sigma), \tag{11.59}$$

and select any threshold $\lambda \in (0, \varepsilon(n-1)/(q-1))$. Lastly, fix an aperture parameter $\kappa \in (0,\infty)$ and consider the Dirichlet problem for L in Ω with data in a Muckenhoupt-weighted Block space (defined in relation to these choices of q, ω, ε, and λ as in Definition 6.7*), formulated as*

$$\begin{cases} u \in [\mathscr{C}^\infty(\Omega)]^M, \\ Lu = 0 \quad \text{in } \Omega, \\ \mathcal{N}_\kappa u \in \mathcal{B}^{q,\lambda}(\partial\Omega,\omega), \\ u\big|_{\partial\Omega}^{\kappa-\text{n.t.}} = f \in [\mathcal{B}^{q,\lambda}(\partial\Omega,\omega)]^M. \end{cases} \tag{11.60}$$

Then there exists $\delta \in (0,1)$, depending on the ADR *character of $\partial\Omega$, n, q, ε, λ, $[\omega]_{A_{q-\varepsilon}}$, and L such that whenever*

$$\|\nu\|_{[\mathrm{BMO}(\partial\Omega,\sigma)]^n} < \delta, \tag{11.61}$$

*i. e., whenever Ω is a δ-*AR *domain, it follows that Ω is actually a* UR *domain and all claims below are true:*

(a) [Existence, estimates, and integral representation]. *Suppose $\mathfrak{A}_L^{\mathrm{dis}} \neq \varnothing$ and $A \in \mathfrak{A}_L^{\mathrm{dis}}$, then the operator*

$$\frac{1}{2}I + K_A : [\mathcal{B}^{q,\lambda}(\partial\Omega,\omega)]^M \to [\mathcal{B}^{q,\lambda}(\partial\Omega,\omega)]^M \tag{11.62}$$

is an isomorphism, and the function $u : \Omega \to \mathbb{C}^M$ defined as

$$u(x) := \left(\mathcal{D}_A\left[\left(\frac{1}{2}I + K_A\right)^{-1} f\right]\right)(x) \quad \text{at each } x \in \Omega \tag{11.63}$$

(with $\mathcal{D}_A$ associated to A as in (3.18)*) is a solution of* (11.60) *satisfying the estimate*

$$\|f\|_{[\mathcal{B}^{q,\lambda}(\partial\Omega,\omega)]^M} \leq \|\mathcal{N}_\kappa u\|_{\mathcal{B}^{q,\lambda}(\partial\Omega,\omega)} \leq C\|f\|_{[\mathcal{B}^{q,\lambda}(\partial\Omega,\omega)]^M}, \tag{11.64}$$

for some constant $C \in (0,\infty)$, depending only on the ADR *character of $\partial\Omega$, n, q, ε, λ, $[\omega]_{A_{q-\varepsilon}}$, and κ.*

(b) [Uniqueness]. *Strengthen the original hypothesis in* (11.59) *and choose a parameter*[2] $\varepsilon := (q-1)/p'$, *where* $p' := p/(p-1) \in (1,\infty)$ *and* $p \in (1,\infty)$ *is such that the conjugate weight satisfies*

$$\omega' := \omega^{-q'/q} \in A_{q'}(\partial\Omega, \sigma) \cap \mathrm{RH}_p(\partial\Omega, \sigma). \tag{11.65}$$

Assume $\mathfrak{A}^{\mathrm{dis}}_{L^\top} \neq \varnothing$. *Then the Dirichlet problem* (11.60) *has at most one solution.*

(c) [Well posedness]. *Strengthen the hypothesis in* (11.59) *and choose a parameter*[3] $\varepsilon := (q-1)/p'$, *where* $p' := p/(p-1) \in (1,\infty)$ *and* $p \in (1,\infty)$ *is such that the conjugate weight satisfies*

$$\omega' := \omega^{-q'/q} \in A_{q'}(\partial\Omega, \sigma) \cap \mathrm{RH}_p(\partial\Omega, \sigma). \tag{11.66}$$

If $\mathfrak{A}^{\mathrm{dis}}_{L} \neq \varnothing$ *and* $\mathfrak{A}^{\mathrm{dis}}_{L^\top} \neq \varnothing$, *the Dirichlet problem* (11.60) *is well posed in the sense of Hadamard. That is,* (11.60) *is uniquely solvable with naturally accompanying estimates as in* (11.64).

(d) [Sharpness]. *If* $\mathfrak{A}^{\mathrm{dis}}_{L} = \varnothing$, *the Dirichlet problem* (11.60) *may fail to be solvable. In fact, solvability may fail for boundary data belonging to an infinite- dimensional subspace of* $[\mathcal{B}^{q,\lambda}(\partial\Omega,\omega)]^M$, *even when* $\Omega = \mathbb{R}^n_+$ *(in which scenario the* BMO*-seminorm of* v *is zero). Also, if* $\mathfrak{A}^{\mathrm{dis}}_{L^\top} = \varnothing$, *then the Dirichlet problem in* (11.60) *may have more than one solution. In fact, the vector space of null-solutions may be infinite-dimensional even when* $\Omega = \mathbb{R}^n_+$.

Remark 11.13. Similar comments as in Remark 11.10 concerning the precise timing of Theorem 11.12 continue to hold. For completeness, we present a precise version of Theorem 11.12.

Fix $n \in \mathbb{N}$ with $n \geq 2$ and $M \in \mathbb{N}$. Suppose L is a weakly elliptic, second-order, homogeneous, constant complex coefficient, $M \times M$ system in $\mathbb{R}^n$. Fix two constants $C_{\mathrm{ADR}}, C_W \in [1,\infty)$. Next, select any integrability exponent $q \in (1,\infty)$ and choose $\varepsilon \in (0, q-1)$. Then for every threshold $\lambda \in (0, \varepsilon(n-1)/(q-1))$, there is some small

$$\delta = \delta(n, C_{\mathrm{ADR}}, C_W, L, q, \varepsilon, \lambda) \in (0,1) \tag{11.67}$$

with the following significance. Assume $\Omega \subseteq \mathbb{R}^n$ is a δ-AR domain whose topological boundary has ADR character controlled by C_{ADR}. Abbreviate $\sigma := \mathcal{H}^{n-1}\lfloor\partial\Omega$ and select any Muckenhoupt weight $\omega \in A_q(\partial\Omega,\sigma)$ satisfying $\omega \in A_{q-\varepsilon}(\partial\Omega,\sigma)$ with $[\omega]_{A_{q-\varepsilon}} \leq C_W$. Finally, fix an arbitrary aperture parameter $\kappa \in (0,\infty)$ and consider the Dirichlet problem:

2 That such a choice of ε is permissible in the context of (11.59) is ensured by Proposition 6.35.

3 Again, Proposition 6.35 ensures that this choice of ε is permissible in the context of (11.59).

$$\begin{cases} u \in [\mathscr{C}^\infty(\Omega)]^M, \\ Lu = 0 \quad \text{in } \Omega, \\ \mathcal{N}_\kappa u \in \mathcal{B}^{q,\lambda}(\partial\Omega, \omega), \\ u\big|_{\partial\Omega}^{\kappa-\text{n.t.}} = f \in [\mathcal{B}^{q,\lambda}(\partial\Omega, \omega)]^M. \end{cases} \tag{11.68}$$

Then Ω is actually a UR domain and the following claims are true:

(a) [Existence, estimates, and integral representation]. Suppose $\mathfrak{A}_L^{\text{dis}} \neq \varnothing$ and $A \in \mathfrak{A}_L^{\text{dis}}$, then with K_A associated with A as in (3.19), the operator

$$\frac{1}{2}I + K_A : [\mathcal{B}^{q,\lambda}(\partial\Omega, \omega)]^M \to [\mathcal{B}^{q,\lambda}(\partial\Omega, \omega)]^M \tag{11.69}$$

is an isomorphism, and the function $u : \partial\Omega \to \mathbb{C}^M$ defined by

$$u(x) := \left(\mathcal{D}_A\left[\left(\frac{1}{2}I + K_A\right)^{-1} f\right]\right)(x) \quad \text{at every } x \in \Omega \tag{11.70}$$

(with $\mathcal{D}_A$ associated to A as in (3.18)) is a solution to the Dirichlet problem (11.68) and satisfies the estimate

$$\|f\|_{[\mathcal{B}^{q,\lambda}(\partial\Omega,\omega)]^M} \leq \|\mathcal{N}_\kappa u\|_{\mathcal{B}^{q,\lambda}(\partial\Omega,\omega)} \leq C\|f\|_{[\mathcal{B}^{q,\lambda}(\partial\Omega,\omega)]^M}, \tag{11.71}$$

for some constant $C = C(n, C_{\text{ADR}}, C_W, L, q, \varepsilon, \lambda, \kappa) \in (0, \infty)$.

(b) [Uniqueness]. Suppose also $\omega' := \omega^{-q'/q} \in A_{q'}(\partial\Omega, \sigma) \cap \text{RH}_p(\partial\Omega, \sigma)$ for some $p \in (1, \infty)$ and define $\varepsilon := (q-1)/p'$ (where $p' := p/(p-1)$), so that $\omega \in A_q(\partial\Omega, \sigma)$ satisfies $\omega \in A_{q-\varepsilon}(\partial\Omega, \sigma)$ by Proposition 6.35. If $\mathfrak{A}_{L^\top}^{\text{dis}} \neq \varnothing$, then the Dirichlet problem (11.68) has at most one solution.

(c) [Well posedness]. Suppose also $\omega' := \omega^{-q'/q} \in A_{q'}(\partial\Omega, \sigma) \cap \text{RH}_p(\partial\Omega, \sigma)$ for some $p \in (1, \infty)$ and define $\varepsilon := (q-1)/p'$ (where $p' := p/(p-1)$), so that $\omega \in A_q(\partial\Omega, \sigma)$ satisfies $\omega \in A_{q-\varepsilon}(\partial\Omega, \sigma)$ by Proposition 6.35. If $\mathfrak{A}_L^{\text{dis}} \neq \varnothing$ and $\mathfrak{A}_{L^\top}^{\text{dis}} \neq \varnothing$, then the Dirichlet problem (11.68) is well posed in the sense of Hadamard. That is, (11.68) is uniquely solvable with the naturally accompanying estimates in (11.71).

We now turn to the task of proving Theorem 11.12.

Proof of Theorem 11.12. The proof of this theorem follows closely that of Theorem 11.9. For the proof of (a), we now we must appeal to Theorem 9.23 (while also keeping in mind Proposition 6.11 in our use of Lemma 11.2) to conclude that $\frac{1}{2}I + K_A$ is invertible on $[\mathcal{B}^{q,\lambda}(\partial\Omega, \omega)]^M$ provided $\|\nu\|_{[\text{BMO}(\partial\Omega,\sigma)]^n} < \delta$ and δ is sufficiently small. Then Proposition 9.16 and Proposition 9.18 ensure that $u := \mathcal{D}_A[(\frac{1}{2}I + K_A)^{-1}f]$ satisfies (11.60) with appropriate nontangential maximal operator estimates as in (11.64). To treat (b), we follow the same scheme as was done in the proof of (b) of Theorem 11.9, but with the roles of Morrey and Block spaces reversed. That is, in our construction of a Green function, we must solve a regularity boundary value problem on $M^{q',\lambda}(\partial\Omega,\omega')$, where we remind the reader that q' is the Hölder conjugate of q, and $\omega' := \omega^{-q'/q} \in A_{q'}(\partial\Omega,\sigma)$ is the conjugate weight to ω. The proof that G in this case exists and is a Green function in the sense of Proposition 11.4 now makes use of Theorem 11.29, Lemma 11.7, Theorem 11.20, Lemma 4.12, and Proposition 4.7. Note that $\omega' \in A_{q'}(\partial\Omega,\sigma) \cap \mathrm{RH}_p(\partial\Omega,\sigma)$ by assumption and that $\lambda \in (0,\varepsilon(n-1)/(q-1)) = (0,(n-1)/p')$ by our choice of ε, ensuring that the Muckenhoupt-weighted Morrey space $M^{q',\lambda}(\partial\Omega,\omega')$ is considered in a regime for which the preceding results indeed apply. We use Theorem 6.28 again to show that the integrability requirement in (11.11) is satisfied, so that by Proposition 11.4 we may conclude that uniqueness holds. Note also that we could have alternatively argued for uniqueness of solutions using Lemma 10.57 and the uniqueness of the Dirichlet problem with data in Muckenhoupt-weighted Lebesgue spaces, cf. [79, Theorem 6.2, pp. 367–369]. The claim in (c) then follows, and we can show that the claims in item (d) are true by arguing similarly as in the proof of item (d) of Theorem 11.9, but using Remark 6.33 and Remark 7.11 instead of Corollary 6.32 and Corollary 7.10. In this case, we also use Lemma 6.26 instead of Lemma 4.12. □

We now present several corollaries to Theorem 11.9, exploring different choices for the system and domains to which our theory applies.

The first corollary we have is the Dirichlet problem for the Laplacian in the upper half-space with boundary datum arbitrarily prescribed in a Muckenhoupt-weighted Morrey space. Despite the fact that the geometry of the domain is as simple as one can have, and that we are dealing with the Laplacian, this result is still new and nontrivial because of our functional analytic setting.

Corollary 11.14. *Fix $n \in \mathbb{N}$ with $n \geq 2$. Select any two integrability exponents $p, q \in (1,\infty)$, along with a Muckenhoupt weight w belonging to $A_p(\mathbb{R}^{n-1},\mathcal{L}^{n-1}) \cap \mathrm{RH}_q(\mathbb{R}^{n-1},\mathcal{L}^{n-1})$, and some threshold $\lambda \in (0,(n-1)/q')$, where $q' := q/(q-1)$. For a fixed aperture parameter $\kappa \in (0,\infty)$, consider the following Dirichlet boundary value problem for the Laplacian in the upper half-space, $\mathbb{R}^n_+$, identifying $\partial\mathbb{R}^n_+ \equiv \mathbb{R}^{n-1}$:*

$$\begin{cases} u \in \mathscr{C}^\infty(\mathbb{R}^n_+), \\ \Delta u = 0 \quad \text{in } \mathbb{R}^n_+, \\ \mathcal{N}_\kappa u \in M^{p,\lambda}(\mathbb{R}^{n-1},w), \\ u\big|_{\partial\Omega}^{\kappa-\mathrm{n.t.}} = f \in M^{p,\lambda}(\mathbb{R}^{n-1},w). \end{cases} \tag{11.72}$$

The Dirichlet problem (11.72) *is uniquely solvable and the solution, u, is given by*

$$u(x',t) := \frac{2}{\omega_{n-1}} \int_{\mathbb{R}^{n-1}} \frac{t}{(|x'-y'|^2+t^2)^{n/2}} f(y')\,dy', \tag{11.73}$$

at each point $(x',t) \in \mathbb{R}^n_+$ and satisfies

$$\|f\|_{M^{p,\lambda}(\mathbb{R}^{n-1},w)} \leq \|\mathcal{N}_\kappa u\|_{M^{p,\lambda}(\mathbb{R}^{n-1},w)} \leq C\|f\|_{M^{p,\lambda}(\mathbb{R}^{n-1},w)} \tag{11.74}$$

for some constant $C \in (0,\infty)$, depending only on n, p, q, λ, $[w]_{A_p}$, $[w]_{\mathrm{RH}_q}$, and κ.

Proof. This is a consequence of Theorem 11.9, under the identifications, $\Omega := \mathbb{R}^n_+$ and $L := \Delta$. Note in view of Proposition 3.16 that Δ has a distinguished coefficient tensor (as does its transpose, since Δ is symmetric), given that K_Δ, the boundary-to-boundary harmonic double layer as defined in (3.25), has a chord-dot-normal structure. Note also that $\mathbb{R}^n_+$ is a δ-AR domain for any choice of $\delta \in (0,1)$, and that K_Δ is actually the zero operator (cf. item (ii) of Proposition 3.16), so that (11.73) follows from the integral representation in (11.34) and the definition of $\mathcal{D}_\Delta$ in (3.24), where we have identified $x = (x', t) \in \mathbb{R}^n_+$ and $y = (y', 0) \in \partial\mathbb{R}^n_+$. □

Another new and nontrivial result that follows from our general theorem is the solvability of the Lamé system of elasticity in the upper half-space with boundary datum in a weighted Morrey space.

Corollary 11.15. *Fix $n \in \mathbb{N}$ with $n \geq 2$. Select any two arbitrary integrability exponents $p, q \in (1,\infty)$ along with a Muckenhoupt weight w belonging to $A_p(\mathbb{R}^{n-1}, \mathcal{L}^{n-1}) \cap \mathrm{RH}_q(\mathbb{R}^{n-1}, \mathcal{L}^{n-1})$, and some threshold $\lambda \in (0,(n-1)/q')$. Choose two parameters $\mu, \lambda \in \mathbb{C}$ with $\mu \neq 0$ and $2\mu + \lambda \neq 0$. For a fixed aperture parameter $\kappa \in (0,\infty)$, consider the following Dirichlet boundary value problem for the Lamé system of elasticity in the upper half-space, $\mathbb{R}^n_+$, identifying $\partial\mathbb{R}^n_+ \equiv \mathbb{R}^{n-1}$:*

$$\begin{cases} u \in [\mathscr{C}^\infty(\mathbb{R}^n_+)]^n, \\ \mu\Delta u + (\mu+\lambda)\nabla \mathrm{div}\, u = 0 \quad \text{in } \mathbb{R}^n_+, \\ \mathcal{N}_\kappa u \in M^{p,\lambda}(\mathbb{R}^{n-1}, w), \\ u\big|^{\kappa-\mathrm{n.t.}}_{\partial\Omega} = f = (f_\alpha)_{1\leq\alpha\leq n} \in [M^{p,\lambda}(\mathbb{R}^{n-1}, w)]^n. \end{cases} \tag{11.75}$$

In the case that $3\mu + \lambda \neq 0$ the boundary value problem (11.75) is uniquely solvable, and the solution, $u = (u_\alpha)_{1\leq\alpha\leq n}$, is given by

$$u_\alpha(x',t) = \frac{4\mu}{3\mu+\lambda}\frac{1}{\omega_{n-1}}\int_{\mathbb{R}^{n-1}} \frac{t}{(|x'-y'|^2+t^2)^{\frac{n}{2}}} f_\alpha(y')\,dy' \tag{11.76}$$

$$+ \frac{\mu+\lambda}{3\mu+\lambda}\frac{2n}{\omega_{n-1}}\int_{\mathbb{R}^{n-1}} \frac{t(x'-y',t)_\alpha(x'-y',t)_\beta}{(|x'-y'|^2+t^2)^{\frac{n+2}{2}}} f_\beta(y')\,dy', \tag{11.77}$$

at each point $(x',t) \in \mathbb{R}^n_+$ and satisfies

$$\|f\|_{[M^{p,\lambda}(\mathbb{R}^{n-1},w)]^n} \|\mathcal{N}_\kappa u\|_{M^{p,\lambda}(\mathbb{R}^{n-1},w)} \leq \|f\|_{[M^{p,\lambda}(\mathbb{R}^{n-1},w)]^n}. \tag{11.78}$$

Proof. This follows from Theorem 11.9, identifying

$$L := L_{\mu,\lambda} := \mu\Delta + (\mu+\lambda)\nabla\,\mathrm{div}, \tag{11.79}$$

and $\Omega := \mathbb{R}^n_+$. Note that the assumption that $\mu \neq 0$ and $2\mu + \lambda \neq 0$ ensure the weak ellipticity of the operator L (cf. [85, Proposition 10.14, p. 366]), while the requirement that $3\mu + \lambda \neq 0$ ensures that L has a distinguished coefficient tensor (cf. [59]). The upper half-space is a δ-AR domain for any choice of $\delta \in (0,1)$, so that such a domain falls within the scope of Theorem 11.9. Moreover, from Proposition 3.16, we know that the boundary-to-boundary harmonic double layer (associated to the distinguished coefficient tensor for the Lamé system) is the zero operator. In order to see why (11.76) holds, note that if $A = (a^{\alpha\beta}_{jk})_{\substack{1\le\alpha,\beta\le n\\ 1\le j,k\le n}} \in \mathfrak{A}^{\mathrm{dis}}_{L_{\mu,\lambda}}$, then necessarily (cf. [59]) the entries of A are given by

$$a^{\alpha\beta}_{jk} = \mu\delta_{jk}\delta_{\alpha\beta} + \frac{(\lambda+\mu)(2\mu+\lambda)}{3\mu+\lambda}\delta_{j\alpha}\delta_{k\beta} + \frac{\mu(\lambda+\mu)}{3\mu+\lambda}\delta_{j\beta}\delta_{k\alpha} \tag{11.80}$$

for each $\alpha, \beta, j, k \in \{1, \dots, n\}$. It is also known (cf. [85, Theorem 10.15, p. 369]) that when $n \geq 3$ the fundamental solution for $L_{\mu,\lambda}$ as in Proposition 3.4 is defined for each $j, k \in \{1, \dots, n\}$ according to

$$E_{jk}(x) = \frac{-1}{2\mu(2\mu+\lambda)\omega_{n-1}}\left[\frac{3\mu+\lambda}{n-2}\frac{\delta_{jk}}{|x|^{n-2}} + \frac{(\mu+\lambda)x_j x_k}{|x|^n}\right] \tag{11.81}$$

at each $x \in \mathbb{R}^n \setminus \{0\}$. When $n = 2$, the fundamental solution for $L_{\mu,\lambda}$ is defined for each $j, k \in \{1, \dots, n\}$ by

$$E_{jk}(x) = \frac{1}{4\pi\mu(2\mu+\lambda)}\left[(3\mu+\lambda)\delta_{jk}\ln|x| - \frac{(\mu+\lambda)x_j x_k}{|x|^2}\right] \tag{11.82}$$

at each $x \in \mathbb{R}^n \setminus \{0\}$. Since A as in (11.80) satisfies $A \in \mathfrak{A}^{\mathrm{dis}}_{L_{\mu,\lambda}}$, item (f) of Proposition 3.12 guarantees the existence of a matrix-valued function $k = (k_{\alpha\beta})_{1\le\alpha,\beta\le n}$ with the property that for each $\gamma, \alpha, \ell \in \{1, \dots, n\}$, we have

$$a^{\beta\alpha}_{j\ell}(\partial_j E_{\gamma\beta})(x) = x_\ell k_{\gamma\alpha}(x) \quad \text{for all } x \in \mathbb{R}^n \setminus \{0\}. \tag{11.83}$$

A straightforward computation using (11.80) and (11.81)–(11.82) then shows that the matrix valued function $k = (k_{\alpha\beta})_{1\le\alpha,\beta\le n}$, as in (11.83) is given by

$$k_{\alpha\beta}(x) := \frac{2\mu}{3\mu+\lambda}\frac{\delta_{\alpha\beta}}{\omega_{n-1}}\frac{1}{|x|^n} + \frac{\mu+\lambda}{3\mu+\lambda}\frac{n}{\omega_{n-1}}\frac{x_\alpha x_\beta}{|x|^{n+2}} \tag{11.84}$$

for each $\alpha, \beta \in \{1, \dots, n\}$ at every point $x \in \mathbb{R}^n \setminus \{0\}$. As such, the integral representation in (11.34) (bearing in mind that K_A is the zero operator since $A \in \mathfrak{A}^{\mathrm{dis}}_{L_{\mu,\lambda}}$), (3.18) (with $x = (x', t) \in \mathbb{R}^n_+$ and $y = (y', 0) \in \partial\mathbb{R}^n_+$), the fact that $\nu(y) = (0, \dots, 0, -1)$ for all $y \in \partial\mathbb{R}^n_+$, (11.84), and (11.83) can be used to ascertain (11.76). □

The geometry of the upper half-space is special, because when the boundary-to-boundary double layer, K_A, is of chord-dot-normal type, we see that K_A is the zero operator. This is cohesive with the kind of estimate achieved in (9.24), which implies that $K_A = 0$ whenever $\|\nu\|_{[\mathrm{BMO}(\partial\Omega,\sigma)]^n} = 0$, something that occurs when $\Omega = \mathbb{R}^n_+$. In this regime, the operator $\frac{1}{2}I + K_A$ is always invertible with inverse $2I$. This yields the following corollary, which actually encompasses the results recorded in Corollaries 11.14 and 11.15.

Corollary 11.16. *Fix $n \in \mathbb{N}$ with $n \geq 2$. Select any two integrability exponents $p, q \in (1, \infty)$ along with a Muckenhoupt weight w belonging to $A_p(\mathbb{R}^{n-1}, \mathcal{L}^{n-1}) \cap \mathrm{RH}_q(\mathbb{R}^{n-1}, \mathcal{L}^{n-1})$, and some threshold $\lambda \in (0, (n-1)/q')$, where $q' := q/(q-1)$. Suppose L is a weakly elliptic, second-order, homogeneous, constant complex coefficient, $M \times M$ system in $\mathbb{R}^n$ with $\mathfrak{A}^{\mathrm{dis}}_L \neq \varnothing$ and $\mathfrak{A}^{\mathrm{dis}}_{L^\top} \neq \varnothing$. For an aperture parameter $\kappa \in (0, \infty)$ consider the following Dirichlet boundary value problem for L in the upper half-space, $\mathbb{R}^n_+$, and identify $\partial\mathbb{R}^n_+ \equiv \mathbb{R}^{n-1}$:*

$$\begin{cases} u \in [\mathscr{C}^\infty(\mathbb{R}^n_+)]^M, \\ Lu = 0 \quad \text{in } \mathbb{R}^n_+, \\ \mathcal{N}_\kappa u \in M^{p,\lambda}(\mathbb{R}^{n-1}, w), \\ u\big|^{\kappa-\mathrm{n.t.}}_{\partial\Omega} = f \in [M^{p,\lambda}(\mathbb{R}^{n-1}, w)]^M. \end{cases} \tag{11.85}$$

Then (11.85) is uniquely solvable and the solution, u, is given by

$$u(x', t) = \left(-2 \int_{\mathbb{R}^{n-1}} a^{\beta\alpha}_{jn} (\partial_j E_{\gamma\beta})(x' - y', t) f_\gamma(y') \, dy' \right)_{1 \leq \alpha \leq M} \tag{11.86}$$

for each $(x', t) \in \mathbb{R}^n_+$ and satisfies for some constant $C \in (0, \infty)$, independent of f,

$$\|f\|_{[M^{p,\lambda}(\mathbb{R}^{n-1},w)]^M} \leq \|\mathcal{N}_\kappa u\|_{M^{p,\lambda}(\mathbb{R}^{n-1},w)} \leq C\|f\|_{[M^{p,\lambda}(\mathbb{R}^{n-1},w)]^M}, \tag{11.87}$$

where $A = (a^{\alpha\beta}_{jk})_{\substack{1\leq j,k\leq n \\ 1\leq\alpha,\beta\leq M}} \in \mathfrak{A}^{\mathrm{dis}}_L$ and $E = (E_{\alpha\beta})_{1\leq\alpha,\beta\leq M}$ is the fundamental solution to L in the sense of Proposition 3.4.

Proof. This is a direct consequence of Theorem 11.9 identifying $\Omega := \mathbb{R}^n_+$, which is a δ-AR domain for any choice of $\delta \in (0,1)$. The fact that the integral formula in (11.86) holds follows from (11.34), and the fact that K_A is the zero operator when $A \in \mathfrak{A}^{\mathrm{dis}}_L$ as stated in item (ii) of Proposition 3.16. □

In the next corollary, we explore the well posedness of a boundary value problem when a specific boundary condition is given. For example, working in $\mathbb{R}^n_+$, we would like to know for what power b does the boundary datum $|\cdot|^b$, defined on $\mathbb{R}^{n-1}$, lead to a well-posed Dirichlet boundary value problem for the Laplacian? Obtaining a well-posedness result for this boundary datum is hopeless when using a theory with a singular focus on Lebesgue spaces for the simple fact that $|\cdot|^b$ does not belong to $L^p(\mathbb{R}^{n-1}, \mathcal{L}^{n-1})$ for any choice of $p \in (1, \infty)$ and $b \in \mathbb{R}$. This next result speaks to the usefulness of exploring functions spaces other than ordinary Lebesgue spaces in relation to partial differential equations. In practice, one does not get to choose their boundary datum; it is usually suggested by empirical data. As such, it is important to develop a theory of boundary value problems in different functional analytic settings in the hopes of being as accommodating as possible to the types of boundary datum one might come across.

Corollary 11.17. *Let $\Omega \subseteq \mathbb{R}^n$ be an* AR *domain and write $\sigma := \mathcal{H}^{n-1}\lfloor\partial\Omega$. Fix an arbitrary integrability exponent $p \in (1,\infty)$, a reference point $x_0 \in \partial\Omega$, any power $a \in (1-n,(p-1)(n-1))$, and some threshold $\lambda \in (0,(n-1)/q')$, where $q' := q/(q-1)$ and $q \in (1,\infty)$ is chosen so as to have $aq > 1-n$. Finally, fix an aperture parameter $\kappa \in (0,\infty)$, and consider the following Dirichlet problem for some $b \in \mathbb{R}$:*

$$\begin{cases} u \in \mathscr{C}^\infty(\Omega), \\ \Delta u = 0 \quad \text{in } \Omega, \\ \mathcal{N}_\kappa u \in M^{p,\lambda}(\partial\Omega, |\cdot - x_0|^a\sigma), \\ u\big|_{\partial\Omega}^{\kappa-\mathrm{n.t.}} = |\cdot - x_0|^b. \end{cases} \tag{11.88}$$

Then corresponding to the case that $b = -(a+n-1-\lambda)/p$, there exists $\delta \in (0,1)$, depending on the ADR *character of $\partial\Omega$, n, p, q, λ, $[|\cdot - x_0|^a]_{A_p}$, and $[|\cdot - x_0|^a]_{\mathrm{RH}_q}$, such that when Ω is a δ-*AR *domain, the Dirichlet problem* (11.88) *is uniquely solvable with*

$$\begin{aligned} \big\||\cdot - x_0|^b\big\|_{M^{p,\lambda}(\partial\Omega,|\cdot-x_0|^a\sigma)} &\le \|\mathcal{N}_\kappa u\|_{M^{p,\lambda}(\partial\Omega,|\cdot-x_0|^a\sigma)} \\ &\le C\big\||\cdot - x_0|^b\big\|_{M^{p,\lambda}(\partial\Omega,|\cdot-x_0|^a\sigma)} \end{aligned} \tag{11.89}$$

for some constant $C \in (0,\infty)$, depending only on the ADR *character of $\partial\Omega$, n, p, q, λ, $[|\cdot - x_0|^a]_{A_p}$, $[|\cdot - x_0|^a]_{\mathrm{RH}_q}$, and κ.*

Proof. Since $a \in (1-n,(n-1)(p-1))$, we have $|\cdot - x_0|^a \in A_p(\partial\Omega,\sigma)$ because of the fact mentioned in (1.41). The discussion in (1.44) also ensures that $|\cdot - x_0|^a \in \mathrm{RH}_q(\partial\Omega,\sigma)$. Note that (cf. [79, Example 7.1, p. 434]) the function $f_{x_0} : \partial\Omega \to \mathbb{R}$, defined according to $f_{x_0}(x) := |x - x_0|^c$ for all $x \in \partial\Omega$, belongs to $M^{p,\lambda}(\partial\Omega,\sigma)$ when $c := -(n-1-\lambda)/p$. We can then see with $b := -(a+n-1-\lambda)/p$ that $|\cdot - x_0|^b$ belongs to $M^{p,\lambda}(\partial\Omega,|\cdot - x_0|^a\sigma)$, since

$$\begin{aligned} \big\||\cdot - x_0|^b\big\|_{M^{p,\lambda}(\partial\Omega,|\cdot-x_0|^a\sigma)} &= \sup_{\Delta\subseteq\partial\Omega}\bigg(\frac{1}{r^\lambda}\int_\Delta |x - x_0|^{bp}|x - x_0|^a\,d\sigma(x)\bigg)^{1/p} \\ &= \sup_{\Delta\subseteq\partial\Omega}\bigg(\frac{1}{r^\lambda}\int_\Delta |x - x_0|^{-(n-1-\lambda)/p}\,d\sigma(x)\bigg)^{1/p} \\ &< +\infty, \end{aligned} \tag{11.90}$$

where the supremum is taken over all surface balls and finiteness in the last line above is a consequence of the membership of $|\cdot - x_0|^{-(n-1-\lambda)/p}$ to $M^{p,\lambda}(\partial\Omega,\sigma)$. Since the Laplacian has a distinguished coefficient tensor (e. g., this can be seen in light of (iii) of Proposition 3.16 via the chord-dot-normal structure of the boundary-to-boundary harmonic double layer, (3.25)), we find that (11.88) is a boundary value problem covered by Theorem 11.9, concluding the proof. □

Remark 11.18. We can rephrase the result in Corollary 11.17 as follows. Let $\Omega \subseteq \mathbb{R}^n$ be an AR domain and write $\sigma := \mathcal{H}^{n-1}\lfloor\partial\Omega$. Fix an arbitrary integrability exponent $p \in (1,\infty)$ along with any exponent $b \in (0, n-1)$.

Then for a fixed aperture parameter $\kappa \in (0,\infty)$, there exists a power $a \in (1-n,(p-1)(n-1))$, an integrability exponent $q \in (1,\infty)$, which satisfies $a > (n-1)/q$, and a threshold $\lambda \in (0,(n-1)/q')$, where $q' := q/(q-1)$ with the following significance. When $\delta \in (0,1)$ is sufficiently small and Ω is a δ-AR domain, the Dirichlet problem for the Laplacian on Ω in (11.88) is uniquely solvable with estimates as in (11.89).

Indeed, there exist such a, q, and λ with the property that

$$\begin{aligned} &|\cdot - x_0|^{-b} \in M^{p,\lambda}(\partial\Omega, |\cdot - x_0|^a\sigma) \quad \text{and} \\ &|\cdot - x_0|^a \in A_p(\partial\Omega,\sigma) \cap \mathrm{RH}_q(\partial\Omega,\sigma). \end{aligned} \tag{11.91}$$

The first line in (11.91) follows from a continuity/connectivity argument allowing us to write $b = (a+n-1-\lambda)/p$, paired with the analysis leading up to (11.90). For the second line of (11.91), the membership of $|\cdot - x_0|^a$ to $A_p(\partial\Omega,\sigma)$ follows from the fact that $a \in (1-n,(p-1)(n-1))$ and (1.41), while the membership of $|\cdot - x_0|^a$ to $\mathrm{RH}_q(\partial\Omega,\sigma)$ follows from the result recorded in (1.44). Subsequently, this confirms that (11.88) is a boundary value problem with a functional analytic setting that falls within the scope of Theorem 11.9.

Additionally, it is worth remarking that the interval $(0, n-1)$ is the optimal range for the exponent b in light of Proposition 4.15, which yields that any Muckenhoupt-weighted Morrey space on $\partial\Omega$ continuously embeds into $L^1(\partial\Omega, \frac{\sigma(x)}{1+|x|^{n-1}})$.

11.2 The Neumann problem on Morrey and Block spaces

The following proposition, a more general version of which appears in [88, Theorem 1.5.10, pp. 156–157], will help with the problem of uniqueness (modulo constants) for the Neumann problem. It is a kind of integral representation for derivatives of solutions to the kinds of weakly elliptic systems we consider.

Proposition 11.19. *Let $\Omega \subseteq \mathbb{R}^n$, where $n \geq 2$, be an AR domain with $\partial\Omega$ unbounded. Write $\sigma := \mathcal{H}^{n-1}\lfloor\partial\Omega$ and $\nu = (\nu_1,\dots,\nu_n)$ for the geometric measure theoretic outward unit normal to Ω. Next, let $L = (a_{rs}^{\alpha\beta}\partial_r\partial_s)_{1\leq\alpha,\beta\leq M}$ be a homogeneous, weakly elliptic, second-order $M \times M$ system in $\mathbb{R}^n$, with complex constant coefficients, and denote by $E = (E_{\gamma\beta})_{1\leq\gamma,\beta\leq M}$ the matrix-valued fundamental solution associated with L as in Proposition 3.4.*

In this setting, assume $u = (u_\beta)_{1\leq\beta\leq M} \in [\mathscr{C}^\infty(\Omega)]^M$ is a vector-valued function which, for some $\kappa > 0$, satisfies

$$
\begin{aligned}
Lu = 0 \quad in\ \Omega, \quad (\nabla u)\big|_{\partial\Omega}^{\kappa-\mathrm{n.t.}} \quad & exists\ at\ \sigma\text{-}a.\,e.\ point\ on\ \partial\Omega, \\
and \quad \mathcal{N}_\kappa(\nabla u) \in L^1\Big(\partial\Omega, \frac{\sigma(x)}{1+|x|^{n-1}}\Big).&
\end{aligned}
\tag{11.92}
$$

Then for each $\ell \in \{1,\dots,n\}$ *and each* $\gamma \in \{1,\dots,M\}$ *one has*

$$
\begin{aligned}
(\partial_\ell u_\gamma)(x) = \int_{\partial\Omega} & a_{rs}^{\beta\alpha}(\partial_r E_{\gamma\beta})(x-y)\{\nu_\ell(y)((\partial_s u_\alpha)\big|_{\partial\Omega}^{\kappa-\mathrm{n.t.}})(y) \\
& - \nu_s(y)((\partial_\ell u_\alpha)\big|_{\partial\Omega}^{\kappa-\mathrm{n.t.}})(y)\}\, d\sigma(y) \\
& - \int_{\partial\Omega} (\partial_\ell E_{\gamma\alpha})(x-y)\nu_r(y) a_{rs}^{\alpha\beta}((\partial_s u_\beta)\big|_{\partial\Omega}^{\kappa-\mathrm{n.t.}})(y)\, d\sigma(y)
\end{aligned}
\tag{11.93}
$$

at every point $x \in \Omega$, *and*

$$
\begin{aligned}
0 = \int_{\partial\Omega} & a_{rs}^{\beta\alpha}(\partial_r E_{\gamma\beta})(x-y)\{\nu_\ell(y)((\partial_s u_\alpha)\big|_{\partial\Omega}^{\kappa-\mathrm{n.t.}})(y) \\
& - \nu_s(y)((\partial_\ell u_\alpha)\big|_{\partial\Omega}^{\kappa-\mathrm{n.t.}})(y)\}\, d\sigma(y) \\
& - \int_{\partial\Omega} (\partial_\ell E_{\gamma\alpha})(x-y)\nu_r(y) a_{rs}^{\alpha\beta}((\partial_s u_\beta)\big|_{\partial\Omega}^{\kappa-\mathrm{n.t.}})(y)\, d\sigma(y)
\end{aligned}
\tag{11.94}
$$

at every point $x \in \mathbb{R}^n \setminus \overline{\Omega}$.

We now record a useful Fatou-type theorem at the level of Muckenhoupt-weighted Morrey spaces, which serves to show that the space of boundary datum we consider is a natural environment given the imposed size condition.

Theorem 11.20. *Let* $\Omega \subseteq \mathbb{R}^n$ *be a* NTA *domain such that* $\partial\Omega$ *is an unbounded, Ahlfors regular set. Abbreviate* $\sigma := \mathcal{H}^{n-1}\lfloor\partial\Omega$ *and let* ν *denote the geometric measure theoretic outward unit normal to* Ω. *Suppose* L *is a weakly elliptic, second-order, homogeneous, constant complex coefficient,* $M \times M$ *system in* $\mathbb{R}^n$. *Now, fix two integrability exponents* $p, q \in (1,\infty)$, *a Muckenhoupt weight* $w \in A_p(\partial\Omega,\sigma) \cap \mathrm{RH}_q(\partial\Omega,\sigma)$, *and a threshold* $\lambda \in (0,(n-1)/q')$, *where* $q' := q/(q-1)$. *Finally, consider a function* $u : \Omega \to \mathbb{C}^M$, *which satisfies*

$$
u \in \big[\mathscr{C}^\infty(\Omega)\big]^M \quad and \quad Lu = 0 \quad in\ \Omega.
\tag{11.95}
$$

Then for any aperture parameter $\kappa \in (0,\infty)$ *and any coefficient tensor* $A \in \mathfrak{A}_L$, *if* $\mathcal{N}_\kappa(\nabla u)$ *belongs to* $M^{p,\lambda}(\partial\Omega, w)$ *it follows that*

$$
\begin{aligned}
u\big|_{\partial\Omega}^{\kappa-\mathrm{n.t.}} \quad & exists\ \sigma\text{-}a.\,e.\ on\ \partial\Omega\ and\ \partial_{\tau_{jk}}\big(u\big|_{\partial\Omega}^{\kappa-\mathrm{n.t.}}\big) \in \big[M^{p,\lambda}(\partial\Omega,w)\big]^M \\
& for\ each\ j,k \in \{1,\dots,n\},\ and \\
(\nabla u)\big|_{\partial\Omega}^{\kappa-\mathrm{n.t.}} \quad & exists\ \sigma\text{-}a.\,e.\ on\ \partial\Omega\ and\ \partial_\nu^A u \in \big[M^{p,\lambda}(\partial\Omega,w)\big]^M.
\end{aligned}
\tag{11.96}
$$

Moreover, analogous results hold when $M^{p,\lambda}(\partial\Omega, w)$ is replaced by $\mathring{M}^{p,\lambda}(\partial\Omega, w)$.

Proof. Fix $\kappa \in (0,\infty)$ and $A \in \mathfrak{A}_L$. Assuming $\mathcal{N}_\kappa(\nabla u) \in M^{p,\lambda}(\partial\Omega, w)$ allows us to use the embedding in Proposition 4.16 to say that $\mathcal{N}_\kappa(\nabla u)$ is in $L^{p_0}_{\rm loc}(\partial\Omega,\sigma)$ for some $p_0 \in (1,\infty)$. Therefore, $u\big|^{\kappa-{\rm n.t.}}_{\partial\Omega}$ exists as a σ-measurable function on $\partial\Omega$ by [79, Proposition 2.24, p. 152]. Note that under our current geometric hypotheses that Ω satisfies a two-sided corkscrew condition (cf. Definition 2.10), and hence having an unbounded Ahlfors regular boundary ensures that Ω is actually a UR domain (cf. (2.9)). Thus, [79, Theorem 3.4, p. 178] is applicable and gives that $(\nabla u)\big|^{\kappa-{\rm n.t.}}_{\partial\Omega}$ exists (in $\mathbb{C}^{M\times n}$) at σ-a. e. point on $\partial\Omega$ and belongs to $[L^{p_0}_{\rm loc}(\partial\Omega,\sigma)]^{M\times n}$ with

$$\big|(\nabla u)\big|^{\kappa-{\rm n.t.}}_{\partial\Omega}\big| \leq \mathcal{N}_\kappa(\nabla u) \quad \text{at } \sigma\text{-a. e. point on } \partial\Omega. \tag{11.97}$$

By Proposition 4.7, we see that $(\nabla u)\big|^{\kappa-{\rm n.t.}}_{\partial\Omega} \in [M^{p,\lambda}(\partial\Omega, w)]^{M\times n}$. This simultaneously shows for any $j,k \in \{1,\dots,n\}$ that $\partial_{\tau_{jk}} u \in [M^{p,\lambda}(\partial\Omega, w)]^M$, again by Proposition 4.7, since

$$|\partial_{\tau_{jk}} u| \leq 2\mathcal{N}_\kappa(\nabla u) \quad \text{at } \sigma\text{-a. e. point on } \partial\Omega \tag{11.98}$$

according to [79, Proposition 2.22, p. 148]. Finally, (11.97) allows us to estimate

$$\big|(\partial_k u)\big|^{\kappa-{\rm n.t.}}_{\partial\Omega}\big| \leq \big|(\nabla u)\big|^{\kappa-{\rm n.t.}}_{\partial\Omega}\big| \leq \mathcal{N}_\kappa(\nabla u) \tag{11.99}$$

for each index $k \in \{1,\dots,n\}$, so it is apparent from (3.18), and Proposition 4.7 that $\partial^A_\nu u \in [M^{p,\lambda}(\partial\Omega, w)]^M$ as well.

Addressing the final claim involving $\mathring{M}^{p,\lambda}(\partial\Omega, w)$, we can argue in much the same way in view of Proposition 10.26. □

We also have a Fatou-type result at the level of Muckenhoupt-weighted Block spaces.

Theorem 11.21. *Let $\Omega \subsetneq \mathbb{R}^n$ be a* NTA *domain such that $\partial\Omega$ is an unbounded, Ahlfors regular set. Abbreviate $\sigma := \mathcal{H}^{n-1}\lfloor\partial\Omega$ and let ν denote the geometric measure theoretic outward unit normal to Ω. Suppose L is a weakly elliptic, second-order, homogeneous, constant complex coefficient, $M\times M$ system in $\mathbb{R}^n$. Now, fix an integrability exponent $q \in (1,\infty)$ and a Muckenhoupt weight $\omega \in A_q(\partial\Omega,\sigma)$. Choose $\varepsilon \in (0, q-1)$ such that $\omega \in A_{p-\varepsilon}(\partial\Omega,\sigma)$ and set a threshold $\lambda \in (0, \varepsilon(n-1)/(q-1))$. Finally, consider a function $u : \Omega \to \mathbb{C}^M$, which satisfies*

$$u \in [\mathscr{C}^\infty(\Omega)]^M \quad \text{and} \quad Lu = 0 \quad \text{in } \Omega. \tag{11.100}$$

Then for any fixed aperture parameter $\kappa \in (0,\infty)$ and coefficient tensor $A \in \mathfrak{A}_L$, if $\mathcal{N}_\kappa(\nabla u)$ belongs to $\mathcal{B}^{q,\lambda}(\partial\Omega,\omega)$, it follows that

$$\begin{aligned} u\big|^{\kappa-{\rm n.t.}}_{\partial\Omega} \quad & \text{exists } \sigma\text{-a. e. on } \partial\Omega \text{ and } \partial_{\tau_{jk}}\big(u\big|^{\kappa-{\rm n.t.}}_{\partial\Omega}\big) \in [\mathcal{B}^{q,\lambda}(\partial\Omega,\omega)]^M \\ & \text{for each } j,k \in \{1,\dots,n\}, \text{ and} \\ (\nabla u)\big|^{\kappa-{\rm n.t.}}_{\partial\Omega} \quad & \text{exists } \sigma\text{-a. e. on } \partial\Omega \text{ and } \partial^A_\nu u \in [\mathcal{B}^{q,\lambda}(\partial\Omega,\omega)]^M. \end{aligned} \tag{11.101}$$

Proof. The proof proceeds along the same argument as that of Theorem 11.20, but replacing $M^{p,\lambda}(\partial\Omega, w)$ with $\mathcal{B}^{q,\lambda}(\partial\Omega, \omega)$ and using Proposition 6.17 and Proposition 6.14 in place of Proposition 4.16 and Proposition 4.7, respectively. □

We now present a well-posedness result for the Neumann problem on rough domains with boundary datum arbitrarily prescribed in a Muckenhoupt-weighted Morrey space, as well as in a Muckenhoupt-weighted vanishing Morrey space. Notably, because of the functional analytic setting, this result is new even in the case of the Laplacian in the upper half-space.

Theorem 11.22 (The weighted Morrey Neumann BVP). *Choose some $n \in \mathbb{N}$ with $n \geq 2$ and $M \in \mathbb{N}$. Let $\Omega \subseteq \mathbb{R}^n$ be an* AR *domain, write $\sigma := \mathcal{H}^{n-1}\lfloor\partial\Omega$, and denote by ν the geometric measure theoretic outward unit normal to Ω. Suppose L is a weakly elliptic, second-order, homogeneous, constant complex coefficient, $M \times M$ system in $\mathbb{R}^n$. Next, fix any two integrability exponents $p, q \in (1,\infty)$, an arbitrary Muckenhoupt weight $w \in A_p(\partial\Omega,\sigma) \cap \mathrm{RH}_q(\partial\Omega,\sigma)$, and a threshold $\lambda \in (0,(n-1)/q')$, where $q' = q/(q-1)$. Finally, fix an aperture parameter $\kappa \in (0,\infty)$ and a coefficient tensor $A \in \mathfrak{A}_L$, and consider the Neumann problem for L in Ω with data in Muckenhoupt-weighted Morrey spaces, formulated below:*

$$\begin{cases} u \in [\mathscr{C}^\infty(\Omega)]^M, \\ Lu = 0 \quad \text{in } \Omega, \\ \mathcal{N}_\kappa(\nabla u) \in M^{p,\lambda}(\partial\Omega, w), \\ \partial_\nu^A u = f \in [M^{p,\lambda}(\partial\Omega, w)]^M. \end{cases} \tag{11.102}$$

Then there exists $\delta \in (0,1)$, depending on the ADR *character of $\partial\Omega$, n, p, q, λ, $[w]_{A_p}$, $[w]_{\mathrm{RH}_q}$, and L, such that whenever*

$$\|\nu\|_{[\mathrm{BMO}(\partial\Omega,\sigma)]^n} < \delta, \tag{11.103}$$

a scenario which renders Ω a δ-AR domain, it follows that Ω is actually a UR *domain and all the claims below are true:*

(a) [Existence, estimates, and integral representation]. *If $\mathfrak{A}_{L^\top}^{\mathrm{dis}} \neq \varnothing$ and $A \in \mathfrak{A}_L$ is such that $A^\top \in \mathfrak{A}_{L^\top}^{\mathrm{dis}}$, then (with $K_{A^\top}^\#$ associated to $A^\top$ as in* (3.20)*) the operator*

$$-\frac{1}{2}I + K_{A^\top}^\# : [M^{p,\lambda}(\partial\Omega, w)]^M \to [M^{p,\lambda}(\partial\Omega, w)]^M \tag{11.104}$$

is an isomorphism and the function $u : \Omega \to \mathbb{C}^M$, defined via

$$u(x) := \left(\mathscr{S}_{\mathrm{mod}}\left[\left(-\frac{1}{2}I + K_{A^\top}^\#\right)^{-1} f\right]\right)(x) \quad \text{at each } x \in \Omega \tag{11.105}$$

(where $\mathscr{S}_{\mathrm{mod}}$ is as in (3.17)*), solves the Neumann problem* (11.102)*, and satisfies the estimate*

$$\|\mathcal{N}_\kappa(\nabla u)\|_{M^{p,\lambda}(\partial\Omega,w)} \leq C\|f\|_{[M^{p,\lambda}(\partial\Omega,w)]^M} \tag{11.106}$$

for some constant $C \in (0,\infty)$*, depending only on the* ADR *character of* $\partial\Omega$*,* n*,* p*,* q*,* λ*,* $[w]_{A_p}$*,* $[w]_{\mathrm{RH}_q}$*,* L*, and* κ*.*

(b) [Uniqueness (modulo constants)]. *If* $\mathfrak{A}_L^{\mathrm{dis}} \neq \varnothing$*, then any two solutions to the Neumann Problem* (11.102) *differ only up to some constant in* $\mathbb{C}^M$*.*

(c) [Well posedness]. *If both* $\mathfrak{A}_L^{\mathrm{dis}} \neq \varnothing$ *and* $\mathfrak{A}_{L^\top}^{\mathrm{dis}} \neq \varnothing$*, then the Neumann problem* (11.102) *is well posed in the sense of Hadamard. That is,* (11.102) *is solvable with any two solutions differing by only a constant (in* $\mathbb{C}^M$*), and any solution satisfies the size estimate in* (11.106)*.*

(d) [Sharpness]. *If* $\mathfrak{A}_{L^\top}^{\mathrm{dis}} = \varnothing$*, then the Neumann problem* (11.102) *may fail to be solvable. In fact, solvability may fail for boundary data belonging to an infinite-dimensional subspace of* $[M^{p,\lambda}(\partial\Omega, w)]^M$ *even when* $\Omega = \mathbb{R}^n_+$ *(in which case the* BMO *seminorm of* ν *is zero). Also, if* $\mathfrak{A}_L^{\mathrm{dis}} = \varnothing$*, then the Neumann problem* (11.102) *may have more than one solution, and in fact the vector space of null solutions may be infinite-dimensional.*

(e) [Weighted vanishing Morrey spaces]. *Analogous results to those found in items (a)–(d) continue to hold when* $M^{p,\lambda}(\partial\Omega, w)$ *is replaced by the Muckenhoupt-weighted vanishing Morrey space,* $\mathring{M}^{p,\lambda}(\partial\Omega, w)$*, i. e., concerning the Neumann problem below:*

$$\begin{cases} u \in [\mathscr{C}^\infty(\Omega)]^M, \\ Lu = 0 \quad \text{in } \Omega, \\ \mathcal{N}_\kappa(\nabla u) \in \mathring{M}^{p,\lambda}(\partial\Omega, w), \\ \partial_\nu^A u = f \in [\mathring{M}^{p,\lambda}(\partial\Omega, w)]^M. \end{cases} \tag{11.107}$$

Remark 11.23. Thanks to Theorem 11.20, we know that $\partial_\nu^A u$ is well-defined in the context of the Neumann problem (11.102), i. e., $\partial_\nu^A u$ necessarily belongs to $[M^{p,\lambda}(\partial\Omega, w)]^M$ whenever u satisfies the first three lines in (11.102). This continues to be true in the context of Muckenhoupt-weighted vanishing Morrey spaces and, as witnessed by Theorem 11.21, Muckenhoupt-weighted Block spaces.

Remark 11.24. Regarding item (d) in Theorem 11.22, note that in the context of (11.102), that each choice of $A \in \mathfrak{A}_L$ leads to a completely different problem, since the conormal derivative (cf. Definition 3.18) is sensitive to the choice of coefficient tensor. It can happen (cf. [79, Theorem 7.22, pp. 492–493]) that even corresponding to the simple case $n = 2$, $L := \Delta$, $\Omega := \mathbb{R}^2_+$, and $w \equiv 1$, matters may be arranged (namely by writing $\Delta = \operatorname{div} A\nabla$ for a certain choice of $A \notin \mathfrak{A}_\Delta^{\mathrm{dis}} = \{I_{2\times 2}\}$) so that the Neumann problem (11.102) fails even to be Fredholm solvable. A similar phenomenon is also true when changing the space of boundary datum to either weighted vanishing Morrey spaces or weighted Block spaces.

Remark 11.25. Note that similar comments as in Remark 11.10, which elucidate the precise timing of the geometric and analytic quantities appearing in the statement of the Dirichlet problem (Theorem 11.9), are also relevant for Theorem 11.22.

The interested reader may contrast Theorem 11.22 with that of Z. Shen's result in [137, Theorem 1.12], which provides a well-posedness result for the Neumann problem for elliptic systems, which are symmetric, "super strongly" elliptic, with real coefficients, in bounded Lipschitz domains Ω, when the boundary datum belongs to the (unweighted) Block space $\mathcal{B}^{2,\lambda}(\partial\Omega, \mathcal{H}^{n-1}\lfloor\partial\Omega)$ (where λ lies in a suitable subinterval of $(0, n-1)$) and integrates to zero along $\partial\Omega$ with respect to the surface measure. See also [79, Theorem 7.22, pp. 492–493] and [90, Theorem 8.4.16, pp. 827–829], which provide a similar result to that of Theorem 11.22, but corresponding to the special case that $w \equiv 1$, i. e., the unweighted case.

Remark 11.26. One feature of the proof of the well posedness of the Neumann problem (11.102), which will follow shortly is that it is "duality-free." That is to say, when considering boundary datum in a Muckenhoupt-weighted Morrey space, the proof makes no mention of the weighted Block space pre-dual to the weighted Morrey space in question. Likewise, when treating the Neumann problem with data in a Muckenhoupt-weighted Block space in Theorem 11.27, the dual of the given weighted Block space is not an aspect of the proof. This is in contrast to the Dirichlet problem (cf. Theorem 11.9), the homogeneous regularity problem (cf. Theorem 11.32), and the transmission problem (cf. Theorem 11.35). Subsequently, when considering boundary data in a Muckenhoupt-weighted Block space the ranges for the indices present in the definition (cf. Definition 6.7) are less restrictive. Specifically, in the context of Theorem 11.27 the conjugate weight to ω does not need to satisfy any reverse Hölder condition and the parameter ε does not need to be chosen so that the corresponding range for the threshold λ coincides with the range for λ arising in the definition of the Muckenhoupt-weighted Morrey space dual to the weighted Block space in question.

Finally, we are ready to present a proof of Theorem 11.22.

Proof of Theorem 11.22. To treat item (a), we start by using [79, Theorem 2.5, p. 86] to conclude that Ω is a UR domain (with UR character controlled in terms of the ADR character of $\partial\Omega$) whenever $\delta \in (0,1)$ is taken to be sufficiently small relative to the ADR character of $\partial\Omega$. To demonstrate solvability, and eventually nontangential estimates, note that since $[M^{p,\lambda}(\partial\Omega, w)]^M$ is a Banach space (cf. Proposition 4.5), we can appeal to Lemma 11.2 to say that $-\frac{1}{2}I + K^{\#}_{A^\top}$ is invertible (and has bounded inverse by the open mapping theorem) on $[M^{p,\lambda}(\partial\Omega, w)]^M$ provided

$$\|K^{\#}_{A^\top}\|_{[M^{p,\lambda}(\partial\Omega,w)]^M \to [M^{p,\lambda}(\partial\Omega,w)]^M} < \frac{1}{2}. \tag{11.108}$$

Since we are presently assuming $A^\top \in \mathfrak{A}^{\mathrm{dis}}_{L^\top}$, this is indeed achievable thanks to Theorem 9.7, which states that for every $m \in \mathbb{N}$ there exists a constant $C_m \in (0, \infty)$, depending only on m, the ADR character of $\partial\Omega$, n, p, q, λ, $[w]_{\mathrm{RH}_q}$, $[w]_{A_p}$, and L such that

$$\|K^{\#}_{A^\top}\|_{[M^{p,\lambda}(\partial\Omega,w)]^M \to [M^{p,\lambda}(\partial\Omega,w)]^M} \le C_m \|\nu\|^{\langle m\rangle}_{[\mathrm{BMO}(\partial\Omega,\sigma)]^n}. \tag{11.109}$$

As such, bearing in mind Definition 2.37 and (5.10), we can take $\delta \in (0,1)$ sufficiently small so that (11.108) is satisfied whenever Ω is a δ-AR domain. From here on out, we assume that Ω is a δ-AR domain for this choice of δ. Consequently, a solution to (11.102) is given by (11.105). Indeed, since our current geometric assumptions on Ω ensure (11.104) is invertible, we see that $(-\frac{1}{2}I + K^{\#}_{A^{\top}})^{-1}f$ is a well-defined function belonging to $[M^{p,\lambda}(\partial\Omega, w)]^M$, and hence Theorem 9.2 further shows that u as in (11.105) is a well-defined function, which satisfies $u \in [\mathscr{C}^{\infty}(\Omega)]^M$ and $Lu = 0$ in Ω. Additionally, the trace result in Theorem 9.2 shows that $\partial^A_\nu u = f$ at σ-a. e. point on $\partial\Omega$. Lastly, Proposition 9.5 and the boundedness of $-\frac{1}{2}I + K^{\#}_{A^{\top}}$ on $[M^{p,\lambda}(\partial\Omega, w)]^M$ yields

$$\begin{aligned}\|\mathcal{N}_\kappa(\nabla u)\|_{M^{p,\lambda}(\partial\Omega,w)} &\leq C\left\|\left(-\frac{1}{2}I + K^{\#}_{A^{\top}}\right)^{-1} f\right\|_{[M^{p,\lambda}(\partial\Omega,w)]^M} \\ &\leq C\|f\|_{[M^{p,\lambda}(\partial\Omega,w)]^M},\end{aligned} \tag{11.110}$$

establishing the inequality in (11.106). This takes care of all claims made in (a).

Turning now to item (b), to demonstrate uniqueness modulo constants we will show that if u satisfies

$$\begin{cases} u \in [\mathscr{C}^{\infty}(\Omega)]^M, \\ Lu = 0 \quad \text{in } \Omega, \\ \mathcal{N}_\kappa(\nabla u) \in M^{p,\lambda}(\partial\Omega, w), \\ \partial^A_\nu u = 0 \in \mathbb{C}^M, \end{cases} \tag{11.111}$$

then u is necessarily constant on Ω, which given the linearity of L will imply the desired uniqueness result. To this end, we will appeal to the integral representation for derivatives given in Proposition 11.19. Note that $(\nabla u)\big|^{\kappa-\text{n.t.}}_{\partial\Omega}$ exists σ-a. e. on $\partial\Omega$ by Theorem 11.20. Since $\mathcal{N}_\kappa(\nabla u)$ belongs to $M^{p,\lambda}(\partial\Omega, w)$ by assumption, the embedding in Proposition 4.15 ensures $\mathcal{N}_\kappa(\nabla u) \in L^1(\partial\Omega, \frac{\sigma(x)}{1+|x|^{n-1}})$. Therefore, the hypotheses of Proposition 11.19 are satisfied, and for each $\ell \in \{1,\dots n\}$ and $\gamma \in \{1,\dots,M\}$ we find (11.93) holds. But note that the second integral of (11.93) actually contains $\nu_r a^{\alpha\beta}_{rs}(\partial_s u_\beta)\big|^{\kappa-\text{n.t.}}_{\partial\Omega} = (\partial^A_\nu u)_\alpha$, which is identically zero by our assumption in the last line of (11.111). Thus, we are left with

$$\begin{aligned}(\partial_\ell u_\gamma)(x) = \int_{\partial\Omega} a^{\beta\alpha}_{rs}(\partial_r E_{\gamma\beta})(x-y)\{&\nu_\ell(y)((\partial_s u_\alpha)\big|^{\kappa-\text{n.t.}}_{\partial\Omega})(y) \\ &- \nu_s(y)((\partial_\ell u_\alpha)\big|^{\kappa-\text{n.t.}}_{\partial\Omega})(y)\}\, d\sigma(y)\end{aligned} \tag{11.112}$$

for each $x \in \Omega$. Observe in view of Proposition 3.4, that the right-hand side of (11.112) defines a boundary-to-domain integral operator of the type under consideration in Theorem 5.7 (cf. (5.37)). Additionally, since $(\nabla u)\big|^{\kappa-\text{n.t.}}_{\partial\Omega}$ exists σ-a. e. on $\partial\Omega$, (2.18), Proposition 4.7, and Remark 4.4 ensure that $\nu_j(\partial_k u_\beta)\big|^{\kappa-\text{n.t.}}_{\partial\Omega} \in M^{p,\lambda}(\partial\Omega, w)$ for every $j,k \in \{1,\dots,n\}$ and $\beta \in \{1,\dots,M\}$. Taking the nontangential boundary trace of (11.112), we arrive at

$$\begin{aligned}((\partial_\ell u_\gamma)\big|_{\partial\Omega}^{\kappa-\text{n.t.}})(x) &= \frac{1}{2\sqrt{-1}} a_{rs}^{\beta\alpha} \widehat{\partial_r E_{\gamma\beta}}(\nu(x))\{\nu_\ell(x)((\partial_s u_\alpha)\big|_{\partial\Omega}^{\kappa-\text{n.t.}})(x) \\ &\quad - \nu_s(x)((\partial_\ell u_\alpha)\big|_{\partial\Omega}^{\kappa-\text{n.t.}})(x)\}\, d\sigma(y) \\ &\quad + \lim_{\varepsilon\to 0^+} \int\limits_{\substack{y\in\partial\Omega \\ |x-y|>\varepsilon}} a_{rs}^{\beta\alpha}(\partial_r E_{\gamma\beta})(x-y)\{\nu_\ell(y)((\partial_s u_\alpha)\big|_{\partial\Omega}^{\kappa-\text{n.t.}})(y) \\ &\quad - \nu_s(y)((\partial_\ell u_\alpha)\big|_{\partial\Omega}^{\kappa-\text{n.t.}})(y)\}\, d\sigma(y),\end{aligned} \tag{11.113}$$

by making use of (5.39). We then use item (e) of Proposition 3.4 to evaluate

$$\begin{aligned}\frac{1}{2\sqrt{-1}}\widehat{\partial_r E_{\gamma\beta}}(\nu(x)) &= -\frac{\sqrt{-1}\nu_r(x)}{2\sqrt{-1}}[L(\nu(x))]_{\gamma\beta}^{-1} \\ &= -\frac{1}{2}\nu_r(x)[L(\nu(x))]_{\gamma\beta}^{-1}.\end{aligned} \tag{11.114}$$

We also introduce the following piece of suggestive notation:

$$\begin{aligned}&(\partial_{T_{jk}} u_\alpha)(x) := \nu_j(x)((\partial_k u_\alpha)\big|_{\partial\Omega}^{\kappa-\text{n.t.}})(x) - \nu_k(x)((\partial_j u_\alpha)\big|_{\partial\Omega}^{\kappa-\text{n.t.}})(x) \\ &\quad \text{for each } j,k\in\{1,\dots,n\},\ \alpha\in\{1,\dots,M\},\ \text{and } \sigma\text{-a. e. } x\in\partial\Omega.\end{aligned} \tag{11.115}$$

Since $(\partial_{T_{jk}} u_\alpha) = -(\partial_{T_{kj}} u_\alpha)$, we can thus rewrite (11.113) (bearing in mind (11.114)) as

$$\begin{aligned}((\partial_\ell u_\gamma)\big|_{\partial\Omega}^{\kappa-\text{n.t.}})(x) &= \frac{1}{2} a_{rs}^{\beta\alpha}\nu_r(x)[L(\nu(x))]_{\gamma\beta}^{-1}(\partial_{T_{s\ell}} u_\alpha)(x) \\ &\quad + \lim_{\varepsilon\to 0^+} \int\limits_{\substack{y\in\partial\Omega \\ |x-y|>\varepsilon}} a_{rs}^{\beta\alpha}(\partial_r E_{\gamma\beta})(x-y)(\partial_{T_{\ell s}} u_\alpha)(y)\, d\sigma(y).\end{aligned} \tag{11.116}$$

Introducing another index $t\in\{1,\dots,n\}$ and keeping in mind (11.115) and (11.116), we find for σ-a. e. point $x\in\partial\Omega$ that

$$\begin{aligned}(\partial_{T_{t\ell}} u_\gamma)(x) &= \nu_t(x)((\partial_\ell u_\gamma)\big|_{\partial\Omega}^{\kappa-\text{n.t.}})(x) - \nu_\ell(x)((\partial_t u_\gamma)\big|_{\partial\Omega}^{\kappa-\text{n.t.}})(x) \\ &= \frac{1}{2} a_{rs}^{\beta\alpha}\nu_r(x)[L(\nu(x))]_{\gamma\beta}^{-1}\{\nu_t(x)(\partial_{T_{s\ell}} u_\alpha)(x) \\ &\quad - \nu_\ell(x)(\partial_{T_{st}} u_\alpha)(x)\} \\ &\quad + \lim_{\varepsilon\to 0^+} \int\limits_{\substack{y\in\partial\Omega \\ |x-y|>\varepsilon}} a_{rs}^{\beta\alpha}(\partial_r E_{\gamma\beta})(x-y)\{\nu_t(x)(\partial_{T_{\ell s}} u_\alpha)(y) \\ &\quad - \nu_\ell(x)(\partial_{T_{ts}} u_\alpha)(y)\}\, d\sigma(y).\end{aligned} \tag{11.117}$$

We now add and subtract both $\nu_\ell(y)(\partial_{T_{st}} u_\alpha)(y)$ and $\nu_t(y)(\partial_{T_{\ell s}} u_\alpha)(y)$ within the second pair of curly brackets in (11.117) and regroup terms to ultimately yield (in view of (11.115)) the identity

$$
\begin{aligned}
(\partial_{T_{\ell t}} u_\gamma)(x) = {} & \frac{1}{2} a^{\alpha\beta}_{rs} \nu_r(x)\nu_s(x)[L(\nu(x))]^{-1}_{\gamma\beta}(\partial_{T_{\ell t}} u_\alpha)(x) \\
& - \lim_{\varepsilon\to 0^+} \int\limits_{\substack{y\in\partial\Omega \\ |x-y|>\varepsilon}} \nu_s(y) a^{\beta\alpha}_{rs}(\partial_r E_{\gamma\beta})(x-y)(\partial_{T_{\ell t}} u_\alpha)(y)\,d\sigma(y) \\
& + \lim_{\varepsilon\to 0^+} \int\limits_{\substack{y\in\partial\Omega \\ |x-y|>\varepsilon}} (\nu_t(x)-\nu_t(y)) a^{\beta\alpha}_{rs}(\partial_r E_{\gamma\beta})(x-y)(\partial_{T_{es}} u_\alpha)(y)\,d\sigma(y) \\
& + \lim_{\varepsilon\to 0^+} \int\limits_{\substack{y\in\partial\Omega \\ |x-y|>\varepsilon}} (\nu_\ell(x)-\nu_\ell(y)) a^{\beta\alpha}_{rs}(\partial_r E_{\gamma\beta})(x-y)(\partial_{T_{st}} u_\alpha)(y)\,d\sigma(y).
\end{aligned} \tag{11.118}
$$

Note that we have made use of the following formula:

$$
\nu_k(x)(\partial_{T_{ij}} u_\alpha)(x) + \nu_i(x)(\partial_{T_{jk}} u_\alpha)(x) = \nu_j(x)(\partial_{T_{ik}} u_\alpha)(x) \tag{11.119}
$$

at σ-a.e. point $x \in \partial\Omega$, which holds for any three indices $i, j, k \in \{1, \dots, n\}$ and $\alpha \in \{1, \dots, M\}$.

Now, observe that the first term on the right-hand side of (11.118) can be written as

$$
\begin{aligned}
& \frac{1}{2}[L(\nu(x))]_{\beta\alpha}[L(\nu(x))]^{-1}_{\gamma\beta}(\partial_{T_{\ell t}} u_\alpha)(x) \\
& \quad = \frac{1}{2}\delta_{\gamma\alpha}(\partial_{T_{\ell t}} u_\alpha)(x) = \frac{1}{2}(\partial_{T_{\ell t}} u_\alpha)(x),
\end{aligned} \tag{11.120}
$$

which can subsequently be absorbed into the left-hand side of (11.118). Recalling (3.19), it is apparent that the second term in (11.118) is

$$
(K_A(\partial_{T_{\ell t}} u))_\gamma(x) \quad \text{where } \partial_{T_{\ell t}} u := (\partial_{T_{\ell t}} u_\alpha)_{1\le\alpha\le M}. \tag{11.121}
$$

Finally, note that the third and fourth terms in (11.118) are sums of commutators of the form

$$
[M_\nu, T](\partial_T u), \tag{11.122}
$$

where M_ν is the operation of pointwise multiplication by generic scalar components of ν, and T is a principal value singular integral operator of formal convolution type with a suitable matrix-valued integral kernel whose entries are linear combinations of first-order partial derivatives of entries of E with coefficients which are entries from A. Also, $\partial_T u$ is any function of the form $\partial_{\tau_{jk}} u_\alpha$, where $j, k \in \{1, \dots, n\}$ and $\alpha \in \{1, \dots, M\}$. With the observations of (11.120)–(11.122) in hand, (11.118) may be reinterpreted as

$$
\begin{aligned}
& \frac{1}{2}\partial_{T_{\ell t}} u_\gamma = K_A(\partial_{T_{\ell t}} u) + [M_\nu, T](\partial_T u), \\
& \text{at } \sigma\text{-a.e. point on } \partial\Omega, \text{ for each } \ell, t \in \{1, \dots, n\}.
\end{aligned} \tag{11.123}
$$

Since we assume that $A \in \mathfrak{A}_L^{\rm dis}$, we may appeal to Theorem 9.7 to estimate

$$\|K_A(\partial_{T_{t\ell}}u)\|_{[M^{p,\lambda}(\partial\Omega,w)]^M} \le C\|\nu\|^{\langle m\rangle}_{[\mathrm{BMO}(\partial\Omega,\sigma)]^n}\|\partial_{T_{t\ell}}u\|_{[M^{p,\lambda}(\partial\Omega,w)]^M}. \tag{11.124}$$

Note that due to the structure of T specified above, and the fact that the integral kernel of T is odd, and positive homogeneous of degree $1-n$, as ensured by Proposition 3.4, we are in a position to appeal to Theorem 5.18 to estimate

$$\|[M_\nu,T](\partial_T u)\|_{[M^{p,\lambda}(\partial\Omega,w)]^M} \le C\|\nu\|_{[\mathrm{BMO}(\partial\Omega,\sigma)]^n}\sum_{t,\ell=1}^{n}\|\partial_{T_{t\ell}}u\|_{[M^{p,\lambda}(\partial\Omega,w)]^M}. \tag{11.125}$$

for some constant $C \in (0,\infty)$, independent of $\partial_T u$. Since $\|\nu\|_{[\mathrm{BMO}(\partial\Omega,\sigma)]^n}$ is controlled by $\delta \in (0,1)$, the nature of the estimates in (11.124) (bearing in mind Remark 5.10) and (11.125) together with (11.123) show upon summing up in t and ℓ that

$$\frac{1}{2}\sum_{t,\ell=1}^{n}\|\partial_{T_{t\ell}}u_\gamma\|_{[M^{p,\lambda}(\partial\Omega,w)]^M} \le C_\delta\sum_{t,\ell=1}^{n}\|\partial_{T_{t\ell}}u\|_{[M^{p,\lambda}(\partial\Omega,w)]^M}, \tag{11.126}$$

where $C_\delta = o(1)$ as $\delta \to 0^+$. Thus, initially assuming that $\delta \in (0,1)$ is sufficiently small, (11.126) implies it must be the case that

$$\sum_{t,\ell=1}^{n}\|\partial_{T_{t\ell}}u_\gamma\|_{[M^{p,\lambda}(\partial\Omega,w)]^M} \le 0, \tag{11.127}$$

so that necessarily,

$$\partial_{T_{t\ell}}u_\alpha = 0 \quad \text{for each } t,\ell \in \{1,\dots,n\} \text{ and } \alpha \in \{1,\dots,M\}. \tag{11.128}$$

Recalling the notation in (11.115) and the integral representation in (11.112), it follows that

$$\partial_\ell u_\gamma = 0 \quad \text{for each } \ell \in \{1,\dots,n\} \text{ and } \gamma \in \{1,\dots,M\}, \tag{11.129}$$

so that in fact the solution u to (11.111) must be locally constant in Ω. However, thanks to [79, Theorem 2.5, p. 86], we see that for $\delta \in (0,1)$ sufficiently small that Ω is actually a connected set. Ultimately, we find there exists $c \in \mathbb{C}^M$ such that $u \equiv c$ in Ω. This takes care of item (b).

The claim in item (c) follows naturally from (a) and (b). Finally, for the claims made in (d), in view of the Fatou-type result in Theorem 11.20, we may argue analogously as in the proof of to [79, Theorem 6.11, pp. 396–397], which will show that if we write $\Delta = a_{jk}\partial_j\partial_k$ with coefficient tensor $A = (a_{jk})_{1\le j,k\le 2}$ given by

$$A := \begin{pmatrix} 1 & -i \\ i & 1 \end{pmatrix}, \tag{11.130}$$

then

$$\begin{aligned}&\{\partial_\nu^A u : u \in \mathscr{C}^\infty(\mathbb{R}^2_+),\ \Delta u = 0 \text{ in } \mathbb{R}^2_+,\ \mathcal{N}_\kappa(\nabla u) \in M^{p,\lambda}(\mathbb{R}, w)\}\\ &\quad = \{f \in M^{p,\lambda}(\mathbb{R}, w) : Hf = -if\}.\end{aligned} \tag{11.131}$$

This has infinite codimension in $M^{p,\lambda}(\mathbb{R}, w)$, given that $H^2 = -I$ on $M^{p,\lambda}(\mathbb{R}, w)$ (e. g., this fact can be seen from knowing that $H^2 = -I$ on $L^{p_0}(\mathbb{R}, w_0)$ for any $p_0 \in (1,\infty)$ and $w_0 \in A_{p_0}(\mathbb{R}, \mathcal{L}^1)$ coupled with the embedding in Lemma 5.17). Additionally, a similar argument as in the proof of [79, Theorem 6.11, pp. 396–397] will also show that

$$\begin{aligned}&\{u \in \mathscr{C}^\infty(\mathbb{R}^2_+) : \Delta u = 0 \text{ in } \mathbb{R}^2_+\ \mathcal{N}_\kappa(\nabla u) \in M^{p,\lambda}(\mathbb{R}, w),\ \partial_\nu^A u = 0\}\\ &\quad = \{\overline{U} : U \text{ holomorphic in } \mathbb{R}^2_+ \text{ with } \mathcal{N}_\kappa(\nabla U) \in M^{p,\lambda}(\mathbb{R}, w)\},\end{aligned} \tag{11.132}$$

which demonstrates failure of uniqueness in this case since the space on the right-hand side is infinite-dimensional, as can be seen from the fact that the function $v_m : \mathbb{C} \to \mathbb{C}$ given by $v_m(z) := (\bar{z} - i)^{-m}$ for $z \in \mathbb{C}$ belongs to this space for any $m \in \mathbb{N}$ (cf. Lemma 4.12). Note that demonstrating the validity of both (11.131) and (11.132) relies on a suitable version of Cauchy's integral representation formula. For this, we turn to [88, Theorem 1.1.1, pp. 3–4], whose present applicability is ensured by Proposition 4.15.

Lastly, we address (e). We observe that the proof of existence is largely unchanged, thanks to Theorem 10.33. Uniqueness then follows immediately from the case for $M^{p,\lambda}(\partial\Omega, w)$ because $\mathring{M}^{p,\lambda}(\partial\Omega, w)$ is a subspace of $M^{p,\lambda}(\partial\Omega, w)$. This takes care of (a) and (b), and the claim in item (c) then follows naturally. For part (d), we find that arguing similarly as before we can also show that (11.131) and (11.132) hold when $M^{p,\lambda}(\mathbb{R}, w)$ is replaced by $\mathring{M}^{p,\lambda}(\mathbb{R}, w)$. This finally completes the proof of Theorem 11.22. □

At this point, we also present a formulation of the Neumann problem for weakly elliptic systems with data in a Muckenhoupt-weighted Block space.

Theorem 11.27 (The weighted Block Neumann BVP). *Choose $n \in \mathbb{N}$ with $n \geq 2$ and $M \in \mathbb{N}$. Let $\Omega \subseteq \mathbb{R}^n$ be an AR domain, abbreviate $\sigma := \mathcal{H}^{n-1}\lfloor\partial\Omega$, and denote by ν the geometric measure theoretic outward unit normal to Ω. Suppose L is a weakly elliptic, second-order, homogeneous, constant complex coefficient, $M \times M$ system in $\mathbb{R}^n$. Fix an integrability exponent $q \in (1,\infty)$ and an arbitrary Muckenhoupt weight $\omega \in A_q(\partial\Omega, \sigma)$. Choose $\varepsilon \in (0, q-1)$ so that $\omega \in A_{q-\varepsilon}(\partial\Omega, \sigma)$ and select any threshold $\lambda \in (0, \varepsilon(n-1)/(q-1))$. Finally, fix an aperture parameter $\kappa \in (0,\infty)$ and a coefficient tensor $A \in \mathfrak{A}_L$, and consider the Neumann problem for L in Ω with data in a Muckenhoupt-weighted Block space, formulated below:*

$$\begin{cases} u \in [\mathscr{C}^\infty(\Omega)]^M, \\ Lu = 0 \quad \text{in } \Omega, \\ \mathcal{N}_\kappa(\nabla u) \in \mathcal{B}^{q,\lambda}(\partial\Omega, \omega), \\ \partial_\nu^A u = f \in [\mathcal{B}^{q,\lambda}(\partial\Omega, \omega)]^M. \end{cases} \tag{11.133}$$

Then there exists $\delta \in (0,1)$, *depending on the* ADR *character of* $\partial\Omega$, n, q, ε, λ, $[\omega]_{A_{q-\varepsilon}}$, *and* L, *such that whenever*

$$\|\nu\|_{[\mathrm{BMO}(\partial\Omega,\sigma)]^n} < \delta, \tag{11.134}$$

a scenario which renders Ω *a* δ-AR *domain, it follows that* Ω *is actually a* UR *domain and all the claims below are true:*

(a) [Existence, estimates, and integral representation]. *If* $\mathfrak{A}^{\mathrm{dis}}_{L^\top} \neq \varnothing$ *and* $A \in \mathfrak{A}_L$ *is such that* $A^\top \in \mathfrak{A}^{\mathrm{dis}}_{L^\top}$, *then (with* $K^{\#}_{A^\top}$ *associated to* $A^\top$ *as in* (3.20)) *the operator*

$$-\frac{1}{2}I + K^{\#}_{A^\top} : \left[\mathcal{B}^{q,\lambda}(\partial\Omega,\omega)\right]^M \to \left[\mathcal{B}^{q,\lambda}(\partial\Omega,\omega)\right]^M \tag{11.135}$$

is an isomorphism and the function $u : \Omega \to \mathbb{C}^M$, *defined via*

$$u(x) := \left(\mathscr{S}_{\mathrm{mod}}\left[\left(-\frac{1}{2}I + K^{\#}_{A^\top}\right)^{-1} f\right]\right)(x) \quad \text{at each } x \in \Omega \tag{11.136}$$

(where $\mathscr{S}_{\mathrm{mod}}$ *is as in* (3.17)), *solves the Neumann problem* (11.133) *and satisfies the estimate*

$$\|\mathcal{N}_\kappa(\nabla u)\|_{\mathcal{B}^{q,\lambda}(\partial\Omega,\omega)} \leq C\|f\|_{[\mathcal{B}^{q,\lambda}(\partial\Omega,\omega)]^M} \tag{11.137}$$

for some constant $C \in (0,\infty)$, *depending only on the* ADR *character of* $\partial\Omega$, n, q, ε, λ, $[\omega]_{A_{q-\varepsilon}}$, L, *and* κ.

(b) [Uniqueness (modulo constants)]. *If* $\mathfrak{A}^{\mathrm{dis}}_{L} \neq \varnothing$ *then any two solutions to the Neumann problem* (11.133) *differ only up to some constant in* $\mathbb{C}^M$.

(c) [Well posedness]. *If both* $\mathfrak{A}^{\mathrm{dis}}_{L} \neq \varnothing$ *and* $\mathfrak{A}^{\mathrm{dis}}_{L^\top} \neq \varnothing$, *then the Neumann problem* (11.133) *is well posed in the sense of Hadamard. That is,* (11.133) *is solvable with any two solutions differing by only a constant (in* $\mathbb{C}^M$), *and any solution satisfies the size estimate in* (11.137).

(d) [Sharpness]. *If* $\mathfrak{A}^{\mathrm{dis}}_{L^\top} = \varnothing$, *then the Neumann problem* (11.133) *may fail to be solvable. In fact, solvability may fail for boundary data belonging to an infinite- dimensional subspace of* $[\mathcal{B}^{q,\lambda}(\partial\Omega,\omega)]^M$ *even when* $\Omega = \mathbb{R}^n_+$ *(in which case the* BMO *seminorm of* ν *is zero). Also, if* $\mathfrak{A}^{\mathrm{dis}}_{L} = \varnothing$, *then the Neumann problem* (11.133) *may have more than one solution, and in fact the vector space of null solutions may be infinite-dimensional.*

Remark 11.28. Similar comments as in Remark 11.10 regarding the precise timing of Theorem 11.27 continue to hold.

Proof of Theorem 11.27. The general structure of the proof remains similar to that of the proof of Theorem 11.22, but with a few key differences. Notably, we require Theorem 9.23 (and Proposition 6.11 when applying Lemma 11.2) in order to conclude that the operator $-\frac{1}{2}I + K^{\#}_A$ is invertible on $[\mathcal{B}^{q,\lambda}(\partial\Omega,\omega)]^M$ provided $\|\nu\|_{[\mathrm{BMO}(\partial\Omega,\sigma)]^n}$ is sufficiently small. Then

Theorem 9.17 and Proposition 9.20 ensure $u := \mathscr{S}_{\rm mod}[(-\frac{1}{2}I+K_A^{\#})^{-1}f]$ satisfies (11.102) with the appropriate nontangential maximal operator estimates. This takes care of (a). The treatment of (b) follows by now appealing to Theorem 11.21 to say that $(\nabla u)\big|_{\partial\Omega}^{\kappa-{\rm n.t.}}$ exists σ-a. e. on $\partial\Omega$ as well as Proposition 6.17 to ensure the proper membership of $\mathcal{N}_\kappa(\nabla u)$ to $L^1(\partial\Omega, \frac{\sigma(x)}{1+|x|^{n-1}})$ so that we may invoke Proposition 11.19. We also use Theorem 9.23 and Theorem 7.13 when necessary to get the appropriate small-norm estimates, which eventually establish the claim in (b). Item (c) then follows naturally. To establish the sharpness result in (d), we find that the equalities in (11.131) and (11.132) continue to hold when $M^{p,\lambda}(\partial\Omega, w)$ is replaced by $\mathcal{B}^{q,\lambda}(\mathbb{R}, \omega)$. Note that for this to work we use the Fatou-type result in Theorem 11.21. Also, the embedding in Proposition 6.17 (specifically $q_0 = 1$) is relevant for the application of Cauchy's integral representation formula presented in [88, Theorem 1.1.1, pp. 3–4]. □

11.3 Regularity problems on Morrey and Block spaces

In this section, we consider both inhomogeneous and homogeneous versions of the regularity problem with boundary datum in both Muckenhoupt-weighted Morrey and Block spaces.

We begin with the formulation of the inhomogeneous regularity problem for weakly elliptic systems with data in Muckenhoupt-weighted Morrey spaces.

Theorem 11.29 (The weighted Morrey inhomogeneous regularity BVP). *Select $n \in \mathbb{N}$ with $n \geq 2$ and $M \in \mathbb{N}$. Let $\Omega \subseteq \mathbb{R}^n$ be an* AR *domain and abbreviate $\sigma := \mathcal{H}^{n-1}\lfloor\partial\Omega$. Suppose L is a weakly elliptic, second-order, homogeneous, constant complex coefficient, $M\times M$ system in $\mathbb{R}^n$. Next, fix any two integrability exponents $p, q \in (1,\infty)$, an arbitrary Muckenhoupt weight $w \in A_p(\partial\Omega,\sigma)\cap{\rm RH}_q(\partial\Omega,\sigma)$, and a threshold $\lambda \in (0,(n-1)/q')$, where $q' := q/(q-1)$. Finally, for a fixed aperture parameter $\kappa \in (0,\infty)$, consider the inhomogeneous regularity problem for L in Ω with data in Muckenhoupt-weighted Morrey-based Sobolev spaces, formulated below:*

$$\begin{cases} u \in [\mathscr{C}^\infty(\Omega)]^M, \\ Lu = 0 \quad in\ \Omega, \\ \mathcal{N}_\kappa u,\ \mathcal{N}_\kappa(\nabla u) \in M^{p,\lambda}(\partial\Omega, w), \\ u\big|_{\partial\Omega}^{\kappa-{\rm n.t.}} = f \in [M_1^{p,\lambda}(\partial\Omega, w)]^M. \end{cases} \tag{11.138}$$

Then there exists $\delta \in (0,1)$, depending on the ADR *character of $\partial\Omega$, n, p, q, λ, $[w]_{A_p}$, and $[w]_{{\rm RH}_q}$ and L, such that whenever*

$$\|\nu\|_{[{\rm BMO}(\partial\Omega,\sigma)]^n} < \delta, \tag{11.139}$$

*i. e., whenever Ω is a δ-*AR *domain, it follows that Ω is actually a* UR *domain and the following claims are all true:*

(a) [Existence, estimates, and integral representation]. *Suppose* $\mathfrak{A}_L^{\mathrm{dis}} \neq \varnothing$ *and* $A \in \mathfrak{A}_L^{\mathrm{dis}}$, *then (with* K_A *associated to* A *as in* (3.19)) *the operator*

$$\frac{1}{2}I + K_A : \left[M_1^{p,\lambda}(\partial\Omega, w)\right]^M \to \left[M_1^{p,\lambda}(\partial\Omega, w)\right]^M \tag{11.140}$$

is an isomorphism, and the function $u : \Omega \to \mathbb{C}^M$, *defined according to*

$$u(x) := \left(\mathcal{D}_A\left[\left(\frac{1}{2}I + K_A\right)^{-1} f\right]\right)(x) \quad \text{at each } x \in \Omega \tag{11.141}$$

(where $\mathcal{D}_A$ *is associated to* A *as in* (3.18)), *solves the regularity problem* (11.138) *and satisfies the size estimate*

$$\|\mathcal{N}_\kappa u\|_{M^{p,\lambda}(\partial\Omega,w)} + \|\mathcal{N}_\kappa(\nabla u)\|_{M^{p,\lambda}(\partial\Omega,w)} \leq C\|f\|_{[M_1^{p,\lambda}(\partial\Omega,w)]^M}, \tag{11.142}$$

for some constant $C \in (0,\infty)$, *depending exclusively on the* ADR *character of* $\partial\Omega$, n, p, q, λ, $[w]_{A_p}$, $[w]_{\mathrm{RH}_q}$, L, *and* κ. *Also,* $(\nabla u)\big|_{\partial\Omega}^{\kappa-\mathrm{n.t.}}$ *exists* σ*-a. e. on* $\partial\Omega$.

(b) [Uniqueness]. *If* $\mathfrak{A}_{L^\top}^{\mathrm{dis}} \neq \varnothing$, *then the regularity problem* (11.138) *has no more than one solution.*

(c) [Well posedness]. *If* $\mathfrak{A}_L^{\mathrm{dis}} \neq \varnothing$ *and* $\mathfrak{A}_{L^\top}^{\mathrm{dis}} \neq \varnothing$, *then the regularity problem* (11.138) *is well posed in the sense of Hadamard. Specifically,* (11.138) *is uniquely solvable with the naturally accompanying size estimate in* (11.142).

(d) [Sharpness]. *If* $\mathfrak{A}_L^{\mathrm{dis}} = \varnothing$, *then the regularity problem* (11.138) *may fail to be solvable. In fact, solvability may fail for boundary data belonging to an infinite- dimensional subspace of* $[M_1^{p,\lambda}(\partial\Omega, w)]^M$, *even in the simple case* $\Omega = \mathbb{R}^n_+$ *(where the* BMO *seminorm of* ν *is zero). If* $\mathfrak{A}_{L^\top}^{\mathrm{dis}} = \varnothing$, *then the regularity problem* (11.138) *may have more than one solution, and in fact, the linear space of null solutions may be infinite-dimensional.*

(e) [Weighted vanishing Morrey spaces]. *Analogous results to those found in items (a)–(d) continue to hold when* $M^{p,\lambda}(\partial\Omega, w)$ *is replaced by the Muckenhoupt-weighted vanishing Morrey space* $\mathring{M}^{p,\lambda}(\partial\Omega, w)$, *i. e., concerning the following regularity problem:*

$$\begin{cases} u \in [\mathscr{C}^\infty(\Omega)]^M, \\ Lu = 0 \quad \text{in } \Omega, \\ \mathcal{N}_\kappa u,\ \mathcal{N}_\kappa(\nabla u) \in \mathring{M}^{p,\lambda}(\partial\Omega, w), \\ u\big|_{\partial\Omega}^{\kappa-\mathrm{n.t.}} = f \in [\mathring{M}_1^{p,\lambda}(\partial\Omega, w)]^M. \end{cases} \tag{11.143}$$

Note that Theorem 11.29 recovers work done in [79, Theorem 7.20, pp. 488–489] and [90] for the unweighted case, corresponding to $w \equiv 1$.

Proof of Theorem 11.29. We begin by observing in light of [79, Theorem 2.5, p. 86] that when $\delta \in (0,1)$ is sufficiently small (relative to the ADR character of $\partial\Omega$), then Ω is a UR domain with UR character controlled by the ADR character of $\partial\Omega$. Turning to the rest of the claims in (a), note that since $M_1^{p,\lambda}(\partial\Omega, w)$ is a Banach space (cf. Proposition 8.11), we can appeal to Lemma 11.2 to say that $\frac{1}{2}I + K_A$ is invertible on $[M_1^{p,\lambda}(\partial\Omega, w)]^M$ via a Neumann series (with bounded inverse via the open mapping theorem) provided

$$\|K_A\|_{[M_1^{p,\lambda}(\partial\Omega,w)]^M \to [M_1^{p,\lambda}(\partial\Omega,w)]^M} < \frac{1}{2}. \tag{11.144}$$

Thankfully, this is achievable under our present assumption that $A \in \mathfrak{A}_L^{\mathrm{dis}}$, since this permits us to use Theorem 9.25 to say for each $m \in \mathbb{N}$ there exists some constant $C_m \in (0, \infty)$, depending only on m, the ADR character of $\partial\Omega$, n, p, q, λ, $[w]_{\mathrm{RH}_q}$, $[w]_{A_p}$, and L, such that (recalling the notation (5.9)),

$$\|K_A\|_{[M_1^{p,\lambda}(\partial\Omega,w)]^M \to [M_1^{p,\lambda}(\partial\Omega,w)]^M} \le C_m \|\nu\|^{\langle m\rangle}_{[\mathrm{BMO}(\partial\Omega,\sigma)]^n}, \tag{11.145}$$

where ν denotes the geometric measure theoretic outward unit normal to Ω. Subsequently, bearing in mind (2.37) and Remark 5.10, there exists $\delta \in (0,1)$ sufficiently small so as to ensure (11.144) is satisfied. From this point on, we assume that Ω is a δ-AR domain for this choice of δ. Consequently, a solution to (11.138) is given by (11.141). Indeed, the invertibility of (11.140) ensures that $(\frac{1}{2}I + K_A)^{-1} f$ is a function in $[M_1^{p,\lambda}(\partial\Omega, w)]^M$, and thus also belongs to $[M^{p,\lambda}(\partial\Omega, w)]^M$. So, in view of Theorem 9.1, we see that u as in (11.141) is a well-defined function for which $u \in [\mathscr{C}^\infty(\Omega)]^M$ and $Lu = 0$ in Ω. Moreover, Theorem 9.1 shows that the boundary trace in (11.138) is satisfied by this particular choice of u. Next, Proposition 9.3 and the boundedness of $(\frac{1}{2}I + K_A)^{-1}$ on $[M^{p,\lambda}(\partial\Omega, w)]^M$ ensure that $\mathcal{N}_\kappa u \in M^{p,\lambda}(\partial\Omega, w)$ with

$$\begin{aligned} \|\mathcal{N}_\kappa u\|_{M^{p,\lambda}(\partial\Omega,w)} &\le C \left\| \left(\frac{1}{2}I + K_A\right)^{-1} f \right\|_{[M^{p,\lambda}(\partial\Omega,w)]^M} \\ &\le C \left\| \left(\frac{1}{2}I + K_A\right)^{-1} f \right\|_{[M_1^{p,\lambda}(\partial\Omega,w)]^M} \\ &\le C \|f\|_{[M^{p,\lambda}(\partial\Omega,w)]^M}. \end{aligned} \tag{11.146}$$

Similarly, Proposition 9.4 gives that $\mathcal{N}_\kappa(\nabla u) \in M_1^{p,\lambda}(\partial\Omega, w)$ with

$$\begin{aligned} \|\mathcal{N}_\kappa(\nabla u)\|_{M_1^{p,\lambda}(\partial\Omega,w)} &\le C \left\| \left(\frac{1}{2}I + K_A\right)^{-1} f \right\|_{[M_1^{p,\lambda}(\partial\Omega,w)]^M} \\ &\le C \|f\|_{[M_1^{p,\lambda}(\partial\Omega,w)]^M}. \end{aligned} \tag{11.147}$$

Together, (11.146) and (11.147) show (bearing in mind (8.18)) that the estimate in (11.142) is satisfied. Lastly, Proposition 9.4 additionally guarantees the σ-a. e. existence of $(\nabla u)\big|_{\partial\Omega}^{\kappa-\mathrm{n.t.}}$ on $\partial\Omega$. This takes care of all claims pertaining to item (a).

Moving on to the claim made in item (b), uniqueness for (11.138) actually follows from item (b) of Theorem 11.9. Alternatively, we can use the integral representation in Proposition 3.20. Indeed, suppose u satisfies (11.138) with $f = 0$ at σ-a. e. point on $\partial\Omega$. The assumption that $\mathcal{N}_\kappa u \in M^{p,\lambda}(\partial\Omega, w)$ ensures that $\mathcal{N}_\kappa u \in L^1_{\mathrm{loc}}(\partial\Omega, \sigma)$ in view of Proposition 4.16. Also, since $\mathcal{N}_\kappa u \in M^{p,\lambda}(\partial\Omega, w)$, we see that $\mathcal{N}_\kappa(\nabla u) \in L^1(\partial\Omega, \frac{\sigma(x)}{1+|x|^{n-1}})$ thanks to Proposition 4.15. Thus, u satisfies (3.79), which permits us to write (bearing in mind that $u\big|^{\kappa-\mathrm{n.t.}}_{\partial\Omega} = 0$ at σ-a. e. point on $\partial\Omega$)

$$u = -\mathscr{S}_{\mathrm{mod}}(\partial^A_\nu u) + c_u \quad \text{in } \Omega, \tag{11.148}$$

for some $\mathbb{C}^M$-valued locally constant function c_u. Note that $\partial^A_\nu u$ is in $[M^{p,\lambda}(\partial\Omega, w)]^M$ by (2.18), Proposition 4.7, and since $\mathcal{N}_\kappa(\nabla u)$ belongs to $M^{p,\lambda}(\partial\Omega, w)$. Taking the conormal derivative of both sides in (11.148), we obtain thanks to Theorem 9.2,

$$\partial^A_\nu u = -\Big(-\frac{1}{2}I + K^\#_{A^\top}\Big)(\partial^A_\nu u) \quad \text{or equivalently} \quad 0 = \Big(\frac{1}{2}I + K^\#_{A^\top}\Big)(\partial^A_\nu u). \tag{11.149}$$

Assuming that $\mathfrak{A}^{\mathrm{dis}}_{L^\top} \neq \varnothing$ and that $A \in \mathfrak{A}_L$ is such that $A^\top \in \mathfrak{A}^{\mathrm{dis}}_{L^\top}$, we find in view of the geometric estimates concerning $K^\#$ contained in Theorem 9.7 that the operator $\frac{1}{2}I + K^\#_{A^\top}$ is invertible on $[M^{p,\lambda}(\partial\Omega, w)]^M$ via a Neumann series (cf. Lemma 11.2) provided $\|\nu\|_{[\mathrm{BMO}(\partial\Omega,\sigma)]^n}$ is sufficiently small. This is enough to conclude that $\partial^A_\nu u = 0$ on $\partial\Omega$, which feeding back into (11.148) shows that u is locally constant in Ω. However, Ω being a δ-AR domain for sufficiently small $\delta \in (0,1)$ ensures that Ω is actually connected (cf. [79, Theorem 2.5, p. 86]), subsequently implying that u is constant in Ω. The fact that $u\big|^{\kappa-\mathrm{n.t.}}_{\partial\Omega} = 0$ at σ-a. e. point on $\partial\Omega$ then guarantees u is identically zero on Ω, as wanted. Yet a third proof for uniqueness of solutions to (11.138) may be given based on the integral representation formula in Proposition 11.19. See the proof of item (b) in Theorem 11.22 to see how to run such an argument.

The claim in item (c) follows naturally from the contents of (a) and (b). The first claim in (d) may be handled essentially using [79, Proposition 3.11, p. 228], though naturally formulated for $M^{p,\lambda}(\mathbb{R}, w)$. The proof of that result goes through virtually the same, however, we note that in order to appeal to [79, Proposition 2.22, p. 148], we employ Proposition 4.16, and mark the relevance of Proposition 4.7. Additionally, for any $f \in M^{p,\lambda}(\mathbb{R}, w)$ the function

$$U(z) := \frac{1}{2\pi i}\int_{\mathbb{R}} \frac{(f + iHf)(y)}{y - z}\, dy \quad \text{for each } z \in \mathbb{R}^2_+, \tag{11.150}$$

(where H denotes the Hilbert transform on $\mathbb{R}$), which appears in the proof of [79, Proposition 3.11, p. 228], is also seen to be a holomorphic function in $\mathbb{R}^2_+$ with $\mathcal{N}_\kappa U$ and $\mathcal{N}_\kappa(\nabla U)$ belonging to $M^{p,\lambda}(\mathbb{R}, w)$. These facts concerning U follow from the observation that the Cauchy integral operator can be viewed as a boundary-to-domain double layer potential for the Laplacian (cf. [89, Example 1.4.1, pp. 43–44]) along with Theorem 9.1, Proposition 9.3, and Proposition 9.4. We can also compute

$$U\big|_{\partial\mathbb{R}^2_+}^{\kappa-\text{n.t.}} = \mathscr{C}(f + iHf)\big|_{\partial\mathbb{R}^2_+}^{\kappa-\text{n.t.}} = \Big(\frac{1}{2}I + \frac{i}{2}H\Big)(f + iHf) \tag{11.151}$$
$$= \frac{1}{2}f + \frac{i}{2}Hf + \frac{i}{2}Hf - \frac{1}{2}H^2 f = f + iHf,$$

at σ-a. e. point in $\mathbb{R}$, where we used that $H^2 = -I$ on $M^{p,\lambda}(\mathbb{R}, w)$, a fact that can be seen using the inclusion in Lemma 5.17 and that $H^2 = -I$ on $L^p(\mathbb{R}, w)$ for any $p \in (1, \infty)$ and any $w \in A_p(\mathbb{R}, \mathcal{L}^1)$. Note also that $Hf \in M^{p,\lambda}(\mathbb{R}, w)$, a consequence of Proposition 9.6 and viewing the Hilbert transform as a scalar multiple of the boundary-to-boundary Cauchy integral operator, which itself can be interpreted as a boundary-to-boundary double layer potential for the Laplacian (cf. [89, Example 1.4.1, pp. 43–44]). The second claim in item (d) is handled by adapting the argument in [79, Example 3.5, p. 210] to accommodate $M^{p,\lambda}(\partial\Omega, w)$; cf. the proof of item (d) in Theorem 11.9. A similar result in higher dimensions can be obtained by considering the operator $L_D := \Delta - 2\nabla\,\mathrm{div}$, cf. [79, Proposition 3.13, pp. 233–234].

Concerning the treatment of item (e), note that item (a) in the case of $\mathring{M}^{p,\lambda}(\partial\Omega, w)$ can be handled with minimal changes, thanks to Theorem 10.38. We observe that uniqueness holds as a consequence of uniqueness of solutions in the case of $M^{p,\lambda}(\partial\Omega, w)$, because $\mathring{M}^{p,\lambda}(\partial\Omega, w)$ is a subspace of $M^{p,\lambda}(\partial\Omega, w)$. Item (c) in this case then immediately follows. The claims in item (d), formulated for $\mathring{M}^{p,\lambda}(\partial\Omega, w)$, are handled much as before. This finally completes the proof of Theorem 11.29. □

Now we present the inhomogeneous regularity problem formulated for Muckenhoupt-weighted Block-based Sobolev spaces.

Theorem 11.30 (The weighted Block inhomogeneous regularity BVP). *Pick $n \in \mathbb{N}$ with $n \geq 2$ and $M \in \mathbb{N}$. Let $\Omega \subseteq \mathbb{R}^n$ be an AR domain and abbreviate $\sigma := \mathcal{H}^{n-1}\lfloor\partial\Omega$. Suppose L is a weakly elliptic, second-order, homogeneous, constant complex coefficient, $M \times M$ system in $\mathbb{R}^n$. Next, fix an integrability exponent $q \in (1, \infty)$, and an arbitrary Muckenhoupt weight $\omega \in A_q(\partial\Omega, \sigma)$. Choose a parameter $\varepsilon \in (0, q-1)$ such that ω belongs to $A_{q-\varepsilon}(\partial\Omega, \sigma)$ and select a threshold $\lambda \in (0, \varepsilon(n-1)/(q-1))$. Finally, for a fixed aperture parameter $\kappa \in (0, \infty)$, consider the inhomogeneous regularity problem for L in Ω with data in a Muckenhoupt-weighted Block-based Sobolev space, formulated below:*

$$\begin{cases} u \in [\mathscr{C}^\infty(\Omega)]^M, \\ Lu = 0 \quad \text{in } \Omega, \\ \mathcal{N}_\kappa u,\ \mathcal{N}_\kappa(\nabla u) \in \mathcal{B}^{q,\lambda}(\partial\Omega, \omega), \\ u\big|_{\partial\Omega}^{\kappa-\text{n.t.}} = f \in [\mathcal{B}_1^{q,\lambda}(\partial\Omega, \omega)]^M. \end{cases} \tag{11.152}$$

Then there exists $\delta \in (0,1)$, depending on the ADR character of $\partial\Omega$, n, q, ε, λ, and $[\omega]_{A_{q-\varepsilon}}$ and L, such that whenever

$$\|\nu\|_{[\mathrm{BMO}(\partial\Omega,\sigma)]^n} < \delta, \tag{11.153}$$

i. e., whenever Ω *is a* δ-AR *domain, it follows that* Ω *is actually a* UR *domain and the following claims are all true:*

(a) [Existence, estimates, and integral representation]. *Suppose* $\mathfrak{A}_L^{\mathrm{dis}} \neq \varnothing$ *and* $A \in \mathfrak{A}_L^{\mathrm{dis}}$, *then (with* K_A *associated to* A *as in* (3.19)) *the operator*

$$\frac{1}{2}I + K_A : [\mathcal{B}_1^{q,\lambda}(\partial\Omega, \omega)]^M \to [\mathcal{B}_1^{q,\lambda}(\partial\Omega, \omega)]^M \tag{11.154}$$

is an isomorphism, and the function $u : \Omega \to \mathbb{C}^M$, *defined according to*

$$u(x) := \left(\mathcal{D}_A\left[\left(\frac{1}{2}I + K_A\right)^{-1} f\right]\right)(x) \quad \textit{at each } x \in \Omega \tag{11.155}$$

(where $\mathcal{D}_A$ *is associated to* A *as in* (3.18)), *solves the regularity problem* (11.152) *and satisfies the size estimate*

$$\|\mathcal{N}_\kappa u\|_{\mathcal{B}^{q,\lambda}(\partial\Omega,\omega)} + \|\mathcal{N}_\kappa(\nabla u)\|_{\mathcal{B}^{q,\lambda}(\partial\Omega,\omega)} \leq C\|f\|_{[\mathcal{B}_1^{q,\lambda}(\partial\Omega,\omega)]^M}, \tag{11.156}$$

for some constant $C \in (0,\infty)$, *depending exclusively on the* ADR *character of* $\partial\Omega$, n, q, ε, λ, $[\omega]_{A_{q-\varepsilon}}$, L, *and* κ. *Also,* $(\nabla u)\big|_{\partial\Omega}^{\kappa-\mathrm{n.t.}}$ *exists* σ*-a. e. on* $\partial\Omega$.

(b) [Uniqueness]. *If* $\mathfrak{A}_{L^\top}^{\mathrm{dis}} \neq \varnothing$, *then the regularity problem* (11.152) *has no more than one solution.*

(c) [Well posedness]. *If* $\mathfrak{A}_L^{\mathrm{dis}} \neq \varnothing$ *and* $\mathfrak{A}_{L^\top}^{\mathrm{dis}} \neq \varnothing$, *then the regularity problem* (11.152) *is well posed in the sense of Hadamard. Specifically,* (11.152) *is uniquely solvable with the naturally accompanying size estimate in* (11.156).

(d) [Sharpness]. *If* $\mathfrak{A}_L^{\mathrm{dis}} = \varnothing$, *then the regularity problem* (11.152) *may fail to be solvable. In fact, solvability may fail for boundary data belonging to an infinite- dimensional subspace of* $[\mathcal{B}_1^{q,\lambda}(\partial\Omega, \omega)]^M$, *even in the simple case* $\Omega = \mathbb{R}_+^n$ *(where the* BMO *seminorm of* v *is zero). If* $\mathfrak{A}_{L^\top}^{\mathrm{dis}} = \varnothing$, *then the regularity problem* (11.152) *may have more than one solution, and in fact, the linear space of null solutions may be infinite-dimensional.*

Remark 11.31. Contrasting the above result with the Dirichlet problem in Theorem 11.12, the proofs of existence and uniqueness that we provide make no use of the condition that the conjugate weight $\omega' := \omega^{-q'/q}$ belongs to the reverse Hölder class $\mathrm{RH}_p(\partial\Omega, \sigma)$ for some integrability exponent $p \in (1, \infty)$ (see also Remark 11.26 in this regard). This is why Theorem 11.30 holds with data in a Muckenhoupt-weighted Block space, where the ranges for the indices that define such a Block space (cf. Definition 6.7) are less restrictive than what is present in Theorem 11.12.

Proof of Theorem 11.30. We use similar techniques as in the proof of Theorem 11.29, but make the following key alterations. First, to treat (a), we rely on Theorem 9.25 as well as Proposition 8.7 in our application of Lemma 11.2 to say that the operator $\frac{1}{2}I + K_A$ is invertible on $[\mathcal{B}_1^{q,\lambda}(\partial\Omega, \omega)]^M$ provided $\|v\|_{[\mathrm{BMO}(\partial\Omega,\sigma)]^n} < \delta$ and δ is sufficiently small. Then Theorem 9.16, Proposition 9.18, and Proposition 9.19 ensure that the function $u := \mathcal{D}_A[(\frac{1}{2}I + K_A)^{-1}f]$ on Ω satisfies (11.152) with appropriate nontangential maximal operator estimates as in (11.156). To finish the treatment of (a), we again appeal to Proposition 9.19 now to conclude that $(\nabla u)\big|_{\partial\Omega}^{\kappa-n.t.}$ exists σ-a. e. on $\partial\Omega$. The claim in item (b) is handled much as before using the integral representation formula in Proposition 3.20, though now we find Proposition 6.17, Theorem 9.17, Theorem 9.23, and Proposition 6.14 of particular relevance in this case. Item (c) then follows naturally, and the claim in item (d) can be handled in a similar way as above, and we note the relevance of Theorem 9.16, Proposition 9.18, Proposition 9.24, and Proposition 9.21 in adapting the argument. Note that the fact that $H^2 = -I$ on $\mathcal{B}^{q,\lambda}(\mathbb{R},\omega)$ can be seen from recalling that $H^2 = -I$ on $L^q(\partial\Omega,\omega)$ coupled with the continuity of H on $\mathcal{B}^{q,\lambda}(\mathbb{R},\omega)$ (cf. Corollary 9.22) and the density of $L^q(\mathbb{R},\omega)\cap\mathcal{B}^{q,\lambda}(\mathbb{R},\omega)$ in $\mathcal{B}^{q,\lambda}(\mathbb{R},\omega)$, a fact witnessed by Proposition 6.24 and the observation that $L^q(\mathbb{R},\omega)$ contains all finite linear combinations of (q,λ,ω)-blocks. This completes the treatment of all claims in Theorem 11.30. □

We would now like to treat a homogeneous version of the regularity problem formulated in (11.138).

Theorem 11.32 (The weighted Morrey homogeneous regularity BVP). *Pick $n \in \mathbb{N}$ with $n \geq 2$ and $M \in \mathbb{N}$. Let $\Omega \subseteq \mathbb{R}^n$ be an* AR *domain and write $\sigma := \mathcal{H}^{n-1}\lfloor\partial\Omega$. Suppose L is a weakly elliptic, second-order, homogeneous, constant complex coefficient, $M\times M$ system in $\mathbb{R}^n$. Fix two integrability exponents $p,q \in (1,\infty)$, an arbitrary Muckenhoupt weight $w \in A_p(\partial\Omega,\sigma)\cap \mathrm{RH}_q(\partial\Omega,\sigma)$, and a threshold $\lambda \in (0,(n-1)/q')$, where q' is the Hölder conjugate of q. For a fixed aperture parameter $\kappa \in (0,\infty)$, consider the homogeneous regularity problem for L in Ω with data belonging to homogeneous Muckenhoupt-weighted Morrey-based Sobolev spaces, formulated below:*

$$\begin{cases} u \in [\mathscr{C}^\infty(\Omega)]^M, \\ Lu = 0 \quad \text{in } \Omega, \\ \mathcal{N}_\kappa(\nabla u) \in M^{p,\lambda}(\partial\Omega, w), \\ u\big|_{\partial\Omega}^{\kappa-\text{n.t.}} = f \in [\dot{M}_1^{p,\lambda}(\partial\Omega,w)]^M. \end{cases} \tag{11.157}$$

Then there exists $\delta \in (0,1)$ depending on the ADR *character of $\partial\Omega$, n, p, q, λ, $[w]_{\mathrm{RH}_q}$, $[w]_{A_p}$, and L such that whenever*

$$\|\nu\|_{[\mathrm{BMO}(\partial\Omega,\sigma)]^n} < \delta, \tag{11.158}$$

*a scenario which renders Ω a δ-*AR *domain; it follows that Ω is actually a* UR *domain and the following claims are true:*

(a) [Existence, integral representation, and estimates]. *If $\mathfrak{A}_L^{\mathrm{dis}} \neq \varnothing$, then (recalling (3.76)) the operator*

$$[S_{\mathrm{mod}}] : [M^{p,\lambda}(\partial\Omega,w)]^M \longrightarrow [\dot{M}_1^{p,\lambda}(\partial\Omega,w)/\sim]^M \tag{11.159}$$

is surjective. Moreover, the regularity problem (11.157) *is solvable with the function* $u : \Omega \to \mathbb{C}^M$*, defined by*

$$u(x) := (\mathscr{S}_{\text{mod}} g)(x) + c \quad \text{at each } x \in \Omega, \tag{11.160}$$

where $c \in \mathbb{C}^M$ *is a suitable constant and* $g \in [M^{p,\lambda}(\partial\Omega, w)]^M$ *is such that* $[S_{\text{mod}}]g = [f] \in [\dot{M}_1^{p,\lambda}(\partial\Omega, w)/\sim]^M$*. Additionally, the solution* (11.160) *satisfies*

$$\|\mathcal{N}_\kappa(\nabla u)\|_{M^{p,\lambda}(\partial\Omega,w)} \approx \|\nabla_{\tan} f\|_{[M^{p,\lambda}(\partial\Omega,w)]^{n \cdot M}} \tag{11.161}$$

with implicit constants independent of the boundary datum, f.

Alternatively, corresponding to the case that $A \in \mathfrak{A}_L^{\text{dis}}$*, and with* $K_{A,\text{mod}}$ *as in* (3.47)*, the operator*

$$\frac{1}{2}I + [K_{A,\text{mod}}] : [\dot{M}_1^{p,\lambda}(\partial\Omega, w)/\sim]^M \longmapsto [\dot{M}_1^{p,\lambda}(\partial\Omega, w)/\sim]^M \tag{11.162}$$

is invertible. Moreover, the regularity problem (11.157) *may be solved with the function* $u : \Omega \to \mathbb{C}^M$*, defined (recall* (3.34)*) according to*

$$u(x) := (\mathcal{D}_{A,\text{mod}} g)(x) + c \quad \text{at each } x \in \Omega, \tag{11.163}$$

where $c \in \mathbb{C}^M$ *is a suitable constant and* $g \in [\dot{M}_1^{p,\lambda}(\partial\Omega, w)/\sim]^M$ *is such that*

$$[g] = \left(\frac{1}{2}I + [K_{A,\text{mod}}]\right)^{-1} [f]. \tag{11.164}$$

Additionally, the solution in (11.163) *continues to satisfy the size estimates recorded in* (11.161).

(b) [Uniqueness modulo constants]. *If* $\mathfrak{A}_{L^\top}^{\text{dis}} \neq \varnothing$*, then any two solutions of* (11.157) *differ by only a constant.*

(c) [Well posedness]. *Whenever both* $\mathfrak{A}_L^{\text{dis}} \neq \varnothing$ *and* $\mathfrak{A}_{L^\top}^{\text{dis}} \neq \varnothing$*, then the regularity problem* (11.157) *is well posed in the sense of Hadamard. That is, there exists a unique solution to* (11.157)*, which also satisfies the size estimates in* (11.161).

(d) [Sharpness]. *If* $\mathfrak{A}_L^{\text{dis}} = \varnothing$*, then* (11.157) *may not be solvable. In fact, solvability may fail for boundary data belonging to an infinite-dimensional subspace of* $[\dot{M}_1^{p,\lambda}(\partial\Omega, w)]^M$*, even when* $\Omega = \mathbb{R}_+^n$ *(a setting in which the* BMO *seminorm of* ν *is zero). Also, if* $\mathfrak{A}_{L^\top}^{\text{dis}} = \varnothing$*, then there may be more than one solution to* (11.157)*, and it may be the case that the linear space of null solutions is infinite-dimensional.*

(e) [Weighted vanishing Morrey spaces]. *Similar statements as in items (a)–(d) continue to hold when* $\dot{M}_1^{p,\lambda}(\partial\Omega, w)$ *is replaced by the homogeneous Muckenhoupt-weighted vanishing Morrey-based Sobolev space* $\mathbb{M}_1^{p,\lambda}(\partial\Omega, w)$*. That is, for the regularity problem below:*

$$\begin{cases} u \in [\mathscr{C}^\infty(\Omega)]^M, \\ Lu = 0 \quad in\ \Omega, \\ \mathcal{N}_\kappa(\nabla u) \in \mathring{M}^{p,\lambda}(\partial\Omega, w), \\ u\big|_{\partial\Omega}^{\kappa-\text{n.t.}} = f \in [\mathbb{M}_1^{p,\lambda}(\partial\Omega, w)]^M. \end{cases} \tag{11.165}$$

Theorem 11.32 may be regarded as an extension of the result in [79, Theorem 7.21, pp. 489–491], which establishes a well-posedness result in the case that $w \equiv 1$.

A similar comment as in Remark 11.31 continues to hold in the context of Theorem 11.32.

We next present the proof of Theorem 11.32.

Proof of Theorem 11.32. To treat item (a), we assume that $\mathfrak{A}_L^{\text{dis}} \neq \varnothing$. Item (2) of Theorem 9.36 ensures the existence of some $\delta \in (0,1)$ (whose nature is specified in the statement of the theorem), such that if $\|\nu\|_{[\text{BMO}(\partial\Omega,\sigma)]} < \delta$, then the operator (11.159) is surjective with quantitative control. That is, there exists $g \in [M^{p,\lambda}(\partial\Omega, w)]^M$ such that $[S_{\text{mod}}]g = f$ with

$$\|g\|_{[M^{p,\lambda}(\partial\Omega,w)]^M} \leq C\|\nabla_{\tan} f\|_{[M^{p,\lambda}(\partial\Omega,w)]^{n\cdot M}} \tag{11.166}$$

for some constant C, independent of f. Note that because

$$[S_{\text{mod}}]g = [S_{\text{mod}}g] = [f], \tag{11.167}$$

we find that $c := f - S_{\text{mod}}g \in \mathbb{C}^M$ is a constant (note here we are also using [79, Theorem 2.5, p. 86], which guarantees Ω is connected provided δ is sufficiently small relative to the ADR character of $\partial\Omega$). Now, assuming that Ω is a δ-AR domain for such a choice of δ, we can further glean from [79, Theorem 2.5, p. 86] that Ω is a UR domain. At this point, define $u : \Omega \to \mathbb{C}^M$ as in (11.160) for c as above. To see that this u satisfies (11.157), start by observing that Theorem 9.2 ensures that $u \in [\mathscr{C}^\infty(\Omega)]^M$ with $Lu = 0$ in Ω. Next, we have by Proposition 9.5 that $\mathcal{N}_\kappa(\nabla u)$ belongs to $M^{p,\lambda}(\partial\Omega, w)$. Lastly, the embedding in Proposition 4.15, and (3.78), ensure that $u\big|_{\partial\Omega}^{\kappa-\text{n.t.}} = S_{\text{mod}}g + c = f$ as required. Concerning the quantitative equivalence in (11.161), notice that for any $j, k \in \{1, \dots, n\}$, we may use [79, Proposition 2.22, p. 148] (whose present applicability in ensured by Proposition 4.16) to write

$$|\partial_{\tau_{jk}} f| = \left|\partial_{\tau_{jk}}\big(u\big|_{\partial\Omega}^{\kappa-\text{n.t.}}\big)\right| \leq 2\mathcal{N}_\kappa(\nabla u). \tag{11.168}$$

Subsequently, thanks to (8.2) and Proposition 4.7, the right-to-left inequality in (11.161) follows. The left-to-right inequality in (11.161) follows from (11.160), Proposition 9.5, and (11.166).

Moving on, assume that $A \in \mathfrak{A}_L^{\text{dis}}$, so by Theorem 9.15, we have for each $m \in \mathbb{N}$ that (cf. Definition 5.9)

$$\big\|[K_{A,\text{mod}}]\big\|_{[\dot{M}_1^{p,\lambda}(\partial\Omega,w)/\sim]^M \to [\dot{M}_1^{p,\lambda}(\partial\Omega,w)/\sim]^M} < C_m \|\nu\|_{[\text{BMO}(\partial\Omega,\sigma)]^n}^{\langle m\rangle} \tag{11.169}$$

for some constant $C_m \in (0,\infty)$ depending on m, the ADR character of $\partial\Omega$, n, p, q, λ, $[w]_{A_p}$, $[w]_{\mathrm{RH}_q}$, and L. Thus, assuming that Ω is a δ-AR domain (which entails $\|\nu\|_{[\mathrm{BMO}(\partial\Omega,\sigma)]^n} < \delta$) for sufficiently small $\delta \in (0,1)$, potentially smaller than what was chosen to ensure Ω is a UR domain, we find that we can force the right-hand side of (11.169) to be less than $\frac{1}{2}$. As such, the operator in (11.162) is invertible via a Neumann series (cf. Lemma 11.2), since $[M^{p,\lambda}(\partial\Omega, w)/\sim]^M$ is a Banach space, a fact recorded in Proposition 8.11. Since the defining property of g in (11.164) entails

$$[f] = \Big(\frac{1}{2}I + [K_{A,\mathrm{mod}}]\Big)[g] = \Big[\Big(\frac{1}{2}I + K_{A,\mathrm{mod}}\Big)g\Big], \tag{11.170}$$

we find (bearing in mind the connectivity of Ω) that $c := f - (\frac{1}{2}I + K_{A,\mathrm{mod}})g$ is a constant in $\mathbb{C}^M$. Now, the fact that u as defined in (11.163) for this choice of c satisfies (11.157) is a consequence of Theorem 9.12. Regarding the size condition, note that the quality of (11.162) being an isomorphism ensures that g belongs to $[\dot{M}_1^{p,\lambda}(\partial\Omega, w)]^M$, and hence (9.55) and Proposition 4.7 imply the desired membership of $\mathcal{N}_\kappa(\nabla u)$ to $M^{p,\lambda}(\partial\Omega, w)$. To see that this choice of u also satisfies the quantitative equivalence in (11.161), start by observing thanks to Proposition 4.7 that

$$\begin{aligned}\|\mathcal{N}_\kappa(\nabla u)\|_{M^{p,\lambda}(\partial\Omega,w)} &\le \|[g]\|_{[\dot{M}_1^{p,\lambda}(\partial\Omega,w)/\sim]^M} \\ &= \Big\|\Big(\frac{1}{2}I + [K_{A,\mathrm{mod}}]\Big)^{-1}[f]\Big\|_{[\dot{M}_1^{p,\lambda}(\partial\Omega,w)/\sim]^M} \\ &\le C\|[f]\|_{[\dot{M}_1^{p,\lambda}(\partial\Omega,w)/\sim]^M} \le C\|\nabla_{\tan} f\|_{[M^{p,\lambda}(\partial\Omega,w)]^{n\cdot M}}.\end{aligned} \tag{11.171}$$

The first inequality is due to (9.55), (8.21), and (8.22), the equality is due to (11.164), the penultimate inequality is a consequence of the fact that (11.162) has a bounded inverse (by the open mapping theorem), and the final inequality is due to (8.21) and (8.8), whose present applicability is guaranteed by the geometry of Ω, (8.19), and Proposition 4.16. This gives the left-to-right inequality in (11.161). For the right-to-left inequality in (11.161), notice that

$$\begin{aligned}\|\nabla_{\tan} f\|_{[M^{p,\lambda}(\partial\Omega,w)]^{n\cdot M}} &\le C\|f\|_{[\dot{M}_1^{p,\lambda}(\partial\Omega,w)]^M} \\ &= C\sum_{j,k=1}^{n} \|\partial_{\tau_{jk}}(u\big|_{\partial\Omega}^{\kappa-\mathrm{n.t.}})\|_{M^{p,\lambda}(\partial\Omega,w)} \\ &\le C\|\mathcal{N}_\kappa(\nabla u)\|_{M^{p,\lambda}(\partial\Omega,w)}.\end{aligned} \tag{11.172}$$

Note that Proposition 4.7 is used throughout the above string of inequalities. Additionally, the first inequality follows from (8.2) and (8.21), while the final inequality is a consequence of [79, Proposition 2.22, p. 148], which provides the necessary bounds on tangential derivatives of the boundary trace of u in terms of the nontangential maximal operator acting on the gradient of u. This finishes the justification of all claims made in item (a).

Turning our attention to the question of uniqueness addressed in (b), suppose that $u : \Omega \to \mathbb{C}^M$ satisfies

$$\begin{cases} u \in [\mathscr{C}^\infty(\Omega)]^M, \\ Lu = 0 \quad \text{in } \Omega, \\ \mathcal{N}_\kappa(\nabla u) \in M^{p,\lambda}(\partial\Omega, w), \\ u\big|_{\partial\Omega}^{\kappa-\text{n.t.}} = 0. \end{cases} \tag{11.173}$$

Our goal is to show that such a u must be constant. Since we are presently assuming $\mathfrak{A}^{\text{dis}}_{L^\top} \neq \varnothing$, choose some $A \in \mathfrak{A}_L$ for which $A^\top \in \mathfrak{A}^{\text{dis}}_{L^\top}$. Also, select $\delta \in (0,1)$ small enough (relative to the ADR character of Ω) so that Ω being a δ-AR domain entails Ω is a two-sided NTA domain with unbounded boundary. That this is achievable is due to [79, Theorem 2.3, pp. 64–65]. Then Theorem 11.20 applies, and we have that $(\nabla u)\big|_{\partial\Omega}^{\kappa-\text{n.t.}}$ exists σ-a. e. on $\partial\Omega$ and $\partial^A_\nu u$ belongs to $[M^{p,\lambda}(\partial\Omega, w)]^M$. Appealing to the integral representation formula in (3.80) of Proposition 3.20, keeping in mind our present assumption that $u\big|_{\partial\Omega}^{\kappa-\text{n.t.}} = 0$, we have

$$u = -\mathscr{S}_{\text{mod}}(\partial^A_\nu u) + c \tag{11.174}$$

for some constant $c \in \mathbb{C}^M$. Note that Proposition 3.20 indeed applies since $\mathcal{N}_\kappa u$ belongs to $L^1_{\text{loc}}(\partial\Omega, \sigma)$. This may be gleaned from [86, Proposition 8.9.17, pp. 816–817],[4] which implies $\mathcal{N}_\kappa u \leq C\mathcal{N}_\kappa(\nabla u)$ at σ-a. e. point on $\partial\Omega$ for some constant $C \in (0, \infty)$. The assumed membership of $\mathcal{N}_\kappa(\nabla u)$ to $M^{p,\lambda}(\partial\Omega, w)$ then places $\mathcal{N}_\kappa u$ in $L^1_{\text{loc}}(\partial\Omega, \sigma)$ thanks to Proposition 4.7 and Proposition 4.16. The fact that c is a genuine constant and not just a locally constant function (as it appears in (3.80)) is due to the fact that our current geometric assumption on Ω (which includes $\|\nu\|_{[\text{BMO}(\partial\Omega,\sigma)]^n} < \delta$) ensures that Ω is actually connected (cf. [79, Theorem 2.5, p. 86]). At this point, we take the conormal derivative of both sides of (11.174), bearing in mind the jump formula in Theorem 9.2, whose present applicability is ensured by the fact that $\partial^A_\nu u \in [M^{p,\lambda}(\partial\Omega, w)]^M$, and that our smallness assumption on δ implies that Ω is a UR domain (cf. [79, Theorem 2.5, p. 86]). We then arrive at

$$\partial^A_\nu u = -\Big(-\frac{1}{2}I + K^\#_{A^\top}\Big)(\partial^A_\nu u), \tag{11.175}$$

or equivalently

$$\Big(\frac{1}{2}I + K^\#_{A^\top}\Big)(\partial^A_\nu u) = 0. \tag{11.176}$$

4 To apply this result, we are also using that Ω is globally (and hence locally) pathwise nontangentially accessible (in the sense of Definition 11.3) thanks to (11.9) and [79, Theorem 2.5, p. 86], which says when Ω is a δ-AR domain for sufficiently small δ, then Ω is a two-sided NTA domain.

However, because $A^\top \in \mathfrak{A}_L^{\mathrm{dis}}$ we may invoke Theorem 9.7, specifically (9.25), to see that for each $m \in \mathbb{N}$,

$$\|K^{\#}_{A^\top}\|_{[M^{p,\lambda}(\partial\Omega,w)]^M \to [M^{p,\lambda}(\partial\Omega,w)]^M} < C_m \|\nu\|^{\langle m\rangle}_{[\mathrm{BMO}(\partial\Omega,\sigma)]^n} \tag{11.177}$$

for some constant $C_m \in (0,\infty)$, depending on m, the ADR character of $\partial\Omega$, n, p, q, λ, $[w]_{A_p}$, $[w]_{\mathrm{RH}_q}$, and L. Since we are currently assuming Ω is a δ-AR domain (note we may need to take δ smaller than initially chosen so as to fully compensate for the constant C_m), we find that the left-hand side of (11.177) is bounded above by $\frac{1}{2}$. Subsequently, the operator $\frac{1}{2}I + K^{\#}_{A^\top}$ is invertible on the Banach space $[M^{p,\lambda}(\partial\Omega, w)]^M$ (cf. Proposition 4.5) via a Neumann series (cf. Lemma 11.2). This fact, and (11.176) allow us to conclude that $\partial^A_\nu u = 0$ at σ-a. e. point on $\partial\Omega$. Recalling (11.174), this ultimately shows that the function u, described in (11.173), is necessarily constant as wanted.

The claim in item (c) now follows directly from items (a) and (b). As for the claim in item (d), we can argue much as in the proof of [79, Proposition 3.12, pp. 229–230]. In adapting that proof to accommodate $\dot{M}^{p,\lambda}_1(\mathbb{R}, w)$, we rely on Theorem 11.20, and the fact that the modified Hilbert transform H_{mod} on $\mathbb{R}$, defined for each $f \in \dot{M}^{p,\lambda}_1(\mathbb{R}, w)$ according to

$$(H_{\mathrm{mod}}f)(x) := \lim_{\varepsilon\to 0^+} \frac{1}{\pi} \int\limits_{y\in\mathbb{R}} \left\{ \frac{\mathbf{1}_{\mathbb{R}\setminus[x-\varepsilon,x+\varepsilon]}(y)}{x-y} - \frac{\mathbf{1}_{\mathbb{R}\setminus[-1,1]}}{-y} \right\} f(y)\, d\sigma(y) \tag{11.178}$$

at $\mathcal{L}^1$-a. e. point $x \in \mathbb{R}$, maps $\dot{M}^{p,\lambda}_1(\mathbb{R}, w)$ boundedly into itself. Note that this fact is a corollary of Proposition 9.14, given that the Hilbert transform on $\mathbb{R}$ can be realized as scalar multiple of a modified double layer potential for the Laplacian for a suitable choice of coefficient tensor (cf. [89, Example 1.4.1, pp. 43–44]). Regarding the potential failure of uniqueness, this is seen from arguing as in [79, Example 3.5, p. 210] to accommodate $M^{p,\lambda}(\mathbb{R}, w)$; cf. the proof of item (d) in Theorem 11.9 for details.

We now focus on item (e). In the case of weighted vanishing Morrey spaces, essentially the same proof goes through thanks to Theorem 10.38, and we can obtain the desired estimates because of Proposition 10.26. Uniqueness in this case follows because $\mathbb{M}^{p,\lambda}_1(\partial\Omega, w)$ is a subspace of $\dot{M}^{p,\lambda}_1(\partial\Omega, w)$. Well posedness then follows immediately. The analogous sharpness results are handled much as before because of Theorem 10.38. This completes the proof of Theorem 11.32. □

We also have a well-posedness result for the homogeneous regularity problem for weakly elliptic systems with data in Muckenhoupt-weighted Block-based (homogeneous) Sobolev spaces.

Theorem 11.33 (The weighted Block homogeneous regularity BVP). *Select $n \in \mathbb{N}$ with $n \geq 2$ and $M \in \mathbb{N}$. Let $\Omega \subseteq \mathbb{R}^n$ be an AR domain and write $\sigma := \mathcal{H}^{n-1}\lfloor\partial\Omega$. Suppose L is a weakly elliptic, second-order, homogeneous, constant complex coefficient, $M \times M$ system in $\mathbb{R}^n$. Fix an integrability exponent $q \in (1,\infty)$ with $q' := q/(q-1)$. Also, choose an arbitrary Muckenhoupt weight $\omega \in A_q(\partial\Omega, \sigma)$, which also satisfies*

$$\omega' := \omega^{-q'/q} \in A_{q'}(\partial\Omega, \sigma) \cap \mathrm{RH}_p(\partial\Omega, \sigma) \tag{11.179}$$

for some $p \in (1, \infty)$. *Define* $\varepsilon := (q-1)/p'$, *which ensures* $\omega \in A_{q-\varepsilon}(\partial\Omega, \sigma)$ *by Proposition 6.35, and select any threshold* $\lambda \in (0, \varepsilon(n-1)/(q-1))$. *For a fixed aperture parameter* $\kappa \in (0, \infty)$, *consider the Homogeneous regularity problem for L in* Ω *with data belonging to a homogeneous Muckenhoupt-weighted Block-based Sobolev space, formulated below:*

$$\begin{cases} u \in [\mathscr{C}^\infty(\Omega)]^M, \\ Lu = 0 \quad \text{in } \Omega, \\ \mathcal{N}_\kappa(\nabla u) \in \mathcal{B}^{q,\lambda}(\partial\Omega, \omega), \\ u\big|_{\partial\Omega}^{\kappa-\mathrm{n.t.}} = f \in [\dot{\mathcal{B}}_1^{q,\lambda}(\partial\Omega, \omega)]^M. \end{cases} \tag{11.180}$$

Then there exists $\delta \in (0,1)$ *depending on the* ADR *character of* $\partial\Omega$, n, q, ε, λ, $[\omega]_{A_{q-\varepsilon}}$, *and L such that whenever*

$$\|\nu\|_{[\mathrm{BMO}(\partial\Omega,\sigma)]^n} < \delta, \tag{11.181}$$

a scenario which renders Ω *a* δ-AR *domain; it follows that* Ω *is actually a* UR *domain and the following claims are true:*

(a) [Existence, integral representation, and estimates]. *If* $\mathfrak{A}_L^{\mathrm{dis}} \neq \varnothing$, *then (recalling* (3.76)) *the operator*

$$[S_{\mathrm{mod}}] : [\mathcal{B}^{q,\lambda}(\partial\Omega, \omega)]^M \longrightarrow [\dot{\mathcal{B}}_1^{q,\lambda}(\partial\Omega, \omega)/\sim]^M \tag{11.182}$$

is surjective. Moreover, the regularity problem (11.180) *is solvable with the function* $u : \Omega \to \mathbb{C}^M$, *defined by*

$$u(x) := (\mathscr{S}_{\mathrm{mod}} g)(x) + c \quad \text{at each } x \in \Omega, \tag{11.183}$$

where $c \in \mathbb{C}^M$ *is a suitable constant and* $g \in [\mathcal{B}^{q,\lambda}(\partial\Omega, \omega)]^M$ *is such that*

$$[S_{\mathrm{mod}}]g = [f] \in [\dot{\mathcal{B}}_1^{q,\lambda}(\partial\Omega, \omega)/\sim]^M. \tag{11.184}$$

Additionally, the solution (11.183) *satisfies*

$$\|\mathcal{N}_\kappa(\nabla u)\|_{\mathcal{B}^{q,\lambda}(\partial\Omega,\omega)} \approx \|\nabla_{\tan} f\|_{[\mathcal{B}^{q,\lambda}(\partial\Omega,\omega)]^{n\cdot M}} \tag{11.185}$$

with implicit constants independent of the boundary datum, f.

Alternatively, corresponding to the case that $A \in \mathfrak{A}_L^{\mathrm{dis}}$, *and with* $K_{A,\mathrm{mod}}$ *as in* (3.47), *the operator*

$$\frac{1}{2}I + [K_{A,\mathrm{mod}}] : [\dot{\mathcal{B}}_1^{q,\lambda}(\partial\Omega, \omega)/\sim]^M \longmapsto [\dot{\mathcal{B}}_1^{q,\lambda}(\partial\Omega, \omega)/\sim]^M \tag{11.186}$$

is invertible. Moreover, the regularity problem (11.180) *may be solved with the function* $u : \Omega \to \mathbb{C}^M$, *defined (recall* (3.34)) *according to*

$$u(x) := (\mathcal{D}_{A,\mathrm{mod}}g)(x) + c \quad \text{at each } x \in \Omega, \tag{11.187}$$

where $c \in \mathbb{C}^M$ *is a suitable constant and* $g \in [\dot{\mathcal{B}}_1^{q,\lambda}(\partial\Omega, \omega)/\sim]^M$ *is such that*

$$[g] = \left(\frac{1}{2}I + [K_{A,\mathrm{mod}}]\right)^{-1}[f]. \tag{11.188}$$

Additionally, the solution in (11.187) *continues to satisfy the size estimates recorded in* (11.185).

(b) [Uniqueness modulo constants]. *If* $\mathfrak{A}_{L^\top}^{\mathrm{dis}} \neq \varnothing$, *then any two solutions of* (11.180) *differ by only a constant.*

(c) [Well posedness]. *Whenever both* $\mathfrak{A}_L^{\mathrm{dis}} \neq \varnothing$ *and* $\mathfrak{A}_{L^\top}^{\mathrm{dis}} \neq \varnothing$, *then the regularity problem* (11.180) *is well posed in the sense of Hadamard. That is, there exists a unique solution to* (11.180), *which also satisfies the size estimates in* (11.185).

(d) [Sharpness]. *If* $\mathfrak{A}_L^{\mathrm{dis}} = \varnothing$, *then* (11.180) *may not be solvable. In fact, solvability may fail for boundary data belonging to an infinite-dimensional subspace of* $[\dot{\mathcal{B}}_1^{q,\lambda}(\partial\Omega, \omega)]^M$, *even when* $\Omega = \mathbb{R}^n_+$ *(a setting in which the* BMO *seminorm of* v *is zero). Also, if* $\mathfrak{A}_{L^\top}^{\mathrm{dis}} = \varnothing$, *then there may be more than one solution to* (11.180), *and it may be the case that the linear space of null solutions is infinite-dimensional.*

Remark 11.34. If we compare the assumptions on the Block space in Theorem 11.33 with those in Theorem 11.30, we will see that the homogeneous regularity problem requires more structure on the Block space in question. Namely, we ask that we have information about the reverse Hölder class to which ω' belongs and that $\varepsilon := (q-1)/p'$. This is in contrast to the assumptions in Theorem 11.30 in which we do not need to know the reverse Hölder class to which ω' belongs and allowing ε to be chosen as anything from the interval $(0, q-1)$ so that $\omega \in A_{q-\varepsilon}(\partial\Omega, \sigma)$. These extra assumptions on the weight ω and parameter ε are necessary for the proof of existence we employ both when using the modified single layer and the modified double later. The information we have about the derivatives of the solution allow us to prove uniqueness in a way that does not make use of the reverse Hölder condition on ω', nor the specific value for ε, much as in the proofs of uniqueness in Theorems 11.27 and 11.30.

Proof of Theorem 11.33. We can argue similarly as in the proof of Theorem 11.32. To prove solvability when working with the modified single layer, we rely on Proposition 9.33 and item (5) of Theorem 9.36 (here is where we use the assumption that $\omega' \in \mathrm{RH}_p(\partial\Omega, \sigma)$ for some $p \in (1, \infty)$). We also appeal to Theorem 9.17, Proposition 9.20, Proposition 6.17, and Proposition 6.14 to ensure that u as defined in (11.183) satisfies (11.180) and (11.185). Alternatively, if we wish to demonstrate solvability using the modified double layer, we turn to Theorem 9.30, Proposition 8.19 (note this necessitates that

$\omega' \in \mathrm{RH}_p(\partial\Omega, \sigma)$ for some $p \in (1, \infty)$), Theorem 9.27, and Proposition 6.14. The question of uniqueness of solutions to (11.180) addressed in item (b) now requires us to appeal to Theorem 11.21, Theorem 9.17, Theorem 9.23, and Proposition 6.11. Well posedness naturally follows, establishing (c). The sharpness results in (d) follow from similar reasoning as before, and we highlight the relevance of Theorem 11.21 and Proposition 9.29, which implies that H_{mod} induces a well-defined and bounded operator on homogeneous Muckenhoupt-weighted Block-based Sobolev spaces. This concludes the treatment of all claims in Theorem 11.33. □

11.4 The transmission problem in Morrey and Block spaces

We are now ready to formulate the transmission problem on Muckenhoupt-weighted-Morrey and Block spaces and present a proof of well posedness. This problem is in some ways a blend between a Neumann problem and a homogeneous regularity problem. For this reason, it is a difficult problem, but our well-posedness result highlights the potency of the method of layer potentials in solving boundary value problems.

Theorem 11.35 (The weighted Morrey transmission BVP). *Fix $n \in \mathbb{N}$ with $n \geq 2$ along with $M \in \mathbb{N}$. Let $\Omega \subseteq \mathbb{R}^n$ be a two-sided* NTA *domain with unbounded, Ahlfors regular boundary, abbreviate $\sigma := \mathcal{H}^{n-1}\lfloor\partial\Omega$, define*

$$\Omega_+ := \Omega, \quad \Omega_- := \mathbb{R}^n \setminus \overline{\Omega}, \tag{11.189}$$

and let ν denote the geometric measure theoretic outward unit normal to Ω. Next, suppose L is a weakly elliptic, second-order, homogeneous, constant complex coefficient, $M \times M$ system in $\mathbb{R}^n$. Now, choose two arbitrary integrability exponents $p, q \in (1, \infty)$ along with a Muckenhoupt weight $w \in A_p(\partial\Omega, \sigma) \cap \mathrm{RH}_q(\partial\Omega, \sigma)$, and some threshold $\lambda \in (0, (n-1)/q')$, where $q' := q/(q-1)$. Having picked some coefficient tensor $A \in \mathfrak{A}_L$, an aperture parameter $\kappa \in (0, \infty)$, and a transmission (or coupling) parameter $\eta \in \mathbb{C}$, consider the transmission problem for L in $\Omega_\pm$ with boundary data in Muckenhoupt-weighted Morrey and Morrey-based Sobolev spaces, formulated below:

$$\begin{cases} u^\pm \in [\mathscr{C}^\infty(\Omega_\pm)]^M, \\ Lu^\pm = 0 \quad \text{in } \Omega_\pm, \\ \mathcal{N}_\kappa(\nabla u^\pm) \in M^{p,\lambda}(\partial\Omega, w), \\ u^+\big|_{\partial\Omega}^{\kappa-\mathrm{n.t.}} - u^-\big|_{\partial\Omega}^{\kappa-\mathrm{n.t.}} = g \in [\dot{M}_1^{p,\lambda}(\partial\Omega, w)]^M, \\ \partial_\nu^A u^+ - \eta \cdot \partial_\nu^A u^- = f \in [M^{p,\lambda}(\partial\Omega, w)]^M. \end{cases} \tag{11.190}$$

Finally, assume one of the following two scenarios:

(i) *Suppose either*

$$A^\top \in \mathfrak{A}_{L^\top}^{\mathrm{dis}} \quad \text{and} \quad \eta \in \mathbb{C} \setminus \{-1\}, \tag{11.191}$$

or

$$A \in \mathfrak{A}_L^{\text{dis}} \quad \textit{and} \quad \eta \in \mathbb{C} \setminus \{-1, 0\}, \tag{11.192}$$

and that

$$\left|\frac{1-\eta}{1+\eta}\right| \|\nu\|_{[\text{BMO}(\partial\Omega,\sigma)]^n}^{\langle m \rangle} < \delta, \tag{11.193}$$

for some $\delta \in (0,1)$, *depending on the* ADR *character of* $\partial\Omega$, n, p, q, λ, $[w]_{\text{RH}_q}$, $[w]_{A_p}$, *and* L.

(ii) *Suppose*

$$|\eta - 1| < \delta_* \tag{11.194}$$

for some $\delta_* \in (0,1)$ *depending only on the* ADR *character of* $\partial\Omega$, *the corkscrew constants of* Ω, n, p, q, λ, $[w]_{\text{RH}_q}$, $[w]_{A_p}$, *and* L.

Then all claims below are true.

(a) [Uniqueness modulo constants]. *Any two solutions to the transmission problem* (11.190) *differ by only a constant (in* $\mathbb{C}^M$*).*

(b) [Well posedness and integral representation]. *In the case that either* (11.191) *and* (11.193) *hold or* (11.194) *holds, the operator*

$$-\frac{1}{2}I + \frac{1-\eta}{1+\eta}K_{A^\top}^{\#} : [M^{p,\lambda}(\partial\Omega, w)]^M \to [M^{p,\lambda}(\partial\Omega, w)]^M \tag{11.195}$$

is an isomorphism. Denote by $\mathcal{D}_{A,\text{mod}}^{\pm}$ *the modified double layer potential operators associated with the sets* $\Omega_{\pm}$ *(cf.* (3.34)*), and likewise* $\mathscr{S}_{\text{mod}}^{\pm}$ *for the modified single layer potential operators (cf.* (3.76)*). Then the functions* $u^{\pm} : \Omega_{\pm} \to \mathbb{C}^M$ *defined according to*

$$u^{\pm} := \pm\mathcal{D}_{A,\text{mod}}^{\pm} g + \mathscr{S}_{\text{mod}}^{\pm} h \quad \textit{in } \Omega_{\pm}, \tag{11.196}$$

where $h \in [M^{p,\lambda}(\partial\Omega, w)]^M$ *is given by*

$$h := \left(-\frac{1}{2}(1+\eta)I + (1-\eta)K_{A^\top}^{\#}\right)^{-1} (f - (1-\eta)\partial_\nu^A(\mathcal{D}_{A,\text{mod}} g)) \tag{11.197}$$

at σ*-a. e. point on* $\partial\Omega$, *solve the transmission problem* (11.190)*. Moreover, there exists a constant* $C \in (0, \infty)$, *independent of* f *and* g, *with the property that*

$$\|\mathcal{N}_\kappa(\nabla u^{\pm})\|_{M^{p,\lambda}(\partial\Omega,w)} \le C(\|f\|_{[M^{p,\lambda}(\partial\Omega,w)]^M} + \|g\|_{[\dot{M}_1^{p,\lambda}(\partial\Omega,w)/\sim]^M}). \tag{11.198}$$

Suppose instead that either (11.192) *and* (11.193) *hold or* (11.194) *holds. Then (cf.* (3.47)*) the operator*

$$\frac{1}{2}I - \frac{1-\eta}{1+\eta}[K_{A,\mathrm{mod}}] : [\dot{M}_1^{p,\lambda}(\partial\Omega, w)/\sim]^M \to [\dot{M}_1^{p,\lambda}(\partial\Omega, w)/\sim]^M \tag{11.199}$$

is an isomorphism. Moreover, the functions $u^\pm : \Omega_\pm \to \mathbb{C}^M$ *defined according to*

$$\begin{aligned} u^+ &:= \mathcal{D}^+_{A,\mathrm{mod}} h - \mathscr{S}^+_{\mathrm{mod}} f - c && \text{in } \Omega_+, \\ u^- &:= -\eta^{-1}\mathcal{D}^-_{A,\mathrm{mod}} h - \eta^{-1}\mathscr{S}^-_{\mathrm{mod}} f && \text{in } \Omega_-, \end{aligned} \tag{11.200}$$

where $h \in [\dot{M}_1^{p,\lambda}(\partial\Omega, w)]^M$ *and* $c \in \mathbb{C}^M$ *are such that*

$$\Big(\frac{1}{2}(1+\eta^{-1})I + (1-\eta^{-1})K_{A,\mathrm{mod}}\Big)h = g + (1-\eta^{-1})S_{\mathrm{mod}} f + c \tag{11.201}$$

at σ*-a. e. point on* $\partial\Omega$*, solve the transmission problem* (11.190)*. Additionally, there exists a constant* $C \in (0,\infty)$*, independent of* f *and* g*, such that the size estimate* (11.198) *holds for* $u^\pm$ *as in* (11.200)*.*

In summary, in the case that either (i) or (ii) holds, the transmission problem (11.190) *is solvable and any two solutions differ by only a constant and satisfy the size estimate* (11.198)*.*

(c) [Other Spaces]. *Analogous results to those contained in (a) and (b) continue to be true when* $M^{p,\lambda}(\partial\Omega, w)$ *and* $\dot{M}_1^{p,\lambda}(\partial\Omega, w)$ *are replaced by* $\mathring{M}^{p,\lambda}(\partial\Omega, w)$ *and* $\mathbb{M}_1^{p,\lambda}(\partial\Omega, w)$ *respectively. That is, similar well-posedness results hold for the transmission problem below:*

$$\begin{cases} u^\pm \in [\mathscr{C}^\infty(\Omega_\pm)]^M, \\ Lu^\pm = 0 \quad \text{in } \Omega_\pm, \\ \mathcal{N}_\kappa(\nabla u^\pm) \in \mathring{M}^{p,\lambda}(\partial\Omega, w), \\ u^+\big|^{\kappa-\mathrm{n.t.}}_{\partial\Omega} - u^-\big|^{\kappa-\mathrm{n.t.}}_{\partial\Omega} = g \in [\mathbb{M}_1^{p,\lambda}(\partial\Omega, w)]^M, \\ \partial^A_\nu u^+ - \eta\cdot\partial^A_\nu u^- = f \in [\mathring{M}^{p,\lambda}(\partial\Omega, w)]^M. \end{cases} \tag{11.202}$$

Before presenting the proof, a few comments are warranted.

Remark 11.36. First, because Ω is assumed to be a two-sided NTA domain, we know that Ω satisfies a two-sided corkscrew condition, which implies $\partial\Omega = \partial_*\Omega$, and hence $\mathcal{H}^{n-1}(\partial\Omega \setminus \partial_*\Omega) = 0$. This means that Ω is an AR domain in the sense of Definition 2.5, because $\partial\Omega$ is assumed to be an Ahlfors regular set. In particular, ν exists σ-a. e. on $\partial\Omega$. Moreover, Ω is a UR domain on account of (2.9). Also, [79, Lemma 2.3, p. 37] ensures that

$$\partial\Omega_+ = \partial\Omega_- \quad \text{and} \quad \partial_*\Omega_+ = \partial_*\Omega_-, \tag{11.203}$$

and in fact the outward unit normal to Ω_- is given by $-\nu$ as one might expect. Thus, it makes sense to write $u^-\big|^{\kappa-\mathrm{n.t.}}_{\partial\Omega}$ with the understanding that this is the same as $u^-\big|^{\kappa-\mathrm{n.t.}}_{\partial\Omega_-}$.

Remark 11.37. Second, we notice that the assumption that Ω be a two-sided NTA domain with unbounded boundary ensures that $\Omega_\pm$ are connected. This follows from the definition of a NTA domain (cf. Definition 2.10), which entails that $\Omega_\pm$ satisfy a (∞, N)-Harnack chain condition for some $N \in \mathbb{N}$, which from Definition 2.9 implies $\Omega_\pm$ are pathwise connected, and hence connected.

Remark 11.38. Third, the σ-a. e. existence of $u^\pm|_{\partial\Omega}^{\kappa-\mathrm{n.t.}}$ on $\partial\Omega$ is an implicit demand in the formulation of the transmission problem (11.190). In view of Theorem 11.20, not only do $u^\pm|_{\partial\Omega}^{\kappa-\mathrm{n.t.}}$ exist σ-a.e on $\partial\Omega$, but also $u^\pm|_{\partial\Omega}^{\kappa-\mathrm{n.t.}}$ belong to the correct space of boundary data, i. e., $[\mathring{M}_1^{p,\lambda}(\partial\Omega, w)]^M$. This is also the case for Muckenhoupt-weighted vanishing Morrey spaces and Muckenhoupt-weighted Block spaces thanks to Theorem 11.21.

Remark 11.39. Fourth, with the fact that the outward unit normal to Ω_- is just $-\nu$ (cf. Remark 11.36) and from the definition of the conormal derivative (cf. (3.18)), we understand $\partial_\nu^A u^-$ via

$$\partial_\nu^A u^- := -\partial_{-\nu}^A u^-. \tag{11.204}$$

Moreover, the Fatou-type result in Theorem 11.20 shows that the conormal derivatives $\partial_\nu^A u^\pm$ are well-defined objects in the context of (11.190); i. e., $\partial_\nu^A u^\pm$ necessarily belong to $[M^{p,\lambda}(\partial\Omega, w)]^M$. This is likewise the case when considering Muckenhoupt-weighted vanishing Morrey spaces. Similarly, for the transmission problem (11.255) formulated in terms of Muckenhoupt-weighted Block spaces (cf. Theorem 11.41), the Fatou-type result in Theorem 11.21 will ensure that $\partial_\nu^A u^\pm$ necessarily belong to $[\mathcal{B}^{q,\lambda}(\partial\Omega, \omega)]^M$.

Remark 11.40. Lastly, corresponding to the case $\eta = -1$, we find that the transmission problem in Muckenhoupt-weighted Morrey Spaces is ill-posed. Even making the choices $\Omega := \mathbb{R}_n^+$, $L := \Delta$, and $A := I_{n\times n}$ in the formulation of (11.190), we find that uniqueness fails. Indeed, if in (11.190) we assume $f = 0$ and $g = 0$ on $\mathbb{R}^{n-1} \equiv \partial\mathbb{R}_\pm^n$, then for any $\phi \in \mathcal{M}^{p,\lambda}(\mathbb{R}^{n-1}, w)$, we find that the functions

$$u^\pm := \mathscr{S}_{\mathrm{mod}}^\pm \phi \quad \text{in } \mathbb{R}_\pm^n \tag{11.205}$$

are harmonic in $\mathbb{R}_\pm^n$, where $\mathscr{S}_{\mathrm{mod}}^\pm$ denotes the modified single layer potential operator for the Laplacian in $\mathbb{R}_\pm^n$. Additionally, according to Theorem 9.2, we see that $\mathcal{N}_\kappa u^\pm$ belongs to $M^{p,\lambda}(\mathbb{R}^{n-1}, w)$. Then the fact that the modified single layer does not jump upon taking nontangential traces (cf. (3.78)), we see that $u^\pm$ as defined in (11.205) satisfies the penultimate condition in (11.190), recalling that we have taken $g = 0$ on $\partial\Omega$. The fact that $u^\pm$ satisfies the final condition in (11.190) with $f = 0$, can be seen from (9.4) and the fact that the (transpose) boundary-to-boundary harmonic double layer potential in the upper half-space is zero. Since $M^{p,\lambda}(\mathbb{R}^{n-1}, w)$ is infinite-dimensional (this is true, e. g., since all characteristic functions of balls belong to a Muckenhoupt-weighted Morrey space thanks to Lemma 4.9), for any $N \in \mathbb{N}$ there exists a collection of linearly independent functions $\{\phi_j\}_{j\in\{1,\dots,N\}} \subseteq M^{p,\lambda}(\mathbb{R}^{n-1}, w)$. Suppose $\{c_j\}_{j\in\{1,\dots,N\}} \subseteq \mathbb{C}$ is a collection of constants such that

$$\sum_{j=1}^{N} c_j \mathscr{S}^{\pm}_{\mathrm{mod}} \phi_j = 0 \quad \text{in } \mathbb{R}^n_{\pm}. \tag{11.206}$$

Then using the linearity of $\mathscr{S}^{\pm}_{\mathrm{mod}}$ and taking the conormal derivative on both sides of (11.206) (bearing in mind (9.4) and the fact that since we are working in the upper-half space $K^{\#}_{\Delta} = 0$ on $\partial\mathbb{R}^n_{\pm}$) we obtain

$$\frac{1}{2}\sum_{j=1}^{N} c_j \phi_j = 0 \implies c_j = 0 \quad \text{for all } j \in \{1, \dots, N\}, \tag{11.207}$$

given the linear independence of $\{\phi_j\}_{j\in\{1,\dots,N\}}$. Ultimately, we see that the space of null solutions to (11.190) is also infinite-dimensional.

The result in Theorem 11.35 encompasses the case $w \equiv 1$, the unweighted case, which was first treated in [79, Theorem 7.23, pp. 494–496].

Proof of Theorem 11.35. Starting with the claims made in (a), to establish uniqueness modulo constants, we will show under either assumption in (i) and (ii) if $u^{\pm}$ satisfies

$$\begin{cases} u^{\pm} \in [\mathscr{C}^{\infty}(\Omega_{\pm})]^M, \\ Lu^{\pm} = 0 \quad \text{in } \Omega_{\pm}, \\ \mathcal{N}_{\kappa}(\nabla u^{\pm}) \in M^{p,\lambda}(\partial\Omega, w), \\ u^{+}\big|^{\kappa-\mathrm{n.t.}}_{\partial\Omega} - u^{-}\big|^{\kappa-\mathrm{n.t.}}_{\partial\Omega} = 0 \quad \text{at } \sigma\text{-a. e. point on } \partial\Omega, \\ \partial^A_{\nu} u^{+} - \eta \cdot \partial^A_{\nu} u^{-} = 0 \quad \text{at } \sigma\text{-a. e. point on } \partial\Omega, \end{cases} \tag{11.208}$$

then in fact there exists $c \in \mathbb{C}^M$ such that $u^{\pm} = c$ in $\Omega_{\pm}$.

Consider the case that (11.191) and (11.193) hold. Suppose $u^{\pm}$ is a solution to (11.208). From Theorem 11.20, we ascertain the σ-a. e. existence and membership of $u^{\pm}\big|^{\kappa-\mathrm{n.t.}}_{\partial\Omega}$ to $[\dot{M}^{p,\lambda}_1(\partial\Omega, w)]^M$ just from knowing that $u^{\pm}$ are null solutions for L in $\Omega_{\pm}$ with the property that $\mathcal{N}_{\kappa}(\nabla u^{\pm}) \in M^{p,\lambda}(\partial\Omega, w)$. Furthermore, bearing in mind the continuous embedding given in Proposition 4.16, we are permitted to use [79, Proposition 2.22, p. 148] to write

$$\begin{aligned} &\nu_j\big((\partial_k u^{+})\big|^{\kappa-\mathrm{n.t.}}_{\partial\Omega}\big) - \nu_k\big((\partial_j u^{+})\big|^{\kappa-\mathrm{n.t.}}_{\partial\Omega}\big) \\ &= \partial_{\tau_{jk}}\big(u^{+}\big|^{\kappa-\mathrm{n.t.}}_{\partial\Omega}\big) = \partial_{\tau_{jk}}\big(u^{-}\big|^{\kappa-\mathrm{n.t.}}_{\partial\Omega}\big) \\ &= \nu_j\big((\partial_k u^{-})\big|^{\kappa-\mathrm{n.t.}}_{\partial\Omega}\big) - \nu_k\big((\partial_j u^{-})\big|^{\kappa-\mathrm{n.t.}}_{\partial\Omega}\big), \end{aligned} \tag{11.209}$$

at σ-a. e. point on $\partial\Omega$ and for each $j, k \in \{1, \dots, n\}$, noting that the second equality follows from the fourth line in (11.208). Recall the notation introduced in (11.115), so that (11.209) may be recast as

$$\partial_{T_{jk}} u^+ = \partial_{T_{jk}} u^- \tag{11.210}$$

at σ-a. e. point on $\partial\Omega$ and for each $j, k \in \{1, \dots, n\}$.

Pressing forward, note that the fifth condition in (11.208) allows us to conclude that $\partial_\nu^A u^+ = \eta \cdot \partial_\nu^A u^-$ at σ-a. e. point on $\partial\Omega$. As such, applying (11.93) of Proposition 11.19 to u^+ and Ω_+, as well as (11.94) of Proposition 11.19 to u^- and Ω_-, we obtain

$$\begin{aligned}(\partial_\ell u_\gamma^+)(x) &= \int_{\partial\Omega} a_{rs}^{\beta\alpha} (\partial_r E_{\gamma\beta})(x-y)(\partial_{T_{\ell s}} u_\alpha^+)(y)\, d\sigma(y) \\ &\quad - \int_{\partial\Omega} (\partial_\ell E_{\gamma\alpha})(x-y)(\partial_\nu^A u^+)_\alpha(y)\, d\sigma(y) \\ &= \int_{\partial\Omega} a_{rs}^{\beta\alpha} (\partial_r E_{\gamma\beta})(x-y)(\partial_{T_{\ell s}} u_\alpha^-)(y)\, d\sigma(y) \\ &\quad - \eta \int_{\partial\Omega} (\partial_\ell E_{\gamma\alpha})(x-y)(\partial_\nu^A u^-)_\alpha(y)\, d\sigma(y) \\ &= (1-\eta) \int_{\partial\Omega} (\partial_\ell E_{\gamma\alpha})(x-y)(\partial_\nu^A u^-)_\alpha(y)\, d\sigma(y)\end{aligned} \tag{11.211}$$

at every point $x \in \Omega_+$. Because of (5.39) in Theorem 5.7, restricting (11.211) nontangentially to the boundary gives at σ-a. e. point $x \in \partial\Omega$ that

$$\begin{aligned}\big((\partial_\ell u_\gamma^+)\big|_{\partial\Omega}^{\kappa-\text{n.t.}}\big)(x) &= (1-\eta)\frac{1}{2i}\widehat{\partial_\ell E_{\gamma\alpha}}(\nu(x))(\partial_\nu^A u^-)_\alpha(x) \\ &\quad + (1-\eta) \lim_{\varepsilon\to 0^+} \int_{\substack{y\in\partial\Omega \\ |x-y|>\varepsilon}} (\partial_\ell E_{\gamma\alpha})(x-y)(\partial_\nu^A u^-)_\alpha(y)\, d\sigma(y)\end{aligned} \tag{11.212}$$

for each $\ell \in \{1, \dots, n\}$ and $\gamma \in \{1, \dots, M\}$. Subsequently, due to the definition of the conormal derivative (3.18), at σ-a. e. point $x \in \partial\Omega$ we find that

$$\begin{aligned}(\partial_\nu^A u^+)_\mu(x) &= \nu_r(x) a_{rs}^{\mu\beta} \big((\partial_s u_\beta^-)\big|_{\partial\Omega}^{\kappa-\text{n.t.}}\big)(x) \\ &= (1-\eta)\frac{1}{2i}\widehat{\partial_s E_{\beta\alpha}}(\nu(x))(\partial_\nu^A u^-)_\alpha(x)\nu_r(x) a_{rs}^{\mu\beta} \\ &\quad + (1-\eta) \lim_{\varepsilon\to 0^+} \int_{\substack{y\in\partial\Omega \\ |x-y|>\varepsilon}} \nu_r(x) a_{rs}^{\mu\beta} (\partial_s E_{\beta\alpha})(x-y)(\partial_\nu^A u^-)_\alpha(y)\, d\sigma(y)\end{aligned} \tag{11.213}$$

for each $\mu \in \{1, \dots, M\}$. Now, because of Proposition 3.4, specifically (3.14), we obtain

$$\begin{aligned}&(1-\eta)\frac{1}{2i}\widehat{\partial_s E_{\beta\alpha}}(\nu(x))(\partial_\nu^A u^-)_\alpha(x)\nu_r(x) a_{rs}^{\mu\beta} \\ &\quad = (1-\eta)\frac{1}{2}\big(a_{rs}^{\mu\beta}\nu_r(x)\nu_s(x)\big)\widehat{E_{\beta\alpha}}(\nu(x))(\partial_\nu^A u^-)_\alpha(x)\end{aligned} \tag{11.214}$$

$$= (\eta - 1)\frac{1}{2}[L(\nu(x))]_{\mu\beta}[L(\nu(x))]^{-1}_{\beta\alpha}(\partial^A_\nu u^-)_\alpha(x)$$
$$= (\eta - 1)\frac{1}{2}\delta_{\mu\alpha}(\partial^A_\nu u^-)_\alpha(x) = \frac{1}{2}(\eta - 1)(\partial^A_\nu u^-)_\mu(x),$$

at σ-a. e. point $x \in \partial\Omega$ and each $\mu \in \{1,\dots,M\}$. Additionally, the definition of $K^\#_A$ in (3.20) and the fact that $(E_L)^\top = E_{L^\top}$ (cf. (3.17) of Proposition 3.4), we may write

$$\lim_{\varepsilon\to 0^+} \int_{\substack{y\in\partial\Omega \\ |x-y|>\varepsilon}} \nu_r(x)a^{\mu\beta}_{rs}(\partial_s E_{\beta\alpha})(x-y)(\partial^A_\nu u^-)_\alpha(y)\,d\sigma(y) = (K^\#_{A^\top}(\partial^A_\nu u^-)(x))_\mu \tag{11.215}$$

at σ-a. e. point $x \in \partial\Omega$, and each $\mu \in \{1,\dots,M\}$. Combining the results in (11.213)–(11.215), we determine that

$$\partial^A_\nu u^+ = (1-\eta)\Big(-\frac{1}{2}I + K^\#_{A^\top}\Big)(\partial^A_\nu u^-) \quad \text{at } \sigma\text{-a. e. point on } \partial\Omega. \tag{11.216}$$

But again, since our current assumptions imply $\partial^A_\nu u^+ = \eta \cdot \partial^A_\nu u^-$, from (11.216) we obtain

$$\Big(-\frac{1}{2}I + \frac{1-\eta}{1+\eta}K^\#_{A^\top}\Big)(\partial^A_\nu u^-) = 0. \tag{11.217}$$

Since we are presently working under the assumption that $A^\top \in \mathfrak{A}^{\rm dis}_{L^\top}$ and $\eta \in \mathbb{C}\setminus\{-1\}$, Theorem 9.7 implies for each $m \in \mathbb{N}$ that there exists some $C_m \in (0,\infty)$ (whose nature is specified in the statement of Theorem 9.7) such that

$$\Big\|\frac{1-\eta}{1+\eta}K^\#_{A^\top}\Big\|_{[M^{p,\lambda}(\partial\Omega,w)]^M\to[M^{p,\lambda}(\partial\Omega,w)]^M} \leq C_m\Big|\frac{1-\eta}{1+\eta}\Big|\,\|\nu\|^{\langle m\rangle}_{[\mathrm{BMO}(\partial\Omega,\sigma)]^n}. \tag{11.218}$$

Thus, if (11.193) holds for $\delta \in (0,1)$ sufficiently small so as to force the right-hand side of (11.218) to be less than $\frac{1}{2}$, then the operator

$$-\frac{1}{2}I + \frac{1-\eta}{1+\eta}K^\#_{A^\top} : [M^{p,\lambda}(\partial\Omega,w)]^M \to [M^{p,\lambda}(\partial\Omega,w)]^M \tag{11.219}$$

is invertible via a Neumann series (cf. Lemma 11.2, which applies thanks to Proposition 4.5), and hence (11.217) yields that

$$\partial^A_\nu u^- = 0. \tag{11.220}$$

Retracing this back to (11.211), we can conclude that $\nabla u^+ = 0$ in Ω_+. Recall our geometric assumptions on Ω actually imply that Ω_+ is connected (cf. Remark 11.37), and hence we find that u^+ is necessarily constant in Ω_+, say $u^+ = c \in \mathbb{C}^M$ in Ω_+. Since $u^\pm$ satisfies the fourth line in (11.208) by assumption, we can see that

$$u^-\big|^{\kappa-\mathrm{n.t.}}_{\partial\Omega} = u^+\big|^{\kappa-\mathrm{n.t.}}_{\partial\Omega} = c \quad \text{at } \sigma\text{-a. e. point on } \partial\Omega. \tag{11.221}$$

Consequently, with the help of (11.209) we can conclude for each pair of indices $j, k \in \{1, \ldots, n\}$ that

$$\nu_j\big((\partial_k u^-)\big|_{\partial\Omega}^{\kappa-\text{n.t.}}\big) - \nu_k\big((\partial_j u^-)\big|_{\partial\Omega}^{\kappa-\text{n.t.}}\big) = \partial_{\tau_{jk}}\big(u^-\big|_{\partial\Omega}^{\kappa-\text{n.t.}}\big) = 0. \tag{11.222}$$

Thus, writing (11.93) for u_- and Ω_-, bearing in mind (11.220) and (11.222), we find that $\nabla u^- = 0$ in Ω_-. Again, our current geometric assumptions on Ω ensure Ω_- is connected (cf. Remark 11.37), so it must be that u^- is constant in Ω_-, and (11.221) actually forces $u^- = c$ on Ω_- as well. This completes the proof of uniqueness modulo constants under the assumptions in (11.191).

We now assume that (11.192) and (11.193) hold. Suppose $u^\pm$ is a solution to (11.208) and note that (11.209) and (11.210) continue to be true. Also, since we are assuming that $u^\pm$ satisfies (11.208), we continue to have that $\partial_\nu^A u^+ = \eta \partial_\nu^A u^-$. Because of these facts, using Proposition 11.19 (specifically (11.93) with $u := u^+$ and $\Omega := \Omega_+$, and (11.94) with $u := u^-$ and $\Omega := \Omega_-$), for each pair of indices $\ell \in \{1, \ldots, n\}$ and $\gamma \in \{1, \ldots, M\}$, and each $x \in \Omega$ we obtain

$$\begin{aligned}
(\partial_\ell u_\gamma^+)(x) &= \int_{\partial\Omega} a_{rs}^{\beta\alpha}(\partial_r E_{\gamma\beta})(x-y)(\partial_{\tau_{\ell s}} u_\alpha^+)(y)\,d\sigma(y) \\
&\quad - \int_{\partial\Omega} (\partial_\ell E_{\gamma\alpha})(x-y)(\partial_\nu^A u^+)_\alpha(y)\,d\sigma(y) \\
&= \int_{\partial\Omega} a_{rs}^{\beta\alpha}(\partial_r E_{\gamma\beta})(x-y)(\partial_{\tau_{\ell s}} u_\alpha^-)(y)\,d\sigma(y) \\
&\quad - \eta \int_{\partial\Omega} (\partial_\ell E_{\gamma\alpha})(x-y)(\partial_\nu^A u^-)_\alpha(y)\,d\sigma(y) \\
&= (1-\eta) \int_{\partial\Omega} a_{rs}^{\beta\alpha}(\partial_r E_{\gamma\beta})(x-y)(\partial_{\tau_{\ell s}} u_\alpha^-)(y)\,d\sigma(y) \\
&= (1-\eta) \int_{\partial\Omega} a_{rs}^{\beta\alpha}(\partial_r E_{\gamma\beta})(x-y)(\partial_{\tau_{\ell s}} u_\alpha^+)(y)\,d\sigma(y).
\end{aligned} \tag{11.223}$$

From this point, we can argue much as in (11.112)–(11.123) to yield (bearing in mind our notation conventions in (11.115) and (11.122))

$$\begin{aligned}
&\partial_{\tau_{t\ell}} u^+ = (1-\eta)\Big(\tfrac{1}{2}I + K_A\Big)(\partial_{\tau_{t\ell}} u^+) + (1-\eta)[M_\nu, T](\partial_T u^+) \\
&\text{at } \sigma\text{-a.e. point on } \partial\Omega, \text{ for each } t, \ell \in \{1, \ldots, n\}.
\end{aligned} \tag{11.224}$$

Therefore, rearranging (11.224), we find for each $\ell, t \in \{1, \ldots, n\}$ that

$$\frac{1}{2}\partial_{\tau_{t\ell}} u^+ = \frac{1-\eta}{1+\eta} K_A(\partial_{\tau_{t\ell}} u^+) + \frac{1-\eta}{1+\eta}[M_\nu, T](\partial_T u^+) \quad \text{on } \partial\Omega. \tag{11.225}$$

Since our present assumptions include $\eta \neq -1$, we may argue similarly as in (11.124)–(11.128) (keeping in mind the extra factor of $\frac{1-\eta}{1+\eta}$) to say that

$$\partial_{T_{t\ell}} u_\alpha^+ = 0 \quad \text{for each } t, \ell \in \{1, \dots, n\} \text{ and } \alpha \in \{1, \dots, M\} \tag{11.226}$$

provided (11.193) holds and $\delta \in (0,1)$ is sufficiently small. We highlight the use of Theorem 9.7 and Theorem 5.18 in such an argument. Inserting the result from (11.226) back into the equation in (11.223), we see it must be the case that $\nabla u^+ = 0$ in Ω_+. Given the connectedness of Ω_+ (cf. Remark 11.37), it follows that in fact u^+ is constant in Ω_+, say $u^+ = c^+ \in \mathbb{C}^M$. Then bearing in mind that $\partial_\nu^A u^+ = \eta \partial_\nu^A u^-$, (11.209), and our present assumption that $\eta \neq 0$, we see that

$$\begin{aligned} &\partial_\nu^A u^- = \eta^{-1} \cdot \partial_\nu^A u^+ = 0 \quad \text{at } \sigma\text{-a. e. point on } \partial\Omega, \\ &\text{and} \quad \partial_{T_{jk}} u_\alpha^- = \partial_{T_{jk}} u_\alpha^+ = 0 \quad \text{at } \sigma\text{-a. e. point on } \partial\Omega, \\ &\text{for each } j, k \in \{1, \dots, n\} \text{ and } \alpha \in \{1, \dots, M\}. \end{aligned} \tag{11.227}$$

Thus, returning to (11.93) of Proposition 11.19, now with $u := u_-$ and $\Omega := \Omega_-$, we can conclude that necessarily $\nabla u^- = 0$ in Ω_-. Thanks to the connectedness of Ω_- (cf. Remark 11.37) we can actually confirm that u^- is also constant in Ω_-, say with $u^- = c^- \in \mathbb{C}^M$. Since $u^\pm$ is assumed to satisfy the fourth line in (11.208), this implies $u^+\big|_{\partial\Omega}^{\kappa-\text{n.t.}} = u^-\big|_{\partial\Omega}^{\kappa-\text{n.t.}}$, and hence

$$c^+ = u^+\big|_{\partial\Omega}^{\kappa-\text{n.t.}} = u^-\big|_{\partial\Omega}^{\kappa-\text{n.t.}} = c^-, \tag{11.228}$$

which now finishes the treatment of (a) under all hypotheses in (i).

Concerning uniqueness under the hypothesis in (ii), note that we may reason as before to obtain (11.217). Similarly, we may invert the operator in (11.219) via a Neumann series argument whenever

$$\left|\frac{1-\eta}{1+\eta}\right| \|K_{A^\top}^\#\|_{[\mathcal{M}^{p,\lambda}(\partial\Omega,w)]^M \to [\mathcal{M}^{p,\lambda}(\partial\Omega,w)]^M} < \frac{1}{2}. \tag{11.229}$$

In view of Proposition 9.6, the above estimate is achievable by assuming $|\eta - 1| < \delta_*$ and that δ_* is sufficiently small. This completes the proof of item (a) under either hypotheses (i) or (ii).

We now move on to address the claims made in (b) and we will treat (b) under the assumptions in (i) first. To begin, let us assume (11.191) and (11.193) hold. To show that the operator (11.195) is an isomorphism, we turn to Theorem 9.7, which if (11.193) is satisfied for small enough $\delta \in (0,1)$ (a regime in which we work from this point on), then

$$\left\|\frac{1-\eta}{1+\eta} K_{A^\top}^\#\right\|_{[\mathcal{M}^{p,\lambda}(\partial\Omega,w)]^M \to [\mathcal{M}^{p,\lambda}(\partial\Omega,w)]^M} < \frac{1}{2}, \tag{11.230}$$

and hence (11.195) is invertible via a Neumann series (cf. Lemma 11.2 which is applicable thanks to Proposition 4.5) and has a bounded inverse by the open mapping theorem. Multiplying the operator in (11.195) by the quantity $1+\eta \in \mathbb{C} \setminus \{0\}$ (recall our present assumptions include $\eta \neq -1$), it follows that the operator

$$-\frac{1}{2}(1+\eta)I + (1-\eta)K^{\#}_{A^{\top}} : \left[M^{p,\lambda}(\partial\Omega, w)\right]^{M} \to \left[M^{p,\lambda}(\partial\Omega, w)\right]^{M} \tag{11.231}$$

is also invertible, and hence h as in (11.197) is a well-defined function belonging to $[M^{p,\lambda}(\partial\Omega, w)]^M$. Indeed, the fact that g belongs to $[M_1^{p,\lambda}(\partial\Omega, w)]^M$ implies $\partial_\nu^A(\mathcal{D}_{A,\mathrm{mod}}g)$ belongs to $[M^{p,\lambda}(\partial\Omega, w)]^M$ (cf. Proposition 9.13), and hence $f-(1-\eta)\partial_\nu^A(\mathcal{D}_{A,\mathrm{mod}}g)$ belongs to $[M^{p,\lambda}(\partial\Omega, w)]^M$. This membership places h in $[M^{p,\lambda}(\partial\Omega, w)]^M$, thanks to the invertibility of the operator in (11.231) and (11.197). To see that $u^\pm$ satisfy (11.190), first recall Theorem 9.12 and Proposition 9.2, which given (11.196) ensure $u^\pm \in [\mathscr{C}^\infty(\Omega_\pm)]^M$ with $Lu^\pm = 0 \in \mathbb{C}^M$ in $\Omega_\pm$. Keeping in mind our notation convention regarding $\mathcal{D}^\pm_{A,\mathrm{mod}}$ adopted in the statement of (b), we find thanks to (9.54) of Theorem 9.12 that

$$(\pm\mathcal{D}^\pm_{A,\mathrm{mod}}g)\Big|^{\kappa-\mathrm{n.t.}}_{\partial\Omega} = \left(\pm\frac{1}{2}I + K_{A,\mathrm{mod}}\right)g \quad \text{at } \sigma\text{-a. e. point on } \partial\Omega. \tag{11.232}$$

Also, thanks to Proposition 4.15 and (3.78) (whose present applicability is ensured by the membership of h to $[M^{p,\lambda}(\partial\Omega, w)]^M$ and Proposition 4.15), we find that

$$(\mathscr{S}^\pm_{\mathrm{mod}}h)\Big|^{\kappa-\mathrm{n.t.}}_{\partial\Omega} = S_{\mathrm{mod}}h \quad \text{at } \sigma\text{-a. e. point on } \partial\Omega. \tag{11.233}$$

Consequently, (11.232), (11.233), and (11.196) lead to

$$\begin{aligned} u^+\Big|^{\kappa-\mathrm{n.t.}}_{\partial\Omega} - u^-\Big|^{\kappa-\mathrm{n.t.}}_{\partial\Omega} &= \left(\frac{1}{2}I + K_{A,\mathrm{mod}}\right)g + S_{\mathrm{mod}}h \\ &\quad - \left[\left(-\frac{1}{2}I + K_{A,\mathrm{mod}}\right)g + S_{\mathrm{mod}}h\right] \\ &= \left(\frac{1}{2}I + K_{A,\mathrm{mod}}\right)g + \left(\frac{1}{2}I - K_{A,\mathrm{mod}}\right)g = g, \end{aligned} \tag{11.234}$$

at σ-a. e. point on $\partial\Omega$ as wanted. Next, recall (11.204) and our notation convention for $\mathscr{S}^\pm_{\mathrm{mod}}$ adopted in the statement of (b). We can then use Proposition 9.2 to say that

$$\partial_\nu^A(\mathscr{S}^\pm_{\mathrm{mod}}h) = \left(\mp\frac{1}{2}I + K^{\#}_{A^{\top}}\right)h \quad \text{at } \sigma\text{-a. e. point on } \partial\Omega. \tag{11.235}$$

Also, bearing in mind (11.204) and our notation convention regarding $\mathcal{D}^\pm_{A,\mathrm{mod}}$, we can say that

$$\partial_\nu^A(\pm\mathcal{D}^\pm_{A,\mathrm{mod}}g) = \partial_\nu^A(\mathcal{D}_{A,\mathrm{mod}}g) \quad \text{at } \sigma\text{-a. e. point on } \partial\Omega. \tag{11.236}$$

Thus, (11.196), (11.235), (11.236), and (11.197) show that

$$\begin{aligned} \partial_\nu^A u^+ - \eta\partial_\nu^A u^- &= \Big[\partial_\nu^A(\mathcal{D}_{A,\mathrm{mod}}g) + \Big(-\frac{1}{2}I + K^\#_{A^\top}\Big)h\Big] \\ &\quad - \eta\Big[\partial_\nu^A(\mathcal{D}_{A,\mathrm{mod}}g) + \Big(\frac{1}{2}I + K^\#_{A^\top}\Big)h\Big] \\ &= (1-\eta)\partial_\nu^A(\mathcal{D}_{A,\mathrm{mod}}g) + \Big(-\frac{1}{2}(1+\eta)I + (1-\eta)K^\#_{A^\top}\Big)h \\ &= (1-\eta)\partial_\nu^A(\mathcal{D}_{A,\mathrm{mod}}g) + f - (1-\eta)\partial_\nu^A(\mathcal{D}_{A,\mathrm{mod}}g) = f, \end{aligned} \tag{11.237}$$

at σ-a. e. point on $\partial\Omega$ as required.

Then, since $g \in [\dot{M}_1^{p,\lambda}(\partial\Omega, w)]^M$, we may use (9.55) of Theorem 9.12 to estimate

$$\|\mathcal{N}_\kappa(\nabla(\mathcal{D}^\pm_{A,\mathrm{mod}}g))\|_{M^{p,\lambda}(\partial\Omega,w)} \le C\|g\|_{[\dot{M}_1^{p,\lambda}(\partial\Omega,w)]^M} \tag{11.238}$$

for some constant $C \in (0,\infty)$, independent of g. Also, since h belongs to $[M^{p,\lambda}(\partial\Omega, w)]^M$, we can use Proposition 9.5, (11.197), the fact that (11.195) has a bounded inverse, and Proposition 9.13 to see that

$$\begin{aligned} \|\mathcal{N}_\kappa(\nabla(\mathscr{S}^\pm_{\mathrm{mod}}h))\|_{M^{p,\lambda}(\partial\Omega,w)} &\le C\|h\|_{[M^{p,\lambda}(\partial\Omega,w)]^M} \\ &\le C\big(\|f\|_{[M^{p,\lambda}(\partial\Omega,w)]^M} + \|g\|_{[\dot{M}_1^{p,\lambda}(\partial\Omega,w)]^M}\big) \end{aligned} \tag{11.239}$$

for some constant $C \in (0,\infty)$, independent of f and g. Thus, thanks to (11.196), (11.238), (11.239), the sublinearity of $\mathcal{N}_\kappa$, and Proposition 4.7, we can confirm that $\mathcal{N}_\kappa(\nabla u^\pm)$ belongs to $M^{p,\lambda}(\partial\Omega, w)$ with the naturally accompanying estimate as given in (11.198). This takes care of existence and estimates in the case that (11.191) holds.

At this point, we now assume (11.192) and (11.193) hold. We may appeal to Theorem 9.15 to ensure that when $\delta \in (0,1)$ is small enough (something we now assume from this point on), then

$$\Big\|\frac{1-\eta}{1+\eta}[K_{A,\mathrm{mod}}]\Big\|_{[\dot{M}_1^{p,\lambda}(\partial\Omega,w)/\sim]^M \to [\dot{M}_1^{p,\lambda}(\partial\Omega,w)/\sim]^M} < \frac{1}{2} \tag{11.240}$$

as long as (11.193) is satisfied. As such, the operator (11.199) is invertible via a Neumann series (cf. Lemma 11.2 and Proposition 8.15) with bounded inverse by the open mapping theorem. Consequently, multiplying the operator in (11.199) by $\frac{1+\eta}{\eta} \in \mathbb{C}\setminus\{0\}$ (recall our present assumption that $\eta \in \mathbb{C}\setminus\{-1,0\}$), we see that the operator

$$\frac{1}{2}\Big(1+\frac{1}{\eta}\Big)I + \Big(1-\frac{1}{\eta}\Big)K_{A,\mathrm{mod}} : [\dot{M}_1^{p,\lambda}(\partial\Omega, w)/\sim]^M \to [\dot{M}_1^{p,\lambda}(\partial\Omega, w)/\sim]^M \tag{11.241}$$

is also invertible. Recall Proposition 9.32, which since $f \in [M^{p,\lambda}(\partial\Omega, w)]^M$ ensures that $S_{\text{mod}} f \in [\dot{M}_1^{p,\lambda}(\partial\Omega, w)]^M$. The fact that (11.241) is a well-defined, invertible operator means that there exists $h \in [\dot{M}_1^{p,\lambda}(\partial\Omega, w)/\sim]^M$ such that

$$\Big(\frac{1}{2}(1+\eta^{-1})I + (1-\eta^{-1})[K_{A,\text{mod}}]\Big)[h] = [g + (1-\eta^{-1})S_{\text{mod}} f], \tag{11.242}$$

or equivalently that there exists some $c \in \mathbb{C}^M$ such that (11.201) holds.

Pressing forward, to see that $u^\pm$ as defined in (11.200) solve (11.190), recall Theorem 9.12 and Proposition 9.2, which ensure that $u^\pm \in [\mathscr{C}^\infty(\Omega_\pm)]^M$ and $Lu^\pm = 0$ in $\Omega_\pm$. Next, bearing in mind the notation convention for $\mathcal{D}^\pm_{A,\text{mod}}$ and that $h \in [\dot{M}_1^{p,\lambda}(\partial\Omega, w)]^M$, we may use (9.54) of Theorem 9.12 to say that

$$(\pm\mathcal{D}^\pm_{A,\text{mod}} h)\big|_{\partial\Omega}^{\kappa-\text{n.t.}} = \Big(\pm\frac{1}{2}I + K_{A,\text{mod}}\Big)h \quad \text{at } \sigma\text{-a. e. point on } \partial\Omega. \tag{11.243}$$

Also, recalling the notation convention for $\mathscr{S}^\pm_{\text{mod}}$, since $f \in [M^{p,\lambda}(\partial\Omega, w)]^M$, from Proposition 4.15 and (3.78), we can conclude that

$$(\mathscr{S}^\pm_{\text{mod}} f)\big|_{\partial\Omega}^{\kappa-\text{n.t.}} = S_{\text{mod}} f \quad \text{at } \sigma\text{-a. e. point on } \partial\Omega. \tag{11.244}$$

It follows from (11.200), (11.243), (11.244), and (11.201) that

$$\begin{aligned} u^+\big|_{\partial\Omega}^{\kappa-\text{n.t.}} - u^-\big|_{\partial\Omega}^{\kappa-\text{n.t.}} &= \Big[\Big(\frac{1}{2}I + K_{A,\text{mod}}\Big)h - S_{\text{mod}} f - c\Big] \\ &\quad - \Big[\eta^{-1}\Big(-\frac{1}{2}I + K_{A,\text{mod}}\Big)h - \eta^{-1}(S_{\text{mod}} f)\Big] \\ &= \Big(\frac{1}{2}(1+\eta^{-1})I + (1-\eta^{-1})K_{A,\text{mod}}\Big)h - (1-\eta^{-1})S_{\text{mod}} f - c \\ &= g + (1-\eta^{-1})S_{\text{mod}} f + c - (1-\eta^{-1})S_{\text{mod}} f - c \\ &= g \end{aligned} \tag{11.245}$$

at σ-a. e. point on $\partial\Omega$ as desired. Moving on, recall (11.204) and our notation convention regarding $\mathscr{S}^\pm_{\text{mod}}$. Since $f \in [M^{p,\lambda}(\partial\Omega, w)]^M$, we can appeal to Proposition 9.2 to say that

$$\partial^A_\nu(\mathscr{S}^\pm_{\text{mod}} f) = \Big(\mp\frac{1}{2}I + K^\#_{A^\top}\Big)f \quad \text{at } \sigma\text{-a. e. point on } \partial\Omega. \tag{11.246}$$

Likewise, keeping in mind the notation convention for $\mathcal{D}^\pm_{A,\text{mod}}$ we find that

$$\partial^A_\nu(\pm\mathcal{D}^\pm_{A,\text{mod}} h) = \partial^A_\nu(\mathcal{D}_{A,\text{mod}} h) \quad \text{at } \sigma\text{-a. e. point on } \partial\Omega. \tag{11.247}$$

Thus, (11.200), (11.246), and (11.247) show that

$$\begin{aligned}\partial_\nu^A u^+ - \eta\partial_\nu^A u^- &= \Big[\partial_\nu^A(\mathcal{D}_{A,\mathrm{mod}}h) - \Big(-\tfrac{1}{2}I + K^\#_{A^\top}\Big)f\Big] \\ &\quad - \eta\Big[\eta^{-1}\partial_\nu^A(\mathcal{D}_{A,\mathrm{mod}}h) - \eta^{-1}\Big(\tfrac{1}{2}I + K^\#_{A^\top}\Big)f\Big] \\ &= -\Big(-\tfrac{1}{2}I + K^\#_{A^\top}\Big)f + \Big(\tfrac{1}{2}I + K^\#_{A^\top}\Big)f = f\end{aligned} \tag{11.248}$$

at σ-a. e. point on $\partial\Omega$ as wanted.

Lastly, to check membership of $\mathcal{N}_\kappa(\nabla u^\pm)$ to $M^{p,\lambda}(\partial\Omega)$ and the size estimates in (11.198), start by observing that the membership of h to $[\dot{M}_1^{p,\lambda}(\partial\Omega, w)]^M$ along with (9.55) of Theorem 9.12 entails the existence of some constant $C \in (0,\infty)$, independent of h, for which

$$\|\mathcal{N}_\kappa(\nabla(\mathcal{D}^\pm_{A,\mathrm{mod}}h))\|_{M^{p,\lambda}(\partial\Omega,w)} \le C\|h\|_{[\dot{M}_1^{p,\lambda}(\partial\Omega,w)]^M}. \tag{11.249}$$

We can then use the fact that (11.241) is invertible with bounded inverse, (11.201), the membership of f to $[M^{p,\lambda}(\partial\Omega, w)]^M$, and Proposition 9.32 to see that

$$\|h\|_{[\dot{M}_1^{p,\lambda}(\partial\Omega,w)]^M} \le C(\|g\|_{[\dot{M}_1^{p,\lambda}(\partial\Omega,w)]^M} + \|f\|_{[M^{p,\lambda}(\partial\Omega,w)]^M}) \tag{11.250}$$

for some constant $C \in (0,\infty)$ independent of f and g. Together, (11.249) and (11.250) show that there exists $C \in (0,\infty)$, independent of f and g such that

$$\|\mathcal{N}_\kappa(\nabla(\mathcal{D}^\pm_{A,\mathrm{mod}}h))\|_{M^{p,\lambda}(\partial\Omega,w)} \le C(\|g\|_{[\dot{M}_1^{p,\lambda}(\partial\Omega,w)]^M} + \|f\|_{[M^{p,\lambda}(\partial\Omega,w)]^M}). \tag{11.251}$$

Since $f \in [M^{p,\lambda}(\partial\Omega, w)]^M$, we can also use Proposition 9.5 to say that there exists a constant $C \in (0,\infty)$, independent of f for which

$$\|\mathcal{N}_\kappa(\nabla(\mathscr{S}^\pm_{\mathrm{mod}}f))\|_{M^{p,\lambda}(\partial\Omega,w)} \le C\|f\|_{[M^{p,\lambda}(\partial\Omega,w)]^M}. \tag{11.252}$$

Ultimately, (11.200), the sublinearity of $\mathcal{N}_\kappa$, (11.251), (11.252), and Proposition 4.7 confirm that $\mathcal{N}_\kappa(\nabla u^\pm)$ belongs to $M^{p,\lambda}(\partial\Omega, w)$ with an estimate as in (11.198).

Finally, in the case that we work with the assumptions in (ii), either proof of existence works with the exception of how (11.230) and (11.240) are obtained. In this case, Proposition 9.6 and Proposition 9.14 paired with (11.194) yield the desired operator norm estimates whenever $\delta_* \in (0,1)$ is small enough. This completes the treatment of item (b) under either hypotheses (i) or (ii).

We now address item (c). Regarding Muckenhoupt-weighted vanishing Morrey spaces, the proof of (a) and (b) is largely unchanged thanks to combinations of Theorem 10.33 and Theorem 10.38. Note also the utility of Proposition 10.26 to obtain the analogous size estimates. This completes the proof of Theorem 11.35. □

Below is a well-posedness result for the transmission problem with boundary data in Mucknehoupt-weighted Block spaces.

Theorem 11.41 (The weighted Block transmission BVP). *Fix $n \in \mathbb{N}$ with $n \geq 2$ along with $M \in \mathbb{N}$. Let $\Omega \subseteq \mathbb{R}^n$ be a two-sided* NTA *domain with unbounded, Ahlfors regular boundary, abbreviate $\sigma := \mathcal{H}^{n-1}\lfloor\partial\Omega$, define*

$$\Omega_+ := \Omega, \quad \Omega_- := \mathbb{R}^n \setminus \overline{\Omega}, \tag{11.253}$$

and let ν denote the geometric measure theoretic outward unit normal to Ω. Next, suppose L is a weakly elliptic, second-order, homogeneous, constant complex coefficient, $M \times M$ system in $\mathbb{R}^n$. Now, pick an arbitrary integrability exponent $q \in (1,\infty)$ and write $q' := q/(q-1)$. Fix a Muckenhoupt weight $\omega \in A_q(\partial\Omega,\sigma)$,

$$\text{choose } \varepsilon \in (0, q-1) \text{ such that } \omega \in A_{q-\varepsilon}(\partial\Omega,\sigma), \tag{11.254}$$

and select any threshold $\lambda \in (0, \varepsilon(n-1)/(q-1))$. Having picked some coefficient tensor $A \in \mathfrak{A}_L$, an aperture parameter $\kappa \in (0,\infty)$, and a transmission (or coupling) parameter $\eta \in \mathbb{C}$, consider the transmission problem for L in $\Omega_\pm$ with boundary data in a Muckenhoupt-weighted Block space and a Block-based Sobolev space, formulated below:

$$\begin{cases} u^\pm \in [\mathscr{C}^\infty(\Omega_\pm)]^M, \\ Lu^\pm = 0 \quad \text{in } \Omega_\pm, \\ \mathcal{N}_\kappa(\nabla u^\pm) \in \mathcal{B}^{q,\lambda}(\partial\Omega,\omega), \\ u^+\big|_{\partial\Omega}^{\kappa-\text{n.t.}} - u^-\big|_{\partial\Omega}^{\kappa-\text{n.t.}} = g \in [\dot{\mathcal{B}}_1^{q,\lambda}(\partial\Omega,\omega)]^M, \\ \partial_\nu^A u^+ - \eta \cdot \partial_\nu^A u^- = f \in [\mathcal{B}^{q,\lambda}(\partial\Omega,\omega)]^M. \end{cases} \tag{11.255}$$

Finally, assume one of the following two scenarios:

(i) *Suppose either*

$$A^\top \in \mathfrak{A}_{L^\top}^{\text{dis}} \quad \text{and} \quad \eta \in \mathbb{C} \setminus \{-1\}, \tag{11.256}$$

or

$$A \in \mathfrak{A}_L^{\text{dis}} \quad \text{and} \quad \eta \in \mathbb{C} \setminus \{-1, 0\}, \tag{11.257}$$

and that

$$\left|\frac{1-\eta}{1+\eta}\right| \|\nu\|_{[\text{BMO}(\partial\Omega,\sigma)]^n}^{\langle m\rangle} < \delta, \tag{11.258}$$

for some $\delta \in (0,1)$, depending on the ADR *character of $\partial\Omega$, n, q, ε, λ, $[\omega]_{A_{q-\varepsilon}}$, and L.*

(ii) *Suppose*

$$|\eta - 1| < \delta_* \tag{11.259}$$

for some $\delta_* \in (0,1)$ depending only on the ADR *character of $\partial\Omega$, the corkscrew constants of Ω, n, q, ε, λ, $[\omega]_{A_{q-\varepsilon}}$, and L.*

Then all claims below are true:

(a) [Uniqueness modulo constants]. *Any two solutions to the transmission problem* (11.255) *differ by only a constant (in* $\mathbb{C}^M$*).*

(b) [Well posedness and integral representation]. *In the case that either* (11.256) *and* (11.258) *hold or* (11.259) *holds, the operator*

$$-\frac{1}{2}I + \frac{1-\eta}{1+\eta}K^{\#}_{A^\top} : [\mathcal{B}^{q,\lambda}(\partial\Omega,\omega)]^M \to [\mathcal{B}^{q,\lambda}(\partial\Omega,\omega)]^M \tag{11.260}$$

is an isomorphism. Denote by $\mathcal{D}^{\pm}_{A,\mathrm{mod}}$ *the modified double layer potential operators associated with the sets* $\Omega_\pm$ *(cf.* (3.34)*), and likewise* $\mathscr{S}^{\pm}_{\mathrm{mod}}$ *for the modified single layer potential operators (cf.* (3.76)*). Then the functions* $u^\pm : \Omega_\pm \to \mathbb{C}^M$ *defined according to*

$$u^\pm := \pm\mathcal{D}^{\pm}_{A,\mathrm{mod}}g + \mathscr{S}^{\pm}_{\mathrm{mod}}h \quad \text{in } \Omega_\pm, \tag{11.261}$$

where $h \in [\mathcal{B}^{q,\lambda}(\partial\Omega,\omega)]^M$ *is given by*

$$h := \Big(-\frac{1}{2}(1+\eta)I + (1-\eta)K^{\#}_{A^\top}\Big)^{-1}\big(f - (1-\eta)\partial^A_\nu(\mathcal{D}_{A,\mathrm{mod}}g)\big) \tag{11.262}$$

at σ*-a. e. point on* $\partial\Omega$*, solve the transmission problem* (11.255)*. Moreover, there exists a constant* $C \in (0,\infty)$*, independent of* f *and* g*, with the property that*

$$\|\mathcal{N}_\kappa(\nabla u^\pm)\|_{\mathcal{B}^{q,\lambda}(\partial\Omega,\omega)} \le C\big(\|f\|_{[\mathcal{B}^{q,\lambda}(\partial\Omega,\omega)]^M} + \|g\|_{[\dot{\mathcal{B}}^{q,\lambda}_1(\partial\Omega,\omega)/\sim]^M}\big). \tag{11.263}$$

Suppose instead that either (11.257) *and* (11.258) *hold or* (11.259) *holds. Additionally, strengthen the original hypotheses in* (11.254) *by assuming that the threshold* ε *is chosen to be*[5]

$$\begin{gathered}\varepsilon := (q-1)/p' \in (0,q-1), \quad \text{where } p' := p/(p-1) \in (1,\infty)\\ \text{and } p \in (1,\infty) \text{ is such that } \omega' \in A_{q'}(\partial\Omega,\sigma) \cap \mathrm{RH}_p(\partial\Omega,\sigma).\end{gathered} \tag{11.264}$$

Then (cf. (3.47)*) the operator*

$$\frac{1}{2}I - \frac{1-\eta}{1+\eta}[K_{A,\mathrm{mod}}] : [\dot{\mathcal{B}}^{q,\lambda}_1(\partial\Omega,\omega)/\sim]^M \to [\dot{\mathcal{B}}^{q,\lambda}_1(\partial\Omega,\omega)/\sim]^M \tag{11.265}$$

is an isomorphism. Moreover, the functions $u^\pm : \Omega_\pm \to \mathbb{C}^M$ *defined according to*

$$\begin{aligned} u^+ &:= \mathcal{D}^{+}_{A,\mathrm{mod}}h - \mathscr{S}^{+}_{\mathrm{mod}}f - c && \text{in } \Omega_+,\\ u^- &:= -\eta^{-1}\mathcal{D}^{-}_{A,\mathrm{mod}}h + \eta^{-1}\mathscr{S}^{-}_{\mathrm{mod}}f && \text{in } \Omega_-,\end{aligned} \tag{11.266}$$

5 That this choice of ε is permissible in the context of (11.254) is ensured by Proposition 6.35.

where $h \in [\dot{\mathcal{B}}_1^{q,\lambda}(\partial\Omega,\omega)]^M$ *and* $c \in \mathbb{C}^M$ *are such that*

$$\Big(\frac{1}{2}(1+\eta^{-1})I + (1-\eta^{-1})K_{A,\mathrm{mod}}\Big)h = g + (1-\eta^{-1})S_{\mathrm{mod}}f + c \tag{11.267}$$

at σ*-a. e. point on* $\partial\Omega$*, solve the transmission problem* (11.255)*. Additionally, there exists a constant* $C \in (0,\infty)$*, independent of* f *and* g*, such that the size estimate* (11.263) *holds for* $u^{\pm}$ *as in* (11.266)*.*

Finally, in the case that (11.256) *holds, the transmission problem* (11.255) *is solvable and any two solutions differ by only a constant and satisfy the size estimate* (11.263)*. If instead* (11.257) *holds, and additionally* ε *is chosen to be the quantity* $(q-1)/p' \in (0,q-1)$*, where* $p' := p/(p-1) \in (1,\infty)$ *and* $p \in (1,\infty)$ *is such that* $\omega' := \omega^{-q'/q} \in A_{q'}(\partial\Omega,\sigma) \cap \mathrm{RH}_p(\partial\Omega,\sigma)$*, then the transmission problem* (11.255) *is solvable and any two solutions differ by only a constant and satisfy the size estimate* (11.263)*.*

Remark 11.42. In the case that $\eta = -1$, the transmission problem (11.255) is ill posed, and the space of null solutions to (11.255) when $\eta = -1$ is infinite-dimensional. This can seen this through a similar argument as in Remark (11.40). Note for such an argument the relevance of Theorem 9.17, and the fact that Muckenhoupt-weighted Block spaces are infinite-dimensional. This can be seen from Proposition 6.25, which says that all characteristic functions of balls belong to a Muckenhoupt-weighted Block space. Alternatively, since the dual of a Muckenhoupt-weighted Block space is a Muckenhoupt-weighted Morrey space, which itself is an infinite-dimensional vector space, we see that a Block space cannot possibly be finite-dimensional.

Proof. The structure of this proof follows closely the proof of Theorem 11.35, though we highlight the key differences. In the case of Muckenhoupt-weighted Block spaces, to tackle uniqueness (modulo constants) under the hypotheses in (11.256) and (11.258), we now rely on Theorem 11.21, Proposition 6.17, (7.90) of Theorem 7.6, Theorem 9.23, and Proposition 6.11. To prove uniqueness (modulo constants) under the hypotheses in (11.257), we now appeal to Theorem 9.23 and Theorem 7.13. If instead we assume (ii), then uniqueness follows much as before, though now thanks to Proposition 9.21.

To address the claims in item (b), we start with the proof of existence under the assumptions in (11.256) and (11.258). Invertibility of the operator

$$-\frac{1}{2}I + \frac{1-\eta}{1+\eta}K^{\#}_{A^{\top}} : [\mathcal{B}^{q,\lambda}(\partial\Omega,\omega)]^M \to [\mathcal{B}^{q,\lambda}(\partial\Omega,\omega)]^M \tag{11.268}$$

comes from Theorem 9.23, Lemma 11.2, and Proposition 6.11. To demonstrate solvability we also rely on Proposition 9.28, Theorem 9.27, Theorem 9.17, and (9.108), Proposition 6.17 with (3.78). As for the estimates in (11.263), we can use (9.109) of Theorem 9.27, Proposition 9.20, Proposition 9.28, and Proposition 6.14.

Concerning the proof of existence under the hypotheses in (11.257) and (11.258), we achieve the invertibility of the operator

$$\frac{1}{2}I - \frac{1-\eta}{1+\eta}[K_{A,\mathrm{mod}}] : [\mathring{\mathcal{B}}_1^{q,\lambda}(\partial\Omega, \omega)/\sim]^M \to [\mathring{\mathcal{B}}_1^{q,\lambda}(\partial\Omega, \omega)/\sim]^M \tag{11.269}$$

thanks to Theorem 9.30, Lemma 11.2, and Proposition 8.19. To demonstrate solvability, we now use Proposition 9.33, Theorem 9.27, (9.108), Proposition 6.17 with (3.78), and the trace result in Theorem 9.17. To get the required estimates in (11.263), we turn to (9.109) of Theorem 9.27, Proposition 9.28, and Proposition 9.20, and Proposition 6.14. Finally, the case of existence under the hypotheses in (ii) can be handled as before using either method thanks to Proposition 9.21 and Proposition 9.29. □

One interesting feature of our well-posedness result for the transmission problem in Theorems 11.35 and 11.41 is that in contrast to the other boundary value problems (cf. Theorems 11.9, 11.12, 11.22, 11.27, 11.29, 11.30, 11.32, and 11.33), we do not necessarily require any flatness condition on the domain Ω. Specifically, we require only that (11.193)/(11.258) holds. In the extreme case that the transmission parameter is equal to 1, then (11.194)/(11.258) is always true. In this case, the transmission problem (11.190)/(11.255) is solvable for *any* two-sided NTA domain with unbounded Ahlfors regular boundary. In particular, in the case that $\eta = 1$, *any* upper-graph Lipschitz domain is a perfectly fine geometric environment in which to seek a well-posedness result for the transmission problem. This is not necessarily the case for our other boundary value problems (Dirichlet, Neumann, and homogeneous/inhomogeneous regularity) in which only those upper-graph Lipschitz domains with sufficiently small Lipschitz constants are counted among the class of δ-AR domains accommodated by the theory to yield a well-posed problem.

Another interesting feature of the transmission problem also concerns the exceptional case $\eta = 1$ and values of η near one. In this regime, we find that the transmission problem in Muckenhoupt-weighted Morrey and Block spaces is well-posed for *any* weakly elliptic system. This is in stark contrast to the Dirichlet, Neumann, and regularity problems, which all require that the system in question possesses a distinguished coefficient tensor. Interestingly, in this exceptional case the boundary conditions actually compensate for a "poorly behaved" system, which otherwise may not have a structure that leads to a well-posedness result. This is yet another remarkable instance of how geometry, analysis, and PDE tightly interface and affect the solvability of a boundary value problem as a whole.

12 Boundary problems in domains with compact boundary

Up to this point, the method we have employed for solving boundary problems has forced the boundary of the underlying domain to be unbounded. More specifically, our well-posedness results throughout Chapter 11 involve the demand that the BMO seminorm of the outward unit normal be small, and in particular less than one. This geometric constraint on the domain then forces the topological boundary of the domain to be unbounded as witnessed in [79, Lemma 2.8, p. 52]. This suggests that in order to study domains with compact boundaries, we need a different notion of flatness. We turn to the following notion of flatness introduced in [90, Section 3.4].

Before properly defining this different way to characterize flatness, we recall the Sarason space of functions of vanishing mean oscillations (VMO). Namely, given a closed Ahlfors regular set $\Sigma \subseteq \mathbb{R}^n$ with surface measure $\sigma := \mathcal{H}^{n-1}\lfloor\Sigma$, the space $\mathrm{VMO}(\Sigma,\sigma)$ is the closed subspace of $\mathrm{BMO}(\Sigma,\sigma)$ consisting of functions $f \in \mathrm{BMO}(\Sigma,\sigma)$ for which

$$\lim_{R\to 0^+} \sup_{\substack{x\in\Sigma \\ r\in(0,R)}} \fint_{\Delta(x,r)} \left| f - \fint_{\Delta(x,r)} f \, d\sigma \right| d\sigma = 0. \tag{12.1}$$

We refer the reader to [87, Section 3.1] for a more detailed treatment of this function space in the context of closed Ahlfors regular sets.

Definition 12.1. Fix $n \in \mathbb{N}$ with $n \geq 2$, and let $\Omega \subseteq \mathbb{R}^n$ be an AR domain with compact boundary. Abbreviate $\sigma := \mathcal{H}^{n-1}\lfloor\partial\Omega$ and denote by ν the geometric measure theoretic outward unit normal to Ω.

1. Given $\delta \in (0,1)$, call Ω a δ-`oscillating` AR `domain` provided

$$\operatorname{dist}\big(\nu, [\mathrm{VMO}\,(\partial\Omega,\sigma)]^n\big) < \delta, \tag{12.2}$$

   where the distance is measured in $[\mathrm{BMO}(\partial\Omega,\sigma)]^n$.
2. Call Ω an `infinitesimally flat` AR `domain` provided ν belongs to the space $[\mathrm{VMO}(\partial\Omega,\sigma)]^n$.

It is clear from the definitions that any infinitesimally flat AR domain is a δ-oscillating domain for any choice of $\delta \in (0,1)$.

While δ-AR domains (cf. Definition 2.5) can be thought of as the geometric measure theoretically sharp version of upper-graph Lipschitz domains, the class of infinitesimally flat AR domains can be thought of as the geometric measure theoretically sharp version of $\mathscr{C}^1$ domains with compact boundary.

With this notion of flatness in Definition 12.1 in hand, we are able to state some well-posedness results in domains with compact boundary where the boundary datum comes from Muckenhoupt-weighted Morrey, vanishing Morrey, or Block spaces. These

https://doi.org/10.1515/9783111458274-012

results follow from the work done in Chapter 10, which fits weighted Morrey and Block spaces into the framework of generalized Banach function Spaces, and the work in [90], which considers boundary problems with data in generalized Banach function spaces.

Our first result of this section regards the Dirichlet problem with data in a weighted Morrey space. Due to the nature of the function space under consideration, this result is new even in the class of $\mathscr{C}^1$ domains with compact boundary (and even when L is merely the Laplacian).

Theorem 12.2 (The weighted Morrey Dirichlet BVP). *Select $n, M \in \mathbb{N}$ with $n \geq 2$, and suppose $\Omega \subseteq \mathbb{R}^n$ is an AR domain with compact boundary. Abbreviate $\sigma := \mathcal{H}^{n-1}\lfloor\partial\Omega$ and denote by ν the geometric measure theoretic outward unit normal to Ω. Suppose L is a homogeneous, second-order, constant complex coefficient, $M \times M$ system in $\mathbb{R}^n$. Fix two integrability exponents $p, q \in (1, \infty)$, an arbitrary Muckenhoupt weight $w \in A_p(\partial\Omega, \sigma) \cap \mathrm{RH}_q(\partial\Omega, \sigma)$, and some threshold $\lambda \in (0, (n-1)/q')$, where $q' := q/(q-1)$. For a fixed aperture parameter $\kappa \in (0, \infty)$, consider the Dirichlet problem for L in Ω with data in a Muckenhoupt-weighted Morrey space, formulated below:*

$$\begin{cases} u \in [\mathscr{C}^\infty(\Omega)]^M, \\ Lu = 0 \quad \text{in } \Omega, \\ \mathcal{N}_\kappa u \in M^{p,\lambda}(\partial\Omega, w), \\ u\big|_{\partial\Omega}^{\kappa-\mathrm{n.t.}} = f \in [M^{p,\lambda}(\partial\Omega, w)]^M, \\ u(x) = O(|x|^{2-n}) \quad \text{as } |x| \to \infty \text{ if } \Omega \text{ is unbounded.} \end{cases} \tag{12.3}$$

Then there exists some threshold $\delta \in (0,1)$, which depends only on n, p, q, λ, $[w]_{A_p}$, $[w]_{\mathrm{RH}_q}$, the system L, the ADR character of $\partial\Omega$, and the relative flatness ratio $\wp(\Omega)/\operatorname{diam}\partial\Omega$ (cf. [90, Definition 3.1.9, p. 123]*), with the property that when*

$$\operatorname{dist}(\nu, [\mathrm{VMO}\,(\partial\Omega, \sigma)]^n) < \delta \tag{12.4}$$

where the distance is considered in the space $[\mathrm{BMO}\,(\partial\Omega, \sigma)]^n$, hence if Ω is a δ-oscillating AR domain[1] *(cf. Definition 12.1), the following statements are true:*

(a) [Existence]. *If the system L is weakly elliptic with $\mathfrak{A}_L^{\mathrm{dis}} \neq \varnothing$, the space of admissible boundary data for the Dirichlet problem* (12.3), *i. e.,*

$$\begin{aligned} \{u\big|_{\partial\Omega}^{\kappa-\mathrm{n.t.}} : u \in [\mathscr{C}^\infty(\Omega)]^M,\ Lu = 0 \text{ in } \Omega, \\ \mathcal{N}_\kappa u \in M^{p,\lambda}(\partial\Omega, w),\ u\big|_{\partial\Omega}^{\kappa-\mathrm{n.t.}} \text{ exists } \sigma\text{-a. e., and} \\ u(x) = O(|x|^{2-n}) \text{ as } |x| \to \infty \text{ if } \Omega \text{ is unbounded}\}, \end{aligned} \tag{12.5}$$

is a closed linear subspace of finite codimension in $[M^{p,\lambda}(\partial\Omega, w)]^M$, and a solution u of (12.3) *exists if and only if the boundary datum f belongs to* (12.5).

1 in particular, if Ω is an infinitesimally flat AR domain.

(b) [Uniqueness]. *If L is weakly elliptic with* $\mathfrak{A}^{\mathrm{dis}}_{L^\top} \neq \varnothing$, *the space of null solutions of the Dirichlet problem* (12.3), *i. e.,*

$$\begin{aligned}\{u \in [\mathscr{C}^\infty(\Omega)]^M : {} & Lu = 0 \text{ in } \Omega, \\ & \mathcal{N}_\kappa u \in M^{p,\lambda}(\partial\Omega, w),\ u\big|^{\kappa-\mathrm{n.t.}}_{\partial\Omega} = 0 \text{ and}, \\ & u(x) = O(|x|^{2-n}) \text{ as } |x| \to \infty \text{ if } \Omega \text{ is unbounded}\},\end{aligned} \tag{12.6}$$

is finite-dimensional; in addition,

for any null-solution u of the boundary problem (12.3) (*i. e., for any function u belonging to the space described in* (12.6)), $\mathcal{N}_\kappa(\nabla u)$ *belongs to* $M^{p,\lambda}(\partial\Omega, w)$ *and the nontangential trace* $(\nabla u)\big|^{\kappa-\mathrm{n.t.}}_{\partial\Omega}$ *exists at σ-a. e. point on* $\partial\Omega$, (12.7)

so, in particular,

the space of null solutions for the Dirichlet problem (12.3) (*cf.* (12.6)) *coincides with that of the regularity problem* (12.32) (*cf.* (12.36)). (12.8)

(c) [Fredholm solvability]. *If one assumes that*

$$L \text{ is weakly elliptic with } \mathfrak{A}^{\mathrm{dis}}_{L} \neq \varnothing \text{ and } \mathfrak{A}^{\mathrm{dis}}_{L^\top} \neq \varnothing, \tag{12.9}$$

then the Dirichlet problem (12.3) *is Fredholm solvable in the sense that the space of admissible boundary data (i. e.,* (12.5)) *has finite codimension in the space* $[M^{p,\lambda}(\partial\Omega, w)]^M$, *and the space of null solutions to* (12.3) (*i. e.,* (12.6)) *is finite-dimensional.*

(d) [Sharpness]. *In the class of weakly elliptic operators L for which* $\mathfrak{A}^{\mathrm{dis}}_{L} = \varnothing$ *or* $\mathfrak{A}^{\mathrm{dis}}_{L^\top} = \varnothing$, *it may happen that the Dirichlet problem* (12.3) *fails to be Fredholm solvable even when the underlying domain Ω is the unit disk in the plane (or the complement of its closure). In fact, the aforementioned failure of Fredholm solvability may be total, in the sense that for the problem* (12.3), *the space of admissible boundary data may have infinite codimension in* $[M^{p,\lambda}(\partial\Omega, w)]^M$, *and the space of null solutions may be infinite-dimensional.*

(e) [Strong ellipticity]. *If* (12.9) *is strengthened to*[2] *assuming that*

L satisfies the Legendre–Hadamard strong ellipticity condition (*cf.* (3.3)) *and one has* $\mathfrak{A}^{\mathrm{dis}}_{L} \neq \varnothing$, (12.10)

then the Dirichlet problem (12.3) *is actually well posed. More specifically, the boundary problem* (12.3) *has a unique solution, and this solution enjoys the additional property that there exists a constant* $C \in (0, \infty)$, *independent of f, such that*

2 According to [90, Proposition 1.4.1, p. 25], whenever L is as in (12.10) we have $\mathfrak{A}^{\mathrm{dis}}_{L} \neq \varnothing$ and $\mathfrak{A}^{\mathrm{dis}}_{L^\top} \neq \varnothing$.

$$\|f\|_{[M^{p,\lambda}(\partial\Omega,w)]^M} \le \|\mathcal{N}_\kappa u\|_{M^{p,\lambda}(\partial\Omega,w)} \le C\|f\|_{[M^{p,\lambda}(\partial\Omega,w)]^M}. \tag{12.11}$$

Proof. Since Proposition 10.41 ensures $M^{p,\lambda}(\partial\Omega, w)$ is a generalized Banach function space and Proposition 5.1, Proposition 7.1, and Theorem 10.46 demonstrate that the Hardy–Littlewood maximal operator on $\partial\Omega$ is bounded on both $M^{p,\lambda}(\partial\Omega, w)$ and it associated space $(M^{p,\lambda}(\partial\Omega, w))'$, we see that Theorem 12.2 is a consequence of [90, Theorem 8.1.9, pp. 614–616]. □

Here is now the Dirichlet problem on domains with compact boundary at the level of Muckenhoupt-weighted Block spaces. Again, because of the nature of the function space, this result is new even in the class of $\mathscr{C}^1$ domains with compact boundary (and when L is simply the Laplacian).

Theorem 12.3 (The weighted Block Dirichlet BVP). *Select $n, M \in \mathbb{N}$ with $n \ge 2$, and suppose $\Omega \subseteq \mathbb{R}^n$ is an* AR *domain with compact boundary. Abbreviate $\sigma := \mathcal{H}^{n-1}\lfloor\partial\Omega$ and denote by ν the geometric measure theoretic outward unit normal to Ω. Suppose L is a homogeneous, second-order, constant complex coefficient, $M \times M$ system in $\mathbb{R}^n$. Fix an integrability exponent $q \in (1, \infty)$ with $q' := q/(q-1)$. Also, choose an arbitrary Muckenhoupt weight $\omega \in A_q(\partial\Omega, \sigma)$, which also satisfies*

$$\omega' := \omega^{-q'/q} \in A_{q'}(\partial\Omega, \sigma) \cap \mathrm{RH}_p(\partial\Omega, \sigma) \tag{12.12}$$

for some $p \in (1, \infty)$. Define $\varepsilon := (q-1)/p'$, which ensures $\omega \in A_{q-\varepsilon}(\partial\Omega, \sigma)$ by Proposition 6.35, and select any threshold

$$\lambda \in (0, \varepsilon(n-1)/(q-1)) = (0, (n-1)/p'). \tag{12.13}$$

For a fixed aperture parameter $\kappa \in (0, \infty)$, consider the Dirichlet problem for L in Ω with data in a Muckenhoupt-weighted Block space, formulated below:

$$\begin{cases} u \in [\mathscr{C}^\infty(\Omega)]^M, \\ Lu = 0 \quad \text{in } \Omega, \\ \mathcal{N}_\kappa u \in \mathcal{B}^{q,\lambda}(\partial\Omega, \omega), \\ u\big|_{\partial\Omega}^{\kappa-\text{n.t.}} = f \in [\mathcal{B}^{q,\lambda}(\partial\Omega, \omega)]^M, \\ u(x) = O(|x|^{2-n}) \quad \text{as } |x| \to \infty \text{ if } \Omega \text{ is unbounded.} \end{cases} \tag{12.14}$$

Then there exists some threshold $\delta \in (0, 1)$, which depends only on n, q, ε, λ, $[\omega]_{A_{q-\varepsilon}}$, the system L, the ADR *character of $\partial\Omega$, and the relative flatness ratio $\wp(\Omega)/\operatorname{diam}\partial\Omega$ (cf.* [90, Definition 3.1.9, p. 123]), *with the property that when*

$$\operatorname{dist}(\nu, [\mathrm{VMO}(\partial\Omega, \sigma)]^n) < \delta \tag{12.15}$$

where the distance is considered in the space $[\mathrm{BMO}(\partial\Omega, \sigma)]^n$, *hence if Ω is a δ-oscillating*

AR *domain*[3] (*cf. Definition* 12.1)*, the following statements are true:*

(a) [Existence]. *If the system L is weakly elliptic with* $\mathfrak{A}_L^{\mathrm{dis}} \neq \varnothing$*, the space of admissible boundary data for the Dirichlet problem* (12.14)*, i. e.,*

$$\begin{aligned}\{u\big|_{\partial\Omega}^{\kappa-\mathrm{n.t.}} : u \in [\mathscr{C}^\infty(\Omega)]^M,\ Lu = 0 \text{ in } \Omega, \\ \mathcal{N}_\kappa u \in \mathcal{B}^{q,\lambda}(\partial\Omega,\omega),\ u\big|_{\partial\Omega}^{\kappa-\mathrm{n.t.}} \text{ exists } \sigma\text{-a. e., and} \\ u(x) = O(|x|^{2-n}) \text{ as } |x| \to \infty \text{ if } \Omega \text{ is unbounded}\},\end{aligned} \tag{12.16}$$

is a closed linear subspace of finite codimension in $[\mathcal{B}^{q,\lambda}(\partial\Omega,\omega)]^M$*, and a solution u of* (12.14) *exists if and only if the boundary datum f belongs to* (12.16).

(b) [Uniqueness]. *If L is weakly elliptic with* $\mathfrak{A}_{L^\top}^{\mathrm{dis}} \neq \varnothing$*, the space of null solutions of the Dirichlet problem* (12.14)*, i. e.,*

$$\begin{aligned}\{u \in [\mathscr{C}^\infty(\Omega)]^M : Lu = 0 \text{ in } \Omega, \\ \mathcal{N}_\kappa u \in \mathcal{B}^{q,\lambda}(\partial\Omega,\omega),\ u\big|_{\partial\Omega}^{\kappa-\mathrm{n.t.}} = 0, \text{ and} \\ u(x) = O(|x|^{2-n}) \text{ as } |x| \to \infty \text{ if } \Omega \text{ is unbounded}\},\end{aligned} \tag{12.17}$$

is finite-dimensional; in addition,

> *for any null-solution u of the boundary problem* (12.14) (*i. e., for any function u belonging to the space described in* (12.17))*,* $\mathcal{N}_\kappa(\nabla u)$ *belongs to* $\mathcal{B}^{q,\lambda}(\partial\Omega,\omega)$ *and the nontangential trace* $(\nabla u)\big|_{\partial\Omega}^{\kappa-\mathrm{n.t.}}$ *exists at σ-a. e. point on ∂Ω,* (12.18)

so, in particular,

> *the space of null solutions for the Dirichlet problem* (12.14) (*cf.* (12.17)) *coincides with that of the regularity problem* (12.40) (*cf.* (12.44)). (12.19)

(c) [Fredholm solvability]. *If one assumes that*

$$L \text{ is weakly elliptic with } \mathfrak{A}_L^{\mathrm{dis}} \neq \varnothing \text{ and } \mathfrak{A}_{L^\top}^{\mathrm{dis}} \neq \varnothing, \tag{12.20}$$

then the Dirichlet problem (12.14) *is Fredholm solvable in the sense that the space of admissible boundary data* (*i. e.,* (12.16)) *has finite codimension in the space* $[\mathcal{B}^{q,\lambda}(\partial\Omega,\omega)]^M$*, and the space of null solutions to* (12.14) (*i. e.,* (12.17)) *is finite-dimensional.*

(d) [Sharpness]. *In the class of weakly elliptic operators L for which* $\mathfrak{A}_L^{\mathrm{dis}} = \varnothing$ *or* $\mathfrak{A}_{L^\top}^{\mathrm{dis}} = \varnothing$*, it may happen that the Dirichlet problem* (12.14) *fails to be Fredholm solvable even when the underlying domain Ω is the unit disk in the plane (or the*

3 in particular, if Ω is an infinitesimally flat AR domain.

complement of its closure). In fact, the aforementioned failure of Fredholm solvability may be total, in the sense that for the problem (12.14)*, the space of admissible boundary data may have infinite codimension in* $[\mathcal{B}^{q,\lambda}(\partial\Omega,\omega)]^M$*, and the space of null solutions may be infinite-dimensional.*

(e) [Strong ellipticity]. *If* (12.20) *is strengthened to*[4] *assuming that*

$$\begin{aligned}&L \text{ satisfies the Legendre–Hadamard strong ellipticity condition}\\&(cf.\ (3.3)) \text{ and one has } \mathfrak{A}_L^{\mathrm{dis}} \neq \varnothing,\end{aligned}\tag{12.21}$$

then the Dirichlet problem (12.14) *is actually well posed. More specifically, the boundary problem* (12.14) *has a unique solution, and this solution enjoys the additional property that there exists* $C \in (0,\infty)$*, independent of* f*, such that*

$$\|f\|_{[\mathcal{B}^{q,\lambda}(\partial\Omega,\omega)]^M} \leq \|\mathcal{N}_\kappa u\|_{\mathcal{B}^{q,\lambda}(\partial\Omega,\omega)} \leq C\|f\|_{[\mathcal{B}^{q,\lambda}(\partial\Omega,\omega)]^M}. \tag{12.22}$$

Proof. Theorem 12.3 is a consequence of [90, Theorem 8.1.9, pp. 614–616] upon making the identification $\mathbb{X} := M^{q',\lambda}(\partial\Omega,\omega')$. Indeed, Proposition 10.41 shows that $M^{q',\lambda}(\partial\Omega,\omega')$ is a generalized Banach function space, Theorem 10.46 ensures that $\mathbb{X}' = \mathcal{B}^{q,\lambda}(\partial\Omega,\omega)$, and the Hardy–Littlewood maximal operator on $\partial\Omega$ is bounded on both $\mathbb{X}$ and $\mathbb{X}'$ thanks to Proposition 5.1 and Proposition 7.1, respectively. □

Here, we present a Dirichlet problem on a domain with a compact boundary, but now the boundary datum belongs to a Muckenhoupt-weighted vanishing Morrey space. This is a new result even in such a mundane setting as the class of $\mathscr{C}^1$ domains with compact boundary (even when L is the Laplacian).

Theorem 12.4 (The weighted vanishing Morrey Dirichlet BVP). *Choose two integers* $n, M \in \mathbb{N}$ *with* $n \geq 2$ *and suppose* $\Omega \subseteq \mathbb{R}^n$ *is an Ahlfors regular domain with compact boundary. Abbreviate* $\sigma := \mathcal{H}^{n-1}\lfloor\partial\Omega$ *and denote by* ν *the geometric measure theoretic outward unit normal to* Ω*. Suppose* L *is a homogeneous, second-order, constant complex coefficient,* $M\times M$ *system in* $\mathbb{R}^n$*. Fix two integrability exponents* $p, q \in (1,\infty)$*, an arbitrary Muckenhoupt weight* $w \in A_p(\partial\Omega,\sigma) \cap \mathrm{RH}_q(\partial\Omega,\sigma)$*, and some threshold* $\lambda \in (0,(n-1)/q')$*, where* $q' := q/(q-1)$*. For a fixed aperture parameter* $\kappa \in (0,\infty)$*, consider the Dirichlet problem for* L *in* Ω *with data in a Muckenhoupt-weighted vanishing Morrey space, formulated below:*

$$\begin{cases} u \in [\mathscr{C}^\infty(\Omega)]^M,\\ Lu = 0 \quad \text{in } \Omega,\\ \mathcal{N}_\kappa u \in \mathring{M}^{p,\lambda}(\partial\Omega,w),\\ u\big|_{\partial\Omega}^{\kappa-\mathrm{n.t.}} = f \in [\mathring{M}^{p,\lambda}(\partial\Omega,w)]^M,\\ u(x) = O(|x|^{2-n}) \quad \text{as } |x|\to\infty \text{ if } \Omega \text{ is unbounded.}\end{cases}\tag{12.23}$$

4 According to [90, Proposition 1.4.1, p. 25], whenever L is as in (12.21) we have $\mathfrak{A}_L^{\mathrm{dis}} \neq \varnothing$ and $\mathfrak{A}_{L^\top}^{\mathrm{dis}} \neq \varnothing$.

Then there exists some threshold $\delta \in (0,1)$, *which depends only on* n, p, q, λ, $[w]_{A_p}$, $[w]_{\mathrm{RH}_q}$, *the system* L, *the* ADR *character of* $\partial\Omega$, *and the relative flatness ratio* $\wp(\Omega)/\operatorname{diam}\partial\Omega$ (*cf.* [90, Definition 3.1.9, p. 123]), *with the property that when*

$$\operatorname{dist}(\nu, [\mathrm{VMO}\,(\partial\Omega,\sigma)]^n) < \delta \tag{12.24}$$

where the distance is considered in the space $[\mathrm{BMO}\,(\partial\Omega,\sigma)]^n$, *hence if* Ω *is a* δ*-oscillating* AR *domain*[5] (*cf. Definition* 12.1), *the following statements are true:*

(a) [Existence]. *If the system* L *is weakly elliptic with* $\mathfrak{A}_L^{\mathrm{dis}} \neq \varnothing$, *the space of admissible boundary data for the Dirichlet problem* (12.23), *i. e.,*

$$\begin{aligned}\{u\big|_{\partial\Omega}^{\kappa-\mathrm{n.t.}} : u \in [\mathscr{C}^\infty(\Omega)]^M,\ & Lu = 0 \text{ in } \Omega, \\ & \mathcal{N}_\kappa u \in \mathring{M}^{p,\lambda}(\partial\Omega, w),\ u\big|_{\partial\Omega}^{\kappa-\mathrm{n.t.}} \text{ exists } \sigma\text{-a. e., and} \\ & u(x) = O(|x|^{2-n}) \text{ as } |x| \to \infty \text{ if } \Omega \text{ is unbounded}\},\end{aligned} \tag{12.25}$$

is a closed linear subspace of finite codimension in $[\mathring{M}^{p,\lambda}(\partial\Omega, w)]^M$, *and a solution* u *of* (12.23) *exists if and only if the boundary datum* f *belongs to* (12.25).

(b) [Uniqueness]. *If* L *is weakly elliptic with* $\mathfrak{A}_{L^\top}^{\mathrm{dis}} \neq \varnothing$, *the space of null solutions of the Dirichlet problem* (12.23), *i. e.,*

$$\begin{aligned}\{u \in [\mathscr{C}^\infty(\Omega)]^M :\ & Lu = 0 \text{ in } \Omega, \\ & \mathcal{N}_\kappa u \in \mathring{M}^{p,\lambda}(\partial\Omega, w),\ u\big|_{\partial\Omega}^{\kappa-\mathrm{n.t.}} = 0, \text{ and} \\ & u(x) = O(|x|^{2-n}) \text{ as } |x| \to \infty \text{ if } \Omega \text{ is unbounded}\},\end{aligned} \tag{12.26}$$

is finite-dimensional; in addition,

for any null-solution u *of the boundary problem* (12.23) (*i. e., for any function* u *belonging to the space described in* (12.26)), $\mathcal{N}_\kappa(\nabla u)$ *belongs to* $\mathring{M}^{p,\lambda}(\partial\Omega, w)$ *and the nontangential trace* $(\nabla u)\big|_{\partial\Omega}^{\kappa-\mathrm{n.t.}}$ *exists at* σ*-a. e. point on* $\partial\Omega$, (12.27)

so, in particular,

the space of null solutions for the Dirichlet problem (12.23) (*cf.* (12.26)) *coincides with that of the regularity problem* (12.47) (*cf.* (12.51)). (12.28)

(c) [Fredholm solvability]. *If one assumes that*

$$L \text{ is weakly elliptic with } \mathfrak{A}_L^{\mathrm{dis}} \neq \varnothing \text{ and } \mathfrak{A}_{L^\top}^{\mathrm{dis}} \neq \varnothing, \tag{12.29}$$

5 in particular, if Ω is an infinitesimally flat AR domain.

then the Dirichlet problem (12.23) *is Fredholm solvable in the sense that the space of admissible boundary data* (*i. e.,* (12.25)) *has finite codimension in the space* $[\mathring{M}^{p,\lambda}(\partial\Omega, w)]^M$, *and the space of null solutions to* (12.23) (*i. e.,* (12.26)) *is finite-dimensional.*

(d) [Sharpness]. *In the class of weakly elliptic operators L for which $\mathfrak{A}_L^{\mathrm{dis}} = \varnothing$ or $\mathfrak{A}_{L^\top}^{\mathrm{dis}} = \varnothing$, it may happen that the Dirichlet problem* (12.23) *fails to be Fredholm solvable even when the underlying domain Ω is the unit disk in the plane (or the complement of its closure). In fact, the aforementioned failure of Fredholm solvability may be total, in the sense that for the problem* (12.23), *the space of admissible boundary data may have infinite codimension in* $[\mathring{M}^{p,\lambda}(\partial\Omega, w)]^M$, *and the space of null solutions may be infinite-dimensional.*

(e) [Strong ellipticity]. *If* (12.29) *is strengthened to*[6] *assuming that*

$$\begin{aligned}&L \text{ satisfies the Legendre–Hadamard strong ellipticity condition}\\&(\text{cf. } (3.3)) \text{ and one has } \mathfrak{A}_L^{\mathrm{dis}} \neq \varnothing,\end{aligned} \tag{12.30}$$

then the Dirichlet problem (12.23) *is actually well posed. More specifically, the boundary problem* (12.23) *has a unique solution, and this solution enjoys the additional property that there exists $C \in (0, \infty)$, independent of f, such that*

$$\|f\|_{[M^{p,\lambda}(\partial\Omega,w)]^M} \leq \|\mathcal{N}_\kappa u\|_{M^{p,\lambda}(\partial\Omega,w)} \leq C\|f\|_{[M^{p,\lambda}(\partial\Omega,w)]^M}. \tag{12.31}$$

Proof. Theorem 12.4 is a consequence of [90, Theorem 8.1.9, pp. 614–616] upon making the identification $\mathbb{X} := M^{p,\lambda}(\partial\Omega, w)$. Indeed, such a choice entails $\mathring{\mathbb{X}} = \mathring{M}^{p,\lambda}(\partial\Omega, w)$ in view of the definition of $\mathring{\mathbb{X}}$ found in [87, Definition 5.2.6, p. 247] and item (c) of Proposition 10.21. Proposition 10.41 shows that $M^{p,\lambda}(\partial\Omega, w)$ is a generalized Banach function space, Theorem 10.46 ensures that $\mathbb{X}' = \mathcal{B}^{p',\lambda}(\partial\Omega, w')$ (where $p' := p/(p-1)$ and $w' := w^{-p'/p} \in A_{p'}(\partial\Omega, \sigma)$), and the Hardy–Littlewood maximal operator on $\partial\Omega$ is bounded on $\mathbb{X}$ and $\mathbb{X}'$ thanks to Proposition 5.1 and Proposition 7.1, respectively. □

We now present the regularity problem on domains with compact boundary and data in a weighted Morrey space.

Theorem 12.5 (The weighted Morrey regularity BVP). *Choose some $n \in \mathbb{N}$ with $n \geq 2$ and $M \in \mathbb{N}$. Suppose $\Omega \subseteq \mathbb{R}^n$ is an AR domain with compact boundary. Abbreviate $\sigma := \mathcal{H}^{n-1}\lfloor\partial\Omega$ and denote by ν the geometric measure theoretic outward unit normal to Ω. Suppose L is a homogeneous, second-order, constant complex coefficient, $M \times M$ system in $\mathbb{R}^n$. Fix two integrability exponents $p, q \in (1, \infty)$, an arbitrary Muckenhoupt weight $w \in A_p(\partial\Omega, \sigma) \cap \mathrm{RH}_q(\partial\Omega, \sigma)$, and a threshold $\lambda \in (0, (n-1)/q')$, where $q' := q/(q-1)$. For a fixed aperture parameter $\kappa > 0$, consider the regularity problem for L in Ω with data in a Muckenhoupt-weighted Morrey-based Sobolev space, formulated below:*

6 According to [90, Proposition 1.4.1, p. 25], whenever L is as in (12.30) we have $\mathfrak{A}_L^{\mathrm{dis}} \neq \varnothing$ and $\mathfrak{A}_{L^\top}^{\mathrm{dis}} \neq \varnothing$.

$$\begin{cases} u \in [\mathscr{C}^\infty(\Omega)]^M, \\ Lu = 0 \quad \text{in } \Omega, \\ \mathcal{N}_\kappa u,\ \mathcal{N}_\kappa(\nabla u) \in M^{p,\lambda}(\partial\Omega, w), \\ u\big|_{\partial\Omega}^{\kappa-\mathrm{n.t.}} = f \in [M_1^{p,\lambda}(\partial\Omega, w)]^M, \\ u(x) = O(|x|^{2-n}) \quad \text{as } |x| \to \infty \text{ if } \Omega \text{ is unbounded.} \end{cases} \tag{12.32}$$

Then there exists some threshold $\delta \in (0,1)$, *which depends only on* n, p, q, λ, $[w]_{A_p}$, $[w]_{\mathrm{RH}_q}$, *the system* L, *the* ADR *character of* $\partial\Omega$, *and the relative flatness ratio* $\wp(\Omega)/\operatorname{diam}\partial\Omega$ (*cf.* [90, Definition 3.1.9, p. 123]), *with the property that whenever*

$$\operatorname{dist}\big(\nu, [\mathrm{VMO}(\partial\Omega, \sigma)]^n\big) < \delta \tag{12.33}$$

where the distance is considered in the space $[\mathrm{BMO}(\partial\Omega, \sigma)]^n$, *hence if* Ω *is a* δ*-oscillating* AR *domain*[7] (*cf. Definition* 12.1), *the following results are true:*

(a) [Existence]. *If* L *is weakly elliptic with* $\mathfrak{A}_L^{\mathrm{dis}} \neq \varnothing$, *the space of admissible boundary data for the boundary problem* (12.32), *i. e.,*

$$\begin{aligned} \{u\big|_{\partial\Omega}^{\kappa-\mathrm{n.t.}} : {} & u \in [\mathscr{C}^\infty(\Omega)]^M,\ Lu = 0 \text{ in } \Omega, \\ & \mathcal{N}_\kappa u,\ \mathcal{N}_\kappa(\nabla u) \in M^{p,\lambda}(\partial\Omega, w), \text{ and} \\ & u(x) = O(|x|^{2-n}) \text{ as } |x| \to \infty \text{ if } \Omega \text{ is unbounded}\}, \end{aligned} \tag{12.34}$$

is a closed linear subspace of finite codimension in $[M_1^{p,\lambda}(\partial\Omega, w)]^M$. *Additionally, a solution* u *of* (12.32) *exists if and only if the boundary datum* f *belongs to* (12.34), *in which scenario a solution* u *may be found such that*

$$\|\mathcal{N}_\kappa u\|_{M^{p,\lambda}(\partial\Omega,w)} + \|\mathcal{N}_\kappa(\nabla u)\|_{M^{p,\lambda}(\partial\Omega,w)} \approx \|f\|_{[M_1^{p,\lambda}(\partial\Omega,w)]^M}. \tag{12.35}$$

(b) [Uniqueness]. *If* L *is weakly elliptic with* $\mathfrak{A}_{L^\top}^{\mathrm{dis}} \neq \varnothing$, *then the space of null-solutions of the boundary problem* (12.32), *i. e.,*

$$\begin{aligned} \{u \in [\mathscr{C}^\infty(\Omega)]^M : {} & Lu = 0 \text{ in } \Omega, \\ & \mathcal{N}_\kappa u,\ \mathcal{N}_\kappa(\nabla u) \in M^{p,\lambda}(\partial\Omega, w),\ u\big|_{\partial\Omega}^{\kappa-\mathrm{n.t.}} = 0, \text{ and} \\ & u(x) = O(|x|^{2-n}) \text{ as } |x| \to \infty \text{ if } \Omega \text{ is unbounded}\}, \end{aligned} \tag{12.36}$$

is finite-dimensional.

(c) [Fredholm solvability]. *If it holds that*

$$L \text{ is weakly elliptic with } \mathfrak{A}_L^{\mathrm{dis}} \neq \varnothing \text{ and } \mathfrak{A}_{L^\top}^{\mathrm{dis}} \neq \varnothing, \tag{12.37}$$

7 in particular, if Ω is an infinitesimally flat AR domain.

then the regularity problem is Fredholm solvable in the sense that the space of admissible boundary data (i. e., (12.34)) *has finite codimension in the space* $[M_1^{p,\lambda}(\partial\Omega, w)]^M$, *and the space of null solutions to* (12.32) *(i. e.,* (12.36)) *is finite-dimensional.*

(d) [Sharpness]. *In the case that* $\mathfrak{A}_L^{\text{dis}} = \varnothing$ *or* $\mathfrak{A}_{L^\top}^{\text{dis}} = \varnothing$, *it may happen that the regularity problem* (12.32) *fails to be Fredholm solvable even when the underlying domain* Ω *is the unit disk in the plane (or the complement of its closure). Moreover, this failure of Fredholm solvability may be total, in the sense that the space of admissible boundary data may have infinite codimension in* $[M_1^{p,\lambda}(\partial\Omega, w)]^M$ *and the space of null solutions may be infinite-dimensional.*

(e) [Strong ellipticity]. *If* (12.37) *is strengthened to*

$$L \text{ satisfies the Legendre–Hadamard strong ellipticity condition (cf. (3.3)) and one has } \mathfrak{A}_L^{\text{dis}} \neq \varnothing, \tag{12.38}$$

then the regularity problem (12.32) *is actually well posed. More specifically, the boundary value problem* (12.32) *has a unique solution, and this solution enjoys the additional property that* (12.35) *holds.*

Proof. Observe that Proposition 10.41 ensures $M^{p,\lambda}(\partial\Omega, w)$ is a generalized Banach function space. Proposition 5.1, Proposition 7.1, and Theorem 10.46 demonstrate that the Hardy–Littlewood maximal operator on $\partial\Omega$ is bounded on both $M^{p,\lambda}(\partial\Omega, w)$ and $(M^{p,\lambda}(\partial\Omega, w))'$. As such, we find that Theorem 12.5 is a consequence of [90, Theorem 8.1.5, pp. 582–585]. □

Moving on, we have the following result regarding the regularity problem on domains with compact boundary and boundary data coming from a Muckenhoupt-weighted Block-based Sobolev space.

Theorem 12.6 (The weighted Block regularity BVP). *Select* $n, M \in \mathbb{N}$ *with* $n \geq 2$, *and suppose* $\Omega \subseteq \mathbb{R}^n$ *is an AR domain with compact boundary. Abbreviate* $\sigma := \mathcal{H}^{n-1}\lfloor\partial\Omega$ *and denote by* ν *the geometric measure theoretic outward unit normal to* Ω. *Suppose* L *is a homogeneous, second-order, constant complex coefficient,* $M \times M$ *system in* $\mathbb{R}^n$. *Fix an integrability exponent* $q \in (1, \infty)$ *with* $q' := q/(q-1)$. *Also, choose an arbitrary Muckenhoupt weight* $\omega \in A_q(\partial\Omega, \sigma)$, *which also satisfies*

$$\omega' := \omega^{-q'/q} \in A_{q'}(\partial\Omega, \sigma) \cap \mathrm{RH}_p(\partial\Omega, \sigma) \tag{12.39}$$

for some $p \in (1, \infty)$. *Define* $\varepsilon := (q-1)/p'$, *which ensures* $\omega \in A_{q-\varepsilon}(\partial\Omega, \sigma)$ *by Proposition 6.35, and select any threshold* $\lambda \in (0, \varepsilon(n-1)/(q-1))$. *For a fixed aperture parameter* $\kappa > 0$, *consider the regularity problem for* L *in* Ω *with data in a Muckenhoupt-weighted Block-based Sobolev space, formulated below:*

$$\begin{cases} u \in [\mathscr{C}^\infty(\Omega)]^M, \\ Lu = 0 \quad \text{in } \Omega, \\ \mathcal{N}_\kappa u,\ \mathcal{N}_\kappa(\nabla u) \in \mathcal{B}^{q,\lambda}(\partial\Omega, \omega), \\ u\big|_{\partial\Omega}^{\kappa-\mathrm{n.t.}} = f \in [\mathcal{B}_1^{q,\lambda}(\partial\Omega, \omega)]^M, \\ u(x) = O(|x|^{2-n}) \quad \text{as } |x| \to \infty \text{ if } \Omega \text{ is unbounded.} \end{cases} \tag{12.40}$$

Then there exists some threshold $\delta \in (0,1)$, which depends only on n, q, ε, λ, $[\omega]_{A_{q-\varepsilon}}$, the system L, the ADR *character of $\partial\Omega$, and the relative flatness ratio $\wp(\Omega)/\mathrm{diam}\,\partial\Omega$ (cf.* [90, Definition 3.1.9, p. 123]), *with the property that whenever*

$$\mathrm{dist}\big(\nu, [\mathrm{VMO}\,(\partial\Omega, \sigma)]^n\big) < \delta \tag{12.41}$$

where the distance is considered in the space $[\mathrm{BMO}\,(\partial\Omega, \sigma)]^n$, *hence if Ω is a δ-oscillating* AR *domain*[8] *(cf. Definition* 12.1), *the following results are true:*

(a) [Existence]. *If L is weakly elliptic with $\mathfrak{A}_L^{\mathrm{dis}} \neq \varnothing$, the space of admissible boundary data for the boundary problem* (12.40), *i. e.,*

$$\begin{aligned} \big\{u\big|_{\partial\Omega}^{\kappa-\mathrm{n.t.}} : u \in [\mathscr{C}^\infty(\Omega)]^M,\ &Lu = 0 \text{ in } \Omega,\ \mathcal{N}_\kappa u, \\ &\mathcal{N}_\kappa(\nabla u) \in \mathcal{B}^{q,\lambda}(\partial\Omega, \omega), \text{ and} \\ &u(x) = O(|x|^{2-n}) \text{ as } |x| \to \infty \text{ if } \Omega \text{ is unbounded}\big\}, \end{aligned} \tag{12.42}$$

is a closed linear subspace of finite codimension in $[\mathcal{B}_1^{q,\lambda}(\partial\Omega, \omega)]^M$. Additionally, a solution u of (12.40) *exists if and only if the boundary datum f belongs to* (12.42), *in which scenario a solution u may be found such that*

$$\|\mathcal{N}_\kappa u\|_{\mathcal{B}^{q,\lambda}(\partial\Omega,\omega)} + \|\mathcal{N}_\kappa(\nabla u)\|_{\mathcal{B}^{q,\lambda}(\partial\Omega,\omega)} \approx \|f\|_{[\mathcal{B}_1^{q,\lambda}(\partial\Omega,\omega)]^M}. \tag{12.43}$$

(b) [Uniqueness]. *If L is weakly elliptic with $\mathfrak{A}_{L^\top}^{\mathrm{dis}} \neq \varnothing$, then the space of null-solutions of the boundary problem* (12.40), *i. e.,*

$$\begin{aligned} \big\{u \in [\mathscr{C}^\infty(\Omega)]^M :\ &Lu = 0 \text{ in } \Omega, \\ &\mathcal{N}_\kappa u,\ \mathcal{N}_\kappa(\nabla u) \in \mathcal{B}^{q,\lambda}(\partial\Omega, \omega),\ u\big|_{\partial\Omega}^{\kappa-\mathrm{n.t.}} = 0, \text{ and} \\ &u(x) = O(|x|^{2-n}) \text{ as } |x| \to \infty \text{ if } \Omega \text{ is unbounded}\big\}, \end{aligned} \tag{12.44}$$

is finite-dimensional.

(c) [Fredholm solvability]. *If it holds that*

$$L \text{ is weakly elliptic with } \mathfrak{A}_L^{\mathrm{dis}} \neq \varnothing \text{ and } \mathfrak{A}_{L^\top}^{\mathrm{dis}} \neq \varnothing, \tag{12.45}$$

8 in particular, if Ω is an infinitesimally flat AR domain.

then the regularity problem is Fredholm solvable in the sense that the space of admissible boundary data (i. e., (12.42)) *has finite codimension in the space* $[\mathcal{B}_1^{q,\lambda}(\partial\Omega,\omega)]^M$*, and the space of null solutions to* (12.40) *(i. e.,* (12.44)) *is finite-dimensional.*

(d) [Sharpness]. *In the case that* $\mathfrak{A}_L^{\mathrm{dis}} = \varnothing$ *or* $\mathfrak{A}_{L^\top}^{\mathrm{dis}} = \varnothing$*, it may happen that the regularity problem* (12.40) *fails to be Fredholm solvable even when the underlying domain* Ω *is the unit disk in the plane (or the complement of its closure). Moreover, this failure of Fredholm solvability may be total, in the sense that the space of admissible boundary data may have infinite codimension in* $[\mathcal{B}_1^{q,\lambda}(\partial\Omega,\omega)]^M$ *and the space of null solutions may be infinite-dimensional.*

(e) [Strong ellipticity]. *If* (12.45) *is strengthened to*

$$\begin{gathered}L \text{ satisfies the Legendre–Hadamard strong ellipticity condition}\\ \text{(cf. (3.3)) and one has } \mathfrak{A}_L^{\mathrm{dis}} \neq \varnothing,\end{gathered} \tag{12.46}$$

then the regularity problem (12.40) *is actually well posed. More specifically, the boundary value problem* (12.40) *has a unique solution, and this solution enjoys the additional property that* (12.43) *holds.*

Proof. Theorem 12.6 is a consequence of [90, Theorem 8.1.5, pp. 582–585] upon making the identification $\mathbb{X} := M^{q',\lambda}(\partial\Omega,\omega')$. Indeed, Proposition 10.41 shows that $M^{q',\lambda}(\partial\Omega,\omega')$ is a generalized Banach function space, Theorem 10.46 ensures that $\mathbb{X}' = \mathcal{B}^{q,\lambda}(\partial\Omega,\omega)$, and according to Proposition 5.1 and Proposition 7.1 the Hardy–Littlewood maximal operator on $\partial\Omega$ is bounded on $\mathbb{X}$ and $\mathbb{X}'$, respectively. □

This next result is a regularity problem for domains with compact boundary with boundary data in a Muckenhoupt-weighted vanishing Morrey-based Sobolev space.

Theorem 12.7 (The weighted vanishing Morrey regularity BVP). *Select two integers* $n, M \in \mathbb{N}$ *with* $n \geq 2$ *and suppose* $\Omega \subseteq \mathbb{R}^n$ *is an* AR *domain with compact boundary. Abbreviate* $\sigma := \mathcal{H}^{n-1}\lfloor\partial\Omega$ *and denote by* ν *the geometric measure theoretic outward unit normal to* Ω*. Suppose* L *is a homogeneous, second-order, constant complex coefficient,* $M \times M$ *system in* $\mathbb{R}^n$*. Fix any two integrability exponents* $p, q \in (1,\infty)$ *along with an arbitrary Muckenhoupt weight* $w \in A_p(\partial\Omega,\sigma) \cap \mathrm{RH}_q(\partial\Omega,\sigma)$*, and some threshold* $\lambda \in (0,(n-1)/q')$*, where* $q' := q/(q-1)$*. For a fixed aperture parameter* $\kappa > 0$*, consider the regularity problem for* L *in* Ω *with data in a Muckenhoupt-weighted vanishing Morrey-based Sobolev space, formulated below:*

$$\begin{cases} u \in [\mathscr{C}^\infty(\Omega)]^M, \\ Lu = 0 \quad \text{in } \Omega, \\ \mathcal{N}_\kappa u,\ \mathcal{N}_\kappa(\nabla u) \in \mathring{M}^{p,\lambda}(\partial\Omega, w), \\ u\big|_{\partial\Omega}^{\kappa-\mathrm{n.t.}} = f \in [\mathring{M}_1^{p,\lambda}(\partial\Omega,w)]^M, \\ u(x) = O(|x|^{2-n}) \quad \text{as } |x| \to \infty \text{ if } \Omega \text{ is unbounded.} \end{cases} \tag{12.47}$$

Then there exists some threshold $\delta \in (0,1)$, *which depends only on* n, p, q, λ, $[w]_{A_p}$, $[w]_{\mathrm{RH}_q}$, *the system* L, *the* ADR *character of* $\partial\Omega$, *and the relative flatness ratio* $\wp(\Omega)/\operatorname{diam}\partial\Omega$ (*cf.* [90, Definition 3.1.9, p. 123]), *with the property that whenever*

$$\operatorname{dist}(\nu, [\mathrm{VMO}\,(\partial\Omega,\sigma)]^n) < \delta \tag{12.48}$$

where the distance is considered in the space $[\mathrm{BMO}\,(\partial\Omega,\sigma)]^n$, *hence if* Ω *is a* δ*-oscillating* AR *domain*[9] (*cf. Definition* 12.1), *the following results are true:*

(a) [Existence]. *If* L *is weakly elliptic with* $\mathfrak{A}_L^{\mathrm{dis}} \neq \varnothing$, *the space of admissible boundary data for the boundary problem* (12.47), *i. e.,*

$$\begin{aligned} \{u\big|_{\partial\Omega}^{\kappa-\mathrm{n.t.}} : u \in [\mathscr{C}^\infty(\Omega)]^M,\ Lu = 0 \text{ in } \Omega,\\ \mathcal{N}_\kappa u,\ \mathcal{N}_\kappa(\nabla u) \in \mathring{M}^{p,\lambda}(\partial\Omega, w), \text{ and}\\ u(x) = O(|x|^{2-n}) \text{ as } |x| \to \infty \text{ if } \Omega \text{ is unbounded}\}, \end{aligned} \tag{12.49}$$

is a closed linear subspace of finite codimension in $[M_1^{p,\lambda}(\partial\Omega, w)]^M$. *Additionally, a solution* u *of* (12.47) *exists if and only if the boundary datum* f *belongs to* (12.49), *in which scenario a solution* u *may be found such that*

$$\|\mathcal{N}_\kappa u\|_{M^{p,\lambda}(\partial\Omega,w)} + \|\mathcal{N}_\kappa(\nabla u)\|_{M^{p,\lambda}(\partial\Omega,w)} \approx \|f\|_{[M_1^{p,\lambda}(\partial\Omega,w)]^M}. \tag{12.50}$$

(b) [Uniqueness]. *If* L *is weakly elliptic with* $\mathfrak{A}_{L^\top}^{\mathrm{dis}} \neq \varnothing$, *then the space of null-solutions of the boundary problem* (12.47), *i. e.,*

$$\begin{aligned} \{u \in [\mathscr{C}^\infty(\Omega)]^M : Lu = 0 \text{ in } \Omega,\\ \mathcal{N}_\kappa u,\ \mathcal{N}_\kappa(\nabla u) \in \mathring{M}^{p,\lambda}(\partial\Omega, w),\ u\big|_{\partial\Omega}^{\kappa-\mathrm{n.t.}} = 0, \text{ and}\\ u(x) = O(|x|^{2-n}) \text{ as } |x| \to \infty \text{ if } \Omega \text{ is unbounded}\}, \end{aligned} \tag{12.51}$$

is finite-dimensional.

(c) [Fredholm solvability]. *If it holds that*

$$L \text{ is weakly elliptic with } \mathfrak{A}_L^{\mathrm{dis}} \neq \varnothing \text{ and } \mathfrak{A}_{L^\top}^{\mathrm{dis}} \neq \varnothing, \tag{12.52}$$

then the regularity problem is Fredholm solvable in the sense that the space of admissible boundary data (i. e., (12.49)) *has finite codimension in the space* $[\mathring{M}_1^{p,\lambda}(\partial\Omega, w)]^M$, *and the space of null solutions to* (12.47) (*i. e.,* (12.51)) *is finite-dimensional.*

(d) [Sharpness]. *In the case that* $\mathfrak{A}_L^{\mathrm{dis}} = \varnothing$ *or* $\mathfrak{A}_{L^\top}^{\mathrm{dis}} = \varnothing$, *it may happen that the regularity problem* (12.47) *fails to be Fredholm solvable even when the underlying domain* Ω *is the unit disk in the plane (or the complement of its closure). Moreover, this failure of*

9 in particular, if Ω is an infinitesimally flat AR domain.

Fredholm solvability may be total, in the sense that the space of admissible boundary data may have infinite codimension in $[\mathring{M}_1^{p,\lambda}(\partial\Omega, w)]^M$ *and the space of null solutions may be infinite-dimensional.*

(e) [Strong ellipticity]. *If* (12.52) *is strengthened to*

$$\begin{aligned}&L \text{ satisfies the Legendre–Hadamard strong ellipticity condition}\\&\text{(cf. (3.3)) and one has } \mathfrak{A}_L^{\text{dis}} \neq \varnothing,\end{aligned} \tag{12.53}$$

then the regularity problem (12.47) *is actually well posed. More specifically, the boundary value problem* (12.47) *has a unique solution, and this solution enjoys the additional property that* (12.50) *holds.*

Proof. Theorem 12.7 follows from [90, Theorem 8.1.5, pp. 582–585] upon making the identification $\mathbb{X} := M^{p,\lambda}(\partial\Omega, w)$. Indeed, such a choice entails $\mathring{\mathbb{X}} = \mathring{M}^{p,\lambda}(\partial\Omega, w)$ in view of the definition of $\mathring{\mathbb{X}}$ found in [87, Definition 5.2.6, p. 247] and item (c) of Proposition 10.21. Proposition 10.41 shows that $M^{p,\lambda}(\partial\Omega, w)$ is a generalized Banach function space, Theorem 10.46 ensures that $\mathbb{X}' = \mathcal{B}^{p',\lambda}(\partial\Omega, w')$ (where p' is the Hölder conjugate of p and $w' := w^{-p'/p} \in A_{p'}(\partial\Omega, \sigma)$), and the Hardy–Littlewood maximal operator on $\partial\Omega$ is bounded on $\mathbb{X}$ and $\mathbb{X}'$, thanks to Proposition 5.1 and Proposition 7.1, respectively. □

Our next set of results will concern the Neumann problem on domains with compact boundary. Before presenting these results, it is of relevance to introduce the following definitions of a topological nature.

Given an arbitrary open set $\Omega \subseteq \mathbb{R}^n$, define

$$\Omega_+ := \Omega \quad \text{and} \quad \Omega_- := \mathbb{R}^n \setminus \overline{\Omega}. \tag{12.54}$$

In relation to these sets, the following function spaces arise naturally:

$$\begin{aligned}\mathbb{C}_{\Omega_\pm} &:= \operatorname{span}_{\mathbb{C}} \{\mathbf{1}_{\mathcal{O}} : \mathcal{O} \text{ bounded connected component of } \Omega_\pm\},\\ \mathbb{C}_{\partial\Omega_\pm} &:= \operatorname{span}_{\mathbb{C}} \{\mathbf{1}_{\partial\mathcal{O}} : \mathcal{O} \text{ bounded connected component of } \Omega_\pm\},\\ \mathbb{C}_{\partial\Omega} &:= \operatorname{span}_{\mathbb{C}} \{\mathbf{1}_{\Sigma} : \Sigma \text{ connected component of } \partial\Omega\}.\end{aligned} \tag{12.55}$$

Also, for any space X of distributions on $\partial\Omega$ we agree to abbreviate

$$X \cap (\mathbb{C}_{\partial\Omega_\pm})^\perp := \{f \in X : \langle f, \chi\rangle = 0 \text{ for each } \chi \in \mathbb{C}_{\partial\Omega_\pm}\}, \tag{12.56}$$

$$X \cap (\mathbb{C}_{\partial\Omega})^\perp := \{f \in X : \langle f, \chi\rangle = 0 \text{ for each } \chi \in \mathbb{C}_{\partial\Omega}\}. \tag{12.57}$$

In addition, we shall consider

$$\begin{aligned}\mathbb{C}^\infty_{\Omega_\pm} &:= \operatorname{span}_{\mathbb{C}} \{\mathbf{1}_{\mathcal{O}} : \mathcal{O} \text{ connected component of } \Omega_\pm\},\\ \mathbb{C}^\infty_{\partial\Omega_\pm} &:= \operatorname{span}_{\mathbb{C}} \{\mathbf{1}_{\partial\mathcal{O}} : \mathcal{O} \text{ connected component of } \Omega_\pm\},\end{aligned} \tag{12.58}$$

and, for any given space X of distributions on $\partial\Omega$, set

$$X \cap (\mathbb{C}^{\infty}_{\partial\Omega_{\pm}})^{\perp} := \{f \in X : \langle f, \chi\rangle = 0 \text{ for each } \chi \in \mathbb{C}^{\infty}_{\partial\Omega_{\pm}}\}, \tag{12.59}$$

as well as

$$X \cap \mathbb{C}^{\perp} := \{f \in X : \langle f, c\rangle = 0 \text{ for each } c \in \mathbb{C}\}. \tag{12.60}$$

We are now ready to present the statement of the Neumann problem in the context of weighted Morrey spaces. Interestingly, we will see that when more structure is assumed of the underlying system, then more can be said about the nature of the space of admissible boundary data and the space of null solutions.

Theorem 12.8 (The weighted Morrey Neumann BVP). *Select $n, M \in \mathbb{N}$ with $n \geq 2$, and suppose $\Omega \subseteq \mathbb{R}^n$ is an* AR *domain with compact boundary. Abbreviate $\sigma := \mathcal{H}^{n-1}\lfloor\partial\Omega$ and denote by ν the geometric measure theoretic outward unit normal to Ω. Suppose L is a homogeneous, second-order, constant complex coefficient, $M \times M$ system in $\mathbb{R}^n$. Fix two integrability exponents $p, q \in (1,\infty)$, an arbitrary Muckenhoupt weight $w \in A_p(\partial\Omega,\sigma) \cap \mathrm{RH}_q(\partial\Omega,\sigma)$, and a threshold $\lambda \in (0,(n-1)/q')$, where $q' := q/(q-1)$. For a fixed aperture parameter $\kappa > 0$, and choice of coefficient tensor $A \in \mathfrak{A}_L$, consider the Neumann problem for L in Ω with data in a Muckenhoupt-weighted Morrey space, formulated below:*

$$\begin{cases} u \in [\mathscr{C}^{\infty}(\Omega)]^M, \\ Lu = 0 \quad \text{in } \Omega, \\ \mathcal{N}_{\kappa}(\nabla u) \in M^{p,\lambda}(\partial\Omega, w), \\ \partial^A_{\nu} u = f \in [M^{p,\lambda}(\partial\Omega, w)]^M, \\ u(x) = O(|x|^{2-n}) \quad \text{as } |x| \to \infty \text{ if } \Omega \text{ is unbounded}, \end{cases} \tag{12.61}$$

where the conormal derivative operator ∂^A_{ν} is defined as in Definition 3.18.

Then there exists some threshold $\delta \in (0,1)$, which depends only on n, p, q, λ, $[w]_{A_p}$, $[w]_{\mathrm{RH}_q}$, A, the ADR *character of $\partial\Omega$, and the relative flatness ratio $\wp(\Omega)/\operatorname{diam}\partial\Omega$ (cf.* [90, Definition 3.1.9, p. 123]), *with the property that whenever*

$$\operatorname{dist}(\nu, [\mathrm{VMO}(\partial\Omega,\sigma)]^n) < \delta \tag{12.62}$$

where the distance is considered in the space $[\mathrm{BMO}(\partial\Omega,\sigma)]^n$, hence whenever Ω is a δ-oscillating AR *domain*[10] *(cf. Definition 12.1), the set Ω becomes a* UR *domain and the following statements are true:*

10 in particular, if Ω is an infinitesimally flat AR domain.

(a) [Existence]. *In the case that L is weakly elliptic,* $\mathfrak{A}^{\text{dis}}_{L^\top} \neq \varnothing$, *and* $A \in \mathfrak{A}_L$ *is such that* $A^\top \in \mathfrak{A}^{\text{dis}}_{L^\top}$ *then the space of admissible boundary data for the boundary problem* (12.61), *i. e.,*

$$\begin{aligned}\{\partial^A_\nu u : u \in [\mathscr{C}^\infty(\Omega)]^M,\ Lu = 0 \text{ in } \Omega,\ \mathcal{N}_\kappa(\nabla u) \in M^{p,\lambda}(\partial\Omega, w),\\ \text{and } u(x) = O(|x|^{2-n}) \text{ as } |x| \to \infty \text{ if } \Omega \text{ is unbounded}\}\end{aligned} \tag{12.63}$$

is a closed linear subspace of finite codimension in $[M^{p,\lambda}(\partial\Omega, w)]^M$. *Also, a solution u of* (12.61) *exists if and only if the boundary datum f belongs to* (12.63), *in which scenario a solution u may be found such that*

$$\|\mathcal{N}_\kappa(\nabla u)\|_{M^{p,\lambda}(\partial\Omega,w)} \leq C\|f\|_{[M^{p,\lambda}(\partial\Omega,w)]^M}, \tag{12.64}$$

for some constant $C \in (0, \infty)$ *independent of f.*

(b) [Uniqueness]. *In the case that L is weakly elliptic with* $\mathfrak{A}^{\text{dis}}_L \neq \varnothing$ *and A is in* $\mathfrak{A}^{\text{dis}}_L$, *then the space of null-solutions of the boundary problem* (12.61), *i. e.,*

$$\begin{aligned}\{u \in [\mathscr{C}^\infty(\Omega)]^M &: Lu = 0 \text{ in } \Omega,\\ &\mathcal{N}_\kappa(\nabla u) \in M^{p,\lambda}(\partial\Omega, w),\ \partial^A_\nu u = 0, \text{ and}\\ &u(x) = O(|x|^{2-n}) \text{ as } |x| \to \infty \text{ if } \Omega \text{ is unbounded}\}\end{aligned} \tag{12.65}$$

is finite-dimensional.

(c) [Fredholm solvability]. *If it happens that*

$$L \text{ is weakly elliptic with } \mathfrak{A}^{\text{dis}}_L \neq \varnothing \text{ and } \mathfrak{A}^{\text{dis}}_{L^\top} \neq \varnothing, \tag{12.66}$$

then whenever $A \in \mathfrak{A}^{\text{dis}}_L$ *it follows that the boundary value problem* (12.61) *is Fredholm solvable in the sense that the space of admissible boundary data for* (12.61) *(i. e.,* (12.63)*) has finite codimension in* $[M^{p,\lambda}(\partial\Omega, w)]^M$, *and the space of null solutions (i. e.,* (12.65)*) is finite-dimensional.*

(d) [Sharpness]. *Even in the case that* $L := \Delta$, *the Laplace operator (which clearly satisfies* (12.66)*), if* $A \notin \mathfrak{A}^{\text{dis}}_L$ *then it may happen that the Neumann problem* (12.61) *fails to be Fredholm solvable even when the underlying domain* Ω *is the unit disk in the plane (or the complement of its closure). In fact, this failure of Fredholm solvability may be total in the sense that the space of admissible boundary data (i. e.,* (12.63)*) may have infinite codimension in* $[M^{p,\lambda}(\partial\Omega, w)]^M$, *and the space of null solutions (i. e.,* (12.65)*) may be infinite-dimensional. A similar lack of Fredholm solvability also occurs in higher dimensions.*

(e) [Strong ellipticity]. *Suppose* (12.66) *is strengthened to the demand that*

$$\begin{aligned}&L \text{ satisfies the Legendre–Hadamard strong ellipticity condition}\\ &(\text{cf. } (3.3)) \text{ and one has } \mathfrak{A}^{\text{dis}}_L \neq \varnothing.\end{aligned} \tag{12.67}$$

Then for each coefficient tensor $A \in \mathfrak{A}_L^{\mathrm{dis}}$ the index of the Neumann problem (12.61), *i. e., the dimension of* (12.65) *minus the codimension of* (12.63) *in the space* $[M^{p,\lambda}(\partial\Omega, w)]^M$, *is equal to zero. Moreover, the space of admissible boundary data for the Neumann problem* (12.61), (*i. e.,* (12.63)) *is equal to*

$$\mathrm{Im}\Big(-\frac{1}{2}I + K^{\#}_{A^{\top}};\ [M^{p,\lambda}(\partial\Omega, w)]^M\Big) \tag{12.68}$$

if either $n \geq 3$, or $n = 2$ and Ω is bounded,

and

$$\Big\{\Big(-\frac{1}{2}I + K^{\#}_{A^{\top}}\Big)g : g \in [M^{p,\lambda}(\partial\Omega, w)]^M \text{ with } \int_{\partial\Omega} g\, d\sigma = 0\Big\} \tag{12.69}$$

if $n = 2$ and Ω is an exterior domain.

Furthermore, there exists a constant $C \in (0, \infty)$ such that whenever the boundary datum f belongs to the space of admissible boundary data, (*i. e.,* (12.63)) *then a solution u of the boundary problem* (12.61) *may be found so that* (12.64) *holds.*
Next, with $\mathcal{D}_A$ denoting the boundary-to-domain double layer potential operator associated with A and Ω as in (3.18), *when either $n \geq 3$, or $n = 2$ and the set Ω is bounded, it follows that the space of null solutions of the Neumann problem* (12.61) (*i. e.,* (12.65)) *is equal to the space*

$$\Big\{\mathcal{D}_A\phi : \phi \in \mathrm{Ker}\Big(-\frac{1}{2}I + K_A;\ [M_1^{p,\lambda}(\partial\Omega, w)]^M\Big)\Big\}. \tag{12.70}$$

Also, the mapping

$$u \mapsto u\big|^{\kappa-\mathrm{n.t.}}_{\partial\Omega}, \tag{12.71}$$

whose domain is the space of null solutions of the Neumann problem (12.61) (*i. e.,* (12.65)), *is an isomorphism onto the finite-dimensional space*

$$\mathrm{Ker}\Big(-\frac{1}{2}I + K_A;\ [M_1^{p,\lambda}(\partial\Omega, w)]^M\Big), \tag{12.72}$$

and the inverse of the isomorphism described in (12.71) *is the boundary-to-domain double layer potential operator $\mathcal{D}_A$ mapping the null space*

$$\mathrm{Ker}\Big(-\frac{1}{2}I + K_A;\ [M_1^{p,\lambda}(\partial\Omega, w)]^M\Big) \tag{12.73}$$

isomorphically onto the space of null solutions of the Neumann problem (12.61) (*i. e.,* (12.65)).

Finally, in the case when $n = 2$ and Ω is an exterior domain it follows that the space of null solutions of the Neumann problem (12.61) (*i. e.*, (12.65)) *is equal to*

$$\Big\{\mathcal{D}_A\phi + c : \phi \in [M_1^{p,\lambda}(\partial\Omega, w)]^M,\ c \in \mathbb{C}^M,\ \Big(-\frac{1}{2}I + K_A\Big)\phi = -c\Big\}. \tag{12.74}$$

Also, the mapping

$$u \mapsto \Big(u\big|_{\partial\Omega}^{\kappa-\text{n.t.}},\ \lim_{|x|\to\infty} u(x)\Big), \tag{12.75}$$

whose domain is the space of null-solutions of the Neumann problem (12.61) (*i. e.*, (12.65)), *is a well-defined isomorphism onto the space*

$$\Big\{(\phi, c) \in [M_1^{p,\lambda}(\partial\Omega, w)]^M \oplus \mathbb{C}^M : \Big(-\frac{1}{2}I + K_A\Big)\phi = -c\Big\}, \tag{12.76}$$

and the inverse of the linear isomorphism described in (12.75) *is the assignment* $(\phi, c) \longmapsto \mathcal{D}_A\phi + c$ *mapping the space in* (12.76) *isomorphically onto the space of null solutions of the Neumann problem* (12.61) (*i. e.*, (12.65)).

(f) [Positive definiteness]. *If* (12.67) *is further strengthened to*[11]

$$\begin{gathered} A := (a_{jk}^{\alpha\beta})_{\substack{1\le\alpha,\beta\le M\\ 1\le j,k\le n}} \in \mathfrak{A}_L^{\text{dis}} \quad \text{is strictly positive definite,} \\ \text{i. e., there exists some } c > 0 \text{ such that} \\ \operatorname{Re}\langle A\zeta, \overline{\zeta}\rangle = \operatorname{Re}(a_{jk}^{\alpha\beta}\zeta_k^\beta\overline{\zeta_j^\alpha}) \ge c|\zeta|^2 \quad \text{for all } \zeta = (\zeta_j^\alpha)_{\substack{1\le\alpha\le M\\ 1\le j\le n}} \in \mathbb{C}^{M\times n} \end{gathered} \tag{12.77}$$

then the Neumann problem (12.61) *has the following additional properties* (*with the spaces* $\mathbb{C}_{\Omega_\pm}$, $\mathbb{C}_{\partial\Omega_\pm}$, $\mathbb{C}_{\Omega_\pm}^\infty$, $\mathbb{C}_{\partial\Omega_\pm}^\infty$ *defined as in* (12.55), (12.58), *and with the conventions* (12.56), (12.59) *in effect*).

First, if either $n \ge 3$, or $n = 2$ and the set Ω is bounded, then the space of admissible boundary data for the Neumann problem (12.61) (*cf.* (12.63)) *is equal to* $[M^{p,\lambda}(\partial\Omega, w) \cap (\mathbb{C}_{\partial\Omega_+})^\perp]^M$, *and the space of null solutions of the Neumann problem* (12.61) (*cf.* (12.65)) *is equal to* $(\mathbb{C}_{\Omega_+})^M$.

Second, if $n = 2$ and Ω is an exterior domain, then the space of admissible boundary data for the Neumann problem (12.61) (*cf.* (12.63)) *is equal to* $[M^{p,\lambda}(\partial\Omega, w)\cap(\mathbb{C}_{\partial\Omega_+}^\infty)^\perp]^M$, *and the space of null solutions of the Neumann problem* (12.61) (*cf.* (12.65)) *is equal to* $(\mathbb{C}_{\Omega_+}^\infty)^M$.

11 According to [90, (1.1.28), p. 9], whenever L has a positive definite coefficient tensor it follows that L satisfies the Legendre–Hadamard strong ellipticity condition.

Remark 12.9. In light of the comments made directly after [90, Theorem 8.1.17, pp. 680–684], we see that the boundary condition in (12.61) is naturally formulated in the sense that the following Fatou-type result holds. Namely, any function satisfying

$$u \in [\mathscr{C}^\infty(\Omega)]^M \quad \text{with } Lu = 0 \text{ in } \Omega \text{ and } \mathcal{N}_\kappa(\nabla u) \in M^{p,\lambda}(\partial\Omega, w) \tag{12.78}$$

has the property that the nontangential boundary trace

$$\begin{aligned} &(\nabla u)\big|_{\partial\Omega}^{\kappa-\text{n.t.}} \quad \text{exists (in } \mathbb{C}^{n\cdot M}) \text{ at } \sigma\text{-a. e. point on } \partial\Omega \text{ and} \\ &\text{the function } (\nabla u)\big|_{\partial\Omega}^{\kappa-\text{n.t.}} \text{ belongs to } [M^{p,\lambda}(\partial\Omega, w)]^{n\cdot M}. \end{aligned} \tag{12.79}$$

In particular, $\partial_\nu^A u$ is meaningfully defined σ-a. e. on $\partial\Omega$ and

$$\partial_\nu^A u \in [M^{p,\lambda}(\partial\Omega, w)]^M. \tag{12.80}$$

Additionally, for any function u as in (12.78), the nontangential boundary trace

$$u\big|_{\partial\Omega}^{\kappa-\text{n.t.}} \quad \text{exists } \sigma\text{-a. e. on } \partial\Omega, \text{ and belongs to } [M_1^{p,\lambda}(\partial\Omega, w)]^M. \tag{12.81}$$

Proof of Theorem 12.8. In view of Proposition 10.41, which identifies the weighted Morrey space $M^{p,\lambda}(\partial\Omega, w)$ as a generalized Banach function space, as well as Proposition 5.1, Proposition 7.1, and Theorem 10.46, which demonstrate that the Hardy–Littlewood maximal operator on $\partial\Omega$ is bounded on both $M^{p,\lambda}(\partial\Omega, w)$ and $(M^{p,\lambda}(\partial\Omega, w))'$, we see that Theorem 12.8 is a direct consequence of [90, Theorem 8.1.17, pp. 680–684] with $\mathbb{X} := M^{p,\lambda}(\partial\Omega, w)$. □

Here is a result concerning the Neumann problem for domains with compact boundary with data belonging to a Muckenhoupt-weighted Block spaces. Note that similar comments as in Remark 12.9 continue to hold in this case.

Theorem 12.10 (The weighted Block Neumann BVP). *Select $n, M \in \mathbb{N}$ with $n \geq 2$, and suppose $\Omega \subseteq \mathbb{R}^n$ is an* AR *domain with compact boundary. Abbreviate $\sigma := \mathcal{H}^{n-1}\lfloor\partial\Omega$ and denote by ν the geometric measure theoretic outward unit normal to Ω. Suppose L is a homogeneous, second-order, constant complex coefficient, $M \times M$ system in $\mathbb{R}^n$. Fix an integrability exponent $q \in (1, \infty)$ with $q' := q/(q-1)$. Also, choose an arbitrary Muckenhoupt weight $\omega \in A_q(\partial\Omega, \sigma)$, which also satisfies*

$$\omega' := \omega^{-q'/q} \in A_{q'}(\partial\Omega, \sigma) \cap \mathrm{RH}_p(\partial\Omega, \sigma) \tag{12.82}$$

for some $p \in (1, \infty)$. Define $\varepsilon := (q-1)/p'$, which ensures $\omega \in A_{q-\varepsilon}(\partial\Omega, \sigma)$ by Proposition 6.35, and select any threshold $\lambda \in (0, \varepsilon(n-1)/(q-1))$. For a fixed aperture parameter $\kappa > 0$, and choice of coefficient tensor $A \in \mathfrak{A}_L$, consider the Neumann problem for L in Ω with data in a Muckenhoupt-weighted Block space, formulated below:

$$\begin{cases} u \in [\mathscr{C}^\infty(\Omega)]^M, \\ Lu = 0 \quad \textit{in } \Omega, \\ \mathcal{N}_\kappa(\nabla u) \in \mathcal{B}^{q,\lambda}(\partial\Omega, \omega), \\ \partial_\nu^A u = f \in [\mathcal{B}^{q,\lambda}(\partial\Omega, \omega)]^M, \\ u(x) = O(|x|^{2-n}) \quad \textit{as } |x| \to \infty \textit{ if } \Omega \textit{ is unbounded}, \end{cases} \tag{12.83}$$

where the conormal derivative operator ∂_ν^A *is defined as in Definition* 3.18.

Then there exists some threshold $\delta \in (0,1)$, *which depends only on* n, q, ε, λ, $[\omega]_{A_q-\varepsilon}$, A, *the* ADR *character of* $\partial\Omega$, *and the relative flatness ratio* $\wp(\Omega)/\operatorname{diam}\partial\Omega$ (*cf.* [90, Definition 3.1.9, p. 123]), *with the property that whenever*

$$\operatorname{dist}(\nu, [\mathrm{VMO}\,(\partial\Omega,\sigma)]^n) < \delta \tag{12.84}$$

where the distance is considered in the space $[\mathrm{BMO}\,(\partial\Omega,\sigma)]^n$, *hence whenever* Ω *is a* δ*-oscillating* AR *domain*[12] (*cf. Definition* 12.1), *the set* Ω *becomes a* UR *domain and the following statements are true:*

(a) [Existence]. *In the case that* L *is weakly elliptic,* $\mathfrak{A}_{L^\top}^{\mathrm{dis}} \neq \varnothing$, *and* $A \in \mathfrak{A}_L$ *is such that* $A^\top \in \mathfrak{A}_{L^\top}^{\mathrm{dis}}$, *then the space of admissible boundary data for the boundary problem* (12.83), *i. e.,*

$$\begin{aligned} \{\partial_\nu^A u : u \in [\mathscr{C}^\infty(\Omega)]^M,\ Lu = 0 \textit{ in } \Omega,\ \mathcal{N}_\kappa(\nabla u) \in \mathcal{B}^{q,\lambda}(\partial\Omega, \omega), \\ \textit{and } u(x) = O(|x|^{2-n}) \textit{ as } |x| \to \infty \textit{ if } \Omega \textit{ is unbounded}\} \end{aligned} \tag{12.85}$$

is a closed linear subspace of finite codimension in $[\mathcal{B}^{q,\lambda}(\partial\Omega,\omega)]^M$. *Also, a solution* u *of* (12.83) *exists if and only if the boundary datum* f *belongs to* (12.85), *in which scenario a solution* u *may be found such that*

$$\|\mathcal{N}_\kappa(\nabla u)\|_{\mathcal{B}^{q,\lambda}(\partial\Omega,\omega)} \leq C\|f\|_{[\mathcal{B}^{q,\lambda}(\partial\Omega,\omega)]^M}, \tag{12.86}$$

for some constant $C \in (0,\infty)$ *independent of* f.

(b) [Uniqueness]. *In the case that* L *is weakly elliptic with* $\mathfrak{A}_L^{\mathrm{dis}} \neq \varnothing$ *and* A *is in* $\mathfrak{A}_L^{\mathrm{dis}}$, *then the space of null-solutions of the boundary problem* (12.83), *i. e.,*

$$\begin{aligned} \{u \in [\mathscr{C}^\infty(\Omega)]^M : Lu = 0 \textit{ in } \Omega, \\ \mathcal{N}_\kappa(\nabla u) \in \mathcal{B}^{q,\lambda}(\partial\Omega,\omega),\ \partial_\nu^A u = 0, \textit{ and} \\ u(x) = O(|x|^{2-n}) \textit{ as } |x| \to \infty \textit{ if } \Omega \textit{ is unbounded}\} \end{aligned} \tag{12.87}$$

is finite-dimensional.

12 in particular, if Ω is an infinitesimally flat AR domain.

(c) [Fredholm Solvability]. *If it happens that*

$$L \text{ is weakly elliptic with } \mathfrak{A}_L^{\mathrm{dis}} \neq \varnothing \text{ and } \mathfrak{A}_{L^\top}^{\mathrm{dis}} \neq \varnothing, \tag{12.88}$$

then whenever $A \in \mathfrak{A}_L^{\mathrm{dis}}$ *it follows that the boundary value problem* (12.83) *is Fredholm solvable in the sense that the space of admissible boundary data for* (12.83) *(i. e.,* (12.85)*) has finite codimension in* $[\mathcal{B}^{q,\lambda}(\partial\Omega,\omega)]^M$, *and the space of null-solutions (i. e.,* (12.87)*) is finite-dimensional.*

(d) [Sharpness]. *Even in the case that* $L := \Delta$, *the Laplace operator (which clearly satisfies* (12.88)*), if* $A \notin \mathfrak{A}_L^{\mathrm{dis}}$ *then it may happen that the Neumann problem* (12.83) *fails to be Fredholm solvable even when the underlying domain* Ω *is the unit disk in the plane (or the complement of its closure). In fact, this failure of Fredholm solvability may be total in the sense that the space of admissible boundary data (i. e.,* (12.85)*) may have infinite codimension in* $[\mathcal{B}^{q,\lambda}(\partial\Omega,\omega)]^M$, *and the space of null solutions (i. e.,* (12.87)*) may be infinite- dimensional. A similar lack of Fredholm solvability also occurs in higher dimensions.*

(e) [Strong ellipticity]. *Suppose* (12.88) *is strengthened to the demand that*

$$\begin{gathered}L \text{ satisfies the Legendre–Hadamard strong ellipticity condition} \\ \text{(cf. (3.3)) and one has } \mathfrak{A}_L^{\mathrm{dis}} \neq \varnothing.\end{gathered} \tag{12.89}$$

Then for each coefficient tensor $A \in \mathfrak{A}_L^{\mathrm{dis}}$ *the index of the Neumann problem* (12.83)*, i. e., the dimension of* (12.87) *minus the codimension of* (12.85) *in the space* $[\mathcal{B}^{q,\lambda}(\partial\Omega,\omega)]^M$, *is equal to zero. Moreover, the space of admissible boundary data for the Neumann problem* (12.83)*, (i. e.,* (12.85)*) is equal to*

$$\begin{gathered}\mathrm{Im}\Big(-\frac{1}{2}I + K^{\#}_{A^\top};\ [\mathcal{B}^{q,\lambda}(\partial\Omega,\omega)]^M\Big) \\ \text{if either } n \geq 3, \text{ or } n = 2 \text{ and } \Omega \text{ is bounded,}\end{gathered} \tag{12.90}$$

and

$$\begin{gathered}\Big\{\Big(-\frac{1}{2}I + K^{\#}_{A^\top}\Big)g : g \in [\mathcal{B}^{q,\lambda}(\partial\Omega,\omega)]^M \text{ with } \int_{\partial\Omega} g\,d\sigma = 0\Big\} \\ \text{if } n = 2 \text{ and } \Omega \text{ is an exterior domain.}\end{gathered} \tag{12.91}$$

Furthermore, there exists a constant $C \in (0,\infty)$ *such that whenever the boundary datum* f *belongs to the space of admissible boundary data, (i. e.,* (12.85)*) then a solution* u *of the boundary problem* (12.83) *may be found so that* (12.86) *holds.*

Next, with $\mathcal{D}_A$ *denoting the boundary-to-domain double layer potential operator associated with* A *and* Ω *as in* (3.18)*, when either* $n \geq 3$, *or* $n = 2$ *and the set* Ω *is bounded,* *it follows that the space of null solutions of the Neumann problem* (12.83) (*i. e.*, (12.87)) *is equal to the space*

$$\Big\{\mathcal{D}_A\phi : \phi \in \operatorname{Ker}\Big(-\frac{1}{2}I + K_A;\ \big[\mathcal{B}_1^{q,\lambda}(\partial\Omega,\omega)\big]^M\Big)\Big\}. \tag{12.92}$$

Also, the mapping

$$u \mapsto u\big|_{\partial\Omega}^{\kappa-\mathrm{n.t.}}, \tag{12.93}$$

whose domain is the space of null solutions of the Neumann problem (12.83) (*i. e.*, (12.87)), *is an isomorphism onto the finite-dimensional space*

$$\operatorname{Ker}\Big(-\frac{1}{2}I + K_A;\ \big[\mathcal{B}_1^{q,\lambda}(\partial\Omega,\omega)\big]^M\Big), \tag{12.94}$$

and the inverse of the isomorphism described in (12.93) *is the boundary-to-domain double layer potential operator $\mathcal{D}_A$ mapping the null space*

$$\operatorname{Ker}\Big(-\frac{1}{2}I + K_A;\ \big[\mathcal{B}_1^{q,\lambda}(\partial\Omega,\omega)\big]^M\Big) \tag{12.95}$$

isomorphically onto the space of null solutions of the Neumann problem (12.83) (*i. e.*, (12.87)).

Finally, in the case when $n = 2$ and Ω is an exterior domain it follows that the space of null solutions of the Neumann problem (12.83) (*i. e.*, (12.87)) *is equal to*

$$\Big\{\mathcal{D}_A\phi + c : \phi \in \big[\mathcal{B}_1^{q,\lambda}(\partial\Omega,\omega)\big]^M,\ c \in \mathbb{C}^M,\ \Big(-\frac{1}{2}I + K_A\Big)\phi = -c\Big\}. \tag{12.96}$$

Also, the mapping

$$u \mapsto \Big(u\big|_{\partial\Omega}^{\kappa-\mathrm{n.t.}},\ \lim_{|x|\to\infty} u(x)\Big), \tag{12.97}$$

whose domain is the space of null solutions of the Neumann problem (12.83) (*i. e.*, (12.87)), *is a well-defined isomorphism onto the space*

$$\Big\{(\phi, c) \in \big[\mathcal{B}_1^{q,\lambda}(\partial\Omega,\omega)\big]^M \oplus \mathbb{C}^M : \Big(-\frac{1}{2}I + K_A\Big)\phi = -c\Big\}, \tag{12.98}$$

and the inverse of the linear isomorphism described in (12.97) *is the assignment $(\phi, c) \longmapsto \mathcal{D}_A\phi + c$ mapping the space in* (12.98) *isomorphically onto the space of null solutions of the Neumann problem* (12.83) (*i. e.*, (12.87)).

(f) [Positive definiteness]. *If* (12.89) *is further strengthened to*[13]

$$A := (a_{jk}^{\alpha\beta})_{\substack{1\le\alpha,\beta\le M\\ 1\le j,k\le n}} \in \mathfrak{A}_L^{\mathrm{dis}} \quad \text{is strictly positive definite,}$$

$$\text{i. e., there exists some } c > 0 \text{ such that} \tag{12.99}$$

$$\mathrm{Re}\langle A\zeta, \overline{\zeta}\rangle = \mathrm{Re}(a_{jk}^{\alpha\beta}\zeta_k^{\beta}\overline{\zeta_j^{\alpha}}) \ge c|\zeta|^2 \quad \text{for all } \zeta = (\zeta_j^{\alpha})_{\substack{1\le\alpha\le M\\ 1\le j\le n}} \in \mathbb{C}^{M\times n}$$

then the Neumann problem (12.83) *has the following additional properties* (*with the spaces* $\mathbb{C}_{\Omega_\pm}$, $\mathbb{C}_{\partial\Omega_\pm}$, $\mathbb{C}^{\infty}_{\Omega_\pm}$, $\mathbb{C}^{\infty}_{\partial\Omega_\pm}$ *defined as in* (12.55), (12.58), *and with the conventions* (12.56), (12.59) *in effect*).

First, if either $n \ge 3$, *or* $n = 2$ *and the set* Ω *is bounded, then the space of admissible boundary data for the Neumann problem* (12.83) (*cf.* (12.85)) *is equal to* $[\mathcal{B}^{q,\lambda}(\partial\Omega,\omega) \cap (\mathbb{C}_{\partial\Omega_+})^{\perp}]^M$, *and the space of null solutions of the Neumann problem* (12.83) (*cf.* (12.87)) *is equal to* $(\mathbb{C}_{\Omega_+})^M$.

Second, if $n = 2$ *and* Ω *is an exterior domain, then the space of admissible boundary data for the Neumann problem* (12.83) (*cf.* (12.85)) *is equal to* $[\mathcal{B}^{q,\lambda}(\partial\Omega,\omega)\cap(\mathbb{C}^{\infty}_{\partial\Omega_+})^{\perp}]^M$, *and the space of null solutions of the Neumann problem* (12.83) (*cf.* (12.87)) *is equal to* $(\mathbb{C}^{\infty}_{\Omega_+})^M$.

Proof. Theorem 12.10 is a consequence of [90, Theorem 8.1.17, pp. 680–684] upon making the identification $\mathbb{X} := M^{q',\lambda}(\partial\Omega,\omega')$. Indeed, Proposition 10.41 shows that $M^{q',\lambda}(\partial\Omega,\omega')$ is a generalized Banach function space, Theorem 10.46 ensures that $\mathbb{X}' = \mathcal{B}^{q,\lambda}(\partial\Omega,\omega)$, and according to Proposition 5.1 and Proposition 7.1 the Hardy–Littlewood maximal operator on $\partial\Omega$ is bounded on $\mathbb{X}$ and $\mathbb{X}'$, respectively. □

This last result concerns the Neumann problem on domains with compact boundary with data in a Muckenhoupt-weighted vanishing Morrey space. As in the case of Muckenhoupt-weighted Block spaces, analogous results to those described in Remark 12.9 still hold in this context.

Theorem 12.11 (The weighted vanishing Morrey Neumann BVP). *Select any two integers* $n, M \in \mathbb{N}$ *with* $n \ge 2$ *and suppose* $\Omega \subseteq \mathbb{R}^n$ *is an AR domain with compact boundary. Abbreviate* $\sigma := \mathcal{H}^{n-1}\lfloor\partial\Omega$ *and denote by* ν *the geometric measure theoretic outward unit normal to* Ω. *Suppose L is a homogeneous, second-order, constant complex coefficient,* $M \times M$ *system in* $\mathbb{R}^n$. *Fix two integrability exponents* $p, q \in (1,\infty)$, *an arbitrary Muckenhoupt weight* $w \in A_p(\partial\Omega,\sigma) \cap \mathrm{RH}_q(\partial\Omega,\sigma)$, *and some threshold* $\lambda \in (0,(n-1)/q')$, *where* $q' := q/(q-1)$. *For a fixed aperture parameter* $\kappa > 0$, *and choice of coefficient tensor* $A \in \mathfrak{A}_L$, *consider the Neumann problem for L in* Ω *with data in a Muckenhoupt-weighted vanishing Morrey space, formulated below:*

---

13 According to [90, (1.1.28), p. 9], whenever L has a positive definite coefficient tensor it follows that L satisfies the Legendre–Hadamard strong ellipticity condition.

$$\begin{cases} u \in [\mathscr{C}^\infty(\Omega)]^M, \\ Lu = 0 \quad \text{in } \Omega, \\ \mathcal{N}_\kappa(\nabla u) \in \mathring{M}^{p,\lambda}(\partial\Omega, w), \\ \partial_\nu^A u = f \in [\mathring{M}^{p,\lambda}(\partial\Omega, w)]^M, \\ u(x) = O(|x|^{2-n}) \quad \text{as } |x| \to \infty \text{ if } \Omega \text{ is unbounded,} \end{cases} \tag{12.100}$$

where the conormal derivative operator ∂_ν^A *is defined as in Definition* 3.18.

Then there exists some threshold $\delta \in (0,1)$, *which depends only on* n, p, q, λ, $[w]_{A_p}$, $[w]_{\mathrm{RH}_q}$, A, *the* ADR *character of* $\partial\Omega$, *and the relative flatness ratio* $\wp(\Omega)/\operatorname{diam}\partial\Omega$ (*cf.* [90, Definition 3.1.9, p. 123]), *with the property that whenever*

$$\operatorname{dist}(\nu, [\mathrm{VMO}(\partial\Omega, \sigma)]^n) < \delta \tag{12.101}$$

where the distance is considered in the space $[\mathrm{BMO}(\partial\Omega, \sigma)]^n$, *hence whenever* Ω *is a* δ*-oscillating* AR *domain*[14] (*cf. Definition* 12.1), *the set* Ω *becomes a* UR *domain and the following statements are true:*

(a) [Existence]. *In the case that* L *is weakly elliptic,* $\mathfrak{A}_{L^\top}^{\mathrm{dis}} \neq \varnothing$, *and* $A \in \mathfrak{A}_L$ *is such that* $A^\top \in \mathfrak{A}_{L^\top}^{\mathrm{dis}}$, *then the space of admissible boundary data for the boundary problem* (12.100), *i. e.,*

$$\begin{aligned} \{\partial_\nu^A u : u \in [\mathscr{C}^\infty(\Omega)]^M,\ Lu = 0 \text{ in } \Omega,\ \mathcal{N}_\kappa(\nabla u) \in \mathring{M}^{p,\lambda}(\partial\Omega, w), \\ \text{and } u(x) = O(|x|^{2-n}) \text{ as } |x| \to \infty \text{ if } \Omega \text{ is unbounded}\} \end{aligned} \tag{12.102}$$

is a closed linear subspace of finite codimension in $[\mathring{M}^{p,\lambda}(\partial\Omega, w)]^M$. *Also, a solution* u *of* (12.100) *exists if and only if the boundary datum* f *belongs to* (12.102), *in which scenario a solution* u *may be found such that*

$$\|\mathcal{N}_\kappa(\nabla u)\|_{M^{p,\lambda}(\partial\Omega,w)} \leq C\|f\|_{[M^{p,\lambda}(\partial\Omega,w)]^M}, \tag{12.103}$$

for some constant $C \in (0,\infty)$ *independent of* f.

(b) [Uniqueness]. *In the case that* L *is weakly elliptic with* $\mathfrak{A}_L^{\mathrm{dis}} \neq \varnothing$ *and* A *is in* $\mathfrak{A}_L^{\mathrm{dis}}$, *then the space of null solutions of the boundary problem* (12.100), *i. e.,*

$$\begin{aligned} \{u \in [\mathscr{C}^\infty(\Omega)]^M : Lu = 0 \text{ in } \Omega, \\ \mathcal{N}_\kappa(\nabla u) \in \mathring{M}^{p,\lambda}(\partial\Omega, w),\ \partial_\nu^A u = 0, \text{ and} \\ u(x) = O(|x|^{2-n}) \text{ as } |x| \to \infty \text{ if } \Omega \text{ is unbounded}\} \end{aligned} \tag{12.104}$$

is finite-dimensional.

---

14 in particular, if Ω is an infinitesimally flat AR domain.

(c) [Fredholm solvability]. *If it happens that*

$$L \text{ is weakly elliptic with } \mathfrak{A}_L^{\mathrm{dis}} \neq \varnothing \text{ and } \mathfrak{A}_{L^\top}^{\mathrm{dis}} \neq \varnothing, \tag{12.105}$$

then whenever $A \in \mathfrak{A}_L^{\mathrm{dis}}$ it follows that the boundary value problem (12.100) *is Fredholm solvable in the sense that the space of admissible boundary data for* (12.100) *(i. e.,* (12.102)*) has finite codimension in $[\mathring{M}^{p,\lambda}(\partial\Omega, w)]^M$, and the space of null solutions (i. e.,* (12.104)*) is finite-dimensional.*

(d) [Sharpness]. *Even in the case that $L := \Delta$, the Laplace operator (which clearly satisfies* (12.105)*), if $A \notin \mathfrak{A}_L^{\mathrm{dis}}$ then it may happen that the Neumann problem* (12.100) *fails to be Fredholm solvable even when the underlying domain Ω is the unit disk in the plane (or the complement of its closure). In fact, this failure of Fredholm solvability may be total in the sense that the space of admissible boundary data (i. e.,* (12.102)*) may have infinite codimension in $[\mathring{M}^{p,\lambda}(\partial\Omega, w)]^M$, and the space of null solutions (i. e.,* (12.104)*) may be infinite-dimensional. A similar lack of Fredholm solvability also occurs in higher dimensions.*

(e) [Strong ellipticity]. *Suppose* (12.105) *is strengthened to the demand that*

$$\begin{aligned}&L \text{ satisfies the Legendre–Hadamard strong ellipticity condition}\\&\text{(cf. (3.3)) and one has } \mathfrak{A}_L^{\mathrm{dis}} \neq \varnothing.\end{aligned} \tag{12.106}$$

Then for each coefficient tensor $A \in \mathfrak{A}_L^{\mathrm{dis}}$ the index of the Neumann problem (12.100)*, i. e., the dimension of* (12.104) *minus the codimension of* (12.102) *in the space $[\mathring{M}^{p,\lambda}(\partial\Omega, w)]^M$, is equal to zero. Moreover, the space of admissible boundary data for the Neumann problem* (12.100)*, (i. e.,* (12.102)*) is equal to*

$$\begin{aligned}&\mathrm{Im}\Big(-\frac{1}{2}I + K^\#_{A^\top};\ [\mathring{M}^{p,\lambda}(\partial\Omega, w)]^M\Big)\\&\text{if either } n \geq 3, \text{ or } n = 2 \text{ and } \Omega \text{ is bounded,}\end{aligned} \tag{12.107}$$

and

$$\begin{aligned}&\Big\{\Big(-\frac{1}{2}I + K^\#_{A^\top}\Big)g : g \in [\mathring{M}^{p,\lambda}(\partial\Omega, w)]^M \text{ with } \int_{\partial\Omega} g\, d\sigma = 0\Big\}\\&\text{if } n = 2 \text{ and } \Omega \text{ is an exterior domain.}\end{aligned} \tag{12.108}$$

Furthermore, there exists a constant $C \in (0, \infty)$ such that whenever the boundary datum f belongs to the space of admissible boundary data, (i. e., (12.102)*) then a solution u of the boundary problem* (12.100) *may be found so that* (12.103) *holds.*

Next, with $\mathcal{D}_A$ denoting the boundary-to-domain double layer potential operator associated with A and Ω as in (3.18)*, when either $n \geq 3$, or $n = 2$ and the set Ω is bounded, it follows that the space of null solutions of the Neumann problem* (12.100) *(i. e.,* (12.104)*) is equal to the space*

$$\Big\{\mathcal{D}_A\phi : \phi \in \operatorname{Ker}\Big(-\frac{1}{2}I + K_A;\ [\mathring{M}_1^{p,\lambda}(\partial\Omega, w)]^M\Big)\Big\}. \tag{12.109}$$

Also, the mapping

$$u \mapsto u\big|_{\partial\Omega}^{\kappa-\mathrm{n.t.}}, \tag{12.110}$$

whose domain is the space of null solutions of the Neumann problem (12.100) (*i. e.*, (12.104)), *is an isomorphism onto the finite-dimensional space*

$$\operatorname{Ker}\Big(-\frac{1}{2}I + K_A;\ [\mathring{M}_1^{p,\lambda}(\partial\Omega, w)]^M\Big), \tag{12.111}$$

and the inverse of the isomorphism described in (12.110) *is the boundary-to-domain double layer potential operator $\mathcal{D}_A$ mapping the null space*

$$\operatorname{Ker}\Big(-\frac{1}{2}I + K_A;\ [\mathring{M}_1^{p,\lambda}(\partial\Omega, w)]^M\Big) \tag{12.112}$$

isomorphically onto the space of null solutions of the Neumann problem (12.100) (*i. e.*, (12.104)).
Finally, in the case when $n = 2$ and Ω is an exterior domain it follows that the space of null solutions of the Neumann problem (12.100) (*i. e.*, (12.104)) *is equal to*

$$\Big\{\mathcal{D}_A\phi + c : \phi \in [\mathring{M}_1^{p,\lambda}(\partial\Omega, w)]^M,\ c \in \mathbb{C}^M,\ \Big(-\frac{1}{2}I + K_A\Big)\phi = -c\Big\}. \tag{12.113}$$

Also, the mapping

$$u \mapsto \Big(u\big|_{\partial\Omega}^{\kappa-\mathrm{n.t.}},\ \lim_{|x|\to\infty} u(x)\Big), \tag{12.114}$$

whose domain is the space of null solutions of the Neumann problem (12.100) (*i. e.*, (12.104)), *is a well-defined isomorphism onto the space*

$$\Big\{(\phi, c) \in [\mathring{M}_1^{p,\lambda}(\partial\Omega, w)]^M \oplus \mathbb{C}^M : \Big(-\frac{1}{2}I + K_A\Big)\phi = -c\Big\}, \tag{12.115}$$

and the inverse of the linear isomorphism described in (12.114) *is the assignment* $(\phi, c) \longmapsto \mathcal{D}_A\phi + c$ *mapping the space in* (12.115) *isomorphically onto the space of null solutions of the Neumann problem* (12.100) (*i. e.*, (12.104)).

(f) [Positive definiteness]. *If* (12.106) *is further strengthened to*[15]

15 According to [90, (1.1.28), p. 9], whenever L has a positive definite coefficient tensor it follows that L satisfies the Legendre–Hadamard strong ellipticity condition.

$$A := (a_{jk}^{\alpha\beta})_{\substack{1\le\alpha,\beta\le M\\ 1\le j,k\le n}} \in \mathfrak{A}_L^{\text{dis}} \quad \textit{is strictly positive definite,}$$
$$\textit{i. e., there exists some } c > 0 \textit{ such that} \tag{12.116}$$
$$\operatorname{Re}\langle A\zeta, \overline{\zeta}\rangle = \operatorname{Re}(a_{jk}^{\alpha\beta}\zeta_k^\beta\overline{\zeta_j^\alpha}) \ge c|\zeta|^2 \quad \textit{for all } \zeta = (\zeta_j^\alpha)_{\substack{1\le\alpha\le M\\ 1\le j\le n}} \in \mathbb{C}^{M\times n}$$

then the Neumann problem (12.100) *has the following additional properties* (*with the spaces* $\mathbb{C}_{\Omega_\pm}$, $\mathbb{C}_{\partial\Omega_\pm}$, $\mathbb{C}_{\Omega_\pm}^\infty$, $\mathbb{C}_{\partial\Omega_\pm}^\infty$ *defined as in* (12.55), (12.58), *and with the conventions* (12.56), (12.59) *in effect*).
First, if either $n \ge 3$, *or* $n = 2$, *and the set* Ω *is bounded, then the space of admissible boundary data for the Neumann problem* (12.100) (*cf.* (12.102)) *is equal to* $[\mathring{M}^{p,\lambda}(\partial\Omega, w) \cap (\mathbb{C}_{\partial\Omega_+})^\perp]^M$, *and the space of null solutions of the Neumann problem* (12.100) (*cf.* (12.104)) *is equal to* $(\mathbb{C}_{\Omega_+})^M$.
Second, if $n = 2$ *and* Ω *is an exterior domain, then the space of admissible boundary data for the Neumann problem* (12.100) (*cf.* (12.102)) *is equal to* $[\mathring{M}^{p,\lambda}(\partial\Omega, w) \cap (\mathbb{C}_{\partial\Omega_+}^\infty)^\perp]^M$, *and the space of null solutions of the Neumann problem* (12.100) (*cf.* (12.104)) *is equal to* $(\mathbb{C}_{\Omega_+}^\infty)^M$.

Proof. Theorem 12.11 follows from [90, Theorem 8.1.17, pp. 680–684] upon making the identification $\mathbb{X} := M^{p,\lambda}(\partial\Omega, w)$. Indeed, such a choice entails $\mathring{\mathbb{X}} = \mathring{M}^{p,\lambda}(\partial\Omega, w)$ in view of the definition of $\mathring{\mathbb{X}}$ found in [87, Definition 5.2.6, p. 247] and item (c) of Proposition 10.21. Proposition 10.41 shows that $M^{p,\lambda}(\partial\Omega, w)$ is a generalized Banach function space, Theorem 10.46 ensures that $\mathbb{X}' = \mathcal{B}^{p',\lambda}(\partial\Omega, w')$ (where p' is the Hölder conjugate of p and $w' := w^{-p'/p} \in A_{p'}(\partial\Omega, \sigma)$), and the Hardy–Littlewood maximal operator on $\partial\Omega$ is bounded on $\mathbb{X}$ and $\mathbb{X}'$, thanks to Proposition 5.1 and Proposition 7.1, respectively. □

Bibliography

[1] D. Adams, *Lectures on L^p potential theory*, Department of Mathematics: Matematiska Institutionen (Department, Univ., Umeå, 1981).

[2] D. Adams, *Morrey Spaces*, Applied Numerical and Harmonic Analysis (Birkhäuser, Cham, 2015).

[3] D. Adams, J. Lewis, On Morrey-Besov inequalities, Stud. Mat. **74**(2), 169–182 (1982).

[4] S. Agmon, A. Douglis, L. Nirenberg, Estimates near the boundary for solutions of elliptic partial differential equations satisfying general boundary conditions I, Commun. Pure Appl. Math. **12**(4), 623–727 (1959).

[5] S. Agmon, A. Douglis, L. Nirenberg, Estimates near the boundary for solutions of elliptic partial differential equations satisfying general boundary conditions I, Commun. Pure Appl. Math. **17**(1), 35–92 (1964).

[6] A. Almeida, Maximal commutators and commutators of potential operators in new vanishing Morrey spaces, Nonlinear Anal. **192**, Paper No. 111684 (2020).

[7] C. Aykol, V. S. Guliyev, A. Serbecti, Boundedness of the maximal operator in the local Morrey-Lorentz spaces, J. Inequal. Appl. **346**, 1–11 (2013).

[8] K. I. Babenko, On conjugate functions, Dokl. Akad. Nauk SSSR (N.S.) **62**, 157–160 (1948) (Russian).

[9] C. Bennett, R. Sharpley, *Interpolation of Operators*, Pure and Applied Mathematics, vol. 129 (Academic Press, 1988).

[10] A. Bensoussan, J. Frehse, *Regularity Results for Nonlinear Elliptic Systems and Applications*, Applied Mathematical Sciences (Springer, Berlin, 2002).

[11] R. Brown, The method of layer potentials for the heat equation in Lipschitz cylinders, Am. J. Math. **111**(2), 339–379 (1989).

[12] A. P. Calderón, A. Zygmund, On the existence of certain singular integrals, Acta Math. **88**, 85–139 (1952).

[13] S. Campanato, Proprietà di inclusions per spazi di Morrey, Ric. Mat. **12**, 67–86 (1963).

[14] S. Campanato, Proprietà di una famiglia di spazi funzionali, Ann. Sc. Norm. Super. Pisa **18**(1), 137–160 (1964).

[15] D-C. Chang, S. Yang, Z. Yang, A two-weight boundedness criterion on spaces of homogeneous type with its application to some elliptic boundary value problems, Math. Methods Appl. Sci. 1–18 (2021).

[16] F. Chiarenza, M. Frasca, Morrey spaces and Hardy-Littlewood maximal function, Rend. Math. Appl. **7**, 273–279 (1987).

[17] M. Christ, A T(b) theorem with remarks on analytic capacity and the Cauchy integral, Colloq. Math. **60**, 601–628 (1990).

[18] R. R. Coiffman, A. McIntosh, Y. Meyer, Le intégrale de Cauchy definit un operateur borne sur L2 pour les courbs Lipschitziennes, Ann. Math. **116**(2), 361–387 (1982).

[19] D. Cruz-Uribe, J. Martell, C. Perez, Extensions of Rubio de Francia's extrapolation theorem, Collect. Math. **57**, 195–231 (2006).

[20] D. Cruz-Uribe, C. J. Neugebauer, The structure of the reverse Hölder class, Trans. Am. Math. Soc. **347**(8), 2941–2960 (1995).

[21] B. Dahlberg, Estimates of harmonic measure, Arch. Ration. Mech. Anal. **65**, 278–288 (1977).

[22] B. Dahlberg, On the Poisson intgeral for Lipschitz and C1 domains, Stud. Math. **66**, 13–24 (1979).

[23] B. Dahlberg, C. Kenig, Hardy spaces and the Neumann problem in Lp for Laplace's equation in Lipschitz domains, Ann. Math. **125**, 437–465 (1987).

[24] B. Dahlberg, C. Kenig, G. Verchota, The Dirichlet problem for the biharmonic equation in a Lipschitz domain, Ann. Inst. Fourier **36**(3), 109–135 (1986).

[25] B. Dahlberg, C. Kenig, G. Verchota, Boundary value problems for the systems of elastostatics in Lipschitz domains, Duke Math. J. **57**(3), 795–818 (1988).

[26] G. David, D. Jerison, Lipschitz approximation to hypersurfaces, harmonic measure, and singular integrals, Indiana Univ. Math. J. **39**(3), 831–845 (1990).

https://doi.org/10.1515/9783111458274-013

[27] G. David, P. Matilla, Removable sets for Lipschitz harmonic functions in the plane, Rev. Mat. Iberoam. **16**(1), 137–215 (2000).

[28] G. David, S. Semmes, *Analysis of and on Uniformly Rectifiable Sets*, Mathematical Surveys and Monographs, vol. 38 (American Mathematical Society, Providence, RI, 1993).

[29] J. Duoandikoetxea, M. Rosenthal, Boundedness of operators on certain weighted Morrey spaces beyond the Muckenhoupt range, Potential Anal. **53**(4), 1255–1268 (2020).

[30] J. Duoandikoetxea, M. Rosenthal, Boundedness properties in a family of weighted Morrey spaces with emphasis on power weights, J. Funct. Anal. **298**(8), Paper No. 108687 (2020).

[31] J. Duoandikoetxea, M. Rosenthal, Muckenhoupt-type conditions on weighted Morrey spaces, J. Fourier Anal. Appl. **27**(32), 1–33 (2021).

[32] B. Dyda, L. Ihnatsyeva, J. Lehrbäck, H. Tuominen, A. V. Vähäkangas, Muckenhoupt Ap-properties of distance functions and applications to Hardy-Sobolev-type inequalities, Potential Anal. **50**, 83–105 (2017).

[33] D. E. Edmunds, H. Triebel, *Function Spaces, Entropy Numbers, Differential Operators*, Cambridge Tracts in Mathematics, vol. 120 (Cambridge University Press, Cambridge, 1996).

[34] A. Eridani, Y. Sawano, Fractional integral operators in generalized Morrey spaces defined on metric measure spaces, Proc. A. Razmadze Math. Inst. **158**, (2012).

[35] L. Escauriaza, M. Mitrea, Transmission problems and spectral theory for singular integral operators on Lipschitz domains, J. Funct. Anal. **216**(1), 141–171 (2004).

[36] L. C. Evans, R. F. Gariepy, *Measure Theory and Fine Properties of Functions*, Studies in Advanced Mathematics, revised edn. (CRC Press, Boca Raton, 2015).

[37] E. Fabes, M. Jodeit, N. Riviere, Potential techniques for boundary value problems on C1 domains, Acta Math. **141**, 165–186 (1978).

[38] E. Fabes, C. Kenig, G. Verchota, The Dirichlet problem for the Stokes system on Lipschitz domains, Duke Math. J. **57**(3), 769–793 (1988).

[39] E. Fabes, O. Mendez, M. Mitrea, Boundary layers on Sobolev-Besov spaces and Poisson's equation for the Laplacian in Lipschitz domains, J. Funct. Anal. **159**, 323–368 (1998).

[40] D. Fan, S. Lu, D. Yang, Boundedness of operators in Morrey spaces on homogeneous spaces and its applications, Acta Math. Sin. Engl. Ser. **14**, 625–634 (1998).

[41] G. D. Fazio, M. A. Ragusa, Commutators and Morrey spaces, Boll. Unione Mat. Ital. **5**, 323–332 (1991).

[42] G. B. Folland, *Real Analysis: Modern Techniques and Their Applications*, Real and Applied Mathematics, 2nd edn. (John Wiley & Sons, New York, 2011).

[43] J. García-Cuerva, J. Rubio de Francia, *Weighted Norm Inequalities and Related Topics* (North-Holland, Amsterdam, 1985).

[44] I. Genebashvili, A. Gogatishvili, V. Kokilashvil, M. Krbec, *Weight Theory for Integral Transforms on Spaces of Homogeneous Type*, Monographs and Surveys in Pure and Applied Mathematics (CRC Press, 1997).

[45] Y. Giga, T. Miyakawa, Navier-Stokes flow in $\mathbb{R}^3$ with measures as initial vorticity and Morrey spaces, Commun. Partial Differ. Equ. **14**, 577–618 (1989).

[46] D. Gilbarg, N. S. Trudinger, *Elliptic Partial Differential Equations of Second Order*, Grundlehren der mathematischen Wissenschaften, vol. 224 (Springer-Verlag, Berlin, 2001).

[47] V. Guliyev, Boundedness of anisotropic singular integral operators on generalized anisotropic Morrey spaces, *Proceedings of the Second ISAAC Congress*, **2**, pp. 1241–1248 (Kluwer Academic Publishers, 2000).

[48] V. Guliyev, Boundedness of the maximal, potential and singular operators in the generalized Morrey spaces, J. Inequal. Appl. **2009**, Paper No. 503948 (2009).

[49] V. Guliyev, S. Aliyev, T. Karaman, P. Shukurov, Boundedness of sublinear operators and commutators n generealized Morrey spaces, Integral Equ. Oper. Theory **71**(3), 327–355 (2011).

[50] V. Guliyev, J. Hasanov, X. Badalov, Maximal and singular integral operators and their commutators on generalized weighted Morrey spaces with variable exponent, Math. Inequal. Appl. **21**(1), 41–61 (2018).

[51] V. Guliyev, J. Hasanov, S. Samko, Boundedness of the maximal, potential and singular integral operators in generalized variable exponent Morrey spaces, Math. Scand. **197**(2), 285–304 (2010).

[52] V. Guliyev, S. Hasanov, Y. Sawano, T. Noi, Non-smooth atomic decompositions for generalized Orlicz-Morrey spaces of the third kind, Acta Appl. Math. **145**, 133–174 (2016).

[53] D. Hakim, Y. Sawano, Complex interpolation of vanishing Morrey spaces, Ann. Funct. Anal. **11**, 643–661 (2020).

[54] G. H. Hardy, J. E. Littlewood, Some more theorems concerning Fourier series and Fourier power series, Duke Math. J. **2**, 354–382 (1936).

[55] N. Hatano, Fractional operators on Morrey-Lorentz spaces and the Olsen inequality, Math. Notes **107**, 63–79 (2020).

[56] N. Hatano, Atomic decomposition for Morrey-Lorentz spaces, arXiv, 2212.13717, (2022).

[57] K. P. Ho, A unification of Orlicz-Morrey spaces and its applications, Manuscr. Math. **172**, Paper No. 617414 (2022).

[58] S. Hofmann, D. Mitrea, M. Mitrea, A. J. Morris, *Lp-Square Function Estimates on Spaces of Homogeneous Type and on Uniformly Rectifiable Sets*, Memoirs of the American Mathematical Society, vol. 245 (American Mathematical Society, Providence, RI, 2017).

[59] S. Hofmann, M. Mitrea, M. Taylor, Singular integrals and elliptic boundary problems on regular Semmes-Kenig-Toro domains, Int. Math. Res. Not. **2010**, 2567–2865 (2010).

[60] T. Hytönen, A. Kairema, Systems of dyadic cubes in a doubling metric space, Colloq. Math. **126**(1), 1–33 (2012).

[61] B. Jawerth, M. Mitrea, Higher dimensional scattering electromagnetic scattering theory on *C*1 and Lipschitz domains, Am. J. Math. **117**(4), 929–963 (1995).

[62] D. Jerison, C. Kenig, An identity with applications to harmonic measure, Bull. Am. Math. Soc. **2**(3), 447–451 (1980).

[63] D. Jerison, C. Kenig, The Neumann problem on Lipschitz domains, Bull. Am. Math. Soc. **4**(2), 203–207 (1981).

[64] D. Jerison, C. Kenig, The inhomogeneous Dirichlet problem in Lipschitz domains, J. Funct. Anal. **130**(1), 161–219 (1995).

[65] H. Jia, H. Wang, Decomposition of Hardy-Morrey spaces, J. Math. Anal. Appl. **354**(1), 99–110 (2009).

[66] R. Johnson, C. J. Neugebauer. Homeomorphisms preserving A_p, Rev. Mat. Iberoam. **32**(2), 249–273 (1987).

[67] R. Johnson, C. J. Neugebauer, Change of variable results for A_p- and reverse Hölder RH_r-classes, Trans. Am. Math. Soc. **328**(2), 639–666 (1991).

[68] R. Kawasumi, E. Nakai, Weighted boundedness of the Hardy-Littlewood maximal and Calderón-Zygmund operators on Orlicz-Morrey and weak Orlicz-Morrey spaces, Math. Inequal. Appl. **24**(4), 1167–1187 (2021).

[69] C. E. Kenig, *Harmonic Analysis Techniques for Second Order Elliptic Boundary Value Problems*, CBMS Regional Conference Series in Mathematics, vol. 83 (American Mathematical Society, Providence, RI, 1994).

[70] Y. Komori, S. Shirai, Weighted Morrey spaces and a singular integral operator, Math. Nachr. **282**(2), 219–231 (2009).

[71] V. A. Kozlov, V. G. Maz'ya, J. Rossman, *Elliptic Boundary Value Problems in Domains with Point Singularities*, Mathematical Surveys and Monographs, vol. 52 (American Mathematical Society, Providence, RI, 1997).

[72] K. Kurata, S. Nishigaki, S. Sugano, Boundedness of integral operators on generalized Morrey spaces and its application to Shrödinger operators, Proc. Am. Math. Soc. **128**(4), 1125–1134 (2000).

[73] L. Lanzhe, Weighted inequalities in generalized Morrey spaces of maximal and singular integral operators on spaces of homogeneous type, Kyungpook Math. J. **40**, 339–346 (2000).

[74] A. T. Lonseth, Sources and applications of integral equations, SIAM Rev. **19**(2), 241–278 (1977).

[75] E. Lorist, Z. Nieraeth, Banach function spaces done right, arXiv, 2307.01173 (2023).

[76] G. Lu, S. Tao, Generalized Morrey spaces over nonhomogeneous metric measure spaces, J. Aust. Math. Soc. **103**, 268–278 (2017).

[77] Y. Lu, D. Yang, W. Yuan, Morrey-Sobolev spaces on measure metric spaces, Potential Anal. **41**, 215–243 (2013).

[78] W. A. J. Luxemburg, *Banach function spaces*, (Technische Hogeschool de Delft, Deflt, 1955).

[79] J. J. Marín, J. M. Martell, D. Mitrea, I. Mitrea, M. Mitrea, *Singular Integral Operators, Quantitative Flatness, and Boundary Problems* (Birkhäuser, Cham, 2022).

[80] J. Martell, D. Mitrea, I. Mitrea, M. Mitrea, The Dirichlet problem for elliptic systems with data in Köthe function spaces, Rev. Mat. Iberoam. **32**(3), 913–970 (2016).

[81] V. Maz'ya, M. Mitrea, T. Shaposhnikova, The Dirichlet problem in Lipschitz domains for higher order elliptic systems with rough coefficients, J. Anal. Math. **110**, 167–239 (2010).

[82] V. G. Maz'ya, J. Rossman, *Elliptic Equations in Polyhedral Domains*, Mathematical Surveys and Monographs, vol. 162 (Amer. Math. Soc., Providence, RI, 2010).

[83] N. Meyers, Mean oscillation over cubes and Hölder continuity, Proc. Am. Math. Soc. **15**(5), 717–721 (1964).

[84] G. Mingione, Gradient estimates below the duality endpoint, Math. Ann. **346**, 571–627 (2010).

[85] D. Mitrea, *Distributions, Partial Differential Equations, and Harmonic Analysis*, Universitext, second edn. (Springer, New York, 2018).

[86] D. Mitrea, I. Mitrea, M. Mitrea, *Geometric Harmonic Analysis-Volume I: A Sharp Divergence Theorem with Nontangential Pointwise Traces*, Developments in Mathematics, vol. 72 (Springer, Cham, 2022).

[87] D. Mitrea, I. Mitrea, M. Mitrea, *Geometric Harmonic Analysis-Volume II: Function Spaces Measuring Size and Smoothness on Rough Sets*, Developments in Mathematics, vol. 73 (Springer, Cham, 2022).

[88] D. Mitrea, I. Mitrea, M. Mitrea, *Geometric Harmonic Analysis-Volume III: Integral Representations, Calderón-Zygmund Theory, Fatou Theorems, and Applications to Scattering*, Developments in Mathematics, vol. 74 (Springer, Cham, 2022).

[89] D. Mitrea, I. Mitrea, M. Mitrea, *Geometric Harmonic Analysis-Volume IV: Boundary Layer Potentials in Uniformly Rectifiable Domains, and Applications to Complex Analysis*, Developments in Mathematics, vol. 75 (Springer, Cham, 2023).

[90] D. Mitrea, I. Mitrea, M. Mitrea, *Geometric Harmonic Analysis-Volume V: Fredholm Theory and Finer Estimates for Integral Operators, with Applications to Boundary Problems*, Developments in Mathematics, vol. 76 (Springer, Cham, 2023).

[91] D. Mitrea, I. Mitrea, M. Mitrea, M. Taylor, *The Hodge-Laplacian: Boundary Value Problems on Riemannian Manifolds*, Studies in Mathematics, vol. 64 (De Gruyter, 2016).

[92] D. Mitrea, M. Mitrea, Finite energy solutions of Maxwell's equations and constructive Hodge decompositions on nonsmooth Riemannian manifolds, J. Funct. Anal. **190**(2), 339–417 (2002).

[93] D. Mitrea, M. Mitrea, J. Pipher, Vector potential theory on Lipschitz domains in $\mathbb{R}^3$ and applications to electromagnetic scattering, J. Fourier Anal. Appl. **3**(2), 131–192 (1997).

[94] I. Mitrea, M. Mitrea, The Poisson problem with mixed boundary conditions in Sobolev and Besov spaces in non-smooth domains, Trans. Am. Math. Soc. **359**(9), 4143–4182 (2007).

[95] I. Mitrea, M. Mitrea, *Multi-Layer Potentials and Boundary Problems for Higher-Order Elliptic Systems in Lipschitz domains*, Lecture Notes in Mathematics (Springer, Berlin, 2013).

[96] M. Mitrea, The method of layer potentials in electromagnetic scattering theory on nonsmooth domains, Duke Math. J. **77**(1), 111–133 (1995).

[97] M. Mitrea, M. Taylor, Boundary layer methods for Lipschitz domains in Riemannian manifolds, J. Funct. Anal. **163**(2), 181–251 (1999).

[98] M. Mitrea, M. Taylor, Potential theory on Lipschitz domains in Riemannian manifolds: Hölder continuous metric tensors, Commun. Partial Differ. Equ. **25**(7–8), 1487–1536 (2000).

[99] M. Mitrea, M. Taylor, Potential theory on Lipschitz domains in Riemannian manifolds: Sobolev-Besov space results and the Poisson problem, J. Funct. Anal. **176**(1), 1–79 (2000).

[100] M. Mitrea, M. Taylor, Potential theory on Lipschitz domains in Riemannian manifolds: L^p, Hardy, and Hölder space results, Commun. Partial Differ. Equ. **9**(2), 369–421 (2001).

[101] M. Mitrea, M. Taylor, Potential theory for Lipschitz subdomains of Riemannian manifolds equipped with Dini continuous metric tensors, Trans. Am. Math. Soc. **355**(5), 1961–1985 (2003).

[102] M. Mitrea, M. Taylor, The Poisson problem in weighted Sobolev spaces on Lipschitz domains, Indiana Univ. Math. J. **55**(3), 1063–1089 (2006).

[103] M. Mitrea, M. Taylor, Boundary problems for the Navier-Stokes system on non-smooth Riemannian manifolds, Math. Ann. **321**(4), 955–987 (2001).

[104] M. Mitrea, M. Wright, Boundary value problems for the Stokes system in arbitrary Lipschitz domains, Astérisque Soc. Math. Fr. **344**, (2012).

[105] T. Mizuhara, Boundedness of some classical operators on generalized Morrey spaces, *ICM-90 Satellite Conference Proceedings*, pp. 183–189 (Springer-Verlag, Tokyo, 1991).

[106] T. Mizuhara, Commutators of singular integral operators on Morrey spaces with some generalized growth conditions, *Proceedings of the Second ISAAC Congress*, vol. 1, pp. 65–73 (Kluwer Academic Publishers, 2000).

[107] C. B. Morrey, On the solutions of quasi-linear elliptic partial differential equations, Trans. Am. Math. Soc. **43**(1), 126–166 (1938).

[108] M. Mourgoglou, X. Tolsa, The regularity problem for the Laplace equation in rough domains, arXiv, 2110.02205 (2022).

[109] B. Muckenhoupt, Weighted norm inequalities for the Hardy maximal function, Trans. Am. Math. Soc. **165**, 207–226 (1972).

[110] E. Nakai, Hardy-Littlewood maximal operator, singular integral operators, and the Riesz potentials on generalized Morrey spaces, Math. Nachr. **166**, 95–103 (1994).

[111] E. Nakai, Orlicz-Morrey spaces and the Hardy-Littlewood maximal function, Stud. Math. **188**, 193–221 (2008).

[112] S. Nakamura, Y. Sawano, The singular integral operator and its commutator on weighted Morrey spaces, Collect. Math. **68**, 145–174 (2017).

[113] J. Peetre, On the theory of $L_{p,\lambda}$ spaces, J. Funct. Anal. **4**(1), 71–87 (1969).

[114] L.-E. Persson, M. A. Ragusa, N. Samko, P. Wall, Commutators of Hardy operators in vanishing Morrey spaces, AIP Conf. Proc. **1493**(1), 859–866 (2012).

[115] J. Pipher, G. Verchota, The Dirichlet Problem in L^p for the biharmonic equation on Lipschitz domains, Am. J. Math. **114**(5), 923–972 (1992).

[116] H. Rafeiro, S. Samko, Herz spaces meet Morrey type spaces and complementary Morrey type spaces, J. Fourier Anal. Appl. **26**(74), (2020).

[117] M. Ragusa, Commutators of fractional integral operators on Vanishing-Morrey spaces, J. Glob. Optim. **40**, 361–368 (2008).

[118] M. Ragusa, Embeddings for Morrey-Lorentz spaces, J. Optim. Theory Appl. **154**, 491–499 (2012).

[119] M. Rosenthal, H. Triebel, Calderón-Zygmund operators in Morrey spaces, Rev. Mat. Complut. **27**, 1–11 (2014).

[120] M. Rosenthal, H. Triebel, Morrey spaces and their preduals, Rev. Mat. Complut. **28**, 1–30 (2015).

[121] W. Rudin, *Functional Analysis*, 2nd edn. (McGraw-Hill Inc., New York, 1991).

[122] N. Samko, Embeddings of weighted generalized Morrey spaces into Lebesgue spaces on fractal sets, Fract. Calc. Appl. Anal. **22**, 1203–1224 (2019).

[123] N. Samko, Weighted Hardy and singular integral operators in weighted Morrey spaces, J. Math. Anal. Appl. **350**, 56–72 (2009).

[124] N. Samko, Maximal, potential and singular operators in vanishing generalized Morrey spaces, J. Glob. Optim. **57**, 1385–1399 (2013).

[125] N. Samko, On a Muckenhoupt-type condition for Morrey spaces, Mediterr. J. Math. **10**, 941–951 (2013).

[126] N. Samko, Weighted Hardy operators in the local generalized vanishing Morrey spaces, Positivity **17**, 683–706 (2013).

[127] N. Samko, Weighted boundedness of certain sublinear operators in generalized Morrey spaces on quasi-metric measure spaces under the growth condition, J. Fourier Anal. Appl. **28**(27), (2022).

[128] Y. Sawano, A note on Besov-Morrey and Triebel-Lizorkin-Morrey spaces, Acta Math. Sin. Engl. Ser. **25**, 1223–1242 (2009).

[129] Y. Sawano, Generalized Morrey spaces for non-doubling measures, NoDEA Nonlinear Differ. Equ. Appl. **15**(4–5), 413–425 (2008).

[130] Y. Sawano, G. Di Fazio, D. Hakim, *Morrey Spaces: Introduction and Applications to Integral Operators and PDE's Volume I*, Chapman & Hall/CRC Monographs and Research Notes in Mathematics (Chapman and Hall/CRC, 2020).

[131] Y. Sawano, G. Di Fazio, D. Hakim, *Morrey Spaces: Introduction and Applications to Integral Operators and PDE's Volume II*, Chapman & Hall/CRC Monographs and Research Notes in Mathematics (Chapman and Hall/CRC, 2020).

[132] Y. Sawano, K. Ho, D. Yang, S. Yang, Hardy spaces for ball quasi-Banach function spaces, Diss. Math. **525**, 1–102 (2017).

[133] Y. Sawano, T. Sugano, H. Tanaka, Generalized fractional integral operators and fractional maximal operators in the framework of Morrey spaces, Trans. Am. Math. Soc. **363**(12), 6481–6503 (2011).

[134] Y. Sawano, T. Sugano, H. Tanaka, Orlicz-Morrey spaces and fractional operators, Potential Anal. **36**(4), 517–556 (2012).

[135] Y. Sawano, H. Tanaka, Decomposition of Besov-Morrey spaces and Triebel-Lizorkin-Morrey spaces, Math. Z. **257**, 871–905 (2007).

[136] Z. Shen, Boundary value problems for parabolic Lamé systems and a nonstationary linearized system of Navier-Stokes equations in Lipschitz cylinders, Am. J. Math. **113**(2), 293–373 (1991).

[137] Z. Shen, Boundary value problems in Morrey spaces for elliptic systems on Lipschitz domains, Am. J. Math. **125**(5), 1079–1115 (2003).

[138] S. Shi, Z. Fu, F. Zhao, Estimates for operators on weighted Morrey spaces and their applications to nondivergence elliptic equations, J. Inequal. Appl. **390**, (2013).

[139] G. Stampacchia, The spaces $\mathcal{L}^{(p,\lambda)}$, $N^{(p,\lambda)}$ and interpolation, Ann. Sc. Norm. Super. Pisa **19**(3), 443–462 (1964).

[140] E. M. Stein, *Harmonic Analysis: Real-Variable Methods, Orthogonality, and Oscillatory Integrals*, Princeton Mathematical Series, vol. 43 (Princeton University Press, Princeton, NJ, 1993).

[141] E. M. Stein, Note on singular integrals, Proc. Am. Math. Soc. **8**(2), 250–254 (1957).

[142] J. Stömberg, R. Wheedon, Fractional integrals on weighted H^p and L^p spaces, Trans. Am. Math. Soc. **287**(1), (1985).

[143] I. Takeshi, Y. Komoro-Furuya, S. Enji, The Adams inequality on weighted Morrey spaces, Tokyo J. Math. **34**(2), 535–545 (2011).

[144] H. Tanaka, Two-weight norm inequalities on Morrey spaces, Ann. Acad. Sci. Fenn. **40**, 773–791 (2015).

[145] M. Taylor, Analysis on Morrey spaces and applications to Navier-Stokes and other evolutions equations, Commun. Partial Differ. Equ. **17**, 1407–1456 (2007).

[146] T. Toro, Potential analysis meets geometric measure theory, *Proceedings of the ICM 2010*, vol. 3, pp. 1485–1497 (2010).

[147] M.-P. Tran, T.-N. Nguyen, Lorentz-Morrey global bounds for singular quasilinear elliptic equations with measure data, Commun. Contemp. Math. **22**(5), Paper No. 1950033 (2020).

[148] H. Triebel, *Interpolation Theory, Function Spaces, Differential Operators*, (Huethig & Wepf, 1995).

[149] G. Verchota, Layer potentials and regularity for the Dirichlet problem for Laplace's equation in Lipschitz domains, J. Funct. Anal. **59**(3), 572–611 (1984).

[150] G. Verchota, The Dirichlet problem for the polyharmonic equation in Lipschitz domains, Indiana Univ. Math. J. **39**(3), 671–702 (1990).

[151] C. Vitanza, Functions with vanishing Morrey norm and elliptic partial differential equations, *Proceedings of Real Analysis and Methods of Elliptic Partial Differential Equations*, pp. 147–150 (Springer, 1990).

[152] W. Yuan, W. Sickel, D. Yang, *Morrey and Campanato Meet Besov, Lizorkin and Triebel*, Lecture Notes in Mathematics (Springer, New York, 2010).

[153] C. Zorko, Morrey space, Proc. Am. Math. Soc. **98**, 586–592 (1986).

Index

https://doi.org/10.1515/9783111458274-014

Symbol index

https://doi.org/10.1515/9783111458274-015

De Gruyter Studies in Mathematics

Volume 98
Peter J. Brockwell, Alexander M. Lindner
Continuous-Parameter Time Series, 2024
ISBN 978-3-11-132499-9, e-ISBN 978-3-11-132503-3, e-ISBN (ePUB) 978-3-11-132520-0

Volume 97
Ştefan Ovidiu I. Tohăneanu
Commutative Algebra Methods for Coding Theory, 2024
ISBN 978-3-11-121292-0, e-ISBN 978-3-11-121479-5, e-ISBN (ePUB) 978-3-11-121538-9

Volume 25
Karl H. Hofmann, Sidney A. Morris
The Structure of Compact Groups. A Primer for the Student – A Handbook for the Expert, 5th Edition, 2023
ISBN 978-3-11-117163-0, e-ISBN 978-3-11-117260-6, e-ISBN (ePUB) 978-3-11-117405-1

Volume 96
Francesco Aldo Costabile, Maria Italia Gualtieri, Anna Napoli
Polynomial Sequences. Basic Methods, Special Classes, and Computational Applications, 2023
ISBN 978-3-11-075723-1, e-ISBN 978-3-11-075724-8, e-ISBN (ePUB) 978-3-11-075732-3

Volume 54
Yasushi Ishikawa
Stochastic Calculus of Variations, For Jump Processes, 3rd Edition 2023
ISBN 978-3-11-067528-3, e-ISBN 978-3-11-067529-0, e-ISBN (ePUB) 978-3-11-067532-0

Volume 95
Marko Kostić
Metrical Almost Periodicity and Applications to Integro-Differential Equations, 2023
ISBN 978-3-11-123303-1, e-ISBN 978-3-11-123387-1, e-ISBN (ePUB) 978-3-11-123417-5

Volume 94
Nora Doll, Hermann Schulz-Baldes, Nils Waterstraat
Spectral Flow. A Functional Analytic and Index-Theoretic Approach, 2023
ISBN 978-3-11-116989-7, e-ISBN 978-3-11-117247-7, e-ISBN (ePUB) 978-3-11-117308-5

Volume 93
Brian Street
Maximal Subellipticity, 2023
ISBN 978-3-11-108517-3, e-ISBN 978-3-11-108564-7, e-ISBN (ePUB) 978-3-11-108594-4

www.degruyter.com

Printed in the USA
CPSIA information can be obtained
at www.ICGtesting.com
CBHW081512031024
15257CB00046B/189

9 783111 458168